Engineering Geological Mapping

Engineering Geological Mapping

W. R. Dearman PhD

Emeritus Professor of Engineering Geology,
Department of Geotechnical Engineering,
University of Newcastle upon Tyne, UK

BUTTERWORTH
HEINEMANN

Butterworth–Heinemann Ltd
Halley Court, Jordan Hill, Oxford OX2 8EJ

PART OF REED INTERNATIONAL P.L.C.

OXFORD LONDON GUILDFORD BOSTON
MUNICH NEW DELHI SINGAPORE SYDNEY
TOKYO TORONTO WELLINGTON

First published 1991

British Library Cataloguing in Publication Data

Dearman, W. R.
Engineering geological mapping.
1. Geological maps
I. Title
551.8

ISBN 0–7506–1010–7

Library of Congress Cataloging-in-Publication Data

Dearman, W. R.
Engineering geological mapping/W. R. Dearman.
p. cm.–(Butterworths advanced series in geotechnical engineering)
Includes bibliographical references.
ISBN 0–7506–1010–7:
1. Engineering geology. 2. Geological mapping. I. Title.
II. Series.
TA705.D33 1990
624.1′51′0223–dc20 90–1816

Filmset by Bath Typesetting Ltd, London Road, Bath, Avon
Printed and bound in Great Britain by Courier International Ltd, Tiptree, Essex

Preface

This book grew from a perceived need to combine and expand two reports on the preparation of engineering geological maps. Work on the first, 'The Preparation of Maps and Plans in Terms of Engineering Geology' was begun in 1968 by a working party set up by the Engineering Group of the Geological Society (of London). The final report was published in late 1972 as Volume 5, No. 4, of the *Quarterly Journal of Engineering Geology*. It was intended as a guide to engineers and engineering geologists as well as to geologists, but not a Code of Practice on the subject since it was recognized by the working party that with the possible rapid evolution of techniques of engineering geological mapping the report would require revision from time to time. No revision has been undertaken, although a foothold for engineering geological mapping has been gained for the first time in the 'Code of Practice for Site Investigations': BS 5930:1981 published by the British Standards Institution. A short section on geological mapping for ground investigation is supplemented by a section dealing with a 'Legend for engineering geological maps and plans', with recommended symbols for soils and rocks, general planar structures, and geological structures and boundaries.

A second report, 'Engineering Geological Maps. A Guide to Their Preparation', was published by Unesco in 1976. For many years Unesco has been concerned with the preparation and publication of small-scale geological maps of various kinds. This new booklet was devoted to a particular aspect of this programme, namely engineering geological mapping. The purpose of engineering geological maps is to show the distribution of specific geological phenomena and characteristics of rocks and soils affecting engineering use of different terrains. The ever-growing demand for such maps has revealed the need for the standardization of principles, systems and methods. This is an urgent but at the same time a difficult problem which can best be solved by international co-operation – sensible comments for the preface to the guidebook, which had been prepared by the Commission on Engineering Geological Maps of the International Association of Engineering Geology (IAEG).

The IAEG, conceived in December 1964 at the International Geological Congress in New Delhi, was officially constituted at Unesco headquarters in January 1967. The First Congress of the IAEG was held in Paris in September 1970, with one of the themes 'Engineering geological mapping'. In his Presidential Address to the General Assembly, Professor Quido Zaruba emphasized that 'the necessity of submitting results to engineers in an understandable form prompted engineering geologists to develop appropriate mapping techniques. The methods of presentation are diverse, but all aim at making each map carry the maximum amount of information in the most intelligible way. Attempts are now being made to standardize the methods of presentation of information on engineering geological maps'. This was a reference to the setting up by the Association of a working group (later commission) for engineering geological maps.

I had been invited by Mr Rudolph Glossop, then Chairman of the Engineering Group of the Geological Society, to be the UK member of the working group and attend the first full meeting at the Congress. In fact, the working group had been established in 1968 at the General Assembly of the IAEG on the occasion of the International Geological Congress in Prague. The Unesco guidebook was the first accomplishment of the commission in response to its stated aims.

Members of the IAEG commission who took part in the preparation of the guide were: Professor Milan Matula (Chairman), Czechoslovakia; Professor G.A. Golodkovskaja, USSR; A. Peter, France; A. Pahl, West Germany; Mrs Dorothy H. Radbruch-Hall (Secretary), USA; and myself as Editor. We were driven and motivated by the energy and enthusiasm of our Chairman, Milan Matula, at the first full working meeting arranged by him in Bratislava, with the preparation of the guidebook quickly becoming our main aim. Later publications by the commission in 1981 gave recommended symbols and rock and soil description and classification for engineering geological mapping. After that the commission disbanded itself, a decision reluctantly accepted by the IAEG as possibly a temporary measure.

Just how far-sighted Professor Zaruba was in 1970 is

borne out by his observation that 'Finally, there is the last group (field of study), very important nowadays, which is connected with the study of the human environment and the planning of land-use, called forth by the increasing pressure of population on space. The so-called environmental geology likewise includes the examination of the effects of human works on existing geological conditions so as to prevent any disturbance of the balance of nature. It also embraces the problem of protection of the natural environment. The speed with which man is changing his environment often makes him forget that natural forces are still at work, and it is a great challenge to engineering geologists, among other natural scientists, to harness them to the profit of man'.

After a decade of relative quiescence, a reformed Mapping Commission of the IAEG is to consider new developments in the field of environmental engineering geology, including hazards and the risks associated with them, some of which are touched on in the later chapters of this book.

Work on the book was begun in 1980, but for a variety of personal reasons was stopped until I was induced to start again in 1988 when a completely new manuscript was prepared and revised within the year. Remote sensing, close-up photogrammetry and the computer manipulation of data and images are now being used more extensively to prepare environmental and engineering geological maps and plans. Those developments would appear to provide the most fruitful practical applications for the future: at large scale during engineering construction, especially as a recording medium for the design of remedial measures and the as-built conditions at engineering works, and at small scales for environmental assessments.

Units

Throughout the book, units are given in their original published form, but where there has been a deliberate change, as for example when a diagram or plan has been redrawn, preference has usually been given to the use of SI units.

W.R.D.

Contents

Colour plates 1 to 20: between pages 136 and 137

1

Introduction

In 1968 the Engineering Geological Mapping Commission of the International Association of Engineering Geology was initiated. In 1970 guidelines for the work of the commission were proposed:

- to clarify the present situation in engineering geological mapping;
- to analyse the various types of maps called engineering geological maps – maps which are to serve for building construction and land-use planning;
- to outline the trends for the future development of engineering geological cartography and to present general recommendations on the information to be provided by a complex engineering geological map;
- to contribute to international exchange of information on this subject.

A long-term programme of topics for discussion was established, which effectively guided the work of the commission for a decade. Selected topics were:

1. What is an engineering geological map? This would involve consideration of: the basic concepts and methodological background of engineering geological mapping; classification of the various kinds of engineering geological maps according to their purpose, scale and content; and the position of engineering geological maps among other geological maps, including environmental geological maps.
2. What are the problems involved in three-dimensional representation of subsurface conditions on engineering geological maps?

Six other topics were also selected, including the use of computers.

To what extent this aim was achieved can be judged from the UNESCO publication prepared by the commission (Anon., 1976): *Engineering Geological Maps. A Guide to their Preparation.*

In the early 1970s a Working Party of the Geological Society of London Engineering Group had been set up to prepare a report on 'The preparation of maps and plans in terms of engineering geology' (Anon., 1972). The theoretical approach of the commission report is very effectively complemented by the more pragmatic working party report.

The present book was started in 1980 by two members of the international commission; active efforts towards completion, after a long period of quiescence, began late in 1987. The aim was both to expand the rather condensed exposition in the guide, and to consider later developments in mapping for both engineering and environmental purposes.

1.1 Definition of engineering geology

There is no difficulty in defining engineering geology. It is one branch of applied geology which, broadly, is the application of geology to industry – not some special type of geology but the whole spectrum of the science. Engineering geology is the discipline of geology applied to civil engineering, particularly to the design, construction and performance of engineering structures interacting with the ground in, for example, foundations, cuttings and other surface excavations, and tunnels.

1.2 Recording the early applications of geology in engineering

It is worth while recalling that engineering geology has had a very long history, even though only in the past two decades has it acquired a degree of sophistication and now stands as an independent subject. In the following brief review of the development of engineering geological mapping, attention is focused selectively on European practice. A similar review could be written for North America.

1.2.1 Architects and foundation mapping

John Smeaton (1724–1792), engineer and mechanic, was engaged as an architect to reconstruct the lighthouse on the Eddystone in the English Channel off Plymouth. Two former lighthouses, constructed mainly of wood, had been destroyed by storms. Smeaton decided that stone was the proper material with which to rebuild the lighthouse. Having formulated detailed ideas on how to construct the building, he then visited the rock for the first time on 26 April 1756, on his sixth attempt to land (Smiles, 1861, p. 32). During this and subsequent visits, Smeaton spent 15 hours on the rock, taking dimensions of all its parts to enable him to construct an accurate model of the foundation of the proposed building. He paid three more visits to the rock for the purpose of correcting his measurements, before constructing a complete model of the lighthouse with his own hands.

Smeaton had devised a simple method of surveying, whereby he could determine the position and elevation of any point on the rock within a circle 32 ft in diameter, centred on the highest point. The surface of the rock was a stepped combination determined by bedding in the slate and jointing; the bedding dip was 26°W. He contrived a robust, three-legged wooden stool with a circular seat. On this he mounted the base of a 12 in diameter theodolite to which was fixed a 16 ft long calibrated wooden stick. With 35 primary locations marked on the rock with the point of a jumper, each could be accurately located using a graduated vertical stick to intersect the horizontal pole, the bearing being determined by the theodolite base (Figure 1.1). Intermediate points between any two locations could be fixed by offset measurements taken from a string line stretched between the two.

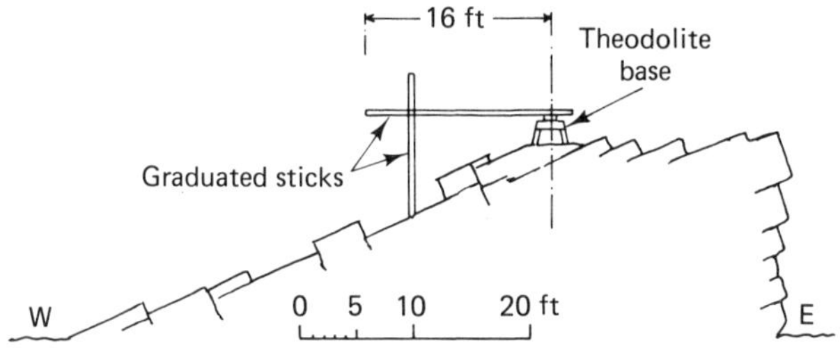

Figure 1.1 Diagram of the method used for large-scale mapping of Eddystone rock, Plymouth, UK (After Smeaton, 1793)

The whole surface of the rock could be modelled by drilling down vertically from the plane upper surface of the wood block to guide the depth for carving; this is the method used in carving. Two copies were made, and on the second the rock excavation (carving might be more appropriate even in the slate) required to suit the building was indicated. The intention was to remove as little of the rock as possible, while ensuring that the flat base of each building block was protected by at least a 3 in upstand in the rock.

Early in the nineteenth century, much of what would now be regarded as major civil engineering construction was still in the hands of architects. Occasionally their plans indicated ground conditions determined by boring. An example from North-east England is the 'General Plan for the Rebuilding and Extension of the Quay, Newcastle Upon Tyne', an ink and watercolour architectural drawing prepared by John Dobson on 1 January 1836. The plan shows the location of five boreholes just on the river side of the line of the new quay wall. Each borehole is marked by a red dot and numbered, with a note alongside of the materials met with in boring. As examples:

No. 1. 8 ft sand, gravel below.
No. 2. 4 ft sand into gravel 2 ft.

At a scale of 1 in = 100 ft, the plan is accompanied by 13 cross-sections at 1 in = 20 ft. The sections show the construction of the wall, profile of the river bed with the sand layer, ground level and fill behind the wall, and the tidal range in the river.

Part of the plan, including the location of Section No. 5 and the positions of Boreholes No. 3 and 4, is shown in Figure 1.2(a); Cross-section No. 5 forms Figure 1.2(b).

1.2.2 Nineteenth century recording of ground conditions at dam sites

The early association of geology with practical problems, although maintained in both metalliferous and coal mining, soon died out in the field of civil engineering, and from 1850 onwards geology became more and more neglected in civil engineering practice in the UK (Glossop, 1969). Despite this, dam engineers and water engineers continued to be concerned with, and to record, geological details of ground conditions encountered in major engineering works; brief details of these were often published.

Vyrnwy dam, Wales, UK

Deacon (1896) described the foundation conditions at the Vyrnwy dam in central Wales. Undertaking as an engineer his own geological investigations, he concluded from his assessments of the form of the valley to be dammed that a rock bar limited on the downstream side the infilled glacial basin that was to form the Vyrnwy reservoir. In a classic site investigation to determine the highest part of the bar, hidden by alluvium, 177 borings and probings, and 13 shafts, were sunk. By means of these, actual contours were drawn (Figure 1.3) and a model made of the rock surface. In the course of excavation, Deacon records that watertight rock was found to be 6–7 ft lower than the rock surface which the contours had shown. Commenting on the foundation conditions, he states that rock masses weighing hundreds of tons, broken from their beds and moved some distance down the valley by the former glacier, or only just detached, were met with.

It is apparent that early papers on major engineering projects occasionally dealt with geological aspects affecting the works. In the late nineteenth century these accounts would be illustrated by steel engravings on which very considerable geological detail could be shown. One such is the somewhat schematic cross-section illustrating the geological conditions associated with driving the tunnel beneath the River Mersey as part of the Vyrnwy scheme for the Liverpool water supply. The diagram (Figure 1.4) comprises four vertical sections through the alluvium and boulder clay overlying New Red Sandstone bedrock. Of particular interest is the disintegrated sandstone (Red Roche) underlying the boulder clay. There is a lack of realism in the representation of the bedding in the sandstone which should be sensibly horizontal, although despite this the section is a realistic geological plan of the subsurface conditions encountered. Incidentally, the tunnel was driven through the water-bearing alluvium rather than at a much greater depth through the sandstone in order to reduce water pressure in the siphon.

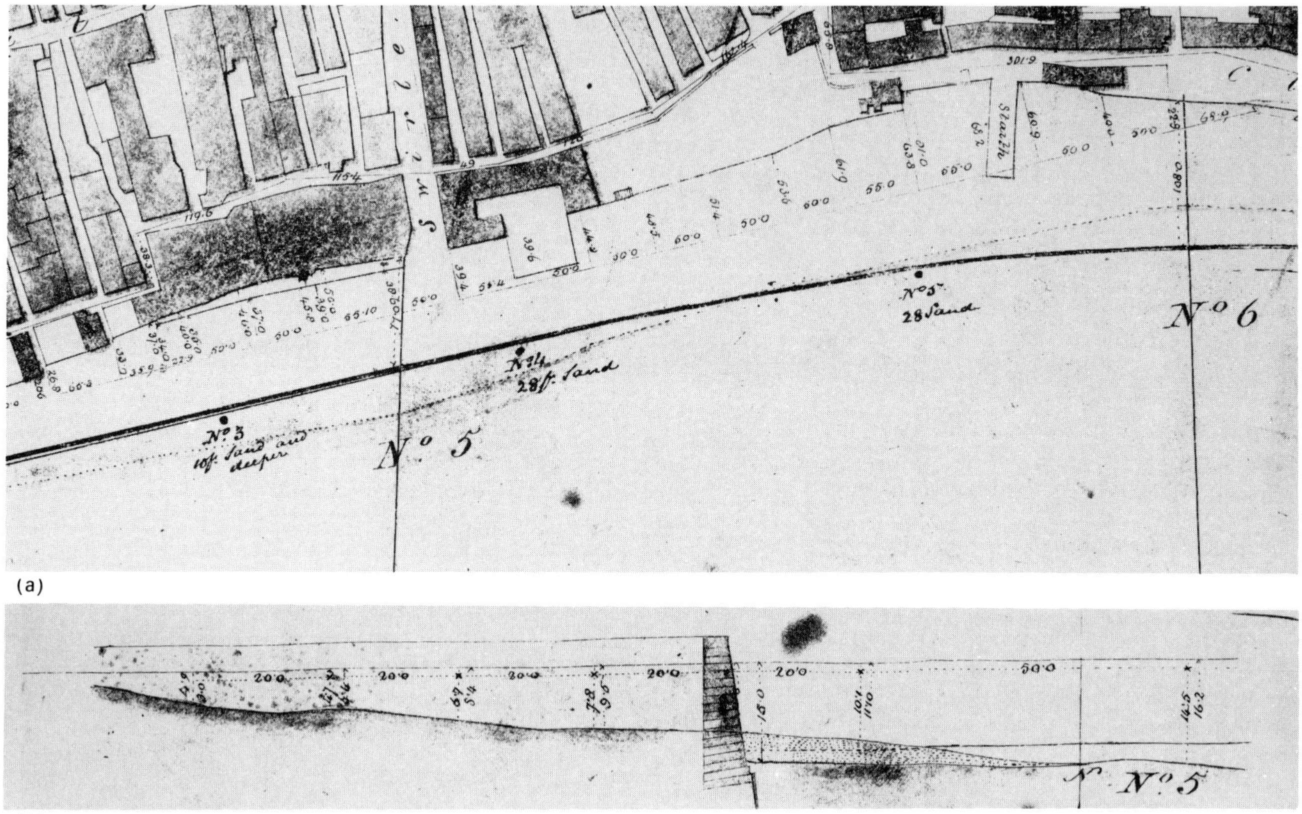

Figure 1.2 Extract from 'General Plan for the Rebuilding and Extension of the Quay, Newcastle Upon Tyne', UK, by John Dobson, showing the location of investigatory boreholes and a cross-section through the proposed quay wall and the river bed (From Dobson, 1836)

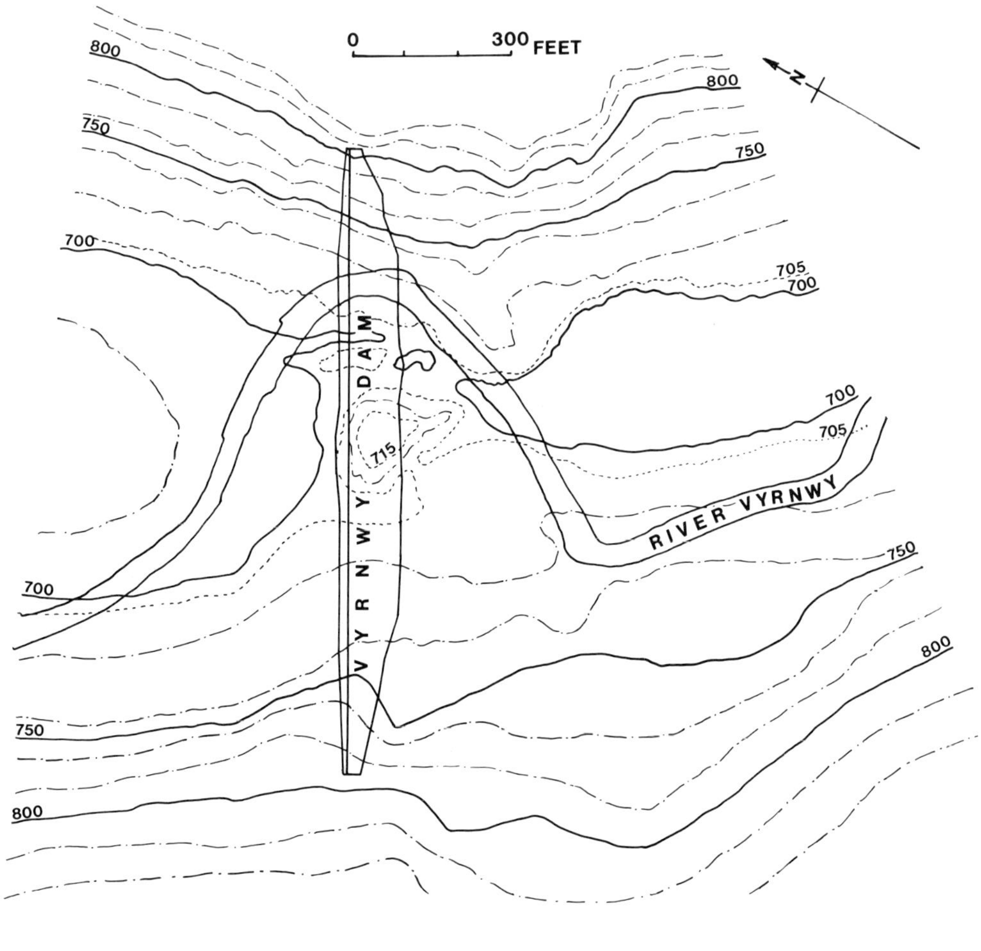

Figure 1.3 Plan drawn at an original scale of 1:1800 showing contours on the rockhead bar below the alluvium at the Vyrnwy dam site, Wales, as proved by borings and probings (Redrawn from Lapworth, 1911, Fig. 5)

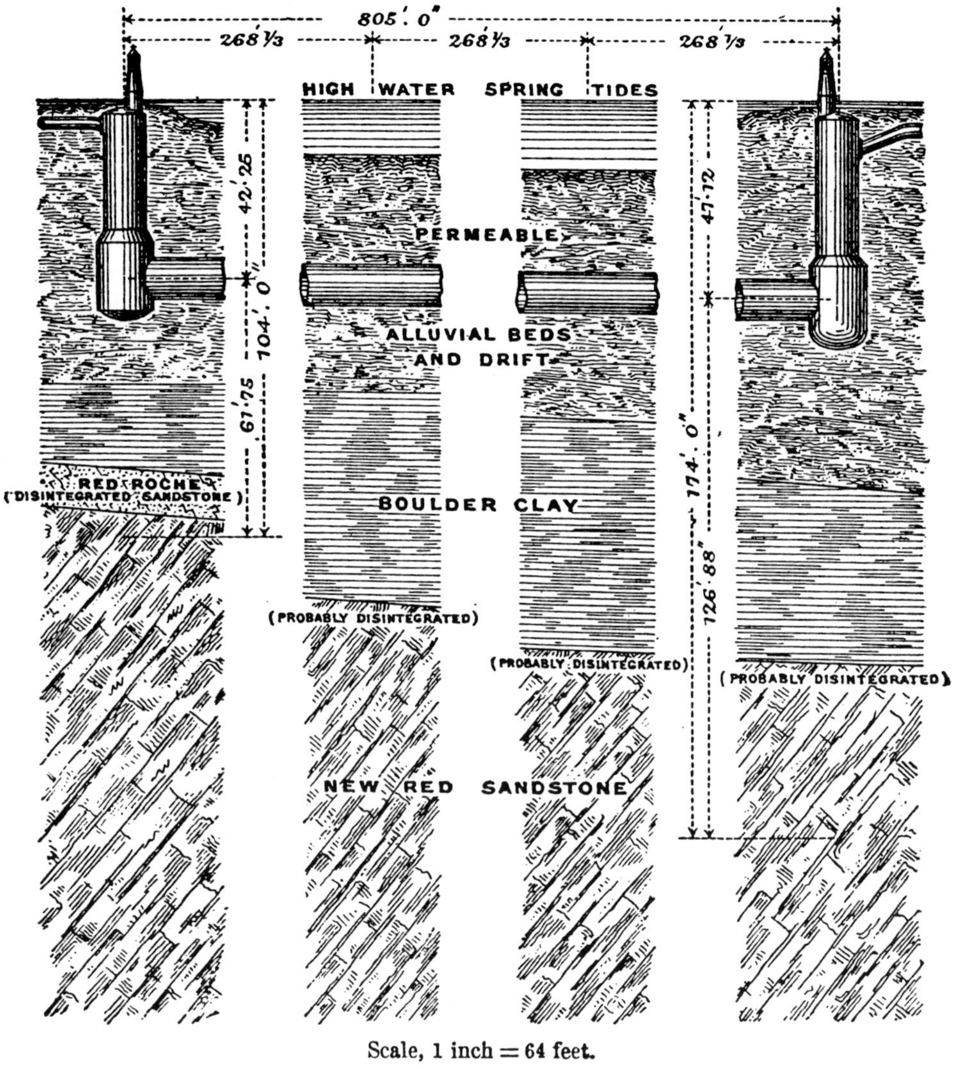

Figure 1.4 Geological cross-section of the ground conditions at the Vyrnwy tunnel-siphon beneath the River Mersey, UK (From Deacon, 1896, Fig. 10)

Burrator reservoir, Devon, UK

At the Burrator Works on Dartmoor, for the water supply of Plymouth in Devon, the site for the main masonry dam proved to be entirely satisfactory on massive fresh granite. Such was not the case, however, for the small earthwork dam across a subsidiary minor col. The cut-off trench for the Sheepstor embankment (Sandeman, 1901) was 680 ft long, whereas the embankment is only 470 ft long. Unexpected ground conditions account for the discrepancy (Figure 1.5).

Several trial pits were sunk to ascertain the most advantageous line for the trench. In two of the trenches, rock was exposed at a depth of about 14 ft, in the event a delusion because on opening the trench they were found to have penetrated to pinnacles of rock which shelved away rapidly to considerable depths. The trench, which at the centre was 105 ft deep, was cut through an extensive layer of decomposed granite crossed by quartz–tourmaline veins and minor intrusions of decomposed fine-grained granite (elvan). 'The veins ranged from 1 inch to 3 feet in thickness, and the larger had the appearance which would be exhibited by the section of a dry rubble wall of irregularly shaped stones fitting one into another . . . these veins were the main reason for so deep a trench being sunk.' Eventually, at depth the rotten granite disappeared, the veins became one with the granite they traversed (so far as engineering purposes were concerned) and the trench ended in solid granite rock.

The geological plan of the trench wall and bottom was plotted as excavation proceeded, and affords a valuable record of the as-found engineering geological conditions on this part of the site. Until quite recently some idea of ground conditions in the trench could be gained from the road-cutting at the western end of the dam. Exposures left after road-widening (Figure 1.6a), now nearly completely grown over, were reasonably well exposed in 1959. Corestones, measuring 2 × 1 m, are only moderately weathered (grade III) internally with the feldspars unaffected, but the outer skin some 15 cm thick is highly weathered and the feldspar megacrysts are kaolinized. There is an abrupt transition to the loose friable completely weathered granite (grade V) with kaolinized feldspars which forms the matrix to the corestones. The line drawing (Figure 1.6b), with details of discontinuities in the corestones and veins in the grade V granite still preserved, was made by drawing directly on the photograph in Indian ink and then bleaching out the photographic image (Fookes, Dearman and Franklin, 1971).

Pennine dams, Northern England

In 1911, Lapworth published many cross-sections of the foundation excavations, particularly the cut-off

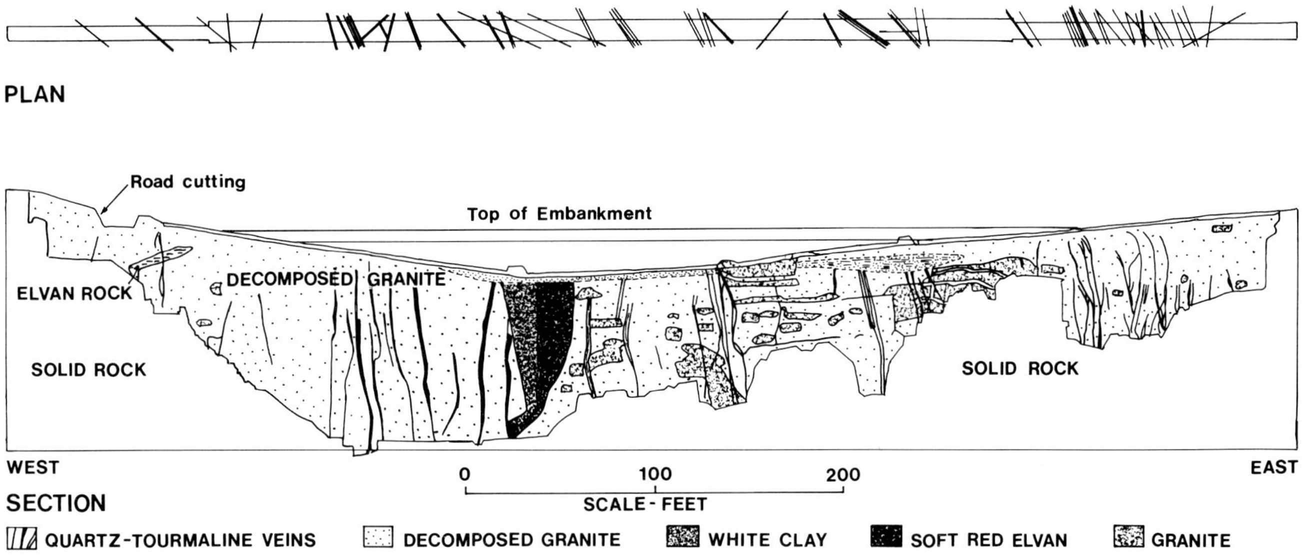

Figure 1.5 Geological plan and cross-section of the cut-off trench for the Sheepstor embankment of the Burrator reservoir, Devon, UK (Redrawn from Sandeman, 1901, Plate 1, Fig. 7)

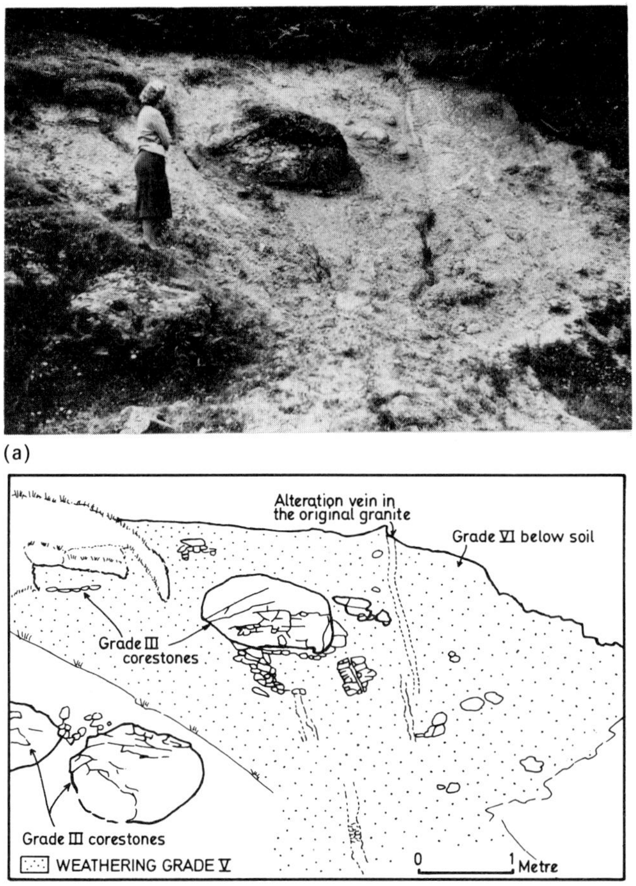

Figure 1.6 Roadside cutting, Burrator reservoir, Sheepstor embankment, Devon, UK, in weathered Dartmoor granite: (a) photograph of the exposure in 1959; (b) drawing from the photograph showing the distribution of weathering grades in the granite (From Fookes, Dearman and Franklin, 1971, Plate IV, by permission)

trenches, for dams. Many of the examples were from the Pennines of Northern England. The one chosen for illustration (Figure 1.7) is the trench for the Yarrow reservoir of the Liverpool Waterworks at Rivington. Sensibly, horizontal alternations of beds of shale and sandstone (grit) were affected by landsliding before the deposition of some 60 ft of alluvium partly infilling the old valley (Lapworth, 1911).

Other cross-sections in similar settings, not reproduced here, show examples of landslides, cambering and valley bulging, and water-bearing wide-open joints in the horizontal sandstones. Somewhat simplified from meticulously engraved originals, Lapworth's drawings are true engineering geological plans, affording a permanent record of the as-found conditions in the walls of cut-off trenches, the method favoured at the time for ensuring a positive water barrier beneath a dam.

It is interesting to note Lapworth's recommendation that the plans should not be drawn at a scale smaller than 1 : 2500.

1.2.3 Development of engineering geological maps in Europe

Peter (1966), discussing geotechnical mapping, reviewed the development of maps in Germany. It was at the Exposition Technique de la Construction, held in Leipzig in 1913, that plans were shown for the first time of both the engineering structure and foundation conditions. These geotechnical plans were made for the towns of Erfurt, Frankfurt on Oder, Danzig and several others. Coloured dot patterns and conventional signs showed areas prone to flooding, where the watertable was less than 1 m deep – mines, quarries, etc. A descriptive memoir gave the results of boreholes put down for site investigation and in the search for water.

In 1919, Moldenhauer produced a geotechnical map of Danzig from the geological map of the city, a technique of historical importance that is the basis of more recent mapping methods. Although the maps have not been available for examination, it is understood that the author divided the ground into a number of depth zones: 0–2 m, 2–4 m, 4–6 m and 6–10 m. Results were presented in two sheets: a map of the location of boreholes (600 alone for Danzig) and a geotechnical map.

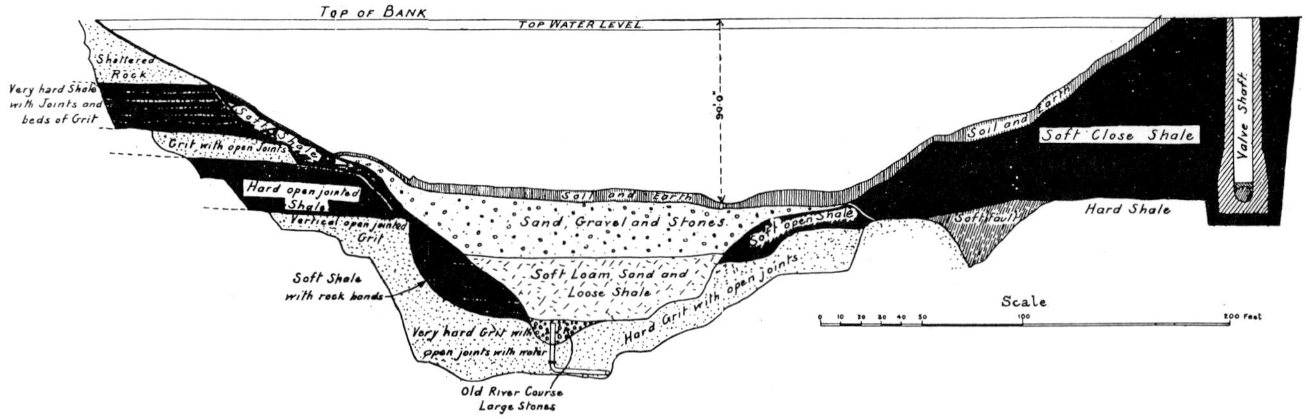

Figure 1.7 Large-scale plan of the wall of the foundation excavation and cut-off trench of the Yarrow reservoir of the Liverpool Waterworks, Rivington, UK (From Lapworth, 1911, Fig. 14)

Stremme in 1932 published a group of maps of Ostendorff dealing with geology and geotechnics. Three geological maps give details of useful rocks and soils, groundwater and surface water, and construction conditions. For the latter, Moldenhauer's method of depth zones was used with regard to the geotechnical map; details were given of estimated allowable bearing pressures, moisture content, and probability of landsliding.

By 1938, Muller had published maps of the parish of Marke with a very modern aspect. The series of maps comprised: an outcrop map of soils; an interpretative map of construction conditions; a hydrological map; a map of treatment for the improvement of soils; a map for use in planning. Suitability for construction of various types of ground was shown by colours:

- green and yellow were used for ground suitable for construction that would permit bearing pressures of not less than 2.5 bar;
- orange indicated moderate ground;
- orange cross-hatching indicated least favourable ground conditions in its present state, where variable hydrological conditions necessitated special attention being paid to foundations – nevertheless, allowable bearing pressures were subject to restriction;
- red corresponded to poor ground requiring costly foundations – it comprised generally fill, marshy ground, steep slopes, etc.

On the soil map, each soil type is shown by a distinctive colour. Black cross-hatching and coloured conventional symbols are used to indicate the characteristic physical and chemical properties of each.

By the 1930s, numerous detailed geological maps, including geotechnical maps, had been prepared essentially for town and country planning in Germany. It seems, however, that many of these maps were, according to Pfannschmidt (1939), pedological soil maps. As such, 'ground maps' at scales of 1:10 000 and 1:25 000 had been prepared for many towns, including as one example Osnabruck. Bruning (1939) also mentions the pedological atlas of Hannover, including a soil type map, an economic/administrative/planning map, and a water map.

Preparation of maps of foundation soils in Warsaw, on which the geological conditions at several levels were plotted, was a pioneering work in urban geology (Sujkowski and Rozycki, 1936).

Early maps purporting to show ground conditions were essentially pedological maps, with the main postwar advance coming from the acceptance of the principle that the maps should be geologically based. The first geological maps for engineering practice were so-called maps of 'soil conditions' or 'ground conditions', representing both the superficial soils and the pre-Quaternary rocks beneath.

In 1947 Zebera, in a publication of the State Geological Institute of the Czechoslovak Republic, outlined a standard form of geotechnical maps for planning that had been developed in Czechoslovakia (Zebera, 1947a). In another report published the same year, dealing with the Bustehrad area, he described the strip method, an ingenious system for showing engineering geological conditions in three dimensions (Zebera, 1947b). These reports discussed the main geological rock units, flagging the factors that might cause difficulties in construction, such as swelling clays and landslides; indicating which soils were most suitable for agriculture and forestry; and noting the places where scenic areas should be preserved. In this respect the maps approach environmental geological maps.

Zebera's description of the methods being used in preparing maps for planning presented a format that had been in the process of development since 1941, when a geological service for the planning of settlements was established under the administration of the State Geological Institute of the Czechoslovak Republic. For this purpose, the geologists of the institute worked out field methods and map standards for the Survey Map of Soil Conditions in the Czechoslovak Republic at the scale of 1:25 000 and Detailed Plans of the Foundation Soils of Settlements in the Czechoslovak Republic at the scale of 1:5000.

Because of its historical importance in engineering geological cartography, the first version of the strip (or stripe) method deserves illustration here, even though it is dealt with in detail in Chapter 6. Originally a 'banding' method (Figure 1.8) was used to show the depth to bedrock beneath the surface. Deep colours represented outcrops of bedrock at the surface, whereas pale colours depicted the soils. Where the depth to bedrock was 4 m or less from the surface, the depth was shown by a band of colour appropriate to the rock. A broad band indicated bedrock down to 1 m depth, a narrow band the depth range 1–2 m, and an interrupted band showed that

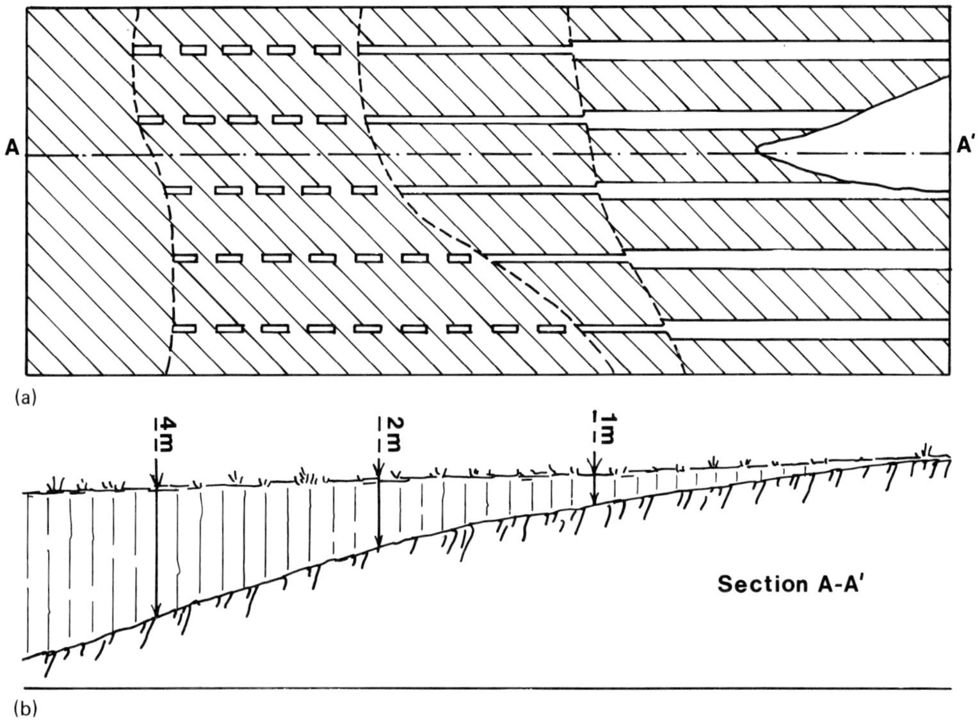

Figure 1.8 The 'banding' method of Zebera (1947b) to show depth of bedrock below the surface, first used in engineering geological maps of Czechoslovakia: (a) plan view; (b) A–A' section showing superficial deposits, lined, above bedrock, plain (After Zaruba and Mencl, 1961, Fig. 2.1)

the bedrock was 2–4 m below the surface. Thicknesses of the superficial deposits were determined by probing to 4 m. Where the deposits were thicker than 4 m, widely spaced hand auger holes were used to determine soil types at depth, but not necessarily to bedrock which was not shown on the plan. Limits of the three types of band on the map obviously outline bedrock contours at 1, 2 and 4 m depth.

Many of the ideas thus generated in Czecholovakia have been modified in the course of time, to spread throughout the international geological community, especially in Eastern Europe. They have appeared in various permutations in numerous publications.

Potentially the most important post-war development in engineering geology was the establishment of the British National Committee of the 1948 International Conference of Soil Mechanics and Foundation Engineering held at Rotterdam. This committee was appointed by the Institution of Civil Engineers. At the same time, with the Geotechnical Society set up, a group of founder members began publication of the journal *Géotechnique*, to encourage the pursuit of engineering geology in addition to soil mechanics. In his foreword to the first number, Terzaghi (1948) stressed that '*Géotechnique* is intended to become a clearing house for significant information in the fields of soil mechanics and engineering geology. Both fields are so inter-dependent that it would be an idle question to ask which one is more important. The fact is aptly expressed by the name of the journal'.

Tracing the history of engineering geology back to William Smith, Terzaghi stated that for too long 'the geological study of the site for proposed constructions furnished an inventory of the materials which would be encountered during excavation, but little or no attention was paid to the factors, such as shearing resistance, permeability and compressibility, which would determine the reaction of the subsurface materials to the changes produced by the excavation or loading'. He also sounded a cautionary note: 'When theory invaded the field of civil engineering our analytical faculties began to develop partly at the expense of our capacity for observation. In the field of geotechnology (at that time soil mechanics and engineering geology), this is a very serious matter.'

It is in the related fields of mapping, as a recording and interpretative tool, and of rock description that engineering geology emerged as a distinct discipline applied by specialist geologists as a practical aid to the civil engineer. As John (1974) later admitted, making geological maps and describing rocks are two basic skills of the geologist that few engineers acquire. All this emerged in the immediate post-war years when geologists – usually mining geologists – began to be employed either by site investigation firms or by consulting engineers, usually those involved with major dam projects.

1.2.4 Engineering geological mapping since 1950

The year 1950, an arbitrary cut-off date, is nevertheless important in the history of engineering geology as it marks the beginning of the world-wide post-war development of engineering geological mapping for both large-scale site-specific engineering plans and smaller scale maps for urban planning. However, within some countries the continuation of large-scale mapping at engineering sites predominated to the almost complete exclusion of the experimental development of small-scale planning maps. Up to three decades elapsed, for example, in the UK before environmental geology maps were commissioned and published.

This is not to deny that geological mapping for engineering purposes, particularly for town planning, continued to evolve very rapidly after the initial impetus from Czechoslovakia, Poland, the Federal Republic of Germany and the USSR, all environments in which the

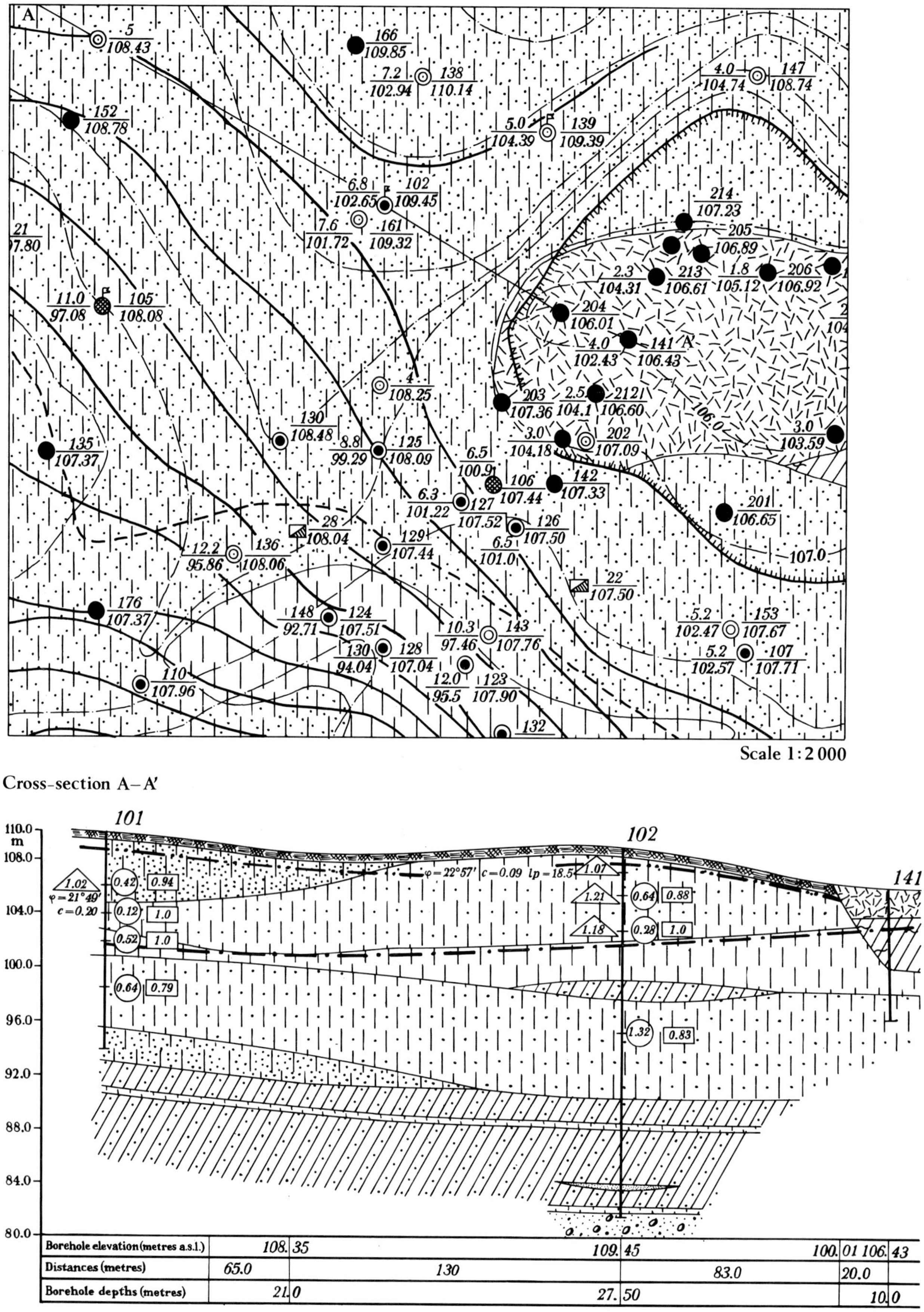

Figure 1.9 Large-scale plan and cross-section of the engineering geological conditions involved in the design of the foundation for an industrial plant in the USSR. Scale as reproduced here is 1:1550 approximately (legend shown opposite)
(Redrawn from Popov *et al.*, 1950)

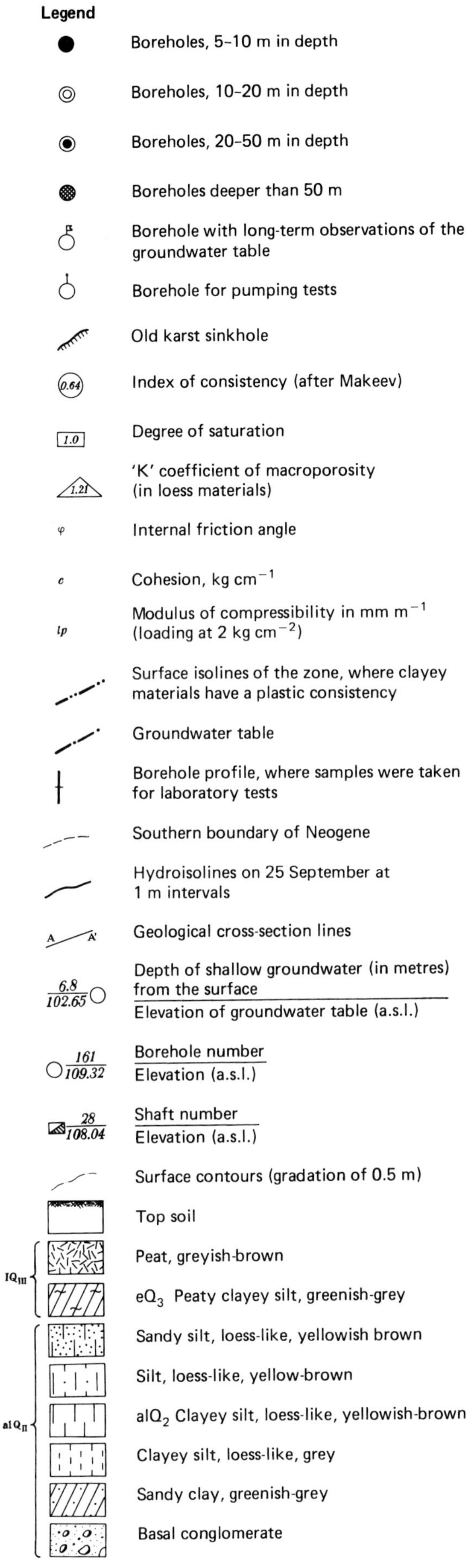

planning concept was paramount. In the early years, many of the cartographic ideas from Eastern Europe were taken up enthusiastically in Western Europe, with the exception of the UK. This early flirtation soon subsided into more conventional approaches to mapping.

As many examples of mapping for environmental planning mainly at medium scales, contrasting with the recording and interpretation of ground conditions and geotechnological treatment at a large scale, are used for illustrations of practice in later chapters, this part of the review will necessarily be in summary form.

The elegant strip or stripe cartographic method for three-dimensional representation was not generally accepted, even in Europe. In France, however, it was used for the first experimental engineering geological map and for a few subsequent maps (Peter, 1966).

In parallel with the stripe method, more or less conventional maps of engineering geological conditions continued to be prepared. Another important development was the concept of zoning maps, representing a major departure from geological maps of rocks and soils in the determination of areas on a map for which engineering geological conditions had a specific uniformity. It has been said, and it is worth emphasizing, that zoning maps are the only true engineering geological maps.

In the USSR, Popov *et al.* (1950) published *The Techniques of Compiling Engineering Geological Maps.* This book contains one of the first published examples of an engineering geological map made for the practical purpose of the engineering design of the foundations for an industrial plant. Their map of engineering geological conditions, reproduced here in black and white (Figure 1.9), had, in the original, a territorial division into engineering geological districts distinguished by colours. A district, in Russian terminology (Anon., 1976), is a zone in which hydrogeological conditions and geodynamic phenomena are uniform. This was a pioneering effort and the map, cross-section and legend are very close in concept to the basic principles of mapping and presentation of data adopted by the Mapping Commission of the International Associaton of Engineering Geology (Anon., 1976, p. 61).

The map is accompanied, in the original, by an axionometric three-dimensional diagram, tables of characteristics of the soil units on the map, and the characteristics of the engineering geological districts.

In the table of soil characteristics each distinguishable engineering type, shown on the map and cross-section (see Figure 1.9), is described in detail, in so far as the lithological character and engineering classification properties are concerned, according to Russian Building Standards. Both physical properties (grain size, specific gravity, bulk density, porosity, natural moisture content, degree of saturation, plasticity limits, consistency, collapsibility), and mechanical properties (angle of internal friction, cohesion, modulus of compressibility at 2 and 4 kg cm^{-2} loading) are listed, as shown in the map legend and on the cross-section.

The table of the characteristics of individual engineering geological districts contains data on the following: description of the lithological profile of foundation soils, hydrogeological conditions, geodynamic phenomena, collapsibility of materials, other special local characteristics, as well as engineering recommendations.

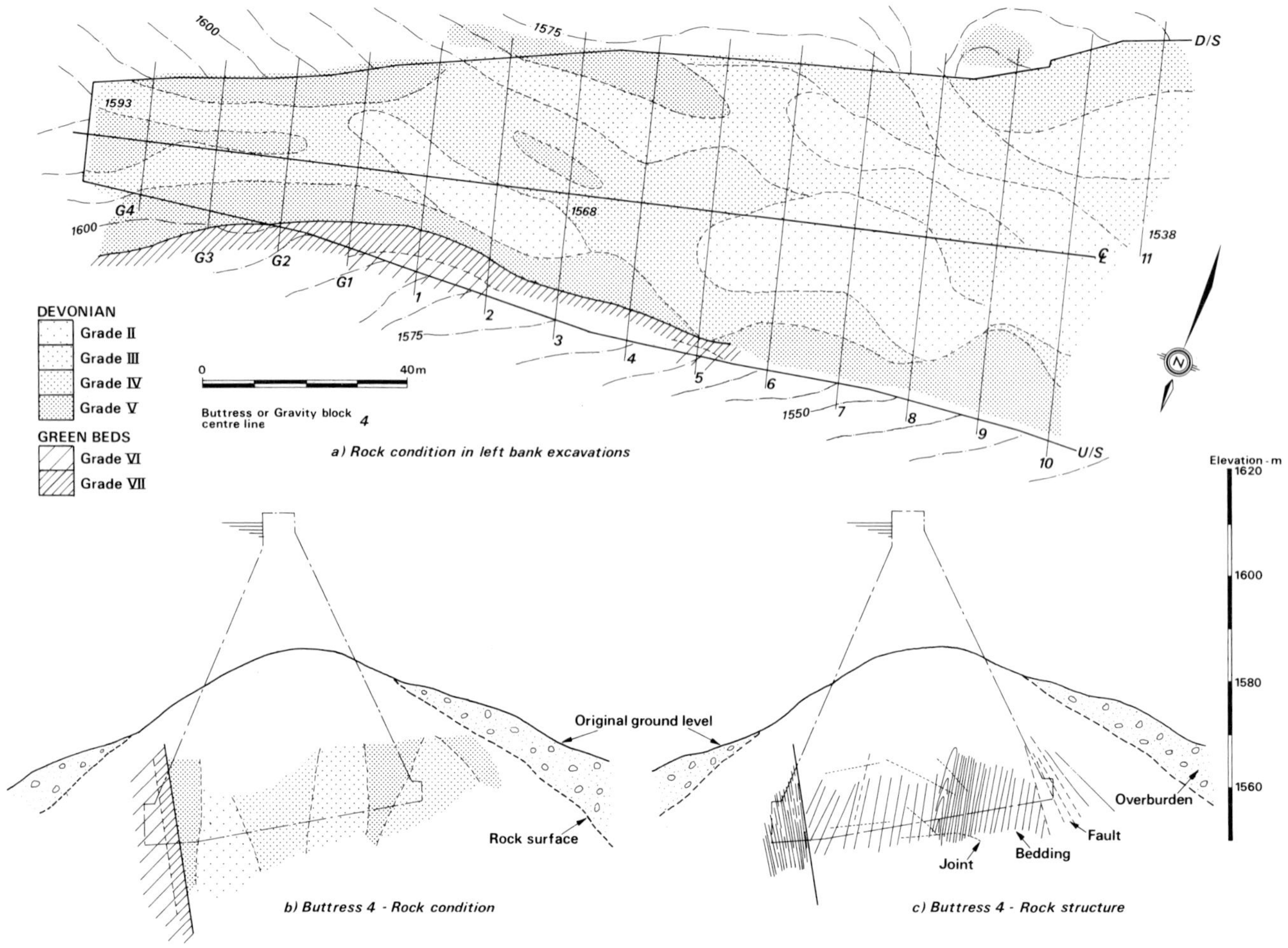

Figure 1.10 Engineering geological plans of the foundations of the left bank of the Latiyan dam, near Tehran: (a) variations in rock condition; (b) section through Buttress 4 illustrating rock condition; (c) rock structure (After Knill and Jones, 1965, Fig. 19)

In any discussion of mapping for engineering purposes in the UK after World War II, it is necessary to refer to 'British Practice' because of the important role that British engineers and engineering geologists played in major overseas engineering projects in addition to those at home. A whole generation of soil mechanicians and engineering geologists received their practical training on the Mangla Project in Pakistan (Binnie *et al.*, 1967), particularly in the detailed large-scale mapping of dam foundations, spillways and tunnels. True engineering geological maps appeared in published papers in Britain for the first time after 1960. A notable collection recording the geological conditions in the foundations of four dams was presented by Knill and Jones (1965). One example will suffice.

For the Latiyan dam near Tehran, Persia, seven grades of rock condition were established for the Devonian and Tertiary beds cropping out in the dam foundations. The grading classification was based on the assessment of a variety of geological characteristics, which may be summarized as:

- state of weathering and loss of cohesion;
- relative compactness of the rock mass;
- intensity and orientation of the various sets of fractures, namely bedding planes, joints and faults;
- relative cleanness of fractures and the rock mass as a whole;
- relative abundance of shale layers.

These characteristics, considered together, controlled the engineering behaviour of the rock. The established grades of rock condition were:

Grade I. Sound massive rock with no weak seams and widely spaced joints.
Grade II. Bedded, relatively solid rock, with some shale layers.
Grade III. Thinly bedded or flaggy rock, with some shale layers.
Grade IV. Blocky, seamy rock with frequent intercalations of shale and clay-silt. Some open joints are present.
Grade V. Broken, faulted rock or weathered rock mixed with shales and/or clay-shales found generally in a loose condition.

Additional grades were applied to Tertiary Green Beds, which were consistently in a poorer condition than the Devonian sandstones:

Grade VI. Thinly bedded rock with thin clay-shale seams.
Grade VII. Friable clay shales.

Distribution of rock grades and the rock structure in the left bank are shown in Figure 1.10. Knill and Jones pointed out that grade IV rock appears to represent grade II rock in a more weathered and fractured state. Because of the form of the excavations, it was possible to predict the distribution of rock grades in depth.

The approach adopted by Knill and Jones had been one of broad rock-mass characterization. This is without doubt a successful attempt at establishing homogeneous, mappable units which could be correlated generally with the results of *in situ* and laboratory design tests (Lane, 1964). In other words, rock groupings, based on engineering geological considerations, correlate with rock mechanics parameters.

By 1969 the full importance of true engineering geological maps had been recognized. Fookes (1969) had shown the shortcomings of the normal general-purpose geological map which, for the engineer, was lacking in quantitative information on the physical properties of rock or soil, the amount and type of discontinuities present, the extent of weathering and groundwater conditions. In addition, details of the superficial deposits, so important in engineering situations, were frequently omitted. He pointed out that the aim of engineering geological mapping should be to produce a map on which the mapped units are defined by engineering properties or behaviour, and the limits of the units are determined by changes of physical and mechanical properties. The boundaries of such mapped units, homogeneous in engineering terms, may well bear no relation to underlying geological structure or to the lithostratigraphical units usually shown on general geological maps.

Assessment of rock quality as a means of delimiting homogeneous mapping units required research into the engineering geological properties of rock material and the *in situ* rock mass. This had claimed attention in many parts of the world (Onadera, 1963; Deere and Miller, 1966; Deere, Merritt and Coon, 1969). At the Jari and Sukian dam sites on the Mangla Dam Project in West Pakistan (Fookes, 1967), examples are given on the plan scale of the correlation between various lithological layers exposed in pits and trenches and a variety of soil index properties.

In the long term it was envisaged that engineering mapping would be undertaken in terms of index properties and perhaps in terms of design parameters. However, in order to cover large mapping areas, rather than restricted construction sites, cheaply determined index properties would have to be accepted as a general guide to the likely range of a design parameter which normally could only be determined by expensive and elaborate *in situ* and laboratory tests.

Mapping in terms of geotechnical properties had already been tried in West Germany between 1951 and 1956, with emphasis on map units determined by settlement characteristics and allowable bearing pressures. For Gottingen, in 1956, Gwinner used map units with constant soil mechanics properties, even though different types of rocks and soils might be present in a single unit.

2

Principles of engineering geological mapping

2.1 Introduction

What should be shown on an engineering geological map depends largely on why the map is being produced, in other words its purpose. In turn, purpose controls the scale of the map, and scale dictates what can be shown and therefore the information that needs to be collected to compile the map. It must always be borne in mind that the purpose of engineering geology is to provide a background for the planning of land-use and for the planning, design, construction and maintenance of civil engineering works. The basic map information is needed to assess the feasibility of a proposed land-use, an environmental assessment or an engineering construction. Engineering geological research is therefore directed towards understanding the relationship of the geological environment to the engineering situation, and the modifications likely to arise from the engineering construction.

It is in the types of information that need to be shown, and in how the information is presented, that an engineering geological map differs from a conventional geological map. The best type of geological map will show, against the background of detailed topography, the distribution of bedrock and superficial deposits, their age, lithology and structure, together with any major tectonic features. A detailed legend, or explanation, with a stratigraphical column, cross-sections and a list of symbols used is an essential aid to map interpretation.

Mapping and map interpretation are basic geological skills that have to be acquired by practice and application. In general these skills are those needed to prepare engineering geological and environmental geological maps, although the emphasis, and therefore the content, of the maps are different. How different depends to a large extent on the scale of the map, with in general an emphasis on the engineering properties of soils and rocks, the nature and distribution of geodynamic processes, and hydrogeology. The amount of detail depends in large part on the scale of the map.

Maps may be made at any scale. Maps of small scale must inevitably deal with broad geological features such as tectonic regions, physiographic provinces, geological age, and the major genetic rock groups such as igneous, metamorphic and sedimentary. At larger scales more attention is paid to detailed lithologies and structures. Similarly there is a change in emphasis in the content of engineering geological maps with progression from small to large scales, that is from the general to the more specific. As a general rule: 'It is desirable to keep in mind the following important aim of engineering geological mapping, namely that the compilation of engineering geological maps should be based on the same general principles regardless of the fact that maps of different scales are designed to solve problems of different kinds. This would make it possible to compare directly maps drawn at the same scale for different areas; it would also facilitate the production of maps of medium to small-scale, for example, from maps of large-scale without the necessity of any radical changes' (Anon., 1976).

Even though the scale of the initial geological mapping may vary from 1:10 000 to 1:50 000 or even 1:250 000, more general geological maps at smaller scales are made by direct reduction and inevitable simplification of the largest scale maps available. There is a natural progression from the specific to the general, from maps of large to small scale. Quite the reverse tends to be the case with engineering geological maps; small-scale engineering geological maps are produced by direct transformation of geological maps of the same scale. For practical applications in engineering, more specific detail is usually required at larger scales. Such information has, in the main, to be acquired by an engineering geological survey directed specifically at the needs of the user, needs which may be perceived by the map-maker or requested by the eventual user.

Standard methods for preparing engineering and environmental geological maps will not be described here, for maps differ widely according to scale, purpose and the geological setting. Constraints of time, money and personnel tend to limit the scope even of maps prepared for research purposes. The way should be left open for developments in mapping techniques both in methods of data acquisition and in final presentation on the map sheet: '. . . creative, constructive, innovative thought . . . can be impaired by standardization of methods no matter how logical the standardization appears to be' (Varnes, 1974).

Despite this, certain broad principles can be established as a framework for the discussion of the general procedures to be followed in preparing engineering geological and environmental geological maps of all kinds:

- the map or group of maps must be suitable for the stated purpose;
- map scale must be matched to purpose;
- the content of the map must be suitable for the purpose and the scale;
- types of data used, and the methods for data collection, must adequately supply the information required by the map user.

Mapping cannot be undertaken without a suitable base map; base maps may be available at an appropriate scale, may be produced for enlargement from existing maps, or may be made specially.

What can be done depends to a large extent on the limitations of time and money already mentioned. There usually are such limitations, even for large government-sponsored projects such as the siting of a nuclear reactor where public safety is an overriding factor in the planning and design processes.

2.2 Purpose

Engineering geological mapping is usually motivated by an engineering purpose: planning for land-use in an urban environment, assessing the distribution of construction materials, selection of motorway alignments, or assessing the environmental impact of mineral developments. In detail, each of these maps will be different but the geological information will be focused on the purpose in mind. As will be illustrated later, the map may show areas where swelling potential in foundation materials may influence housing development; alternatively the map, or plan because of the large scale, may provide the information necessary to solve a problem of stability in a road-cutting. In both examples, the scale of the mapping would have been determined by the purpose of the map. Large scale may be needed for pinpoint assessments; small scale for general assessment of variations in foundation conditions

2.3 Scale

Map scale is more often than not directly related to the range of map scales conventionally used for topographic maps in a country. At larger scales, the scale selected for a particular purpose may be determined by engineering conventions. Some standard range of scales is desirable, and two established conventions need to be considered. The first is the simple distinction between maps and plans (Anon., 1972): (a) engineering geological maps at scales of 1:10 000 and smaller; (b) engineering geological plans at scales of 1:10 000 and larger.

This division conforms to the engineering convention that working drawings are referred to as plans, usually with a specified range: 1:10 000, 1:2500, 1:1250 (all available as Ordnance Survey Plans), and 1:500 and 1:100 for more detailed applications. At the map scale in the UK, the steps are 1:10 000, 1:25 000, 1:50 000 and 1:250 000 – all metricated versions of the old Imperial scales.

As an alternative, an international range of scales that are proposed by the UNESCO guidebook (Anon., 1976) is used here: 1:10 000 and larger (large-scale maps); 1:10 000–1:100 000 (medium-scale maps); 1:100 000 and smaller (small-scale maps).

It has to be accepted that maps at an unnecessarily large scale, i.e. those with empty enlargement, are a waste of time and money. Simple geological conditions can be dealt with at an appropriate small scale; records of engineering foundation conditions are rarely made at scales less than 1:500, and a scale of 1:100 may be necessary to show appropriate detail.

Maps at 1:100 000 scale and smaller are seldom used by engineers except at the feasibility stage of a project where a particular site or alignment has to be selected from a number of possibilities. They are therefore planning maps in which effort is directed towards avoiding problematical areas from the point of view of construction, or where land-use may directly damage the physical environment.

2.4 Classification and description of rocks and soils

The subject of classification and description of rocks and soils for engineering purposes is a large one, and has been exhaustively covered in several publications. Here, the principles of classification and description of rocks and soils are discussed at length in Chapter 4, although it is appropriate at this stage to give an outline.

Classification is based on four attributes: genesis, structure, predominant grain size, and mineralogy of the constituent grains. For soils, ease of disaggregation and simple standard index tests have, to a large extent, formalized classification based on the examination of disturbed samples. Rocks, on the other hand, are more difficult for the non-geologist and there is no substitute for adequate training. A plea is made for using a limited number of rock names (see Table 4.1), although geologists ignore this, and engineers and planners tend to accept esoteric names as distinctive tags for discussion. The names do not necessarily contribute to the understanding of engineering properties.

However, for effective description a rock or soil name is essential. Description has largely been formalized with attributes subdivided into those related to the material and those related to the mass. The order of the descriptive attributes is mainly a matter of personal or corporate preference, and includes for rocks:

ROCK NAME
  Supplementary petrographic properties
Rock material properties
  Colour
  Texture
    Grain size
    Other textural features and fabric
  State of weathering
    State of alteration
  Strength
Rock mass properties
  Structure
  Discontinuities
  Weathering profile

and for soils:

SOIL NAME
Including minor constituents
Genetic type
Soil material properties
Colour
Texture
Particle shape and composition
State of weathering
Strength
Consistency
Undrained shear strength
Moisture condition
Relative density
Compactness
Soil mass properties
Structure
Discontinuities
Weathering profile

Full descriptions of rocks and soils are reserved for the legend of small-scale maps, whereas large-scale plans may show the same rock type subdivided into separate map units on the basis of one or more attributes, for example, degree of weathering or jointing, or both.

The problems of deciding which attributes of soils and rocks should be shown on maps have been discussed by Varnes (1974). Ideally, an engineering geological map should show rock and soil units on the basis of uniform physical properties in terms of a spectrum of specified properties within a class. This is manifestly impracticable, even if only a small number of classified attributes is used. Attributes must be combined if the ground is to be mapped in terms of a small number of quasi-homogeneous units for each of which the range of engineering parameters is shown on the legend.

2.5 Classification of mapping units for engineering geological mapping

In conventional geological mapping the map units are recognizable lithostratigraphical units. Units have a distinctive lithology or association of lithologies and may be diachronous, that is from the point of view of geological age there may be transitions from one lithology to another. Relations of lithostratigraphical mapping units are shown in Figure 2.1 and there is an international hierarchy of terms.

On engineering geological maps, the classification of rocks and soils should be based on the principle that the physical or engineering geological properties of a rock, or soil, in its present state are dependent on the combined effects of mode of origin, subsequent diagenetic, metamorphic and tectonic history, and on weathering processes. This approach makes it possible to determine the reasons not only for their spatial distribution as shown on the map, but also the reasons for the lithological and physical characteristics of the rocks and soils involved. Uniform and continuous rock units, at different map scales, are based on the classification of individual rock samples, field observations and measurements.

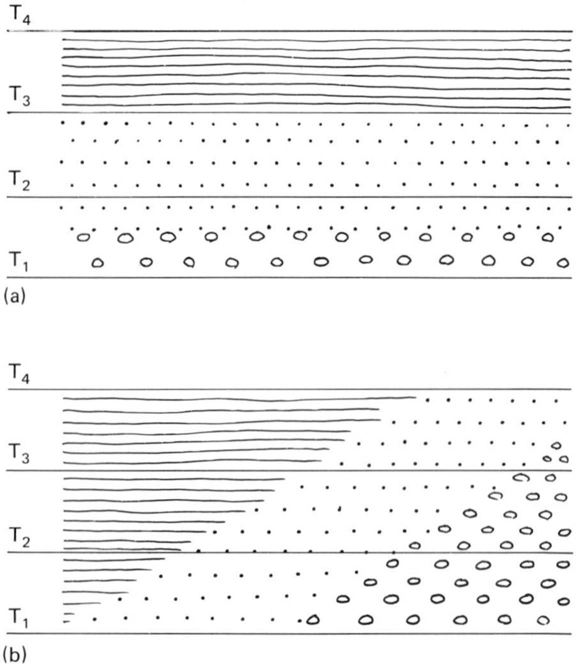

Figure 2.1 Lithostratigraphy and mapping units: (a) synchronous members and lithological types; (b) diachronous members and lithological types (T_1, T_2, T_3, T_4 are time intervals)

2.5.1 *Lithostratigraphical mapping units*

Taking British maps as an example, and for the time being restricting attention to sedimentary rocks, the map units are lithostratigraphical and may be diachronous, that is the age of the mapped units may change across the map. The conventional hierarchy of lithostratigraphical terms (Hedberg, 1972) used in the map explanation for the units shown on the map is as follows:

bed	=	named or unnamed individual layer;
member	=	named or unnamed lithological entity within a formation;
formation	=	fundamental unit of lithostratigraphy;
group	=	two or more formations

Most important is the concept of a formation consisting of a succession of strata useful for mapping or description. Most formations possess certain lithological features that may indicate genetic relationship. Ordinarily, the upper and lower boundaries of a formation are determined lithologically, but they may be unconformities or be determined by the occurrence of guide fossils. The age or time value of a formation is not necessarily the same wherever it is recognized (Anon., 1962). A formation should be conspicuously different from adjacent formations.

2.5.2 *Engineering geological map units*

In the UNESCO guidebook, four schematic maps and plans are used to illustrate the characteristics of mapping units depending on scale. Distinctive names, slightly modified here, are given to the mapping units:

Engineering type (ET)	UNESCO: engineering geological type

Lithological type (LT)	UNESCO: the same
Engineering formation (EF)	UNESCO: lithological complex
Engineering group (EG)	UNESCO: lithological suite

There will be different degrees of homogeneity for each unit, and it will be observed that engineering geological mapping is concerned, mainly, with the recognition of map units with the same engineering properties. Exceptions to this basic rule are to be found in the very detailed plans of engineering sites prepared at large scales, usually to depict and help in the resolution of engineering problems. The general characteristics of each mapping unit may be summarized as shown in the following subsections.

2.5.2.1 Engineering type (ET)

Engineering types can only be mapped at the largest plan scales, and at scales larger than 1:500 individual beds of rock or distinctive soil layers are commonly shown. Indeed, it is possible, and usually it is necessary, to map in even greater detail. An individual sandstone unit, which may be a single bed or a small number of beds, interbedded with mudstones, may be broken by a fault with resultant closer jointing within the fault zone (Figure 2.2a). A fault zone may provide easy access for groundwater with the possibility of localized chemical weathering. Thus a fresh sandstone with widely spaced jointing may be weathered to grade II (see Table 4.12) with closely spaced jointing; locally it could even be weathered to a grade V sand. Instead of sandstone as a single descriptive term, there will now be two or more distinctive types of sandstone each with its own physical and mechanical properties.

If the engineering setting so demands, a plan recording the face of an excavation in the sandstone will show all of the distinctive sandstone engineering types. As another example, where the foundation rock has been strongly affected by tropical chemical weathering, a plan of the weathering profile, for instance in a road-cutting, may show up to six different types of 'granite'. Each recognizable engineering type should be described separately using the schemes proposed (see Chapter 4) for the engineering description of soils and rocks and shown on the plan.

An *engineering type* is thus a subdivision within a lithological type which is uniform in its physical state. As a map unit, it may be discriminated, for example, on the basis of weathering grade, discontinuity frequency and pattern, strength or consistency. Each type will have distinctive geological and engineering properties. Scale of the mapping is greater than 1:5000 and usually much greater; at 1:100 very detailed plans can be made.

2.5.2.2 Lithological type (LT)

As the mapping scale becomes smaller, such geological detail within an individual bed or small group of beds of similar lithology cannot be recorded. For the sandstone described above, individual beds or groups of beds are mapped, regardless of variations in say jointing or weathering, as *lithological types* (Figure 2.2b). Typically the scale of such mapping is 1:5000 to 1:10 000.

A lithological type is homogeneous in terms of mineralogical composition, texture and structure. As a map unit it is discriminated on the basis of petrographic investigation.

2.5.2.3 Engineering formation (EF)

At scales of 1:10 000 and smaller, mapping is more generalized tending towards the production of regional engineering geological maps. Individual map units down to a thickness of 3 m can be plotted at 1 : 50 000 scale together with faults and general structural indications.

An *engineering formation* comprises a set of genetically related lithological types developed under specific palaeogeographical and geotectonic conditions. Within an engineering formation the spatial arrangement of lithological types is uniform in either lithological character or physical state. In consequence, it is not possible to define the physical and mechanical properties of the whole formation, but only to give data on the individual lithological types comprising it and to indicate the general behaviour of the whole. The engineering formation is used as a mapping unit on medium-scale (Figure 2.2c) and some small-scale maps, that is, mainly 1:10 000–1:200 000.

2.5.2.4 Engineering group (EG)

The *engineering group* comprises many engineering formations that developed under generally similar palaeogeographical and tectonic conditions. It has certain common lithological characteristics throughout which impart a general unity and serve to distinguish it from other engineering groups. Only very general engineering geological properties can be defined for the units that are only used on small-scale maps, smaller than 1:200 000 (Figure 2.2d).

2.6 Relation between engineering geological and lithostratigraphical map units

In engineering geology the mapping units do not need to be used with any stratigraphical implications; they are purely based on lithology. At large scales the engineering type, with no stratigraphical equivalent, is the only mapping unit that is exclusively engineering geological. A lithological type is equivalent to a lithostratigraphical bed or member, depending on whether one or more beds are involved. However, at small scales, apart from the omission of any indication of age in engineering geological mapping, there is no significant difference in the map units recognized as formations and groups either when used conventionally or in engineering geology.

A proposed nomenclature for mapping units for use in engineering geology, related to map scale, is given in Table 2.1.

2.7 Other aspects of mapping

Rocks and soils are basic to engineering geology, but, from the point of view of engineering, water has an almost equal importance. Evaluation of hydrological

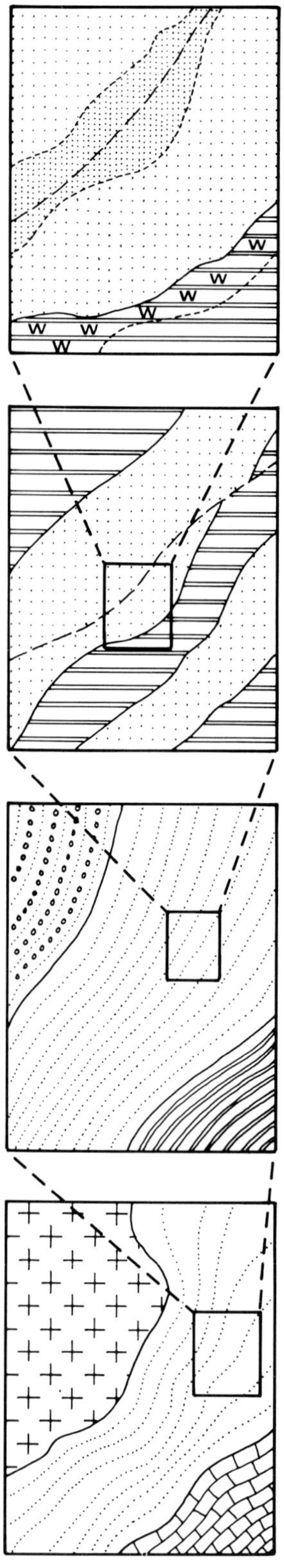

ENGINEERING TYPES

MUDSTONE, light greyish brown, fine to fine grained, thinly bedded, closely jointed, slightly weathered which slakes slowly on exposure, moderately weak

MUDSTONE, light greyish brown, fine to fine grained, extremely closely jointed, moderately to highly weathered, which crumbles in the fingers, very weak

SANDSTONE, dark brown coarse grained, medium to thickly bedded, with widely spaced joints, moderately to slightly weathered, strong

SANDSTONE, yellowish-brown, coarse grained, medium to thickly bedded, with closely to very closely spaced joints, sheared, highly weathered, weak

Scale 1:5000 and larger

Attribute of homogeneity: uniformity of physical state in each engineering type

Methods of delimiting map units: e.g. investigation of degree of weathering, determination of discontinuity frequency and pattern, strength, consistency

Map units characterized by determination of physical and mechanical properties

(a)

LITHOLOGICAL TYPES

MUDSTONE, light greyish brown, fine to very fine grained, thinly bedded

SANDSTONE, dark brown and yellowish brown, coarse grained, medium to thickly bedded

Scale 1:5000 to 1:10 000

Attribute of homogeneity: mineralogical composition, texture and structure

Methods of delimiting map units: petrographic investigation

Map units characterized by boring and sampling, geophysical testing, limited *in situ* testing, systematic laboratory testing

(b)

ENGINEERING FORMATIONS

Conglomerate formation. Littoral conglomerates, sandstones and subordinate mudstones (Lower Triassic)

Sandstone formation. Flyschoid calcareous sandstones, mudstones and subordinate siltstones (Lower Triassic)

Mudstone formation. Shallow neritic mudstones with subordinate sandstones and siltstones

Scale 1:10 000 to 1:200 000

Attribute of homogeneity: specific grouping of certain lithological types and their spatial arrangement

Methods of delimiting map units: areal mapping with facies analysis

Map units characterized by boring and sampling, geophysical and petrographical investigations, laboratory determination of index properties

(c)

ENGINEERING GROUPS

Granite group. Multiple granitic intrusion with remnants of schists comprising granite, amphibolite and gneiss

Clastic group. Conglomerate, sandstone and mudstone formations

Calcareous group. Limestone, dolomite and calcareous mudstone formations

Scale 1:200 000 and smaller

Attribute of homogeneity: specific grouping of certain engineering formations and their spatial arrangement

Methods of delimiting map units: existing geological maps; reconnaisance mapping; aerial photographs

Map units characterized by evaluation of existing data

(d)

Figure 2.2 Effect of scale on the basic units used in engineering geological mapping, and how the units are investigated and characterized (After Anon., 1976, Fig. I)

conditions is therefore paramount. Geomorphological mapping has been adopted with enthusiasm as a means of prediction of ground conditions during site investigation, including the assessment of materials for construction.

2.7.1 Hydrological conditions

Hydrological conditions, including surface hydrology, affect land-use, planning, site selection and the cost, durability and even the safety of engineering structures.

Table 2.1 Proposed nomenclature for engineering geological map units and their lithostratigraphical equivalents

Scale		*Nomenclature for engineering geology*	*Lithostratigraphical nomenclature*
Large	1:5000 and larger	Engineering type (ET)	No equivalent
	1:5000–1:10 000	Lithological type (LT)	Bed Member
Medium	1:10 000–1:100 000	Engineering formation (EF)	Formation
Small	1:100 000 and smaller	Engineering group (EG)	Group

Groundwater and surface water play an important part in present-day weathering, slope movements, active development of karst, volume changes by shrinking and swelling of susceptible soils, and the tunnel erosion and general collapse of loess. Rock and soil properties are largely dependent on their moisture content. Groundwater may influence methods of excavation and construction by flowing into excavations, and by producing seepage forces and uplift pressures. Both surface and underground waste disposal may be influenced by hydrogeological conditions.

Provision of hydrological information on maps facilitates prediction of undesirable changes in the hydrological regime and the recommendation of methods to avoid them. Therefore, on engineering geological maps it is necessary to assess and represent the following:

- the distribution of surface and subsurface water;
- infiltration conditions;
- water content;
- direction and velocity of groundwater flow;
- springs and seepages from water-bearing horizons;
- depth to watertable and its fluctuation range;
- regions of confined water and piezometric levels;
- hydrochemical properties such as pH, salinity, corrosiveness;
- presence of bacterial and other pollutants.

On small-scale maps, hydrogeological information is represented by point symbols and numbers. On medium-scale maps, the watertable can be represented by contours and its range of fluctuation indicated by numbers. In mountainous areas this is not feasible, and depth to watertable and other features can only be shown by numbers. On large-scale plans, hydrogeological conditions are represented by isohypses, isobaths and isopiestic lines, with known fluctuations shown numerically.

If much data are available, a separate map or plan is obviously desirable. Hydrological maps are produced by government agencies as a specialist product, and an international range of symbols has been published by UNESCO (Anon., 1970).

2.7.2 Geomorphological conditions

Geomorphological mapping is an essential part of engineering geological mapping, and indeed of conventional geological mapping, which can be carried out cheaply and quickly by suitably trained people. It is often a decisive factor in planning a site investigation.

Evaluation of geomorphological conditions is more than a simple description of surface topography. It should provide the basis for an explanation of:

- the relation between surface conditions and geology;
- the origin, age and development of individual geomorphological elements;
- the influence of geomorphological conditions on hydrology and geodynamic processes;
- the impending development of geomorphological features such as the erosion of river and lake banks, movement of dunes, collapse in karstic or undermined areas.

On medium- and large-scale maps and plans the actual boundaries and morphology of geomorphological features can be mapped. On small-scale maps, point symbols are used to show significant geomorphological elements. The base maps are contoured to show the surface topography.

2.7.3 Active geomorphological processes

Surface geological processes that may be active at the present time, or the legacy of formerly active processes, include:

- those due to erosion and deposition;
- aeolian processes;
- slope movements;
- permafrost;
- formation of karstic conditions;
- suffusion;
- volume changes in soils.

All these features, important in planning and engineering construction, can be shown on special-purpose or general-purpose comprehensive maps, and on topical maps. The amount of detail depends on the scale of the map.

2.7.4 Geodynamic conditions

Geodynamic phenomena result from so-called endogenetic processes operating in the Earth's crust at the present time that have a surface expression. Chief among these are seismic and volcanic activity, which may be associated, and neo-tectonic movements resulting in uplift or depression of the surface.

Seismic activity may be mapped in terms of the intensity of the event, using isoseismal lines, or surface expression of present-day and historical events. Mappable features include: offset streams, terraces and man-made structures; sag ponds; lines of springs; linear trenches and fault scarps.

The frequency and intensity of volcanic activity, together with the nature, location and extent of the volcanic products, are of importance in engineering geology.

2.7.5 Principles of engineering geological zoning

Zoning maps show individual areas which are approximately homogeneous in terms of engineering geological conditions. They are one type of derivative map in which the area covered by the map sheet is divided into a number of distinctive zoning units. 'Engineering geological conditions' is an all-embracing phrase for the geological system of rocks and soils, water, both passive and active geomorphological conditions, and geodynamic processes that may occur at an individual engineering site or in an area.

The concept was first formulated in Eastern Europe, and formalized into a classification system in which the following natural territorial units were recognized: *regions* – based on the uniformity of geotectonic structural elements; *areas* – on the basis of the uniformity of individual regional geomorphological units; *zones* – based on homogeneity of lithology and structural arrangement of rocks and soils; *districts* – in which hydrogeological conditions and geodynamic phenomena, including active geomorphological processes, are uniform.

Regions, areas and zones can be accepted in terms of homogeneity, whereas the term 'district' combines two distinct, unrelated geological aspects – hydrogeology and geodynamics – into what should by definition be a single natural territorial unit. Five units, and not four, are called for. At small map scales, and it has to be remembered that the classification has been proposed for general-purpose zoning on small-scale engineering geological maps, the divisions are firmly embedded in geological and geographical classifications of the Earth's surface. For example, physiographic provinces in geography are equivalent to the 'areas' of the proposed classification.

It would be preferable to retain well-established descriptive terms for the proposed divisions, if only because at small scales the differences between the geological and engineering geological approach are minimal, if they exist at all. An alternative would be to call all the units zones, or engineering zones if such emphasis is felt to be necessary, for example geotectonic zone, lithological zone. Zone when applied to rocks and soils has, of course, very precise geological implications, making the use of 'engineering zone' preferable.

Nevertheless, the zoning concept is both important and useful in engineering geological mapping, and environmental geology mapping, at both small and large scales. A map of special-purpose engineering geological zoning might be prepared with a particular type of engineering activity in mind, for example, roads, dams, tunnels. On such a map the zoning units would be based on the analysis of geological phenomena and on geotechnical parameters, and evaluated in terms of a particular engineering purpose.

Examples of zoning maps, of various kinds, are given in later chapters.

2.7.6 Symbols for engineering geological maps

Apart from the usual cartographic symbols, the following may be shown by distinctive symbols on engineering geological maps. A general list will suffice here, as full details are given in Chapter 5. The aspects are: geology; hydrogeology; geomorphology and active processes; seismicity and volcanicity; boreholes; site investigations and tests; mines and quarries.

2.8 Types of map

For engineering and environmental purposes the types of map that may be prepared are many and varied. It is desirable to categorize them in some way, as in the UNESCO guidebook (Anon., 1976) where kinds of map are determined by the criteria of purpose, content and scale. With some slight modifications to the nomenclature, the interrelationships are as shown in Figure 2.3.

Comprehensive maps, as the name implies, attempt to show everything of relevance in the engineering geological environment, which is only possible at small and medium scales. Maps dealing with only one component of the geological environment are specialized in content and here are called 'thematic' because they cover only one aspect such as landslides. Specialized maps are likely to be compiled at medium and large scales.

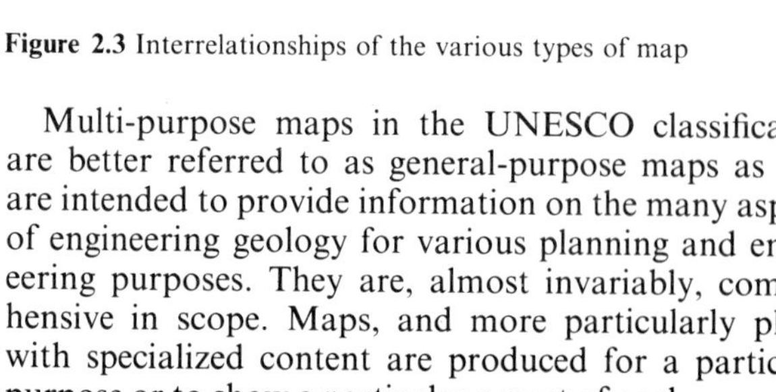

Figure 2.3 Interrelationships of the various types of map

Multi-purpose maps in the UNESCO classification are better referred to as general-purpose maps as they are intended to provide information on the many aspects of engineering geology for various planning and engineering purposes. They are, almost invariably, comprehensive in scope. Maps, and more particularly plans, with specialized content are produced for a particular purpose or to show a particular aspect of geology – they may be thematic.

3

How to make engineering geological maps

3.1 Introduction

In the previous chapter, four principles were laid down for compiling engineering geological and environmental geological maps (see page 13); these can be used as a checklist for the steps that have to be followed. Of the utmost importance is the initial decision on what types of data should be collected and what methods should be used in collecting them.

The lithological units that can be used on maps of different scales have already been illustrated schematically in Figure 2.2. For each map, a summary is given in the diagram of: (a) what determines the lithological map unit, that is what factors are used to characterize each map unit which is supposedly homogeneous or at best quasi-homogeneous; (b) how the boundaries are drawn; (c) how the map units are characterized.

Information on rock and soil types, whether they be engineering or lithological types, or engineering formations or groups, depending on map scale, is only one aspect of the engineering geological environment shown on maps and plans for engineering or planning purposes.

The purpose of the map or maps must be held firmly in mind while making decisions regarding scale, type and number of data to be gathered. If the purpose of the map does not guide these decisions, the result will be wasted effort, time and money, and more than likely an inadequate and unsatisfactory final product.

3.1.1 Scale

Selection of an appropriate scale will depend on the purpose for which the map is intended and the amount of detail that has to be shown. Plan scales follow the conventions of the civil engineering industry, with 1:500 commonly used for working layouts of a scheme, and 1:100 or 1:50 for critical detailed design and recording. Practice over many decades has firmly established these scales as the most convenient for purpose, although 1:10 000 does set a practical limit for site work. The same is not true for general geological mapping from which engineering and environmental geological maps are usually derived. At one limit of the medium-scale range, 1:10 000 base maps, if available, are ideal for basic mapping in as much detail as is usually afforded by the ground. From maps of this scale, successive reductions and consequent generalization can encompass the medium- and small-scale range of maps. The eventual scale selected for a map will be related to the purpose for which the map is intended, and what base maps are available.

Topographical maps may be enlarged, although it is obvious that no additional detail appears on the map. Such change of scale is sometimes referred to as 'empty enlargement'; however, the enlargement will have been done for a practical purpose. Usually this would have been to provide a suitable base for recording local site detail such as the location of site investigation boreholes, pits and trenches. Enlargement, or reduction for that matter, is usually available on a commercial basis to predetermined scales or special scales on request. Whatever scale is most useful and appropriate should be selected for each individual project.

3.1.2 Type and number of maps

Before primary field mapping is undertaken to compile an engineering geological map, or as is more usually the case an existing geological map is to be interpreted, and possibly supplemented by additional field work, a decision has to be made on the number of maps to be prepared (see page 47). This is a very important initial step that will save both time and money, and should ensure that the content of the maps, or plans, is suitable for the purpose specified by the client. The client for the mapping may be a firm of consulting engineers which has to work within financial constraints in turn imposed by their client, or planners in local government, and so on.

The methods of compiling engineering geological maps on base maps do not differ from the methods used for conventional geological mapping. There is no need to describe the methods here as they are well known. A summary account appears in section 3.2.2.

3.2 Data collection

3.2.1 Introduction

Data collection is an essential part of what is usually called a 'desk study'. Such initial investigations should reveal whether or not an area has been mapped before, and if so what maps, memoirs and references in the literature are available. If a conventional geological map has been published at an appropriate scale, then all that is required is to collect necessary engineering geological data. The latter almost certainly would involve some field work and probably laboratory testing. However, in many parts of the world, geological maps are not available or are not complete enough to provide the required background data for an engineering geological evaluation. Then, of course, primary mapping is needed but may be focused quite sharply on the engineering geological requirements for a regional assessment or an actual construction site.

3.2.2 Methods of conventional geological mapping

Methods of conventional geological mapping, at various scales, are dealt with by Lahee (1961), Low (1957) and Barnes (1981). Techniques are best learned by training in the field given by an experienced geological mapper. Lahee gives a very comprehensive account to which reference can be made to advantage; here only some significant points will be mentioned. Geological field work usually involves:

- the study and interpretation of rocks and soils, topographic forms, and significant geomorphological and other processes;
- the determination of the location of points or outcrops where observations are made;
- the plotting of these points or outcrops, and of other geological data, on a map.

From the map it should be possible to infer:

- the relations between superficial deposits and bedrock;
- the succession of different engineering soil types in the superficial deposits;
- the succession and structural relations of the lithological types forming the bedrock.

These can be illustrated by:

- field sketches or associated photographs of significant features of the landscape, or individual exposures;
- carefully constructed horizontal sections, drawn with the same horizontal and vertical scales, or with the minimum of vertical exaggeration needed for drafting purposes;
- block diagrams, using the isometric or other suitable projections.

This is not the place for a detailed account of geological mapping, which is exhaustively discussed, for example, by Lahee (1961).

3.2.2.1 Field mapping in the UK

Official mapping, and the training in geological mapping given by universities and colleges, is based on the highly detailed plans at a scale of 1:10 000 produced by the Ordnance Survey. There is usually no difficulty in locating points on the map, except in upland or mountainous areas where man-made features are limited in number.

Mapping is by exposure supplemented by surface soils, topographic features, vegetation changes and springs and seepages. The extent of each exposure is marked to scale on the plan – remembering that at 1:10 000 a 10 m square will be a 1 mm square on the plan, in fact a large dot. Groups of small exposures separated by areas with thin soil cover can be mapped as a large outcrop. Exposures are outlined by a fine coloured line and the rock or soil type marked by coloured pencil.

An exposure map provides the factual information on which the interpretation of the geology is based (Figure 3.1). The completed field plan should show the boundaries between superficial and solid formations, and boundaries in the solid formations. Where a boundary is exposed, a continuous line is used; elsewhere, a dashed line is the convention. There are also conventional ways of marking boundaries, and these as well as structural, geomorphological and other symbols are listed and illustrated in Chapter 5. They should, as far as possible, be adhered to and illustrated in the legend to the plan or map.

3.2.3 Three-dimensional representation

From basic geological information shown on the map, in conjunction with surface topographic contours, the geologist at an early state in his training learns to draw horizontal sections across the map. The outcrop pattern and its interaction with topography, structural data measured in exposures, thickness of formations either estimated or measured, are used to construct the sections at either natural or exaggerated vertical scale. Sections are usually drawn at right angles to the general strike of sedimentary formations, and the positions of all sections are shown on the map with each end clearly indicated.

Instructions for constructing cross-sections, and other methods of three-dimensional representation – columnar sections, isometric and oblique projections, block diagrams and fence diagrams – are given by Barnes (1981) and in greater detail by Lahee (1961).

3.2.4 Types of data to be collected for engineering geological maps

After the purpose, scale, type and number of maps and base maps have been decided upon for a given project, it is then necessary to decide what types of data must be collected. The data required will be outlined below for maps at different scales, starting with general-purpose medium-scale maps as they embody most of the principles to be followed in compiling engineering geological maps.

Ideally, information on the following aspects has to be assembled for a general-purpose engineering geological

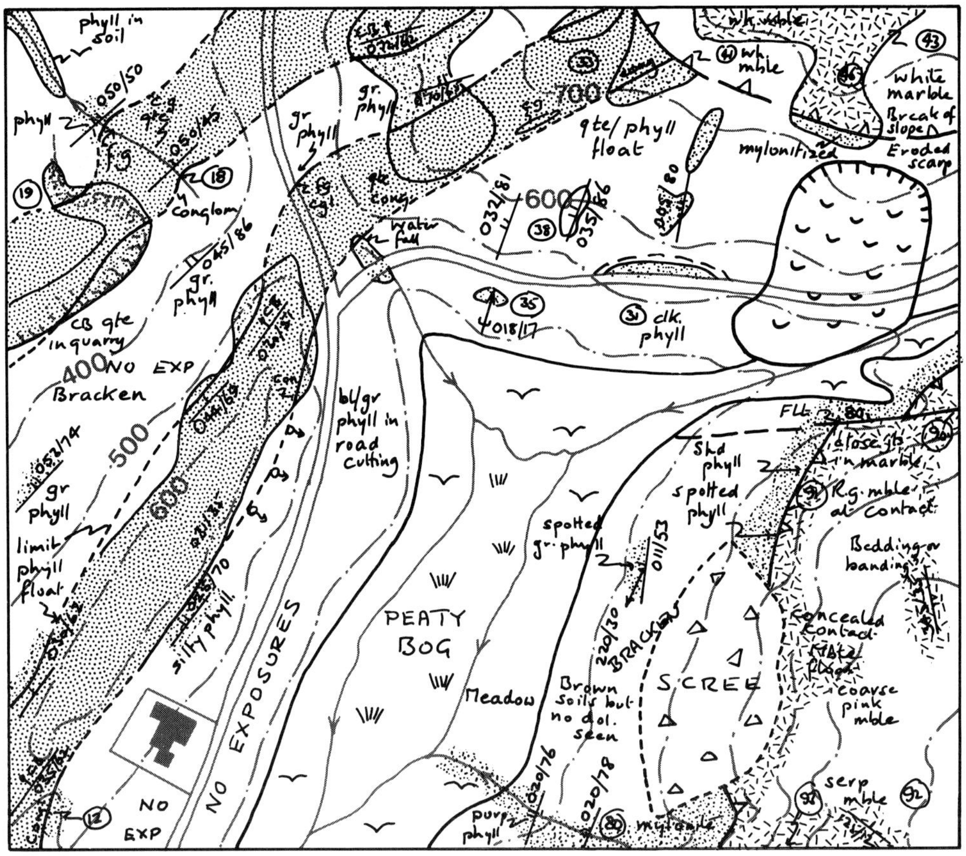

Figure 3.1 Part of an inked-in field slip showing various methods of geological mapping at the 1:10 000 scale (Redrawn from Barnes, 1981, Plate 4.1)

map (see subsection 3.2.4.1 for explanation of * and + symbols):

Geological details and location of boreholes and mine shafts
* Lithology, rocks and soils
Physical properties of rocks and soils
* Structure
Subsurface conditions
Discontinuities
Evidence of active geomorphological processes and geodynamic processes
Karst
\+ Landslides
Subsidence
\+ Active faults
\+ Active volcanism
Seismicity and evidence of recent seismicity
Erosion
Weathering
Evidence of permafrost
Geomorphological features
Hydrogeological features

3.2.4.1 Medium-scale general-purpose maps

As a first step in assembling the necessary data for the map, all pertinent literature, whether geological or not, should be examined to recover as much usable information as possible. General lithology and structure of an area may be on existing geological maps. Historical records and other public documents may give additional information useful for compiling an engineering geological map, such as information on historic earthquakes or landslides, locations of now concealed fortifications, locations of former shorelines or former breakwaters. Construction records may provide data on physical properties of rocks and soils, subsurface data, excavation methods for various rock types, or support methods in tunnels.

If there is no geological map of the area, then one must be compiled, using standard geological mapping methods (see section 3.2.2). One of the quickest ways to obtain preliminary information, as well as data on easily recognizable features such as landslides, erosion patterns and physiographic characteristics, is by examination of stereo-pairs of aerial photographs, generally at scales of between 1:10 000 and 1:50 000. High resolution high altitude photographs can also be useful.

Not all of the aspects listed are likely to occur in the area of a single map sheet, and many of them that do are best dealt with in a descriptive memoir. At the median scale of say 1:50 000 for medium-scale maps, those marked with an asterisk (*) are essential, for without them the map would not be geological. Features which it would be desirable to show are large landslides, active faults to supplement the structural information and active volcanism. These are denoted by a cross (+). In some circumstances, isopachytes on selected superficial deposits and contours on rockhead can be shown without cluttering up the map. Most of the remaining aspects usually demand a separate map, additional small maps in the legend, or description in the map explanation or accompanying memoir. Examples are given in Chapter 10.

Medium-scale maps, therefore, are prepared by field investigations supported by archival studies. The detail that can be shown is determined by the cartographical problems associated with map scale. In deciding on the depth to which the engineering geological conditions should be shown, the main purposes of the map as well as the complexity of the geology will have to be taken into account.

Collection of data involves representing information on the base map. At medium scale, this implies mapping in terms of engineering formations and drawing on the map their boundaries using the conventional methods of indicating degrees of certainty with which the boundaries have been determined. Continuity of map units may be broken or displaced by faults which can be mapped. Sufficient structural data should be plotted on the map to enable interpretation of the general structure. Usually, landslides, active or dormant, are outlined and the area marked by a distinctive symbol. Significant areas of man-made ground, including infilled quarries, embankments and reclaimed areas along major rivers and estuaries, mine dumps and reclaimed opencast mines, are usually shown.

Lithology, together with physical properties of rocks and soils of the various mapped formations, are indicated on the legend. Descriptions of lithological types in the approved form would have been made in the course of the field study, backed up by simple field tests for strength and consistency.

Since 1860, 1:10 560 maps of economically important areas were produced by the basic geological survey of the UK. Where these maps were available, medium-scale 1:63 360 (now 1:50 000) geological maps were compiled by direct reduction. In other countries, for example the USA, the basic mapping scale is 1:24 000. Maps of selected areas in the UK, usually of classic geological localities and scenic areas important for recreation, are published at 1:25 000; these are also direct reductions from the 1:10 000 maps, as indeed are the background topographic sheets.

3.2.4.2 Large-scale general-purpose maps

Here emphasis is on maps, usually at a scale of 1:10 000, in other words at the lower limit of the scale range for this category, made of a whole area of interest and not site specific. Large-scale maps are prepared by detailed field investigations and mapping using conventional geological survey techniques, except that the engineering geological descriptions will be extended to both lithological and engineering rock and soil types, as appropriate. Detailed field investigations and mapping are supplemented by using all existing archival material, systematic subsurface exploration by boring, pitting and trenching, geophysical investigation, and field and laboratory testing. In urban areas, sufficient information may be available in site investigation reports to form the main basis for mapping, and consequently only selective field work may be necessary to complete the mapping.

At a large scale, all those aspects of the geological environment listed in section 3.2.4 can be mapped in considerable detail. For example, all the morphological aspects of landslides can be shown by distinctive symbols, and indications given of relative activity within a landslide complex.

3.2.4.3 Small-scale general-purpose maps

Small-scale maps of areas in which the geology is well known can be compiled from available maps, literature and archival sources. For poorly investigated areas, photogeological interpretation and other aspects of terrain analysis may be supplemented by reconnaissance mapping.

Map units are usually engineering groups of rocks and soils and, apart from a general absence of stratigraphical information, a small-scale engineering geological map hardly differs in any relevant aspect from the usual geological map. Surface topography is shown by contours, and significant geomorphological features by point symbols. Point symbols may also be used to show the location of significant geodynamic processes, and seismic activity by isoseisms.

3.2.5 Documentation maps

A documentation map provides a record of the sources of information used in compiling an engineering geological map. The documentation map should be drawn at the same scale as the associated engineering geological map, and only if there is no danger of overcrowding the geological map should the two be combined.

The information on a documentation map includes symbols showing, for example, the position, depth and type of individual boreholes; observed or investigated exposures, pits, trenches, quarries and other excavations; and wells and springs. Drill holes should be indicated in which *in situ* tests and geophysical logging have been done; from which samples for laboratory testing were taken; which were tested by pumping in or pumping out tests, as well as those used for observation of the underground water regime. The location of sites used for engineering field experiments and geophysical experiments should also be indicated.

Information related to past investigations, frequently available in considerable detail from site investigation reports, photogeology and archives, should be outlined.

All the symbols used (see Chapter 5) should be explained on the map legend and the symbols standardized as far as possible.

3.2.6 Methods of data collection

Apart from the general methods of geological survey in the field, acquisition of data may be aided by the following techniques: photogeology and other methods of remote sensing; terrain evaluation; geomorphological mapping; geophysical methods on land and on the sea bed; boring, drilling and sampling; laboratory and field testing; hydrogeological investigations, including instrumented drill holes; detailed mapping of excavations; economic evaluations; and land-use capability analyses based on geological data.

These methods are best illustrated by examples of current practice as given in subsequent chapters. It is not the intention here to discuss, for example, how boring, drilling and sampling are done, or the theoretical background to photogeology or geophysical techniques. A guide to appropriate literature will be given later; however, the results obtainable by applying simple or sophisticated techniques are important, particularly to show the situations in which they can be of use, often considerable use, in engineering geological mapping.

3.3 Derivative maps and plans

In practical applications, engineering geological data are required for a specific purpose. Such purposes may include: prediction of potential hazard and assessment of risk; economic feasibility of projected developments. Potential hazards depend on the nature of the terrain and the background geological conditions. They may be summarized as being genetically associated with landslides, seismic activity, volcanic activity, karst and subsidence, collapsible soils, shrinking and swelling soils, and permafrost.

Economic feasibility can generally be held to be associated with land capability in one form or another. Analyses of ground conditions are usually associated with estimates of go, caution, and no go situations, and zoning maps with these three, or more, subdivisions may be prepared. Examples are given in the appropriate later chapters.

3.4 Mapping techniques for site plans

Plans may be drawn at the site investigation stage or the construction stage of a project, and there is an important difference in the mapping techniques available at each stage. During site investigation every attempt is made to interpret the ground conditions in as much detail as is thought necessary. Within financial constraints that are usually imposed, if boreholes, down-the-hole investigations, pits, trenches and other surface investigations are needed to expand information from surface mapping – geological or geomorphological or both – then they are provided, mapped and logged. A balance has to be achieved on the effort expended to understand the site sufficiently to proceed from the feasibility to the design stage of a project.

The situation is quite different at the construction stage, where complete exposure is achieved, albeit not necessarily all at the same time. As much detail can be recorded as is felt necessary, the construction programme permitting.

Mapping is at large scale, using techniques only rarely applied in pure geological mapping. The main requisite is a formal base on which to carry out the mapping; the base may be a photograph, a scaled baseline or a surveyed grid. With these, it is possible to record in considerable detail: the walls and floor of a site investigation pit or trench; a rock face, either natural or excavated; the base and walls of a major excavation; a tunnel or a large underground cavern. Soils and rocks can be mapped with equal ease.

Here, concern is only with techniques. However, apart from the general statement above, techniques are best discussed and illustrated when plan-scale mapping is dealt with (see Chapter 11).

4

Rock and soil description and classification for engineering geological mapping

4.1 Introduction

In the ordinary sense of the word, 'rock' refers to those consolidated masses of mineral material which enter into the composition of the Earth's crust. 'The term is used by geologists not only for... hard substances... but also for sands, clays, etc.' (De la Beche, 1832). The reason for this is clear: '... there is often so insensible a passage from a soft and incoherent state to that of stone, that geologists of all countries have found it indispensable to have one technical term to include both' (Lyell, 1852).

A modern definition (Gary, McAfee and Wolfe, 1972) of rock is: 'Any naturally formed, consolidated or unconsolidated material (but not soil) composed of two or more minerals, or occasionally of one mineral, and having some degree of chemical or mineralogic constancy.' Here 'soil' is evidently used in the pedological sense.

Application of the term 'soil' in the engineering sense is related to the development of the science of soil mechanics: the uncemented sediments such as sands, clays and silts are known to the engineer as 'soils' or 'earth', in contrast to the more rigid limestones, sandstones, granites and other 'rocks'. It is only recently that our knowledge of their behaviour in relation to engineering problems has been placed on a reasonably scientific basis. This is due in large part to Terzaghi, who brought laboratory experiments, theory and field observations into an integral philosophy which was first expressed in the celebrated *Erdbaumechanik*, published in 1925.

Since that time, rocks have been studied in the same way in relation to engineering problems, and 'rock mechanics' is now firmly established. The two disciplines, sharing similar theoretical and experimental approaches, meet in the transition from soil to rock, where the geological materials may be called hard soils or soft (semi-solid) rocks.

Apart from this overlap, in engineering there is a clear distinction to be made between the behaviour of the two materials. The following quotation (Anon., 1972) may be used as a guide to the distinction between soil and rock; 'Soil is an aggregate of mineral grains that can be separated by such gentle means as agitation in water. Rock, on the other hand, is a natural aggregate of minerals connected by strong and permanent cohesive forces. Since the terms "strong" and "permanent" are subject to different interpretations, the boundary between soil and rock is necessarily an arbitrary one' (Terzaghi and Peck, 1967, p. 4).

Apart from the uncemented sediments already mentioned, hard rocks may be partially or completely converted to engineering soils by weathering. For engineering appraisals, such intact mixtures of rock and soil, in which the original geological textures and structures are largely retained, may be classified either as rock or soil depending on whether the one or the other material determines engineering behaviour. Degree and openness of jointing is another factor.

Particularly relevant to a practical definition of soil and rock are assessments of ease of excavation: rock which can be removed by heavy earth-moving equipment tends to be defined by engineers and particularly contractors as 'soil' and they bid on this basis.

Description of geological materials, both in terms of those aspects which can be determined in the hand specimen or small sample and those properties which can only be assessed in a natural exposure or in the walls and floor of an excavation, is a necessary prerequisite to assigning to it a geological name. A general name, for example 'granite' rather than say 'unakite' – a variety of pink and green granite from the Unaka range in North Carolina – places the material in the correct genetic group of a geological classification, with implications for mode of origin, grain size and other textural aspects, and structure.

Although it is clear that a sandstone and sand will have different physical and mechanical properties, the values for these properties cannot be deduced from the name alone. A sand may be loose or dense; a sandstone may be weakly or strongly cemented with important consequences in terms of strength or durability. Sandstone weathers to sand, so that the two may be intimately associated. In just the same way, a granite may be fresh and unequivocally rock, or extremely weathered when equally unequivocally it would be classified as an engineering soil. Rock properties in engineering terms cannot be inferred either from a geological name or from the usual type of geological description.

On conventional geological maps, where the map units have been determined on a lithostratigraphical basis, the name of a unit may be misleading if a standard soil name is used. An often quoted example from the UK is 'London Clay' which is principally clay but includes some sands and silts (Burnett and Fookes, 1974).

What is needed for rock, long available for engineering soils, is a method of description that discriminates between two types, or two states, of the same rock while at the same time allowing probable engineering properties to be inferred. This type of approach is unlike the usually accepted geological description found, for example, on 1:50 000 scale maps.

On the explanation, or legend, of a geological map, a clear distinction is made between map units of different ages. In detail, treatment is different from country to country: there may be a stratigraphic column with, for each formation or member, a detailed description including lithological details and minor variations, mode of origin, thickness. A typical example (Williams and Hackman, 1971) of U.S. Geological Survey practice is:

Morrison Formation

Variegated fluvial and lacustrine mudstone, siltstone and sandstone; local thin limestone beds. 0 to 750 feet thick.

Jmb, Brushy Basin Shale Member; light red, purple, and greyish-green bentonitic lacustrine mudstone; minor limestone lenses and concretions; minor white, grey and buff crossbedded sandstone lenses; minor but distinctive conglomerate lenses containing multicoloured chert pebbles. 0 to 400 feet thick, increasing from west to northeast.

Jms, Salt Wash Sandstone Member; light grey, tan and white fluvial, thickly crossbedded fine- to coarse-grained lenticular sandstone containing scattered pebbles and thin conglomerate beds; contains interbeds of variegated sandy mudstone. Northeast of the Henry Mountains as much as 50 feet of massive gypsum occurs at the base. 0 to 400 feet thick, increasing from west to southeast.

When the geological map is used as the basis for a series of derivative maps, a map showing types of bedrock and superficial deposits may be prepared on which lithology and not lithostratigraphy determines the selection of map units. As an example, the two members of the Morrison Formation described above, which in places cannot be distinguished from immediately overlying Cedar Mountain Formation, are described as:

SANDSTONE AND CONGLOMERATIC SANDSTONE—
Massive to thin-bedded; contains 10–35 percent shale, siltstone and limestone.

The combined formations are:

SHALE CONTAINING 10–35 PERCENT SANDSTONE AND CONGLOMERATIC SANDSTONE—
In some places amounts of siltstone, gypsum and limestone.

In the UK, bedrock (the so-called 'Solid' formations) and superficial deposits (Drift) are clearly differentiated. This division has a background in the early history of geology. Murchison (1839) introduced 'drift' for what was then called 'diluvium', preferring it to the latter term, which implied a flood, if not The Flood, as he thought all this material had drifted in marine currents and accumulated under the sea in comparatively recent times.

It is now known that practically the whole of Murchison's drift was formed as a result of glacial conditions; glacial drift includes all glacial and glaciofluvial deposits which are referable directly or indirectly to ice. The 'Solid' formations end with the deposits of the Pliocene System. All the later deposits, whether belonging to the Pleistocene System or even later, are called 'Drifts'.

The 'drift' editions of many geological survey maps include river alluvium and other non-glacial superficial deposits not shown on the corresponding 'solid' editions.

In many parts of the world, superficial deposits in the sense used above are absent and there is a transition downwards from a pedological soil, through soil formed in the weathering zone, into fresh rock beneath. On geological maps, the soil which has developed in place as a result of weathering is usually taken as part of that rock; for engineering purposes the distinction is of the greatest importance and both soil and rock would be indicated on maps.

4.2 Early developments in rock description and classification

In the early days of rock mechanics the need for a geological, or petrographic, name for rock material was largely discounted. Emphasis was placed on rock properties as a basis for classification (Coates, 1964), although it had earlier been argued that a rock name, provided that it was simple but accurate, gave useful information for the engineer. Once it becomes apparent that a 'strong, elastic, brittle, layered, blocky rock' is in fact a limestone and not a sandstone, that simple field name adds relevant information (Coates and Parsons, 1966) and the guidance of a name indicating general composition clearly is useful.

The original impetus to standardize rock description appears to have come from Deere (1963) in his prescription of a method of describing, technically, rock cores for engineering purposes. This was followed by work on the classification of intact rock, the rock material (Deere and Miller, 1966) and later *in situ* rock, the rock mass (Deere, Merritt and Coon, 1969).

A geological rock name is assigned on the basis of mineralogical composition, texture and mode of origin, and how to do it can only be learned by training and experience. Deere (1963) suggested that this lithological description should be supplemented by descriptive terms for grain size, colour and minor constituents. He also went further than this and suggested that, for rocks, an 'index property', possibly correlative with other more important mechanical properties, might be provided by a quantitative, numerical 'hardness' value in conjunction with a qualitative textural classification.

Texture may be equated with fabric – the way in which mineral grains in the rock are arranged and bound together. Deere recognized three textural types:

- *interlocking texture* shown by granite, basalt and other igneous rocks, massive metamorphic rocks including quartzite for example, and among the sediments, rock salt, limestone and dolostone;
- *cemented texture* of sandstone, conglomerate, some limestone, tuffs;
- *laminated–foliated texture* for rocks which contain minerals with a preferred orientation, such as shale, slate, phylitte, schist and gneiss. Such rocks have an obvious anisotropy even in hand specimens.

Diagrammatically shown in Figure 4.1, the three types of texture are also illustrated by photomicrographs of thin slices of rock in Figure 4.2.

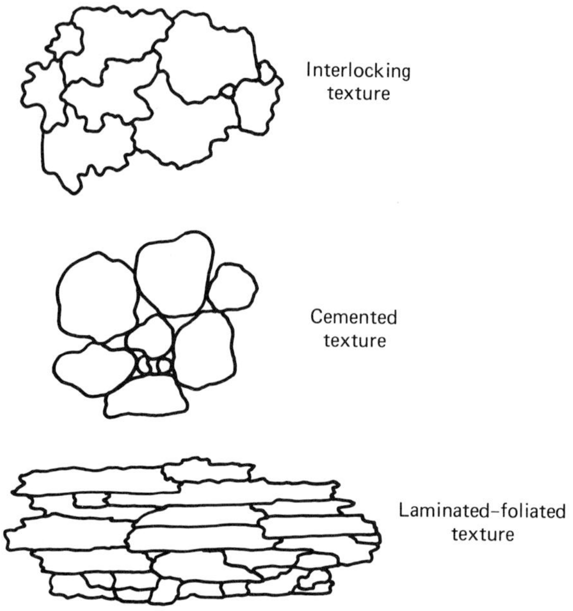

Figure 4.1 The three types of rock texture

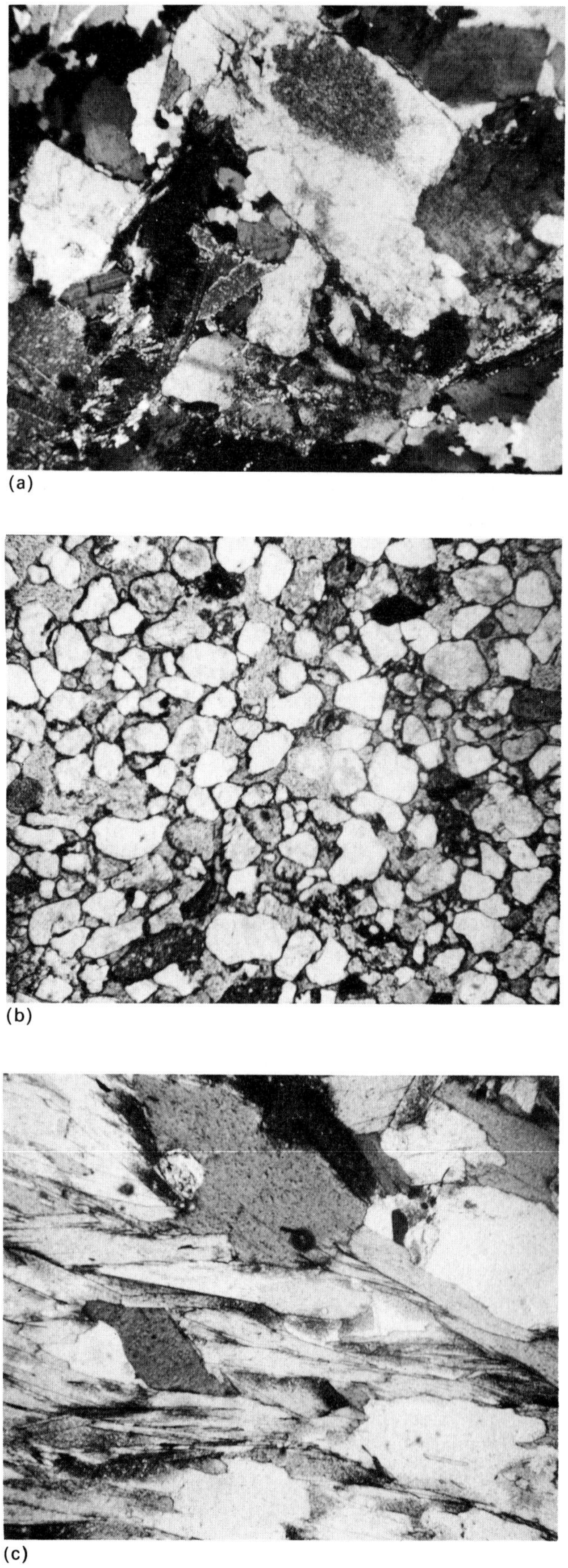

Figure 4.2 Photomicrographs illustrating the three types of rock texture. Field of view 3.3 mm wide. (a) The interlocking texture of granite (crossed polars); (b) The cemented texture in sandstone; (c) Laminated–foliated texture in biotite schist

There was an obvious need for a simple quantitative field test which could be carried out on rock core or irregular hand specimens or on the rock face to give an index value as a means of classification. The Schmidt concrete hammer (Figure 4.3), giving a rebound number, was used, but has now largely been superseded by the point load test (Figure 4.4).

A logical way to describe a granite would then be:

coarse-grained, light grey, biotite GRANITE with interlocking texture, hard.

- *hard* could be quantified as strong, one class of a rock strength classification, or specified as: unconfined compression strength 70 MPa;
- *interlocking* – one of the textural types could be quantified by study of thin sections of the rock.
- *light grey* could be quantified using a colour chart;
- *coarse-grained* is a specified grain-size class, but could be quantified.

The only reason for retaining the rock name is that it provides so much more information about the geological setting and associations than does the use of the non-specific term 'rock'. For this reason, geological names should be retained, but they must be simple.

Figure 4.3 Schmidt concrete hammer being used to determine the rebound number on a rock sample

Figure 4.4 Point load test machine

Recalling the earlier works of Terzaghi (1929, 1946), Stini (1950), Muller (1958, 1959, 1960), Pacher (1957), Casagrande (1962) and John (1962), Deere paid particular attention to the physical discontinuities in a rock mass that Terzaghi had referred to as mechanical defects. Such defects are very important because they influence deformability, shear strength and permeability. He described the types of discontinuity, quantifying the spacing of joints and bedding planes in English and approximate equivalent metric units. The metric classification is now largely enshrined in accepted rock mechanics terminology, although a different range of spacing intervals is still used in engineering geology. Essential to the approach is the recognized need, wherever possible, to quantify classes to which descriptive terms are applied.

Emphasizing the importance of rock discontinuities which have a direct bearing on the homogeneity of the rock mass, Deere proposed that the lengths of pieces of core obtained during each coring run should be measured and recorded. This approach foreshadowed rock quality designation (RQD) and although the lengths of the core pieces are influenced by drilling method and technique, they were still thought to be of sufficient importance to warrant describing.

Most of the descriptive terms are classified and described later, but even so it is worth while briefly to list Deere's proposals and some of his comments.

Attitude of discontinuities

Strike or, in modern times, dip direction cannot be measured on cores and recourse had to be made to direct observation down the borehole, geological mapping at the surface or in shafts and tunnels. Intersecting joints in core may give an indication of conjugate joint sets.

Character of surfaces and filling materials

The engineering behaviour of a jointed rock mass is probably more influenced by the character and infilling of the discontinuities than their mere presence. Tightness, irregularity and kind and amount of infilling material should be described. Whatever the process responsible for the infilling, the end product and not the infilling process is the important aspect. Alteration may result in material that grades inwards from the discontinuity from material with soil-like properties to material with rock-like properties. All filling or alteration products may be described as a soil or rock in terms of thickness, type and hardness.

If all possible descriptive terms are quantified, or when necessary illustrated, the result will be a modern version of the technical description of rocks for engineering purposes. It is to be regretted that there is a divergence of views on the numerical steps to be used in the classification of, say, bedding plane spacing as between rock mechanics and engineering geology. In rock mechanics, the progression is in steps 1–3–10, etc., whereas in engineering geology 2–6–20, etc., steps are used. Unfortunately, no proposals have been made, and there would have to be international agreement, to rationalize the approach by recommending one or the other system and the engineering geology intervals will continue to be used.

The scheme in which a lithological rock name is supplemented to give a descriptive rock name from which engineering properties may be inferred has been adopted and refined by a number of Working Parties. The first to be published was in the report on 'The logging of rock cores for engineering purposes' (Anon., 1970b).

A slightly different version appears in the report on 'The preparation of maps and plans in terms of engineering geology' (Anon., 1972), further modified for international use in Anon. (1979) and again in Anon. (1981a). What should be regarded as the definitive scheme for UK application appeared in the 'Code of practice for site investigations' (Anon., 1981b).

The descriptive scheme recommended for use in mapping is the version prepared by the Commission of the International Association of Engineering Geology on Engineering Geological Mapping (Anon., 1981a). The same group earlier published a guide to the preparation of engineering geological maps (UNESCO/IAEG, 1976) as well as recommended symbols for engineering geological mapping (Anon., 1981c).

4.3 Principles of classification and description of rocks and soils

Classification of rocks and soils on engineering geological maps should be based on the principle that the physical or engineering geological properties of a rock in its present state are dependent on the combined effects of mode of origin, subsequent diagenetic, metamorphic and tectonic history, and on weathering processes. This principle of classification makes it possible not only to determine the reasons for the lithological and physical characteristic of soils and rocks, but also for their spatial distribution. This is a basic principle of engineering geological mapping as of other geological mapping and implies not only the classification of individual rock samples but also the use of many individual rock samples, field observations and measurements to delineate uniform and continuous rock units (UNESCO-IAEG, 1976).

A full description of a soil or rock for engineering purposes involves giving an appropriate geological name together with as much detailed information as can be gained from the examination of a sample or exposure. The description may be systematic if a limited range of standard descriptive terms is used, and for ease of application should place the soil or rock in a limited number of groups with definable properties, thus providing a good guide to engineering performance.

At present not enough is known of the relations between engineering properties and lithological characteristics to establish a unified classification system for both soil and rock units on engineering geological maps. In such a classification each class would represent a soil or rock unit with a narrow range of engineering properties. Consequently, a three-part approach to engineering geological description and classification has to be adopted, involving:

1. A lithological classification.
2. A description of rock and soil material and mass and the assignment of descriptive names for rock and soil types.
3. Classification of rock and soil units delineated on engineering geological maps at various scales.

There is an obvious need to make the classifications and descriptions simple so that they can be readily understood and applied, even by those with limited geological knowledge. It should be emphasized that although the classifications proposed for rock and soil are very different, the suggested methods for the description of rock and soil are the same, both as materials and in the mass.

4.4 Bases for classification of rocks and soils for engineering purposes

In geological terms, a *rock* is a naturally occurring aggregate of minerals, and so includes loose aggregates, which would be called soils by an engineer, and the harder, solid materials which would be stone in layman's terms or rock to the engineer. The geological name often reflects the physical state of the aggregate, for example *sand*, with the equivalent lithified state indicated by adding 'stone' to the name of the loose aggregate: for example, *sandstone*. This distinction only applies to rocks of one group, namely the sedimentary rocks. Both igneous and metamorphic rocks can only exist in the solid state, except some pyroclastics. If weathering has progressed far enough, then rocks from both of these groups may be converted entirely to soils; for example, a highly weathered granite is by definition a soil and hence will have properties similar to those of the appropriate engineering soil. In the same way, a sandstone can revert to sand.

Although it is logical to classify rocks in geological terms regardless of physical state, it has to be accepted that there is a clear difference in the engineering properties, and hence engineering behaviour, of those rocks which the engineer would call engineering soils, and other natural materials. This distinction has both a practical and an historic basis: a quantitative approach to the behaviour of soils as foundation materials and in excavations, and also when used as construction materials, led to the development of the science of soil mechanics, in which theory and experimental data were applied to practical engineering design (Terzaghi, 1925).

A similar quantitative approach to the design of engineering structures in rock – the science of rock mechanics – has developed rapidly in the last two decades. There is a clear distinction to be made between the engineering behaviour of soil and rock. Division may be made, for example, on the basis of hardness or strength (Piteau, 1970), porosity and density (Duncan, 1969), permeability and compressibility, all of which are important engineering properties. But there is the problem, in attempting too precise a definition, of the gradation in properties that is known to exist between soil and rock, making a more qualitative definition preferable.

Rocks, using the term in the geological sense, may be divided into three major genetic groups – igneous, sedimentary and metamorphic – depending upon the processes active in their formation. Engineering soils, of various genetic types, for example alluvial, aeolian and glacial, occupy an important position in the sedimentary group. To these three groups must be added a fourth – residual soils formed by the weathering of rocks – and a fifth – made ground or fill not involving natural materials.

A number of criteria are used in classification, including chemical and related mineralogical composition, grain size, texture and fabric, which also provide a firm basis for the identification of rocks for engineering purposes. The most important criterion is grain size because of the ease with which it can be quantified in engineering soils, or estimated for rocks. In rocks, the predominant grain size, or range of grain sizes, is used as a means of subdivision of each genetic group. Contrast this with engineering soils, where as little as 35% of fine constituents, the clay and silt sizes, determines the engineering classification and much less may have an important influence on engineering properties. It is this aspect, more than any other, coupled with the ability to determine grading readily, that necessitates and justifies the separate methods for the description and classification of soils and rocks as adopted here.

4.5 A lithological classification of rocks

The classification of rocks (Tables 4.1a to 4.1c) is based on the determination of:

1. Genetic group
   (a) Sedimentary: detrital, chemical/organic
   (b) Metamorphic
   (c) Igneous
2. Structure
   (a) Bedded
   (b) Foliated
   (c) Massive
3. Predominant grain size
   (a) Very coarse-grained: greater than 60 mm
   (b) Coarse-grained : 2–60 mm
   (c) Medium-grained : 0.06–2 mm
   (d) Fine-grained : 0.002–0.06 mm
   (e) Very fine-grained : less than 0.002 mm
   (f) Glassy/amorphous
4. Composition (mineralogical)
   (a) Rock grains
   (b) Quartz
   (c) Feldspars and felspathoids
   (d) Mafic (dark-coloured) and related minerals
   (e) Clay minerals
   (f) Carbonates
   (g) Salts: siliceous and carbonaceous materials
   (h) Glass

Rock names are given to particular combinations of these features and correct naming requires recognition of the four attributes listed above. Accuracy and facility in this sphere can only come with training.

A simple, technically useful classification of rock types in each genetic group is given in Table 4.1. It is unified only with respect to predominant grain size. The selected grain size categories are not those used in geology, but are those used in the engineering classification of soils. This somewhat radical departure from conventional usage was felt to be justified as a means of simplification of rock classification and to give consistency between the rock and soil classifications. The limits of the size ranges in the finer soils define marked changes in the engineering properties of those soils (Glossop and Skempton, 1945) and are therefore well established in engineering usage. Any step towards unification should take account of this, and consequently the conventional size classes used in geological classifications have been slightly adjusted to conform to the soil classification.

Rock names were mainly selected from those used in non-specialist geological textbooks and in many cases are not used strictly, except as general terms for broad groups of related rock types.

4.6 Description of rocks for engineering purposes

The description of rock involves the following steps:

- determination of the rock name – the 'lithological rock name';
- description of the properties of the rock material;
- description of additional properties necessary to describe the features of the rock mass.

The properties of the rock mass are controlled partly by the properties of the rock material, but in many rock masses structural features substantially control the engineering properties. Such features include structures, discontinuities such as joints and bedding plane partings. Distribution of rock and soil materials in the weathering profile is an important engineering feature.

The three steps listed above provide a 'descriptive rock name' from which engineering properties may more readily be inferred than from a 'lithological rock name'.

4.6.1 Descriptive rock name

In a rock description the main characteristics should be given in the following order:

ROCK NAME
  Supplementary petrographic properties
Rock material properties
  Colour
  Texture
    Grain size
    Other textural features and fabric
  State of weathering
    State of alteration
  Strength
Rock mass properties
  Structure
  Discontinuities
  Weathering profile

The descriptive scheme has been modified from that recommended in Anon. (1972). The main differences are in the treatment of state of weathering and the weathering profile, and an expansion in the description of structure. Structural aspects have been dealt with more thoroughly by Anon. (1977) and also in ISRM (1977) and Brown (1981).

Most of the descriptive terms are classified semi-quantitatively. The result is a composite descriptive name involving a comprehensive description of engineering properties as an adjunct to the lithological rock name. It includes the characteristics of descriptive and physical properties (colour and texture), physical state (jointing, weathered and altered state), mechanical properties (strength, deformability), and if necessary permeability and durability.

4.6.2 Lithological name

The lithological name is of primary importance because it indicates the genetic rock group and provides basic information on mineral composition and grain size. Supplementary petrographic properties may be used where necessary to qualify the rock name, signifying for example a relative abundance of a particular mineral – say, biotite granite – or indicating minor admixtures of other lithological types. These supplementary features

Table 4.1 (a) Classification of rock types: igneous

PYROCLASTIC	IGNEOUS				GENETIC GROUP		
	Massive				*Usual structure*		
At least 50% of grains are of igneous rock	Quartz, feldspars, micas, dark minerals		Feldspar, dark minerals	Dark minerals	Composition		
	Acid	Intermediate	Basic	Ultrabasic			*Predominant grain size (mm)*
Rounded grains: AGGLOMERATE	PEGMATITE			Pyroxenite Peridotite	Very coarse-grained		
						60	
Angular grains: VOLCANIC BRECCIA	GRANITE	DIORITE	GABBRO		Coarse-grained		
						2	
TUFF*			DOLERITE		Medium-grained		
						0.06	
Fine-grained TUFF	RHYOLITE	ANDESITE	BASALT		Fine-grained		
						0.002	
Very fine-grained TUFF					Very fine-grained		
	VOLCANIC GLASSES				GLASSY AMORPHOUS		

*A tuff containing both pyroclastic and detrital material, but predominantly pyroclastic, is called TUFFITE.

Table 4.1 (b) Classification of rock types: sedimentary

DETRITAL SEDIMENTARY						CHEMICAL/ ORGANIC	GENETIC GROUP		
			Bedded				*Usual structure*		
Grains of rock, quartz, feldspar and clay minerals			At least 50% of grains are of carbonate			Salts, carbonates, silica, carbonaceous	Composition		
Rudaceous	Grains are of rock fragments Rounded grains: CONGLOMERATE Angular grains: BRECCIA			LIMESTONE (undifferentiated)	Calcirudite	SALINE ROCKS: Halite Anhydrite Gypsum	Very coarse-grained		*Predominant grain size (mm)*
								60	
							Coarse-grained		
								2	
Arenaceous	Grains are mainly mineral fragments SANDSTONE: Grains are mainly mineral fragments				Calcarenite	CALCAREOUS ROCKS	Medium-grained		
								0.06	
Argillaceous or Lutaceous	MUDSTONE SHALE: fissile mudstone	SILTSTONE: 50% fine-grained particles	Marlstone		Calcisiltite	LIMESTONE DOLOMITE	Fine-grained		
					CHALK			0.002	
		CLAYSTONE: 50% very fine-grained particles			Calcilutite		Very fine-grained		
						SILICEOUS ROCKS Chert Flint CARBONACEOUS ROCKS LIGNITE COAL	GLASSY AMORPHOUS		

Table 4.1 (c) Classification of rock types: metamorphic

METAMORPHIC		GENETIC GROUP	
Foliated	*Massive*	*Usual structure*	
Quartz, feldspars, micas, dark minerals	Quartz, feldspars, micas, dark minerals, carbonates	Composition	*Predominant grain size (mm)*
Tectonic breccia	Tectonic breccia	Very coarse-grained	
MIGMATITE			60
GNEISS	HORNFELS: Marble Granulite	Coarse-grained	
			2
SCHIST	QUARTZITE	Medium-grained	
Amphibolite	Amphibolite		
Phyllite			0.06
SLATE		Fine-grained	
			0.002
Mylonite	Mylonite	Very fine-grained	
		GLASSY AMORPHOUS	

may be very useful as a means of discriminating between different rocks that have the same lithological name. Minor constituents may also have an important effect on the mechanical and physical properties of rocks, and should be carefully considered.

The rock name is selected from the classification tables (see Table 4.1) and these are the only rock names that are recommended for use. In arriving at a name for a rock, there is no substitute for either geological knowledge or an aid to identification that is reliable and easy to apply.

4.6.3 Description of rock material

4.6.3.1 Colour

A semi-quantitative colour index, the sum of the dark or coloured minerals expressed as percentages, may be useful in the identification of igneous rocks. For general application to all rocks, a quantitative descriptive scheme in terms of three parameters is available. Colour can be expressed in terms of *hue* which is the basic colour or a mixture of basic colours, the *chroma* or brilliance or intensity of colour, and the *value* or lightness of the colour. A Rock-Color Chart, designed primarily for field use, using this system has been published by the Geological Society of America (1963) and is based on Munsell (1941). Use of this chart is strongly recommended as a standard for rock colour nomenclature.

As an alternative to the Rock-Color Chart, the following simple system (Anon., 1972), which limits the subjectivity of an evaluation, may be used. One term is selected, as required, from each column in Table 4.2, and combined as a colour assessment. Examples of use are: light yellowish brown, dark reddish brown, dark brown, etc. If necessary, colour differences can be emphasized separately by the use of terms such as spotted, dappled, mottled, streaked; for example, light yellowish brown spotted with dark brown.

Table 4.2 Terms for lightness, chroma and hue which may be used in combination for colour description

Lightness	*Chroma*	*Hue*
Light	Pinkish	Pink
Dark	Reddish	Red
	Yellowish	Yellow
	Brownish	Brown
	Greenish	Green
	Bluish	Blue
		White
	Greyish	Grey
		Black

4.6.3.2 Texture

Texture is all-important in naming a rock type since there are examples of igneous, metamorphic and sedimentary rocks which have the same, or at least very similar, mineral assemblages; textural differences serve to distinguish them in samples of rock material. Textural and other terms used later are defined in Gary, McAfee and Wolfe (1972) and other geological dictionaries.

Textural features are as varied as mineral composition and need to be considered in some detail before significant diagnostic textures can be selected for each of the main rock groups and for the discrimination of different classes in each main group.

At this point it is necessary to define certain terms. Texture, in petrological terms (Gary, McAfee and Wolfe, 1972) is: 'The general physical appearance or character of a rock, including the geometric aspects of, and the mutual relations among, the component particles or crystals; e.g. the size, shape, and arrangement of the constituent elements of a sedimentary rock, or the crystallinity, granularity, and fabric of the constituent elements of an igneous rock. The term is applied to the smaller features as seen on a smooth surface of a homogeneous rock or mineral aggregate.'

Of the textural elements used for description and classification, the most important is grain size which, for the predominant size of grain, can be classified semi-quantitatively. From Table 4.1 the relations between rock names and grain sizes can be understood. It will be recalled that the class boundaries have been fixed at limits of grain size grades adopted for engineering soils, i.e. the boundaries between clay, silt and sand sizes that are justified and determined by the differences in the physical behaviour of those soils (Glossop and Skempton, 1945).

The texture of a rock refers to individual grains and the arrangement of grains, referred to as the rock fabric, which may show a preferred orientation. Structure is concerned with the larger scale interrelationship of textural features. Common terms should be used where possible; a separate term may not be necessary if it is implicit in the rock name or is more appropriately referred to as a minor lithological characteristic.

Terms frequently used include sheared, cleaved, foliated, lineated, massive, flow-banded, veined, porphyritic, homogeneous. Sedimentary rocks occur in beds which may be regular, laminated, cross-laminated, graded or show slump-structure; bedding planes may be ripple-marked, sun-cracked or sole-marked.

Because grain size considerably affects the physical properties of a rock it should always be indicated directly in the rock description rather than relying on the grain size implication in the rock name.

It is usually sufficient to estimate grain size by eye, which may be aided by a hand lens in the case of fine-grained and amorphous rocks. The limit of unaided vision is approximately 0.06 mm.

Many other aspects of rock texture may be used to amplify the description, such as are given below.

Relative grain size

This may be described, for example, as uniform, non-uniform or porphyritic.

Grain shape

This may be described by reference to the general form of the particles, their angularity which indicates the degree of rounding at edges and corners, and their surface characteristics (Table 4.3).

Table 4.3 Terms used in description of grain (particle) shape

FORM	Equidimensional Flat Elongated Flat and elongated Irregular
ANGULARITY	Subangular Subrounded Rounded
SURFACE CHARACTERISTICS	Rough Smooth

Fabric

This term, which is used in the definition of texture, is not easily understood. The fabric of an igneous rock is defined as the external appearance or pattern produced by the shapes and orientation of the crystalline and non-crystalline parts of the rock; it is dependent on the relative sizes and shapes of the parts and their positions with respect to one another and to the groundmass when present. The same definition would be acceptable for metamorphic rocks, but a slight modification is needed for sedimentary rocks in which fabric refers to the orientation (or lack of it) in space of the elements (discrete particles, crystals, cement) of which the rock is composed.

The spatial arrangement of grains in the rock may show a preferred orientation (see Figure 4.1) or lack of it, and may produce patterns by non-uniform arrangements of grains, crystals and groundmass.

Porosity

The size, shape and orientation of pore or void spaces should be described under this aspect of rock texture.

4.6.3.3 State of weathering

Description of the state of weathering of rock material is of particular importance in describing engineering rocks because weathering has profound effects on the physical and mechanical properties of rock material. In any description there needs to be a statement as to whether or not the rock material is considered to be either in a fresh state or is weathered. Weathering effects may be described in terms of discoloration, chemical decomposition or physical disintegration.

On the basis of mechanical and chemical changes, a descriptive scheme for weathering grades of rock material may be established (Table 4.4). The extent of particular weathering effects may be subdivided using such qualifying terms, for example, highly decomposed, extremely discoloured, slightly disintegrated, as will aid the description of the material being examined. These descriptive qualifying terms may be quantified if necessary by estimation from drill core or in the natural exposure (Table 4.5).

Table 4.4 Descriptive scheme for weathering grades of rock material

Term	*Description*
Fresh	No visible sign of weathering of rock material
Discoloured	Colour of original fresh rock material changed and is evidence of weathering. Degree of change from original colour should be indicated. If colour change confined to particular mineral constituents, this should be mentioned
Decomposed	Rock weathered to condition of a soil in which original material fabric still intact, but some or all mineral grains decomposed
Disintegrated	Rock weathered to condition of a soil in which original material fabric still intact. Rock friable, but mineral grains not decomposed

Table 4.5 Terms for description of degree of weathering of rock material

Term	*Degree of change* (%)
Fresh	0
Slightly	Over 0–10
Moderately	10–35
Highly	35–75
Extremely	Over 75

Depending on the character and distribution of the weathering changes, and the extent to which a rigid rock framework is retained, the weathered rock material may assume the characteristics of an engineering soil at an early stage.

Extremely weathered rock material will almost certainly be an engineering soil, and may be classed a residual soil if the original rock fabric has collapsed or changed so as to remove most traces of the original fabric.

Examples of use of the terms in Table 4.5 are: fresh rock; slightly decomposed; moderately disintegrated; highly discoloured. Usually combinations occur: highly disintegrated and moderately decomposed, etc.

4.6.3.4 State of alteration

Where appropriate, terms used for weathering of rock material may be used, as in many instances the effects of alteration are not easily distinguishable from those brought about by weathering. Wherever possible, common terms are preferable, e.g. slightly kaolinized, highly mineralized; they may be quantified using the scale in Table 4.5.

4.6.3.5 Strength

The uniaxial compression test gives a reliable indication of the strength of rock material, although the test results are dependent on the moisture content of the specimen, any anisotropy in the material, and the test procedure adopted. A scale of strength is given in Table 4.6.

As an alternative method of strength testing for use in the field, the point load test (Broch and Franklin, 1972; ISRM, 1977; Brown, 1981) is recommended. Provided that individual test results are normalized to a standard specimen thickness of 50 mm, and the recommended test procedures are followed, this test provides a good estimate of unconfined compressive strength. The relation UCS = 14PLS (where UCS is the unconfined compressive strength and PLS is the point load strength) has been demonstrated repeatedly and can be accepted as a reasonably reliable approximation.

Table 4.6 Scale of strength for dry rock material

Term	*Compressive strength* (MPa)
Weak	1.5*–15
Moderately strong	15–50†
Strong	50–120
Very strong	120–230
Extremely strong	Over 230

*Rocks with strength under 1.5 MPa are, as a rule, hard soils and should be tested accordingly.
†Soft rocks are weaker than 50 MPa; strong rocks are stronger than 50 MPa.

4.6.4 Description of rock mass

4.6.4.1 Structure

The structure of the rock mass is related to the larger scale interrelations of textural features and the association of one or more rock types in the mass. Terms used to describe sedimentary rock types include bedded, interbedded, laminated, folded, etc.; metamorphic rocks may be cleaved, foliated, schistose, banded, gneissose, lineated, folded; igneous rocks may be massive, flow-banded, etc.

The problem of the association of more than one rock type in a rock mass is difficult to deal with because, for example, in a sequence of sedimentary rocks one lithological type may be more susceptible to weathering than another, and the presence of a highly weatherable rock in a sedimentary sequence may have an enhancing effect on the susceptibility of the rock types to weathering. In addition, the engineering properties of an interbedded rock mass may depend more on one rock type than on the other.

4.6.4.2 Discontinuities

A discontinuity is a surface within the rock mass that is open or potentially openable under the stress level applicable in engineering because the tensile strength across the surface is lower than that of the rock material. Thus a discontinuity is not necessarily a plane of actual physical separation within the rock mass, for example slaty cleavage.

Discontinuities have many modes of origin, but two main types may be recognized: those that occur in sets, for example bedding planes, joints, cleavages, foliations, and those that are unique, for example individual joints or faults. Of these, bedding planes, cleavages and foliations are discontinuities only where there is a parting or substantial weakening of the rock across them. Similarly joints and fault planes may be healed by introduced minerals or by alteration along them.

Description of discontinuities is aimed at determining their nature, orientation, spacing, persistence, roughness, wall strength, aperture, infilling, seepage, number of sets, and block size and shape (ISRM, 1977; Brown, 1981).

Orientation

A compass with clinometer is used to measure the dip – the maximum inclination from the horizontal – of the planar feature being measured with reference to magnetic or true north. Alternatively, the dip and dip direction of the plane may be determined, and this is commonly preferred.

Individual discontinuities are shown on maps by conventional dip and strike symbols (see Figure 5.12), although frequently it may be more useful to give the orientation of joint sets over the whole map area or for selected smaller areas. Two expedients are used: the joint rosette or a contoured spherical projection.

Joint rosettes

A large number of orientation measurements can be plotted on a simplified compass rose marked from 0° to 360° with radial lines at 10° intervals. Strike readings are grouped into the nearest 10° sectors, with the number of readings given along radial axes using numbered circles representing 5, 10, 15, etc., readings (Figure 4.5). The resultant strike 'petals' have mirror images about the centre of the rosette. The range of dip measurements must be shown outside the circumference of the rosette.

Two methods of representing orientation data are shown in Figure 4.5. Readings grouped in the nearest 10° sectors can be represented either as solid radial sectors, as shown on the left-hand side of the diagram, or their strike values can be averaged resulting in sharp petals (right-hand side).

Rosette diagrams are not suitable for representing sub-horizontal discontinuity sets. In this case, recourse should be made to a spherical projection.

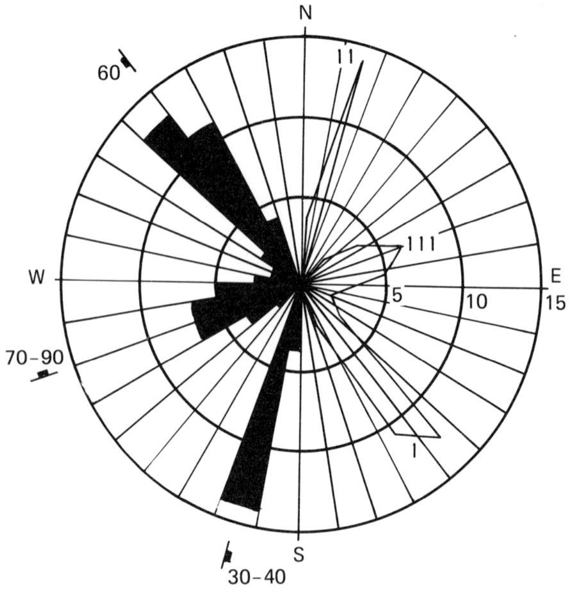

Figure 4.5 Two methods of representing discontinuity orientation data on a joint rosette (After Brown, 1981, Fig. 3)

Spherical projections

The most commonly used spherical projections are the equal area projection and the equal angle projection. In the former, the spatial distribution of data is accurately presented on a Schmidt, or Lambert, equal area net. The equal angle projection, using the Wulff net, accurately represents the angular relationships between planes and lines.

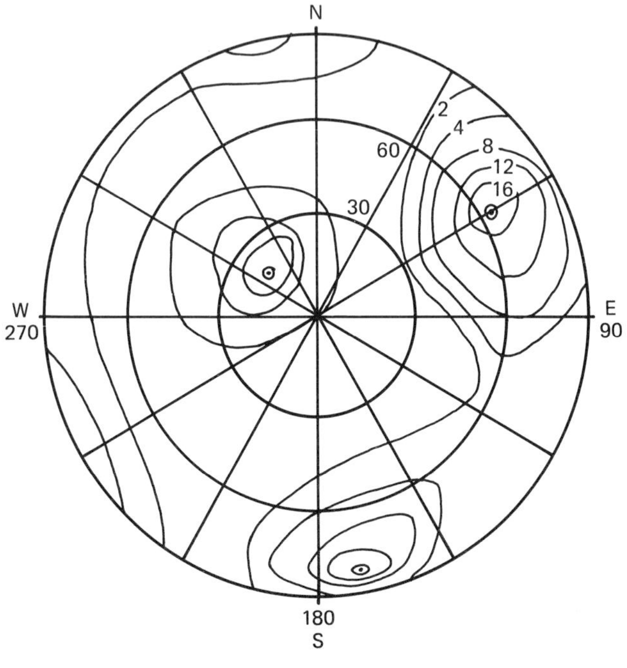

Figure 4.6 Contoured Schmidt diagram representing the orientation of three sets of discontinuities on a polar equal area net

Equal angle projections are most frequently used to analyse the angular relationships between a small number of planes. Each discontinuity plane, or the mean orientation of a set, can be uniquely represented as a great circle, or as a pole on a reference hemisphere.

Handling a large amount of orientation data, such as might be gathered from natural rock outcrops in the field or from the walls of an excavation, requires a different approach. It is usual to plot the pole to each plane on an equal area net (Figure 4.6). Clusters of poles may become evident as plotting proceeds. The Schmidt contouring method is used to determine pole densities, with the central value of each highest concentration representing the mean orientation of that particular set of discontinuities.

This method of plotting can handle planes with any attitude, and the data can be plotted by computer. In many situations rock mechanicians, in particular, prefer to plot planar orientation in terms of dip direction and dip amount. What has been plotted should always be stated, otherwise gross errors of interpretation can arise.

Either the complete contoured diagram, or a diagram showing just the plots, may be drawn on the map or plan.

Spacing

This term refers to the mean or modal spacing of a set of joints, that is the perpendicular distance between adjacent discontinuities. Descriptive terms for discontinuity spacing are given in Table 4.7.

Table 4.7 Discontinuity spacing

Term	*Spacing*
Very widely spaced	Over 2 m
Widely spaced	600 mm–2 m
Medium spaced	200–600 mm
Closely spaced	60–200 mm
Very closely spaced	Under 60 mm

Persistence

This is a measure of the areal extent of a discontinuity from its inception to its termination in solid rock or against another discontinuity. For major joints, the plane may extend beyond the limits of the exposure and then the maximum trace length or area should be recorded.

Roughness

A discontinuity surface may be planar, undulating or stepped, and descriptive terms (Table 4.8) are based on two scales of observation: small scale (several centimetres); large scale (several metres). Typical roughness profiles in each category are illustrated in Figure 4.7.

Wall strength

The shear strength of a discontinuity may be significantly affected by the condition or strength of the rock material forming the walls of the discontinuity, especially where infilling is small or absent and wall roughness is significant. Wall strength may be measured in terms of compressive strength, which may be lower than the strength of the fresh rock material due to weathering or alteration of the walls.

Table 4.8 Roughness categories for discontinuity surfaces (From ISRM, 1977, by permission)

Term	*Category*
Rough (or irregular), stepped	I
Smooth, stepped	II
Slickensided, stepped	III
Rough (or irregular), undulating	IV
Smooth, undulating	V
Slickensided, undulating	VI
Rough (or irregular), planar	VII
Smooth, planar	VIII
Slickensided, planar	IX

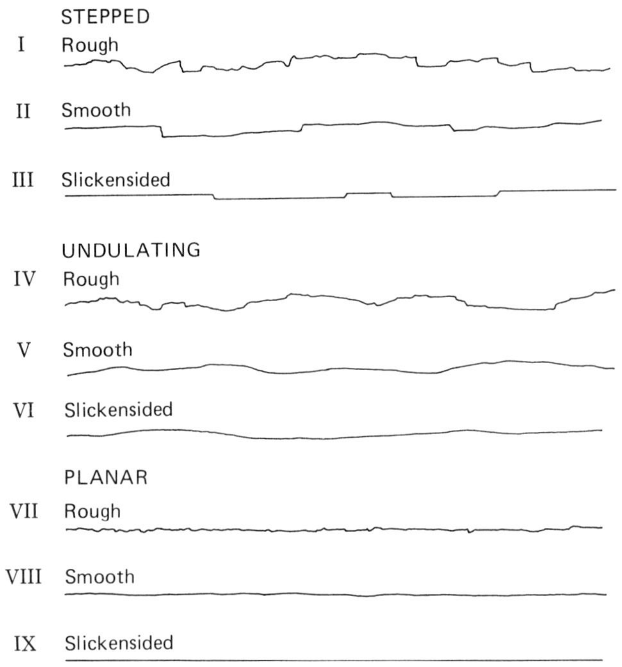

Figure 4.7 Typical profiles for the roughness categories for discontinuity surfaces (After ISRM 1977, Fig. 17)

The stage of weathering of the rock material may be described in terms of degree of discoloration, decomposition and disintegration from the fresh state (Dearman, 1974, 1976). An estimate of unconfined compressive strength may be obtained using the Schmidt hammer value (Hucka, 1965; Deere and Miller, 1966), or a simple manual penetration test (Anon., 1957), as appropriate to the material.

Aperture

The mean perpendicular distance between adjacent walls of a discontinuity, in which the space is infilled with air or water, is referred to as the aperture. The mean separation may be described using the terms in Table 4.9.

Table 4.9 Aperture of discontinuity surfaces

Term	*Aperture* (mm)
Very wide	Over 200
Wide	60–200
Moderately wide	20–60
Moderately narrow	6–20
Narrow	2–6
Very narrow	0–2
Tight	Zero

Infilling

The infilling between discontinuity surfaces may be soil introduced into the opening, minerals such as calcite or, in the case of faults, clay gouge or breccia. The width of an infilled discontinuity – the perpendicular distance from wall to wall – is important in conjunction with the roughness in determining the resistance to shear along the discontinuity.

The infilling material should be identified and described. Strength of the infill may be assessed visually and manually, or should be measured.

Seepage

Water flow and free moisture visible in individual discontinuities should be described and the rate of flow estimated.

Number of sets

Discontinuities may be irregular, of small persistence, without any arrangement into distinct sets: they are then referred to as non-systematic. On the other hand, systematic joints will usually be persistent, with individual joints in each set parallel or sub-parallel. Systematic joints may be present in one or more sets, and influence rock mass and block size.

Block size and shape

The spacing of discontinuities may be described with reference to the size and shape of rock blocks bounded by discontinuities and by the relative persistence of different discontinuity sets. Matula and Holzer (1978) have given models of rock mass blockiness, as illustrated in Figure 4.8. These may be described as follows:

- *Polyhedral blocks*. Irregular discontinuities without arrangement into distinct sets, and of small persistence.
- *Tabular blocks*. One dominant set of parallel discontinuities (1), for example bedding planes, with other non-continuous joints; thickness of blocks much less than length or width.
- *Prismatic blocks*. Two dominant sets of discontinuities (1 and 2), approximately orthogonal and parallel, with a third irregular set; thickness of blocks much less than length or width.
- *Equidimensional blocks*. Three dominant sets of discontinuities (1, 2 and 3), approximately orthogonal, with occasional irregular joints, giving equidimensional blocks.
- *Rhomboidal blocks*. Three (or more) dominant mutually oblique, sets of joints (1, 2 and 3), giving oblique-shaped, equidimensional blocks.
- *Columnar blocks*. Several, usually more than three, sets of continuous, parallel joints (1, 2, 3, 4, 5) crossed usually by irregular joints; length much greater than other dimensions.

On weathering, each of these block forms will be modified by rounding of the sharp edges, and the eventual formation of spheroidal blocks; the original joint pattern usually survives the weathering process.

The size of the blocks bounded by discontinuities may be described using the terms in Table 4.10.

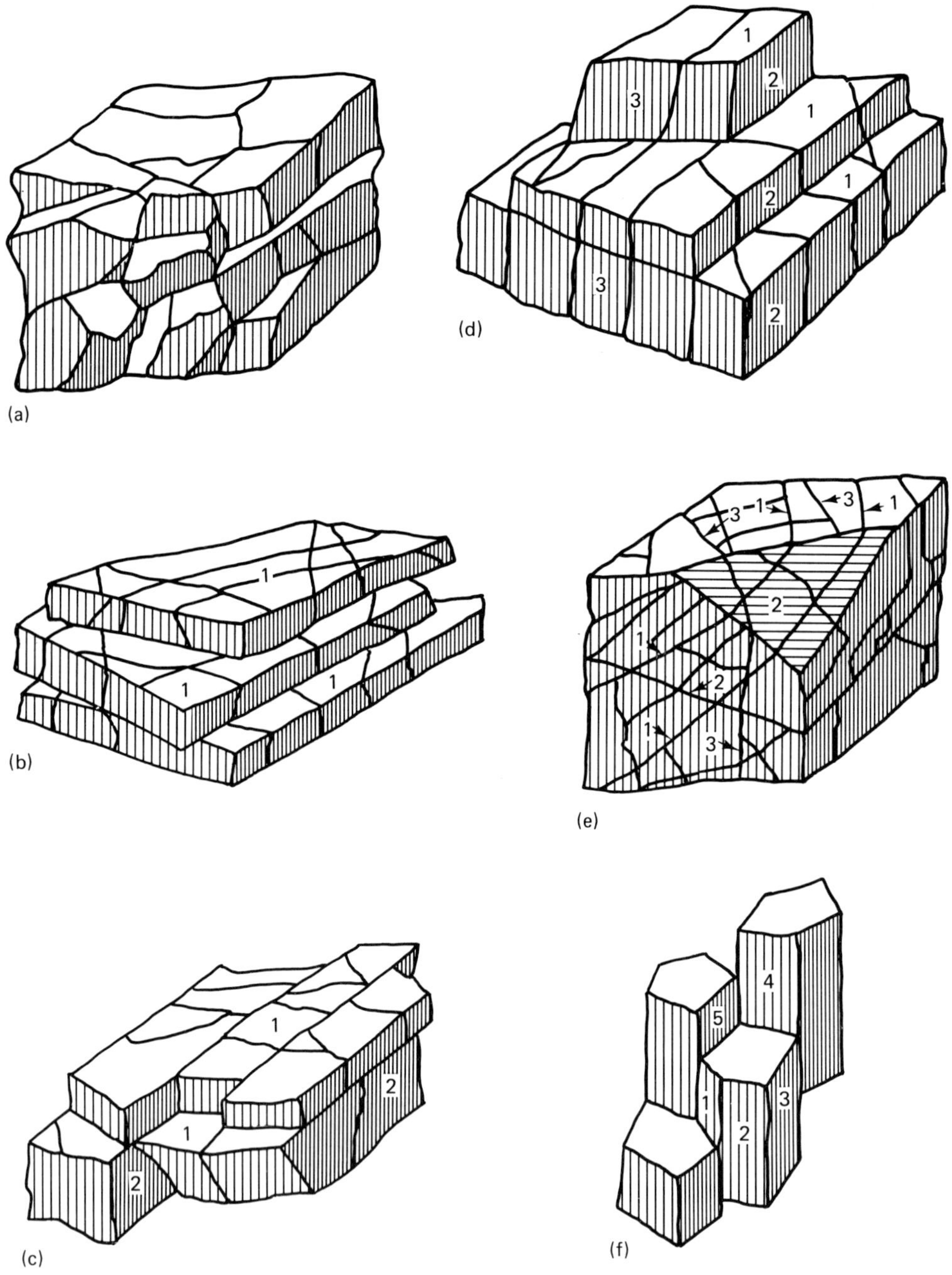

Figure 4.8 Rock mass block shape; a classification based on numbers of sets of persistent discontinuities and their orientation: (a) polyhedral blocks, Po-type; (b) tabular blocks, Ta-type; (c) prismatic blocks, Pr-type; (d) equidimensional blocks, Eq-type; (e) rhomboidal blocks, Rh-type; (f) columnar blocks, Co-type (After Matula and Holzer, 1978, Fig. 1)

Table 4.10 Terms for description of block size in rock mass

Term	*Average dimension*
Very large	Over 2 m
Large	600 mm–2 m
Medium	200–600 mm
Small	60–200 mm
Very small	Under 60 mm

4.7 Description of weathering profile in rock mass

4.7.1 Introduction

The description of the weathering of rock materials is of particular importance in ground investigations as most construction on or in a rock mass is undertaken at shallow depths within the zone of influence of surface weathering. Even for potentially deeper structures weathering is important; for instance, the portals of a tunnel may be weathered, and faults may have permitted very deep penetration to the tunnel line, particularly of chemical weathering effects. There is no doubt that a weathered rock mass can be described, logged or mapped in terms of the distribution of different types or grades of weathered rock material. Such an approach has to be taken in logging drill core, with the distribution of mass rock grades being inferred later from the graphic borehole logs.

Many attempts have been made to devise weathering grade scales for particular types of rock. However, most of the scales, supposedly of general application (Anon., 1970b, 1972, 1977, 1981a,b) derive from those set up for granites. Of the three prominent early workers in this field, Moye (1955) was concerned with consistent description of rock material for dam sites and tunnels on the Snowy Mountains Scheme in Australia. On the other hand, Ruxton and Berry (1957), working on exposures in Hong Kong, classified both the material and the mass in terms that are interchangeable.

On engineering sites, where extensive exposures may be available, the distribution of material types can be logged, and grouped into grades the boundaries of which are usually gradational.

The recommendations given in subsection 4.6.3.3 provide terms for the description of material weathering that have to be taken into account in devising a workable weathering scale for the rock mass.

There are two aspects of weathering: one dominated by mechanical disintegration, the other by chemical decomposition, including solution. Generally, both mechanical and chemical effects act together, but, depending on climatic regime, one or other of these aspects may be dominant. Mechanical weathering results in the opening of discontinuities, the formation of new discontinuities by rock fracture, the opening of grain boundaries, and the fracture or cleavage of individual mineral grains. Chemical weathering leads to the eventual decomposition of silicate minerals to clay minerals; some minerals, notably quartz, resist this action and may survive unchanged. Early stages of chemical weathering result in discoloration of the rock affected. Solution is an ever present aspect of chemical weathering which is particularly important in carbonate and saline minerals.

It should always be borne in mind that the recommendations in section 4.6 for description of the rock mass are intended to be of a general nature, and may be elaborated by the investigator to describe most appropriately the distribution of weathered rock and materials encountered on a project.

Table 4.11 Rock substance: condition terms (After Stapledon, 1976)

Term	*Abbreviation*	*Meaning*
Fresh	(F)	Rock shows no sign of discoloration
Slightly weathered	(SW)	Rock slightly discoloured, but generally shows little or no change of strength from fresh rock
Highly weathered	(HW)	Rock strength visibly altered by weathering. Rock may be highly discoloured, usually by ironstaining. Porosity may be increased by leaching, or may be decreased due to deposition of weathering products in pores
Extremely weathered	(XW)	Rock weathered to such an extent that it has soil properties, i.e. it either disintegrates or can be remoulded in water

At this point it is pertinent to consider one other approach to the description of weathering of rock material. Stapledon (1976), presenting his Australian experience, has defined the four classes:

- fresh;
- slightly discoloured (rock);
- highly discoloured;
- decomposed and/or disintegrated (soil).

justified by Dearman (1986), in an unambiguous and simple way using what he calls 'condition terms'. These are set out in Table 4.11 with one minor change to discoloration, rather than decomposition, in the definition of 'fresh'.

The descriptive condition terms are used with rock type names and descriptive terms for rock material strength, e.g. GRANITE, slightly weathered, very high strength, or GRANITE (SW), which is all that is needed for a rock material description. The other descriptive terms recommended in section 4.6.3 – colour, grain size, etc. – are generally superfluous unless one rock type needs to be distinguished from another of the same type for engineering purposes.

4.7.2 Weathering profile

Weathering of the rock mass is described in terms of the distribution and relative proportions of fresh rock and discoloured, decomposed and disintegrated rock, and the effects on discontinuities.

Weathering eventually converts rock to an engineering soil and the weathering profile may be described in terms of three basic units: rock, rock-and-soil, and soil. In rock masses consisting of one rock type these units will form a distinct succession with rock at depth being overlain in turn by rock-and-soil and soil at the surface if a complete weathering profile has been developed and subsequently preserved.

It may be found convenient to subdivide each unit to give six distinct weathering grades, and the justification for doing so is given in Dearman (1974). A typical scale of weathering grades applicable to a weathering profile in rock is given in Table 4.12; the grades are transitional, and it may be necessary to subdivide a grade for a particular application (Irfan and Dearman, 1978).

Table 4.12 Scale of weathering grades of rock mass (From Anon., 1981b, by permission)

Term	*Description*	*Grade*
Fresh	No visible sign of rock material weathering; perhaps slight discoloration on major discontinuity surfaces	I
Slightly weathered	Discoloration indicates weathering of rock material and discontinuity surfaces. All rock material may be discoloured by weathering	II
Moderately weathered	Less than half of rock material is decomposed or disintegrated to a soil. Fresh or discoloured rock present either as a continuous framework or as corestones	III
Highly weathered	More than half of rock material decomposed or disintegrated to a soil. Fresh or discoloured rock present either as a discontinuous framework or as corestones	IV
Extremely weathered	All rock material decomposed and/or disintegrated to soil. Original mass structure still largely intact	V
Residual soil	All rock material converted to soil. Mass structure and material fabric destroyed. There is a large change in volume, but soil has not been significantly transported	VI

Geological structures, and in particular faults, may have influenced the form of the weathering profile by permitting penetration of weathering agencies deep into otherwise fresh rock. This eventuality had been foreseen in a grading scheme proposed for rock cores (Anon., 1970b) by the recognition of the need for grade 1B: 'Faintly weathered: weathering limited to the surface of major discontinuities.'

Alternation of different rock types in differing proportions in a rock mass may be most conveniently dealt with by ignoring the lithological variations and applying the weathering grade criteria set down in Table 4.12.

4.7.3 Scale of weathering grades for a rock mass

The terms given in Table 4.12 are considered to be of general application to rock masses and may be modified to suit particular situations. Subdivisions of grades may be made if necessary.

It should be understood that all grades of weathering may not be seen in a given rock mass, and that in some cases a particular grade may be present to a very small extent. Distribution of the various weathering grades of rock material in the rock mass may be related to the porosity of the rock material and the presence of open discontinuities of all types in the rock mass. Distribution of weathering grades in a rock mass may be determined by mapping natural and artificial exposures. However, isolated natural exposures of rock, and excavations of limited extent, are not necessarily representative of the whole rock mass.

4.7.4 Diagrammatic representation of a weathering profile in massive rock

A schematic vertical section through a weathering profile, such as that given in Figure 4.9(a), may be translated into reality by considering some of the complexities resulting from structures in the rock mass (Figure 4.9b).

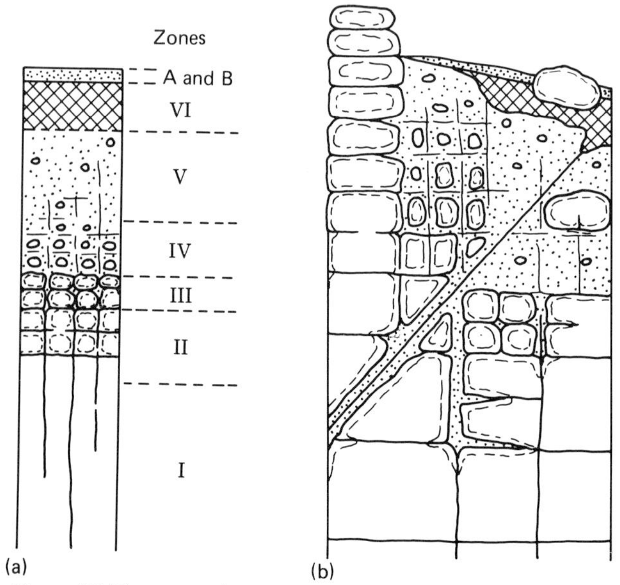

Figure 4.9 Diagrammatic representation of a simplified weathering profile in massive rock: (a) idealized weathering profile; (b) example of a complex profile

4.8 Classification of soils for engineering purposes

The classification of soils into a number of soil groups is based on: (a) grading of the constituent particles, and (b) plasticity of that fraction of the material consisting of particles finer than 0.425 mm – selected maximum size is a convenient BS Test Sieve size within the medium sand particle size range.

Determinations, following existing classifications, are made on disturbed samples, and the characteristics are independent of the condition in which the soil occurs in the ground, and disregard the influence of the structure of the soil mass. The soil classification can give a good guide to how the disturbed soil will behave as a construction material, for instance in embankments, at various moisture contents.

Grading and plasticity may be estimated in the field, or determined under laboratory conditions. The classification is carried out on material nominally finer than 60 mm; boulders and cobbles larger than this size are picked out and recorded as a percentage of the whole sample.

The soil classification given in Table 4.13 is modified from the Unified Soil Classification (USC) devised by Casagrande (Anon., 1957) and the British Soil Classification for Engineering Purposes (BSCS) (Anon., 1981b).

Soils are divided into three main classes – coarse, fine and organic. A coarse soil contains less than 35% of grains finer than 0.06 mm, and coarse soils are additionally subdivided into sand and gravels on the basis of grain size (Table 4.13). A fine soil contains more than 35% of grains finer than 0.06 mm, and fine soils are subdivided into silts and clays with 65–100% fines.

Grading and plasticity are divided into a number of clearly defined ranges, each of which may be referred to by a descriptive name and letter:

Coarse components		
Main Terms	GRAVEL	G
	SAND	S
Qualifying Terms	Well graded	W
	Poorly graded	P
	Uniformly graded	Pu
	Gap graded	Pg
Fine components		
Main Terms	FINE SOIL, FINES – may be differentiated into M or C:	F
	SILT (M – soil) – plots below A-line of plasticity chart	M
	CLAY – plots above A-line (fully plastic)	C
Qualifying Terms	Of low plasticity*	L
	Of intermediate plasticity	I
	Of high plasticity	H
	Of very high plasticity	V
	Of extremely high plasticity	E
Organic components		
Main Term	PEAT	Pt
Qualifying Term	Organic – may be suffixed to any symbol	O

* See Table 4.18 for range of liquid limits. Qualifying terms applied when the fine soil content is significant.

Table 4.13 Soil classification for engineering geological mapping*

Main divisions of soil groups		*Fines* (percentage finer than 0.06 mm)	*Soil groups*	*Group symbols*	*Subgroup symbols*	*Soil name*
COARSE SOILS: More than 35% of material less than 60 mm is larger than 0.06 mm	GRAVEL: More than 50% of coarse material is larger than 2 mm	0–5	GRAVEL	G	GW GP GPu GPg	GRAVEL, well graded GRAVEL, poorly graded GRAVEL, uniformly graded GRAVEL, gap graded
		5–35	GRAVEL, silty GRAVEL WITH FINES GRAVEL, clayey	GM GF GC	GML, etc. GCL, etc.	GRAVEL, silty, of low plasticity GRAVEL, clayey, of low plasticity
	SAND More than 50% of coarse material is smaller than 2 mm	0–5	SAND	S	SW SP SPu SPg	SAND, well graded SAND, poorly graded SAND, uniformly graded SAND, gap graded
		5–35	SAND, silty SAND WITH FINES SAND, clayey	SM SF SC	SML, etc. SCL, etc.	SAND, silty, of low plasticity SAND, clayey, of low plasticity
FINE SOILS: More than 35% of material less than 60 mm is finer than 0.06 mm	SILTS and CLAYS, Gravelly or sandy	35–65	SILT, gravelly FINE SOIL, gravelly CLAY, gravelly	MG FG CG	MLG, etc. CLG, etc.	SILT, gravelly, of low plasticity CLAY, gravelly, of low plasticity
			SILT, sandy FINE SOIL, sandy CLAY, sandy	MS FS CS	MLS, etc. CLS, etc.	SILT, sandy, of low plasticity CLAY, sandy, of low plasticity
	SILTS and CLAYS	65–100	SILT (M-Soil) FINE SOIL CLAY	M F C	ML, etc. CL CI CH CV CE	SILT of low plasticity CLAY of low plasticity CLAY of intermediate plasticity CLAY of high plasticity CLAY of very high plasticity CLAY of extremely high plasticity
ORGANIC SOILS	ORGANIC sand, silt or clay			Descriptive letter O suffixed to any symbol		
PEAT	Predominantly plant remains which may be fibrous or amorphous			Pt		

* Material coarser than 60 mm is removed and recorded as COBBLES (60–200 mm) or BOULDERS (over 200 mm).

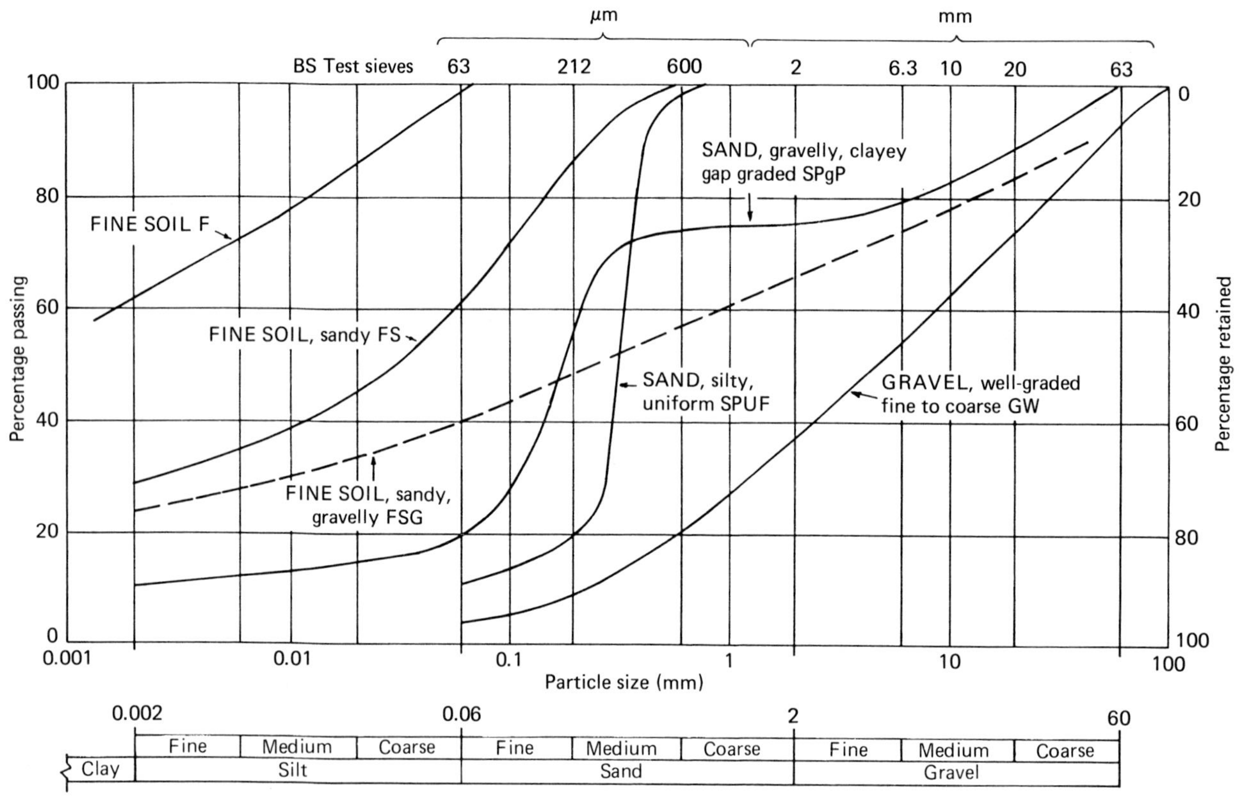

Figure 4.10 Grading chart for soils with grading curves for selected soil types

Particle size distribution, if determined by laboratory tests, may be plotted on a grading chart (Figure 4.10). The grading curve will assist in determining the proportions of different materials present in the soil, and thus the designation of the soil.

4.9 Description of soils

The soil name, which provides information on the grading characteristics of the soil, is usually a combination of a basic soil type with secondary admixtures of other basic soil types. Supplementing the name are descriptive terms for characteristics of the soil material and the soil mass.

In a soil description the main characteristics preferably should be given in the following order:

SOIL NAME
- Including minor constituents
- Genetic type

Soil material properties
- Colour
- Texture
  - Particle shape and composition
- State of weathering
- Strength
  - Consistency
  - Undrained shear strength
  - Moisture condition
  - Relative density
  - Compactness

Soil mass properties
- Structure
- Discontinuities
- Weathering profile

4.9.1 Descriptive soil name

Basic soil types

Basic soil types are named after soil particle sizes falling within specified limits (Table 4.14).

Table 4.14 Particle size distribution for basic, single component, soil types

Soil name	*Qualifying term*	*Particle size* (mm)
Boulders		Over 200
Cobbles		60–200
	Coarse	20–60
	Medium	6–20
	Fine	2–6
	Coarse	0.6–2
Sand	Medium	0.2–0.6
	Fine	0.06–0.2
	Coarse	0.02–0.06
Silt	Medium	0.006–0.02
	Fine	0.002–0.006
Clay		Under 0.002

Composite soil types

Most soils are composite types, containing a range of basic soil types. The proportion by weight of each component is then used as the basis for naming the soil (Table 4.13).

Coarse composite soils
As noted in the classification table (Table 4.13), a coarse soil contains 65% or more of soil in the particle size range 0.06–60 mm: the sand and gravel sizes. Composition of the coarse fraction may be described using the terms in Table 4.15.

The gravel and sand sizes may be subdivided into coarse, medium and fine (Table 4.14) and the type of grading can be specified (soil name in Table 4.13). Fine soils are absent.

Table 4.15 Description of sand and gravel composite soil types

Term	*Description*
GRAVEL	Up to 5%* sand
GRAVEL, sandy	5–35% sand
GRAVEL/SAND	Approximately equal proportions of gravel and sand
SAND, gravelly	5–35% gravel
SAND	Up to 5% gravel

*Percentages are of whole material less boulders and cobbles, and are approximate estimates in a field description.

Very coarse composite soils (boulders and cobbles)
In very coarse soils over 50% of the material is over 60 mm. Table 4.16 gives descriptive terms for these soils.

Mixtures of very coarse soils with finer materials (i.e. coarse and fine soils)
Very coarse-grained mixtures with finer materials may be described on the basis of approximate estimates of composition (Table 4.17).

The name of the finer material, which may be gravel, sand or silt, or mixtures of these, should be given (see Table 4.13), e.g. SAND, gravelly, with some boulders; CLAY, sandy, with some cobbles.

Organic soils
Organic clays, silts and sands contain appreciable quantities of organic vegetable matter, may have a marked odour and a dark grey, dark brown or dark bluish-grey colour. Fine soils with large amounts of organic matter usually plot below the A-line (Figure 4.11) as organic silts, and have high to extremely high liquid limits.

Peat consists predominantly of plant remains which may be fibrous or amorphous.

Plasticity
Clay and silt, alone and in mixtures with coarse material, may be classified in terms of plasticity (Table 4.18). As an alternative, a classification may be made in terms of plasticity index (Table 4.19).

Table 4.16 Descriptive terms for very coarse soils

Main name	*Qualifying name*	*Description*
BOULDERS or BOULDER GRAVEL	Cobbly	Over 50% of very coarse material is over 200 mm: boulder size
COBBLES or COBBLE GRAVEL	Bouldery	Over 50% of very coarse material is cobble size: 60–200 mm
COBBLE/BOULDER GRAVEL		Equal proportions of cobble and boulder sizes

Table 4.17 Descriptive terms for mixed very coarse and finer grained soils

Term	*Description*
BOULDERS (or COBBLES) with some finer material	5–20% finer material
BOULDERS (or COBBLES) with much finer material	20–50% finer material
FINER MATERIAL with many BOULDERS (or COBBLES)	50–20% boulders (or cobbles)
FINER MATERIAL with some BOULDERS (or COBBLES)	20–5% boulders (or cobbles)

Table 4.18 Terms for description of plasticity of fine soils and fine soil fractions in coarse soils

Term	*Range of liquid limit (%)*
Of low plasticity	Under 35
Of intermediate plasticity	35–50
Of high plasticity	50–70
Of very high plasticity	70–90
Of extremely high plasticity	Over 90

Table 4.19 Terms for description of plasticity index classes for fine soils and fine soil fractions in coarse soils

Term	*Plasticity index (%)*
Non-plastic	Under 1
Slightly plastic	1–7
Moderately plastic	7–17
Highly plastic	17–35
Extremely plastic	Over 35*

*For very 'fat' clays.

Genetic type of deposit
Descriptive soil names are usually supplemented by reference to the type of deposit described where this is not evident from the name of the geological formation. Examples are: alluvium, eluvium, colluvium, talus, scree, loess, till, moraine, boulder clay, outwash, aeolian, dune, beach sand, marine, deltaic, tidal swamp, swamp, lagoonal, littoral.

4.9.2 Description of soil material

4.9.2.1 Colour

Soil colour may be evaluated quantitatively using the Munsell Soil Color Chart (Munsell, 1941) or the simple system recommended for rocks (see Table 4.2).

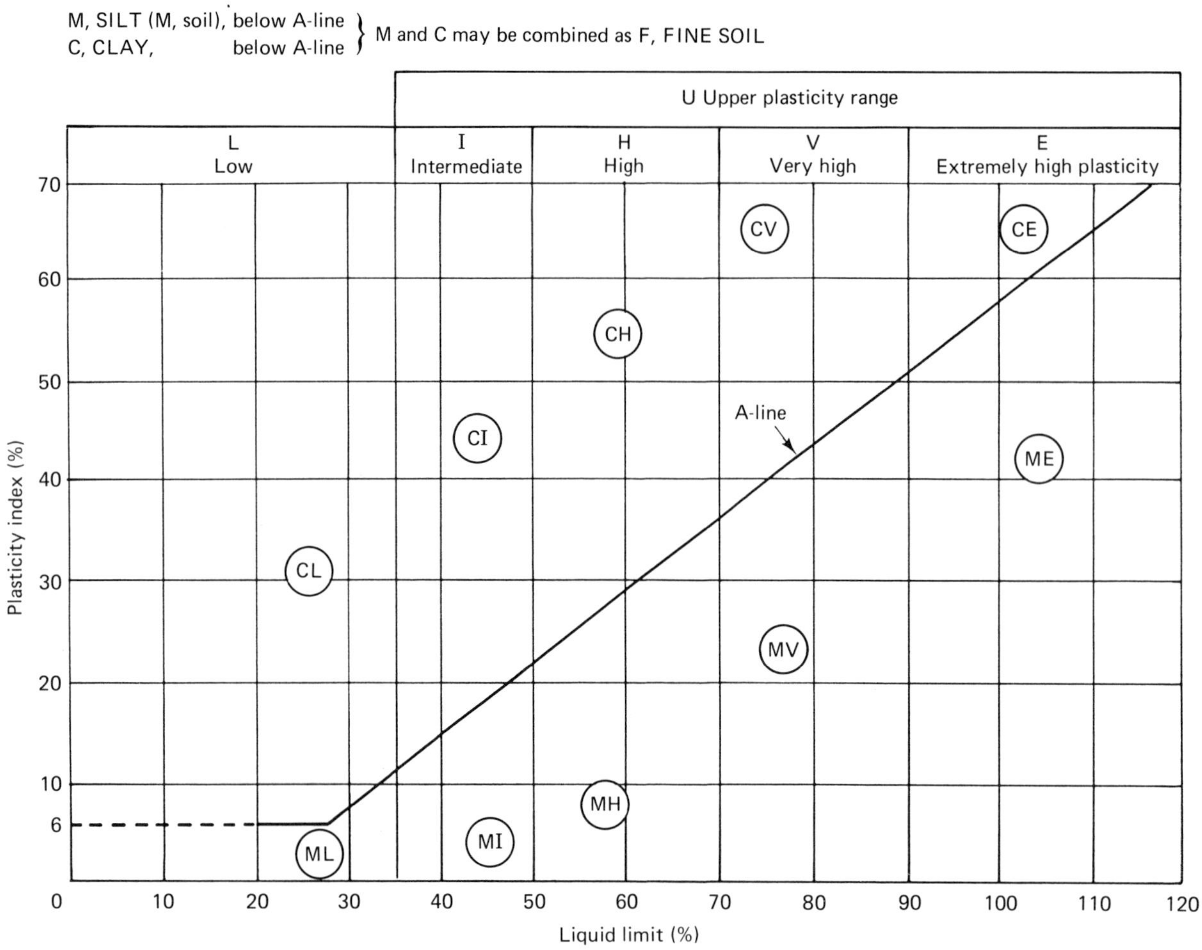

Figure 4.11 Plasticity chart for the classification of soils and the finer part of coarse soils

4.9.2.2 Particle shape and composition

Where appropriate, particle shape may be described by reference to the general form of the particles, their angularity which indicates the degree of rounding at corners and edges, and their surface characteristics (see Table 4.3). The surface of particles may be polished, etched, pitted, honeycombed, as examples of additional descriptive terms.

Composition of particles visible to the naked eye or under a hand lens may be described. Larger particles are usually rock fragments and may be so described (see Table 4.1). Sand and finer grains are usually individual mineral grains, e.g. mica, quartz, feldspar. Gravel and sand particles may be coated with mineral matter, e.g. limonite, calcite. Crystals, often of large size, may be present, e.g. gypsum in clay, pyrite in chalk. Shells may be an important constituent of soils.

4.9.2.3 State of weathering

The same terms applied to rock may be used to describe the weathering of soil material; namely, discoloration, decomposition and disintegration. Descriptive terms, quantified as in Table 4.4, imply changes from fresh soil, but the resultant products of the progressive weathering of soils are also engineering soils that may be described individually in the recommended manner.

Most obvious of the changes will be discoloration in fine soils, and discoloration, colour-banding, cracking, disintegration and decomposition of individual mineral and rock fragments in coarse soils.

4.9.2.4 Strength

Strength may be estimated by simple field tests on disturbed and undisturbed samples or by laboratory tests on undisturbed samples.

For fine soils, with over 35% of silt and clay size, consistency may be estimated using simple manual tests (Table 4.20) on disturbed samples. Undrained shear strength of fine soils may be classified and described as in Table 4.21, and the strength values are directly related to the estimations made by the simple manual tests given in Table 4.20.

Table 4.20 Field tests for consistency of fine soils

Term	*Field test*
Very soft	Exudes between fingers when squeezed in hand
Soft	Can be moulded by light finger pressure
Firm	Can be moulded by strong finger pressure
Stiff	Cannot be moulded by fingers; can be indented by thumb
Very stiff or hard	Can be indented by thumb nail

Table 4.21 Classification of fine soils in terms of undrained shear strength

Term	*Undrained shear strength* ($kN\,m^{-2}$)
Very soft	Under 20
Soft	20–40
Firm	40–75
Stiff	75–150
Very stiff	Over 150

If numerical, laboratory determined values for moisture content (m), liquid limit (LL), and plastic limit (PL) are available, the physical state of fine soils can be closely expressed by the consistency index (CI):

$$CI = \frac{LL - m}{LL - PL}$$

A scale in terms of consistency index is given in Table 4.22.

Table 4.22 Classification of fine soils in terms of consistency index

Term	*Consistency index*
Very soft	Less than 0.05
Soft	0.05–0.25
Firm	0.25–0.75
Stiff	0.75–1.00
Very stiff or hard	Over 1.00

For coarse soils (see Table 4.13), the compactness of the soil material may be estimated using the simple field tests specified in Table 4.23.

Table 4.23 Field test and descriptive terms for compactness of coarse soils (gravels and sands)

Term	*Field test*
Loose	Can be excavated with spade; 50 mm wooden peg* easily driven
Dense	Requires pick for excavation; 50 mm wooden peg hard to drive
Slightly cemented	Visual examination; pick removes lumps which can be abraded

*The peg should be of square cross-section and sharpened to a 100 mm long point.

The relative density of sands and gravels is usually determined by the standard penetration test (SPT). A scale in terms of *N*-values is given in Table 4.24.

Table 4.24 Standard penetration test *N*-values and descriptive terms for coarse soils

Term	*SPT N-values* (blows/300 mm penetration)
Very loose	1–4
Loose	4–10
Medium dense	10–30
Dense	30–50
Very dense	Over 50

Relative density of sands and gravels may also be determined from the void ratio (e_n) of the soil *in situ* and the void ratios corresponding to the loosest (e_{min}) and densest (e_{max}) laboratory states for a given soil. Relative density (RD), classified in Table 4.25, is derived from:

$$RD = \frac{e_{max} - e_n}{e_{max} - e_{min}} \times 100$$

Table 4.25 Relative density of sands and gravels

Term	*Relative density* (%)
Very loose	Less than 20
Loose	20–33
Medium dense	33–66
Dense	66–90
Very dense*	90–100

*Usually weakly cemented.

4.9.3 Description of soil mass

The description of the soil mass requires information additional to the description of the soil as a material.

A soil mass is first described as a soil material, or as a combination or mixture of soil types, followed by information of engineering significance concerning the soil mass. Such information includes:

- the geological structure and the distribution of soil types in the soil mass;
- the characteristics of discontinuities in the soil mass;
- the weathering profile.

Features of the soil mass can only be determined by field observation of pits, trenches and larger excavations; from natural exposures; or from a sufficient number of undisturbed samples.

4.9.3.1 Structure

The larger aspects of structure are related to folding and faulting of stratified and interstratified deposits. Beds, layers and lenses of different soil types may be present, and individual layers may be gradational from one soil type to another. Folding and faulting introduce gross defects into the soil mass, and interstratification causes heterogeneity and anisotropy, all of which may be more important from the viewpoint of engineering properties than the properties of individual types of soil material.

Special bedding characteristics should be described; for example, disturbed bedding, cross-bedding, and structures associated with bedding surfaces such as ripple marks. Thin laminae of silt in clay, or fine sand in silt, as two examples, give rise to partings that separate easily. Intercalations and lenses of soft clayey or organic matter, or of soluble materials, should be described in detail because of their very important influence on engineering properties and the behaviour of the soil mass.

Pedological features, such as roots and root holes penetrating the soil mass, should be described.

4.9.3.2 Discontinuities

Discontinuities in soils, including fissures and shear planes, should be described and quantified using the terms already given for rocks (see subsection 4.6.4.2).

4.9.4 Weathering profile

Weathering of the soil mass should be described in the same way as for rock (see subsection 4.6.3.3), including the gradation to rock if the soil profile has resulted from weathering of rock. However, Table 4.12 has to be modified if it is to be generally applicable to all soils, other than very stiff or hard, jointed and fissured fine soils. Weathering grades for the soil mass are proposed in Table 4.26. Relics of the original soil of weathering grades I and II are present in grades III and IV, corresponding to the corestones in the rock mass weathering grade classification (see Table 4.12).

Table 4.26 Weathering grades for soil mass

Term	*Description*	*Grade*
Fresh	No visible sign of soil material weathering; perhaps slight discoloration on major discontinuity surfaces	I
Slightly weathered	In fine soils, discoloration indicates weathering of soil material and discontinuity surfaces; there is not a marked change in consistency of the discoloured soil. Relics of fresh soil may be present. In coarse soils, individual fragments and discontinuities are discoloured; there is no marked change in relative density	II
Moderately weathered	In fine soils, soil discoloured; less than 35% of soil will show a marked change in consistency; relics of grade I and/or II present. In coarse soils, less than 35% of soil has markedly lower relative density	III
Highly weathered	In fine soils, soil discoloured and more than 35% of soil will show marked change in consistency; relics of grade II present. In coarse soils, more than 35% of soil has markedly lower relative density	IV
Extremely weathered	In fine soils, soil discoloured, relics of grade II absent, soil shows marked change in consistency from fresh soil. In coarse soils, marked decrease in relative density	V

4.10 Additional geological information for rock and soil description

Wherever possible, additional information – as detailed below – should be given.

4.10.1 Geological formation

The name of the geological formation can usually be found published on geological maps, and should be given in the form determined by local custom, for example London Clay, Bannisdale Slates, Morrison Formation.

4.10.2 Age

The age of the formation should be stated in terms of System, Series, Stage or whatever stratigraphical subdivisions are available.

4.11 Made ground

Made ground may be of two types, conveniently referred to as 'engineered fill' and 'non-engineered fill'.

4.11.1 Engineered fill

Excavated natural soils may be used as construction materials particularly for embankments, as general fill in large-scale cut-and-fill grading operations for housing developments, or dumped as ground fill wherever volumes available are in excess of engineering requirements. It is possible to carry out classification tests on such materials and to describe them using the proposals in sections 4.4 and 4.5.

Run of quarry rock material as produced by blasting, or aggregates prepared by controlled crushing and grading, may similarly be used as construction materials. Although difficult to sample if the particle size is very large, these engineering materials may also be classified and described using sections 4.8 and 4.9.

4.11.2 Non-engineered fill

Except where natural soils and rock aggregates have been used as construction materials, it is seldom possible to carry out classification tests on materials in heaps or that have been used to infill excavations. This is particularly the case where urban demolition has taken place, or where excavations or areas of low-lying ground have been used either for garbage disposal or the dumping of industrial wastes of all kinds.

A clear and complete description is therefore of the greatest importance, and in addition to describing normal soil constituents should include the details included in the following subsections.

4.11.2.1 Mode of origin

The fill may have been placed as a single dumped layer or as many thin layers with or without blanketing of each layer by natural soils. Modes of origin include household and industrial garbage, industrial wastes of all kinds, including waste-solids, chemicals and liquids, and general demolition debris from urban renewal. Fill may be layered or 'cross-bedded', depending upon the method of placement.

4.11.2.2 Presence of large objects

Masses of masonry, reinforced concrete and steel girders may, for example, be present.

4.11.2.3 Voids

Urban fill may obscure partly filled or unfilled cellars of previously existing buildings, sewers and other underground pipes, empty oil drums and other hollow objects including tyres, dumped automobiles, refrigerators, etc.

4.11.2.4 Chemical and organic matter

Many areas of non-engineered fill may represent the sites

of disused quarries and clay pits that have been used for the storage and disposal of solid and liquid chemical wastes which may be inorganic or organic in character.

4.11.2.5 Toxic materials including gases

Acid wastes may react with limestone and concrete to produce carbon dioxide. Decomposing plant and animal matter will evolve methane, ammonia and other toxic gases. The dumped materials may themselves be toxic.

4.11.2.6 Age

Most easily dated are old newspapers, magazines and calendars which provide an indication of the maximum age of the deposit.

4.11.3 *Effects of non-engineered fill on adjacent* in situ *rocks and soils*

Industrial liquids, leachates formed by percolation or meteoric water through the fill, and other solutions, may percolate into adjacent and underlying rocks and soils. Resultant reaction with, or precipitation within, the *in situ* rock and soil mass may bring about extensive alteration, masking their original character.

5

How to present data

5.1 Introduction

Understanding, in other words being able to interpret and use, a topographic map is a basic skill that is readily acquired. Training, together with more than a little practical application in the field, is needed if the user is to become a competent map reader. Manuals of map reading, often combined with field sketching, are usually published by the army in different countries (Anon., 1921).

Proficiency in map reading, and this applies equally to topographic as well as to the various types of geological maps, requires an understanding of the many conventions by which the natural and artificial (man-made) features of the country are portrayed. In practice this has to be coupled with the power to visualize accurately the features represented, making full use of all the information which a map is intended to provide. It is therefore necessary to interpret the various cartographic symbols, and for this the guidance of the map legend or explanation is of paramount importance.

Maps and plans must be made so that, as with standard topographic and cultural maps, the end product is readily intelligible to prospective users: planners, engineers, decision-makers and the general public. A group of users may have the background knowledge and training to be able to interpret maps of basic data. Other groups, and the most often quoted example is the planners, may be more at home with interpretative maps, produced with the needs of the user firmly in mind.

Brook and Marker (1987) have very fairly commented that 'Turning to applied earth scientists, whether they be geologists, engineering geologists, hydrogeologists, mining engineers, geochemists, geophysicists, geomorphologists or soil scientists, it is equally the case that their professional languages are as opaque to planners as they are in many cases between the individual disciplines. Similarly, planning terminology is often not fully understood by these specialists. This requires an exercise in communication. The applied earth scientist must appreciate the form in which information is digestible by planners (and engineers). Conversely the planner (and the engineer) must appreciate what to look for and what questions to ask.' There is, fortunately, a common skill between geologists, planners and engineers – familiarity with maps. What to put on a map, and how to present it, are therefore of prime importance.

5.2 Format and content of engineering geological maps

5.2.1 Format

Published in 1972, The Working Party Report on 'The preparation of maps and plans in terms of engineering geology' (Anon., 1972) made specific recommendations on the format and content of engineering geological maps, and these are of general application.

An engineering geological map may be supplemented by notes in the margins, as is done by the United States Geological Survey; on its reverse side, as on maps of the New Zealand Geological Survey; or in an accompanying pamphlet or table, as for maps produced in Eastern European countries, the United States Geological Survey and the Milton Keynes engineering geological map produced in manuscript by the British Geological Survey.

For a single sheet format (Figure 5.1), a main centrally placed base map at a suitable scale which depicts topography and geology may be overprinted with engineering geology lines and symbols in appropriate colours. If desired, interpretation of data can be shown at similar or smaller scales, for the whole or parts of the area covered by the main map, by means of patterns or contoured marginal diagrams. Three-dimensional interpretation of structure is assisted by one or more cross-sections, and the geological succession may be shown by a vertical column with brief notes on relevant rock properties. Lengthy descriptive material should not form part of the map but should be presented separately. It is not entirely satisfactory that important information should be printed on the reverse of the map sheet. Unless two copies of the map can be used, constant turning over of the single sheet may be necessary to relate explanation and map details.

Where the importance of the mapped area or the amount of data available warrants it, the results of an

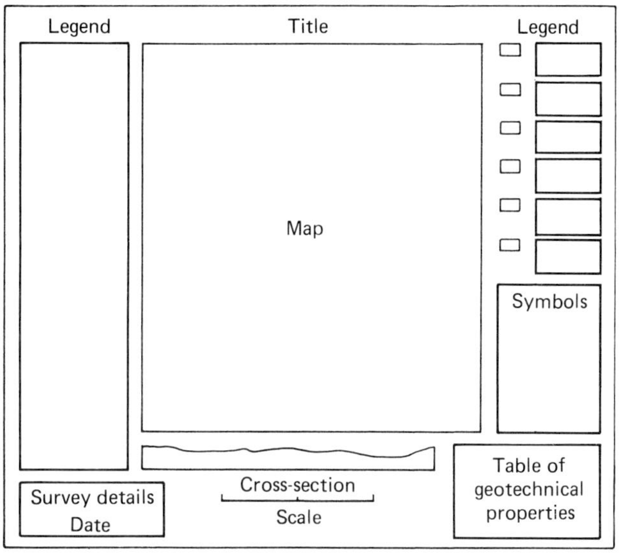

Figure 5.1 Schematic layout of an engineering geological map

engineering geological survey should be presented in a series of maps. For example for general planning of a new town area, a set of maps might comprise:

- geology;
- hydrology;
- geomorphology with slope stability conditions;
- foundation conditions;
- sources of data, boreholes, pits, trenches, geophysical survey, archival material.

Each map should use the same topographic base, usually printed in subdued colour.

Alternatively, when more than one sheet is used, a base geological map may be supplemented by transparent overlays of the other aspects, although this is certain to be more expensive and less convenient than printing separate sheets.

Engineering geological maps can be based on topographic maps of the Ordnance Survey or on maps specially prepared, for example, from aerial survey.

It is desirable that all sheets, cartographic or otherwise, should be the same size, adequately identified and cross-referenced. The use of colour on printed maps is strongly recommended for ease of cartographic representation. In general, colour should be used to emphasize engineering geology features including geomorphology and groundwater conditions, leaving geology and topography relatively subdued. Where possible, the colouring of lines, patterns and symbols should accord with generally accepted or published practice in other related fields. In the UK, the British Geological Survey has its own standard patterns and symbols, and the same applies to other official geological surveys (Berkman, 1982, for Australia, is one example).

The International Standards Organization has yet to complete its work on standard symbols and patterns for geological maps. The International Association of Engineering Geology has published recommendations for international use (Anon., 1981a) which mainly follows Anon. (1972), although there are important differences. For site investigation, the British Standards Institution (Anon., 1981b) also follows Anon. (1972), with the noticeable exception of a symbol for silt and siltstone that is firmly embedded in the mythology of civil engineering site investigation.

5.2.2 Content of engineering geological maps

In the Working Party Report (Anon., 1972) on engineering geological maps, outline recommendations are given for the content of British maps and those prepared following established British practice in other countries. The report is quoted more or less verbatim on various aspects of content.

Geological aspects

Mappable units
Division of rocks and soils into mappable units should be on the basis of descriptive engineering geological terms using either the recommended symbols or distinguishing colours.

Geological boundaries
An indication should be given of the accuracy of geological boundaries between mappable units and of certain structures such as faults, using the recommended convention.

Description of rocks and soils
The general description of rocks and soils should be in the terms set out in Chapter 4. Information should be given on lateral and vertical variations in lithology and about variations in thickness.

Description of exposures
Notes referring to exposures of rocks and soils should give descriptions in the recommended form with structural features marked using the symbols listed.

Description of weathering and alteration
Notes on the distribution, depth and nature of weathering and alteration must be given wherever possible.

Description of joints and other structural planes
As much information as possible must be given about jointing of rock, including frequency, direction, inclination, weathering or staining, whether open or closed or filled, if filled with what, whether or not slickensided and other surface characteristics. Similar detailed information should be given about other structural planes including bedding and cleavage.

Visible faults should be described in detail, including particulars of gouge, slickensiding, width of shatter zone, extent of drag and amount of throw if measured or estimated.

Horizons known to be subject to shearing should be indicated and shear surfaces and shear zones should be particularly noted and described.

Subsurface conditions
Where enough information is available, and where appropriate, isopachytes and contours of selected horizons, for example rock-head, seam contours in worked coal, may be given.

Hydrogeological aspects

Availability of information
Reference should be made to hydrogeological maps and publications if they are available for the area.

General hydrogeological conditions
Where information on hydrogeology is not otherwise available, there should be a note in the legend about the general hydrogeological conditions prevailing in the area. This could include notes on groundwater basins, artesian overflow areas, qualitative permeability, boundaries between saline and potable water.

Hydrogeological conditions should be quantified wherever possible and all relevant hydrogeological information should be given, including piezometric levels, coefficient of permeability based on *in situ* or laboratory determinations, storage coefficient and geochemical parameters (groundwater and formation water chemistry).

Hydrogeological properties of rocks and soils
Explanation in the legend should include information on hydrogeological properties of the rocks and soils. Aquifers, aquitards, aquicludes and aquifuges should be distinguished by a simple convention.

Springs and seepages
Springs, both permanent and intermittent, should be marked by symbols; seepage lines should be indicated by notes or symbols. Periodic streams, or bournes, should be shown wherever possible. Flows should be quantified wherever possible.

Geomorphological aspects

General
Geomorphological features should be shown by notes on the map, for example steep scarp, gentle dip slope, or by the symbols given wherever practicable or where space permits.

Mass movements
Landslips, subsidences, solifluxion features, cambering phenomena and any other mass movement features should be shown using the symbols given.

Seismicity and other geodynamic phenomena
Any recorded seismic activity should be noted.

Boreholes

Boreholes and wells should be classified according to purpose, and details should be given of the rocks and soils penetrated.

Site investigations and tests

Construction and other sites should be marked for which a site investigation has been made. The locations of any isolated samples taken for engineering laboratory tests should be noted.

Relevant geophysical information could include sites and types of individual measurements, for example an expanding electrode resistivity measurement; contours showing measured quantities and derived physical properties, for example resistivity and gravity; geological features interpreted from geophysical measurements, for example the line of a fault.

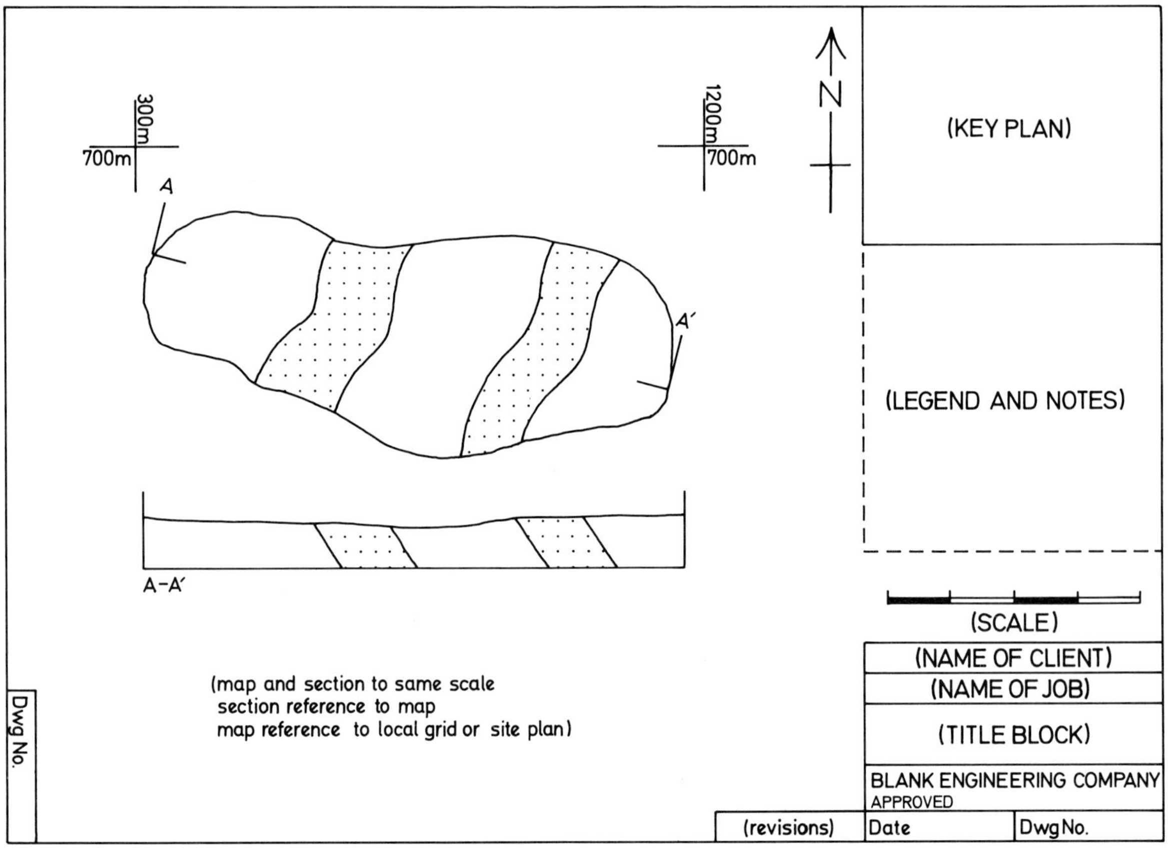

Figure 5.2 Suggested layout for large drawings

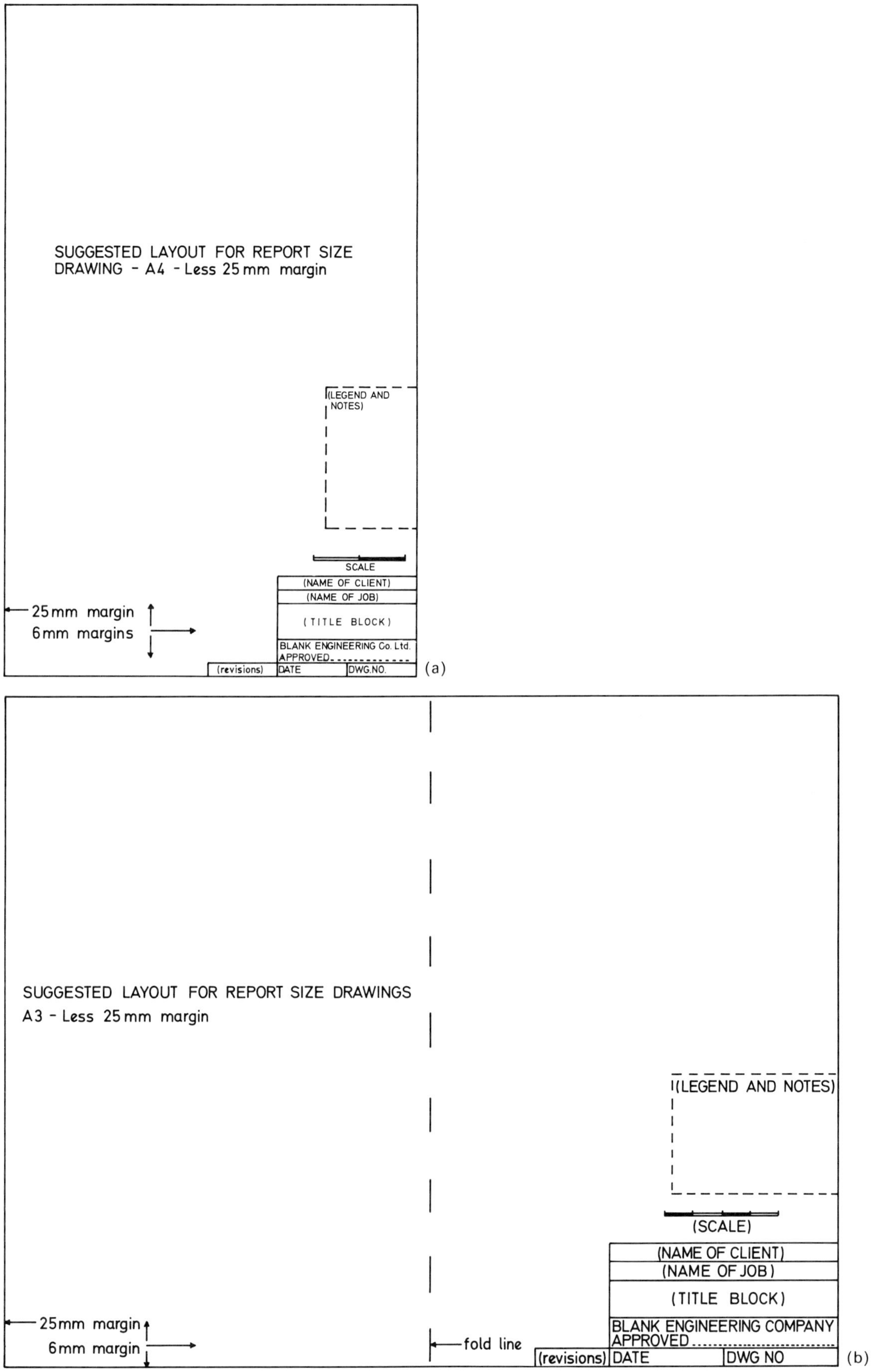

Figure 5.3 Suggested layout for report sized drawings. (a) A4 size drawing, less 25 mm binding margin on the left. An alternative layout is with the long side of the A4 sheet forming the bottom edge of the drawing. (b) A3 size drawing, less 25 mm binding margin on the left. This layout can be extended in width with additional folds

Mines and quarries
All mines and quarries, whether active or abandoned, should be noted together with dates of working, notes on material extracted, and whether plans are available.

Areas of made ground should be shown with notes on the original nature of the material and the date of dumping.

The date of every observation should be recorded.

5.3 Layout and content of engineering geological plans

5.3.1 Format

Engineering geological plans are just one type of engineering drawing and, as such, the layout should conform to standard practices. Individual organizations usually have their own house style, although in general this is some variant of the layout in Figure 5.2 for large drawings. Drawings to be included in a report are most conveniently of the standard A4 paper size (Figure 5.3) or A3, both with a binding margin on the left and ample trimming margins on the other edges.

5.3.2 Content

Engineering geology plans will normally be based on site plans made specially for the location, but for certain purposes may be produced by enlarging existing reliable maps to an appropriate scale.

The notes on the plan should be supplemented by an accompanying detailed report illustrated with relevant photographs, sketches, logs, sections, detailed plans at a larger scale and test data. The various aspects that have to be dealt with are similar to those for maps (see section 5.2.2).

5.4 Supplementation of existing geological maps for engineering purposes

In many countries there is adequate coverage by conventional geological maps, although no engineering geological maps may be available, and indeed may not be deemed necessary. However, should engineering information be required, this may be achieved either by expert interpretation of an existing geological map (see section 10.2) or by supplementation of the appropriate geological map. What can be done depends upon the nature of the project and the scale of the available maps. There are a number of good examples that illustrate what it is possible to achieve either as part of a routine engineering geological survey or for a specific project.

First we may consider general recommendations for the supplementation of existing geological maps. These recommendations apply particularly to the UK, where official geological maps, of varying quality, are now available on the scale of 1:10 560 or 1:10 000 covering at least 85% of the land area. This tremendously valuable geological archive is supplemented by published maps at 1:63 360 or 1:50 000, for many of which descriptive memoirs have been published.

The 1:10 560 maps are published only for those areas of economic importance including London, the main coal-mining areas, and many of the metalliferous mining areas. All others are available as photographic copies. This extensive coverage is itself supplemented by maps published in the proceedings of many learned societies, university theses, etc.

Together with the maps of the Geological Survey, these provide a very valuable source of basic geological map information, generally at a scale marking the lower acceptable limit for engineering geological plans. At the published scale the maps are useful at the feasibility or preliminary planning stage of an engineering undertaking; photographically enlarged they could, and frequently do, form the basis of site investigation or pre-construction plans.

Direct supplementation of geological maps could be made:

- by additional descriptive information in engineering geological terms either as an additional explanation or legend added to the margin of a published sheet or to a photocopy of an archival sheet;
- by separately printed notes;
- by additions to the printed map or a photocopy of an archival sheet giving engineering geological information for a specific project.

Some, or all, of the aspects listed below may be of engineering significance at a particular site. The scale of the map has a direct bearing on the degree of supplementation possible.

5.4.1 Recommendations for supplementation of existing geological maps

The following recommendations apply specifically to maps at scales of 1:10 000 or smaller (Anon., 1972). It is especially important that the final map does not become too cluttered with detail, so causing cartographical problems involved in superimposing additional engineering geological information onto basic geological information. As a generally acceptable rule, 50 symbols are the maximum number that can be used on a map, although it may be possible to exceed this by the use of overprints in colour.

Perhaps the best method of supplementing an existing map is by the use of transparent overlays, when separate overlays may be used for different categories of additional engineering geological information. On a routine basis this method may be preferable because of the ease of updating both the information on the map and any attempted interpretation. By using the dye-line process, copies of supplemented Geologial Survey documents can be provided quickly and cheaply, either singly for individual use or as multiple copies for a client. Supplemented maps may show some or all of the aspects dealt with in section 5.2.2.

5.4.1.1 Examples of map supplementation

Examples of supplemented maps are uncommon in the literature. In consequence, for the Working Party Report (Anon., 1972), part of an unpublished 1:10 560 geological map prepared by a committee member had additional information included, based on his personal experience of the area. Mainly, the additions on Figure 5.4 refer to discontinuity patterns, details of weathering grades and extremely weathered zones, and some geomorphological features including springs, slope changes and a solifluction lobe. The area is covered by solifluction debris, referred to as 'head' deposits on maps of the British Geological Survey, and by widely scattered granite and tuff boulders.

The explanation, of which only a part is reproduced, gives more details on the superficial soils and solid rocks than is usual in the legend to a geological map. Descriptions of soils and rocks are in the style recommended in Chapter 4.

For any supplementation exercise, the intended purpose must be clearly focused; there seems to be little point in undertaking an exercise in supplementation for a general and therefore unspecified purpose. There follows an example of a medium-scale geological map, published in colour, supplemented during the reconnaissance work done on the 110 mile Suva–Nandi Highway, Fiji, by the engineer. The original map was only available on a 1:50 000 scale.

The mapped sheet is the base geology map (Figure 5.5) to which engineering geological observations have been added. Information is included on the potential borrow material (which was in relatively short supply), engineering characteristics especially grade of weathering, slope stability and foundation conditions, and future pit, auger and drill hole locations which were intended to form part of the succeeding site investigation. The reconnaissance was carried out with a Land Rover by an engineering geologist, with the soils engineer directly involved in the design of the road. The reconnaissance traversed the route of the existing old gravel road which approximately coincided with the alignment of the proposed highway.

The reconnaissance started in London with a 5 man-day study of available information on geotechnical observations of residual soils of the area, local geology, topography and climate. This was followed by two weeks of field work: the first week was spent travelling the old road and making field observations; the second week was occupied in writing up, revisiting selected locations, collecting local data and planning the proposed site investigation. The site investigation, which took over a year to complete, elaborated on the reconnaissance observations, and provided detailed data on bridge sites, foundation conditions, slopes and borrow materials. The contract for the site investigation was let after a competitive bid on tender documents developed from the work carried out on the reconnaissance. Preliminary designs went ahead on the results of the reconnaissance mapping and on the detail added and modified during and after completion of the full site investigation (Lovegrove and Fookes, 1972).

5.5 Other aspects of presentation

Although in later chapters the many illustrations deal effectively with what maps and plans do and what they should look like, it is important here to highlight those aspects of general application in engineering geological cartography.

5.5.1 Scale

The importance of avoiding empty enlargement, using a scale that is larger than necessary to show all the required detail, has already been mentioned. Equally important is the use of a small range of 'standardized' scales within each of the classification divisions into small, medium and large scales. For example:

1:50	for plans of pits and trenches
1:500	for site plans
1:2500	for large-scale Ordnance Survey plans
1:10 000	for medium/large-scale Ordnance Survey plans and basic, detailed geological maps
1:25 000 1:50 000 1:250 000	for geological maps at successively smaller scales

and so on.

Another point is to avoid using odd or obsolete scales, if possible; in the UK, this implies using 1:10 000 rather than 1:10 560, and 1:50 000 rather than 1:63 360 – and not 1:21 120 as used on the engineering map of Belfast (see Figure 10.22). This may be a counsel of perfection, as maps at more suitable scales may not be available.

5.5.2 Number of maps and plans

Trying to portray too much information on a single map sheet is counter-productive because the map becomes very difficult, if not practically impossible, to decipher. The recommended four map sheets, each with different content, are ideal for general use. Even with large-scale geotechnical plans (see Figures 10.15, 10.16 *et seq.*), different aspects of ground treatment may be displayed on separate sheets to advantage.

5.5.3 Three-dimensional representation

Interpretation of geological maps requires training, skill and the application of certain rules if the two-dimensional picture presented by the map is to be translated into an accurate three-dimensional view of the structures or the relative dispositions of the various structural units. In solid formations, indications of the dip of strata at different points, rules for the construction of folds from a number of dips, and the intersection of geological boundaries with topography all provide means of projecting the surface information below ground level.

There are many standard texts on the subject. Lahee (1961) has provided a manual of field geology that, in addition to including a wealth of information on techniques for making geological maps (Chapter 3), shows how the maps may be interpreted and the geological history of the area thereby deciphered. Usually the geological structure is portrayed by a geological section akin to cutting a deep vertical slice into the ground on which the structures can be seen in cross-section. Busk (1929) has provided a technique for the accurate geometrical construction of sections through a certain type of fold which, if used with care, is of general application.

Standard methods for geological map interpretation are always concerned more with the so-called solid

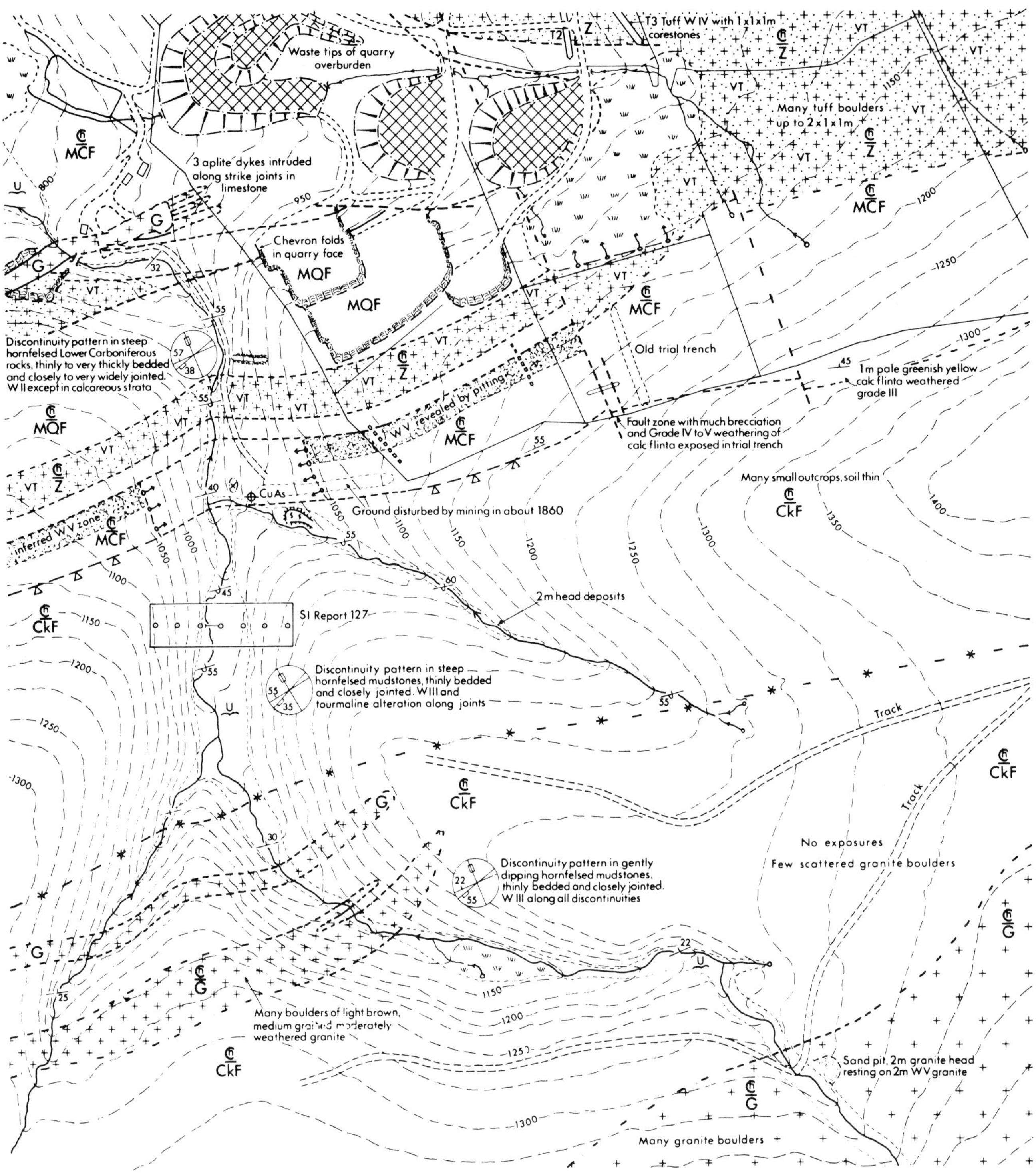

(a)

formations than the superficial deposits. However, in many parts of the world, it is the younger superficial deposits, to depths of a few metres or at the most a few tens of metres, that are the prime concern of the foundation engineer. Such deposits are usually devoid of structures in the usual sense, except for horizontal layering and intercalations of different soil types all lying on solid bedrock that might have a marked hidden topography. In these circumstances, the usual methods of geological interpretation are not applicable, and the necessary data may have to be acquired specially by boring or pitting, or indirectly by geophysical means.

One of the most difficult problems in engineering geological cartography is how to show on a plan the spatial relationships of near surface rock and soil units. Geological sections, fence diagrams or block diagrams provide an easy answer. However, to show on a plan the lateral extent, lithological and thickness variations and depth below surface of a number of geological units, in their correct order, and also to represent their lithological characteristics and some engineering properties, requires both very detailed investigation and careful attention to cartography.

Several methods have been developed, including

EXPLANATION

SUPERFICIAL DEPOSITS (DRIFT) — Thickness in ft.

RECENT AND PLEISTOCENE

ALLUVIUM — up to 8

On the granite, alluvium is a brownish-yellow, loose, sub-angular, coarse gravelly SAND with some peat and rounded boulders of moderately weathered granite up to 3 ft, and pebbles of quartz. Downstream, alluvium is a silty gravelly SAND with rounded granite boulders up to 3 ft and sub-angular cobbles and boulders of the solid rocks. The deposits are moderately to highly permeable. Locally much disturbed by streaming for tin.

U

RIVER TERRACES (UNDIFFERENTIATED) — up to 40

Dark yellowish-brown, loose but locally weakly to strongly cemented in horizontal layers by manganiferous or ferruginous material, sub-angular to rounded, sandy GRAVEL with rounded to sub-angular cobbles and boulders of local rocks. Boulders occasionally up to 3 ft. The deposits are highly permeable except where cemented. Locally much disturbed by streaming for tin.

HEAD — 6-10, locally >40

Almost everywhere present and largely obscures the solid formations. Represents solifluxion debris and grades downslope into alluvium and terrace deposits.

Within the outcrop of the granite, head comprises yellowish-brown, loose, layered, sandy GRAVEL with some clay, and gravelly silty SAND with cobbles and boulders of moderately weathered granite; grades down into moderately to highly weathered granite in situ. On the Upper Carboniferous outcrop next to the granite, head is typically reddish-brown, loose to compact, homogeneous, clayey gravelly SAND with many sub-angular cobbles; on steep slopes fines may be absent and head is then loose, clean COBBLES of the local rocks beneath 6 - 12 in. of humic soil.

On the Lower Carboniferous rocks, head is reddish-brown, loose to compact, homogeneous silty clayey SAND with some cobbles and boulders of local rocks; it may be layered with an upper gray horizon separated by a black cemented layer typically 3 in. thick from reddish-brown head down to bedrock.

SOLID FORMATIONS

CARBONIFEROUS

Upper Carboniferous (Namurian)

CkF

CRACKINGTON FORMATION — ?

Dark to very dark grey, very fine grained, thinly bedded to thinly laminated, very closely jointed, slightly to moderately weathered, poorly cleaved SHALE, weak, impermeable except along open joints. Interbedded with very subordinate grey to dark greenish grey, fine-grained, very thinly bedded, thinly laminated and cross-laminated, closely jointed, slightly to moderately weathered SILTSTONE, moderately strong, and dark greenish grey medium grained, very thinly to medium bedded, with closely to widely spaced joints slightly to moderately weathered, SANDSTONE, strong.

The shale slakes on exposure and is suitable for brick making.

S

SANDSTONE

It has been possible to map groups of beds in which SANDSTONE predominates. Beds are usually less than 12 in. thick and are separated by very thin beds of siltstone and shale.

Sandstones are suitable for aggregate production.

Within the contact metamorphic aureole of the granite, dark grey, very pale orange to dusky yellowish brown, fine to medium grained, thinly bedded, closely jointed, slightly to moderately weathered, hornfelsed SHALE and SANDSTONE, strong, impervious except along open joints. Locally with fine grained black tourmaline developed as selvedges up to 1 in. wide along discontinuities and with irregular quartz veins up to 2 in. wide.

Lower Carboniferous (Dinantian)

MELDON CHERT FORMATION — 240

(b)

Figure 5.4 Example of the supplementation of a large-scale plan as proposed: (a) unpublished 1:10 560 plan of part of the northern border of the metamorphic aureole of the Dartmoor granite, South-west England (reproduced at 1:7300); (b) part of the explanation to the map with expanded rock and soil descriptions (From Anon., 1972, 2.9.1 and 2.9.2, by permission)

simple superposition of symbols; a sequence of letter and number symbols; uncovered maps; striped maps; unitized maps; and the profile-legend type of map.

Superposition of symbols

A simple method of showing that the surface layer overlies another is to overprint the pattern of the lower layer with a halftone colour representing the upper layer, or vice versa.

The method is widely used on the so-called 'Solid and Drift' 1:50 000 maps published in the UK by the Institute of Geological Sciences. The upper or drift layer is shown by a characteristic colour and symbol, whereas only the boundaries of the solid formations are shown.

Identification of the particular formation depends on a combined letter-and-number symbol (Figure 5.6). Except where there are extensive borehole data, sub-drift boundaries are inferred from the relatively few exposures of the solid rocks usually available.

Sequences of letter and number symbols

A codified system is used on official engineering geological maps produced at a scale of 1:25 000 in Czechoslovakia (Matula, 1969). Forty distinct types of ground

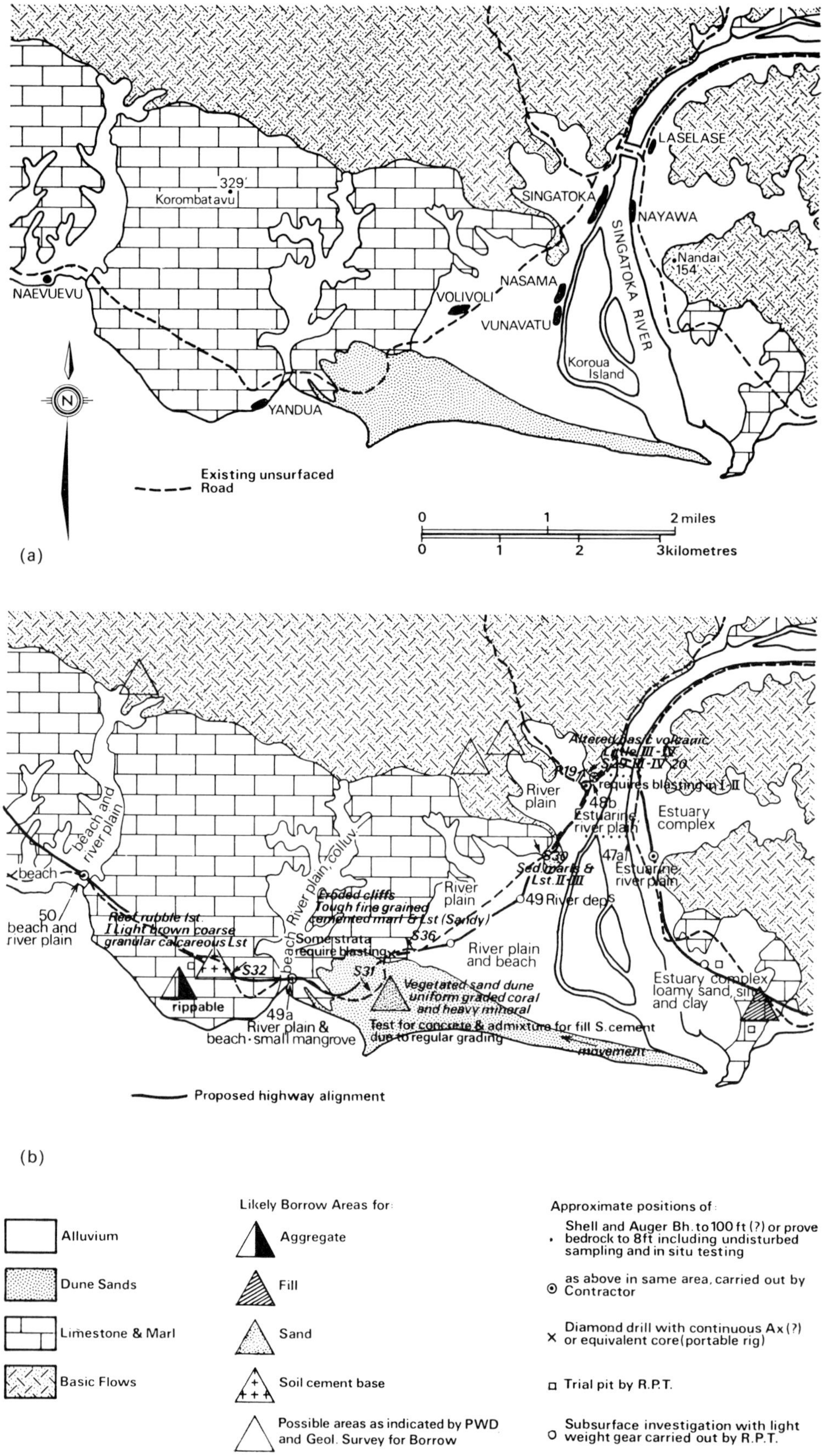

Figure 5.5 Section of one of the maps drawn at the reconnaissance stage of the Suva-Nandi Highway, Fiji, by adding engineering geological detail to the existing 1:50 000 geological map: (a) base geological map showing the route of the existing unsurfaced road; (b) geological map annotated with the results of the soils reconnaissance and the proposed highway alignment

are indicated by symbols – for example Ft, river terrace deposits – on the map units. The symbols have no stratigraphical significance except for the separation of Quaternary and pre-Quaternary rock and soil units. Should it be considered necessary to take into account two superimposed different types of ground down to a maximum depth of 10–15 m, symbols may be combined – for example EsFt, which is a zone of river terrace deposits (Ft) overlain by aeolian loess (Es).

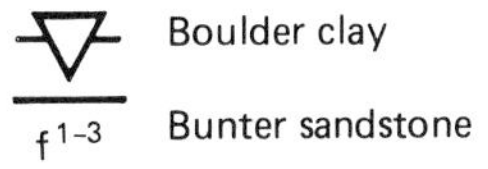

Figure 5.6 British Geological Survey method of symbols for geological maps

The map units are more or less quasi-homogeneous zones of surface geological materials, soils or rocks, each with distinctive engineering properties. If the information is available, each zone may be characterized in terms of soil or rock type and range of thickness, and subzones may be used to show the distribution of different thickness ranges within a zone. Symbols used for zones and subzones are listed in Table 5.1, with abbreviated descriptions of the soil and rock types. Lower case letters are used for the Quaternary soils, upper case for the pre-Quaternary.

As an example of the use of the system, the symbol $h1g2S^1$ means that cohesive soils (h) less than 2 m thick (1) at the surface are underlain by gravels (g) between 2 and 5 m thick (2), that rest on hard rocks (S) of the pre-Quaternary basement at a depth of less than 5 m (^{1}). An advantage of this method is that the map does not have to be printed in colour, while symbols for soil and rock types are unnecessary. Also, it is easy to represent a sequence of more than one layer in the Quaternary soils.

It should be noted that the codification of the map units or zones, for example Ft, is quite distinct from the letter symbols used to show the types and sequences of soils and rocks in Table 5.1.

Uncovered maps

Uncovered maps show the traces of geological boundaries on surfaces other than the ground surface. These maps are of three types:

- the selected surface is horizontal, at a particular altitude – the level map;
- the selected surface is at a constant depth below ground level – the specific-depth map;
- the selected surface is a particular geological horizon – the geological horizon map

Level maps are most often prepared for civil engineering sites at the site investigation stage or during construction where excavation for foundations is completed to a specified level. These maps are thus either predictive or record maps, usually made in conjunction with elevations of the walls of excavations. In large underground openings, a level map of the roof would also be prepared. Apart from this, where the maps are made as large-scale engineering plans, the only other common use of level maps is in the mapping of ore bodies at successive levels in a mine.

Stripe maps

Historically the development of the 'stripe method' to show both distribution, depth and thickness of a second and third layer in Quaternary deposits overlying bedrock, served to emphasize first the importance in engineering geology of the Quaternary cover, and secondly the need for three-dimensional representation of these deposits (see section 1.2.3; Zebera, 1947a,b; Pasek and Rybar, 1961). In essence, a stripe drawn on a map provides a window through the surface layer to the next layer, or layers where two stripes are used. There are several variations of the stripe method which are discussed in more detail, with examples, in Chapter 6.

Unitized maps

On unitized maps, a succession of two or more distinctive rock or soil units, rather than just the surface unit as on conventional geological maps, is represented by a particular single colour or pattern. Boundaries of each unit are vertical and a conventional structural interpretation of the character of the geological boundary is not possible. This very important aspect of engineering geological mapping is discussed in detail in Chapter 7.

Representation of thickness of units on maps

Thickness variations in a single surface unit, Quaternary deposits or a weathering mantle on bedrock are best shown by isopachytes: lines of equal thickness. Alternatively both surface topographic contours and contours

Table 5.1 Classification of rock and soil types, their thickness and depth to bedrock as used in typological zoning maps in Czechoslovakia (From Matula 1969, 1979, by permission)

Quaternary soils	*Pre-Quaternary rocks and soils*
g gravelly soils	S solid (hard) rocks
p sandy soils	B semi-solid (weak) rocks
n alternation of gravelly and sandy soils	F alternation of hard and weak rocks (flyschoid)
h cohesive soils	Z highly weathered rocks
k combinations of cohesive and non-cohesive soils	G gravelly soils
s loess soils	P sandy soils
o organic soils	N alternation of gravels and sands
b bouldery soils	I cohesive soils
	G combinations of cohesive and non-cohesive soils
Strata thickness	*Depth to pre-Quaternary surface**
1 <2 m	1 <5 m
2 2–5 m	2 5–10 m
3 >5 m	3 >10 m

*It is conventional that these numbers are shown superscript (see text).

on bedrock surface (rockhead) may be drawn; thicknesses then have to be estimated by subtracting the rockhead level from the ground surface level directly above. Contours can also be drawn on other interfaces, making estimation of individual thicknesses slightly more complicated.

Representation of engineering properties on maps

Engineering geological maps and plans may be used to display, in addition to the geology, the spatial distribution of units mapped in terms of an engineering grade, which has to be defined, or specific engineering properties – rock deformability, soil plastic limit by way of illustration – either at point locations or over the whole map area.

Representation of surface engineering grades is akin to unitized mapping in so far as a particular colour or pattern on the map represents specified engineering geological conditions divorced from geological details. It is an interpretative map and as such is truly engineering geological.

Variations in individual engineering properties are best shown by isolines of the property values.

Cross-sections and fence diagrams

Drawing a cross-section along a specified line on the map is a standard geological map interpretation technique briefly alluded to above. Some element of vertical exaggeration is usually required for drafting reasons and the user has to be aware of the resultant false dip indications. A series of site investigation borehole logs is usually plotted as cross-sections, for which special symbols are available.

A fence or panel diagram provides a method of showing the relation between the lithological logs of a group of boreholes. Individual plots are connected by vertical cross-sections (Figure 5.7). The diagram may be constructed on a map of borehole locations, or in isometric or perspective projection.

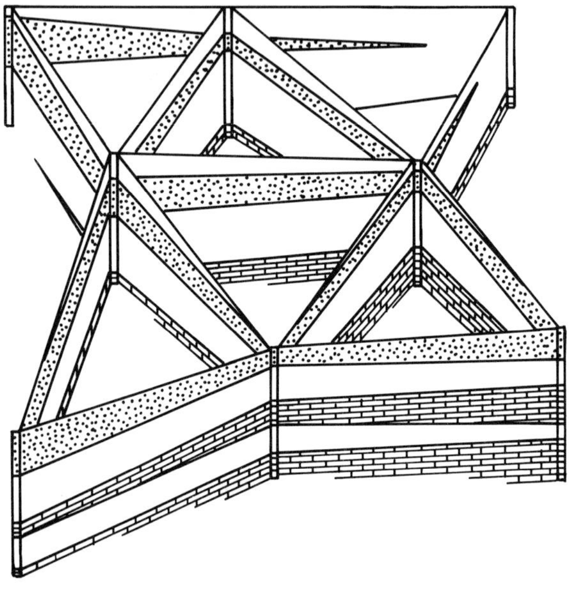

Figure 5.7 Typical fence diagram

Block diagrams

A particularly effective method of geological illustration is the block diagram. This is a view of an imaginary block of the Earth's crust drawn so that two sides and the top are visible. The top of the block may be drawn as if it is a model of the topography, a relief block diagram, on which the geology is displayed. Alternatively, the block may be drawn with a flat top, as a map block. In either case, the block may be separated along a significant geological surface, for example rockhead, to emphasize selected geological features (Figure 5.8) or to make it easier to visualize internal geological aspects of the block.

Methods for construction of isometric, cabinet, and perspective block diagrams are given by Lahee (1961, pp. 715–721).

5.6 Explanation or legend

The explanation or legend on an engineering geological map provides a guide to the symbols used in drawing the map. In addition, the geology may be illustrated by one or more cross-sections, fence diagrams or block diagrams. Inset smaller scale maps could illustrate specialized aspects. Summary accounts may be given of rock and soil properties, hydrogeological conditions, geomorphological features and geodynamic processes, and an evaluation of individual zoning units.

Ideally, a general engineering geological map or group of maps should be accompanied by a descriptive memoir. Plans of engineering sites almost invariably are included in comprehensive reports, in which not only are geological and engineering conditions described but also the geotechnical engineering conditions and solutions to problems are proposed and discussed.

5.7 Legend for engineering geological maps and plans

The legend allows each symbol on the map to be identified by category and type, and is thus an all-important guide to map interpretation. Such symbols, or combinations of symbols, have to be clear, distinctive and informative at all scales.

5.7.1 Symbols for rocks

Symbols for the principal rock types that are likely to be encountered are given in Figure 5.9, and the nomenclature follows that given for the classification for rocks in Tables 4.1(a–c). The general intention of the classification scheme is to simplify petrographical nomenclature for engineering geological purposes, and the examples provided of sedimentary, igneous and metamorphic rock types may be regarded as a minimum working list of rock types. The symbols are simple and distinctive; those symbols for sedimentary rocks combine easily into those for composite types of rocks.

The symbols are from the recommendations given in Anon. (1972), which were modified where considered necessary from the symbols in an early edition of the Code of Practice CP2001: Site investigations (Anon., 1957). One important difference has been carried over into the latest edition of the code (Anon., 1981b); it concerns the symbol for silt and siltstone. Civil engineers

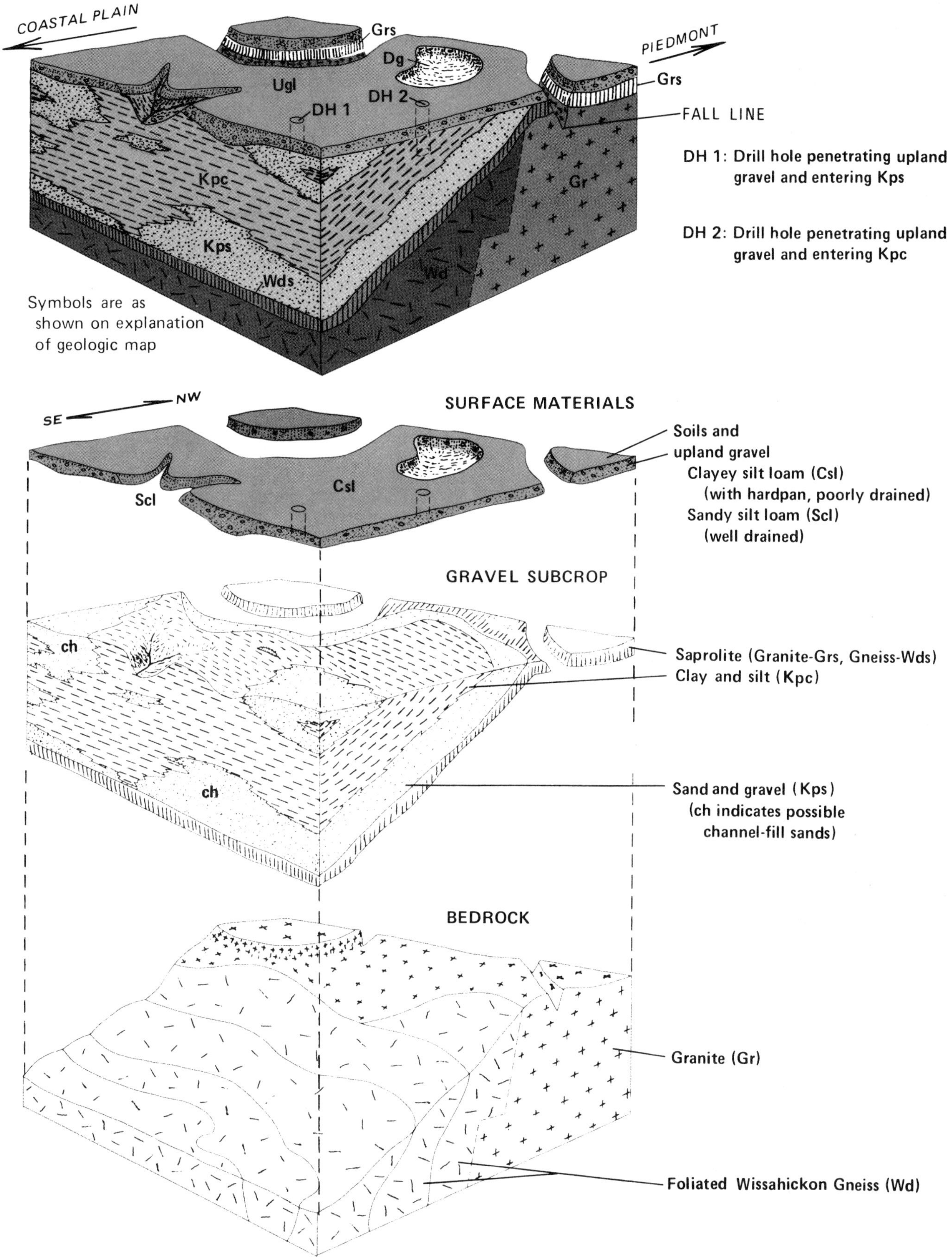

Figure 5.8 Block diagram of superficial soils and solid formations beneath. In the lower part of the diagram, the block is exploded to show the geology at two selected horizons, one rockhead and the deeper one on a prominent unconformity (From Robinson and Spieker, 1978, by permission)

are insistent that in the UK the cross symbol for silt and siltstone should be retained, whereas vertical lines are used here. The idea of using vertical lines for the silt symbol was taken from Hvorslev (1948). This symbol was originally used by the U.S. Corps of Engineers, Vicksburg District, and subsequently has been followed by the Norwegian Geotechnical Unit and the Ontario Department of Highways, among others.

The International Standards Organisation (ISO) has in turn proposed different symbols for silt and siltstone which were adopted in Anon. (1981b).

It appears to be very unlikely that international standard symbols will be generally accepted in the foreseeable future. Consequently some alternative symbols are illustrated in Appendix 1.

5.7.2 Symbols for soils

The symbols are simple and distinctive; they permit the same basic ornament to be used for a soft or clastic rock in its unlithified and lithified states; they combine easily into symbols for composite types of soils and clastic rock.

Although the proposed symbols are primarily intended for use in logs and sections, usually on a large scale, they can also be used for plans and maps. For the latter it may be advantageous to lighten the ornament by spacing it more widely or by using thinner lines.

Symbols are given for the four divisions of soils based on particle size. Each symbol has two variants – one for use when the material is the chief soil constituent, the other for use when it is the secondary constituent. The symbols for the corresponding rocks, in which the particles are cemented, are given here as well as in their appropriate subsection to illustrate the unity of symbolism (Figure 5.10).

5.7.3 Symbols for hydrogeological properties of rocks and soils

Aquifers, aquitards, aquicludes and aquifuges may be distinguished, in rectangles in the explanatory legends to maps and plans, by a simple convention (Figure 5.11).

Relevant symbols will be found in Figure 6.5 and Plate 20.

5.7.4 Symbols for structural features

Two aspects of structural information are provided by geological maps; one comprises precise information on the attitude, direction and location of individual structures, and the other depicts the trace of a structure at the surface or a particular level below ground. Structural traces are also used on interpretative horizontal sections and in recording detail on the vertical walls of excavations.

It is conventional to use three distinctive symbols for each type of geological structure, namely a general symbol and two special symbols – one for a horizontal attitude and the other for a vertical attitude.

All official geological surveys have, or have adopted, standard symbols for geological structures and boundaries. A selection is given in Figure 5.12.

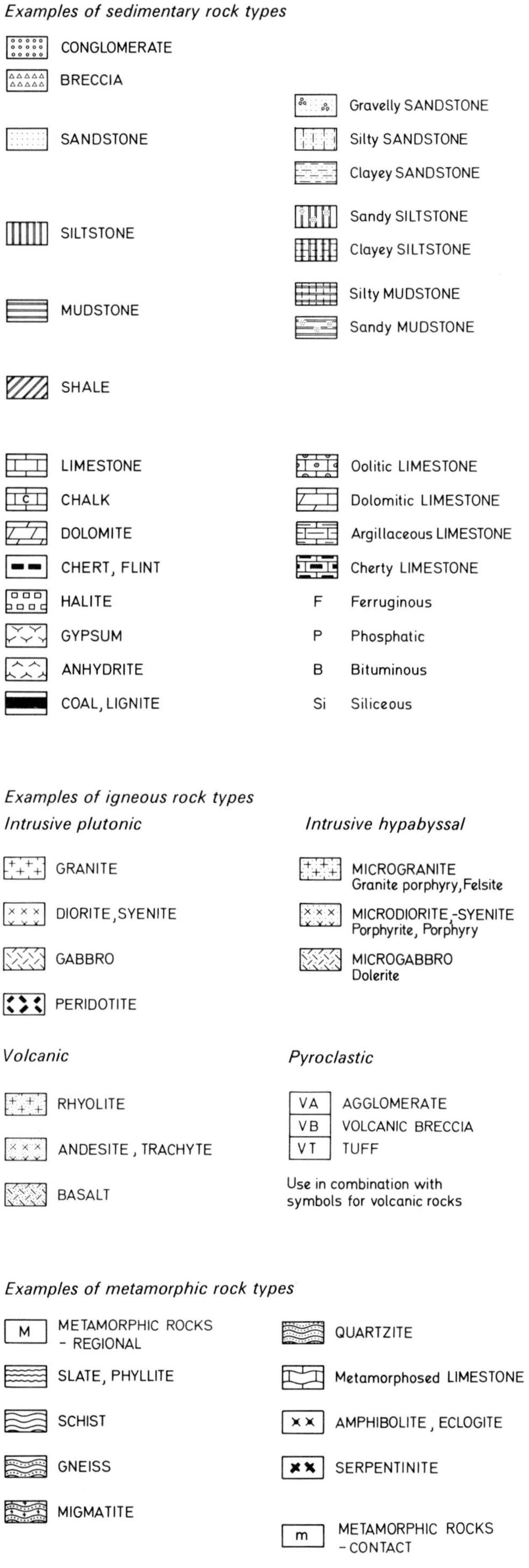

Figure 5.9 Symbols for rocks (From Anon., 1972, by permission)

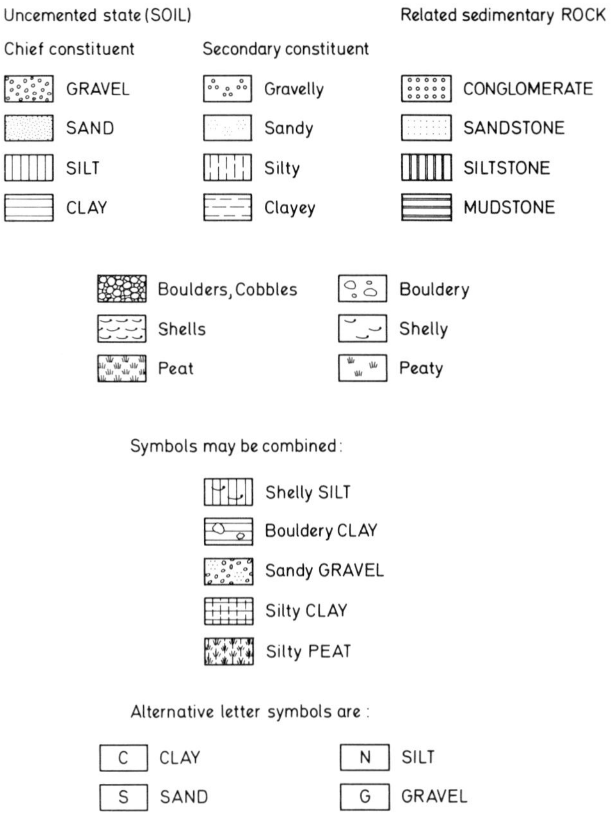

Figure 5.10 Symbols for soils (From Anon., 1972, by permission)

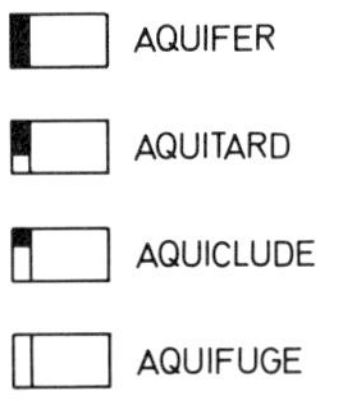

Figure 5.11 Representation of hydrogeological properties on a map legend (From Anon., 1972, by permission)

Planar structures

For each planar structure the long bar of the symbol indicates the strike direction and the short bar the dip direction; the dip amount is given in degrees measured from the horizontal.

Formerly, the dip arrow was used exclusively to indicate the direction and amount of dip of bedding planes. It is still used occasionally and provides an acceptable alternative to the bar symbol.

Bedding, foliation, banding and cleavage in sedimentary and metamorphic rocks may be crumpled, corrugated or undulating, although the general disposition may be horizontal, inclined or vertical. These conditions may be indicated by sinuous strike bars.

Linear structures

Abbreviations may be used in conjunction with lineation symbols to indicate the type of linear structure. A symbol used alone signifies a linear structure of unspecified type:

BD Boudinage
CB Cleavage/bedding intersection
BC Bedding/cleavage intersection
L Mineral lineation. When due to specific minerals, indicated thus: Lp (pyrite), Lq (quartz)
M Mullion structure
P Puckering
R Rodding; may be qualified Rq (quartz rodding)
SL Slickensides; may be qualified SLq (quartz slickensides)
TB Lineation due to top structures on bedding, e.g. ripple mark crests
BB Lineation due to bottom structures on bedding, e.g. flute and groove structures

The types of linear structure present should also be specified on the legend accompanying the map or plan.

Folds of intermediate or large size would usually be obvious from the disposition of the bedding. Small-scale folds may be regarded as interruptions on otherwise evenly inclined strata; for purposes of structural interpretation it may be necessary to record the relative lengths of the limbs of minor folds and their orientation. If the folds are viewed down the plunge, i.e. towards the head of the dip arrow on the symbol for fold plunge, then fold shape may be indicated by a sketch adjoining the symbol. The shape of a fold may also be indicated by abbreviations with the symbols. For example:

IS Isoclinal
T Tight
OP Open

Traces of geological structures and geological boundaries

A distinction is usually made on geological maps between boundaries of drift (or superficial) deposits and boundaries of solid deposits (see Figure 5.12). Some indication is also usually given of the accuracy of boundaries, with broken lines denoting uncertainty in the positions of solid geological boundaries and faults. This principle may be applied to the trend and where appropriate to the position of the traces of other planar structures.

On engineering geology plans, faults and fault zones do not call for distinctive structural symbols, but will usually be mapped as zones of which the margins are plotted and the internal structures and filling materials are shown in detail.

Structural features determined by photogeology

If all the information shown on the map has been obtained by photogeological means, there is no need for a separate set of symbols. On engineering geology maps which combine the results of photogeological and ground surveys, the photogeological symbols (otherwise the same as the symbols already proposed) should carry a single dot (Figure 5.13).

Trace symbols for use on engineering geology sections

Symbols for the main types of discontinuity met with in mapping the walls and floors of excavations are given in Figure 5.14.

Trace symbols for features interpreted by geophysical means

Symbols are needed for contours on bedrock (rockhead) or a particular marker horizon, faults and other boundaries determined primarily by geophysical means. They

General planar structures

- HORIZONTAL STRATA
- INCLINED STRATA, dip in degrees, direction of succession unknown
- INCLINED STRATA, dip in degrees, normal succession
- INCLINED STRATA, dip in degrees, inverted succession
- VERTICAL STRATA, long axis is strike direction
- GENTLY INCLINED STRATA
- MODERATELY INCLINED STRATA
- HIGHLY INCLINED STRATA

- FOLIATION or BANDING, horizontal
- FOLIATION or BANDING, inclined, dip in degrees
- FOLIATION or BANDING, vertical

- CLEAVAGE, horizontal
- CLEAVAGE, inclined, dip in degrees
- CLEAVAGE, vertical

- JOINT, horizontal
- JOINT, inclined, dip in degrees
- JOINT, vertical

- FOLD AXIAL PLANE, horizontal
- FOLD AXIAL PLANE, inclined, dip in degrees
- FOLD AXIAL PLANE, vertical
- FOLD AXIAL PLANE, strike, dip direction unknown
- FLOW FOLIATION in igneous rocks, horizontal
- FLOW FOLIATION in igneous rocks, inclined, dip in degrees
- FLOW FOLIATION in igneous rocks, vertical
- FLOW FOLIATION in igneous rocks, strike, dip direction unknown
- SHEAR ZONE, horizontal
- SHEAR ZONE, inclined, dip in degrees
- SHEAR ZONE, vertical

Planar structures underground

Special symbols should be used if surface and underground information are combined on one map or plan

- HORIZONTAL STRATA
- GENTLY INCLINED STRATA
- INCLINED STRATA, dip in degrees
- VERTICAL STRATA

Symbols for linear structures

- LINEATION, horizontal
- LINEATION, inclined, inclination in degrees
- LINEATION, vertical
- MINOR FOLD AXIS, horizontal
- MINOR FOLD AXIS, inclined, plunge in degrees
- MINOR FOLD AXIS, vertical
- ANTICLINE, axis with plunge in degrees
- SYNCLINE, axis with plunge in degrees

Combinations of structural symbols

In a combined symbol the location is represented by the central point of the symbol for the planar structure. It is convenient for the linear structure to emanate from the same point. When the linear symbol is used alone the arrow head represents the location of the structure.

Some combinations are difficult to represent, for example bedding parallel to cleavage, while others such as axial plane cleavage are indicated by direct superposition of symbols.

- Axial plane cleavage, i.e. cleavage parallel to the axial plane of a fold (horizontal and 5°dip cases shown)
- CB: Lineation on bedding caused by intersection of cleavage with bedding
- FP: Flow foliation in an igneous rock with linear arrangement of feldspar phenocrysts - FP

Geological boundary lines

- DRIFT, certain
- DRIFT, approximate
- SOLID, certain
- SOLID, approximate
- SOLID, assumed (calculated or conjectured)

Traces of planar structures

- STRATIFICATION or BEDDING TRACE
- FOLIATION or BANDING
- FLOW FOLIATION in igneous rocks
- CLEAVAGE
- JOINTS
- AXIAL TRACE OF ANTICLINE
- AXIAL TRACE OF SYNCLINE

Special symbol

- Outer limit of zone of contact metamorphism, crosses inside zone

Faults

- FAULT, crossmark on downthrow side, dip in degrees, throw T in metres
- FAULT, with horizontal component of relative movement
- TERMINATION OF FAULT

Figure 5.12 Symbols for structural features (From Anon., 1972, by permission)

- HORIZONTAL STRATA
- INCLINED STRATA, estimated dip in degrees
- VERTICAL STRATA
- JOINTS
- GEOLOGICAL BOUNDARY, certain, approximate, assumed

Figure 5.13 Symbols for structural features determined by photogeology (From Anon., 1972, by permission)

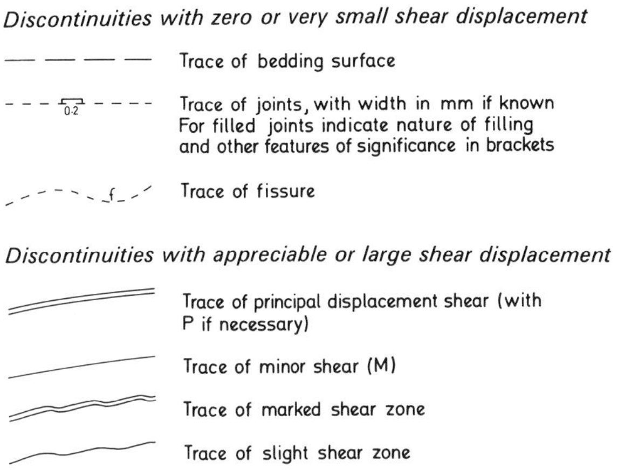

Figure 5.14 Symbols for traces of planar structures mapped in excavations (From Anon., 1972, by permission)

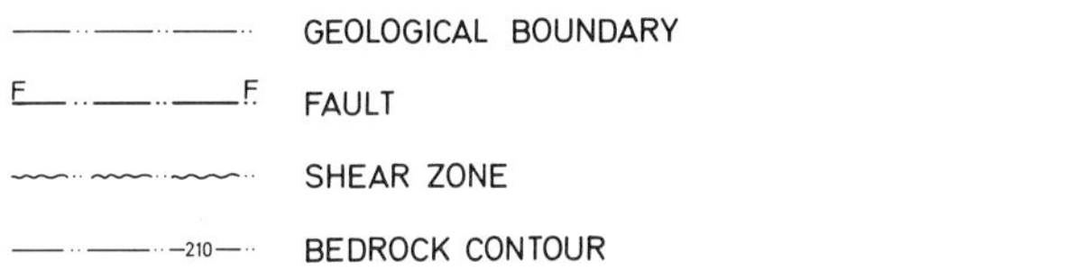

Figure 5.15 Trace symbols for features determined by geophysical means (From Anon., 1972, by permission)

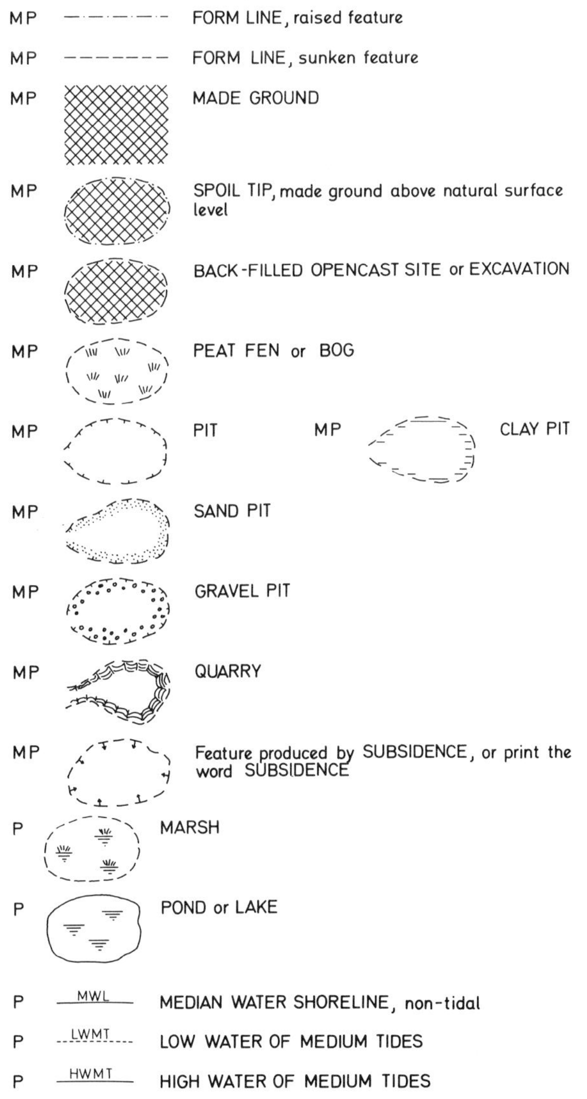

should be close to the normal geological symbols but distinctive (Figure 5.15). The difference here is that double dots are used.

5.7.5 Symbols for general features

Different boundary lines are used for raised and sunken features, and most of the sunken features are either man made or man induced; for example, the various types of pits and quarries. The cross-hatched pattern for made ground or fill (Figure 5.16) can, if necessary, be rotated through 90° or drawn with different spacings or line weights to indicate different types of fill. Although the distinction is usually obvious on the map, for foundation purposes it may be advisable to separate made ground associated with reclamation along a river or estuary

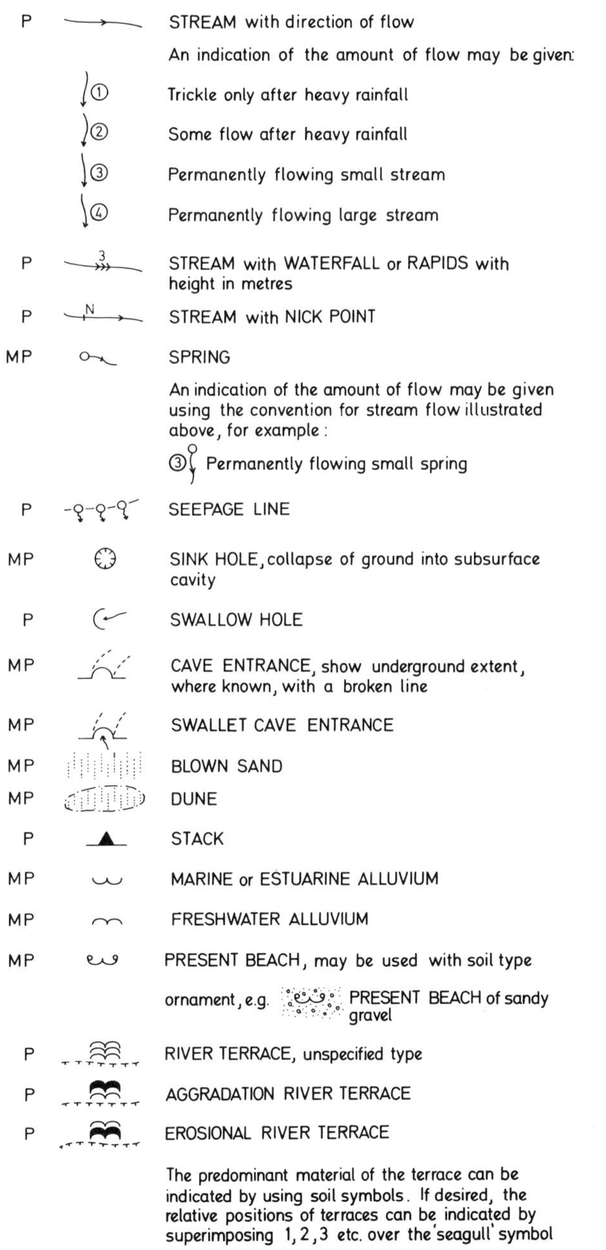

Figure 5.16 Symbols for general features on engineering geological maps and plans (From Anon., 1972, by permission)

from infilled open cast coal workings. The type, or types, of soils comprising the fill may also be pointed out by a note on the plan or map (see Figure 10.22).

Permanent flooding of a pit or quarry, either complete or partial, may be shown by combining the appropriate symbol with that for water.

A selection of symbols is given in Figure 5.16.

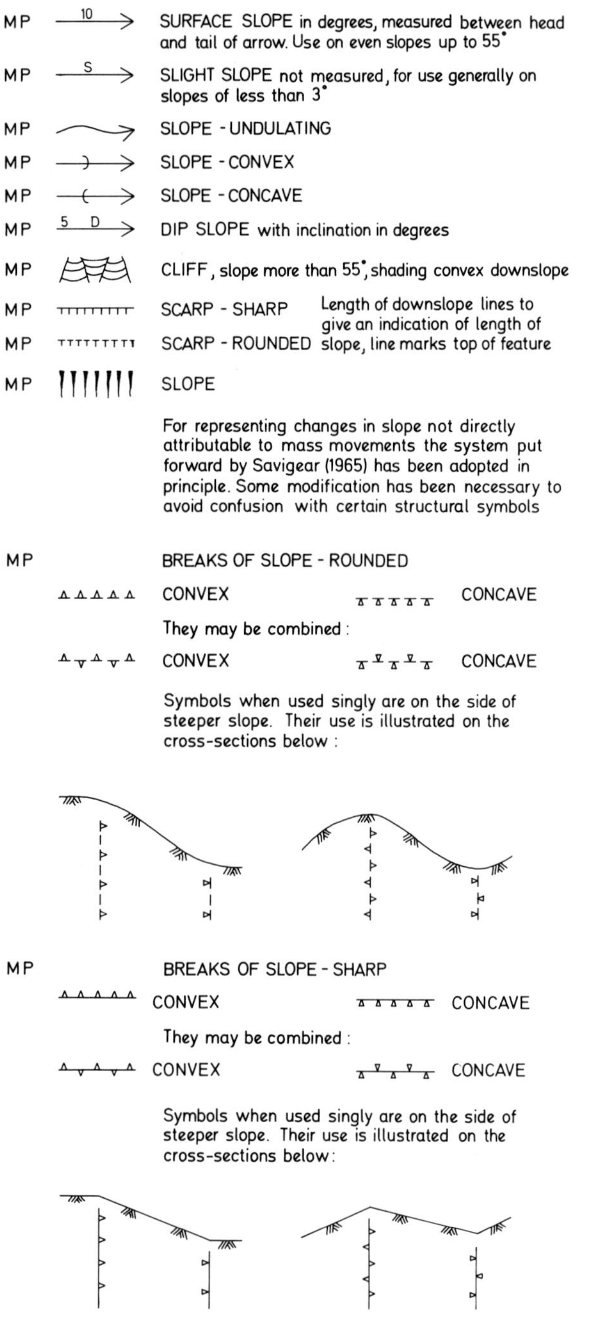

Figure 5.17 Symbols for geomorphological features (From Anon., 1972, by permission)

5.7.6 Symbols for geomorphological features

Comprehensive lists of geomorphological symbols have been published by the International Geographical Union (1968) and only a selection is presented here (Figure 5.17). The symbols were originally chosen for British use (Anon., 1972), and at that time, because of a general lack of interest in small-scale engineering geological mapping, no consideration was given to so-called 'point symbols'. Some examples were later published (Anon., 1981c) and are reproduced separately (see section 5.7.10). Symbols labelled 'M' are intended for use on medium-scale engineering geology maps; symbols labelled 'P' are for large-scale plans. In some cases ('MP') the same symbol is used at both scales.

In the same way as that conventionally adopted for geological boundaries, degrees of uncertainty can be introduced into many of the symbols by using a broken line to indicate uncertainty.

5.7.7 Symbols for glacial features

Only a limited number of glacial features are listed (Figure 5.18), and most mapping aspects can be dealt with quite adequately by using selected geomorphological symbols.

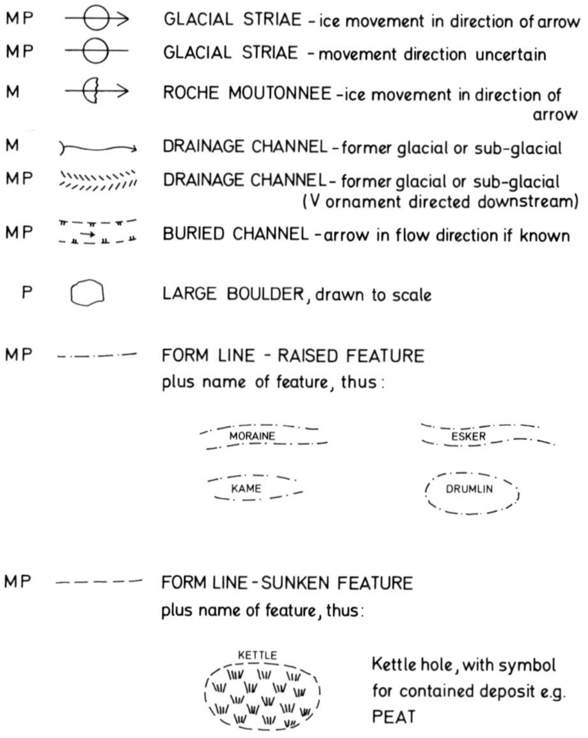

Figure 5.18 Symbols for glacial features (from Anon., 1972, by permission)

5.7.8 Symbols for mass movement features

Erosional and depositional forms produced by mass movements can be mapped at large scale using selected symbols for general features (see Figure 5.16), such as form lines, springs, seepages, marshes, ponds, and symbols for geomorphological features (see Figure 5.17). With the latter, surface topographic forms – slopes, streams and breaks of slope – are all that are needed to highlight the significant aspects of the topography, especially if they are added to a detailed, contoured plan of the area.

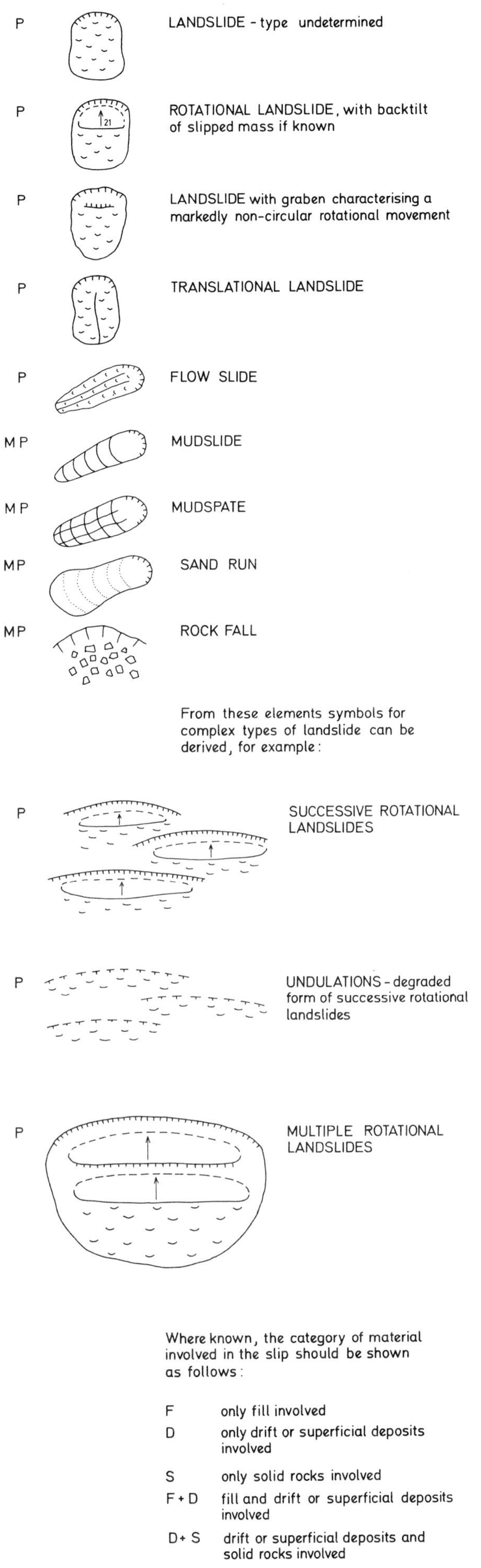

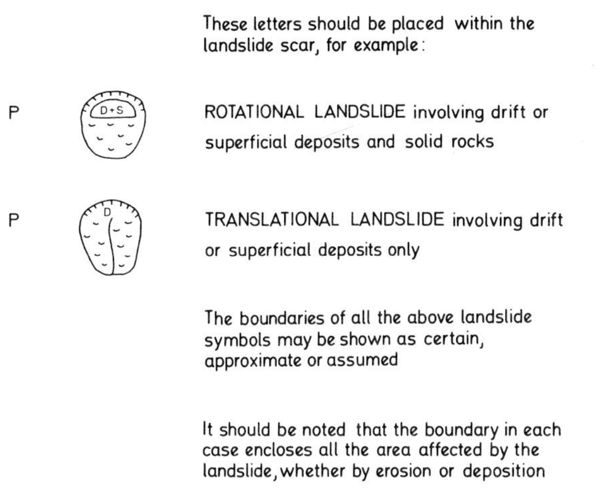

Figure 5.19 Symbols for mass movement features (From Anon., 1972, by permission)

The selected symbols (Figure 5.19) are used in conjunction with the mapped limits of the feature, wherever possible.

5.7.9 Symbols for periglacial features

Periglacial features are of two kinds – those in present-day permafrost areas, and those relict features indicative of past permafrost conditions (Figure 5.20).

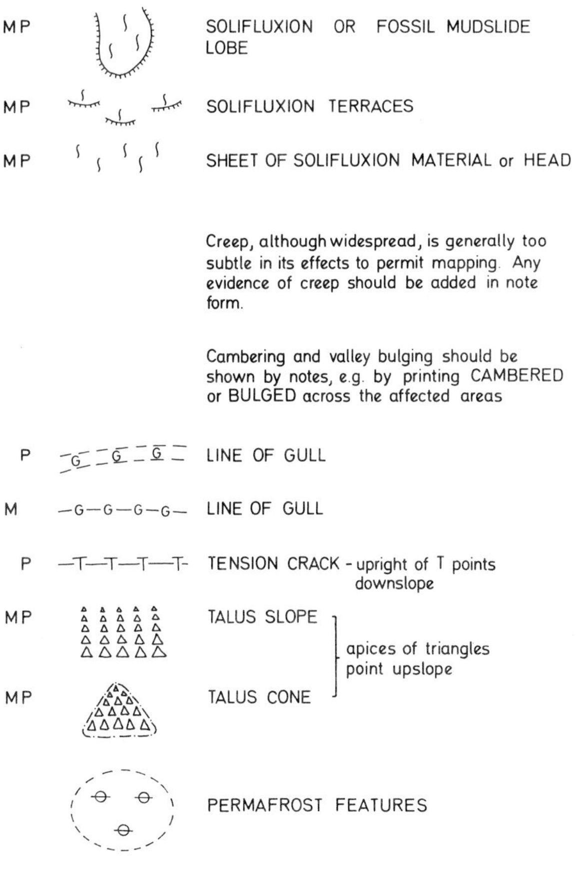

Figure 5.20 Symbols for periglacial features (After Anon., 1972, 1981c)

5.7.10 Point symbols for small-scale maps

On small-scale maps it is not possible to show the actual boundary or form of many geomorphological and other surface features. Point symbols (Figure 5.21) may be used to locate and typify such features.

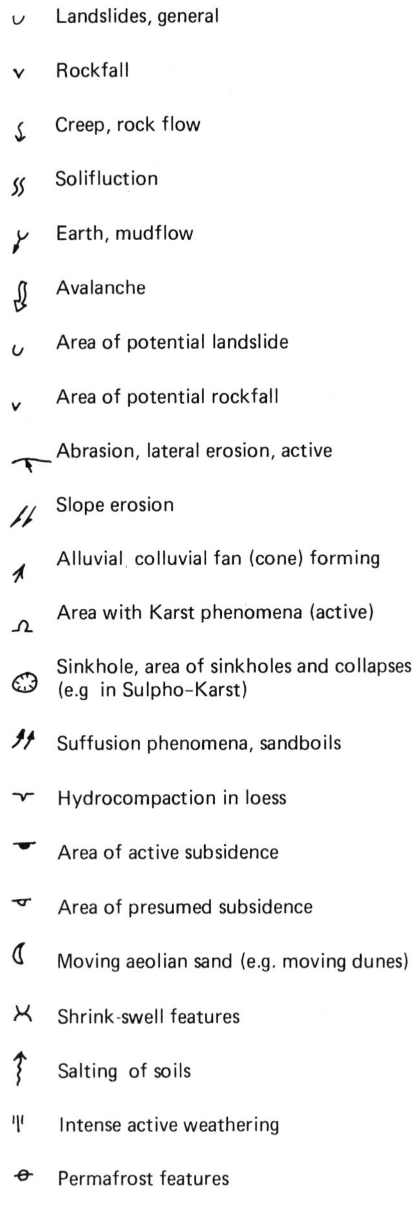

Figure 5.21 Selection of point symbols for use on small-scale engineering geological maps (From Anon., 1981c, by permission)

5.7.11 Symbols for shafts, tunnels, boreholes and wells

Holes in the ground may be accessible for mapping or geological details may have been preserved in an archive or data bank. As such they may be an important source of geological and hydrogeological information, and their location and type need to be recorded by symbols on plans or maps (Figure 5.22).

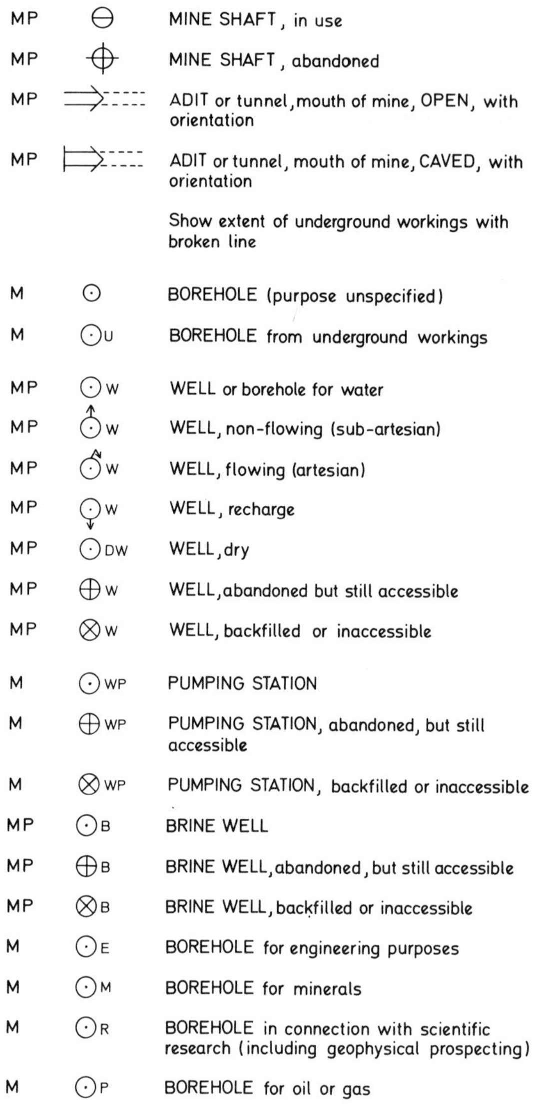

Figure 5.22 Symbols for shafts, tunnels, boreholes and wells (After Anon., 1972, 1981c)

5.7.12 Site investigation symbols

On medium-scale maps, symbols are needed to show the area of a site investigation and the location of pits, trenches and other investigatory excavations.

At large scale, usually on a site plan, the location of different types of borehole, *in situ* tests, photographs and samples can all be shown (Figure 5.23).

5.7.13 Symbols for use on cross-sections for engineering geological plans

Large-scale cross-sections may be of two kinds: those interpreting ground conditions from the results of site investigation boreholes, and factual records of vertical exposures produced by pitting and trenching. Some appropriate additional symbols are given in Figure 5.24.

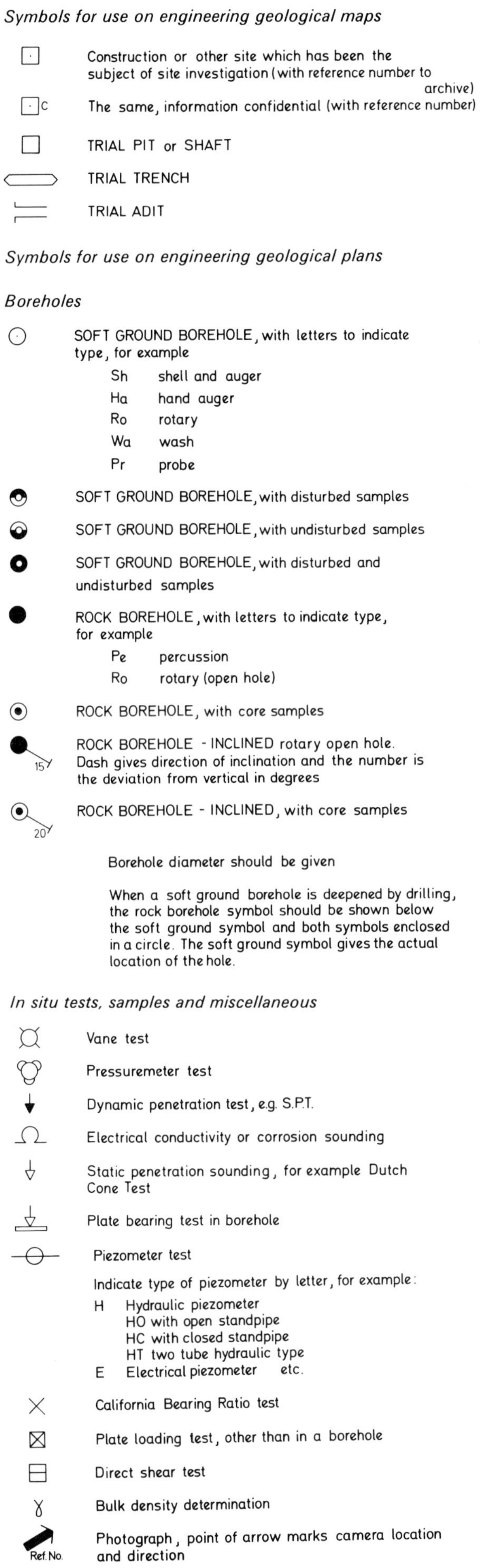

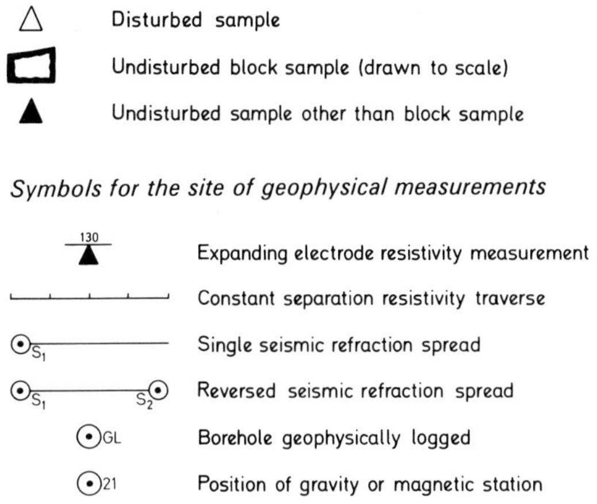

Figure 5.23 Site investigation symbols (From Anon., 1972, by permission)

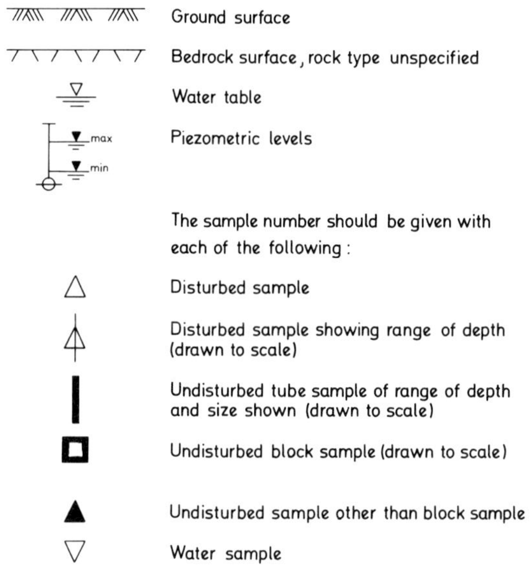

Figure 5.24 Symbols for use on cross-sections on engineering geological plans (From Anon., 1972, by permission)

5.7.14 Colour scheme

Engineering geological plans will often be uncoloured, but colour should be used on engineering geological maps wherever possible. Colour can be used for linework and symbols (Table 5.2), and also as an infill to depict mapped divisions or formations. For the latter it is not necessary to follow a code, although there are advantages in following accepted geological cartographic practice. Soils are best shown by pale shades of colour, or by stipple or shading over the solid colour representing the underlying rock.

Symbols for rocks
All patterns denoting rocks should be brown or in the colour accepted elsewhere for the infill colour of the division or formation.

Symbols for soils
Soil patterns, if used, should be in black. These should not be superimposed on rock patterns, but the latter may be overprinted by the soil letter symbols suggested by

Dumbleton (1968) which should be in black. The symbols for made ground should be in red.

Table 5.2 Recommended colour scheme for linework and symbols

Colour	*Depiction*
Grey	Base topography, except surface contours
Black	Geological contacts, structural symbols, soil group letter symbols
Brown	Areal representation of lithology, unless appropriate stratigraphic colour used
Blue	Surface water features
Violet	Groundwater features
Orange	Geochemical including hydrogeochemical and geophysical features
Green	Surface contours and geomorphological including glacial features
Red	Works of man, including shafts, pits and boreholes, but excluding features such as roads, railways and buildings which are commonly included as base topography; site investigation symbols
Yellow or gold	Mineral veins and ore bodies

Symbols for glacial, mass movement and fossil periglacial, geomorphological and general features
Symbols should generally be shown in green except as otherwise indicated:

- movement features may be shown in black if desired;
- spoil tip and back-filled open cast site or excavation – red;
- peat fen or bog – black or blue;
- pits and quarries – red;
- marsh, pond, shoreline, spring, etc. – blue;
- swallow hole – blue;
- cave – black;
- swallet cave – black and blue;
- shafts, wells and boreholes – red.

A useful addition for boreholes through soil into rock, where the plan colour infill indicates the type of soil, is to omit the central dot in the symbol and to fill the interior of the circle with the colour of the underlying rock division:

- alluvium, etc., and terraces – black.

Symbols for structural features
All structural features should be shown in black.

Symbols for site investigations
All site investigation symbols should be shown in red, except for piezometer readings which should be in violet.

5.8 Another example of a legend

The legends set out in detail in section 5.7 contain symbols recommended for use in British practice. As becomes very apparent when published maps and plans are examined, there is a great deal of variation from these recommendations in the symbols actually used by government agencies, professional engineering practice, and in the research field. No claim has been made that the symbols published here cover every conceivable need; neither is their use mandatory. They simply represent good practice as proposed and agreed by a large group of specialist users in 1972.

In contrast, in 1967 a very detailed system of symbols for engineering geological maps at all scales was published, with map examples in colour, for use in the Council of Mutual Economic Aid (COMECON) group of countries (Bachmann *et al.*, 1967). These are instructions for preparing engineering geological maps, and their use is mandatory in those countries. Some of the instructions are included below.

Colours for representation of genetic groups of rocks
With respect to a taxonomic rock and soil classification, the different genetic types in both Engineering Groups and Engineering Formations (using the nomenclature proposed in section 2.5.2) are shown in standard colours – light tones for engineering soils and dark tones for rocks.

Patterns for rocks and soil types
Distinctive patterns are provided for 57 sedimentary soils and rocks; 7 igneous rocks; 8 volcanic rocks with variations of each symbol for lava, tuff and tuffite; and 13 metamorphic rocks. These are petrographic, i.e. lithological, rock types and the patterns have no stratigraphical implications.

Orange patterns are used for the rock and soil types at the surface. Grey patterns are used for the second formation below the surface.

Thickness ranges in first layer of superficial deposits
Three thickness ranges – less than 2 m, 2–5 m, and more than 5 m – are indicated by increasingly darker tones of the appropriate 'genetic soil type' colour.

Depth ranges in second layer
Three depth ranges are indicated using grey lithological symbols. Dark grey is used for less than 5 m, light grey for 5–10 m. For a depth range greater than 10 m the symbol is omitted because the map is only used to show rocks and soils to a depth of 10 m, and the orange lithological pattern suffices.

Special case of more than one layer in the superficial deposits
Where two thin layers of superficial deposits are present the vertical stripe method is used, in which narrow vertical bands interrupt the colour and orange pattern of the top layer. The vertical bands are given the appropriate genetic-group colour and the orange lithological symbol.

Striped areas may be underlain by either superficial or bedrock deposits; the lithology of the third unit down is then shown by the appropriate grey lithological pattern.

In some cases, where there could be confusion, a different lithological pattern is used for the third unit.

Uniform rock or soil sequence to 10 m
If the rock or soil sequence is uniform throughout the depth shown, only one unit is shown by the appropriate genetic-colour and orange lithological symbol.

Depths greater than 10 m

If appropriate data are available, for example from deep boreholes, deeper strata may be shown on engineering geological cross-sections and in the map explanation.

Hydrogeological conditions

Hydrogeological conditions are represented by conventional symbols in blue.

Depth of saturated strata (depending on the map scale) is expressed numerically, by hydroisobath, hydroisohypses, or hydropiezes. In numerical or hydroisobath presentation it is recommended that the general directions of groundwater flow be shown by a special symbol.

The results of chemical analyses of groundwater are represented by blue circles divided into four sectors. The pH value and the total water hardness are numerically inscribed in these sectors; the corrosiveness is expressed by filling the appropriate sectors.

Geodynamic phenomena

Geodynamic phenomena are represented by red symbols. Some symbols may be used at all map scales. For others, there are two versions of a symbol: one for use on medium- and large-scale maps and another for small-scale maps. Many of the symbols are point symbols for use only on small-scale maps.

Symbols for engineering geological zoning

Regions, areas, zones, subzones and districts distinguished on a map of engineering geological zoning are represented in various colours. Type of line and thickness of line in black are specified for the boundaries of all zoning units.

The hierarchy of zoning units is indicated by numbers and letters in black, e.g. Region : A, Area : a; Zone : II; Sub-zone : 2; District : α.

Documentation

A limited number of black symbols are proposed for various aspects of site investigation, boreholes, quarries, sandpits, etc.

6

Czechoslovakian stripe method and parallel developments

6.1 Introduction

Fundamental to the geological aspects of engineering geological mapping is the perceived need to show all the significant details of the superficial deposits, mainly Quaternary in age, particularly with respect to the thickness ranges of the uppermost two or three units. Conventionally, the accepted method is to draw stratum contours on the boundaries of the various units; then by using surface topographic contours it is possible to estimate thickness at any locality. As a method it requires a certain amount of skill that is readily acquired with patience, but it lacks the possibility of immediate general assessments because there is no visual impact. Three or four squiggly lines, in different colours or patterns, need quite careful inspection and interpretation. What is needed is to take the available raw data on the map, work out the thickness variations in each layer in terms of a small number of thickness ranges of significance in ground engineering, and then present the results as outlined areas on the map, preferably in colour.

An elegant solution, proposed by Zebera (1947b) and subsequently modified by Simek (1962), was the adoption of what was first called the 'strip' method and later more generally the 'stripe' method. The stripes are analogous to trenches through the superficial deposits to the top surface of the next layer beneath which might be bedrock or another layer in the superficial deposits.

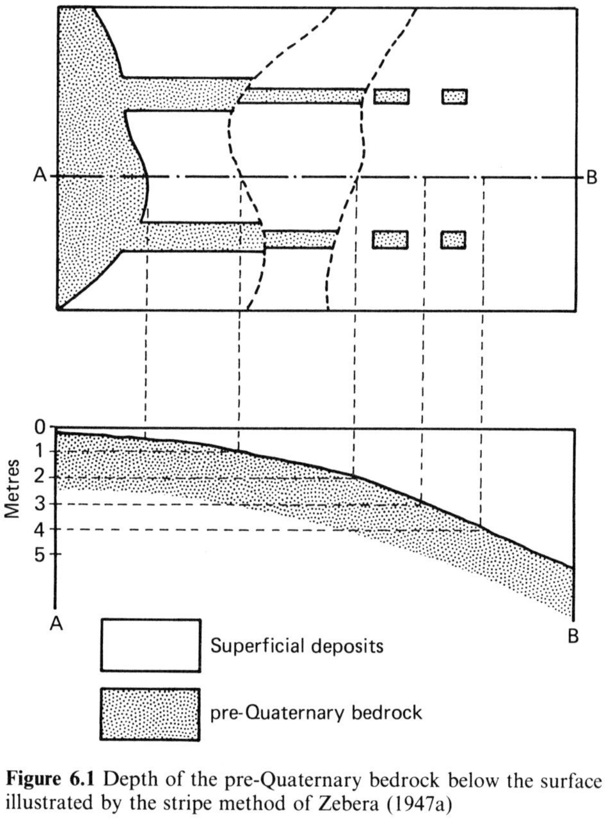

Figure 6.1 Depth of the pre-Quaternary bedrock below the surface illustrated by the stripe method of Zebera (1947a)

6.1.1 Zebera's stripe method of 1947

At first the stripe method only represented the depth to bedrock surface beneath the Quaternary deposits (Zebera, 1947a). Usually depths were shown to 4 m and at greater depths the bedrock symbol was omitted. Down to 0.3 m was accepted as bedrock (Figure 6.1); depth of bedrock to 0.3–1 m was shown by a wide band containing the bedrock symbol or colour, and 1–2 m by a narrow band. For greater depths the narrow stripe was broken into rectangles, long for 2–3 m and short for 3–4 m depth. Rockhead contours were also drawn through the ends of the stripes.

6.1.2 Other applications of the Zebera stripe method

It soon became evident that maps limited to soil thicknesses down to 4 m were too restricted for engineering purposes, and that 10 m would be more appropriate. With this thickness of superficial deposits there was a strong likelihood that more than one soil type might be present, with the certainty also that individual layers would not be of uniform thickness.

The surface layer presents no difficulties. Different shades of the same colour, with the deepest shade reserved for the thickest layer, or the same pattern with different spacings, can be used for up to three thickness ranges and still leave the map visually attractive and easy to understand. It is the next step, representing the second layer, and also possibly a third layer, that is difficult.

Three superimposed layers in the superficial deposits above bedrock can be accommodated using the stripe method proposed by Zebera. Equally adaptable to black and white maps as well as those in colour, the method initially was used to display three layers with three thickness ranges: 0.8–2 m, 2–5 m and 5–10 m.

The first layer, which crops out completely, is divided into the three thickness areas distinguished either by a pattern representing lithology, for which there are variations for spacing, or by shades of the same colour with the deepest shade for the thickest unit (Figure 6.2 and Plate 1). Roman numerals I, II and III are used for the three thickness ranges, and where there is uncertainty this is shown by enclosing Roman numerals in parentheses.

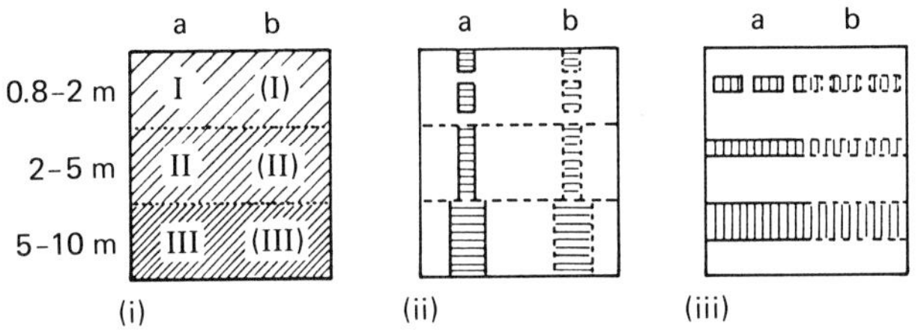

Figure 6.2 Czechoslovakian stripe method as developed by Zebera (1947a). (i) Three thickness ranges in the surface layer: (a) Roman numerals indicate the three thickness ranges; (b) thicknesses in parentheses are approximate. (ii) Three thickness ranges in the second layer with (a) certain, (b) approximate. (iii) Same convention for the third layer (After Pasek and Rybar, 1961, Fig. 1)

A vertical stripe, appropriately coloured in three shades or with a distinctive pattern, designates the second layer which of course can also crop out. Where the second layer lies beneath the top layer, a wide stripe is used for the thickness range 5–10 m, a narrower stripe for 2–5 m and an interrupted stripe for 0.8–2 m. Where thickness ranges are approximate, the boundaries of the stripes and rectangles are drawn with pecked lines.

For a third layer horizontal stripes, interrupted by the vertical stripes, are used with the same conventions.

The method is illustrated by two sketches: in plan (Figure 6.2) and as a coloured cartoon map (Plate 1).

Two maps in Plate 1, intended to be overlain, are here separated for clarity. Bedrock is mapped as it crops out on rockhead. At one locality, two symbols are superimposed; diagonal broken green lines representing tuff and tuffite are associated with the pattern for kaolinized gneiss, here printed in purple to indicate that it is a second layer beneath the tuffs. It must be assumed that the superficial deposits, an alluvial deposit overlain by loess, are less than 10 m thick, thus enabling two layers in the bedrock to be mapped. Another feature of interest is the evident gradational boundary between Miocene clays and sands in the south-east corner of the map; no definite boundary line is drawn.

Where bedrock crops out, colours are omitted on the map of superficial deposits, and consequently only green patterns would appear. Loess is the highest deposit in the succession, with three thickness ranges coloured in red. It blankets in part the alluvial terrace deposits, coloured yellow, represented in vertical stripes for the three thickness ranges. Holocene silts overlie alluvial terrace deposits. Glacial deposits presumably overlie bedrock as a single layer.

6.1.3 Simek stripe method

A sequence of two soil layers on pre-Quaternary bedrock can be shown by parallel stripes of uniform width across the outcrop of the uppermost layer (Figure 6.3). Stripes are coloured or ornamented to represent the second layer, and within the stripe the thickness of that layer is given in Roman numerals, either coded for different thickness ranges or giving an actual thickness. Conditions are realistic in the figure for the geological sequence, although a more rugged bedrock topography or a slightly undulating alluvial surface blanketed by loess could be accommodated. Thickness ranges in the superficials, here in 1 m lifts, are limited by contours and number coded. Graded colours as adopted by Pasek and Rybar (1961) in Plate 1 could also be used.

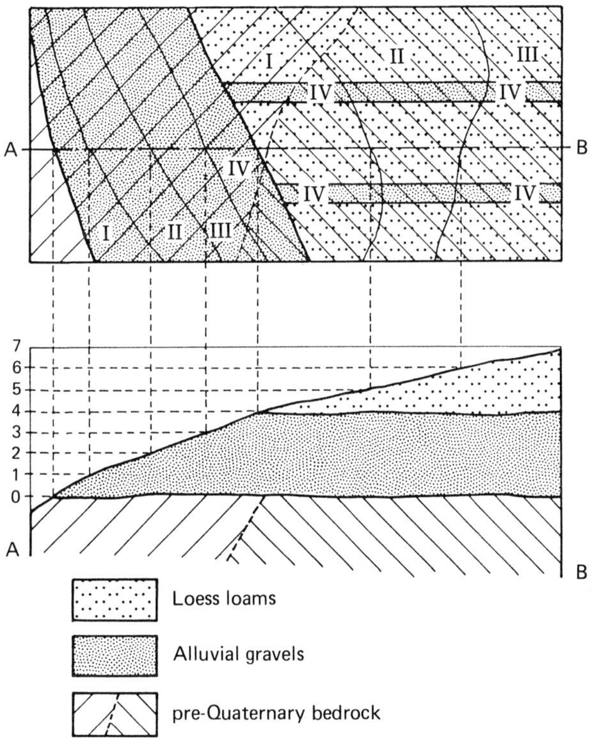

Figure 6.3 Stripe method of Simek (1962) to illustrate the sequence of superficial soils

Two stripes at right angles, that for the second layer interrupting the stripe for the third layer, opened the door for a Union Jack of five layers as the ultimate complexity. This and other developments came later.

6.2 Practical applications of the stripe method

In order to represent one or two layers in the all-important superficial deposits using the stripe method, a great deal of three-dimensional information is needed. From the practical point of view, this implies borehole logs from site investigations, perhaps supplemented by additional holes where data are scanty. It is only in the larger towns that such archival material is likely to be available, or in satellite new towns undergoing development in which the expenses of providing the necessary boreholes can be met.

Except for a general regional overview for planning purposes, when small-scale maps are all that are necessary, urban and urban-fringe geology demands at least medium-scale and preferably large-scale maps to be effective. Maps at scales of 1:5000 and 1:10 000 were originally called 'detailed' maps. Their production in Czechoslovakia has a long history, being mainly concerned with planning in more than 300 land areas in the country, including 50 major towns.

From the beginning, the compilation of detailed maps was standardized (Zebera, 1947b), with the parallel development of uncovered maps and maps using the stripe method. At the same time, pedological mapping was supplanted by engineering soil mapping because emphasis had to be placed on foundation conditions. Surface deposits, both soils and rocks, were placed into 28 groups according to grain size or petrographical character, with due allowance for bearing capacity and other technical properties. Mapping was done regardless of the geological conventions of genesis and stratigraphy. The only exceptions were loess and Recent alluvial soils, which were mapped as the uppermost superficial deposits.

6.2.1 Uncovered maps

In the late 1940s detailed maps were made of certain towns in which the top 2 m, sometimes more, of the superficial deposit were ignored. Geological boundaries were shown as they would appear at shallow foundation level on these 'uncovered' lithological maps. Some were published, at least in summary form, and one such is for the town of Turnov (Zaruba, 1946).

The publication is a short pamphlet illustrated by a large-scale engineering geological map (Figure 6.4) and cross-sections. Zaruba made the point, at that early date, that the results of a geological survey are just one of the important factors affecting town planning. In this case, there were no new problems to solve, but the results of the survey had to be interpreted and presented in such a way that they would be intelligible to specialists in other disciplines in town planning.

Turnov is an example of very favourable ground conditions. The town has developed on broad river terraces which offer good foundation conditions for all sorts of construction. With the prospect of continuing expansion of the town, the geologists had to determine the quality of foundation soils and rocks, water supply conditions, and deposits of natural building materials.

The best foundation conditions are provided by the gravels of the Pleistocene river terraces and Cretaceous sandstones. Cretaceous clays and clayey marls are far less suitable due to their poor bearing capacity and compressibility. Water-saturated alluvial plains in the bottom of the river valley, intermittently flooded at times of high water, are not suitable for any type of foundation. Additionally, unstable steep slopes and landslides cannot be considered as suitable building sites.

The engineering geological map (Figure 6.4) was compiled with special attention to the superficial deposits. In the legend, map units are broadly related to likely foundation conditions.

Gradually, it became apparent that subsoil maps did not meet the demands of engineering practice. Although the geological setting was well displayed to a depth of 2 m, this is just the depth at which the foundation soil proper begins, because in the continental climate of Czechoslovakia light engineering structures are usually founded at a depth of 1.5–2 m. Maps of foundation soils should, therefore, show the soils on which structures are founded because it is at levels below the foundations that the soils have a true supporting role.

Other criticisms were levelled at the maps. Soils were sampled from 2 m deep pits and classified only on grain size. Origin was not considered, so that the geological setting could not be fully appreciated. Foundation properties were based on strength and estimates of bearing capacity. Zeman (1962) modified the approach by dividing the map into areas bordered by green, yellow and red – implying good, conditionally good and unfavourable conditions for foundations.

The stripe method was more adaptable. Surface soils were distinguished by coloured planes, with coloured stripes, as already discussed above, for soils to certain depths, usually 0.3–1 m, 1–2 m and 2–4 m (see Figure 6.2). Standard symbols distinguished swamp areas and those with a watertable 1 m and 2 m below the surface, landslides, quarries, sand pits, etc. Consequently, the maps represented both the horizontal and vertical distribution of foundation soils, with significant hydrogeological data.

6.2.2 Engineering geological plans of Prague at 1:5000

In 1958, Simek led the first attempts to produce detailed engineering geological maps of some parts of Prague undergoing housing development (Simek, 1962). Compilation took account of archival material in the form of site investigation reports, data from test pits, new boreholes and temporary exposures.

The simple stripe method (see Figure 6.3), with two directions of stripe, was used to show the superposition of superficial deposits with thickness expressed both in the plane areas of colour and in the stripes by Roman numerals. With a maximum of seven thickness ranges, the stripe method can cope with, in this instance, a maximum thickness for three beds of more than 30 m, a depth range unlikely to be needed in normal circumstances.

Compiled between 1960 and 1962, the map sheet Kralupy nad Vltarou 5–9 covers an area of 2 × 2.5 km at a scale of 1:5000, in four sheets: geology, hydrogeology, documentation and zoning. The zoning map is discussed later (Chapter 7) and here only the first three will be dealt with.

The first sheet is a detailed geological map that shows the pre-Quaternary bedrock, the superficial deposits and fills as well as their thickness.

Brown pattern distinguishes the pre-Quaternary bedrock; more expressive patterns are used to distinguish rocks with unfavourable technical properties. The pat-

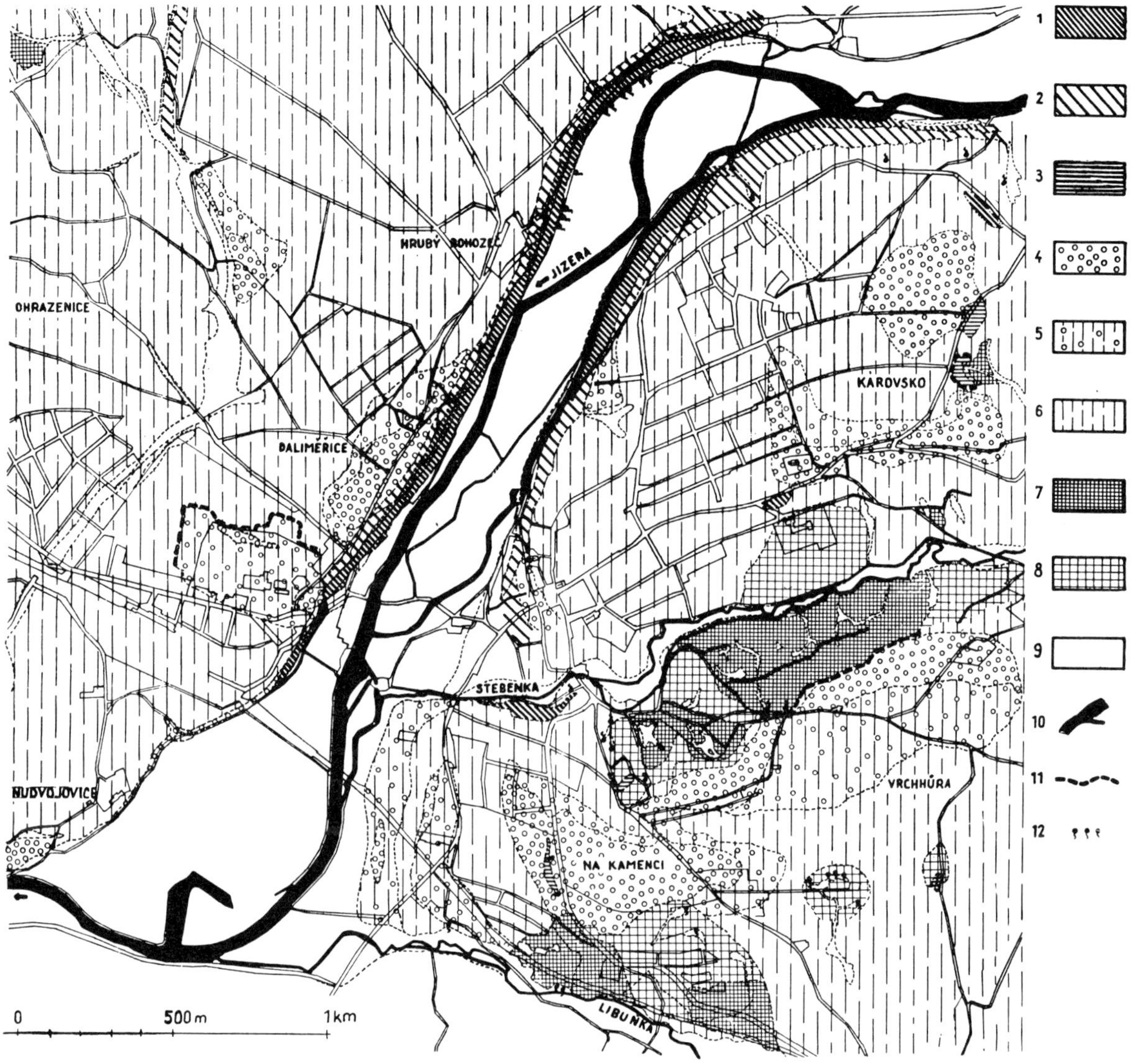

Rock type	*Foundation conditions*
1. Cretaceous sandstones and sandy marls of IX zone	Very good
2. The same rocks, 0.5–2 m below the surface, covered with loams	Very good
3. Cretaceous clays of X zone	Poor
4. Sandy gravels of river terraces	Very good
5. The same gravels 0.5–2 m below the surface, covered with loams	Very good
6. Loess and loams	Highly compressible, suitable for light loads
7. Landslide areas	Not suitable for foundations
8. Area of old earth movements, with tendency to sliding	Not suitable
9. Water saturated soils and areas intermittently flooded	Not suitable
10. Water areas	———
11. Unstable steep slopes and excavations	Not suitable
12. Springs	———

Figure 6.4 Engineering geological plan of Turnov, Czechoslovakia (From Zaruba, 1946, by permission)

terns are dashed if the bedrock is more than 10 m below land surface. Surficial deposits are distinguished by colour if they occur on the surface and by coloured horizontal stripes in the second horizon. In practice, a third horizon occurs, but seldom; if encountered, it may be indicated by vertical stripes. The thickness – from the surface of the area – is indicated in Roman numerals. Isolines of the thickness of surficial deposits are denoted by fainter lines, full or dashed. Fills are shown by a dense red hatching and, if thicker than 2 m, their thickness is given in metres.

Part of the map is reproduced at 1:10 000, with the detailed legend, as Plate 2. The legend, in the absence of a memoir or report – and if one was prepared it is not available – provides the only key to map interpretation and this is a vital part of the map. Not all of the symbols shown may appear on the map extract.

Seven ranges of thickness of the Quaternary deposits

290
285
280
275
270
3
3
5
6
S 34
S 54
725
S 84
S 92
S 127
S 128
S 74
250

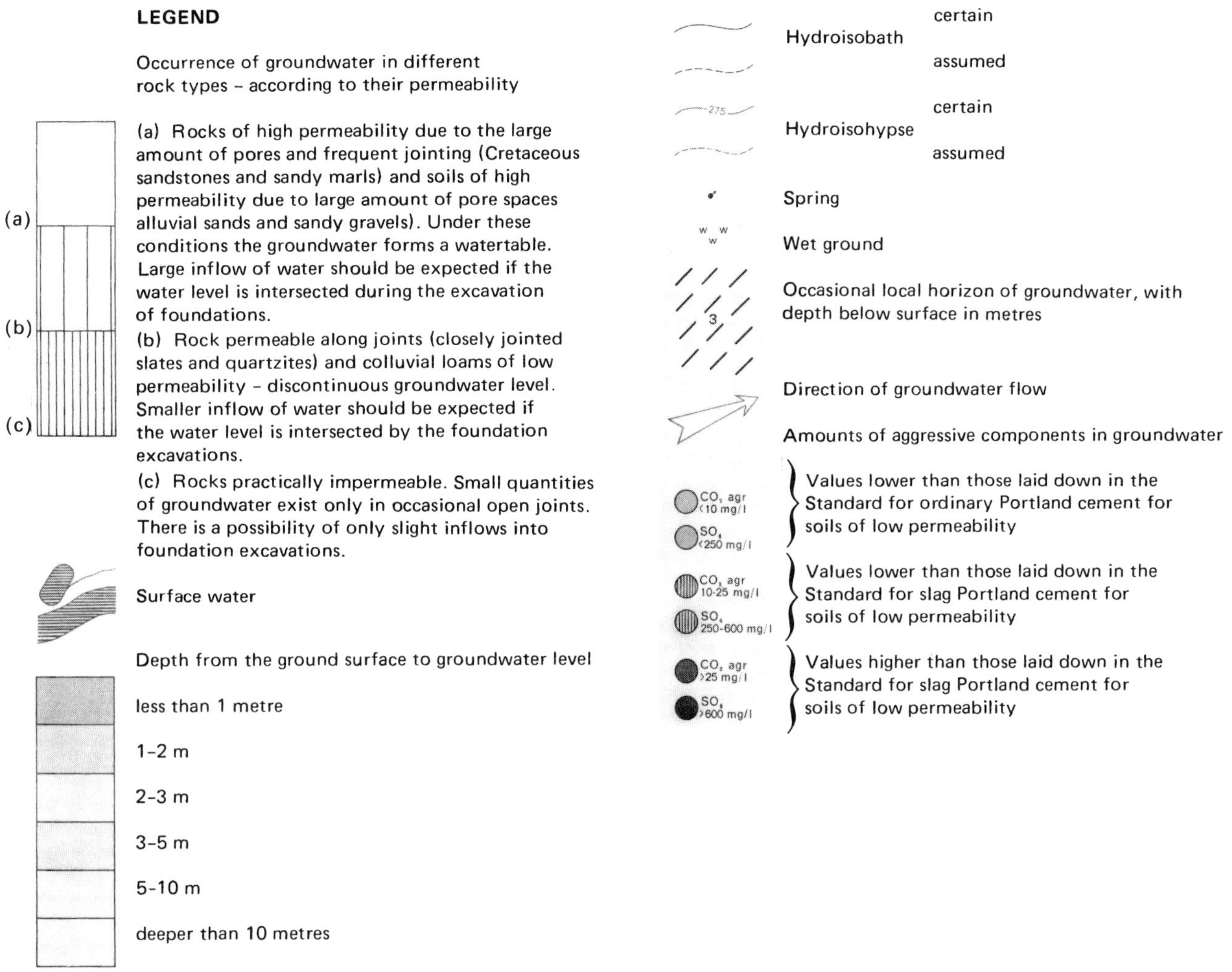

Figure 6.5 Part of the hydrogeological map accompanying the engineering geological map of Prague (Plate 2). Printed at 1:5000, reproduced at 1:10 000

are shown up to a thickness of more than 10 m. Where the thickness of these drift deposits varies irregularly, or when it was not determined exactly, the depth grades are combined. Fills above the original ground surface were not taken into account when the thickness contours (iosopachytes) of the first and second layers of drift deposits were drawn.

Outcrops of pre-Quaternary rocks are uncoloured and marked only by the characteristic pattern of the legend. Of particular interest are the three map units that represent weathered bedrock. These are not particularly easy to distinguish from Quaternary deposits because of the close similarity of the colours used, except for the carmine. Bedrock has been quarried locally; for example, on the south-eastern border of the map, extract quarries are marked in Ordovician slates and Cretaceous kaolinitic and glauconitic sandstones. These areas, and adjacent outcrops of the same rock types, are surrounded by large grey areas representing sands and loams with rock fragments derived by weathering on the parent bedrock.

A large part of the map area is covered by loess deposits, typically 1–2 m thick but reaching 5–10 m in places. The loess blankets the weathered type of bedrock, and where this occurs the weathering crust forms the second layer and is shown in the horizontal stripes.

Over most of the area loess rests on sands and sandy-gravel terrace deposits. Extensive loam pits have stripped the surface layer of loess to expose the terrace gravels beneath; in places the gravels themselves have also been worked. The pits may have active faces and elsewhere the old workings are partly refilled.

In the northern part of the map, outcrops of Algonkian lydites, greywackes and slates have an extensive mantle, up to 3 m thick, derived from the weathering of bedrock. Extensive bouldery screes of lydite have spread southwards over the terrace gravels, and to the east are overlain in turn by loess. The loess outcrop contains occasional large lydite boulders at the surface and within the body of the loess. This is the only area on the map where horizontal and vertical stripes are used for the three surface layers.

Close examination of the map, with the legend to hand, can be most rewarding, giving a very positive impression of the large amount of information that the stripe method can convey very clearly.

The second map sheet (Figure 6.5) provides information on ground and surface water. The rocks in which groundwater occurs are distinguished according to their characteristics into rocks with a porous permeability, and into almost impermeable rocks. The highest groundwater level is indicated by isobaths. In rocks with a

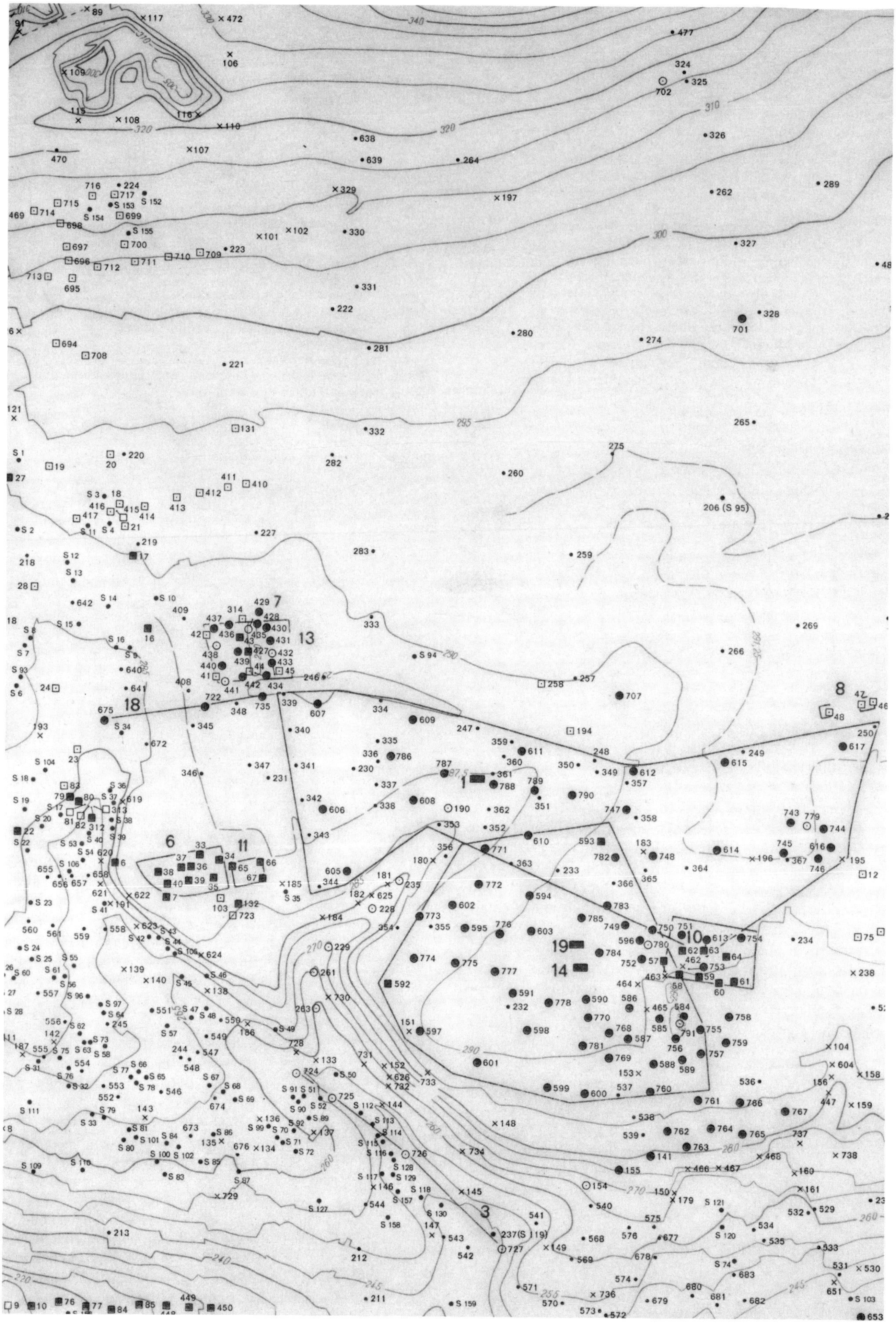

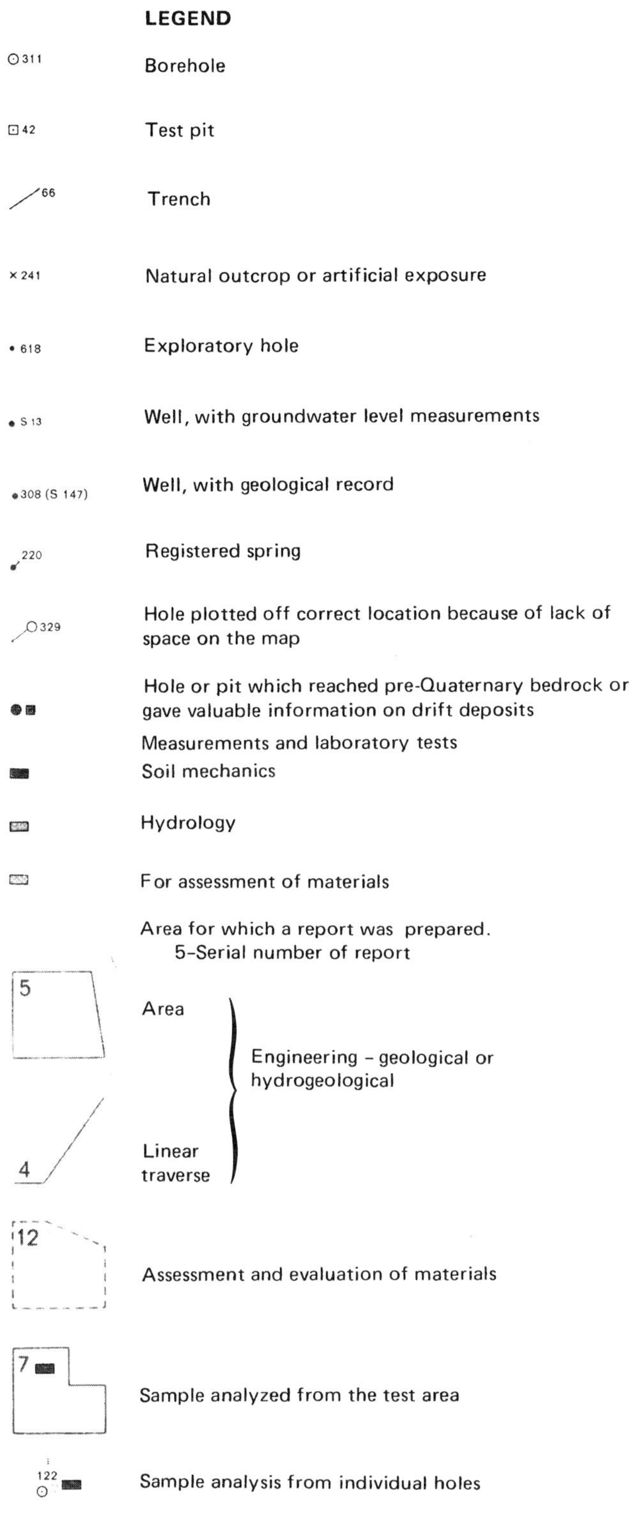

Figure 6.6 Part of the documentation map accompanying the engineering geological map of Prague (Plate 2). Printed at 1:5000, reproduced at 1:10 000

porous permeability, the groundwater level is indicated also by hydroisohypses. The aggressivity of groundwater and other hydrogeological data are expressed by symbols. An occasional local horizon of groundwater is shown by hatching, the number indicating its depth.

Using the map legend, the map can generally be deciphered in the black and white version without too much difficulty. Detail in the southern part of the map is clearly related to the concentration of site investigation data presented in Figure 6.6.

The third map of geological documentation comprises the complete geological documentation, both that produced by the survey and that of previous workers. Test pits reaching the pre-Quaternary bedrock are marked red. Red lines delimit the area under investigation. Symbols denote the places where physical or mechanical analyses of rocks, and loading tests, were made. A general impression of the map can be conveyed in black and white (Figure 6.6).

6.3 Development of medium-scale 1:25 000 engineering geological maps

Between the years 1957 and 1961 the Geological Institute of the Czechoslovak Academy of Sciences (Pasek and Rybar, 1961) compiled engineering geological maps at a scale of 1:25 000 for four adjoining areas in Western Bohemia, using a new cartographic method. In addition, the maps presented a new concept of engineering geological mapping, paying full attention to desk studies and particularly the content of the map. Field surveys checked the initial compilations. The results of the survey were assembled in the now familiar set of four sheets.

6.3.1 1:25 000 maps of Western Bohemia, 1961

Sheet A. This is a geological map with representation of the Quaternary superficial deposits as three superposed layers with indications of thickness variations. The geology is depicted to a depth of 10 m below ground surface. Pre-Quaternary bedrock is distinguished by monochrome patterns and the superficial deposits by plane colours. Bedrock patterns correspond to the character of the rocks according to predetermined standards. A second bedrock layer within the 10 m depth is shown by an appropriate pattern in a different colour. The boundary of the second layer corresponds approximately to a 10 m depth.

Superficial deposits are shown on the map by plane colours spread out over the bedrock patterns. Where bedrock crops out, or is overlain by less than 0.8 m of weathered rock or superficial deposits, the area is left uncoloured on the map. Different thicknesses of superficial deposits, in the ranges 0.8–2 m, 2–5 m and 5–10 m, are coloured with shade of colour deepening as depth range increases. Where superficial deposits are thicker than 10 m, the bedrock pattern is omitted. If an outcropping superficial deposit overlies a second deposit, the lower deposit is marked by a coloured vertical stripe, with the colour appropriate to the second layer. The convention first adopted by Zebera (1947a) is used to indicate the thickness range of the soils in the second layer.

Sheet B. This sheet is referred to as a map of engineering symbols. Red is used as a warning colour for those morphological forms in an area that would adversely influence foundation conditions. Blue is used for hydrogeological symbols; black for places where rocks have some distinct adverse property such as very close cleavage or deep weathering. All quarries, pits, mines and so on are given appropriate symbols. Approved reserves of economic minerals and nature reserves are hatched in red to emphasize that these areas are not be be developed.

Sheet C. This is the geological documentation map. Here are documented, by point symbols, test pits, boreholes, etc., with their exact location and whether or not the thicknesses of the Quaternary soils were determined. Localities where soil samples were taken for laboratory analysis, or where groundwater samples were taken, are marked by distinctive symbols. The same applies to quarries, clay and sand pits, and mines. A card index, today a computer data bank, of documentation points supplements the map, and the results of laboratory and field analyses are documented.

Sheet D. This map gives an engineering geological evaluation of the map area for foundations of specified types of structures. It is a derivative map from the fundamental map sheets A, B and C and is compiled for a specific purpose.

The map area is subdivided into several units. Evaluation of all the engineering geological facts makes it possible to distinguish suitable and usable areas as well as those that might be used only under carefully specified conditions, and those that present entirely unfavourable foundation conditions. Each area is colour coded: green indicates suitable; blue indicates certain conditions; red indicates no-go.

Workings in mineral deposits, and approved economic reserves, are marked. The map is supplemented by a table giving the characteristics of the classified areas.

It is unfortunately the case that these maps are not available for reproduction. However, some maps in a similar series have been published. With these maps a standard four-sheet format was formalized: geological map; hydrogeological map; documentation map; and a map of engineering geological zoning – the true engineering geological map.

6.3.2 Engineering geological 1:25 000 maps of the Zvolen Basin

The first version of this series of maps was compiled in 1961 by the Department of Engineering Geology of the Comenius University in Bratislava (Matula, 1961). New versions were progressively developed between 1963 and 1966, and the final version was adopted as a model by the COMECOM Commission in Engineering Geology (Matula and Pasek, 1966). Used in illustration here is a map of engineering geological conditions that has been widely published as a typical example (Matula, 1969; Varnes, 1974; Anon., 1976) of the Czechoslovak mapping technique.

A desk study of available data was supplemented by systematic field surveys and by soils and other field tests. Four map sheets were produced.

A conventional geological map, in which the mapped geological units are represented by symbols and colours, distinguishes the Quaternary deposits from those that are older.

Outcrops of the pre-Quaternary bedrock are shown by plane colours and patterns. Extension of the pre-Quaternary soils of Pliocene and Miocene age beneath the Quaternary deposits is indicated only by the appropriate patterns in grey. Heavy patterns are used from the bedrock surface to 5 m depth, whereas lighter patterns are used for 5–10 m depth.

Quaternary soils are colour coded and also have a distinctive pattern to indicate lithology. For some deposits, two thickness units are shown, 0–5 m and 5–10 m, and for slopewash there are three thickness units, less than 2 m, 2–5 m and 5–10 m, shown by deepening shades of the same colour. The second layer beneath slopewash is shown using the Simek (1962) convention of vertical stripes with two gradations of colour.

Hydrogeological conditions are shown in blue for hydroisobaths of maximum seasonal groundwater depth at intervals of 2, 5 and 10 m. Specific symbols indicate the corrosiveness of groundwater.

Red is used to show important faults and the distribution, kind and intensity of geodynamic processes, gully erosion and accumulation of alluvial cones.

In this map, rocks and soils are divided into engineering formations which are distinguished on the basis of physical and engineering characteristics. The map, illustrated in colour, and the legend in Plates 3 and 4 are from Matula (1969).

Table 6.1 Engineering classification of map units to accompany map of engineering geological conditions in the Zvolen Basin (Plates 3 and 4)*

Symbol	*Soil classification*	*Building foundation conditions*	*Road construction*		*Ease of excavation*
			Foundation	*Material for embankments*	
${}^{l}_{f}Q_{3-4}$	ML CL	D_2 D_3	MV	MV	1–3 1–3
${}^{s}_{f}Q_{3-4}$	SM	C_2	V	VV	2
${}^{g}_{f}Q_{3-4}$	GW GM	B_1	VV	VV	4
${}_{o}Q_{3-4}$	OH	E_3	N	N	2–3
${}^{l}_{d}Q_{3-4}$	CL	D_2 D_2	MV	MV	2–3
${}^{tr}N_2$–Q_4	—	A_2	N	MV	5
${}^{l}_{lf}N_2$	CL ML	D_2 D_3	MV	MV	2–3
${}^{g}_{lf}N_2$	GV GM	B_1	VV	V	3
${}^{tu}N_1$	—	A_2	V	MV	5–6

*Classification according to ČSN (Czechoslovak Building Standard): A_2 = soft rocks; B_1 = gravels with pebbles in contact; C_2 = medium to fine sands; D_2 = clays of medium plasticity; D_3 = clays of high plasticity; E_3 = organic silky clays; VV = very suitable; V = suitable; MV = barely suitable; N = unsuitable.

Conventionally, the map legend provides the key to the symbols used on the map. In addition, there is a genetic-lithological classification giving summary information on the age, genetic group, a stratigraphical symbol and short lithological description. Colours for the various map units, colour shades for different thicknesses, and red symbols for lithology complete the legend which is essential if the map is to be interpreted.

Engineering aspects are dealt with in separate tables that do not form part of the map legend. The first table is an engineering classification giving codified information for each map unit: a standard soil classification supplements foundation conditions for buildings and roads, use of materials in embankments and ease of excavation (Table 6.1).

A whole range of engineering geological properties with typical values (Table 6.2), usually applied to rock and soil descriptions (Chapter 4), complements the engineering classification. However, there are no extended descriptions of the type familiar in the UK (see section 4.6.1); this is reserved for the comments on the foundations as they occur in the mass, *in situ* (Table 6.3).

Using generalized symbols, rather than actual values, for each map unit, guidance can be found on road construction, soil classification and ease of excavation. Classification is according to the Czechoslovak Building Standard.

The second map, not illustrated, is a typical hydrogeological map. Shallow groundwater in the Quaternary superficial deposits is shown in colours related to permeability. Blue hatching is used for the extent of the groundwater table, and the thickness of the hatched lines is related to the likely groundwater yield. Distinctive symbols, also used in Plate 3, mark the location of springs and their flow, and with mineral springs also the physical and chemical properties of the water.

The third in the series of engineering geological maps is a typical documentation map, with a content similar to that described above; it is not illustrated.

6.4 Use of the stripe method in the Federal Republic of Germany

During the period 1960–1966, a succession of committees of engineering geologists from the various COMECON countries met to consider the preparation of standard engineering geological maps (Kabel, 1968). Published in German as 'Instruktion für die Anfertigung einheitlicher ingenieurgeologischer Gründkarten' (Bachmann *et al.*, 1967), the report defines principles and terminology for engineering geological maps prepared within the group of countries. Model maps were compiled by the USSR at a scale of 1:500 000, by Poland at 1:200 000, by Czechoslovakia at 1:25 000 and by the German Democratic Republic at 1 : 5000. The 'instructions' are divided into three main parts.

The first part contains the general principles and instructions for making geological maps. It deals with:

- nomenclature and classification of engineering geological maps;
- the content of maps with regard to engineering geological conditions, and maps of engineering geological zoning (Rayonierung);
- the methodology of representing natural factors (lithology, tectonics, hydrogeology, exodynamic features) and the taxonomy of regional units;
- engineering geological profiles and tables;
- the working steps in the preparation of maps;
- the basic data needed for their preparation.

The second part contains additional instructions for making maps at different scales, and classifies maps according to scale into:

General maps	1:500 000 and smaller
Small-scale maps	1:200 000 and 1:100 000
Medium-scale maps	1:50 000 and 1:25 000
Large-scale maps	1:10 000 and larger

The third part considers the requirements for the explanatory notes, or legend, on the map sheet.

The guidelines are very comprehensive and include lists of standard symbols, patterns and colours; even the prescribed layout for the report to accompany the map is set out. While the standards were being worked out, mapping continued with, at large scales, the development of interesting variations on the stripe method.

6.4.1 Engineering geological map of Berlin

An interesting variation of the stripe method was used by Robling and Tamm (1961) on the 1:4000 scale engineering geological map of the Central Ward of the City of Berlin. Four sheets were used: a documentation map, a geological map, a groundwater map, and a subsoil or building ground map (Baugrundkarte). Reuter (1968) later expressed the opinion that the subsoil map, based on lithology, could be regarded as the real engineering geological map *sensu stricto*.

The subsoil map is of interest here; the other maps are conventional. On the map extract (Figure 6.7) the surface geology, shown by widely spaced patterns, is simple: mainly sand with localized areas of undifferentiated fill with peat and peaty sand, some of which are infilling ancient fortification moats. The surface distribution holds down to a depth of 3 m. For depths below 3 m, two methods are used to show the ground conditions down to more than 10 m; the resultant map is somewhat confusing.

A modification of the simple stripe method is used to show lithology in four stages: 3–4 m, 4–6 m, 6–10 m and more than 10 m. On the map, the four layers of this 'stratigraphical' succession are mainly in sand; on the other hand, the representation in Figure 6.8 shows the presence at depth, more than 10 m, of areas of glacial till either overlain irregularly by, or passing laterally at, that level into sands. The stripe, a wide band on the map subdivided into four successively narrower bands, represents lithology down to 10 m and below. Closer patterns are used for soil types below 3 m.

Combined in the stripes is the other type of surface outcrop – the peats and fill – locally more than 10 m deep. Sufficient information on their distribution is available from site investigation and other boreholes to enable the base of the deposits to be contoured at 3, 4, 6 and 10 m depths. It is therefore possible to incorporate

Table 6.2 Engineering geological characteristics of rock materials in each map unit shown on map of engineering geological conditions in Zvolen Basin (Plate 3)

Symbol	*Lithological types*	*Specific gravity,* ρ_s (g cm^{-3})	*Dry bulk density,* ρ_d (g cm^{-3})	*Porosity* (%)	*Consistency,* I_c / *Rebound hardness (R. Schmidt)*	*Moisture content* (5)
${}^{l}_{f}Q_{3-4}$	Sandy-loessic loams (in the Hron floodplain)	2.69†	1.51 / 1.42–1.62‡	43 / 39–47	0.6 / 0.1–0.9	30 / 25–32
	Clayey-loessic loams (in older terraces)	2.61	1.57 / 1.30–1.78	40 / 36–51	0.8 / 0.4–1.3	29 / 15–35
${}^{s}_{f}Q_{3-4}$	Loamy sands	2.65	1.30–1.70§	36–47		9–10
${}^{g}_{f}Q_{3-4}$	Sandy gravels (Hron Valley terraces)	2.65–2.68	1.80–1.90			
${}_{o}Q_{3-4}$	Sapropelic muddy silts of abandoned river arms in the floodplain*	2.60	1.37	48	0.13	33
${}^{l}_{d}Q_{3-4}$	Sandy-loessic loam	2.67	1.62 / 1.13–1.77	40 / 32–58	0.7 / 0.5–1.3	26 / 18–43
${}^{tr}N_2$–Q_4	More compact travertine / Strongly porous travertine	2.66 / 2.68	1.93 / 1.60	8 / 16		
${}^{l}_{lf}N_2$	Loessic-clayey soil	2.63	1.76 / 1.15–1.60	45 / 40–55	0.8 / 0.3–1.0	29 / 23–46
	Sandy-loessic soil		1.67 / 1.30–1.70	37 / 30–50	0.9 / 0.7–1.1	20 / 14–39
${}^{g}_{lf}N_2$	Weathered sandy gravels					
	Loamy gravels					
${}^{tu}N_1$	Andesitic arenaceous tuffs	2.68	1.68	36	16	
	Arenaceous tuffites	2.71	1.69	37	26	
	Rudaceous tuffites	2.61	1.85	29	—	

*Mean statistical values.
†Medium and minimum–maximum values.
‡Minimum–maximum values.
§Analogous values from neighbouring area.
‖k_z coefficient of softening (ratio of saturated to dry strength).

the deposits in the stripe system (Figures 6.7 and 6.8) by suitable shading where the deposit crosses the appropriate stripe level, and at the same time show the plan distribution at the various levels on the map.

Down to depths of 10 m, it is very likely that other lithological types, or stratigraphical units, will be present that do not necessarily crop out in the map area. Such deposits can be catered for. In Figure 6.8 the distinctive boundary line at 10 m depth can be used to show the limit of the surface sands at that depth where they overlie glacial till. The outcrop at 10 m depth is shown by the appropriate shading in the more than 10 m stripe.

If the rather confusing depth-contour system is confined to geotechnically highly significant soil types, as for example peat and fill, then the map provides a direct indication not only of areas with problematic foundation soils but also the depth to a suitable foundation soil within the area. The depth zones used are related to the usual building methods and foundation conditions in Berlin according to established standards.

Four aspects need to be emphasized:

1. From the combination of stripes and depth contours on the interface between different soil types – on the map these are sand and fill – the sequence of soil types can be determined at any location down to 10 m depth.
2. Each stripe system gives the sequence, read from top to bottom of the stripe on the map, of soils from the surface.
3. Subdivisions in the main stripe refer only to the locations crossed by each minor stripe.
4. The stripes are particularly useful in highlighting the distribution of soil types that do not crop out; for example, the glacial till at 10 m in Figure 6.7.

Apart from item 4, interpretation of the map is quite easily done from the contours of the interface between different soil units, although it has to be admitted that the stripes have a visual impact that requires a minimum of interpretation. As such the stripes are useful.

w_p(%)	w_l(%)	I_p(%) / Indentation hardness kgf mm^{-2}	Shearing strength (φ, deg)	Cohesion, c (kgf cm^{-2})	Uniaxial compressive strength, σ (kgf cm^{-2})	Modulus of compressibility (kgf cm^{-2}): Modulus of elasticity, E (10^{-4} kp cm^{-2})	Modulus of deformation, E_o	Modulus of dynamic elasticity, E_{dyn}	Others
			25	0.5	—	—	85	—	
24 / 21–30	37 / 32–44	14 / 10–22							
			11	0.8	—	—	—	—	
29 / 18–35	41 / 30–54	16 / 7–24							
—			27	0.1					
			30	36					
21	35	14							
23 / 17–40	39 / 23–52	14 / 6–20	21	0.8	—	—	102–140	—	
					302			84	k_z = 0.87‖
		150 / —							
					118			—	k_z = 0.60
25 / 19–40	40 / 29–60	16 / 12–23	9	1.0			80		
20 / 15–24	36 / 22–48	15 / 6–24	14	0.7			147–150		
—		48			115–171	4	2.8	5	k_z = 0.61–0.71
—		70			225	10.9	4.3	15	k_z = 0.78
—		—			115	—	—	—	k_z = 0.38

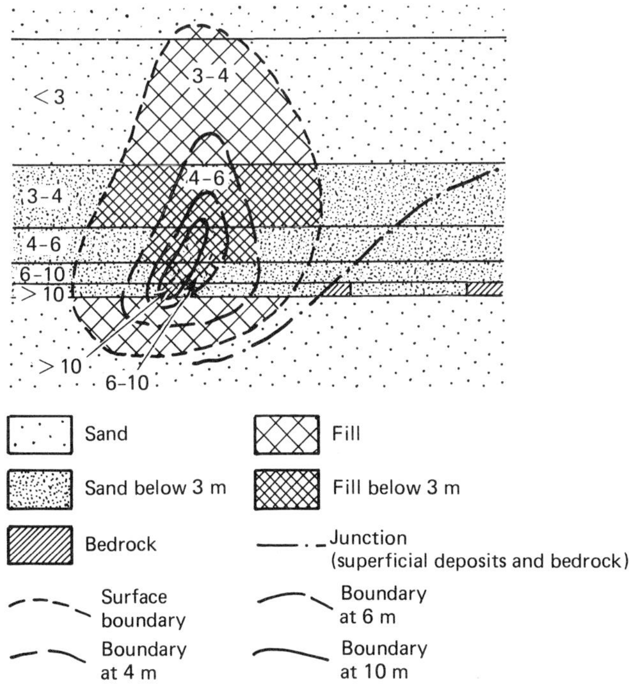

Figure 6.7 Part of the subsoil map for the centre of Berlin (From Robling and Tamm, 1961, Fig. 4, by permission)

6.4.2 Engineering geological map of Frankenfelde

As an illustration of large-scale maps compiled according to the COMECOM instructions, Bachmann and Reuter (1968, Fig. 2) produced a specimen map at a scale of 1:5000 of the engineering geological conditions in the town of Frankenfelde. The work was done during the years 1966 and 1967. The map is printed in colour with two cross-sections and a comprehensive legend.

A conscious decision was taken that the areas of fill should not be regarded as belonging to any geological formation and invariably occur as the uppermost layer. This meant that the two uppermost layers in the succession of superficial soils can be shown by using the stripe method.

The vertical stripe method is used on the Frankenfelde map wherever two Quaternary superficial deposits overlie either the Permian bedrock or the Pleistocene glacial till. Lithological units in the Quaternary are colour

Table 6.3 Engineering geological characteristics of rock mass in each map unit shown on map of engineering geological conditions in Zvolen Basin (Plate 3)

Stratigraphical symbol	*Engineering formations*	*Rock mass characteristics*
${}^{l}_{f}Q_{3-4}$	Sandy-loessic to loessic-clayey alluvial loams	Thickness 0–2 m. Horizontal thin to laminar bedding. Vertical anisotropy. Strong slaking. Soft to solid consistency. Intercalations of organic soils. Slight permeability
${}^{s}_{f}Q_{3-4}$	Alluvial sands predominantly medium-grained with loamy intercalations	Thickness 0–1.5 m. Medium sands, intercalations of silts, cross-bedded. Considerable loam admixture decreases the permeability and increases coherence. Facially very variable. Strongly weathered in older terraces
${}^{g}_{f}Q_{3-4}$	Fine- to medium-grained sandy gravels, loamy in older terraces	Sandy medium gravels with sand intercalations 3–9 m thick. Irregular bedding, vertical anisotropy. Moderate permeability $5.10^{-4}\,m\,s^{-1}$. Water-bearing, locally corrosive. Boulders at the base. In older terraces strongly weathered, loamy, permeability 3.10^{-6}. In lateral valleys coarse, bouldery, heterogeneous
${}_{o}Q_{3-4}$	Sapropelic muds in abandoned river channels, sandy-loessic to clayey	Deposits occupy numerous irregular abandoned river arms and depressions at the edges of alluvial cones. Sandy and loessic sapropelic muds, very soft to soft consistency, water saturated. Very low permeability, slakes readily. No bearing capacity, very strongly compressible
${}^{l}_{d}Q_{3-4}$	Sandy-loessic to clayey slope loams	Irregularly to cryptobedded, predominantly 1.5–3 m thick, at the foot of slopes more than 10 m thick, sandy-loessic brown loams, with fragments of volcanic rocks. On Pliocene deposits more clayey, rust-brown, with gravels. Stiff to very stiff, weather to prismatic fragments. In places slake readily. Very slightly permeable. Used for brick making
${}^{tr}N_2$–Q_4	Travertines, mounds and slope sheets	Irregular bodies, accumulations and sheets adapted to the relief. Variable thickness. Semi-solid rocks, strongly fissured, macroporous and permeable. Soluble, karstified. Hardly suitable for building or decorative stone
${}^{l}_{lf}N_2$	Lacustrine and alluvial clayey-loessic, sandy-loessic, rarely clayey loams	Facially very homogeneous cryptobedded clayey to sandy-loessic soils, up to 10–15 m thick. Brown to rusty-brown, spotted. Stiff to very stiff consistency. Diagenetically consolidated. Prismatic disintegration. Practically impermeable
${}^{g}_{lf}N_2$	Lacustrine and alluvial sandy to loamy gravels with sand beds	Sandy-loamy medium gravels 5–50 m thick. Lenticular intercalations of sands. Strongly weathered, in places weakly cemented. Medium permeability
${}^{tu}N_1$	Stratified andesitic tuffs and tuffites	Coarse-bedded rudaceous–arenaceous tuffs deposited in water environment. They alternate with rudaceous to argillaceous tuffites. Varied colours, facial variability and physical heterogeneity. Weak semi-solid rocks, weather readily, disintegrate in water. Little jointing and low permeability

shaded to represent thickness bands of less than 2 m, 2–5 m and 5–10 m, with these distinctions carried into the stripes wherever appropriate. So far, the approach is conventional and differs from the Simek method only in not having Roman numerals in the stripes to indicate thickness ranges.

A large part of the town centre has significant accumulations of man-made fill marked on the map with a distinctive rectangular grid pattern in grey. Two spacings of the pattern indicate thicknesses of less than 2 m and 2–10 m. The area of fill is left uncoloured to emphasize its distribution, but, as a background, patterns appropriate to the Permian bedrock and Pleistocene glacial till are printed in stronger grey lines.

Where fill rests directly on the glacial till, no colour is used – only the two patterns. If there is an intervening deposit higher in the Quaternary succession, this occupies the stripe with colour and pattern in red. The same convention applies elsewhere on the map wherever there are two superficial deposits overlying the glacial till.

In the areas of fill, wherever there are two units in the superficial Quaternary soils intervening between the fill at the surface and the bedrock glacial till, a double vertical stripe is used. A wide stripe indicates the deposit beneath the fill and a central stripe of standard width the second deposit down resting on the glacial till.

Illustrated in Figure 6.9, the double stripe in areas of fill allows for units to be shown very clearly on the map. Everywhere on the map the local succession is indicated by stratigraphical sequences of the symbols for the lithological units down to and including bedrock; the sequence may be of one unit where bedrock, including glacial till, crops out, with up to four units in the fill areas.

The plan is interpreted as a cross-section in Figure 6.10.

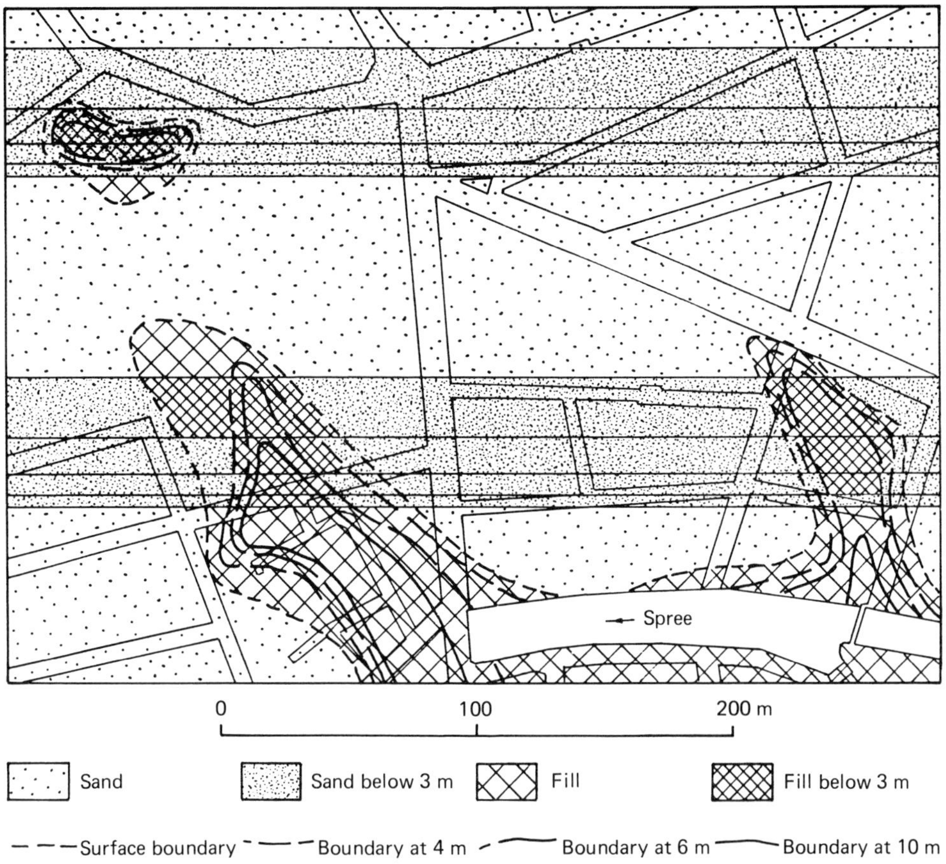

Figure 6.8 Explanation of the stripe method used on the Berlin map (From Reuter, 1968, Fig. 3, by permission)

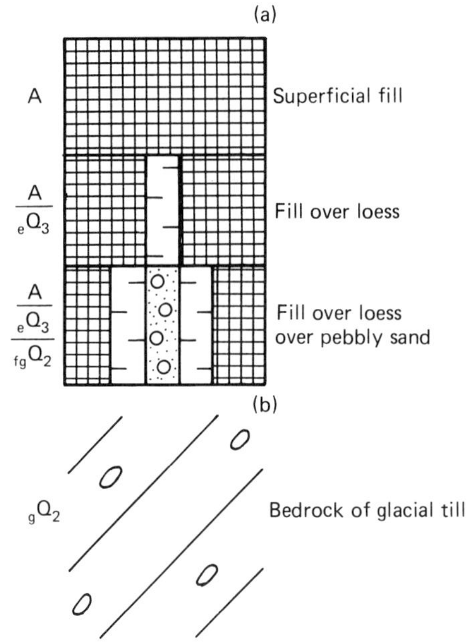

Figure 6.9 Double stripe method used to represent one or two superficial deposits resting on bedrock, all overlain by fill: (a) superficial units; (b) underlying 'bedrock', in this case glacial till. On the map, (a) overprints (b)

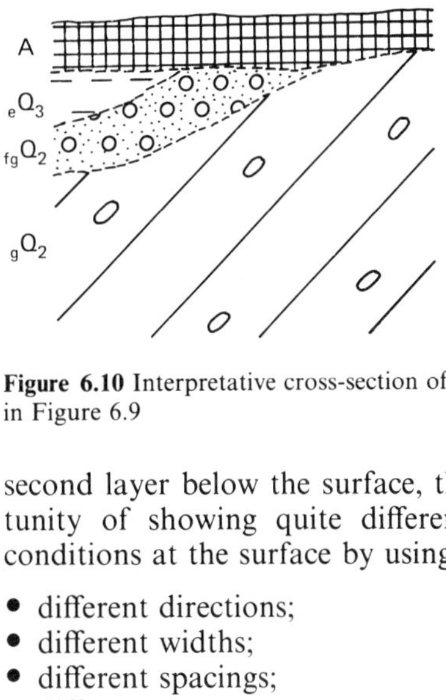

Figure 6.10 Interpretative cross-section of the double stripe diagram in Figure 6.9

6.5 Application of the stripe method for other purposes

Most applications of the stripe method that have been described so far use one or two stripes, each of the same width and spacing. Frankenfelde is an obvious exception. If the stripe is used to show the nature of the second layer below the surface, there is still the opportunity of showing quite different aspects of ground conditions at the surface by using two stripes of:

- different directions;
- different widths;
- different spacings;
- different patterns.

What possibilities are there? For clarity and ease of discrimination they are limited to four directions of stripe (Figure 6.11), which may be used individually or combined to indicate up to four layers plus the surface layer. Probably two widths of stripe are enough and three spacings sufficient. A cross is the only obvious pattern to use, either upright or diagonal.

6.5.1 Map of a permafrost region, USSR

There can be little doubt that some applications of

the stripe method have reached the practical limit of complexity, although if carefully drawn they remain relatively easy to understand and interpret.

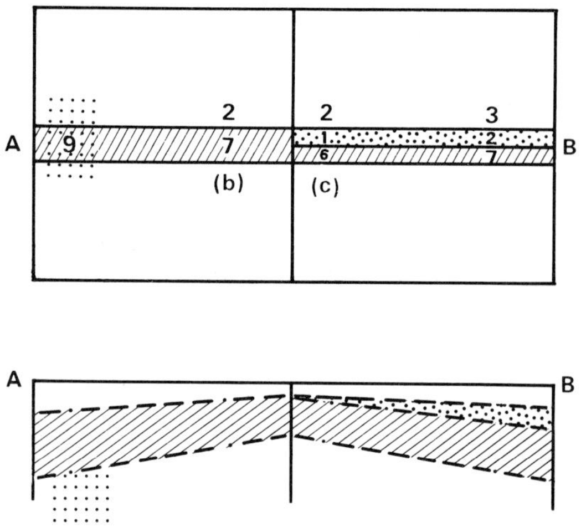

Figure 6.11 Stripe method applied to the discrimination of four layers beneath the surface layer

Golodkovskaja and Demidyuk (1970) have published excerpts from two maps of a mining area, originally drawn at 1:10 000 – the plan scale – and one of them is reproduced here at approximately 1:100 000. The selected map (Figure 6.12A) shows the engineering and geological conditions of the surface strata; the other provides information on the engineering geology of the bedrock.

The map is an actual practical example of an area underlain by a mineral deposit in a permafrost region. It has been used in the design and construction of mining and industrial works, urban, road and other structures needed for the exploitation of the deposit. The original version of this map, reproduced here redrawn and slightly modified, was published as a 'Montage of engineering and geological map of Quaternary deposits with elements of territorial engineering and geological evaluation for ground construction' (Golodkovskaja and Demidyuk, 1970, Fig. 1).

In a permafrost region the stripe method, with horizontal and vertical stripes, crossed stripes and one direction of oblique stripes, shows the age and genesis of the uppermost and immediately underlying soils or rocks, their lithology, thermal condition and frozen water (ice) content. Lithology is given by the patterns in the legend; on the map, symbols are depth of seasonal and permanent frozen ground, depth of frozen and partly frozen water, thickness of Quaternary deposits, and contemporary geological processes and phenomena. There is also an evaluation of the area for construction purposes.

From the point of view of engineering geological cartography, the map illustrates an interesting application of the 'stripe method': the direction and width of the stripes indicates the average annual temperature range of the superficial soils (symbols 9–12), whereas the three degrees of spacing of the stripes (symbol 13) indicate the frozen water (ice) content of the materials. This latter treatment of the stripe symbols is, as shown on the map, applicable to the whole range of symbols 9–12.

Despite the abundance of symbols the map is easily deciphered in this black and white version; colour might possibly make interpretation easier. A great deal of information is packed into the map, but even so the map would need to be accompanied, at least, by complementary topographical and hydrogeological maps.

Obviously, an alternative cartographic approach would have been to produce a set of maps, possibly as overlays to a base topographic map. Some of these are given in Figure 6.12B, and a decision has to be made by the organization or person commissioning the map whether to have a one-sheet presentation that requires a fair amount of application to understand, or several maps each dealing with individual aspects and each readily interpreted.

The first derived map (a) is the distribution of the superficial deposits, mainly Quaternary, that is obvious from the original map sheet. Stripes in various directions reveal the underlying marine Upper Quaternary and Permian–Triassic volcanics, occupying two broad areas on the map (b). These two maps deal with the only lithological aspects of the area.

Areas on three of the remaining maps (c,d,e) all differ in some respect, presenting different geological aspects revealed by the stripes or symbols: map c is of areas with the same average annual range of temperature; map d deals with the frozen water (ice) content; map e gives the distribution of different lithological types.

The final map (f) evaluates the engineering geological conditions into three categories based on the amount of preparation needed before engineering construction can be undertaken. This map is most clearly comparable to map c, suggesting that the most important engineering constraint is a combination of average temperature range and ice content of the surface soils; lithology plays a very subordinate role.

From the point of view of the engineer, there can be little doubt that dividing the map into three types of zone, as in map f, is what the engineer needs. There is one proviso, and that is the qualification of the engineering preparation required would have to be set down in detailed engineering terms. The map then would become a geotechnical map rather than an engineering geological map.

6.6 Engineering geological cartography in France: 1965–1970

In France, engineering geological maps are referred to as 'cartes géotechniques'. Peter (1966) and Gazel and Peter (1969), reviewing the topic, concluded that such maps were not new and could be traced back at least to maps shown at a technical exhibition in Leipzig in 1913 (see section 1.2.3).

It was not until long after World War II that the French enthusiastically took up engineering geological mapping. Mapping techniques, originally elaborated in Czechoslovakia, were the inspiration for the French Geological Survey – the Bureau de Recherches Géologiques et Minières (BRGM) – to produce their own maps. Most of these, with an accompanying memoir, were not published and remained documents for official use only. The first appeared in 1965–66 and it was in the course of the following three years that most of them were produced by BRGM.

Following the Czechoslovak lead, it was accepted that ideally a series of maps would be necessary, including a geological map, a hydrogeological map and a map of rock and soil properties. Reflecting to some extent special aspects of the geological setting, the other maps that should be included were:

1. Geomorphological map, showing actual or potential landslides; solution effects in calcareous and gypsiferous terrains; collapse of aging underground quarries close to the surface.
2. Map of natural construction materials.
3. Other maps showing, for example: the location of mines and underground quarries, working or abandoned, with consequent risk of subsidence and damage to buildings; problems related to the pollution and protection of groundwater; seismicity; pedology, to the extent that it was desirable to conserve agriculture in a region, or the local soil was known to be especially fertile.

A separate documentation map might be produced, although in many cases this would be combined with the geological map.

Sanejouand (1972) also reviewed engineering geological mapping in France. Observing that what had been accomplished depended primarily on the organization for which the work had been done from the point of view of map scale, and also the organization or person involved in producing the map where content was concerned, there appeared to be a need to classify maps by scale, content and purpose. It was also necessary to distinguish maps made for experimental purposes from those produced to solve real problems.

A geotechnical study almost invariably required more than one map, as outlined above. He recognized two kinds of map – analytical and synthetic.

Analytical maps

These are maps on which one or more geological aspects are presented. Good examples are hydrogeology and superficial deposits. These are called 'cartes de facteurs' or 'factor' maps. As a group they may be divided broadly into:

- maps of common factors accumulated in the course of the mapping programme;
- maps of special factors, derived from the information on the former types of maps in order to solve a particular, concrete problem – as an example, a map of contours on the top of a selected bed or on rockhead that was preferred as a foundation level.

It had to be accepted that factor maps were not, as a general rule, of direct use to planners and similar specialists; a certain minimum amount of geological knowledge is required to understand and interpret such maps. This is now held to be commonly true, and interpretative maps (referred to as synthetic maps) have to be prepared for this non-geologist type of user, as explained below.

Synthetic maps

These are maps on which different factors are assessed for a particular use. Examples available at the time were concerned with the engineering properties of soils, or were maps of zones which could be considered homogeneous for a particular application. The most usual types were maps of suitability for foundations, and maps of natural materials.

Foundation suitability maps were of two kinds, dependent upon the rigorousness of the zoning. In the first category, zoning classifies the soils into a small number of categories. Foundation soils may be, for example, good, moderate or bad. The maps are basically of two kinds:

1. Those in which only horizontal variations are considered. Areas on the map with the same properties are given the same colour or pattern.
2. Those in which horizontal variations are coupled with layers of constant thickness, the different combinations having different but specific foundation conditions. In order of decreasing precision, the following may be distinguished:

(a) Maps where layers of soils of constant thickness are classified in terms of up to five distinct foundation conditions (class I is best, whereas class V is worst) and the various classes may be present in any order in the soil layers.
(b) Maps on which different soil types – good, moderate or bad for shallow foundations – are distinctively coloured green, yellow or red. Also on the same sheet would be shown, by point symbols and local patterns, depths at which other suitable foundation conditions might be found.
(c) Maps on which the outcrops of different soils on a geological map are categorized in terms of suitability for foundations.

It is clear that only the latter are based on a geological map as of necessity; in the others, map units are related to predetermined soil categories and not to geological units.

In broad terms, it became clear that the factors that had to be taken into account in the preparation of engineering geological maps were:

- the character and properties of rocks and soils;
- hydrogeology;
- geomorphology
- exogenic geodynamic aspects;
- endogenetic geodynamic aspects;
- natural materials;
- modifications to the natural environment caused by man.

Also very important were:

- documentation aspects.

6.6.1 Engineering geological map of the Creil (Oise) region

Prototype experimental maps were prepared in 1967–1968 of the Creil region near Paris as an essay in how to produce an engineering geological map. Extracts from the maps have been published in colour in Gazel and Peter (1969) and Sanejouand (1972). Covering an area of 56 km^2, there are two maps at a scale of 1:20 000, a 'factor' map and a 'suitability' map, accompanied by two hydrogeological maps at a scale of 1:50 000. This study was based on the 1:50 000 scale geological maps of the solid and superficial deposits available for the area, supplemented by some additional geological surveying made necessary by the larger mapping scale.

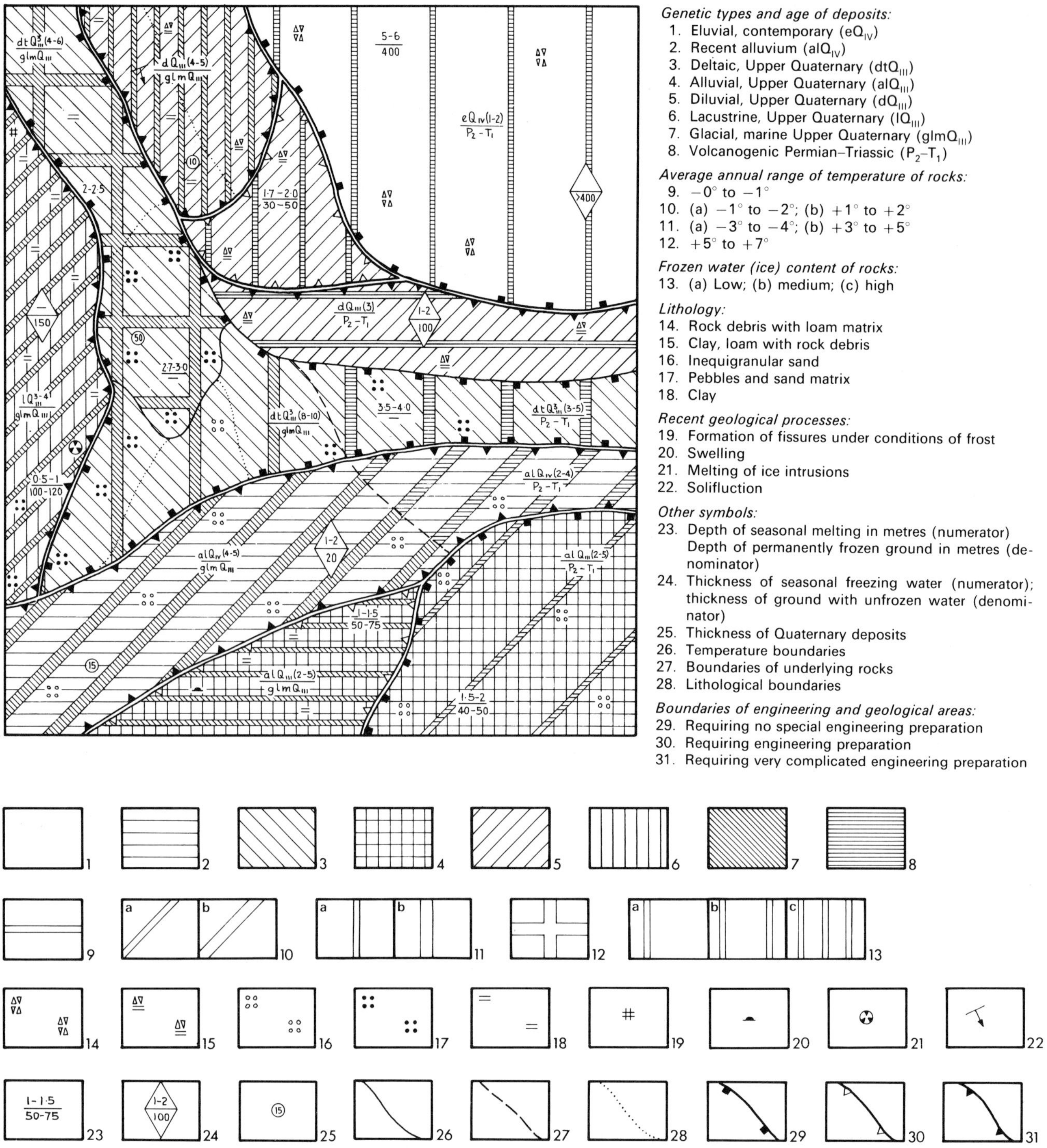

Figure 6.12A Map of a permafrost region in the USSR illustrating not only lithology but some aspects of the physical state of rocks and soils. Originally at 1:10 000, reproduced here at 1:150 000 (From Golodkovskaja and Demidyuk, 1970, by permission)

Geotechnical properties were provided by 149 old borehole records, supplemented by more shallow auger holes and 89 pressuremeter determinations, down to 8 m depths.

Geological conditions are relatively simple. Solid formations are Eocene Lutetien limestones, Cuisian sands and Sparnacian clays and sands. Superficial deposits are loams and gravels on the limestone plateaus, slope loams mainly found on the Cuisian sands, and old alluvial sandy-gravels or clays and peats of the modern alluvium in the Oise and Rhone valleys. In addition there are areas of fill.

Geological and documentation factor map

On this map (Plate 5) are shown the location of all documentation points and their character. Also indicated are the lithology of the superficial and solid deposits with information on the order of superposition and the thicknesses of the superficials.

The different lithologies in the solid deposits at rockhead are represented by patterns printed in green. On the other hand, outcrops of the superficial deposits are in colour. Superposition is shown by the stripe method, using stripes of constant width, by which, it will be recalled, the second layer from the surface appears in

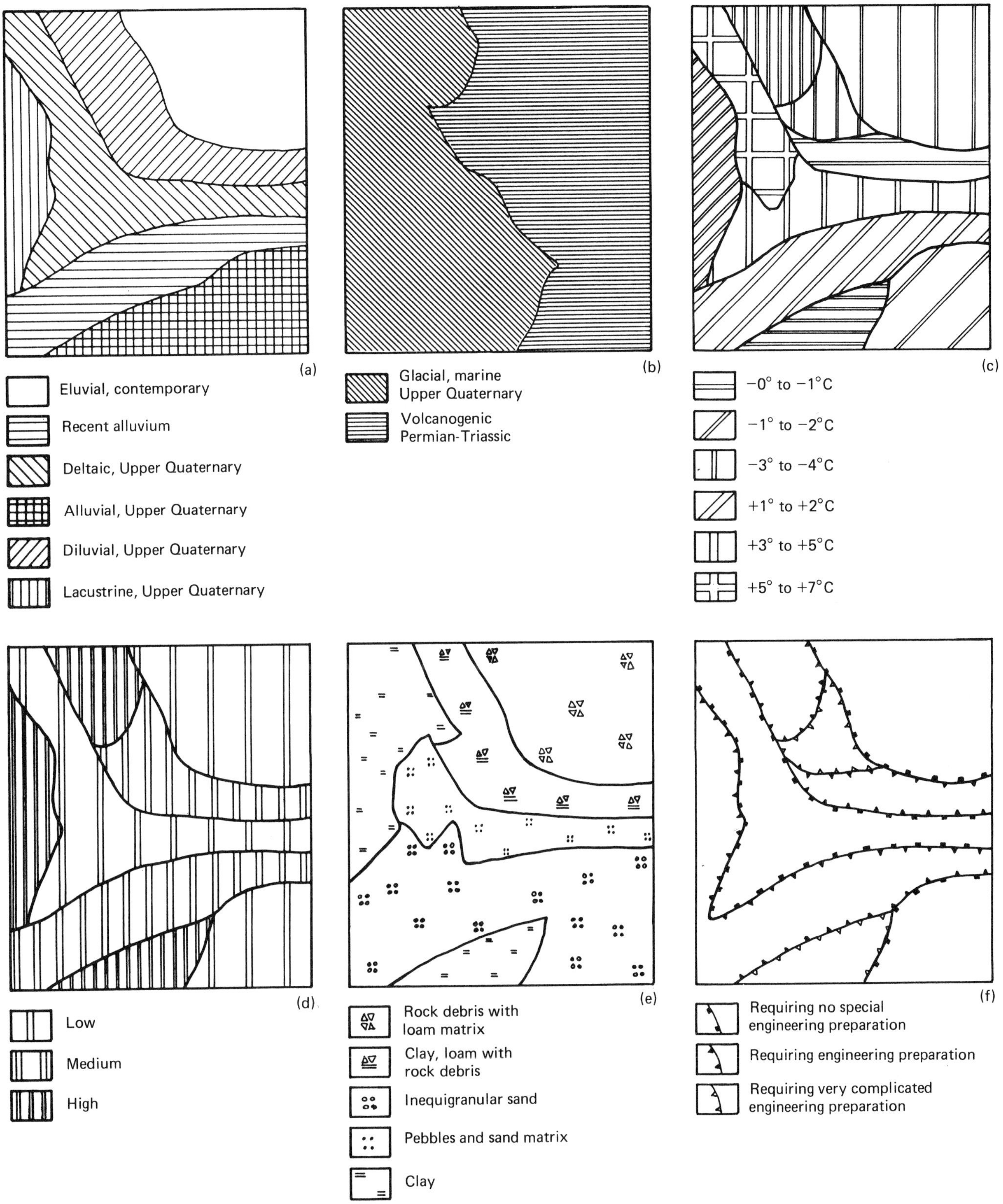

Figure 6.12B Individual maps derived from the composite map: (a) distribution of the superficial deposits on the basis of genetic type and age; (b) distribution of rocks and soils beneath the superficial deposits; (c) average annual range of temperature; (d) ice content; (e) distribution of lithological types in the superficial deposits; (f) three categories of engineering ground conditions

this instance as a vertical band intersecting a third horizontal band where the succession demands. The stripes are coloured with hues of three different lightness values for the three thickness ranges – less than 2 m, 2–5 m and more than 5 m – in the superficial deposits. Boundaries and isopachytes for the three layers have distinctive lines.

Hydrogeological maps

Of the two maps, one is a classical hydrogeological map,

showing the piezometric surface in the alluvial deposits, directions of flow, with the location of wells and artesian bores to the underlying Eocene solid formations. In the case of the latter, the yield in $m^3 h^{-1}$ is also given. The second map reveals the groundwater table, where water is present at the surface in marshes and at depths of less than 1 m, 1–4 m, 4–10 m and more than 10 m. Wells are also located.

Comment

Sanejouand (1972) published an extract of the map at the original 1:20 000 scale, together with the original legend. The same approach has been adopted here, except that the topographic detail has been omitted for clarity. He states, unequivocally, that in his opinion the attempt to put on the same map sheet the superficial deposits, thickness indications and the solid rocks immediately beneath has led to a very complex map that was hardly legible. With this type of map, satisfactory reproduction can only be assured using offset litho printing techniques, and, as Sanejouand points out, this is expensive and cannot in most cases be justified for the short print runs envisaged for limited circulation of the map.

From the viewpoint of clarity, the map published in Gazel and Peter (1969) has a better selection of colours than the Sanejouand version. The slightly simplified map, as reproduced here, has improved colour balance than either of the originals, giving a better impression of this style of engineering geological cartography, albeit still somewhat complex. The part of the full map sheet selected had to be used because the 'suitability' map (Plate 6) covers the same area.

Map of foundation types and construction materials

A suitability map (Plate 6) was based on information provided from each location where geotechnical tests had been carried out. A limited number of types of foundation was considered: strip footings, 1 m wide at 1 m depth; strip footings, 1 m wide at 4 m depth; bored piles, 1.20 m diameter; driven piles, 0.30 × 0.30 m.

A segmented circle around each borehole location on the map (Figure 6.13 and Plate 6) summarizes the geotechnical information. As an example, the top segment, which may be uncoloured or red, gives the allowable bearing pressure for 1 m wide strip footings 1 m deep. When uncoloured, the allowable bearing pressure is less than 0.6 bar, and this type of footing is not recommended; when the colour is light red, the bearing pressure is between 0.6 and 1 bar; when bright red, the value is more than 1 bar.

Three other parts of the circle provide similar information for the other types of foundation. A black ring around the circle, covering one or more of the segments, implies that settlements of more than 2 cm will occur at the allowable bearing pressure for the particular type of foundation.

Limited factual information from boreholes has been extended over the whole of the map area on the basis of known geological conditions. Where there was still uncertainty, the authors of the map were content to give no information on foundation types. Where the symbols, rectangles or circles, are left uncoloured the implication is that the allowable bearing capacity of foundations or the working load of piles had not been established by *in situ* tests. Nevertheless, the recommended type of foundation is indicated, with large areas of the map being suitable for strip footings, whereas others are dominated by piles. The distribution of the four types of foundation, either determined or extrapolated, can be shown on separate maps in black and white (Figure 6.14).

The other aspect of this particular map is the localization and distribution of workable construction materials, either in open quarries or underground workings.

In discussing the map, Sanejouand brings up again the question of the degree of complexity, demanding a quite sophisticated cartographic approach. It would be preferable to use separate maps for foundations and materials. The approach to both would be through general zoning maps rather than the very complicated point symbol method adopted in this case.

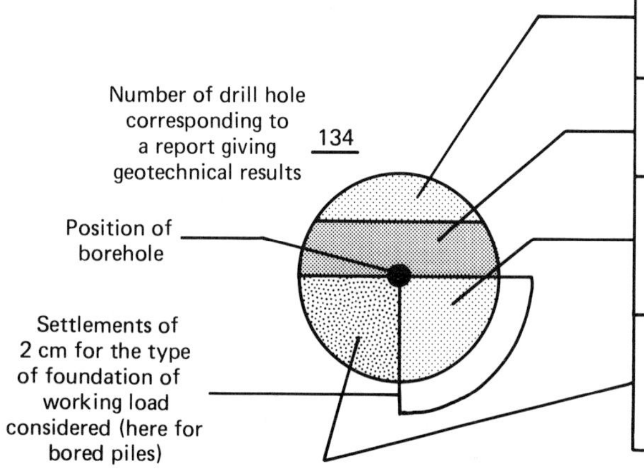

Type of foundation	Significance of colours		
	Colourless	Light colour	Dark colour
Strip foundations 1 m wide by 1 m deep (red)	Not recommended; bearing pressure less than 0.6 bar	Bearing pressure 0.6 to 1 bar	Bearing pressure ⩾ 1 bar
Strip foundations 1 m wide by 4 m deep (blue)	Not recommended	Bearing pressure 1 to 2 bar	Bearing pressure ⩾ 2 bar
Bored piles 1.20 m diameter and 3 to 4 m long (red)	Not recommended or working load less than 40 t	Working load 40 to 100 t	Working load ⩾ 100 t
Driven piles 0.30 X 0.30 m working at a nominal working load of 50 t		Pile length < 7 m	Pile length > 7 m

Figure 6.13 Scheme for the representation of point values for types of foundation (From Sanejouand, 1972, Plate VII, by permission)

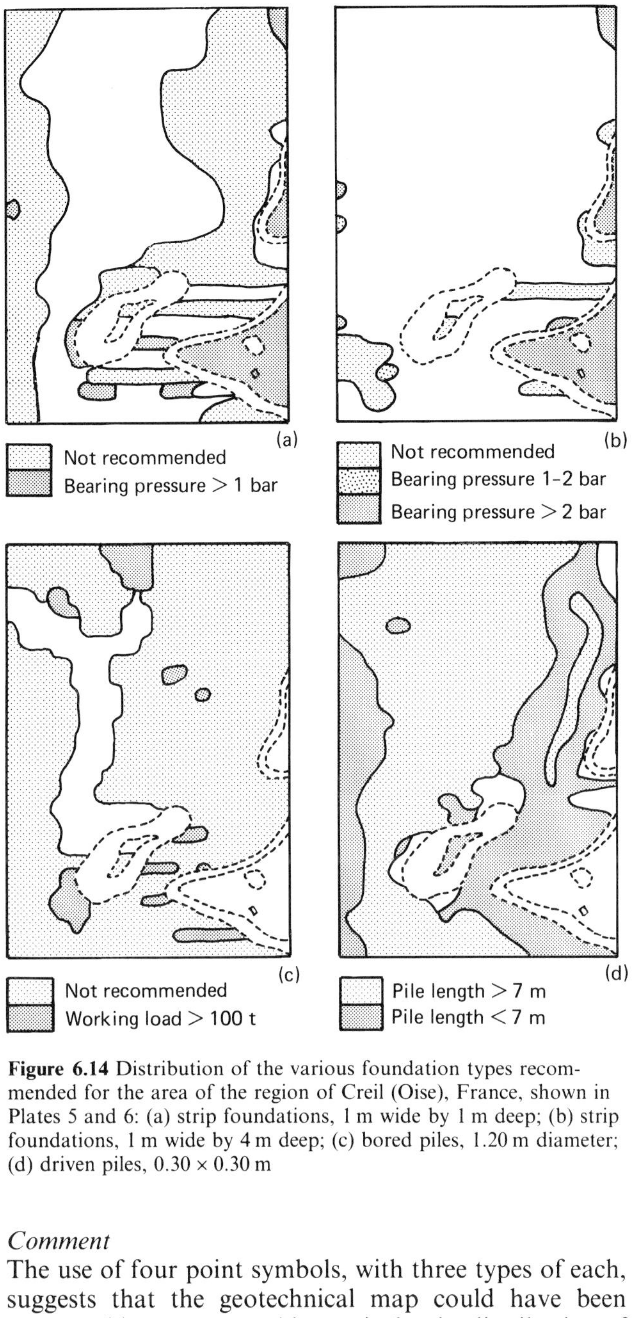

Figure 6.14 Distribution of the various foundation types recommended for the area of the region of Creil (Oise), France, shown in Plates 5 and 6: (a) strip foundations, 1 m wide by 1 m deep; (b) strip foundations, 1 m wide by 4 m deep; (c) bored piles, 1.20 m diameter; (d) driven piles, 0.30 × 0.30 m

Comment

The use of four point symbols, with three types of each, suggests that the geotechnical map could have been generated by computer. Alternatively, the distribution of each foundation suggestion could have been given in separate maps with two or three types of area, as has already been demonstrated in Figure 6.14.

Cartographically the map is too complex, and detailed examination of the original (Sanejouand, 1972, Plate VII) reveals errors of detail that are not easy to see and thus to correct. A zoning scheme would be better, and very much easier to understand and interpret over the mapped area. Nevertheless the map is a true geotechnical map far removed from the original geological map.

6.6.2 Variations of the stripe method developed in Belgium and France

The single and double stripe system is an elegant method that is easy to apply and interpret, and one that does not detract too greatly from the main aim of geological mapping, namely to show the distribution of rocks and soils at the surface. Belgian and French workers developed the method considerably, with interesting variations of the simple Simek theme, even into the representation of foundation conditions.

Engineering geological map of Mons, Belgium

A descriptive memoir is accompanied by four maps: three are factor maps with the titles Map of the Geological Substratum, Map of the Superficial Deposits, Hydrogeological Map, and the fourth is a map of the suitability of soils for foundations with the title 'Carte d'interpretation géotechnique'. All the maps are drawn at a scale of 1:10 000; in the opinion of the authors (Debaille and Ghiste, 1969), the best scale for a geotechnical map is 1:5000 or 1:10 000.

Vertical stripes are used on the solid geology map to show the two uppermost stratigraphical units.

Treatment of the superficial deposits is the more interesting because the superposition of three layers rather than two is shown by a modification of the single stripe method. A wide stripe is reserved for the surface layer (Figure 6.15) and the intervening narrow stripes are subdivided into wide and narrow compartments. The wide compartment represents the second layer down, and the narrow compartment the third layer. For all three layers, thickness categories may be shown by Roman numerals. Areas of fill on the ground surface are shown by overprinting which does not obscure the underlying detail of the superficial deposits, so that four layers can be shown (Figure 6.16).

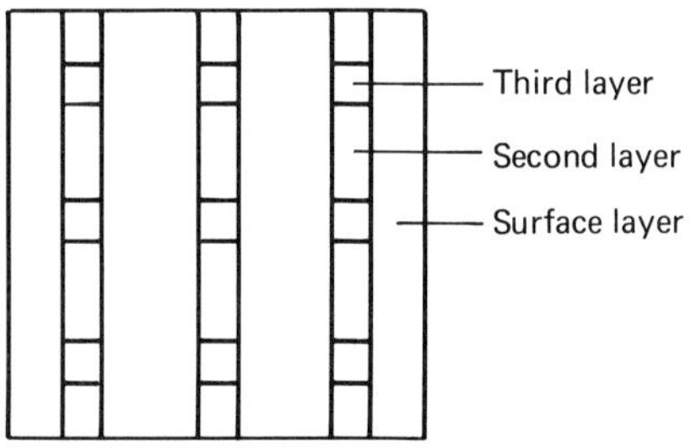

Figure 6.15 Engineering geological map of the superficial deposits of Mons, Belgium. Method of representing the three layers by the stripe method

Map of suitability of soils as foundations in Mons

On the map, three soil layers are recognized – 0–3 m, 3–8 m and 8–15 m – and five classes of soil graduated from good to bad and represented on the map in green and yellow, with red for the worst foundation conditions. Sanejouand (1972) commented that the map had been designed so that at any point on the map the foundation conditions could be determined down to 15 m. Used in this way, and with care, the complexity of the map representation (Plate 7) can be accepted. The map was compiled for use by civil engineers, although the method of representation of variations in ground conditions over the area of the map using three different types of boundary line for the first, second and third beds is only really effective in showing the distribution of the five soil classes in the uppermost, first, layer. It is difficult to imagine that the map would appeal, for example, to town planners, who are concerned with the general spatial variations of ground conditions rather than more detailed information at a particular point on the map.

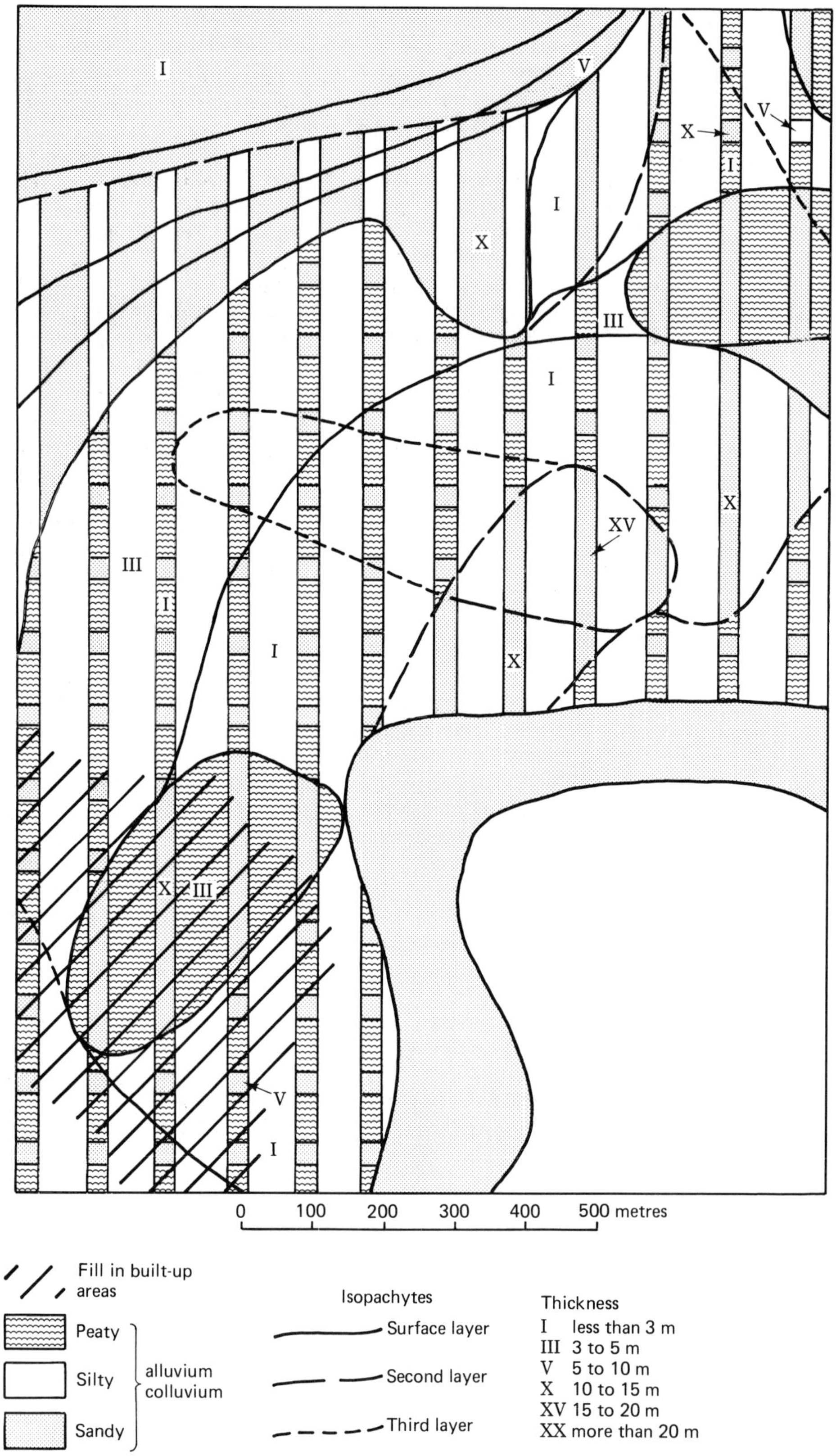

Figure 6.16 Extract from the Mons, Belgium, map of the superficial deposits showing the use of the stripe method (From Debaille and Ghiste, 1969, by permission)

In the memoir, the complete criteria used in the determination of the five soil classes are explained in detail. A great number of borehole records, *in situ* and laboratory tests, and water level records were used in compiling the map.

French maps

At the time that the maps already described were being prepared, there was a great deal of activity in the presentation of both geological and geotechnical maps using variations of the stripe method (BRGM, 1968, 1969). Reading the reports, and studying the maps, there is a distinct impression that the stripe method had served its initial purpose and that there was about to be a major advance in the presentation of 'geotechnical' maps, initially in a quite complicated format.

6.6.3 Map of suitability of soils for foundations

The map selected is of the Bassin de la Crau et de l'Étang de Berre produced by the Laboratoire Régional des Ponts et Chaussées, Marseille, for the Organisation Régionale d'Étude d'Aire Métropolitaine 13, at a scale of 1:50 000. Soils are grouped into a small number of classes based on suitability for foundations for three depth ranges: 0–3 m, 3–6 m, 6–9 m. Soils in the different layers are classified into four categories: good or very good (GVG) in green on the map; good on the whole (G), yellow; bad and locally mediocre (B), red; and very bad and fill (VB), violet.

Two different parts of the map have been published, Plate I in Sanejouand (1972) and Fig. 5.3.1.2 (Anon., 1976), and the latter is reproduced here (Plate 8).

Documentation available in the BRGM and from private companies was used in compiling the map. Hand augering and penetrometer tests were used to define certain zones more precisely. How many borehole records were available, how the soils were classified, and how the three soil thicknesses were chosen was not revealed to Sanejouand. On the complete map sheet the isopiestic lines for the watertable of La Crau are given; these are the only hydrogeological data available, except to note that on the map extract published here the uncoloured areas are all water.

Recognition of three distinct layers and four qualities of soil makes it theoretically possible to show 64 different zones on the map. Although the system is very flexible, it is doubtful whether it is realistic to give regional estimates for layers of constant thickness, and whether the small scale of the map justifies recognition of three depth zones. In addition, the map as printed is not very legible. However, the cartographic solution to the geotechnical problem is not without interest as a method of possible general application in appropriate situations.

Perhaps an explanation of the legend would not be out of place because at first glance the map is very complex, even with topographical details omitted. Good or very good (GVG) soils are coloured green, good on the whole (G) is yellow, red is used for bad and locally mediocre (B), and purple for very bad and fill (VB). This colour coding is applied in four ways. Solid bright colours refer to constant ground conditions down to 9 m; light tones of solid colour record conditions from 3 to 9 m; horizontal bars of bright colours apply to the depth range 0–6 m. In this way, single layer and double layer conditions are dealt with, leaving 3 m thick layers either individually or in combination with 6 m thick layers at the top or bottom of a local succession.

The colour coding, using the same four colours, copes with this by having horizontal lines for the depth range 0–3 m, vertical lines for 3–6 m and diagonal lines for 6–9 m. Horizontal or diagonal lines combined with pale solid colours or bright colour bars take account of the double layer case, where one 3 m layer at the top or bottom is combined with a 6 m layer. For the case of three separate layers, appropriate combinations of horizontal, vertical and diagonal lines are used. For example, an area in which a sequence of very bad from 0 to 3 m, underlain in turn by bad from 3 to 6 m and good from 6 to 9 m is coded by a combination of horizontal purple lines, vertical red lines and diagonal yellow lines. Such a combination, and others, can be found on the fragment of the map published.

6.7 Application of the stripe method in the UK

It has to be accepted that the stripe method, in its various manifestations, is especially attractive as a way of illustrating the distribution of two or more geological units in the superficial, mainly Quaternary, deposits. These uppermost soil layers are of particular importance in geotechnical engineering and, as such, a plan display of the various lithological units and their engineering properties is of prime importance for both engineering and planning applications.

In the UK the stripe method has only been used on manuscript plans of Tyne and Wear County in Northern England, prepared during an engineering geological survey of the county at various scales. A conventional approach was adopted for medium-scale maps of the whole area. The results of the survey are discussed in Chapter 10.

7

The zoning concept in engineering geological mapping

7.1 Introduction

Geological mapping, and this includes engineering geological mapping, always involves drawing lines on a map or plan around areal homogeneous units. Each map unit comprises in general terms a zone – an acceptable misnomer for an enclosed area. Homogeneity, or quasi-homogeneity, is related to particular geological conditions or engineering properties and the concept may be applied to all map scales.

The concept of zoning has been formalized in the COMECON countries, where it was first developed, into a hierarchy – region, area, zone, district, each appropriate to particular map scales. What is involved is the characterization of map units for a particular purpose, and there are a number of ways in which this has been achieved. Some of these will be discussed in the following sections, but first some aspects of classification and identification are discussed.

7.1.1 Classification and identification of map units

As with the more conventional geological map, the process of making an engineering geological map involves defining and then delimiting, for the purpose in hand, either homogeneous or almost or sufficiently homogeneous, that is quasi-homogeneous, units. A basic problem is that of identifying those properties and relations, or in other words those characteristic qualities or attributes, that apply specifically to the units to be mapped and are sufficient to define them. Ideally a mapping unit is defined by a unique attribute or a unique set of attributes.

In geological mapping, four categories of attributes are involved: those related to *time*, those determined by *space*, those associated with *properties*, and those concerned with *relations* between objects. Time is established by the fossil record, geochronology and the general principles of stratigraphy. Space is determined by physical boundaries, and there are a great number of properties that can be used in mapping. Relations between map units are usually defined geometrically.

Varnes (1974) has discussed in detail the general characteristics and processes of classification, and in developing his ideas has distinguished between 'classification' and 'identification', quoting Sokal (1966, p. 108): 'When a set of unordered objects has been grouped on the basis of like properties, biologists call this "classification". Once a classification has been established the allocation of additional unidentified objects to the correct class is generally known as "identification."'

Classification of units on geological maps into distinct formations is based on lithology as the main attribute, but the mapping process is also involved with the establishment of geometrical relations with other units. Genetic and temporal relations have to be determined if all the geological aspects of each map unit are to be understood. As an example, a sandstone B in a sedimentary sequence A–B–C is recognized on the basis of its distinctive lithology, and its position in a lithostratigraphical sequence becomes apparent when the boundaries with equally distinctive lithologies A and C below and above are plotted. Fossils within the sandstone may confirm that the whole of the mapped unit may have accumulated within the same time span. Alternatively the evidence may point to a lack of coincidence between lithological boundaries and time lines within the deposit, the deposit being diachronous. If, for a particular purpose, the time attribute is essential, then one map unit B based on lithology becomes several units, B_1, B_2, B_3, when both typological and temporal attributes are considered. This aspect has already been touched on and illustrated (Figure 2.1) in Chapter 2.

Areas of classes comprising the mapping units do not necessarily touch one another as in the last case. Varnes (1974, Fig. 1) provides the following example. Units on an engineering geological map may be defined by the attributes of:

- lithology, L;
- slope with a specified range of inclinations, S;
- groundwater conditions, W.

The distribution of these attributes may have been mapped, as in Figure 7.1, and the ruled map unit is shown as two adjacent but separate areas determined by the overlap of all three attributes. If within the outcrop

L there are strata of different ages but of essentially the same lithology, e.g. areas L_1, L_2, L_3, rocks in the two ruled areas are non-contiguous in time, although they remain valid mapping units on the basis of the original common attributes L, S and W.

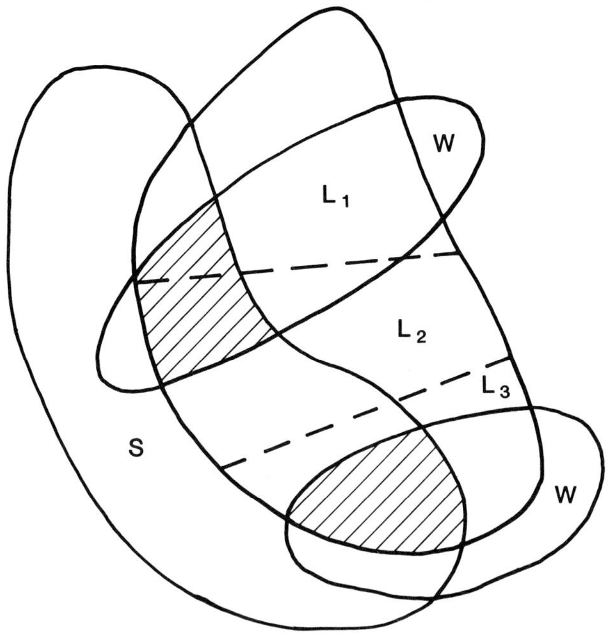

Figure 7.1 Map unit, ruled, defined by overlapping attributes of lithology L, slope S and groundwater condition W (After Varnes, 1974, Fig. 1)

Once the purpose of the mapping unit has been identified, it needs to be very carefully defined. There are three stages involved. First, each unit must be assigned to one or more of the fundamental temporal, spatial, typological or relational categories. Secondly, a clear statement must be made of the essential attributes in each of the fundamental categories that are applicable to the unit. The third stage, perhaps the most important in classification, is to assign the degree of internal heterogeneity that can be accepted in each attribute if the map is to fulfil its purpose.

The whole of the classification of soils and rocks for engineering purposes, discussed in detail in Chapter 4, is concerned with statements of the range of properties that determine the classes of various typological attributes, such as strength, spacing of discontinuities and grain size. Implicit in rock and soil names, granite and silty-clay for instance, are statements on the acceptable range of grain sizes, and for granite there are additional implications on mineralogical composition.

Emphasis is thus on heterogeneity rather than homogeneity, as stressed by Varnes, and mapping is essentially the delineation of areas that may be regarded as 'homogeneous' in that they are acceptably heterogeneous *for the intended purpose of the map*. The resulting map consists of two parts:

- the two-dimensional plan showing the outlines of the mapped areas;
- the legend, or explanation as it is sometimes called, that describes in words and symbols the attributes on which each mapped unit has been classified.

Because mapping involves classifying data and outlining particular class boundaries, the methods of classification are of great importance in constructing new maps and interpreting existing ones.

7.2 Methods of classification

The process of classification involves the formation of classes by grouping objects on the basis of their common properties or attributes. Whether the property is present or not is the simplest relation, but many attributes are gradational and have a range of values or degree that can be specified by some system of measurement.

7.2.1 Use of matrices

A matrix to display data is very helpful in constructing or analysing classifications if more than a very small number of objects and associated attributes are involved. In the matrix in Figure 7.2(a), the relation between the corresponding object and attribute is expressed by symbols a, b, c, etc.; the symbols can be replaced by 1 or 0, a nominal measurement indicating the presence or absence of the relation (Figure 7.2b). If gradational attributes are involved, then class numbers can be assigned to the attributes (Figure 7.2c) or actual numerical values (Figure 7.2d). The matrix can be used in two ways. If the objects in Figure 7.2 are areas on a map, then the attributes of a particular area can be determined from the particular horizontal row on the matrix, noting the measures in each column. On the other hand, areas that exhibit a particular attribute or particular classes of that attribute can be seen by scanning the appropriate vertical column.

The two ways in which a matrix is used therefore reflect the two basic ways in which maps are used, and the design of maps involves methods of displaying one or the other of these forms of matrix.

Another use for matrices is the study of covariance because the rows and columns can be manipulated to establish groupings of either attributes or objects that can be used to define classes. In Figure 7.3(a) the matrix shows established relationships between attributes of places. The rows of Figure 7.3(a) can be regrouped as in Figure 7.3(b) to identify two new classes, or map units, that have similar but not identical attributes, as can be seen by comparing the attributes of rows A, D, I, one new class, or rows B, E, G, the other new class. The number of new map units can thus be reduced from 9(A, B, C, D, E, F, G, H, I) to 5(A + D + I, B + E + G, and C, F and H). An essential proviso is that the new groupings, including the two new ones that are quasi-homogeneous, fit in with the purpose of the map.

Regrouping of the columns of attributes, 1–10, reveals two groupings of attributes that covary: 3 and 7 perfectly, 1 and 9 almost perfectly (Figure 7.3c). Reasons for the covariance can then be studied, and one or other of the two attributes in each of the new groupings may be used to classify the map units.

Specialized statistical methods are available to group objects or attributes, or both, to give the most meaningful units for the purpose in hand. At some stage, groups are determined that have the maximum heterogeneity that is acceptable in defining the map units. As the process of grouping continues, the statements that

(a)

Map units	1	2	3	4	5	6
A	a_1	a_2	a_3	a_4	a_5	a_6
B	b_1	b_2	b_3	b_4	b_5	b_6
C	c_1	c_2	c_3	c_4	c_5	c_6
D	d_1	d_2	d_3	d_4	d_5	d_6
E	e_1	e_2	e_3	e_4	e_5	e_6

(b)

Map units	1	2	3	4	5	6
A	1	1	1	1	1	1
B	1	1	1	1	1	0
C	1	1	0	1	1	1
D	1	1	1	1	0	1
E	1	1	1	1	1	1

(c)

Map units	1	2	3	4	5	6
A	1	2	1	1	2	1
B	2	3	3	2	4	5
C	3	1	5	3	1	3
D	4	5	2	5	5	2
E	5	4	4	4	3	4

(d)

Map units	1	2	3	4	5	6
A	500	16	40	0.73	1500	+6.2
B	400	13	25	0.62	300	n.a.
C	300	17	0.0	0.57	2670	−1.1
D	200	10	30	0.39	0	+3.6
E	100	12	5	0.43	1200	−2.3

Figure 7.2 Matrices relating objects and attributes according to scales of measurement: (a) and (b) nominal; (c) ordinal; (d) interval or ratio (From Varnes, 1974, Fig. 3, by permission)

(a)

Map units	1	2	3	4	5	6	7	8	9	10
A		X	X	X		X	X	X		X
B	X	X	X		X	X	X		X	X
C	X			X		X		X	X	
D			X	X		X	X	X		X
E	X	X	X		X	X	X		X	
F		X		X				X		X
G	X	X	X		X		X		X	X
H		X	X				X			
I		X	X	X		X	X	X	X	

(b)

Map units	1	2	3	4	5	6	7	8	9	10
A		X	X	X		X	X	X		X
D			X	X		X	X	X		X
I		X	X	X		X	X	X	X	
B	X	X	X		X	X	X		X	X
E	X	X	X		X	X	X		X	
G	X	X	X		X		X		X	X
C	X			X		X		X	X	
F		X		X				X		X
H		X	X				X			

(c)

Map units	3	7	1	9	2	4	5	6	8	10
A	X	X			X	X		X	X	X
B	X	X	X	X	X		X	X		X
C			X	X		X		X	X	
D	X	X				X		X	X	X
E	X	X	X	X	X		X	X		
F					X	X			X	X
G	X	X	X	X	X		X			X
H	X	X			X					
I	X	X		X	X	X		X	X	

Figure 7.3 Matrices showing: (a) attributes of places; (b) places grouped by similar attributes; (c) attributes grouped by similar places (From Varnes, 1974, Fig. 6, by permission)

can be made about each group become more and more generalized, until the process is ended.

A three-dimensional matrix may be used to explore the relations between three attributes, and to check for a particular map whether all the possibilities of the classification are discussed or specifically stated to be absent. The category boxes (Figure 7.4) represent possible map units and on the matrix the actual contacts between the map units are represented by adjoining surfaces, edges and corners of the boxes. All possible units, as determined by the matrix, may not be present in the mapped area. Alternatively, for practical purposes it may be necessary to combine units where either lack of information or the restricted extent of the map unit makes it expedient.

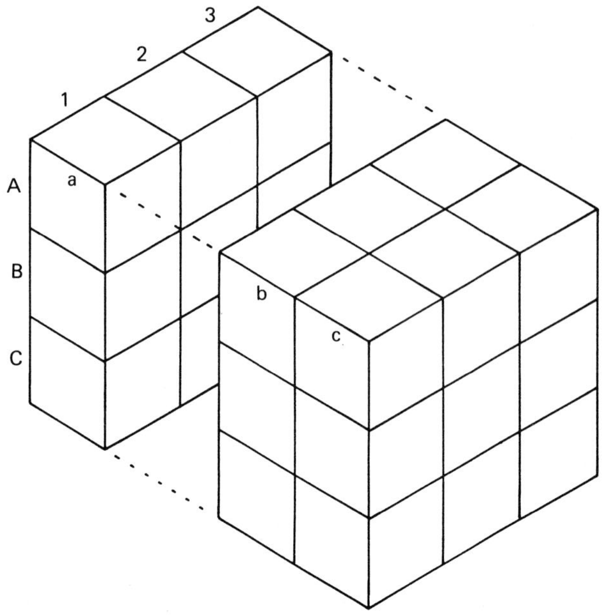

Figure 7.4 Three-dimensional matrix

7.3 Effect of map scale on selection of mapping units

The method of classification discussed so far involves grouping, and the result is the recognition of mapped areas that have specified attributes. Alternatively, instead of this approach that gives general knowledge of map units, classification may proceed by division leading to map units for each of which there is precise knowledge of the attributes.

In the process of mapping, division consists only in the addition of new boundaries that define smaller map units within each of which, as successive division takes place, there is less variance than before. A practical limit is reached either when further subdivision does not reduce the heterogeneity of the chosen attributes or the units become too small to be shown on the map. The importance of map scale as a controlling factor cannot be overemphasized.

The scale at which any geological mapping is carried out is the prime factor in determining both what can be mapped and what should be mapped; that is, the map scale determines the general mapping technique in terms of the amount of detail that can be shown. Can, for example, individual joints be shown, or individual beds in an outcrop of sedimentary rocks, or the distribution of different weathering grades, or does detail of this kind only find a place in the descriptive legend? Are all the topographical and other details of a landslipped area shown, or is the presence of a landslip merely shown by a point symbol?

What is possible cartographically depends upon the nature of the feature being shown: at 1:2500 scale a normal country road will be shown by lines 3 mm apart and all the bends, angles and variations in width can be plotted and dimensions can be accurately scaled from the map. Changing the scale to 1:10 000 has little effect, except on the accuracy of directly scaled distances. However, at 1:50 000 the character of the representation changes, roads are shown by equally spaced lines and the width of the spacing is used to characterize the type of road. Nevertheless all roads are represented at all three scales, whereas at 1:250 000 only selected roads can be shown if the map is to retain a general uncluttered appearance. Individual buildings and groups of buildings are treated by successive generalizations as the scale of the map becomes smaller.

Geological detail is affected in the same way and one illustration will suffice to illustrate the point. In Figure 7.5, which illustrates a mapped succession of tuffs and interbedded shales and quartzites, each bed and some of the lithological variations within beds can be shown, as can joints and small-scale faults, and the distribution of weathering grades on a map of the quarry exposure. At 1:500 scale, that used for the main mapping of the quarry in which the exposure occurred, the main lithological units in the tuff sequence are all that can be shown. With successively decreasing scale, detail decreases with only the general tuff sequence shown at 1:10 000, whereas at 1:50 000 the mapped outcrop is the entire geological formation comprising shales and quartzites including the interbedded tuffs; and at 1:625 000 merely the whole of the Lower Carboniferous series is depicted.

7.4 The zoning concept: examples from Czechoslovakia

7.4.1 Introduction

In following, in Chapter 6, the evolution of cartographic techniques to represent geological conditions in the all-important superficial deposits of an area, the stripe map was seen to be usually one of a set of geological maps prepared for planning and engineering purposes. A minimum set of four maps was called for:

1. The geological map: using the stripe method if appropriate.
2. A hydrogeological map: may be combined with (1).
3. A documentation map: may be combined with (1).
4. A map of engineering geological zoning.

Three maps of this set are not specifically engineering geological in character, and it has been said that the zoning map, evaluating geological and hydrogeological conditions for a variety of purposes, is the only true engineering geological map. Geotechnical aspects do not enter into the zoning concept, although the engineering characteristics of each zone on a map will, at some stage, be assessed and discussed, often for a variety of purposes.

What can be shown on a map – the content in other words – depends upon the scale. The classification of scales accepted in 1976 (see Chapter 2 and Table 2.1) may be modified and extended in a more specific form as:

- General maps : 1:500 000 and smaller;
- Small-scale maps : 1:200 000 and 1:100 000;
- Medium-scale maps : 1:50 000 and 1:25 000;
- Large-scale maps : 1:10 000 and larger.

This provides a convenient classification for zoning maps. Maps at scales of 1:100 000 and smaller are summary or synoptic maps, generally conceded to be of no use for civil engineering construction or for areal planning. Nevertheless they have been compiled and can be accepted as providing a broader view of the regional setting for any medium-scale maps that subsequently might be prepared.

It is pointed out that maps of the pre-Quaternary bedrock and mineral deposits have been prepared for the whole of Czechoslovakia at a scale of 1:200 000. From these maps, experienced geologists should be able, at least, to formulate an opinion on the engineering geological conditions (see section 10.2). This is even more the case if there is a parallel map of the Quaternary deposits.

7.4.2 COMECON 'Instructions'

From the early 1960s, Eastern European engineering geologists were working on the standardization of engineering geological maps at all scales (Kabel, 1969). General maps at 1: 500 000 and smaller were intended for preliminary environmental planning, including the most rational distribution of towns and large industrial centres from the point of view of natural conditions. As a rule both geology and zoning are shown on the same map sheet, with the important proviso that superficial deposits less than 5 m thick are not shown.

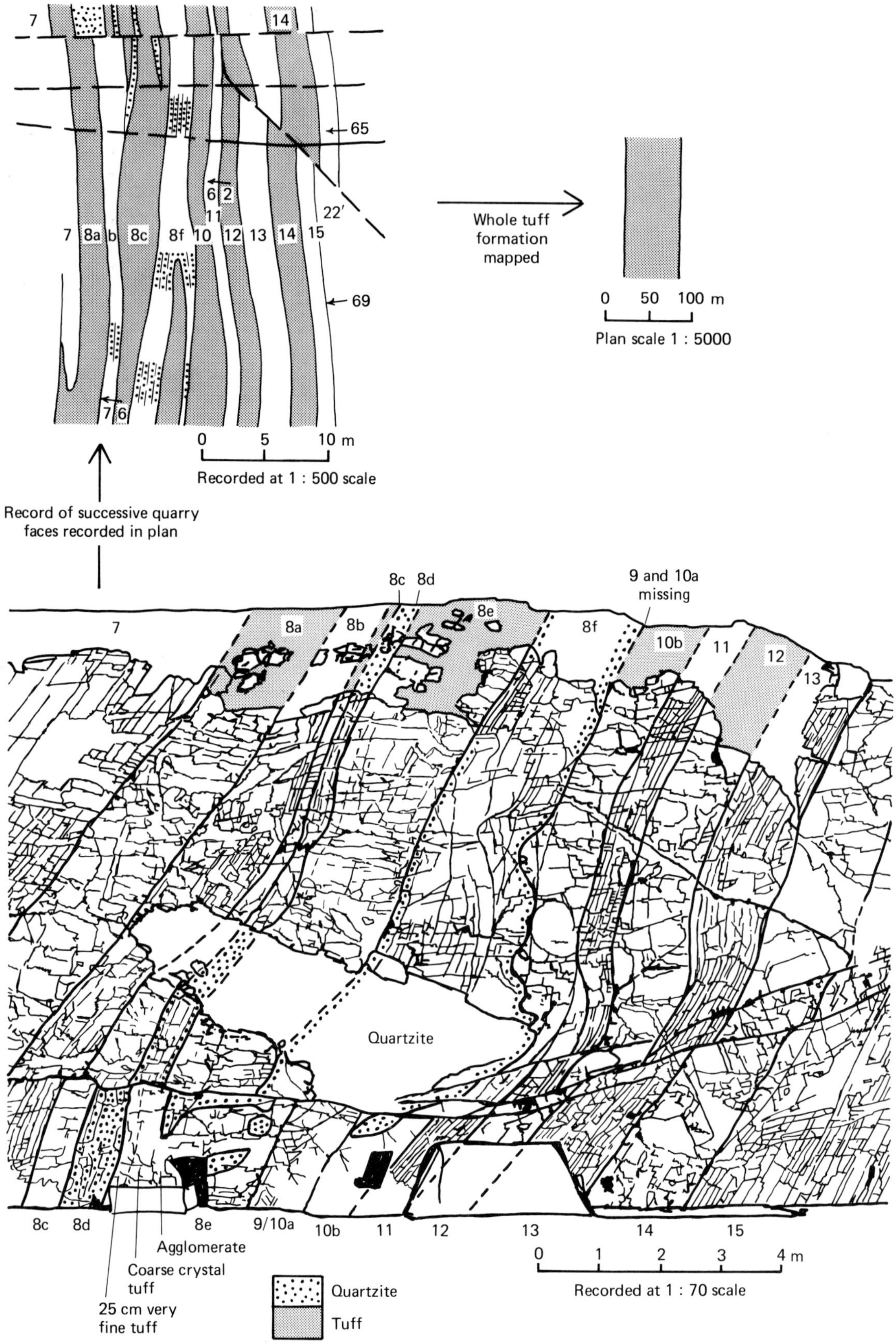

Figure 7.5 Effect of scale on the detail that can be recorded in a sequence of tuffs

Small-scale maps can be closely focused on the actual location of towns and villages, industrial installations, transportation routes, and for planning hydrogeological developments. At 1:200 000 the maps have to be supplemented by medium-scale maps for town planning and for preliminary design of site investigations for diverse civil engineering constructions. At these scales, geology and zoning are usually presented as separate map sheets.

The 'Instructions' (Bachmann *et al.*, 1967) were accompanied by examples, in colour, of maps representative of the four map-scale ranges. Czechoslovakian engineering geologists were commissioned to prepare exemplars for the medium-scale maps that have been published separately (Matula and Pasek, 1966). There is a distinct impression that much of the impetus for standardization came from this source, and their maps are used in the subsequent discussion.

7.5 General zoning maps

General maps were earlier called synoptic or summary maps, and small-scale maps are usually put in the same category.

7.5.1 Summary map of Slovakia at 1 : 500 000

Together with the 1:25 000 maps of the Zvolen Basin, Matula (1969) presented, at a scale of 1:500 000, a summary engineering geological map of the whole of Slovakia. Compiled in 1965, at the time when the COMECON proposals for engineering geology maps were being actively discussed, the map, together with its extended legend and the published account of the regional engineering geology, gives a valuable insight into the Eastern European approach to engineering geological zoning.

Initially there appears to have been a problem in translating 'zoning' into Western European languages. The word appears as 'rayonization' or 'rayonierung' and this might have been necessary because the word 'zone' was used for one of the units. Zoning is the basic method of evaluating engineering geological conditions for engineering purposes. The process consists, simply, of the division of the country into the recognized sequence of homogeneous units (regions, areas, zones and districts), the hierarchical sequence in which the appropriate map scales become progressively larger, whereas the individual mapped units become progressively smaller.

Regions are recognized as first-order units in each of which geotechnic evolution and tectonic structure are the same or at least generally similar.

In the West Carpathians (Figure 7.6) four types of engineering geological region have been recognized. These reflect the evolution, during the general uplift in the Pliocene, of the Carpathian mountains separated by tectonic depressions. Each region is differentiated by the age, structure and lithological nature of the solid rocks, even in the tectonic depressions.

The oldest rocks, Precambrian and Palaeozoic metamorphics intruded by granites, overlain by Upper Permian to Middle Cretaceous sediments, form the main core mountains typical of Region A. Phyllites, schists, gneisses and migmatites are highly deformed, and with the younger sediments – sandstones, shales, marlstones and limestones mainly – are involved in large-scale folds and nappes dislocated by faults. Neo-tectonic movements along bounding faults led to the relative uplift of the mountain ranges.

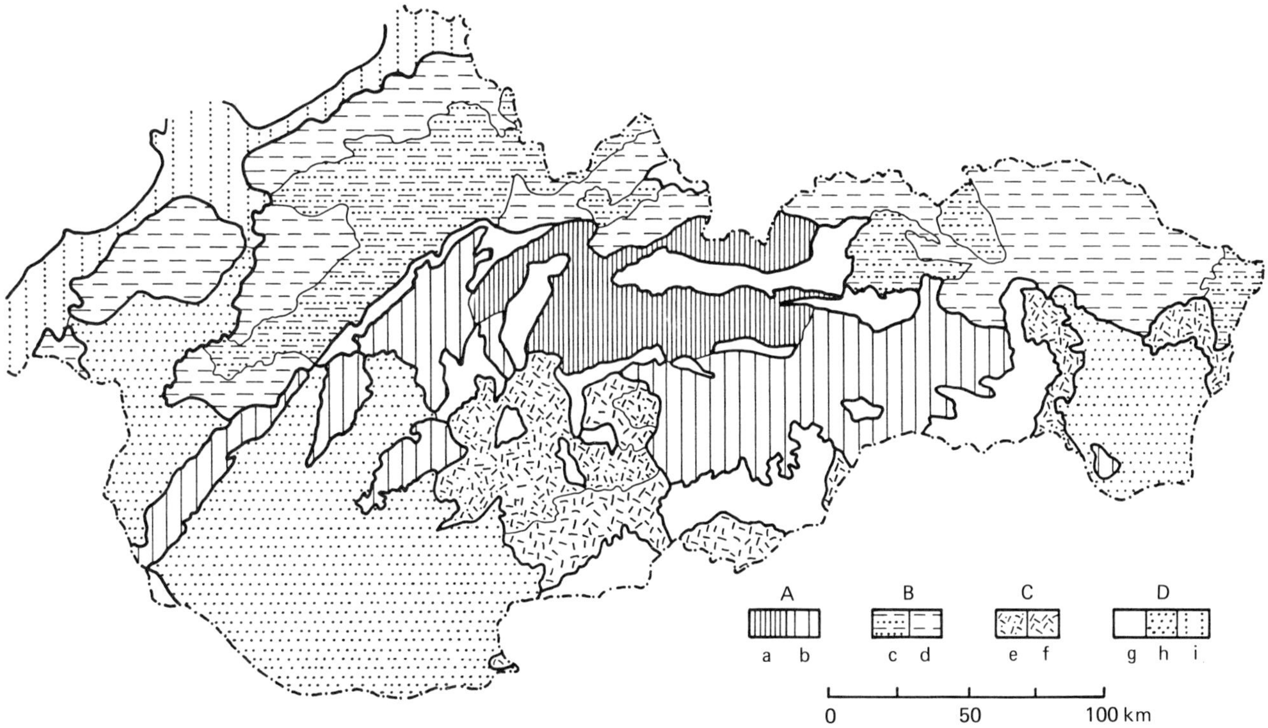

Figure 7.6 Engineering geological regions and areas of the Czechoslovak Carpathians. Regions: A, core mountains; B, Carpathian flysch; C, Neogene volcanics; D, Neogene tectonic depressions. Areas: Aa, high core mountains; Ab, moderately high core mountains; Bc, flysch highlands; Bd, flysch uplands; Ce, volcanic highlands; Cf, volcanic uplands; Dg, intramontane basins; Dh, intra-Carpathian basins; Di, Carpathian foredeep (From Matula, 1969, by permission)

Regions are subdivided into areas on their geomorphology. Two types of area are recognized in the core mountains: areas Aa of high mountains rising to 2500 m, with up to 800 m of relative glacial and bare rock relief; and areas Ab of lower mountains rarely rising above 1500 m, with up to 600 m of relative relief. In Ab areas the relief has mainly been modelled by erosion, karstic processes and periglacial solifluction. On the map legend there is additional summary information – full details are given in the text – on superficial deposits, hydrogeology, the most important active geomorphological processes, and an evaluation of construction conditions.

Distribution of engineering geological regions and areas throughout Slovakia is shown in Figure 7.6, with a summary of the characteristics of each.

Third-order classification units are the *engineering geological zones* into which areas are divided, and in turn each zone may be further divided into *engineering geological districts*. Map scale has an important bearing on how far the zoning process can be taken. On the 1:500 000 map, regions and areas are outlined, and it is not until the scale is increased to 1:200 000 (Plate 9) that areas are broken down into zones. At a map scale of 1:25 000 zones may be divided into subzones.

District zoning is restricted to large-scale construction plans and is mainly concerned with mapping engineering types of rocks and soils (Chapter 11).

As individual map units become progressively smaller in extent, engineering geological conditions in each unit become more homogeneous. A practical limit is reached with the concept of the district when the prerequisite is complete exposure in, for example, a foundation excavation.

7.5.2 Summary map of the Zvolen Basin, Czechoslovakia, at 1:200 000

A typical map at this scale is that published by Matula and Ondrasik (Anon., 1976) for regional and land-use planning. It is, essentially, a geological map that goes beyond the usual confines of a classical engineering geological map in that it includes data on rocks, groundwater and geodynamic phenomena.

At this scale, rocks and soils are divided into engineering groups and formations. Colours are used for the representation of lithological character, and patterns for the lithology of Quaternary superficial soils (in ochre) and the pre-Quaternary basement (in grey). Pre-Quaternary deposits are shown only where they are thicker than 3 m. The age of the rocks and soils is shown, only in the legend, by geological symbols.

Hydrogeological conditions are represented by numerical symbols in blue for each formation of rocks and soils. Point symbols in red show locations of geotechnically important geodynamic phenomena, for example landslides, gully erosion and karst.

The map (Plate 9), as an extension of the legend, has separate tables listing the quantitative and qualitative properties of geological materials (Table 7.1) and of the various rock and soil masses (Table 7.2).

Table 7.1 Engineering geological characteristics of rock and soil materials

Main lithological types	*Dry bulk density* (g cm^{-3})	*Porosity* (%)	*Uniaxial compressive strength* (kgf cm^{-2}. 10^2)	*Indentation hardness* (*Šrejner*) (kgf mm^{-2})	*Modulus of deformation* E_0 (kgf cm^{-2}.10^5)	*Durability*
Sandy to clayey loam	1.53–1.79	35–58	0.0055	—	—	Soften, swell, not frost-resistant
Sandy loam	1.50–1.65	35–45	—	—	—	Not frost-resistant
Sandy gravels	1.80–1.90		—	—	—	
Sandy loams and stony-loamy soils						
Travertine	1.60–1.95	1–10	1–3	150–190	—	Slightly soluble
Loamy gravels	—	—	—	—	—	
Sandy clays, clays	1.50–1.75	30–35	—	—	—	Swell, soften
Basalts	2.85–2.95	1.5–3.0	19–25	700–900	6–8	Very stable
Basaltic tuffs	—	—	—	—	—	Slightly affected by slaking
Andesites	2.67–2.71	1.5–4.5	15–23	500–700	5–7	Stable
Block and agglomeratic tuffs	1.83–2.12	18–33	1.6–3	—	—	Stable
Tuffs	1.30–1.70	30–40	1–2.5	50–80	0.4–0.8	Variable, not frost-resistant, slakes
Tuffites	1.68–1.85	29–37	1–1.15	—	—	—
Limestones	2.70–2.73	0.23–1.39	7–14	200–300	0.03–56	
Dolomites	2.71–2.82	0.68–3.79	6–15	120–400	3–6	Stable
Quartzites	2.53–2.58	3.0–4.0	10–14	380–420	3–3.4	Stable
Granodiorites and diorites	2.62–2.70	1.1–2.7	12–17	270–600	0.4–0.6	Stable

Table 7.2 Engineering geological properties of the main rock and soil units in the mass

Symbol	*Ease of excavation*	*Description of rock and soil masses*
$_{d}Q_{3-4}$	2–4	Thickness 1–9 m, fragmentary debris at the base, very low permeability, without permanent groundwater table, much gully erosion and landsliding used for brick materials, impervious materials for dams
$_{f}Q_{3-4}$	1–3 3–4	(a) Thickness up to 3 m, permeable with permanent continuous groundwater table (b) Thickness up to 11 m, slightly permeable (10^{-4} to 10^{-3} m s^{-1}), without a permanent watertable; suitable as fill material
$_{pr}Q_{2-3}$	4	Conical forms; coarse—alternating with fine—grained materials
$^{t}N_2$–Q	5	Dome-shaped bodies, stratified structure, open fissures, slightly permeable along fissures and cavities, slightly soluble in water; used for construction purposes
$_{l}N_2$	4 2–3	Cross-bedded sandy gravels and sands with alternating silty clays and clays; lateral transitions and wedging out. Sands and gravels are dense amd slightly to moderately permeable; suitable for fills, locally may be suitable for concrete. Lenticles of cohesive soils are well consolidated, slightly permeable to effectively impermeable; used for brick making
β	7 3–5	Outliers and remnants of lava flows several metres to tens of metres thick, predominant prismatic jointing. Homogeneous, except marginally. Intense weathering to several metres, moderately permeable along fissures. Suitable for building stone and crushed aggregates. Irregular bodies of ashy to coarse-clastic tuffs, very heterogeneous, permeable along fissures and through pores. Suitable for light construction material
α	6–7 5–7	(a) Irregular andesitic bodies several to 100 m thick, homogeneous except at the margins, predominantly prismatic jointing, frequent tectonic disturbance, sometimes propylitized. Fissure permeability. Suitable as a construction material (b) Tuffs tens to hundreds of metres thick, very heterogeneous tuffs and agglomerates, variably cemented, permeable through pores and along joints. Of limited use as a construction material
tu_{α}	5–6	Tens to hundreds of metres thick irregular horizons of chaotic and stratified tuffs alternating with tuffites of varied grain size and degree of heterogeneity. Irregular jointing and frequent selective weathering. Permeable through pores and fissures. Of limited use as a construction material
ld_{T2}	5–6 4–6	Thick to massive bedded relatively homogeneous limestones and dolomites. Limestones are highly karstified and highly permeable. Used as a construction material; purer horizons are used for lime production Massive dolomites with intercalations of shale and limestone are tectonically disturbed. Weathered to various depths to dolomite sand containing dolomite fragments. Locally used as fill material
q_{T1}	4–6	Thickly bedded to massive quartzites with horizons of sericitic and chloritic shales, dynamometamorphosed to a varying degree. Weathered to a fine rubble. Permeable in tectonically disturbed zones. A suitable material for crushed aggregates
γ	—	Massive, homogeneous, in places tectonically disturbed and dynamometamorphosed. Polyhedral jointing. Permeable along fissures and disturbed zones. A suitable construction material

Zoning has been applied to delimit different types of engineering geological areas and zones. Discrimination of areas, as has been discussed above, is based on the uniformity of individual geomorphological units. Zones are delimited on the basis of the general character and structural arrangement of engineering formations. Simply as an illustration of the types of information that can accompany a zoning map is a tabular presentation (Table 7.3) giving short descriptions of two contrasted types of zones and three associated areas. Geological, hydrogeological and present geodynamic conditions are listed for each zone, together with general estimations of engineering geological conditions for construction purposes: excavations and cuttings, fill, foundations, roads and building materials. More information could be given in an accompanying descriptive memoir.

7.6 Medium-scale zoning maps

Zoning at medium scale has been attempted in several countries with slightly different approaches that in some cases reflect local ground conditions. In all examples studied, the essence of the method is the comprehensive evaluation of the engineering geological environment – including geological, geomorphological and hydrogeological aspects – into a limited number of uniform map units which evaluate ground conditions down to a specified depth. At this scale, the maps are usually of general application.

7.6.1 Zoning map of the Zvolen Basin, Czechoslovakia, at 1:25 000

The map (Figure 7.7) complements that of engineering geological conditions (see Plate 3) described in Chapter 6, using the stripe method.

A formal, codified approach has been applied to the zoning process. Here the latest published version will be considered as an example (Anon., 1976, Fig. 5.2.2.3), although hydrogeological data on the original are omitted so that the map can be published in black and white.

Engineering geological zones are discriminated on (a) the general uniformity of lithological character; and (b) the arrangement of engineering formations in the upper-

Table 7.3 Engineering geological conditions for selected areas and zones, and for construction purposes

Areas	*Zones*			
	Type	*Geological conditions*	*Hydrogeological conditions*	*Present geodynamic processes*
Ce Areas of the volcanic highlands Young and variegated mountainous landscape, formed by the destruction of original stratovolcanic forms due to differential neotectonic movements and erosion	Dm Zone of slopewash deposits on magmatic and metamorphic rocks	Loamy slopewash soils 2–7 m thick. In the substratum there are solid and semi-solid, slightly compressible volcanic and pyroclastic rocks	Very slightly water-bearing to practically dry	Intense slope erosion in erosive furrows and gullies
Dg Areas of intramountain basins Regional depressions of tectonic origin. Selective erosion and the accumulation of soft Pliocene and Quaternary sediments formed moderate to flat landscape forms with predominant floodplains and terraces of large rivers	Fu Zone of floodplain alluvial deposits	Coarse-grained gravels, sandy gravels, sands and sandy to clayey loams. Large rivers may also have sapropelic fills in abandoned river branches and depressions. Thickness of alluvium 5–11 m	Gravel deposits are highly saturated. The groundwater level is at a depth of 5 m. Large areas often flood at high groundwater levels	Washing out of the banks causes small slope failures in some places
	Nk Zone of pre-Quaternary alternating cohesive and cohesionless sediments	Moderately compacted Pliocene gravels, with silty-clayey bond, and weakly consolidated Pliocene clays of higher plasticity	On the gravelly Pliocene complex abundant small hillside springs may occur on slopes	Intensive slope erosion in the form of deep furrows. Frequent small slides at contacts with tuffites

most 10 m below the ground surface. Engineering geological zones may be subdivided into subzones on the bases of (i) the sequence; and (ii) the thicknesses of individual soil and rock types.

Conditions in the subzones are represented by schematic cross-sections (Table 7.4) of the soils and rocks.

Individual zones are indicated by symbols expressing both the genesis and lithology of the rocks and soils involved. In the generally accepted zoning classification in the CSSR, there are 40 distinct types of zone. A few examples must suffice:

Fu: zone of river valley deposits
Ft: zone of deposits in river terraces
Dm: zone of slope-wash deposits on magmatic and metamorphic rocks
Ep: zone of aeolian sands

If within the prescribed depth there are two superimposed formations, symbols can be combined; a zone of river terrace deposits (Ft) overlain by aeolian sand (Ep) could be represented by EpFt.

Subzones have a sequence of symbols formed by grouping the appropriate symbols for rock and soil types with indications for thickness of the strata and the depth to pre-Quaternary bedrock, in the order of the beds in the local succession.

The symbols are listed in Table 5.1. As an example of their use, $h1g2S^1$ indicates a soil sequence in which cohesive soils (h) less than 2 m thick, are underlain by gravels (g) between 2 and 5 m thick, on hard rocks (S) at a depth of less than 5 m.

All the subzone symbols on the map have only one or two letters, an accepted convention for subzones of small extent. The lithology of the pre-Quaternary basement is shown by grey patterns of different weights, as on the map of engineering geological conditions.

In addition to zones and subzones, information for the user is completed by data on hydrogeological conditions (omitted in Figure 7.7). Parts of the map area affected by specified geodynamic processes, in this case landsliding and gully erosion, are delimited as particular districts. Protected zones, for example spas, are also shown.

Of equal importance in the context of civil engineering, making the map content engineering geological rather than purely geological, are the very detailed tables on the engineering geological and geotechnical characteristics of the subzones. Only three subzones are dealt with here as a sample of the approach adopted (Table 7.5). The table supplements the geological information in Table 7.4, with an extensive range of numerical data on, for example, bearing capacity, pile bearing capacity, slope angles, dewatering of excavations, and more general comments on foundation evaluation, possible use of geomaterials, recommendations for additional investigations, and so on.

On the engineering geological zonation map of part of the Zvolen Basin (Matula, 1969), primary zonation is established on the basis of topography:

Zone I Floodplain of the Hron river and its tributaries.
Zone II Terraces of the Hron river and its tributaries
Zone III Moderate deluvial slopes, up to 15°.

Engineering geological conditions for construction purposes				
Excavations and cuttings	*Fill construction*	*Structural foundations*	*Roads*	*Building materials*
Cuttings and side-slope cuts will mainly be excavated in slopewash deposits and the underlying weathered rocks; groundwater issuing from the slope is easily collected and drained from the site	The construction of larger fills may lead to slope failure. Use of a combination of half fill–half cut would be advantageous	Foundation conditions are very good. The only serious problem is the question of ensuring slope stability during construction	When determining the line of roads and railways difficulties arise because of the effect of the gullies, and the need for good drainage to maintain stability	Brick clays
As a rule, foundation excavations have to be made below groundwater level. In deeper excavations in gravels large influxes of water	Does not cause difficulties, but strongly compressible non-load-bearing sludgy and putrified muddy accumulations in oxbow lakes and the fringes of alluvial cones require attention	In foundations of buildings with cellars difficulties are caused by the high watertable. Necessity of installing waterproofing (uplift pressure), in some cases precautions against corrosiveness. Fluctuating groundwater level and frequent flooding deteriorate the foundation materials	With regard to its very slight relief the zone is convenient. Difficulties occur in districts of marshes and oxbow lakes	Abundant stocks of high quality gravel and sand for the production of concrete. Abundant fill material
Cuttings are excavated mostly in moderate slopes. Eventual groundwater influxes from the slope are easily collected and drained from the site	The construction of bigger fills may lead to slope failure. Use of a combination of half fill–half cut of smaller dimensions would be advantageous	This zone is less suitable for heavy construction of larger settlements and industrial structures. It is mainly suitable for simpler structures	On account of clayey materials and the presence of deep scours, the zone is only conditionally suitable for transport constructions	Possible advantageous large gravel pits and clay pits. Gravels of Pliocene complex are, however, of lesser quality

Table 7.4 Characteristics of three selected zones and subzones on the map of engineering geological zoning, Zvolen Basin, Czechoslovakia (From Anon., 1976, by permission)*

Zones				*Subzones*		
Type	*Geological–geomorphological characteristics*	*Hydrogeological conditions*	*Geodynamic conditions*	*Type*	*Pattern (on the map)*	*Lithological characteristics; vertical lithological composition*
Ng	Pliocene gravels alternating with sand horizons forming a flat rolling landscape with moderate slopes. Slopewash loams and sands with some gravel <1 m thick	Groundwater usually more than 5 m deep. Small springs from gravelly horizons	Extensive gully erosion. Slides of small extent	G3 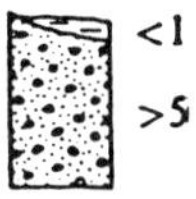	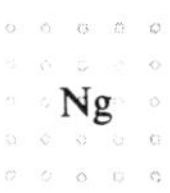	Gravels, sandy gravels, less loamy gravels, with sand and clayey-sandy horizons
Dm	Slopewash loams 5–8 m thick on andesitic agglomerates and tuffs. Moderate slopes, locally intensively dissected by erosion	Very slightly water-bearing slopewash on relatively permeable bedrock	Extensive sheet and gully erosion. Shallow landslides and earth flows	$h2B^1_v$ 	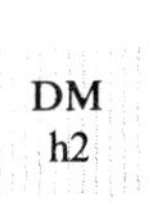	Slopewash loams, clayey, less sandy, stiff to very stiff, 2–5 m thick, on agglomeratic tuffs and agglomerates, interbedded with lapilli tuffs
Fu	Floodplain alluvium formed of gravel beds 5–9 m thick covered by sand, or loam 0.5–2 m thick	Area regularly affected by large floods. Watertable usually <2 m deep. Water locally corrosive (high SO_4 content)	In places the river banks are undercut	$klg3B^2_v$ 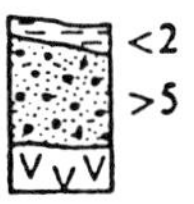	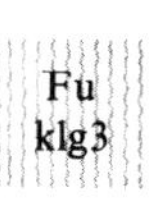	Loam and sand 0.5–2 m thick, underlain with sandy gravels 6–9 m thick. At a depth of 5–9 m substratum as in zone Dm (tuffs)

*The characteristics for only three selected types of subzone of three typical zones are presented as an example.

Zone IV Steep eluvio-deluvial slopes, over 15° and narrow dividing ridges.
Zone V Wide and flat dividing ridges.

Subzones are related particularly to the thickness and type of the Quaternary formations at the surface combined in most cases with the nature of the underlying

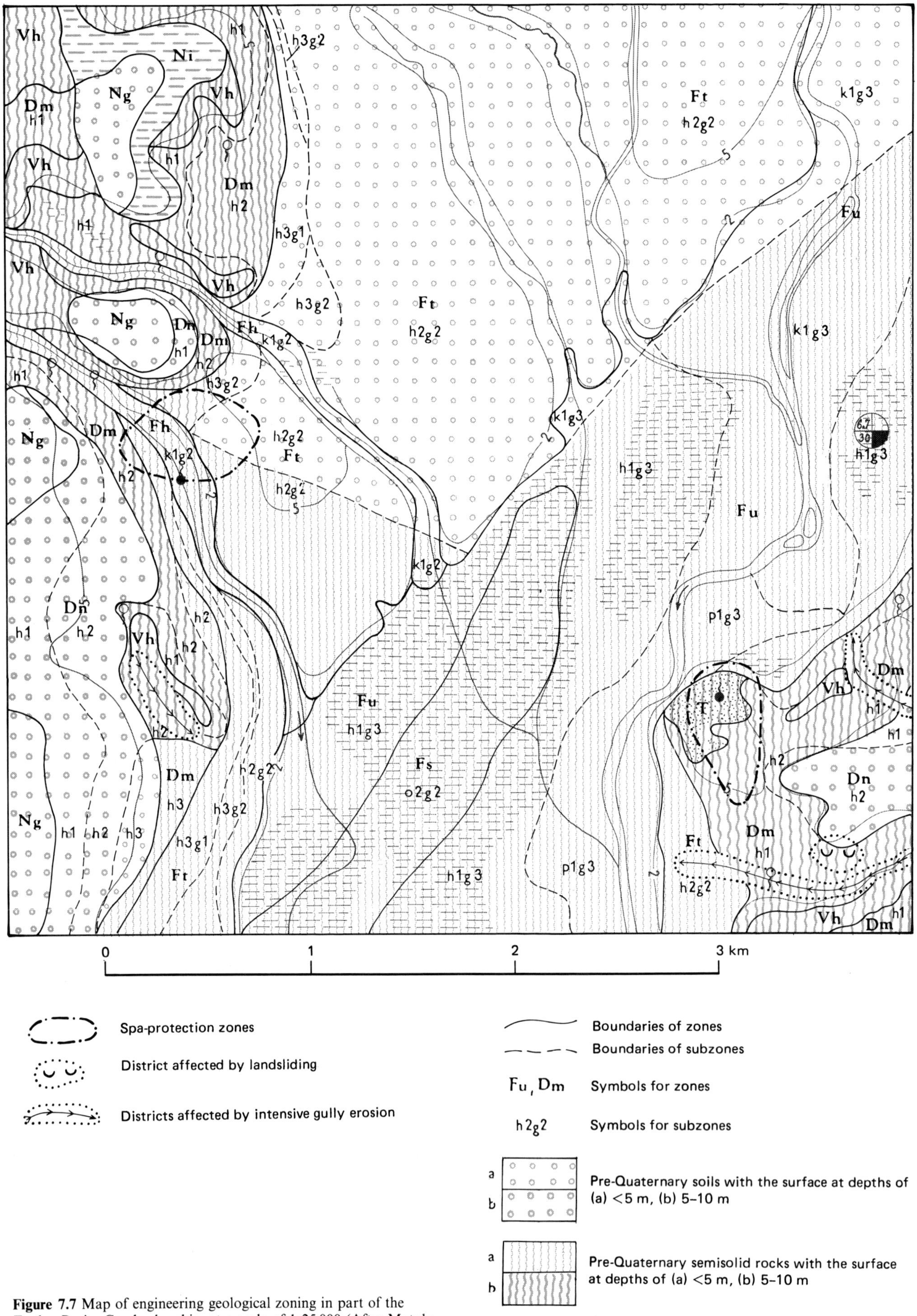

Figure 7.7 Map of engineering geological zoning in part of the Zvolen Basin, Czechoslovakia, at a scale of 1:25 000 (After Matula, 1969, Appendix 3)

pre-Quaternary formations that include Pliocene clays, sands and gravels in addition to older rocks.

Insistence on differentiation between what are essentially superficial deposits of two ages – Pliocene and Quaternary – as is made quite clear on the geological profile accompanying the engineering geological map of the Zvolen Basin, has led to inconsistencies in the definition of zones and subzones that have been discussed by Varnes (Varnes, 1974, pp. 27–29).

Varnes regards the map as an excellent example, in full colour, and generally available outside central and Eastern Europe, largely derived from a more-or-less conventional geological map of a real area. The geological map, using the stripe method (see Plate 3), was prepared knowing that the zoning map was to be derived from it. Planning of this kind makes it more likely that the derivative zoning map will be satisfactory for a specific purpose.

7.7 Zoning plans

Basic geological maps of urban areas are published at scales between 1:10 000 and 1:2500. As far as rocks and soils are concerned, at this range of scales the limit to the detail that can be shown is map units determined as distinct geological types. It is from the geological maps, with a great deal of three-dimensional detail, and associated hydrogeological, geomorphological and geodynamic aspects, that the zoning maps are prepared. If the zoning map is aimed at a particular purpose, then the engineering properties of rock and soil masses will also have to be taken into account.

At the larger scales of 1:500 to 1:100 of direct application, for example, to the foundations of engineering structures, mapping is even more detailed. Distribution of lithological types can be shown as zones and zones can be subdivided into districts on the basis of recognition of engineering types within each lithological type. No examples are available from Czechoslovakia, and in Western Europe and elsewhere the nomenclature is ignored, and plans of the distribution of engineering types (although the name is not admitted) are prepared as necessary. These are discussed in Chapters 10 and 11.

7.7.1 Zoning plan of Prague, Czechoslovakia at 1:5000

The plan of engineering geological zoning is derived from the geological plan (see Plate 2 and section 6.2) and the accompanying hydrogeological plan (Figure 6.5). It is a single purpose plan (Plate 11) evaluating the area in terms of suitability of the ground for the foundations of stated types of housing in the northern parts of Prague. Two types of housing are considered: multi-storey buildings and individual houses. All the complexity of the geological map has been dispensed with, and in its place the plan sheet is divided into a small number of simple units distinguished by different colours.

Five classes of foundation conditions are recognized:

Class I — Very suitable, where the depth to a suitable foundation level is up to 1 m.
Class II — Suitable, where the depth is up to 2 m.
Class III — Not very suitable, where the depth is up to 3 m.
Class IV — Conditionally acceptable, where locally the bearing capacity is lower than in classes I and II.
Class V — Unsuitable for variety of reasons.

These appraisals of suitability relate to multi-storey buildings; foundations for individual houses are dealt with in a similar fashion.

A great deal of information is presented on the map in a clear unequivocal fashion. Large areas of plain distinctive colours make interpretation by direct references to the legend a simple process that, having planners and other non-geologist users in mind, can be carried out quickly and accurately by non-specialists.

7.8 Other types of zoning map

One of the problems in producing zoning maps is that the boundaries of the zones or subzones are drawn on the map as they would appear at the ground surface. On the Czechoslovakian examples (Plate 11), ground conditions are considered down to 10 m, and it has to be accepted that the boundary of the zone is vertical. Whether this is important depends on map scale and the geology. In urban areas where the scale is likely to be 1:5000 or even larger it is manifestly untrue, as can be judged from the stripe map of Berlin (see Figure 6.8), where fill areas tend to be conical. The same is true for the uncovered maps of Warsaw which are discussed later (see section 10.5 and Figure 10.8a,b).

With this proviso firmly in mind, remembering that uncertainty also applies to the boundaries of geological maps, zoning is very useful in presenting information on ground conditions for the non-geologists of the planning fraternity. In a variety of forms, zoning has been used in other parts of the world.

7.8.1 Unitized maps

On a unitized map, a particular colour or pattern is used to represent a succession of two or more distinctive units, rather than just the surface unit. Boundaries of each unit are vertical and a conventional structural interpretation of the character of the geological boundary cannot be applied.

Mapping units on soil survey maps are unitized, each unit representing a pedological soil group with a characteristic profile of three or more horizons grading down into a specified geological parent material. Another unitized system is that adopted for terrain evaluation for engineering purposes involving the recognition of land facets (Dowling, 1968) or terrain units (Aitchison and Grant, 1967, 1968a, 1968b), each of which is characterized in terms of surface form, soils, materials and hydrology, and vegetation. Thus in this system, also, map boundaries do not necessarily coincide with geological boundaries and there is a strong and obvious affinity with engineering geological zoning maps. Some of the map units in the engineering geological zonation map of the Zvolen Basin (section 7.6.1) require that particular materials lie on others.

Table 7.5 Engineering geological and geotechnical characteristics of the three selected subzones in Table 7.4

Subzone	*Estimated bearing capacity (q_0 kgf cm^{-2}) at a 2 m/6 m depth for foundation width 1 m. Depth of water below foundation:*		*Pile bearing capacity* (Mp). *Piles Ø1000 cm^2, length 6 m*	*Foundation evaluation according to the standard for foundation*	*Suitability for road foundation according to standard for transport roads*		*Capillarity*	*Frost susceptibility*	*Evaluation of road construction conditions*
	>1 m	<1 m			*Road on natural subgrade*	*Embankment on natural subgrade*			
G3	7/9	4.7/6	60	Foundation conditions simple. Foundation site suitable. Only conditionally suitable in loamy gravels. Gravels and sandy gravels, classes 8 and 10. Loamy gravels, classes 9 and 11	Very suitable	Very suitable	None (medium)	None (medium)	Territory suitable for roads in cuttings and embankments
h2B[1]	1.2/6	12/6	60	Foundation conditions simple. Loams, classes 20 and 21. Substratum classes 2 to 5. Foundation sites conditionally suitable	Barely suitable	Barely suitable	High	High	Because of soil character and extensive dissection, territory barely suitable for transport road
Klg3B[2]	6/8	4/5	40	Foundation conditions simple. Foundation sites may be unsuitable because of high water or flooding	Stratum I (loamsand) little suitable. Stratum II (gravel) very suitable	Stratum I (loamsand) little suitable. Stratum II (gravel) very suitable	Medium to high	Medium to high	Because of hydrogeological conditions, roads should be built on embankments

Unitized maps of St. Louis County, Missouri, USA

The unitized method, conveniently used to represent the distribution of superficial deposits on a limited number of bedrock types, has been partly used on the engineering geological map of the Creve Coeur Quadrangle, Missouri, USA (Rockaway and Lutzen, 1970). The map at a scale of 1:24 000 is accompanied by three tables and a short illustrated report that describes and evaluates the soil and bedrock characteristics of the area. Used as a general guide, the intention was to provide a convenient means of arriving at the most efficient, economical and environmentally acceptable programme for urban growth and land development. Factors taken into account included the engineering properties of soils and rocks, surface water and groundwater flow, subsurface geological structures, the origin of present-day topography and the stability of that surface.

A second map, at a scale of 1:63 360 (Lutzen and Rockaway, 1971), covers the whole of the county.

Four topographic areas (Figure 7.8) can be recognized in St. Louis County:

1. Flat or very gently sloping floodplains of the major rivers. Floodplain alluvium fills very deep trenches in the bedrock.
2. More than 50% of the county is rolling uplands, blanketed with loess, with relief seldom exceeding 100 ft and slopes of 2–5%.
3. In the southwest of the county, slopes in rugged topography are rarely less than 5% and relief often more than 200 ft. Superficial soils are thin and bedrock exposures are common.
4. The fourth area, in the north east, is a flat lowland underlain by Pleistocene lake-bed deposits, the Florissant Basin.

Groundwater is abundantly available from the alluvial deposits. Fracture water occurs in the limestones, although these sources are subject to pollution because sinkholes and other solution features allow surface water, for example from septic tanks, to enter bedrock.

Floodplains of the Missouri, Mississippi and Meramec rivers (Figure 7.9) are underlain by more than 100 ft of alluvium: stratified gravels, sands, silts, clays and organic deposits. Remnants of older higher level river terrace deposits often remain on the valley slopes.

Bedrock formations are predominantly horizontal limestones overlain by shales with siltstones, sandstones, some limestones and thin coals. The upper surface of the limestone is very irregular and may be masked by residual deposits.

Deposits of windblown silt (loess), derived from the floodplain of the Missouri River during the Pleistocene, completely cover bedrock to depths up to 80 ft, blanketing a bedrock topography similar to that of the present day. The thickest loess deposits, over 50 ft, are found along the bluffs of the Missouri River (Figure 7.9), thinning to 5–10 ft along ridge tops to the south. On the

Maximum slope angle in temporary excavations $\left(Depth \frac{3\,m}{4\,m}\right)$		*Estimated maximum flow in litres/min through 1 m of the circumference of the excavation to lower the water table by:*			*Ease of excavation*	*Possible rock use*	*Recommendations for additional investigations*
Dry	*Under water*	2 m	4 m	6 m			
$\frac{1:1.4}{1:1.8}$	$\frac{1:2.2}{1:2.5}$	5	15	50	2–3	Gravel and loam extraction; gravels of poor quality	Because of the irregular occurrence of sand and clayey-sand intercalations, a dense network of boreholes is required
$\frac{1:0.7}{1:0.9}$	$\frac{1:1.3}{1:1.2}$	—	<1	<1	2–3	—	Very variable petrographic composition and consistency of loams
$\frac{1:1.4}{1:1.8}$	$\frac{1:2.2}{1:2.5}$	40	180	400	$\frac{1–2}{2–3}$	Considerable quantities of a good-quality fluvial gravel for concrete	Variable corrosive groundwater properties

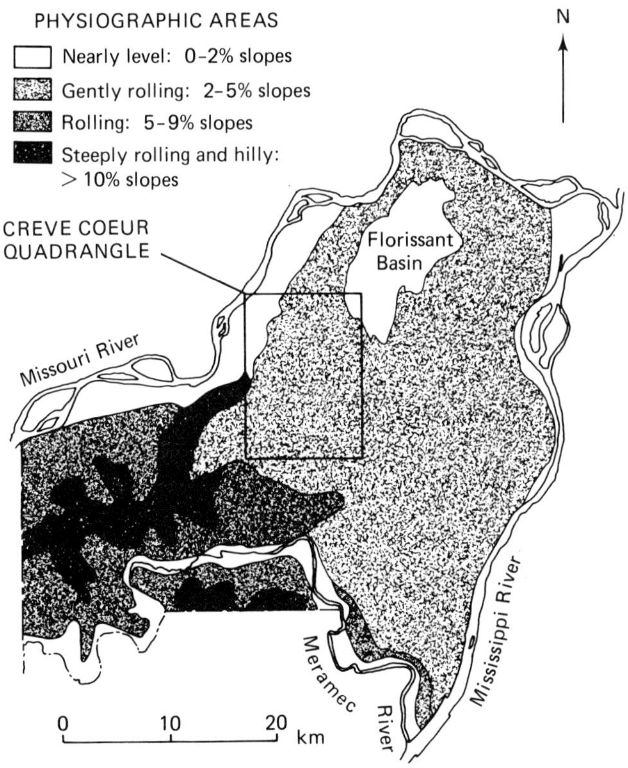

Figure 7.8 Physiographic map of St. Louis County, Missouri, USA showing the four topographic areas. Creve Coeur Quadrangle is outlined (After Lutzen and Rockaway, 1971, Fig. 1)

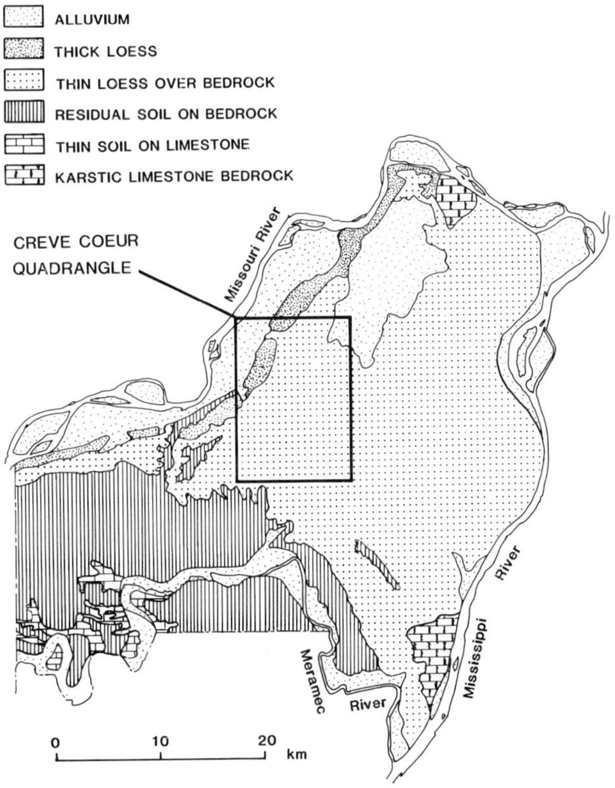

Figure 7.9 Unitized small-scale engineering geological map of St. Louis County, Missouri, USA (After Lutzen and Rockaway, 1971, Plate 1)

adjacent hillsides, loess has either been removed or reworked by surface water.

Engineering geological conditions of the county are governed by soil, regardless of bedrock type, thin soils are varied bedrock types, and karst processes in limestone bedrock. Six distinct mapping units may be recognized:

- alluvium,
- thick loess.

in which engineering is concerned mainly with soil properties;

- thin loess over bedrock,
- thin soil on limestone bedrock,
- residual soils on bedrock,

where both soil and rock, and often the interface between the two, influence ground conditions; and

- karstic limestone bedrock,

where variations are related to solution features in limestones which may be compared with the ground surface or masked or plugged by soil.

Distribution of these six units is shown in Figure 7.9, a small-scale unitized engineering geological map of the county, originally drawn at a scale of approximately 1:360 000 by simplification of the published 1:63 360 map.

There are two types of loess deposit: thick loess comprises an upper and lower loess of different compositions, whereas in areas of thin loess, the upper loess has been largely removed by erosion.

Comparison of the small-scale unitized map and the physiographic map shows the interrelation between the two maps. Alluvium is obviously related to the main river floodplains and the Florrisant Basin filled with Pleistocene Lake deposits. Rolling and hilly country in the south-west of the county is dominated by residual soils masking bedrock. The residual soils are usually thin, in part thin colluvial slope deposits, but may reach 30 ft in thickness. Practically the whole of the eastern and north-eastern part of the county, gently rolling, is underlain by thin loess masking either a limestone or shale bedrock. Thick loess borders the alluvium of the Missouri River floodplain. These areas in the east of the county are markedly karstic, and in the south-west, adjacent to the Meramec river, there are significant areas of limestone covered by thin soils.

It is the distribution of the two types of loess on different bedrock units that the map units used on the medium-scale maps show particularly well. Two different methods are used to show the characteristics of the map units: block diagrams in the report (Figure 7.10) and thumbnail sketches (Figure 7.11) in the explanation of map units.

The St. Louis County, Missouri, USA, 1:63 360 map

At the medium scale of 1 : 63 360 Lutzen and Rockaway (1971) have developed an engineering geological map of the whole of St. Louis County, Missouri, in which Creve Coeur, discussed later, is a single quadrangle. The aim is to provide assistance to urban developers in the definition and evaluation of those geological factors that influence land use. Physical properties of bedrock and unconsolidated materials as well as topographic features are used to define map units which have specific engineering geological properties. The map is then used to show the distribution of rock formations and soils having similar engineering properties. Interpretation is limited only to cross-referencing the appropriate map unit, with the graphic and other descriptions provided by the map legend, the tables and the report.

The rationale of the mapping has been the classification of the county into homogeneous areas based on evaluation of bedrock characteristics, or on the superficial deposits where these blanket bedrock to significant depths thus acting as the controlling geological factors in land-use planning. Primary state-wide map units, not all present in the county, have been classified as:

Unit I	Alluvium, with three subunits.
Units II–V	Limestone bedrock, with 11 subunits.
Units VI and VII	Shale.
Unit IX	Sandstone.
Unit X	Cyclic deposits, predominant shales with some sandstones and limestones.

Units VII and IX are not represented in the map area.

In the report, block diagrams very effectively illustrate the relationships of the mapped units to both topography and geology. Typical associations of karst (sinkhole) areas of Limestone Unit II with other engineering geological units are shown in Figure 7.10. Geological names on the original diagram have been removed in order to conform to the map legend and to emphasize that the units are in fact homogeneous engineering geological mapping units.

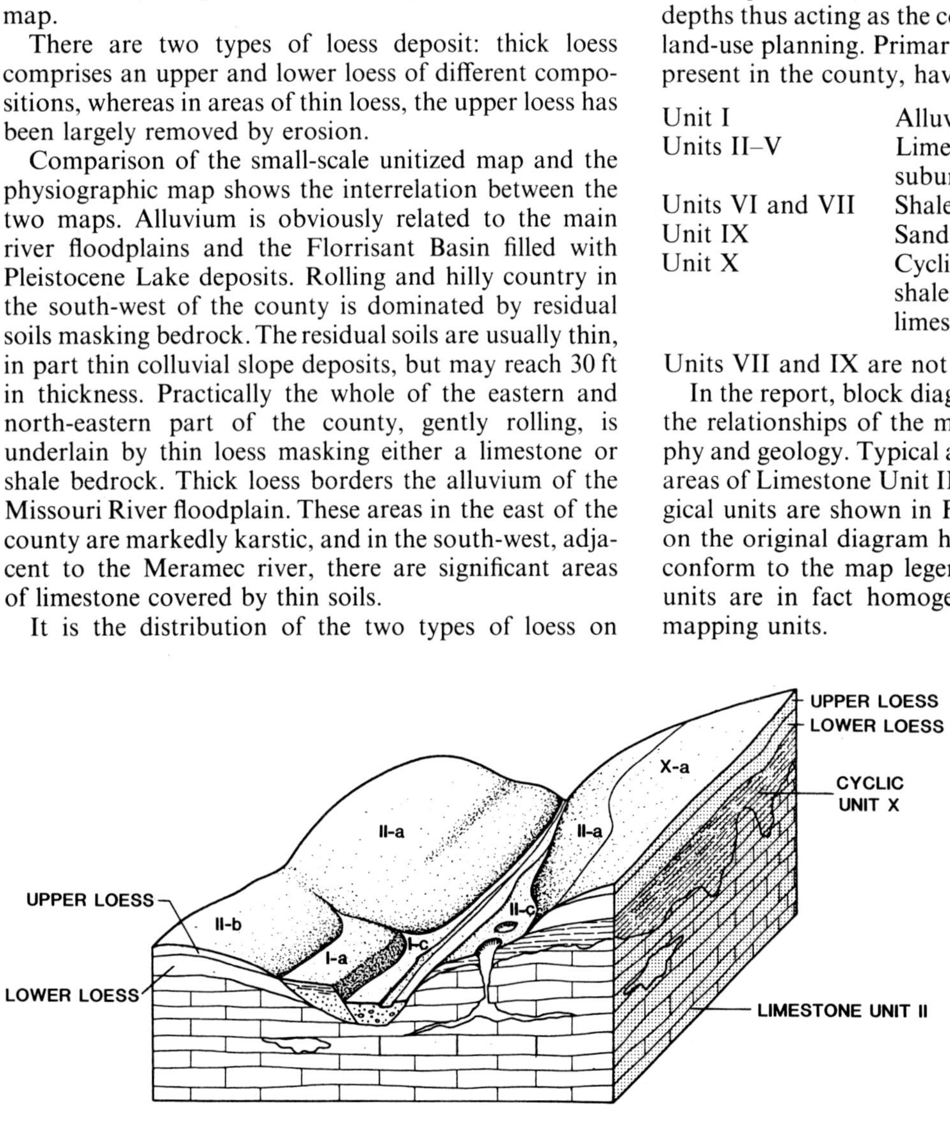

Figure 7.10 Typical relationship of karst areas to other engineering geological units, St. Louis County, Missouri, USA (From Lutzen and Rockaway, 1971, Fig. 4, by permission)

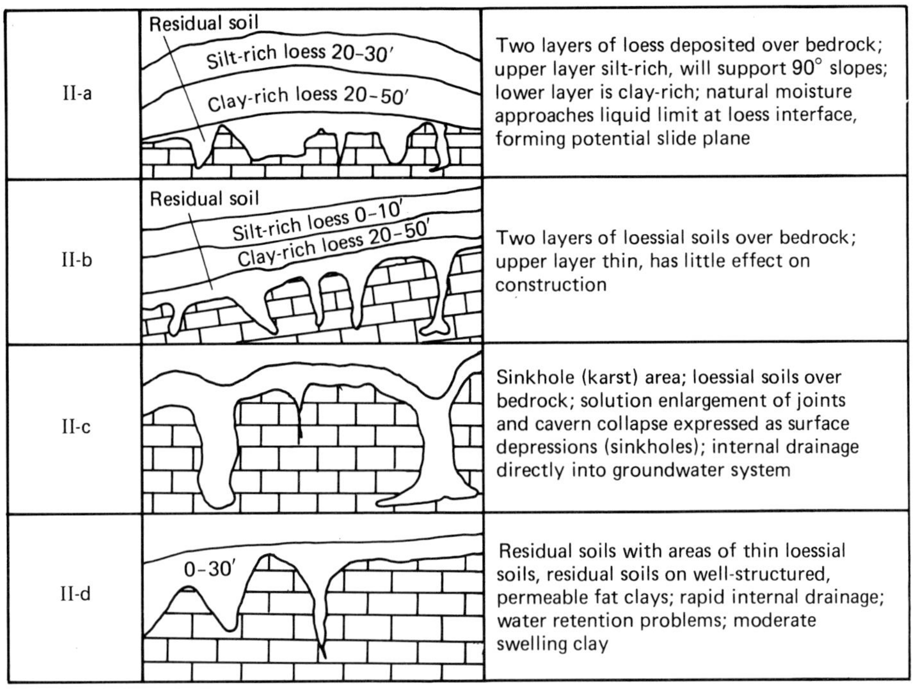

Figure 7.11 Extract from the legend to the St. Louis County, Missouri, USA, map giving details of Unit II, limestone bedrock with extensive near-surface weathering and solution enlargement along discontinuities – all covered by loess. The original is in colour (After Lutzen and Rockaway, 1974, Plate 1)

The Creve Coeur, Missouri, USA, 1:24 000 map

Major map units identified in the Creve Coeur area are:

Unit I   Alluvium, not unitized but could be.
Unit II   Limestone bedrock; unitized with four subunits on the basis of divisions in the superficial deposits.
Unit X   Shales with sandstones, etc., unitized with three subunits.

Only Unit II has been selected as an illustration of unitized map units, but some of the other units inevitably appear on the diagrams.

Apart from alluvium, which is a main mapping unit, the superficial soils are:

1. The upper loess, 20–30 ft thick, composed of uniform non-stratified silt with little or no clay, and a limited development of a pedological soil profile.
2. The lower loess, 20 ft or more thick, with a significantly higher clay content than the upper loess.
3. A residual clay soil and partially decomposed bedrock beneath the lower loess.

Relative proportions of the upper and lower loesses are used to define two subunits, II-a and II-b, on the carbonate bedrock. One other subunit, II-c, includes areas of karstic collapse marked by sinkholes or surface depressions infilled with reworked loess. Another subunit, II-d, marks steep bluffs along the margin of the Missouri River floodplain where limestone is either exposed or covered with a thin soil layer.

It is thus apparent that three quite different attributes are used to define limestone subunits: thickness and type of loess; presence of sinkholes in areas of thin loess; steep rock slopes with thin soil cover. Such contrasted attributes have, it seems, been used as a means of emphasizing problems likely to be associated with urban development. In fact, the subunits could have been defined on the basis of the superficial cover alone:

subunit II-a: thick upper loess over thick lower loess;
subunit II-b: thin to very thin upper loess over thin to medium-thick lower loess;
subunit II-c: thin collapsed lower loess;
subunit II-d: thin colluvium derived from loess.

However, in order to convey a very important characteristic of each unit, it is essential to make a general statement on topography. For subunit II-a, topography is rolling to rugged; II-b, gentle to rolling; II-c, karst (sinkholes); II-d, rugged (bluffs).

Complete unitization of the map would have been preferable to the mixture of main units adopted by Lutzen and Rockaway, based on the distribution of alluvium and types of bedrock, but only from the point of view of a consistent and tidy approach, by referring to the purpose of the map as an aid to the localization and interpretation of problems likely to be associated with urban development. Emphasis is thus clearly on the behaviour of the superficial soils, except where bedrock is near surface or there are particular hazards associated with bedrock.

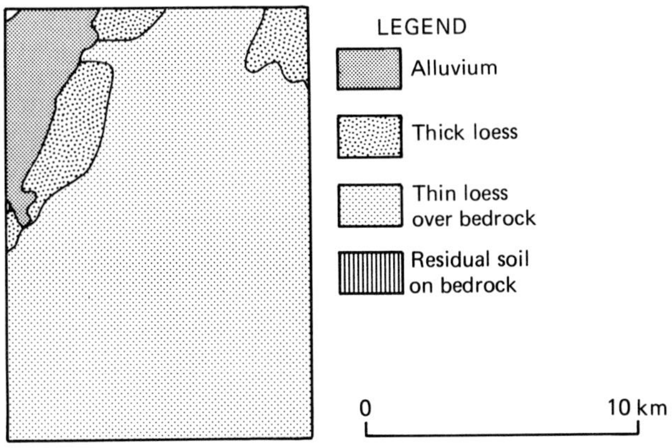

Figure 7.12 Small-scale general unitized map of the Creve Coeur Quadrangle, St. Louis County, Missouri, USA

Description of the 1:24 000 map of Creve Coeur, Missouri, USA

The geology of Creve Coeur Quadrangle is dominated by three units: alluvium, thick loess and thin loess over bedrock (Figure 7.12) when viewed at a small scale. At the 1:24 000 mapping scale, these units are subdivided into:

I ALLUVIUM
- I-a Swamp
- I-b Thick alluvium
- I-c Thin alluvium
- I-d Alluvial terrace
- I-e Lake deposits

II CARBONATE BEDROCK
- II-a Loess
- II-b Loess
- II-c Karst
- II-d Bedrock

X CYCLIC DEPOSITS
- X-a Loess
- X-b Loess
- X-c Loess

and Cyclic Unit X overlies Carbonate Bedrock II (see Figure 7.10).

At the published map scale (Plate 10) the Creve Coeur Quadrangle is dominated by Unit X-b which generally occupies higher ground than adjacent Unit II areas. Unit II therefore occupies the main scarp bordering the floodplain of the Missouri – Unit I-b on the map – and valleys draining to the floodplain and also to the south-east corner of the map.

Unit X: cyclic deposits

Within the area of Unit X-b, Units X-a and X-c play a very subordinate role. Bedrock is covered by loess so that rock exposures are limited to steep slopes or man-made excavations. Boundary lines drawn to separate this unit from Unit II are based on limited data from drilling records and must be considered as approximate.

Unit X is subdivided on the basis of the engineering characteristics of the superficial soils, particularly with respect to stratification and clay content. There are three subunits; the first has a thick cover of loess; the second thin loess; the third has an expansive clay at shallow depths.

Subunit X-a

Loess: loess deposited over bedrock (predominantly shale); two layers – thick upper layer (silt-rich) and thick lower layer (clay-rich); problems with potential slide plane at interface of loess and shale.

Subunit X-b

Loess: loess deposited over bedrock (predominantly shale); two layers – thin silt-rich layer over thick clay-rich layer; potential slide plane at interface of loess and shale.

Subunit X-c

Loess: loess deposited over bedrock (predominantly shale); swelling clays encountered at shallow depth.

The descriptions given above are taken from the map legend, where they are used with thumbnail sketches of each unit from which thickness ranges can be judged.

On the map (Plate 10) Subunit X-a only occurs capping high ground just beyond the valley escarpment, and is surrounded by Subunit X-b. Subunit X-c with swelling clays occupies relatively small areas of higher ground within the area of general thin loess cover of Subunit X-b.

Unit II: carbonate bedrock

The main expanse of this unit is occupied by Unit II-b; Units II-a and II-d occur along the escarpment and Unit II-c sporadically within Unit II-b (Plate 10).

Carbonate bedrock comprises a sequence of three limestones, one slightly argillaceous with some chert and the other two pure, massive limestones with occasional thin shale beds. They are grouped together because of the similarity in their engineering properties.

Solution features accompanying limestone weathering have significantly affected the physical properties of these formations. Although solution features are usually close to the limestone surface, they may extend to considerable depth. Solution of the limestone by surface waters produces an irregular and unpredictable surface between bedrock and overlying soils. One form of this solutioning is called pinnacling, where waters have migrated down the vertical joints, consequently enlarging them and leaving intervening pinnacles of limestone. Vertical and horizontal solution openings may intersect so that unweathered blocks of limestone may be left surrounded by clay washed in from the surface, filling the cavities.

Solution at shallow depths may enlarge cavities until roof collapse and surface settlement leave sinkholes. Although exposed at only a few locations, sinkholes are undoubtedly present under the thick loess cover in many parts of the Creve Coeur Quadrangle.

Unit II is subdivided into four subunits on the engineering characteristics of the overlying loess. Two of three subunits, II-a and II-b, are based on the thickness and types of materials comprising the loess; II-c outlines areas of relatively thin loess and surface expression of karst (sinkhole) topography; II-d outlines the steep limestone bluffs with bedrock near the surface. The difference between II-a and II-b is gradational and consequently the boundaries are somewhat arbitrary.

Subunit II-a

Loess: loess deposited over carbonate bedrock; two layers – upper layer (silt-rich) will stand on 90° slopes, lower layers clay-rich; at interface of loess layers natural moisture content near liquid limit.

Subunit II-b

Loess: loess deposited over carbonate bedrock; two layers – thin silt-rich layer over thick clay-rich loess; high water content at interface of loess layers.

Subunit II-c

Karst (sink area): loess deposited over carbonate bedrock; solution enlargement of joints or cavern collapse expressed as depressions on surface; internal drainage directly to groundwater system.

Subunit II-d

Bedrock: steep slopes (greater than 60° above horizontal); bedrock at or near surface; thin soil zone has tendency to creep down slope.

Unit I: alluvium
Unit I and its subdivisions are alluvial materials ranging from coarse gravel to clay. The deposits generally show stratification and segregation of materials, according to grain size and are associated with water deposition. Alluvium covers the valley floors of all major streams and on some of the smaller creeks. The deepest alluvium occurs along the Missouri River Valley.

There are significant differences in engineering properties of these alluvial deposits, which have been mapped as five subunits: Subunit I-a materials are swamp or marsh-type deposits with high organic content; I-b materials consist of stratified sands, silts and clays – and gravels filling the valleys of larger streams; I-c materials are alluvial or colluvial deposits found along narrow valleys of smaller creeks; I-d materials (probably remnants of Pleistocene terraces) consist of stratified finer sands, silts and clays that are found above the present floodplains of larger streams; and I-e are lake-bottom (lacustrine) deposits.

Typical associations of map units I and II along the Missouri River bluffs are shown in Figure 7.13.

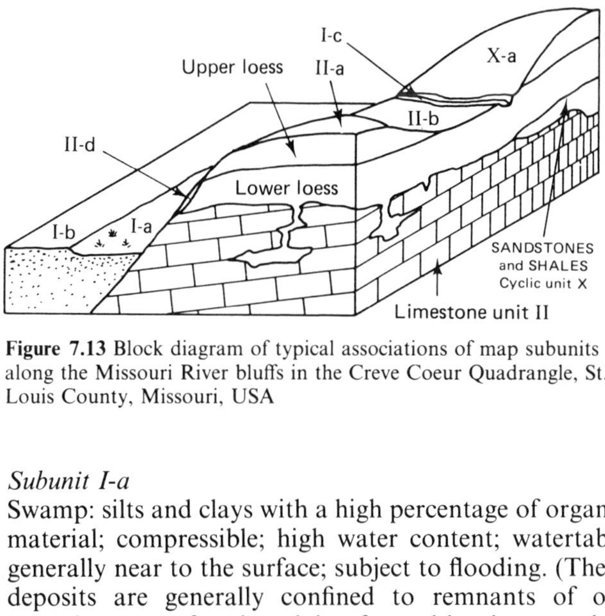

Figure 7.13 Block diagram of typical associations of map subunits along the Missouri River bluffs in the Creve Coeur Quadrangle, St. Louis County, Missouri, USA

Subunit I-a
Swamp: silts and clays with a high percentage of organic material; compressible; high water content; watertable generally near to the surface; subject to flooding. (These deposits are generally confined to remnants of old meander scars of ox-bow lakes formed by the meandering of the Missouri River. Even when free water is not observed, the natural moisture content may range from 60% to 100%, as determined by weight of water to soil.)

Subunit I-b
Thick alluvium: stratified sands, silts and clays with gravel and/or clay lenses; high watertable; subject to flooding.

Subunit I-c
Thin alluvium: stratified sands, silts and clays deposited in tributary stream valley; varying thickness; generally silt or clay over sand and/or gravels; subject to flooding.

Subunit I-d
Alluvial terrace: stratified sands, silts and clays; above level of present floodplain.

Subunit I-e
Lake deposits: stratified silts and clays; compressible; high water content; some organic material.

Engineering assessments for the 1:24 000 and 1:63 360 maps
An important aspect of engineering geological mapping is the inclusion of tables (Tables 7.6–7.8) of geotechnical information and engineering interpretations of ground conditions either as part of the map legend or in the report. For the St. Louis County maps, a separate sheet of tables and the report are used. The data could have been evaluated further using combinations of criteria, as has been done for the Zvolen Basin region (Figure 7.7).

Geotechnical information
A summary of typical engineering properties of the superficial deposits is given in tabular form (Table 7.7) only for the Creve Coeur map. Detail includes classification, grain size distribution, Atterberg limits, shrinkage limit, standard Proctor, permeability, vane shear strength and swelling index.

Engineering interpretation
For both maps a three-level rating system – slight, moderate or severe – is used to classify the severity of potential land-use or engineering problems caused by the geological features of the area (Table 7.8). Each map unit is evaluated with respect to waste disposal, water pollution, slope stability, excavation properties, bearing capacity, conservation of mineral resources, and special problems.

Ratings for the different units are based on the experience of the map compilers and others with experience of working with the materials, and from the results of engineering tests carried out in the laboratory.

A rating of *slight* is an indication that the geological materials will only have a limited effect on a particular development. On the other hand, *moderate* implies that there have been problems associated with a particular unit, either long or short term, that may be overcome by proper design. Particularly high-risk conditions are associated with a *severe* rating. In such a setting, extensive remedial measures will be needed to avoid potential failure.

For each type of urban development, potential problems are discussed in the report for each map unit. Emphasis is given to the special problems associated with loess which is subject to rapid erosion if stripped of vegetative cover in the early stages of urban development. More insidious is piping leading to undetected subsurface erosion under streets and foundations, causing significant loss of foundation support and possible collapse to the surface.

Comparison of some cartographic aspects of both maps
It is self-evident that more detail can be shown on a large-scale map than on one at small scale. As an example, at the 1:25 000 scale Alluvium Unit I is mapped as five subunits instead of three at 1:63 360 and Cyclic Unit X comprises three instead of two subunits. Ideally the same criteria should be used to classify map units regardless of the scale of the map, otherwise confusion may result. This general principle of engineering geological mapping (Anon, 1976, p. 15)

Table 7.6 Description and characteristics of Map Unit II for the Creve Coeur Quadrangle, St Louis, Missouri, USA (From Rockaway and Lutzen, 1970, by permission)

Map unit	*Bedrock*	*Weathering characteristics of bedrock*	*Overburden (soils)*	
			Materials	*Thickness* (ft)
II-a	Mississippian limestone	Irregular surface, solution enlargement of joints and bedding planes	Thick, silty loess over thick, clayey loess	50–100
II-b	Same as Unit II-a	Same as Unit II-a	Thin silty loess over thick clayey loess	20–50
II-c	Same as Unit II-a	Irregular surface, collapse of cavern roof and extensive solution features	Silts and clays	10–50 or more
II-d	Same as Unit II-a	Same as Unit II-a	Silts and clays	0–20

Table 7.7 Summary of typical engineering properties of overburden material for Map Unit II for the Creve Coeur Quadrangle, St. Louis, Missouri, USA (From Rockaway and Lutzen, 1970, by permission)

Map unit	*Parent materials*		*Classification*		*Grain size distribution; % finer than* (mm)				
			Unified	*AASHO*	2.0	0.25	0.074	0.005	0.002
II-a	Peoria and Roxana loess over limestone	Peoria	ML	A-4			90–95	23–30	15–25
		Roxana	ML-CL	A-4 to A-7-6			95–98	25–35	20–30
II-b	Generally Roxana loess over limestone		ML to CL may be borderline (CL-ML)	A-6 to A-7		95–99	92–98	25–50	15–30
II-c	Thin soil over limestone karst topography		ML to ML-CL	A-6 to A-7			95–99	23–35	20–30
II-d	Steep limestone bluffs		No engineering analysis made due to topography and thin soil cover						

Table 7.8 Interpretation of problems associated with urban development in Creve Coeur County, St. Louis, Missouri, USA. Only Map Unit II is given here (From Rockaway and Lutzen, 1970, by permission)

Unit	*Waste disposal*			*Construction*				
	Sanitary landfill	*Sewage lagoon*	*Septic tank*	*Slope stability*	*Slab foundations*	*Footings*	*Excavation*	*Road performance*
II-a	Slight	Moderate	Slight	Slight	Slight	Slight*	Slight, severe**	Slight
II-b	Slight	Slight	Slight*	Slight, moderate*	Slight	Slight	Slight, severe**	Slight
II-c	Severe	Severe	Severe	Moderate	Moderate	Moderate	Severe	Moderate
II-d	No engineering analysis							

Topography	*Groundwater*	*Drainage (overburden—soils)* *Surface (run-off)*	*Subsurface*	*Remarks*
Rolling to rugged	In bedrock	Rapid	Moderate to rapid in upper loess; slow to moderate in lower loess	Upper loess tends to stand in vertical cuts
Gentle to rolling	Same as Unit II-a	Rapid	Slow to moderate	
Karst (sinkhole)	Same as Unit II-a	Rapid	Moderate to rapid	
Steep slopes and bluffs	Same as Unit II-a	Rapid	Rapid	

Atterberg limits		*Shrinkage limit*	*Standard Proctor*		*Permeability* ($cm\ s^{-1} \times 10^{-4}$)	*Shear strength* (Vane, ton ft^{-2})	*Swell index confined uplift (FHA)* (lbf ft^{-2})	*Remarks*
LL	*PI*		*Max. dry density*	*Opt. moisture content*				
25–30	4–8	20–25	106–112	18–20	10–46	0.75–1.0	<775	Two ages of loess deposit cover the bedrock and each is substantially different; the Peoria loess will stand in vertical cuts, but the Roxana will not
30–36	12–15	25–30	101–105	14–18	0.03–1.0	0.75–1.25	1000–1500	
25–52	6–32	25–30	100–114	15–22	1.0–2.8	0.75–1.0		Peoria loess generally covers Roxana loess but is usually less than 5 ft thick. Major foundation problems will be found on the Roxana
30–40	10–15	18–24	105–110	15–20	1.0–2.8	0.75–1.0	<775	Soil is usually reworked with loess filling in cavern collapsed area. Area has typical karst topography
No engineering analysis made due to topography and thin soil cover								

Lake impoundments	*Economic use*	*Special problems*	*Remarks*
Moderate	Good	Subject to severe erosion in cuts and fills when vegetation cover is removed	* Loess loses strength rapidly upon saturation ** Severe if bedrock encountered
Slight	Good	Same as Unit II-a	* Moderate if seepage occurs along some horizons ** Severe if bedrock encountered
Severe	Good	Same as Unit II-a	Sinkholes should be cleaned out and backfilled with suitable material before any improvements attempted
No engineering analysis			

unfortunately has not been upheld, as there is a lack of correspondence between Unit I-a, and also Unit II-d, with change of map scale.

There is also a difference in emphasis in the legend to the maps. The large-scale map legend illustrated the pinnacled surface of limestone Units II-a and II-b, overlain by partially decomposed bedrock and residual clay. On the legend to the 1:63 360 map this important detail is omitted, although it is mentioned in the report.

Varnes (1974, Fig. 29) has assembled a matrix showing the definition of units on the Creve Coeur map. Most of the units are relational, as they are defined by being loess over cyclic deposits or over limestone. In addition, the relation of alluvial deposits to bedrock can be inferred from the map for four out of five of the alluvial deposits; the exception is Subunit I-b, thick Missouri River floodplain alluvium which is over 100 ft thick, in which case the underlying bedrock is hardly likely to be of significance in planning or civil engineering.

Questioning some of the relations, Varnes points out that whereas areas of thin loess are divided into several subunits depending on the underlying solid material (Subunits II-b,c,d; X-b,c), areas of thin alluvium and terrace alluvium are not. However, it is implicit in the map legend that the authors considered alluvium to warrant individual attention because high degrees of lithological variability may cause strength and permeability values to change significantly from layer to layer. This situation requires extensive, detailed subsurface investigations and results in complex design criteria. There is in addition the possibility that organic materials (peat) may require special consideration. Matrix logic is one thing – a map for practical applications may be regarded as more important even if illogical in some aspects.

A major criticism of the map is that, apart from the alluvium units which are all the same colour, differentiated by number and letter coding, visually distinctive map units are based on bedrock differences. Yet with the acknowledged problems associated with the superficial soils, the summary map of Creve Coeur (see Figures 7.9 and 7.12) is more firmly based in planning and engineering reality than the version published in colour (see Plate 11). It is tempting to redraw the 1:24 000 map, with redefined map units, to illustrate the point (see Figure 7.14).

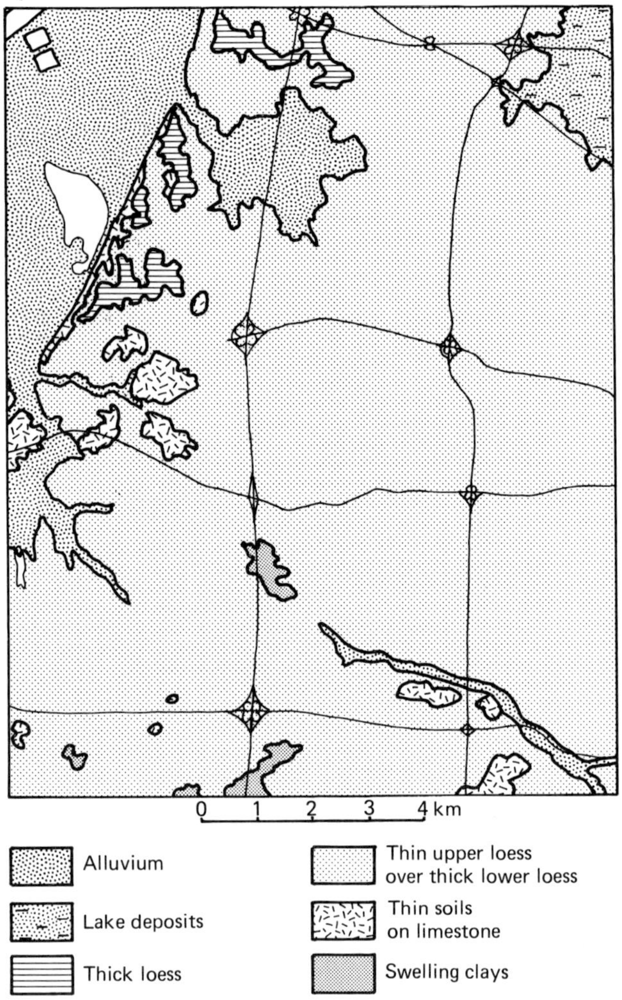

Figure 7.14 Version of the 1:24 000 map of the Creve Coeur Quadrangle, St. Louis County, Missouri, USA, drawn with revised map units

7.8.2 The profile-legend method

Mapping of predominantly alluvial, flat regions with very limited exposure presents problems of representation. An area in which these problems play a prominent part is the Holocene marine clay deposits in the western and northern parts of the Netherlands. Difficulties in satisfactorily recording the results of field work led to the development of a new kind of map legend – the profile-type legend – and with it a new type of unitized map.

The older deposits of the Holocene marine transgression are called the 'Calais deposits'. During the next stage, thick deposits of peat, often more than 2 m thick, were formed. These are the 'Holland deposits'. Finally, this peat cover was buried beneath younger Holocene marine deposits, the 'Dunkirk deposits'.

If at this stage a conventional geological map had been produced, it would only have shown the youngest deposits which over most of the area are the Dunkirk deposits. The extent and facies development of the deeper layers would always be obscured.

With the advent of hand-boring equipment that makes it relatively easy to sample to depths of at least 6 m in sands and 10 m in peat and clay, a great deal of information became available on the distribution in depth of the three types of deposit. To enable this to be shown on the map a new type of legend was devised, the profile-legend type, giving another version of a unitized map.

Three lithological units have to be ditinguished, namely:

- the Dunkirk deposits (marine);
- the Holland deposits (peat);
- the Calais deposits (marine).

A schematic cross-section (Figure 7.15) shows seven possible combinations, the main profile types (Figure 7.16) representing the primary distribution of the lithostratigraphical types.

In detail the cross-section is too simple, because in fact there were minor phases of transgression and regression of the sea leading to interfingering of the deposits. The peats of the Holland deposits interfinger with both the

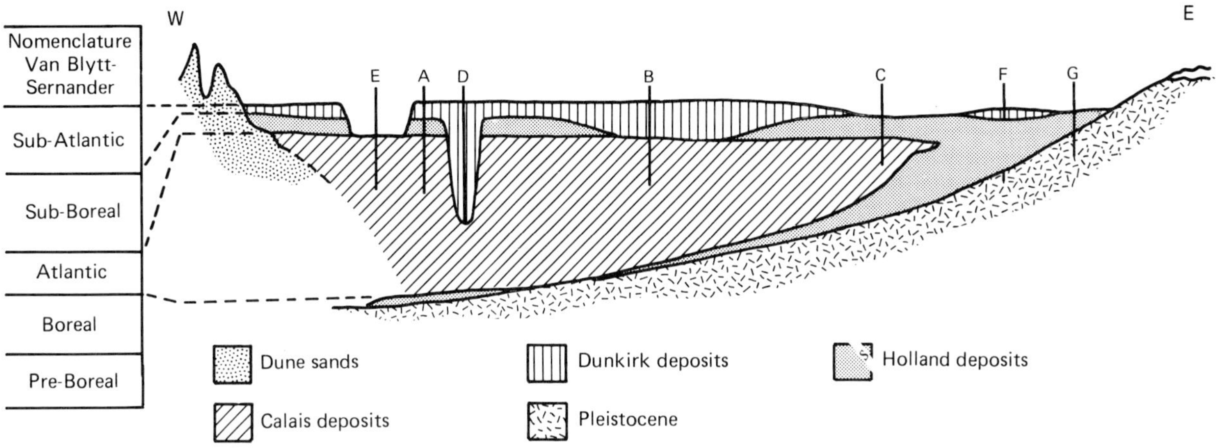

Figure 7.15 General cross-section of the Holocene marine deposits in the Netherlands (After Hageman, 1963, Fig. 2)

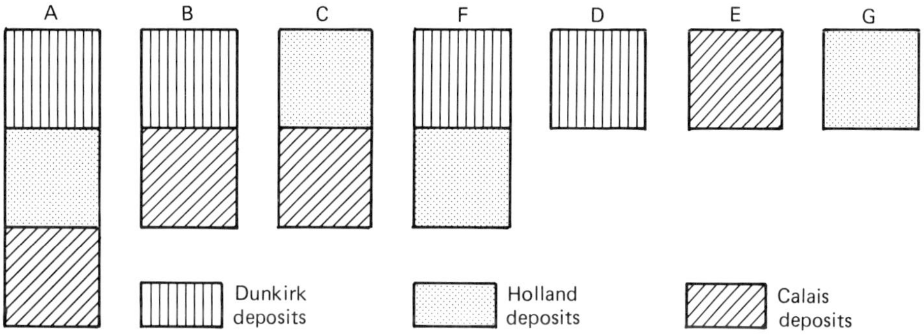

Figure 7.16 The seven main combinations of Dunkirk, Holland and Calais deposits

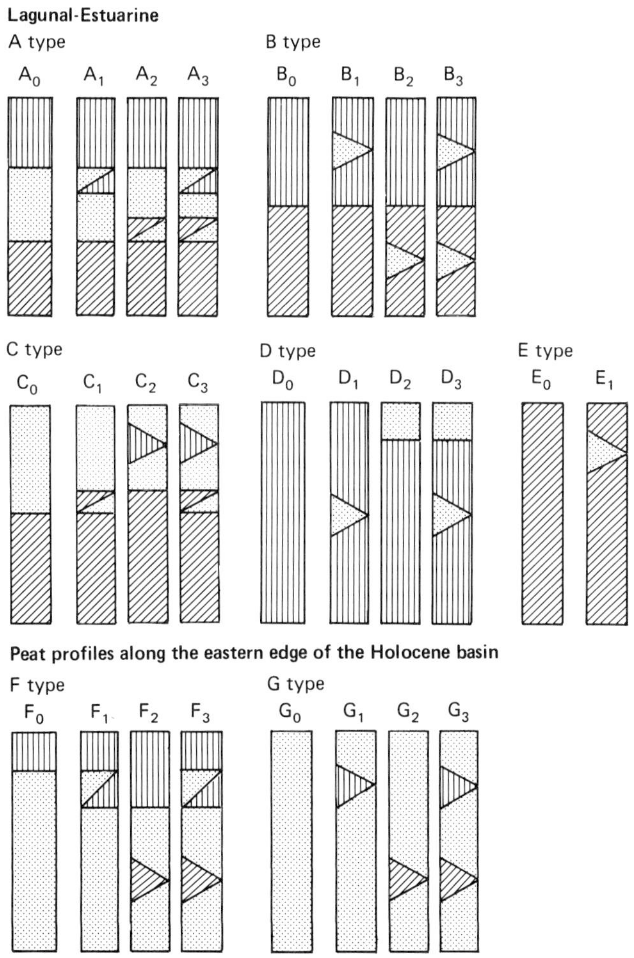

Figure 7.17 The nineteen subtypes of the seven main combinations of Holocene marine deposits (From Hageman, 1963, Fig. 8, by permission)

underlying Calais deposits and the overlying Dunkirk deposits. It is assumed that the Calais and Dunkirk deposits never occur interfingered, and are separated by unconformity.

Hageman (1963) recognized 19 subtypes caused by interfingering of the peats in the seven main types (Figure 7.17). Geological conditions giving rise to the subtypes are illustrated by schematic cross-sections in Figure 7.18.

The geological map of Zeeuwsch-Vlaanderen (Oostblad), published in colour at a scale of 1:50 000, uses the same method of representing complex spatial information (Rummelen, 1965). Each map unit has a distinctive colour-and-pattern combination and a symbol. Varnes (1974, p. 36) has commented on the profile-legend method that the system tends to subordinate attributes concerning the identity of particular deposits at the surface to those related to geometrical relationships. Thus the same unit at the surface may be coloured dark tan, light green or olive-brown depending upon the nature of associated sediments below – a complete departure from geological map conventions.

On this map, the Dunkirk deposits are in three recognizable divisions and, if the two types of Pleistocene sands are included, 39 distinct map units are represented on the legend (Varnes, 1974, Pl.2G) by colours and symbols. Thus a very great deal of information is conveyed by the map on spatial relations, although the map does not show clearly the areal extent of the main units at the surface. An important aspect which is lacking, however, is depths of thicknesses of the units present.

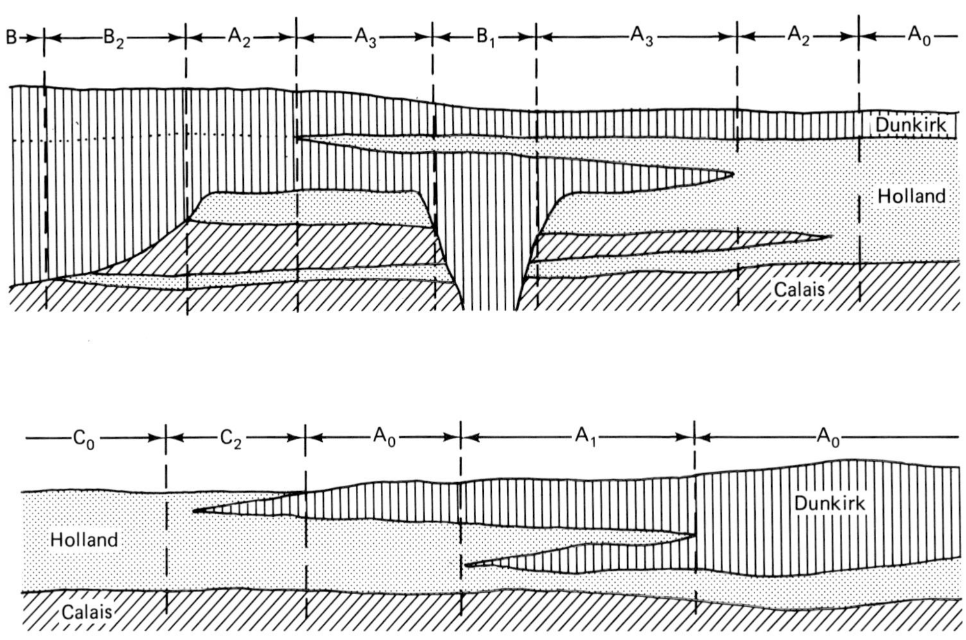

Figure 7.18 Cross-sections illustrating the possible relations between some of the 19 subtypes in Figure 7.17 (From Hageman, 1963, Fig. 8, by permission)

8

General or synoptic engineering geological maps

8.1 Introduction

A conventional geological map shows the distribution, stratigraphical and structural arrangement, age and genesis of soils and rocks. However, the name given to a deposit may give no indication, or at best may not give a correct indication, of the lithology of the deposit over the whole area of the map. Many examples of slightly and even completely misleading or uninformative names could be quoted, but the implications of adopting lithostratigraphical units as the only practical geological mapping units should by now be well understood and the implicit limitations appreciated.

This chapter is concerned with maps at the smallest scales which can be called general or synoptic engineering geological maps, because they have been compiled by engineering geologists for general engineering applications. At map scales of 1:500 000 and smaller, they can only have very general application, as the name implies. Some at least of the examples given later were generated by scientific curiosity, although that might not necessarily be admitted by the authors. It is pertinent to review some published examples of general maps to see how far they have been successful in depicting the four prerequisites of an engineering geological map, namely the representation of rock and soil types, hydrogeology and geodynamic processes. Embracing a range of scales from approximately 1:200 000 000 to 1:7 500,000, the maps depict the world, half a continent, and a part and the whole of a small country. Many more examples could have been found, but these will suffice for the present purpose.

8.2 Engineering geological map of the world

Although the published version is at an approximate scale of 1:200 000 000, the comprehensive engineering geological map of the world by Ershova and Sergeev (1979) was prepared at the larger scale of 1:43 000 000. Map units for the continents, continental shelf and slope, and the oceans, are based on combinations of 14 geostructural units with 10 zonal geological units that take account of the phase state and salinity of water within the ground. Some combinations are mutually exclusive, so that the map involves 82 distinct units in addition to linear symbols for deep trenches, faults and a variety of boundaries. The descriptive text deals very broadly with rock and soil types, hydrogeology, surface forms and surface processes, thus complying with all the distinctive requirements for an engineering geological map.

The published map (Ershova and Sergeev, 1979, Fig. 1) has been simplified, mainly by omission of detail in the oceanic areas, and is presented on three maps:

- a geological map of the distribution of the major geostructural units: cratons, orogens and riftogens;
- a zonal map of the distribution of geological features mainly determined by climate, i.e. heat and humidity: the zonal geological units;
- a superposition of the two maps giving the global distribution of engineering geological types.

8.2.1 Geostructural units

In the evolution of the Earth's surface a primary role has been played by the interaction of endogenetic, that is internal, geological processes that have determined the broad structures of the continents and ocean basins. These have outlined the main geological regions of the globe largely controlled by plate tectonics with associated tectonic and neotectonic activity. Neotectonic refers to the study of post-Miocene structures including volcanic and seismic activity and uplift of the main mountainous areas.

It is relatively easy to subdivide, for example, the continents into cratons, orogenic belts and riftogens, with the possibility of further division of these basic regions.

Cratons

A craton is a part of the Earth's crust that has attained stability and, after the initial episodes of intense sedimentary, tectonic and igneous activity, has now reached a period of quiescence associated with slight continuous uplift, leading to the formation of a plain or upland topography. Extensive central cratons of the continents are composed of shields and platforms.

A shield is a large area of exposed basement rocks in a craton, surrounded by sediment-covered platforms. The rocks of most shield areas are of Precambrian age. In contrast, a platform comprises flat-lying or gently deformed strata, mainly sedimentary, which may be some kilometres thick resting on a basement of older rocks.

Cratons may be 'old', in which the rocks involved are predominantly of Precambrian age, and 'young', composed of phanerozoic rocks.

Orogenic belts

Orogenic belts, or orogens, are linear or arcuate foldbelts that have been subjected to thrusting, folding and faulting in the outer, higher layers and plastic folding, metamorphism and plutonism in the inner and deeper layers. Mountainous topography may be formed on orogenic belts, but only in the youngest mountains is there likely to be a causal relation between rock structure and surface landscape.

Riftogens

Riftogens, or rift valley systems, are long narrow continental troughs bounded by normal faults; they are graben of regional extent. Riftogens are superimposed structures of an individual type and are not subdivided by the nature of their substratum.

8.2.1.1 Types of geostructural unit

The following types of geostructural unit occur:

I	PRECAMBRIAN CRATONS
Ii	Shields
Iii	Platforms
II	PALAEOZOIC CRATONS
IIi	Shields
IIii	Platforms
III	OROGENS AND RIFTOGENS
IIIi	Epicratonic on folded Precambrian and Palaeozoic basement
IIIii	Epicratonic on Mesozoic folded basement
IIIiii	Epigeosynclinal on Mesozoic–Cainozoic folded basement.

There are two types of geostructural region, cratons and orogens-and-riftogens; the latter are grouped because they are both large mobile zones of the lithosphere characterized by intensive, often highly differential neotectonic movements leading to prevailing uplift into medium to high mountain relief. Cratons, on the other hand, are large blocks of lithosphere characterized by a plain or upland topography maintained by slight, predominantly positive, crustal movements.

Each type can be described in considerable detail in terms of rock types present, engineering properties, hydrogeological conditions, relief, geodynamic features including as appropriate seismic and volcanic activity; the list is a long one that demands a descriptive memoir to deal with adequately.

8.2.2 Regional geological map

The three major elements of the structure of the continents are shown in Figure 8.1, in which cratons are not differentiated on the basis of geological age into Precambrian and Palaeozoic. Sergeev and Ershova (1989, pp. 18–39) have recently published a version of this map in colour (their Fig. 17), and versions of maps of each geostructural unit in which, for example, cratons are differentiated into Precambrian and Lower Palaeozoic shields and platforms, and orogenic belts are differentiated into three types of orogen (their Figs 10 and 11).

8.2.3 Zonal geological units

Natural factors that have influenced zonal geological conditions are external, or exogenic, in character. They are mainly the properties of rocks and soils; the phase-state, chemical activity and depth of groundwater; external physical processes such as weathering, erosion, deposition, salinization, geocryological processes, etc.; distribution and types of Quaternary deposits; and relief.

Under the impact of climatic conditions, superficial deposits and soils are formed, relief is modified, swamps develop, river systems generated, rocks and soils are wettened, dry out, or freeze and thaw, glaciers grow or decline, groundwaters and superficial deposits become salinized, and water aggressiveness changes. These processes are of great engineering geological significance, affecting and modifying the upper few tens of metres of the ground that are of particular concern to the engineering geologist.

The following zones and subzones are recognized:

A PERMAFROST AND ICE
- A.1 Glaciers, permafrost without seasonal thawing
- A.2 Continuous permafrost with seasonal thawing
- A.3 Discontinuous permafrost with seasonal thawing

B HIGH MOISTURE CONDITIONS
- B.1 With seasonal freezing
- B.2 Without seasonal freezing

C MODERATE MOISTURE CONDITIONS WITH EVIDENCE OF SALINIZATION
- C.1 With seasonal freezing
- C.2 Without seasonal freezing

D LOW MOISTURE CONDITIONS WITH SALINIZATION
- D.1 With seasonal freezing
- D.2 Without seasonal freezing

Two main zonal types are distinguished: permafrost and unfrozen. In unfrozen rocks there are three zones: a humid zone with abundant moisture, an arid zone with marked water deficiency and salinization, and a transitional zone. Each zone in the belt of unfrozen rocks is divided into two subzones: that with seasonal freezing, occurring chiefly in the temperate climatic zone, and that without seasonal freezing, mainly in the tropics.

Ershova and Sergeev (1979) give examples of the characteristics of three subzones.

8.2.4 Map of zonal geological units

The four major zones are shown in Figure 8.2, with the permafrost-and-ice zone subdivided into the three subzones.

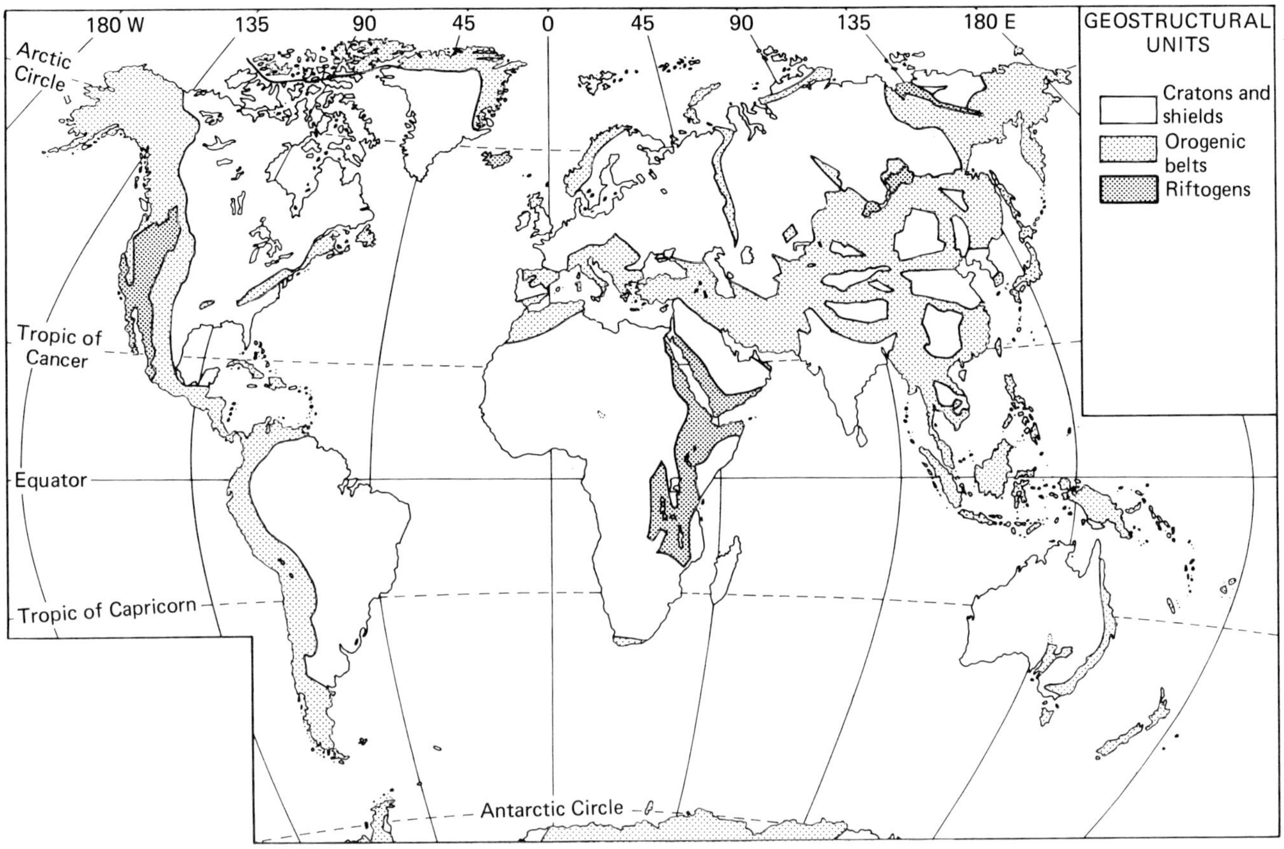

Figure 8.1 General map of geostructural units on the continental land masses of the Earth

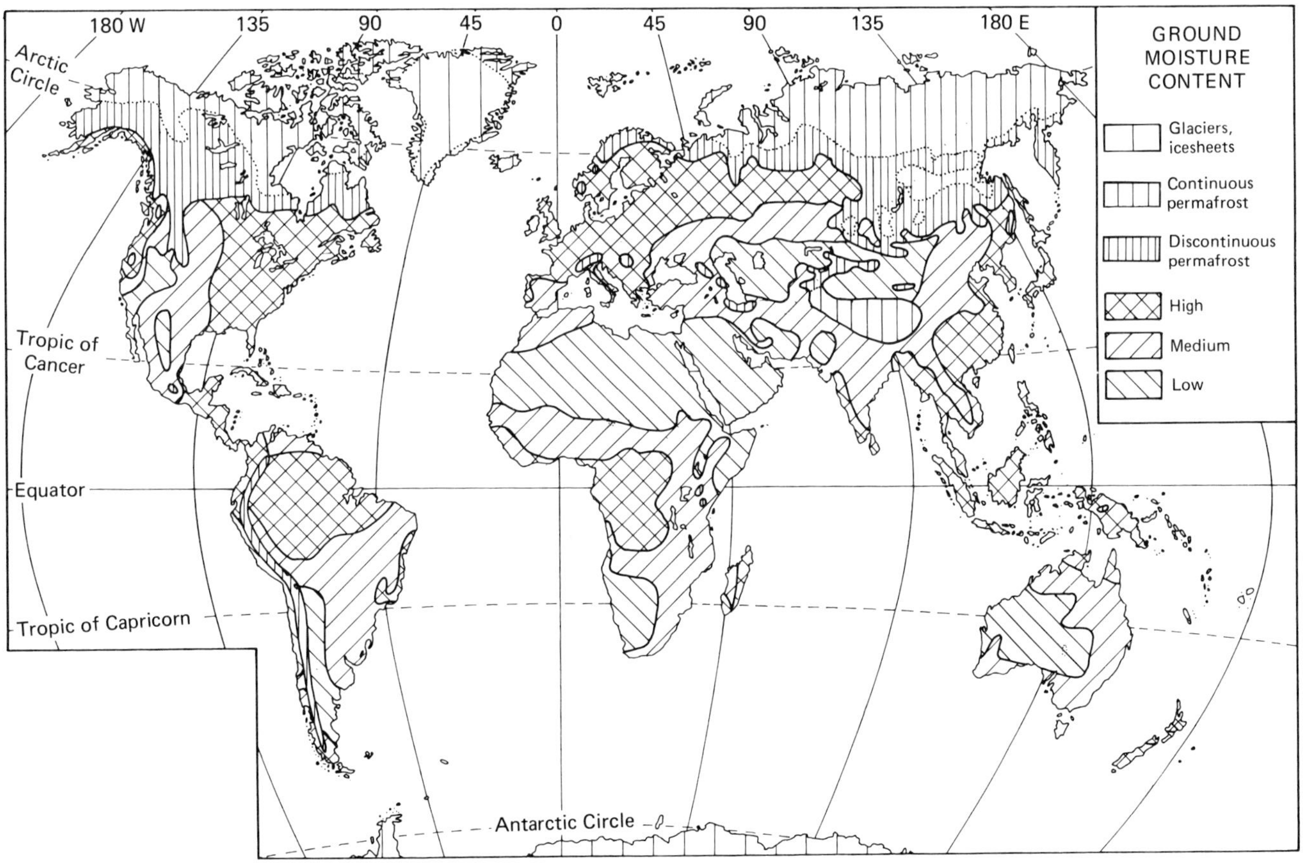

Figure 8.2 General map of zonal geological units of the Earth. Only the zone of permafrost-and-ice is divided into subzones as an example

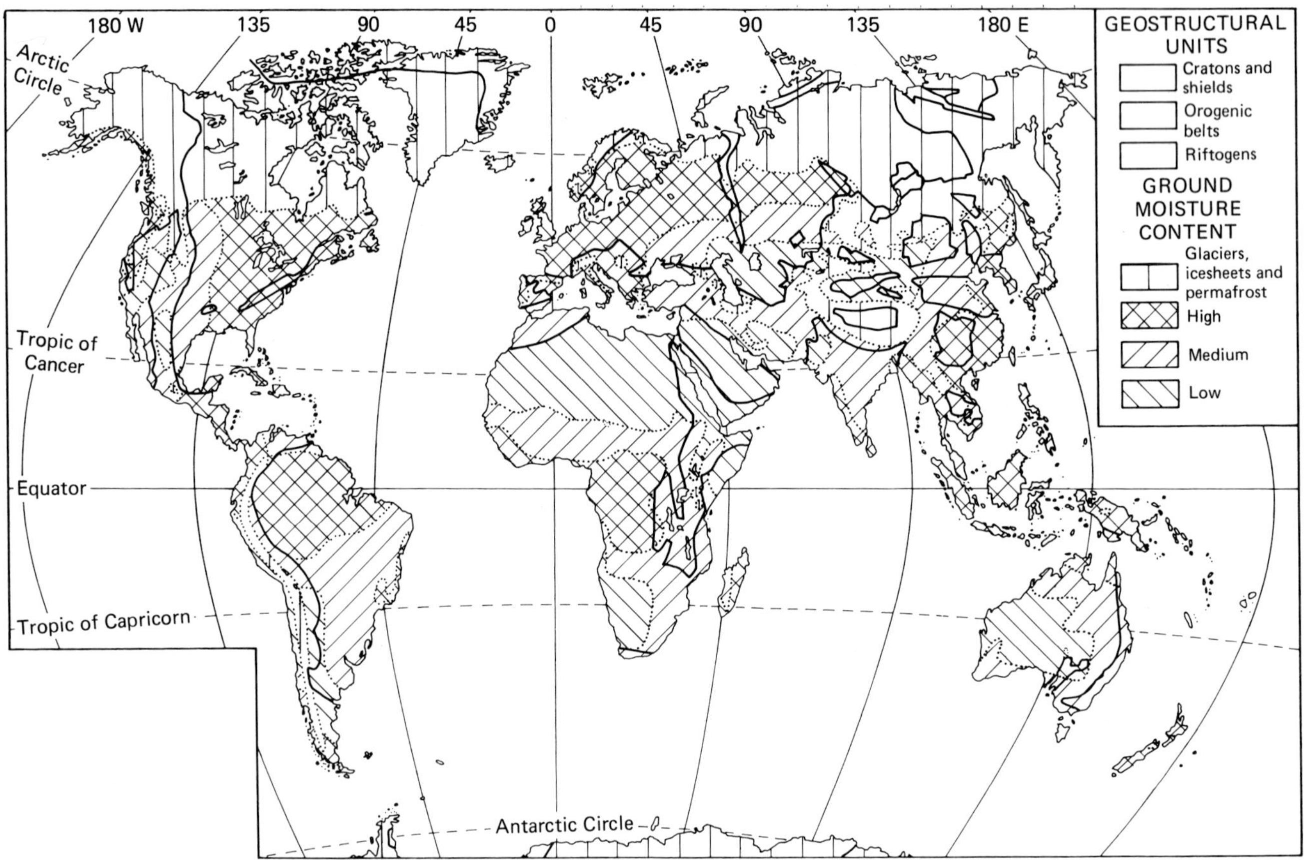

Figure 8.3 Map of selected engineering geological types on the Earth, produced by superimposing Figures 8.1 and 8.2. As printed, the map scale is 1:200 000 000 approx.

8.2.5 Map of engineering geological types on the Earth

Superimpose the two maps in Figures 8.1 and 8.2 and the result is a map (Figure 8.3) of the continents zoned in terms of engineering geological types, displaying 63 of the 82 distinct units on the complete map (Ershova and Sergeev, 1979, Fig. 1). Each engineering geological type is a combination of a specific type of geostructural unit with an equally specific zonal geological unit largely determined by climate.

There is one important proviso. It must be remembered that the zonal distribution of the natural exogenic engineering geological factors over the Earth is particularly marked in areas with flat topography and only slight neotectonic movements, typified by cratonic regions. In these regions the zonal geological factors are particularly important for the development of characteristic engineering geological conditions. This is not the case in regions of intense neotectonic movements, typified by orogens and riftogens, where the zonal distribution is disturbed by vertical zoning.

Besides, mountainous relief is distinguished by intense gravitational mass movement and erosion that rapidly remove chemical weathering crusts in tropical climatic areas. Instead, physical weathering prevails with intense mechanical disintegration.

It might be argued that publication of a map of engineering geological types of the world can have little practical application. However, each type on the map is characterized by specific regional geological and zonal geological features, enabling consistent and objective distinction of engineering geological types featuring similar engineering geological conditions. Wherever a particular type occurs, the engineering conditions will be the same. Thus it should be possible to codify directions for engineering geological mapping and allied investigation techniques, with general standards for construction and planning. This, it is considered, is of particular interest for developing countries since it will enable them, when planning construction, to apply the experience gained elsewhere in areas with similar types of engineering geological conditions.

Engineering geological map of Slovakia

The engineering geological map of Slovakia at a scale of 1:500 000 (Matula, 1969) has already been described in dealing with zoning maps (Chapter 7, Figure 7.6). It will be recalled that it is essentially a geological map with additional information on the hydrogeological situation, contemporary geodynamic processes, and a division into four main geological divisions. Geomorphological characteristics are listed in the legend to the map, and descriptions of engineering geological conditions are exhaustively dealt with in an accompanying memoir.

No further description is needed here, although it may be remarked that the approach adopted is that used for compiling the engineering geological map of the world, as is to be expected.

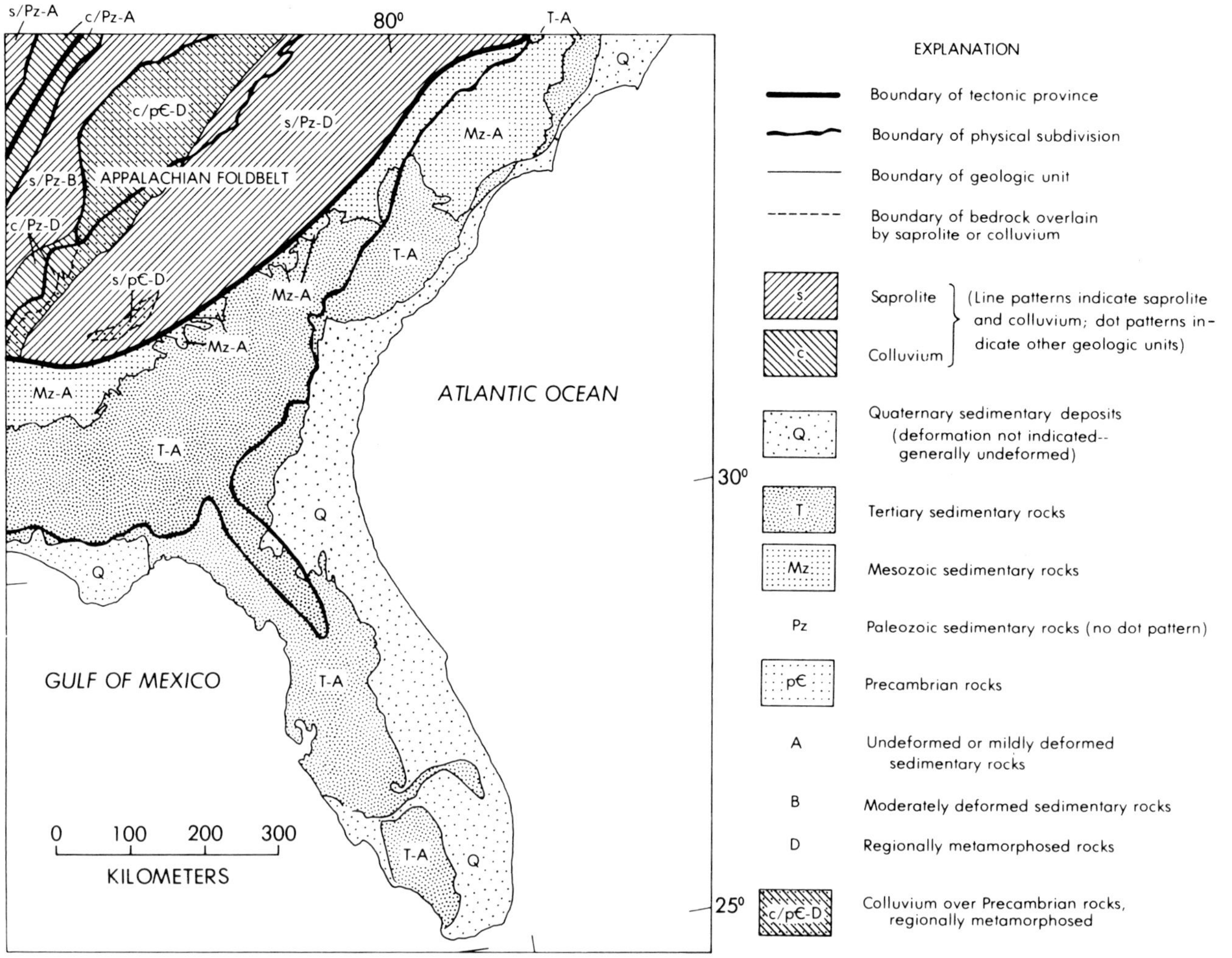

Figure 8.4 Segment of the base geological map compiled for the engineering geological map of the USA, scale 1:7 500 000 (From Radbruch-Hall, Edwards and Batson, 1979, Fig. 1, by permission)

8.3 Engineering geological map of the USA

Much more specific in both purpose and content is the series of experimental engineering geological maps of the conterminous USA prepared by computer methods (Radbruch-Hall, Edwards and Batson, 1979). As conceived, the three maps were at a scale of 1:7 500 000 but they are reproduced in the paper at a scale of approximately 1:21 000 000. The set of maps includes a base geological map also showing the intensity of certain geological conditions, processes and topographic conditions that might pose constraints to construction; a map showing which hazards or combinations of hazards exist at any point on the map; and a map showing the areas where construction or land development may worsen existing hazards and/or cause secondary environmental effects.

Here then are specific criteria that had to be assessed, evaluated and combined to produce the final maps. How was it done?

Because of the small scale, the base geological map (Figure 8.4) only shows the distribution of sedimentary, plutonic and volcanic rocks, together with the Quaternary alluvium. In the restricted region of the Appalachians, colluvium and saprolite are shown because these are very thick superficial deposits, and there the underlying solid rocks are indicated only by symbols. Sedimentary rocks are divided by age, because younger rocks are generally less consolidated and less deformed than the older. Quaternary and Tertiary volcanic rocks are also differentiated. Geological complexity, indicated by a letter symbol, is classified into four grades on the basis of the following criteria:

- undeformed or mildly deformed – dips generally less than 20°;
- moderately deformed – faulting does not obscure the pattern of lithological or major structural units;
- severely deformed – includes areas of closely spaced faults that obliterate lithological and structural continuity;
- regionally metamorphosed.

Boundaries were drawn on the map around geological units, physical subdivisions and tectonic provinces.

Once the geological base map had been prepared, areas were defined that, because of topographic and geological conditions, presented constraints to construction. This engineering geological map took account of the effects of landslides, karst, volcanic hazards, horizontal acceleration caused by seismic activity, standing water and steep slopes. Maps were available for all these

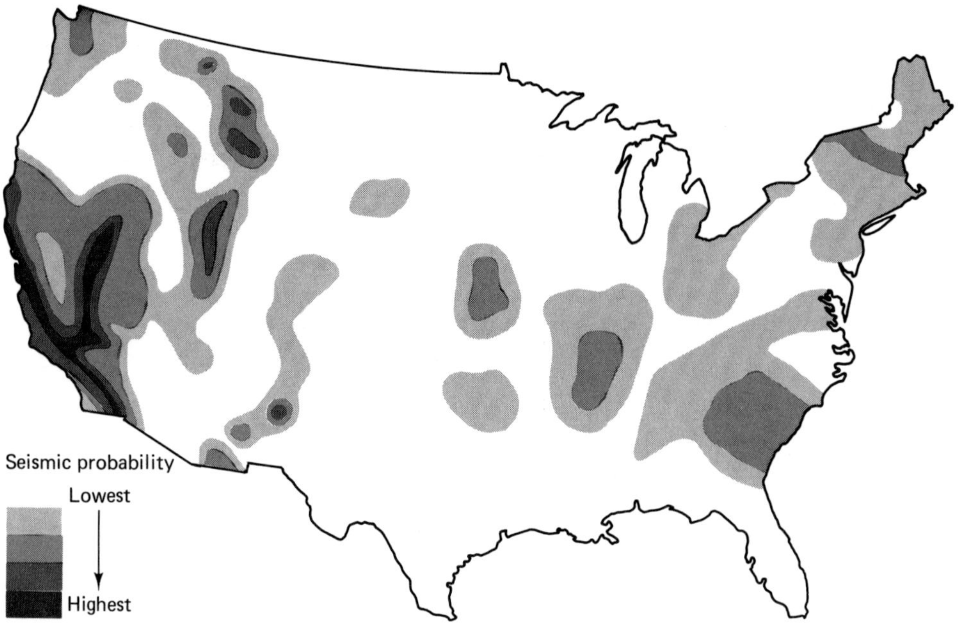

Figure 8.5 Seismic probability map generated by computer. This map is one component used in preparing the engineering geological map of the USA. Areas of greatest probability are indicated by the darker shade. Scale approximately 1:21 000 000 (From Radbruch-Hall, Edwards and Batson, 1979, Fig. 2, by permission)

(e.g. Figure 8.5) and a comprehensive list of other factors was given that would have been taken into account if the data had been available.

How to combine all these features on derivative maps posed the problem of relative weighting. Three criteria were used in the assessment: severity, permanence and recurrence, and difficulty of amelioration.

The basis for evaluating each of these criteria, giving each class specific numerical values, was as follows:

I *Basis for evaluating severity*:
- (a) Can cause some engineering problems; there may be rare or minor damage to property.
- (b) Definite engineering problems: some damage to property, there may rarely be loss of life.
- (c) Can cause severe engineering problems; there may be extensive damage to property; some loss of life.
- (d) and (e) Much extensive property damage and loss of life. Occurs only on map of seismic probability.

II *Basis for evaluating permanence or recurrence*:
- (a) A condition or process considered to be rare or sporadic (not known to occur within historic time, which is arbitrarily set at 4000 years).
- (b) Intermittent (has been known to occur during the last 4000 years).
- (c) Permanent or continuing.

III *Basis for evaluating estimated difficulty of amelioration*:
- (a) Could probably be ameliorated in the course of construction.
- (b) Might be ameliorated but with great difficulty or at great expense.
- (c) Problem can be handled best by avoidance.

Maps were compiled showing:

- the distribution of various geological and topographical factors that pose constraints to construction;
- the various degrees of severity of a number of combined geological and topographical factors that pose constraints to construction
- areas where surface construction may cause direct and indirect damage to the physical environment.

The resultant maps (Radbruch-Hall, Edwards and Batson, 1979, Figs 3–5) show clearly and graphically the areas of the USA where the most restrictive geological conditions exist for construction, within the limits of the data used. Where the combined map can be superimposed on the geological base in colour, it can also show the relation of the restrictive conditions to the tectonic provinces, physical subdivisions, geological complexity of the rocks, and the rock and soil units. In using the map, however, it is not possible to tell which specific problems exist for any given spot. For this purpose, the second map of the set shows in a combination of shades of grey and black and white pattern the extent of the various individual problem conditions and processes as shown on the component maps. In so far as the data from which this map was prepared were stored in the computer, the map was easily generated from the already available data.

A third map was prepared from the same data by evaluating the map areas on the basis of the influence construction might have on the geological and topographical factors being considered. For example, volcanic hazards were not evaluated, as construction on or near a volcano will have no influence on its activity. All categories of landslide amounts and susceptibility were given a high rating for potential environmental damage, inasmuch as construction could both increase the amount of sliding and have secondary effects such as clogging streams with debris. Areas of standing water were given a lower rating because draining such an area for development will not increase the amount of standing water, but may have secondary effects such as causing subsidence or altering groundwater conditions.

After re-evaluation of the map areas on this basis, an environmental map was reproduced from the stored data. It shows where construction might cause damage to the environment, either directly, by worsening the condition or process, or indirectly, by damaging other aspects of the environment.

It was emphasized, in conclusion, that these maps cannot be used for selection of specific engineering sites, but are useful for preliminary planning of large-scale national construction programmes such as major highways, pipelines or hydroelectric projects.

8.4 Content of general engineering geological maps

These examples serve as a guide to what can be shown on engineering geological maps. It is worth while, therefore, to attempt a critical summary of the maps as a basis for formulating the possible contents of the engineering geological map of the UK.

All the maps take account, either completely or in part, of the four basic aspects of the geological environment that are of particular importance to engineering geology. This should be a prerequisite for a general-purpose engineering geological map.

To recapitulate, in particular relation to the UK, those requirements of paramount importance are:

1. *The character of rocks and soils*
   For each mapping unit, rocks and soils should be defined in terms of:
   (a) lithological type or types,
   (b) structural complexity,
   (c) relation between superficial and solid deposits which are in turn related to:
   (d) geological age, and
   (e) genetic group.
2. *The general hydrogeological conditions*
   For each mapping unit, hydrogeology depends on whether the unit can be regarded as:
   (a) an aquifer,
   (b) an aquiclude;
   and in aquifers whether the groundwater is
   (c) unconfined, or
   (d) confined;
   and for groundwaters:
   (e) the chemical character of the water.
   Depending on the prevailing lithological types and structure of each map unit, the general occurrence of:
   (f) springs and seepages,
   (g) absence of surface water, for example in limestone regions,
   (h) presence of surface water or areas with a very high watertable, should be particularly noted.
3. *The general geomorphological conditions*
   Should be defined for each mapping unit, or groups of mapping units as appropriate, and should indicate:
   (a) the character of the landscape which can most readily be achieved by preparing:
   (b) a slope map.
4. *The general geodynamic conditions*
   On the map should be a selection of active geomorphological and other earth processes that are particularly relevant to engineering construction. The following should be included:
   (a) areas of active or dormant landslides,
   (b) areas of active coastal erosion or accretion,
   (c) areas susceptible to flooding,
   (d) karstic areas,
   (e) areas of active or dormant man-induced subsidence.

It should be noted that one of the map series, that of the USA, attempted an analysis of the interplay of these four aspects of general geological conditions with civil engineering construction. But this analytical aspect of geotechnical mapping is surely a necessary sequel to the compilation of factual data, at least for broad-scale planning for which the general engineering geological map is designed (Matula, 1979; Matula and Letko, 1980).

8.5 General engineering geological map of the soils and rocks of the UK

A very broad cover of the geology of the UK is afforded by the two sheets of the solid geology of the whole country published at 1:625 000. Supplementing the latter at the same scale are those dealing with sand and gravel resources (Ordnance Survey, 1965) and with the superficial cover of Quaternary and Recent deposits (Institute of Geological Sciences, 1977). In various published atlases, smaller scale versions are available down to scales as small as 1:5 000 000.

These are conventional geological maps and an attempt has been made to derive from these bases a series of general engineering geological maps.

8.5.1 Basic rock and soil types and combinations of types relevant to the bedrock

Complete lithostratigraphical geological maps of the UK are readily available at suitable scales for the projected engineering geological map. Without considerable background knowledge or by reference to a comprehensive memoir, however, it is usually impossible to determine with certainty the lithological types of rock and soil present in the different map units. Some method of representation has to be devised for this specific purpose, which should enable not only the component rocks and soils but also the relative proportions of each lithological type to be determined.

After much experimentation a simple answer was found already published in the *Atlas of the British Isles* (Bickmore and Shaw, 1963). A simplified geological map at a scale of 1 : 2 000 000 had been overprinted with a diagonal pattern of ruled lines. Excluding superficial deposits, the predominant types of rock were grouped as argillaceous, arenaceous, calcareous, metamorphic and igneous. For each of the sedimentary groups, subgroups were set up in order of increasing consolidation, indicated by a letter and number combination, with the metamorphic representative of each group being the most consolidated. Argillaceous rocks were coloured grey, with slates and greywacke a darker grey; arenaceous were yellow, and calcareous blue. The 16 lithological types are listed in Figure 8.6.

Proportions of the basic lithological types in each map unit were represented by a stripe system by which 11 different proportions of one, two or three lithological types were clearly differentiated. This stripe system should not be confused with the stripe method of three-dimensional representation as developed in Czechoslovakia (Pasek, 1968; see also Chapter 6). By the use of colour very small areas on the map could be quite clearly distinguished and it has to be admitted that the map sheet is very successful in giving an immediate impression of the distribution of lithological types over the country.

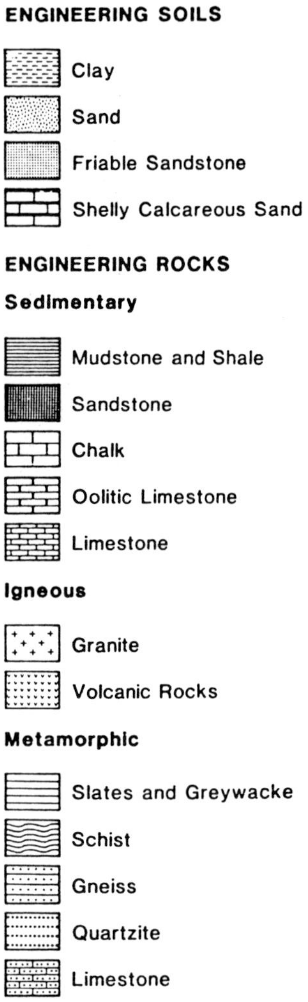

Figure 8.6 Symbols for engineering soils and rocks shown on the engineering geological map

This elegant solution to the representation of the varying proportions of a selected group of lithological types in each unit on a geological map of the UK depends to a large measure for its success on the use of colour. For the present publication, colour had to be translated into a small number of distinctive black and white patterns. An inevitable consequence was that the number of mapped outcrops had to be reduced because the small areas occupied by some lithologies, for example basic igneous rocks, were too small to be effectively represented. Only slight simplification was however necessary in the choice of lithological types, and 17 predominant types were reduced to 16 by grouping hard sandstone with quartzite.

The final selection of lithological types had to take account of an accepted classification of rocks and soils for engineering geological mapping, and lithological types of particular significance in the context of the geology of the UK.

8.5.2 Rock and soil classification adopted for the UK engineering geological map

On the map of bedrock, or the pre-Quaternary or Solid Formations, 16 distinct lithologies have been recognized: four engineering soils, five sedimentary rocks, two igneous rocks and five metamorphic rocks (Figure 8.6). Spanning an age range from Precambrian to Pliocene, the names reflect both the predominant lithologies present and some special types. The latter are used because they are either unique to a particular stratigraphic horizon or give a special flavour to the geology of a part of the UK. Oolitic limestone, so redolent of the buildings and farming industry of the Jurassic belt of England (Clifton-Taylor, 1972), is sufficiently different from chalk and the older limestones to justify special inclusion on the map. Slates and greywackes are characteristic of highly deformed older geosynclinal areas. Shelly calcareous sand typifies the Pliocene to Pleistocene Crags of Norfolk. Friable sandstone as an engineering soil is sufficiently different in its engineering properties from the engineering rock, sandstone, to justify its inclusion.

Thus there is about the map a certain illogicality in the selection of some of the mapping units, rather than a rigid doctrinaire application of accepted principles.

Sixteen distinct lithologies are used in different combinations and different proportions to differentiate many more engineering formations (Figure 8.7). Only 11 of the 16 distinct lithologies define individual map units; 19 other units are combinations of two or three basic types in different proportions (Figure 8.8).

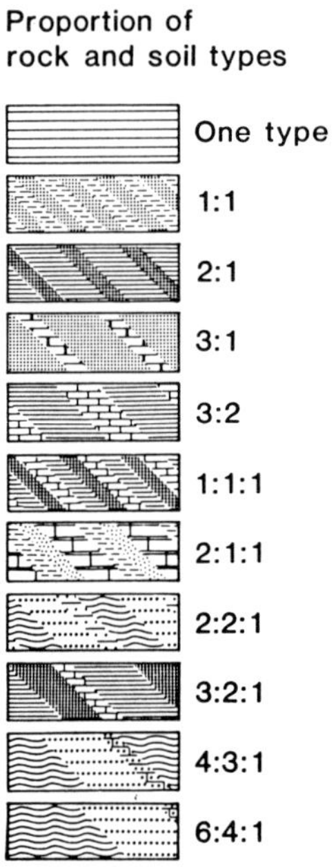

Figure 8.7 Method adopted for showing the proportions of the different rock and soil types in individual units on the engineering geological map

Limitations of the classification

Selection of the 16 lithological types was influenced chiefly by cartographic problems due to the use of a black and white method of reproduction rather than colour.

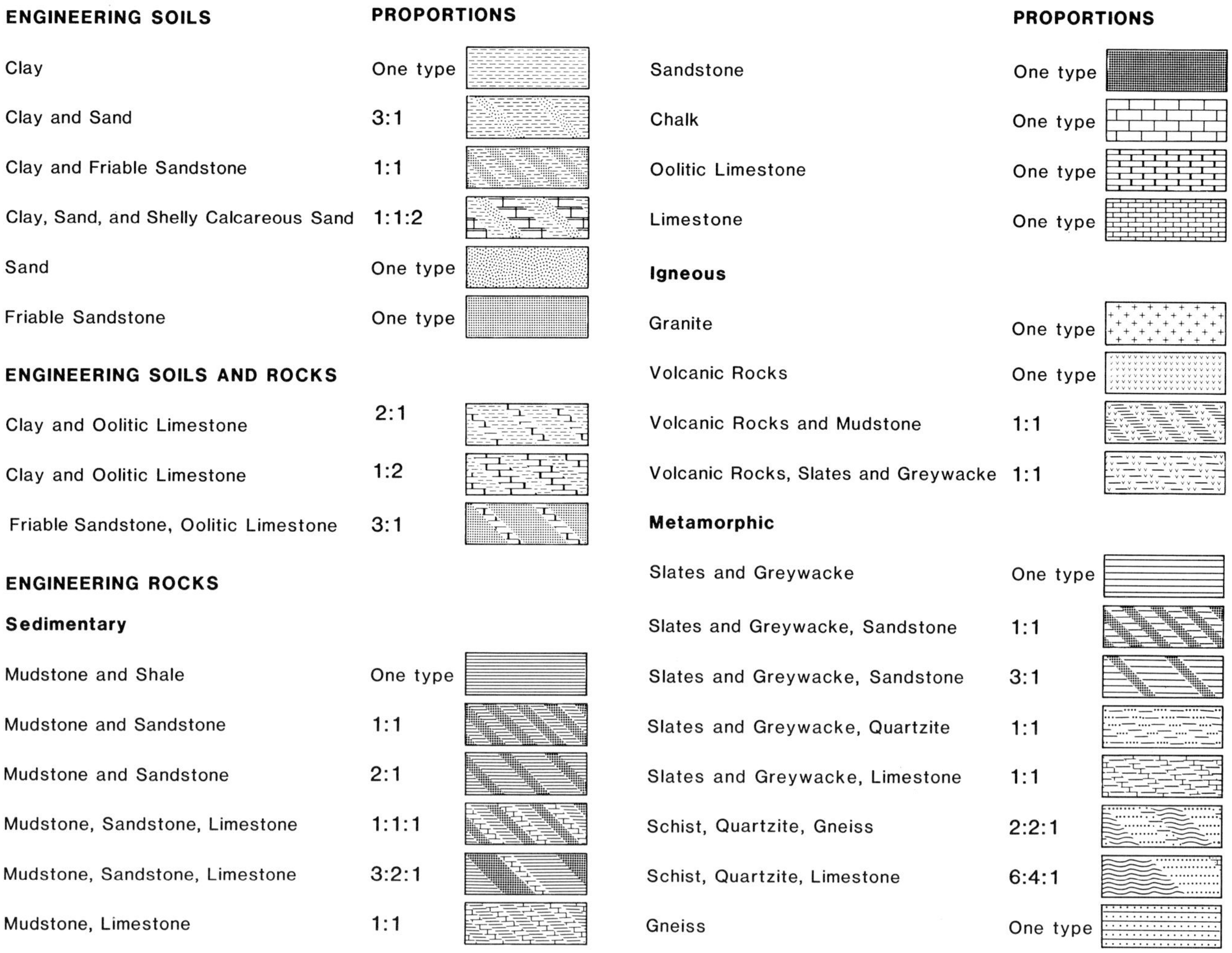

Figure 8.8 Symbols used for the 31 distinct mapping units on the engineering geological maps

Igneous rocks are limited to granite and volcanic rocks; basic intrusions, and there are some economically important ones including the Great Whin Sill – a dolerite sill of very great extent in north-east England that on the map could only be represented by a sinuous line – have been omitted except for a few coarse-grained bodies that are indicated as granite. Pyroclastics are included with volcanic rocks, as are the fine-grained igneous rocks and volcanic glasses.

No contact metamorphic rocks associated with large igneous intrusions are shown, or tectonic breccia or mylonite. Apart from these, the other groups of metamorphic rocks (see Table 4.1) are well, but not completely, represented.

Choice of sedimentary rocks is even more selective. Mudstone and shale, sandstone and three types of limestone are all that are represented. Greywacke, that elsewhere has been included as a sedimentary rock (Pettijohn 1957; Dearman, 1974), is here regarded as a metamorphic rock. Because of its frequent association with slate, one of the mapping units is the engineering formation 'slates and greywacke'.

Engineering soils in the Solid Formations are simply divided into fine and coarse soils. Fine soils are represented as clay, whereas coarse soils are restricted to three types of sand – sand, friable sandstone and shelly calcareous sand.

The very simple classification of rocks and soils used on the engineering geological map of the UK is set out in Figure 8.9.

8.5.3 Engineering geological maps

A series of maps has been prepared and, for such a relatively small land mass, the geological conditions are extraordinarily varied and complex. In some parts of the country, bedrock is almost completely obscured by Quaternary and Recent superficial deposits. Consequently the distribution of lithological types in the bedrock can only be shown effectively if the superficial deposits are imagined to be stripped off. A general impression of bedrock in terms of four classes of structural complexity (Figure 8.10) is supplemented by a detailed presentation of the distribution of lithological types within the mapped engineering formations (Figure 8.11).

Superficial deposits are dealt with in two ways: as a direct overlay obscuring part of the bedrock (Figure 8.12) and as a separate map of superficial engineering soils (Figure 8.13).

Finally a map sheet has been prepared of the distribution of engineering soils regardless of whether the deposits are members of the bedrock or are superficial deposits (Figure 8.14).

Sedimentary			Metamorphic		Igneous	Genetic group		
Soils	Rocks		Foliated	Massive				
SAND Friable sandstone Shelly calcareous sand	SANDSTONE	LIMESTONE Oolitic limestone	GNEISS	QUARTZITE	GRANITE	Coarse-grained	More than 0.06	Grain size (mm)
			SCHIST	LIMESTONE				
CLAY Till	MUDSTONE and SHALE	CHALK	SLATE	Greywacke	VOLCANIC ROCKS Rhyolite Andesite Basalt	Fine-grained	Less than 0.06	

Figure 8.9 Classification of rocks and soils on the engineering geological map of the UK

8.5.4 Map of relative geological complexity in the bedrock

Following Radbruch-Hall (1979) one version of the map (Figure 8.10) shows the distribution of four grades of geological complexity based on these criteria:

1. *Undeformed or mildly deformed.* Very large folds in which the bedding on individual limbs has a regional dip of generally less than 20°. Faults, although present, are not an important element of the structure.
2. *Moderately deformed.* Large-scale folds with wavelengths measured in kilometres. Bedding may dip at more than 20°. Folds are broken by numerous faults that nevertheless do not obscure the pattern of lithological or major structural units.
3. *Highly deformed.* Folds are generally small scale, with amplitudes often less than 30 m; they vary in attitude from upright to recumbent, and the interlimb angle from open to tight. Beds are cleaved, and the folds are broken by faults.
4. *Regionally metamorphosed.* Associations of slates, schists, gneisses, with usually subordinate quartzites and limestones, are often intensely folded on both large and small scales. There may be more than one cleavage. The metamorphic rocks are intruded by granites.

8.5.5 Distribution of lithological types in the bedrock

There is a close relation between the geological complexity of the bedrock, as shown in Figure 8.10 and the distribution of engineering soils and rocks in Figure 8.11.

As is to be expected from the tectonic history of the country, the south-east is occupied by undeformed strata ranging in age from Permian to Pliocene. This is the only bedrock unit in which engineering soils – overconsolidated clays, sands, friable sandstones and shelly calcareous sands – are present as original constituents. Associated strata are limestones, predominantly chalk and oolitic limestone, but with an important Magnesian limestone unit.

Some undeformed strata are much older, including sandstone engineering formations of Precambrian and Devonian age in northern Scotland. But these have been included in the moderately deformed category because the small areas are either faulted or are associated with highly deformed or regionally metamorphosed rocks.

On the map, moderately deformed areas are usually associated with basins and domes many tens of kilometres across. Typically they contain associations of sandstones and shales in basins containing coal measures in which coals, seatearths and shales have been extensively exploited. Flanking the basins are sandstone–shale and sandstone–shale–limestone sequences that in some areas also contain local coals that have been worked. Anticlinal regions between major coal-bearing basins expose some of the major limestones in the UK.

Highly deformed areas are dominated by the slate-and-greywacke association, with volcanics and post-orogenic granites, marking the former positions of Lower Palaeozoic geosynclines. Within each area there is a marked grain to the geological structures with parallel fold axes maintaining a constant trend for many tens of kilometres. However, the plunge of folds varies within this constant trend, as does the attitude of the fold axial plane, the shape, and the complexity of the small-scale folds. Low-grade regional metamorphism has affected the rocks, and there are usually extensive contact metamorphic aureoles around the intrusive granites.

The largest areas of regionally metamorphosed rocks are found in the north of Scotland and in Northern Ireland. Structures, both large and small scale, are very complex with repeated folding and more than one cleavage, but despite this, individual units of limestone and quartzite can be traced for many tens of kilometres. Deformation and metamorphism have been accompanied by the extensive emplacement of both synorogenic and postorogenic granites.

8.5.6 Importance of glacial till

It is generally conceded that engineering soils are, in the context of civil engineering works in the UK, of especial importance. One of the most widespread types of

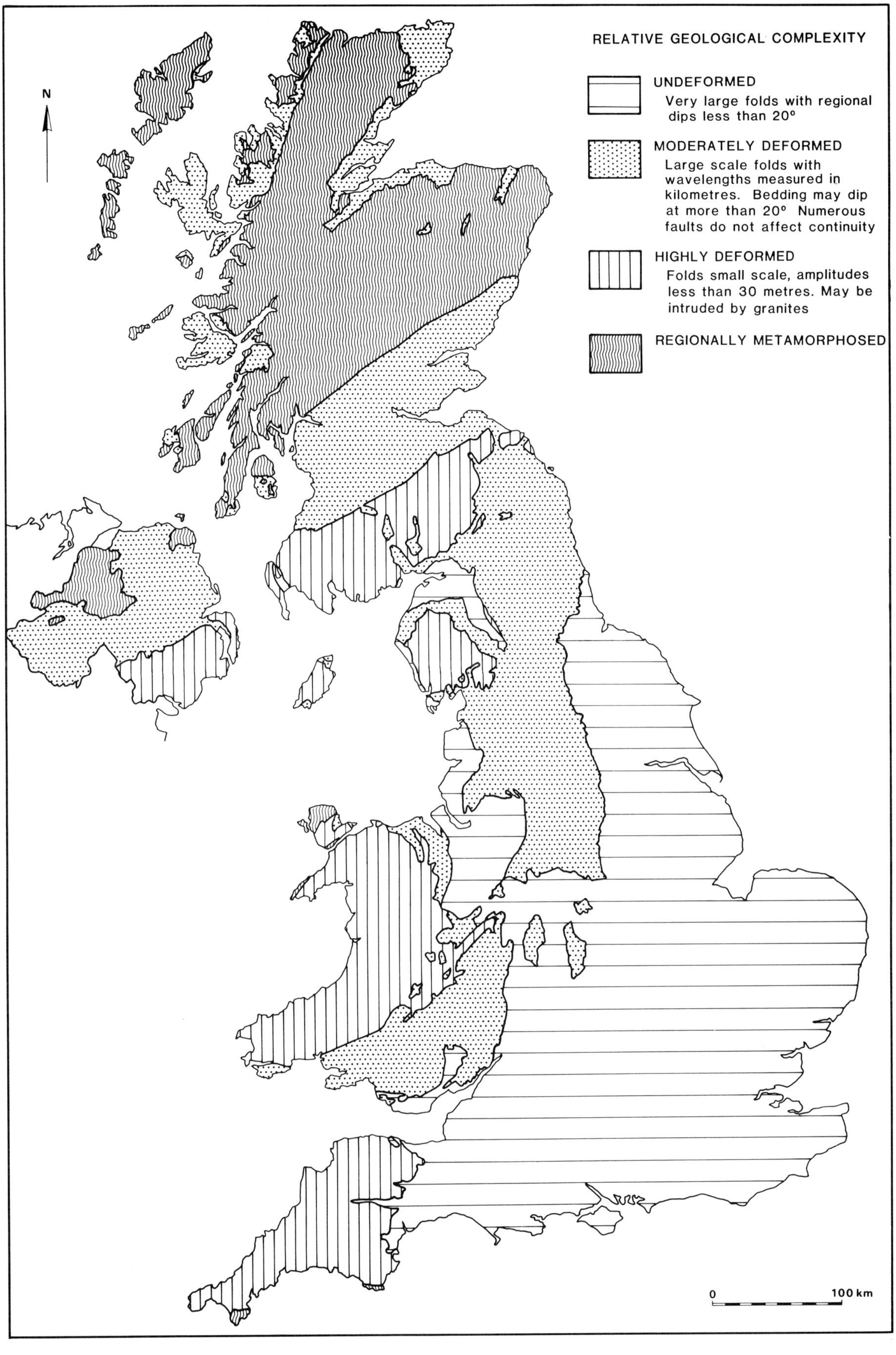

Figure 8.10 Distribution of the four classes of structural complexity in the bedrock

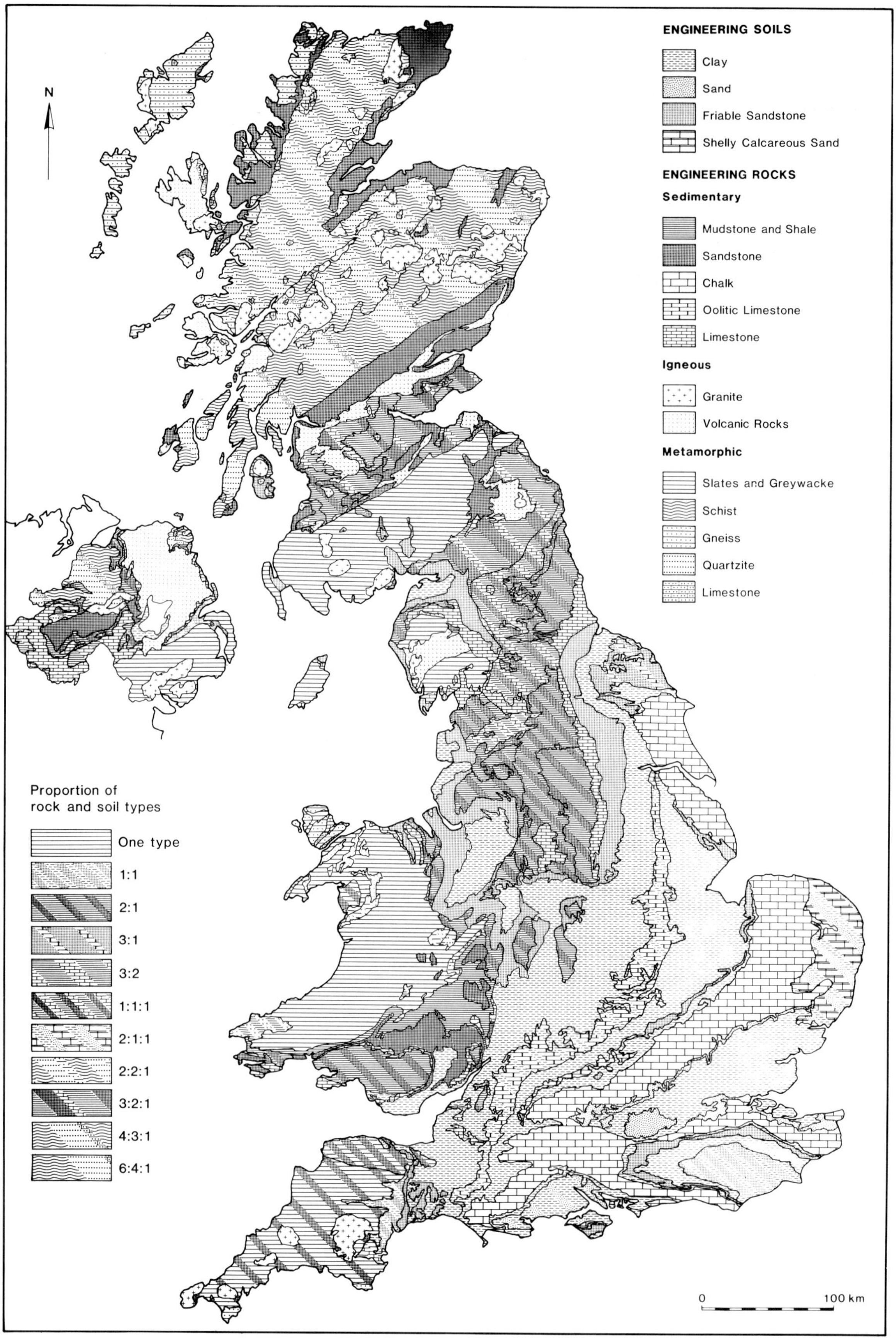

Figure 8.11 Engineering geological map of the solid or bedrock of the UK. Drawn at an original scale of 1:2 000 000, and reproduced here at approximately 1:4 000 000. (After Bickmore and Shaw, 1963: Map of Predominant Types of Rock)

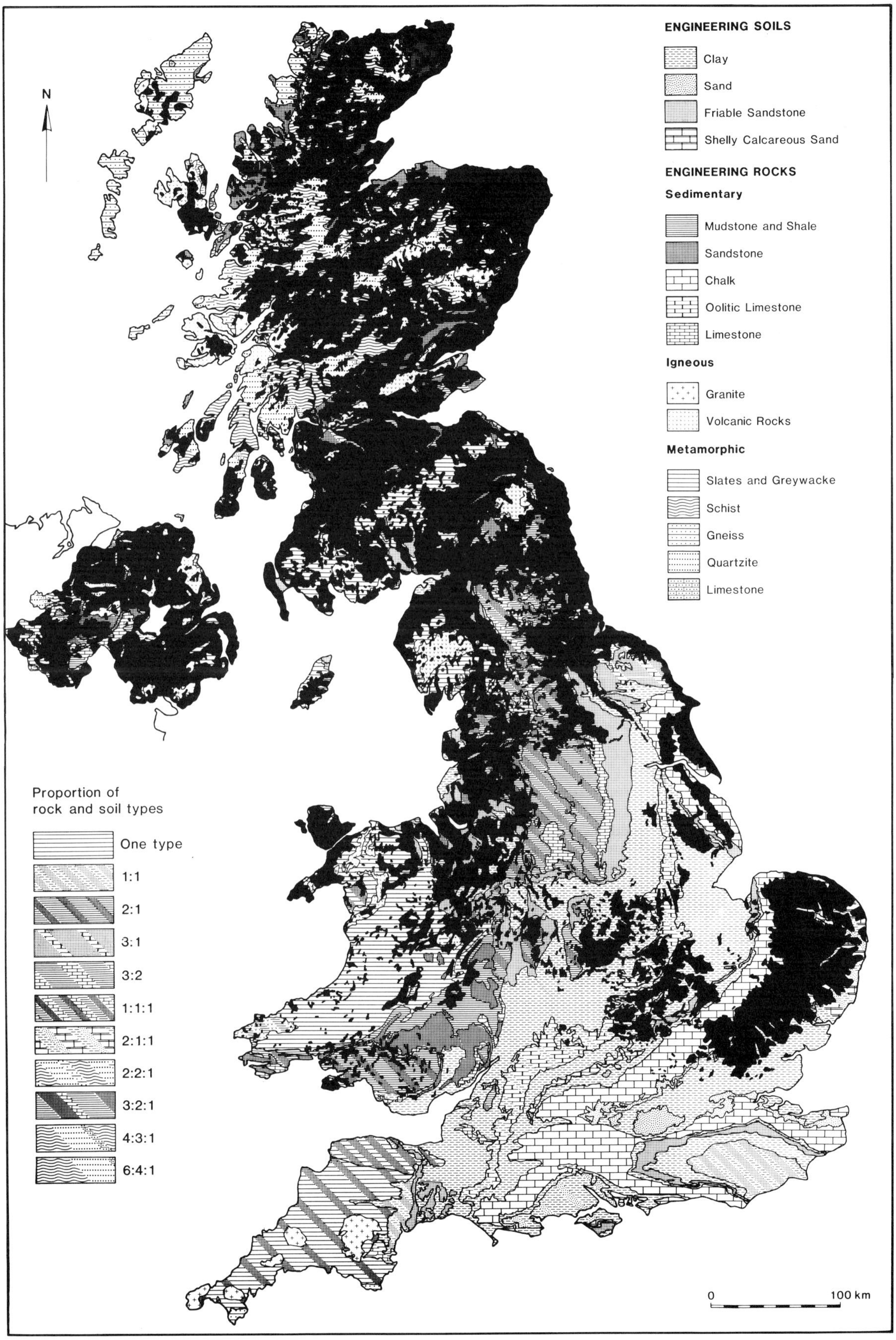

Figure 8.12 Map of the distribution of glacial till in relation to bedrock in the UK

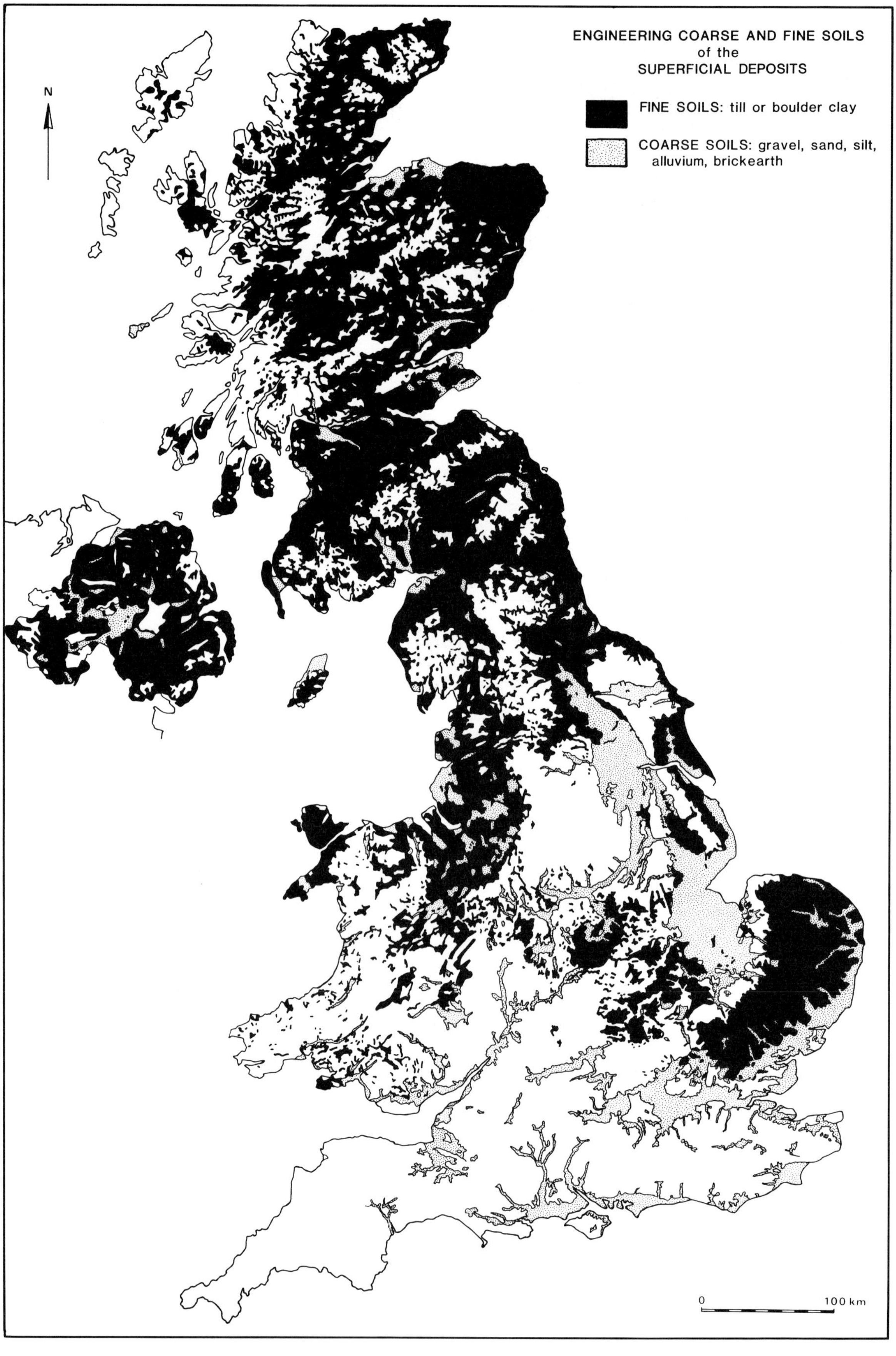

Figure 8.13 Map showing the distribution of engineering soils comprising the superficial deposits of the UK (After Bickmore and Shaw, 1963: Superficial Deposits)

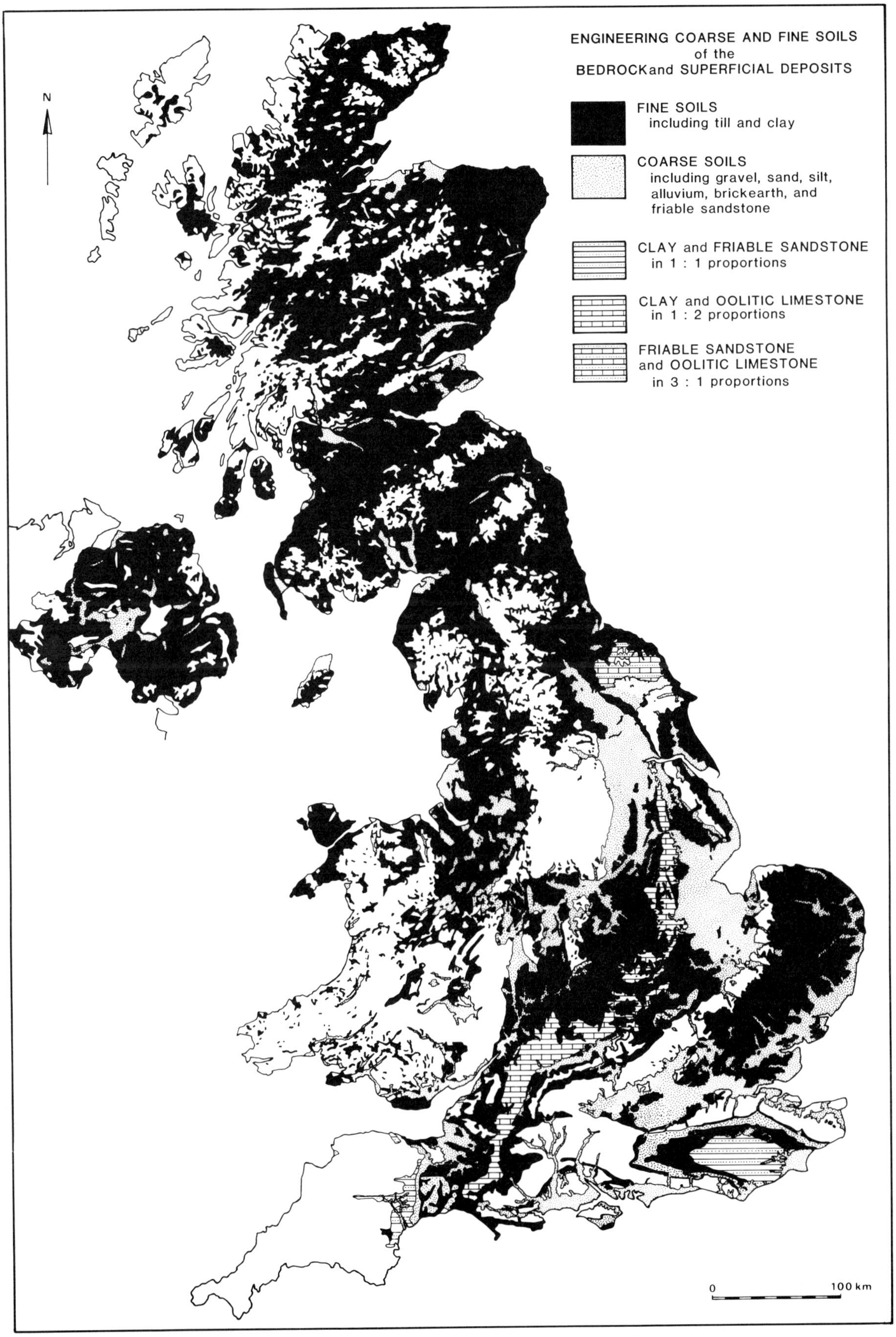

Figure 8.14 Distribution of engineering coarse and fine soils of the bedrock and superficial deposits of the UK

engineering soil is the various glacial tills (Eyles and Dearman, 1981) that mask so much of the bedrock or so-called solid geology of the whole island. Distributed as shown in Figure 8.12, glacial till obscures more than 50% of the bedrock of the northern half of the map. Although the map gives no indication of the thickness of the till, it must be apparent that the engineering properties of till would be of vital significance for a majority of engineering works in the area.

To the south, till deposits of any extent are restricted to lowland areas and in the far south are absent altogether. Extensive, continuous till cover is associated with former ice sheets. Sporadic and localized deposits of till are indicative of former valley glaciers in those generally unglaciated regions of high relief.

8.5.7 Distribution of superficial deposits

If the very extensive spreads of cohesionless soils, represented in the superficial deposits by gravel, sand, silt, alluvium and brickearth, are added to the map of glacial till on bedrock (see Figure 8.12), then the importance of engineering assessments of the superficial deposits is even more apparent. For clarity, in Figure 8.13 bedrock details have been omitted, and the extent of the additional cover of cohesionless soils, obscuring bedrock over large areas of central, eastern and southern England, is very impressive. At the scale of the map, some very large areas of bedrock are completely masked by the superficial deposits. Also, there are very large spreads of cohesionless soils in southern England beyond the limit of the till deposits.

8.5.8 General distribution of coarse and fine engineering soils

The final stage in producing a map showing the distribution of engineering soils has been to add on to Figure 8.13 the engineering soils found in the bedrock. This has been done on the last map (Figure 8.14), reinforcing the importance of engineering soils in the south of England. Account has been taken of two extensive outcrops, one – in the extreme south-east – of friable sandstones and clays, and the other the band of oolitic limestones and clays, clay and friable sandstone and oolitic limestone, running diagonally across the country from the south to the east coast.

8.5.9 Additional maps of engineering geological conditions

The map of the distribution of superficial deposits (Figure 8.13) simply discriminates between coarse and fine soils without any statement on the different types of soil, the relation of the deposits to bedrock, and any superficial structures that may be present.

The importance of those engineering soils of Britain represented by superficial deposits, ranging in age from the Quaternary to Recent, cannot be overemphasized. Soil types vary from consolidated bouldery silty clays: tills, through normally consolidated clays, laminated clays, silts, sands and gravels, to peats and organic soils. In fact the complete spectrum of engineering soils (Table 4.13) is represented in very variable associations resting on bedrock that may be a few metres to many tens of metres below the present land surface.

Associated with the superficial soils are superficial structures, involving both soil and rock, resulting from glacial and periglacial activity. Solifluction deposits with shear-based solifluction sheets and lobes, cambering, valley bulging, landsliding and frost wedging are just a few examples of superficial structures that pose geotechnical problems. Problematic ground conditions arise, for example, because of very rapid transitions from one soil to another contrasted type, the juxtaposition of rock and soil, and the presence of unexpectedly weak horizons.

There is an extensive literature on the geotechnical properties of tills in the UK. A selective review would highlight publications by Boulton and Paul (1976), Eyles and Sladen (1981), Fookes *et al.* (1978), McGown *et al.* (1978), McGown and Derbyshire (1977), Marsland (1977, 1979) and Tarbet (1973). Periglacial effects have been reviewed by Higginbottom and Fookes (1970). Slope processes, including cambering and valley bulging, feature in papers by Horswill and Horton (1976), Skempton and Weeks (1976) and Chandler *et al.* (1976). Weathering effects on the geotechnical properties of some bedrock soils and weak rocks have been quantified for the Keuper Marl by Chandler (1969), for the Lias by Chandler (1972) and for chalk by Ward and Burland (1968), Ward and Burland (1986) and Wakeling (1970). Highly disturbed soils on these three lithologies are all of a solifluction origin and hence are, in a sense, glacigenic in origin.

It appeared that the next step should be how to explore some of these aspects. A possible method involved:

- illustration in block diagrams of associations of glacial landforms and the component sediments;
- showing the significant landsystems so identified on general maps to provide a predictive tool in site investigation.

8.6 Glacial terrain map of Britain

On a new small-scale map of the Quaternary of Britain recurrent terrain types have been identified, based on work at the margins of modern glaciers. It has been possible to portray most of the sedimentary and geomorphological products in the non-marine glacial environment by a few simple block models. These models show, for example, the range of rockhead conditions, the extent of the overburden and the penetration of overburden into rock, and the occurrence of specific soil types and landforms.

The dominant soil-producing Quaternary event in Britain has been that of episodic glaciation depositing offlapping, progressively younger till sheets with a maximum southward extent during the first or Anglian glaciation marked approximately by the Bristol Channel to the Thames line (Figure 8.15). Within the limit of the latest phase of glaciation, the Devensian, freshly constructed, relatively undissected highland and lowland

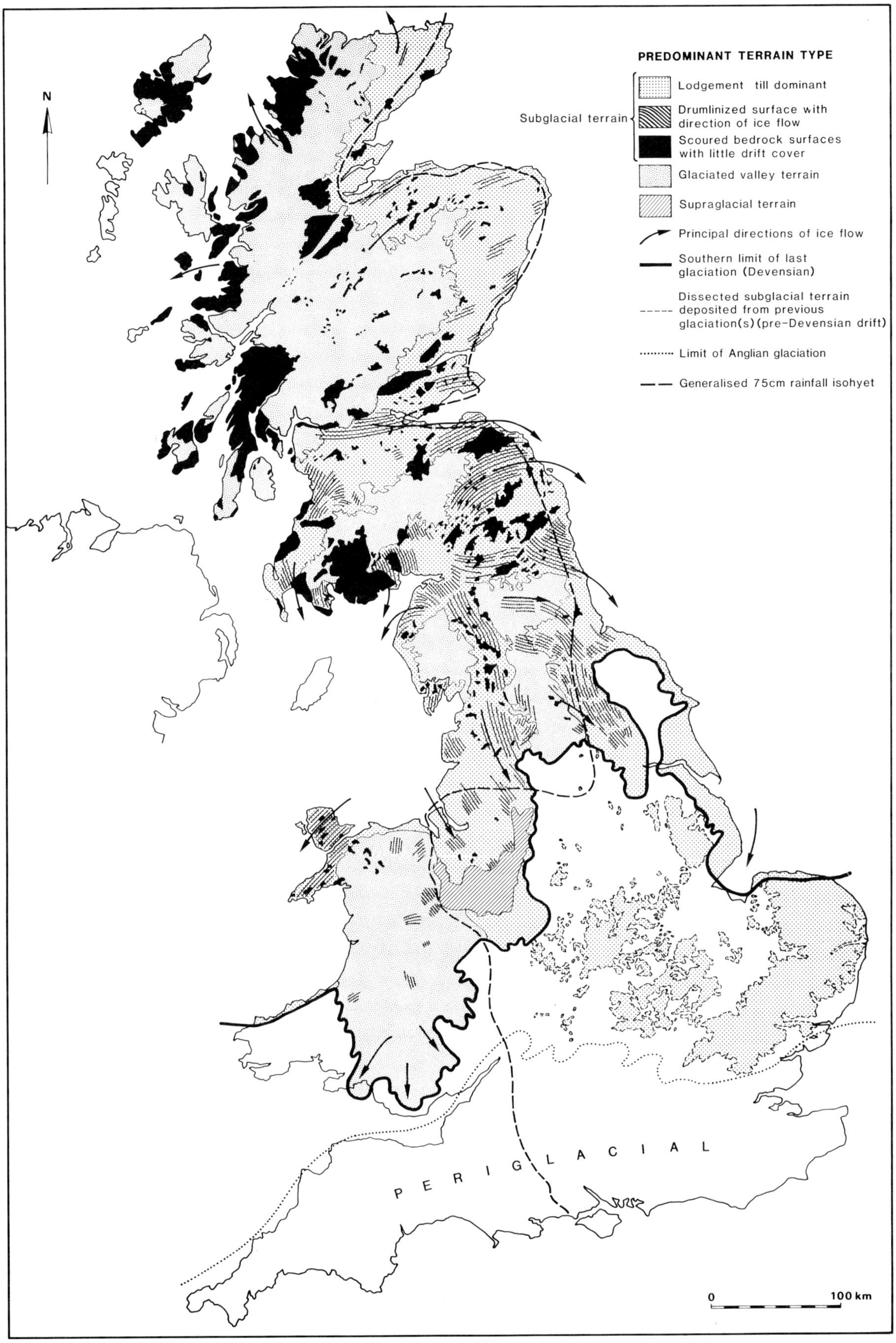

Figure 8.15 Map of the distribution of predominant glacial terrain types in the UK (From Eyles and Dearman, 1981, Fig. 6, by permission)

glaciated terrain contrasts with weathered, fluvially and periglacially eroded glaciated terrain elsewhere.

In these older terrains where glacigenic sequences have been partially eroded, that is, where surface morphology cannot be used as a guide to the occurrence and spatial distribution of subsurface ground conditions, characteristic and particular soil features allow recognition of the former local land system. From this information the occurrence, adjacent engineering soils and overall ground conditions can be predicted and mapped.

Outside the overall glacial limit in Britain, and at well-defined localities inside the limit, bedrock lithology plays a fundamental role in association with fluvial and periglacial activity and *in situ* weathering in determining the predominant engineering soils. Here general ground conditions are dictated by bedrock lithology and may result from repetitive sequences of strongly contrasting lithologies (e.g. the Jurassic and the Lower Cretaceous), or from the presence of thick fairly uniform sequences (e.g. the Upper Cretaceous). These again can be modelled as a series of representative strata as a predictive tool in site investigation and mapping, but only some of the aspects are dealt with in the present account.

8.6.1 Terrain evaluation

Terrain evaluation is aimed at understanding the natural features of the landscape. As a process, evaluation inevitably involves terrain classification in which the landscape is separated into natural units. Each constituent unit, at whatever scale is being considered, must be internally homogeneous and distinct from the others. Recognition of landscape units implies that there is a genetic relationship between landforms and the processes and materials involved in their development. The processes are mainly surface geological processes that have been active in the recent past, and these may be very different from the processes that have been active at the present time. The materials are the superficial and solid deposits that crop out at the surface and immediately underlie it.

At the scale of the present maps that are compilations from the basic 1:625 000 geological map of Britain (Institute of Geological Sciences, 1977), the appropriate landscape unit is the land system.

Each land system can be defined in terms of relief, soil and bedrock, but whereas the term soil originally referred to pedological soil, in the present context soil is concerned with the engineering soil complexes underlying, and at the same time controlling, surface topography.

Although terrain evaluation is said to be concerned only with the uppermost 6 m of the ground (Mitchell, 1973), the land systems which here involve engineering soils are considered to extend to bedrock regardless of depth.

The basic premise of the present approach is that if the land system can be identified, then it is possible to identify in turn the relevant engineering geological conditions. As such the approach has an important role in the desk study phase of site investigation, with the potential for saving time and hence money at this initial planning stage. Moreover, the land system once identified provides a model for the interpretation of the results of site investigation, and focuses attention on those aspects or particular features that should be looked for in examining samples recovered by boring.

8.6.2 Quaternary lithologies

For the purposes of general engineering geological mapping, the Quaternary of Britain can be divided into two basic groups: the non-glacial Quaternary, and the glacial Quaternary.

Non-glacial Quaternary

The non-glacial group comprises a distinctive shallow water, marine, estuarine and freshwater sequence of Lower and Middle Quaternary age (older than 300 000 years before present, YBP). These occupy a basin in the underlying Cretaceous in East Anglia, that deepens in the east below the North Sea where the total thickness of Quaternary soils is over 1 km (McCave, Caston and Fannin, 1977).

The succession comprises the famous 'crag' sequences of shelly calcareous sands found in Norfolk and Suffolk. The non-glacial group includes soils deposited in cold environments, but there is no direct evidence of glaciation. As a whole the group comprises shelly clay to sand sequences, but because it is almost entirely overlain by till, sands and gravels, the distribution of the group cannot be shown in Figure 8.13, but it does appear in Figure 8.11, the map of the bedrock.

Glacial Quaternary

Of fundamental importance in engineering geology are soils of the glacial Quaternary which were deposited during a number of glacial and interglacial cycles post-dating 300 000 YBP. Deposits of three distinct glaciations can be recognized. In order of age these are Anglian (300 000–250 000 YBP), Wolstonian (200 000–150 000 YBP) and Devensian (135 000–10 000 YBP). The southward extent of the Anglian glaciation is the greatest and that of the Devensian the least. As a result a large tract of dissected Anglian and Wolstonian glacial soils preserved on interfluve slopes lies south of the Devensian limit (see Figure 8.15). The limit of the Wolstonian glaciation is poorly defined and in what follows glacial soils of Wolstonian and Anglian age are referred to collectively as pre-Devensian.

Beyond the limit of the Anglian glaciation lies an unglaciated zone where there has been little net sedimentation during the Quaternary and where the drainage basins and their underlying ground conditions have been extensively modified by periglacial cold climate processes. Effects of Tertiary weathering processes are also found. Large tracts of unglaciated southern Britain and the pre-Devensian glacial terrains of the Midlands and East Anglia are fossil periglacial landscapes.

8.6.3 Glacial soils and their classification

A complex range of engineering soils results from glaciation. As a means of classifying and mapping solid sequences and landforms at the margins of modern-day glaciers, it has been found that there is a close genetic association between the subsurface geology and landforms (Boulton and Paul, 1976). It is possible to recognize a number of distinct terrain types each having a characteristic topography and subsurface conditions. A corollary of this is that, by the identification of the landforms and terrain type, the geometry and character of the subsurface can be generally predicted. By recognizing these terrain types the character of the subsurface ground conditions can be generalized for large areas. The value of this approach to ground engineering lies in its use as an engineering geological mapping tool as a rapid guide to likely subsurface conditions.

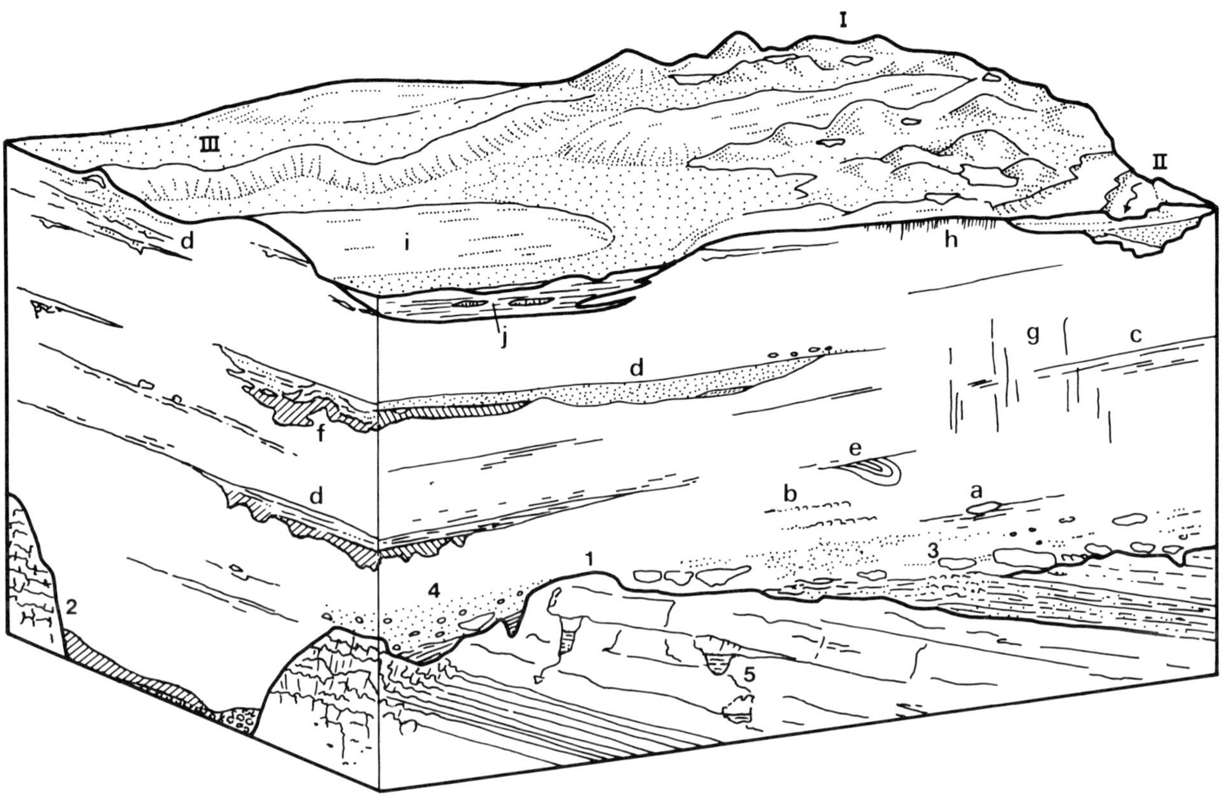

Rockhead
1. Striated rockhead.
2. Buried channel oversteepened by subglacial meltwaters filled with subglacially deposited sands and gravels and lenses of waterlain tills: note frost-shattered bedrock; the latter is frequently 'bulged' upward into channels as a result of stress-release (Price and Knill, 1967).
3. Glacially sheared (glacitectonised) rockhead, rock rafts and boulder pavement: note wedging off of joint-guided blocks along bedding plane (Knill, 1968).
4. Lowermost bouldery till unit composed of local lithologies, thickening in the lee of rock highs - contains scree-like masses of till deposited in subglacial cavities.
5. Sub-till 'cold water' karst from enhanced solution of limestones by subglacial meltwaters: note presence of soil-filled cavities.

Lodgement till
a. Preferentially aligned stones with long axes generally parallel to ice flow direction. The latter is rapidly determined by 1 above and the orientation of glacially-shaped boulders having a flat-iron form (bullets).
b. Crude shear lamination caused by the shearing out of soft lithologies during till deposition (Kruger, 1979).
c. Slickensided horizontal bedding plane shears resulting from glacitectonism producing a distinct fissility (McGown *et al.*, 1978).
d. High permeability gravels, sands and laminated clays deposited in subglacial channels or ponds and truncated by overlying till units: the texture of these tills reflects incorporation of fluvial soils producing clay to cobble enriched till. These fluvial units are elongated in the direction of ice flow, and lensoid in cross-section.
e. Folded and sheared-off channel fill. Note individual till beds ranging in thickness from 10 cm to 20 m, each having distinct erratic suites, colours and geotechnical properties formed by the lateral shifting of ice flow units. Ice flow direction measured by (a) may change from one bed to another.
f. Diapiric intrusion of till squeezed up into subglacial channel fills.
g. Slipped and flowed waterlain masses of till formed either by bank-collapse along subglacial channels or by debris melting out and flowing from the glacial base into subglacial ponds and channels.
h. Near-vertical joints whose orientation is related to ice flow direction or to joints in underlying bedrock (Boulton and Paul, 1976; Cox and Harrison, 1979).
i. Desiccation joints and oxidative surface weathering to 2 m depth in western Britain (McGown *et al.*, 1975) increasing to 8 m in drier eastern Britain (Madgett and Catt, 1978, Eyles and Sladen, 1981).
j. Peat in poorly-drained inter-drumlin depressions. Peat is commonly buried below solifluction debris derived from drumlin side-slopes; often recorded in site investigation logs as peat below till (Dickson *et al.*, 1976).

Outwash
I Hummocky kame and kettle topography obscuring drumlins and composed of poorly-sorted dirty gravels.
II Outwash fan channelled into till and composed of cross-stratified sands and gravels.
III Esker ridge; a subglacial or englacial channel fill that survives as a positive topographic feature not having been sheared off and buried by till as are other subglacial channel fills (d, e).

Scale: The length of section might be tens of metres to tens of kilometres. Height from several metres to a maximum of 100 metres.

Figure 8.16 Block diagram illustrating the ground conditions and engineering soils typical of subglacial terrain

Glacial and periglacial ground conditions can be generalized into four terrain models:

- subglacial terrain;
- supraglacial terrain;
- glaciated valley terrain;
- periglacial terrain.

The first two are characteristic of glaciated lowlands with a subdued rockhead surface almost completely obscured by glacial soils. These soils were deposited in the main by large ice lobes.

Glaciated valley terrain, on the other hand, results where the bedrock relief is so marked, as in highland areas, that bedrock protrudes through the drift cover breaking up the ice lobes into valley glaciers.

Finally, ground conditions and landforms resulting from periglacial conditions may be superimposed to varying degrees on the three terrain types described above, but predominate in the unglaciated zone south of the limit of glaciation.

The four terrain types are described in more detail in Eyles and Dearman (1981).

8.6.4 Distribution of glacial terrain types

Distribution of complexly layered lithologies and the ground conditions encountered in glaciated and periglaciated terrains are too variable to present on a map at the scale of Figure 8.13. Certainly the general distribution of till and fluvioglacial sands and gravels can be depicted together with fine-grained soils (silts, clays and organic complexes) resulting from post-glacial sedimentation along lowland meandering rivers and their estuarine outlets. This, however, gives no information on the subsurface ground conditions and soil types. Similar comments can be directed to the glacial geology maps produced by Clayton (1963), West (1968), Derbyshire (1975) and the compilation produced by the Institute of Geological Sciences (1977).

For each of the terrain types identified above, namely subglacial, supraglacial, glaciated valleys and periglacial, block diagrams can be constructed that show the type and geometry of soil horizons and principal diagnostic landforms. These terrain models provide a representative survey of ground conditions (Figures 8.16–8.19).

The areas in which the four glacial and periglacial terrain types are represented in Britain and where the models of ground conditions afford a generalized indication of subsurface soils and their geometry are shown in Figure 8.15.

Subglacial terrain

Within the area of subglacial terrain, the dominant glacial soil is lodgement till, the 'boulder clay' of countless publications. At many sites, two or more distinct till units can be mapped in each section. Near the margin of the Devensian glaciers, where the depth of glacial erosion was limited compared with the central parts

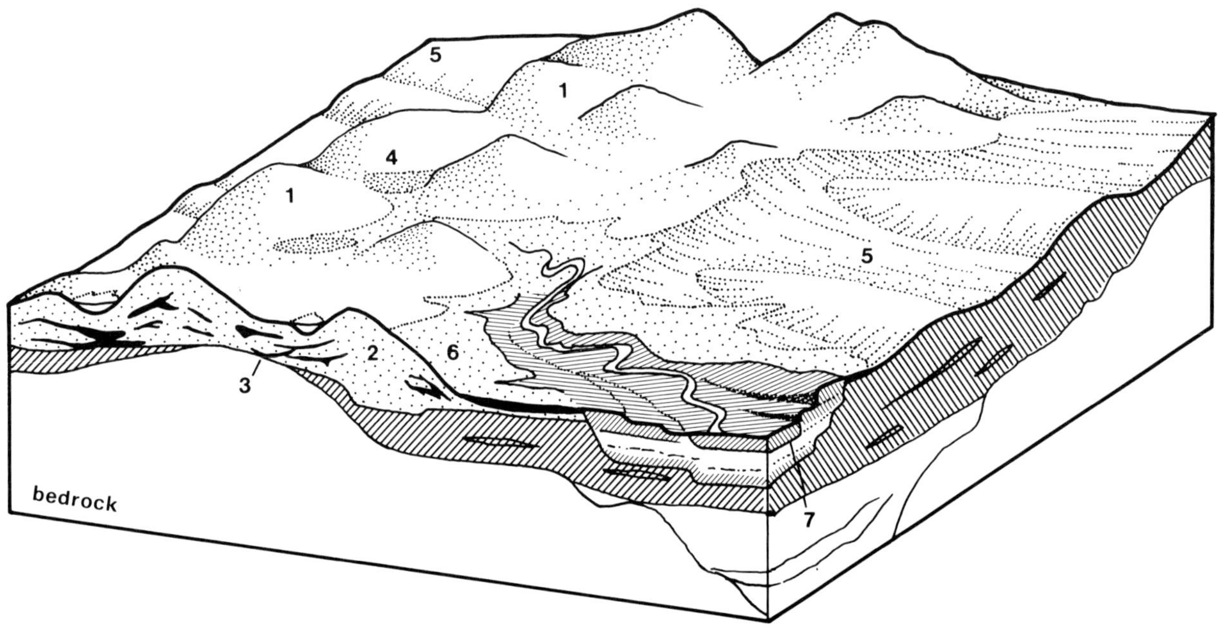

1. Hummocky supraglacial terrain obscuring streamlined subglacial terrain.
2. Poorly sorted glaciofluvial soils with lenses of flowed till masses of irregular and non-repetitive geometry. Hummocks are frequently capped by loose till or soils deposited in former topographic lows (clays, silts) reflecting subsequent topographic inversions following the melt of ice-cores.
3. Synclinally-flexured bedding often parallel to the ground surface and faulted hummock margins both caused by the settlement of soils into subsurface voids created by melting ice-cores.
4. Peat filled kettle hollow between hummocks.
5. Drumlinized lodgement till on which the belt of supraglacial terrain has been superimposed. Supraglacial deposition may be aided by bedrock obstructions to ice flow.
6. Laminated clays resulting from meltwater ponding between the ice front and the belt of hummocky supraglacial terrain.
7. Terraced glaciofluvial soils along course of former meltwater river and modern meandering stream.

Figure 8.17 Ground conditions and engineering soils of supraglacial terrain

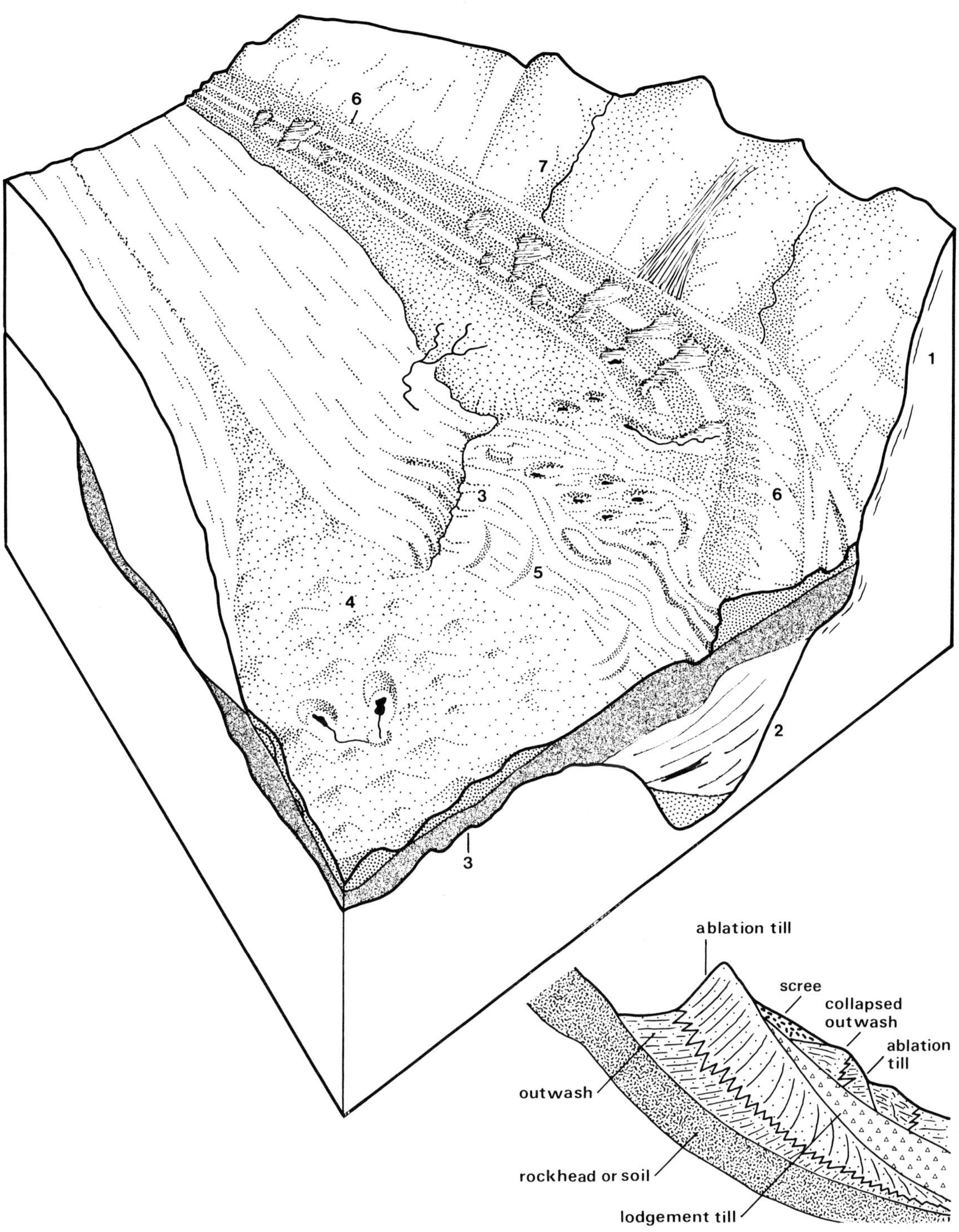

Rockhead

1. Oversteepened failure-prone bedrock disturbed by valley joints related to stress-release, glacitectonism, frost-shattering, bulging and solution superimposed on tectonic structures.
2. Buried valleys filled with layers of glaciofluvial soils and tills.

Tills and outwash

3. Lodgement till with streamlined, drumlinized surface revealed as an inlier through a cover of hummocky ice-cored ablation till and kettled outwash.
4. Ablation till deposited as a 'controlled' medial moraine.
5. Minor ridges of ablation till dumped from crevasses or the icefront.
6. Lateral moraine ridge (with the ground conditions shown in the insert) with kame terraces being gullied by the melt of ice-cores.
7. Valleyside fans which may subsequently breach the lateral moraine ridge resulting in smooth valleyside slopes, as part of post-glacial infill processes with the underlying ground conditions depicted in Figure 8.19.

Figure 8.18 Glaciated valley terrain illustrating ground conditions and engineering soils

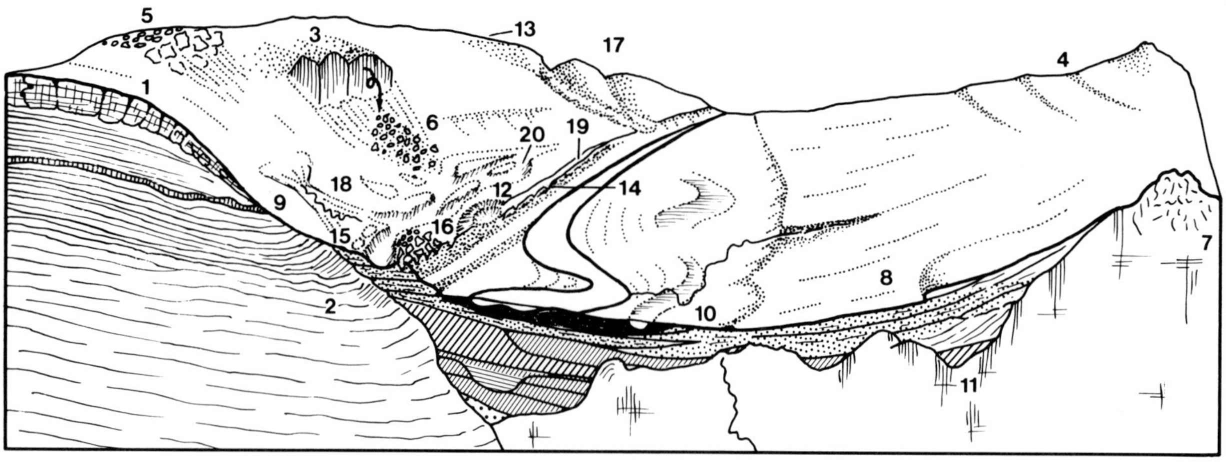

1. Flat-bedded sedimentary rocks cambered into valley – harder beds are jointed allowing individual blocks to founder and move downslope as complex slides or rafts riding on sheared-out weaker rocks or soils. Distinctive fissure fills may be encountered upslope.
2. Strata bulged and displaced upwards along the valley floor and cut across by undisturbed terraced gravels as part of the complex fill of a buried valley. The fill includes lodgement till and outwash.
3. Valley-side tors and cambered 'edges' exposed by slope-stripping and found in association with screes (clitter). Rockfalls common.
4. Summit tors on crystalline bedrock surrounded by blockfield of frost heaved bedrock, and rock-cut terraces cut by freeze-thaw and washing on the margins of snow patches (niveofluvial erosion). Such terraces are well developed on the Dartmoor metamorphic aureole.
5. 'Patterned ground' composed of polygonal nets and stripes.
6. Stratified screes and blockstreams (below the Hythe Beds escarpment in Kent the latter are 4–6 km in length).
7. Deeply weathered bedrock as a result of frost-action or older, non-glacial processes.
8. Solifluction fans composed of downslope thickening, crudely-bedded sheets of rubble ('head') fining downslope into hill-wash on the lower slope foot. On the Chalk periglacial head may be transported to form lobes of coombe rock.
9. Mudflow underlain by low-angled shear planes – common on Jurassic and Cretaceous mudstones.
10. Solifluction mantle interbedded with gravels deposited by periglacial braided streams and overlain by peats and silts comprising the floodplain of the modern meandering stream.
11. Glacial soils lying in pockets in rock head, below solifluction debris.
12. Doughnut-like degraded circular rampart of former pingo; may also affect bedrock surfaces below overlying drift.
13. Clay-with-flints and other remainié soil mantles (plateau gravels or drift and 'taele' gravels) resulting from mechanical churning of pedological soils and bedrock by frost heaving, solifluction and enhanced solution on carbonate rich lithologies by aggressive cold water (cold water karst). Rockhead underlying such mantles is often high, irregular and disturbed by frost action or solution.
14. Soil involutions and contortions resulting from seasonal freezing of thaw-soaked soil.
15. Faulted soil zones as a result of ground contraction during freezing and collapse into subsurface voids as a result of the melt of buried ground ice. The latter may also generate (rarely) a pitted ground surface (thermokarst).
16. Fossil ice-wedge casts penetrating terraced gravels deposited by braided flood-prone periglacial streams. The bedrock surface below these gravels may be diversified by scour-hollows.
17. Dry valley system cut by increased surface run-off as a result of impermeable frozen ground.
18. Alluvial fan deposited by streams from former snow patches on the valleysides. Snowpatch hollows are referred to as nivation hollows.
19. Sheets of windblown coarse-silt (loess) up to 3 m thick, thickening in the lee of river terrace scarps.
20. Windblown 'cover' sands often in the form of parabolic dunes.

Figure 8.19 Ground conditions and engineering soils associated with periglacial terrain

of the ice cap, organic material and non-glacial soils lying between tills indicate either pre-Devensian glacial deposits or readvancing ice lobes. In drumlinized terrains elsewhere, ubiquitous 'tripartite' stratigraphies are composed of lower till(s), middle sands, gravel and laminated clays and upper till(s). They have been used extensively as a mapping unit by the Institute of Geological Sciences and taken to indicate multiple glaciation. The threefold or multiple successions can be explained by the former presence of subglacial melt streams and ponds coupled with shifting ice divides and changing ice flow directions bringing in different erratic types.

The activity of subglacial melt waters is also deduced from overdeepened sections along the bedrock floors of many river valleys. In many lowland areas with a thick drift cover, the amplitude of the bedrock surface has a higher relief than the generally smoothed till surface (Anson and Sharp, 1960; Anderson, 1974). Data on drift thicknesses in these areas form a useful base for a series of thematic engineering geological maps (Dearman *et al.*, 1979).

Drumlinized subglacial terrain is found most widely in lowlands below 300 m in elevation.

The increasing dissection of the till cover south of the limit of the last glaciation should be noted (see Figure 8.13). The loss of original glacial topography prohibits the use of glacial landforms as a guide to subsurface conditions. Nevertheless, the identification of homogeneous till sheets south of the limit suggests that these are the remains of lodgement plains, and they have thus been included in the subglacial terrain group.

Supraglacial terrain

Large areas occupied by supraglacial terrain are limited to arcuate morainic belts marking the positions of Devensian ice lobes.

Drumlinized terrain emerges from under the hummocky cover indicating deposition and superposition of

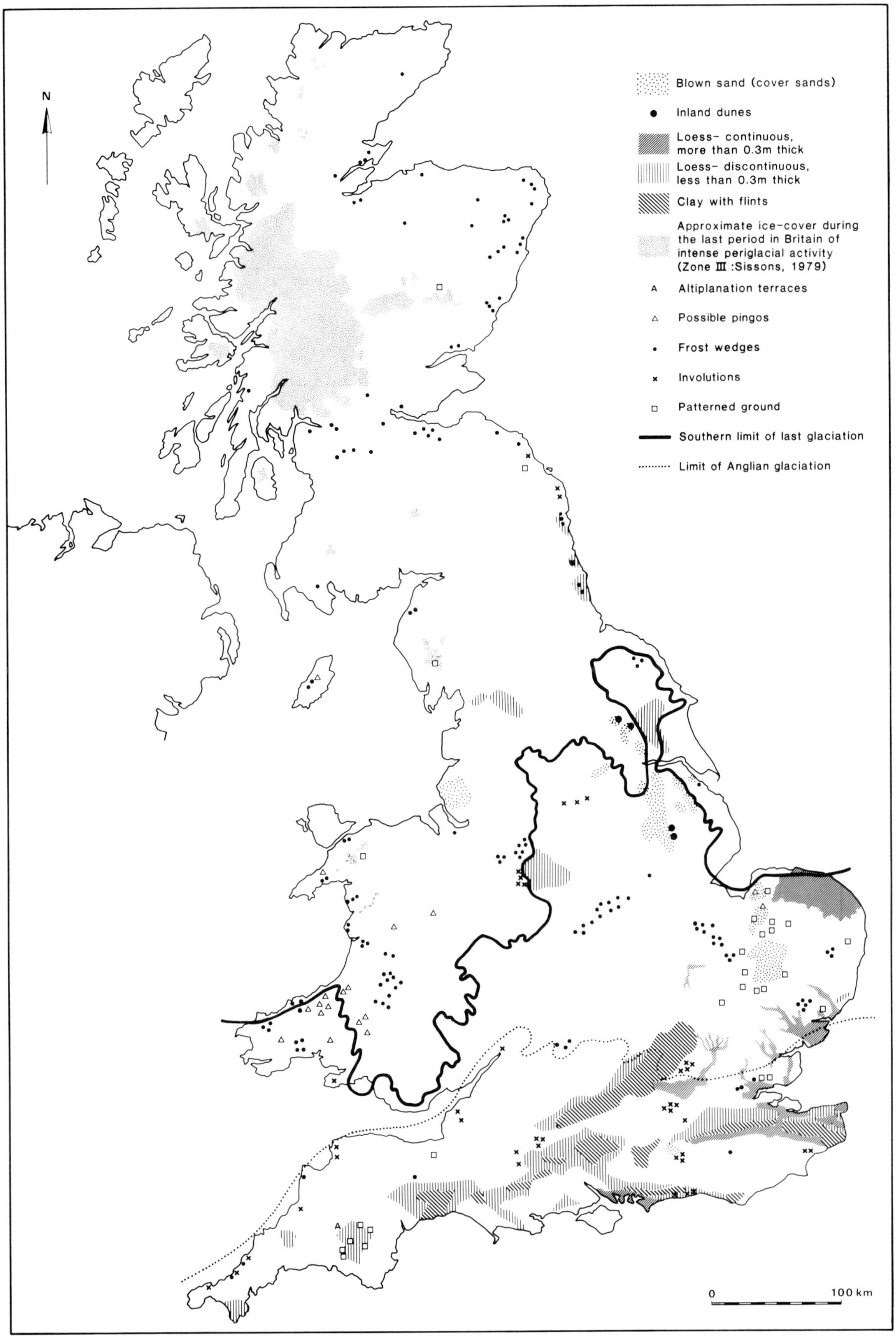

Figure 8.20 Map of the distribution of selected periglacial features in the UK

flow tills and outwash as 'distinct moraine belts' along the former ice margin. Economically such areas are important major aggregate sources. The 'trapping' of meltwater rivers within the belt of ice-covered ridges along the former ice margins accounts for the large volume of glaciofluvial soils associated with this distinctive terrain type.

Glaciated valley terrain

Hummocky morainic topography along valley floor and complex lateral accumulations on valley sides is a ubiquitous terrain type in upland and highland Britain. Deposits of ablation till as medial and lateral moraine ridges frequently overlying lodgement till are most clearly exhibited in those areas where ice survived until the end of Zone III (10 000 YBP approx.). The classic 'hummocky moraine' in Scotland is of this type (Sissons, 1967).

Occurring as it does in wetter upland and highland areas, this terrain type has been modified by postglacial periglacial processes. Following deglaciation, oversteepened and frost-disturbed valley slopes became unstable such that rapid and often spectacular downslope movement ensued. The U-shape of many glaciated valleys results from the formation of valleyside fans (Figure 8.18). Coalescence of fans from opposite valley sides leaves a thick lunate infill composed of a mixture of disaggregated tills, outwash, scree and bedrock walls. Valley infill is often confined to *in situ* till; an excellent geotechnical description is provided by Fookes, Gordon and Higginbottom (1978), where valley infill is referred to as mountain till. A description of the stratigraphy of periglacial valley infill sequences, and criteria for their recognition, is given by Thomas (1976).

Periglacial terrain

In those areas of eastern Britain underlain by bedrock suites that include marked lithological alternations between mudstone, shale, sandstones and limestones, periglacial terrain is well developed. On areas of pre-Devensian glacial soils a large volume of the periglacial slope mantles and valley infill is composed of resedimented till (Baker, 1976) and weathered bedrock lithologies. These may overlie deep-seated non-diastrophic structures that may be due in part to periglacial processes. South of the limit of glaciation, the periglacial mantle is derived entirely from weathered bedrock.

A map showing the distribution of individual periglacial features, reported in publications too numerous to cite, is of limited engineering use. Rather more useful would be to combine these data with the lithological map and highlight prominent scarp-forming lithologies among the sedimentary suites; these are likely to be foci for periglacially distributed ground conditions.

8.6.5 Discussion

The gross distribution of engineering ground conditions resulting from successive glaciations in Britain (Figures 8.13, 8.15 and 8.20) is a reflection of topography and precipitation. The 75 cm isohyet divides a wetter western highland portion of Britain from the drier eastern lowlands. The highland area is dominated by crystalline and metamorphic rocks, and there the ice caps periodically reformed producing heavily dissected terrain. Drier eastern lowlands are underlain by sedimentary rocks where the erosive effects of glaciers have been more limited and where glacial deposition had been dominant. In the east, deeply weathered granites survive in northern England and in eastern Scotland (Phemister and Simpson, 1949) and are found in association with mature drainage basins that have been little affected by glaciation (Linton, 1959). Lodgement tills of even Devensian age are deeply weathered in eastern Britain (Madget and Catt, 1978; Eyles and Sladen, 1981).

It is not appropriate here to give more detail on the glacial history of the UK. The four terrain models are fully annotated, sufficiently to be able to understand the distribution of ground conditions in Figures 8.15 and 8.20. Additional discussion of the geomorphological aspects will be found in Fookes and Vaughan (1986).

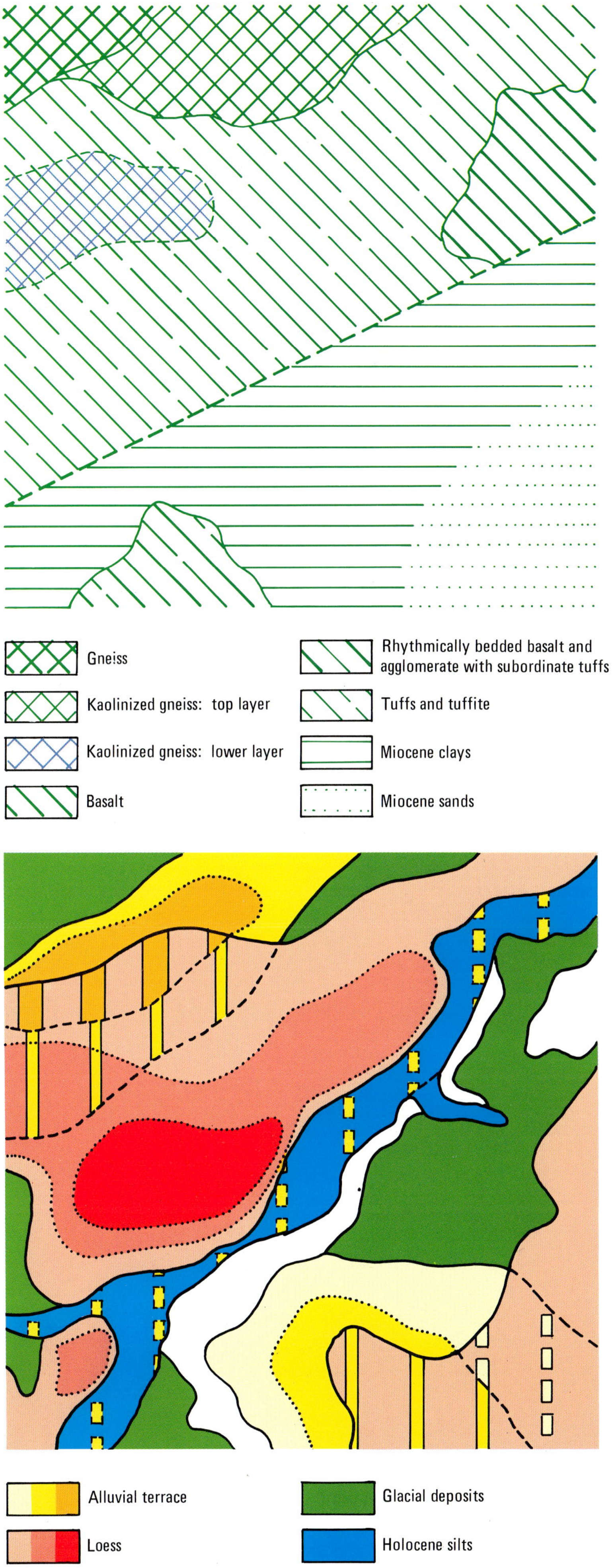

Plate 1 Example of map at a scale of 1:25 000 showing the method of representing the solid formations by patterns (above), and (below) the superficial deposits by colours and stripes (After Pasek and Rybar, 1961, Fig. 4)

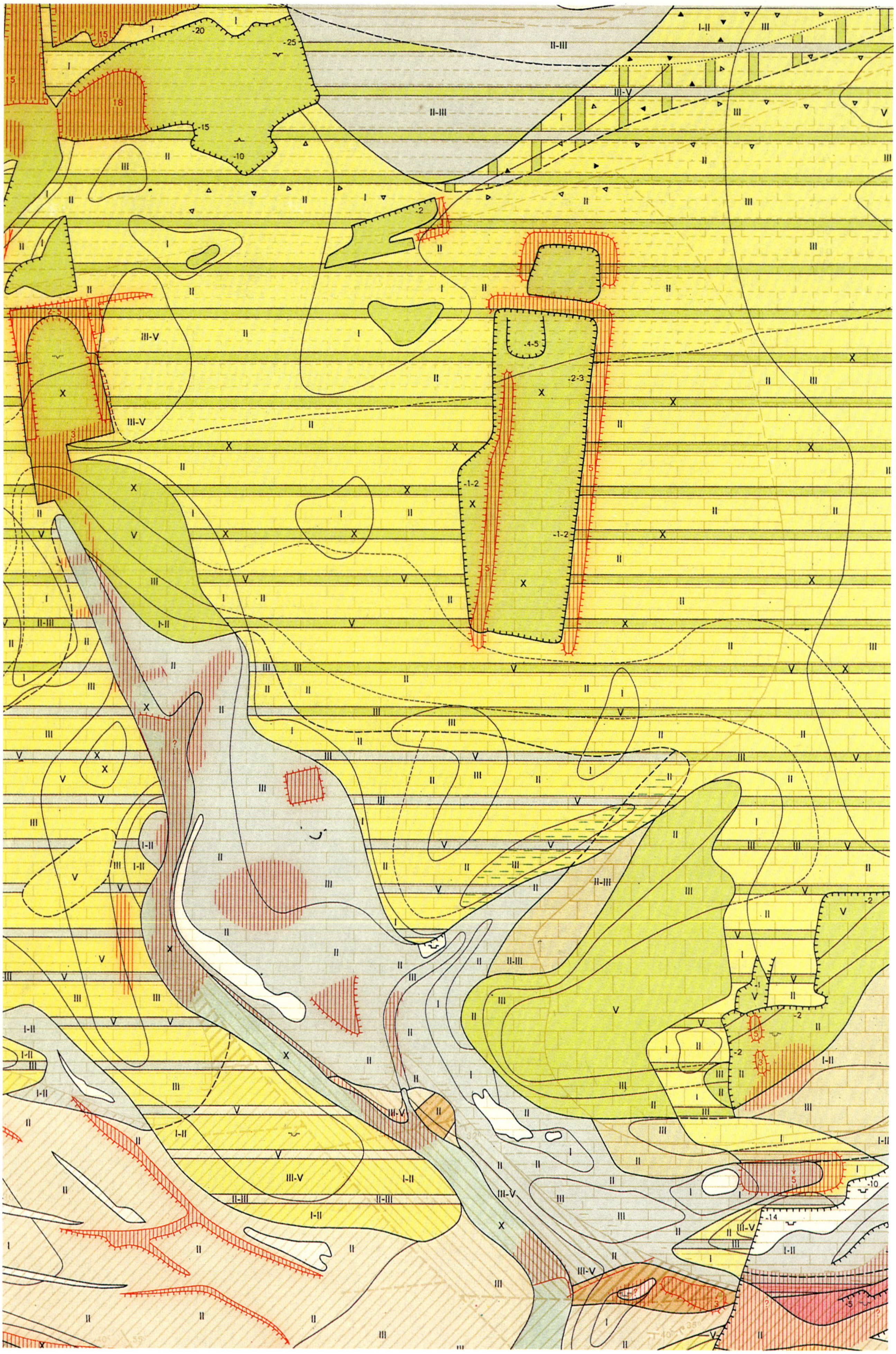

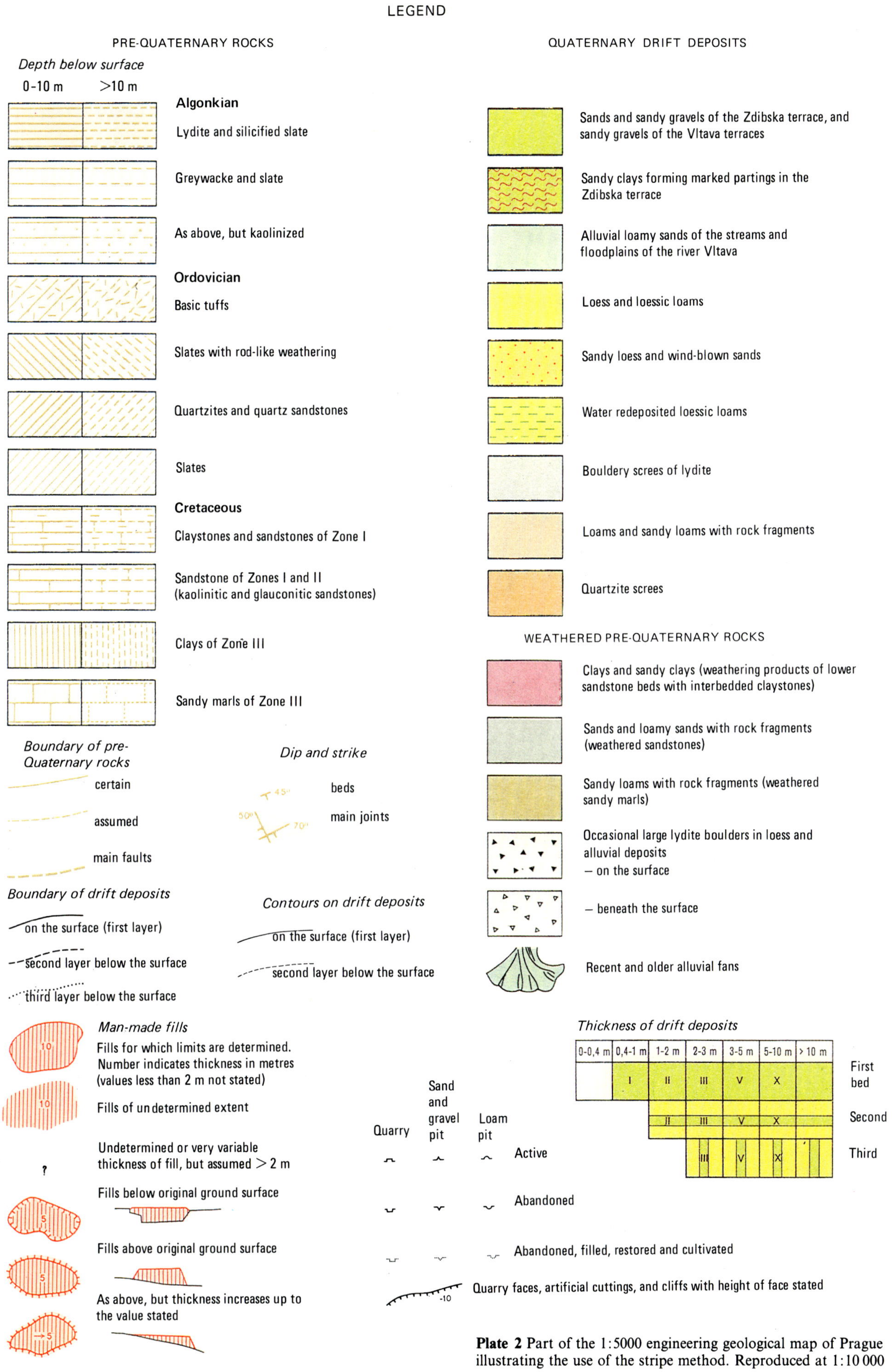

Plate 2 Part of the 1:5000 engineering geological map of Prague illustrating the use of the stripe method. Reproduced at 1:10 000

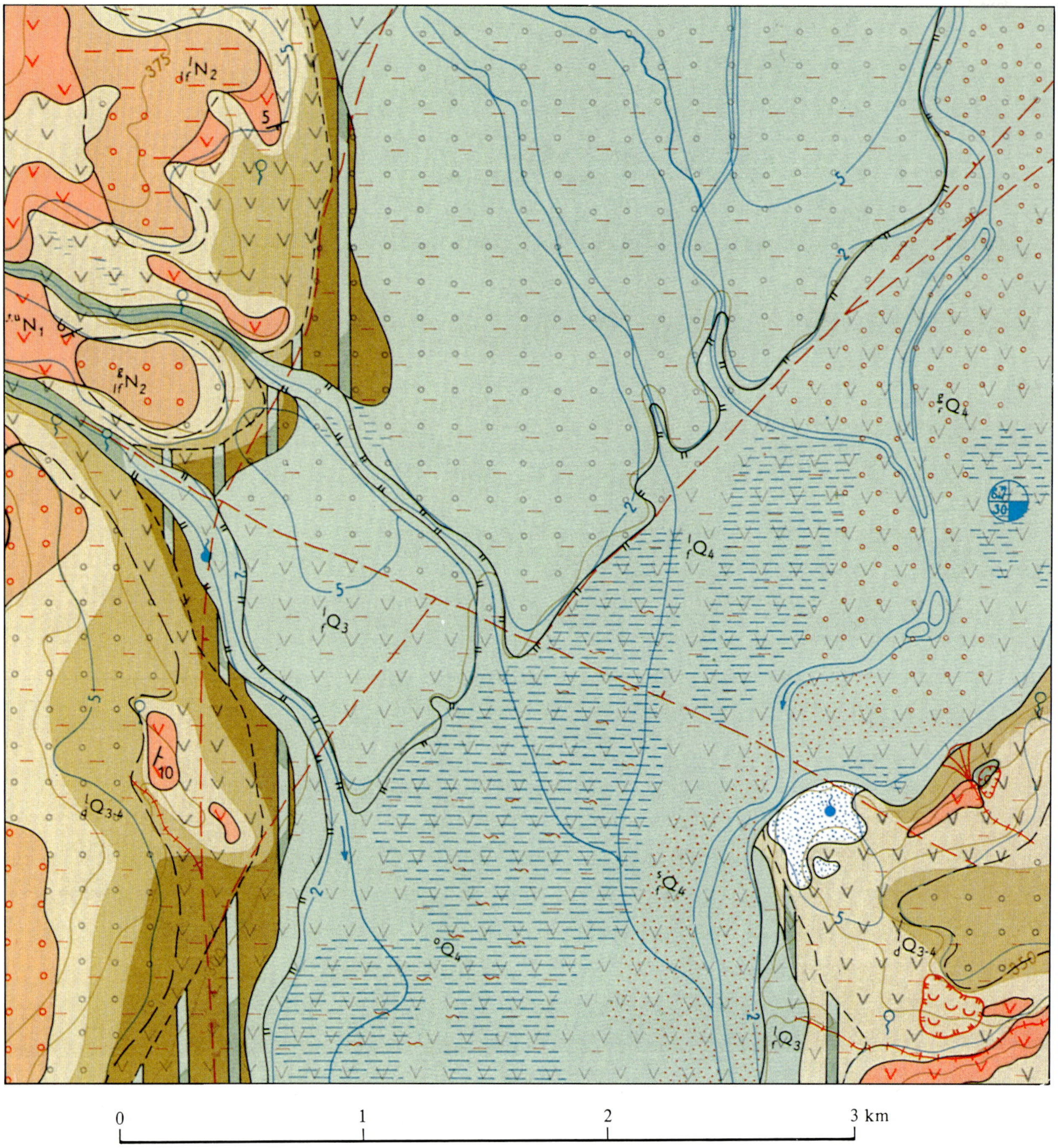

Legend

- Hydroisobaths of the maximum groundwater table
- Springs of fresh and mineral water
- Waterlogged areas
- Corrosiveness of groundwater: pH degree, hardness of water (in German degrees); the full sector indicates the aggressivity of CO_2, and/or SO_4 according to Czechoslovak standards (symbol labels: pH, CO_2, H, SO_4)
- Gully erosion
- Formation of alluvial cones
- Active landslides
- Older (potential) landslides
- Edges of river terraces
- Boundary of engineering formations on the surface
- Boundary of formations under the surface between: (a) Quaternary deposits and (b) Quaternary deposits and pre-Quaternary basement; between pre-Quaternary formations
- 10 Strike and dip of strata
- Important tectonic faults and their dip

Plate 3 Map of engineering geological conditions in part of the Zvolen Basin, Czechoslovakia, at a scale of 1:25 000 (From Matula, 1969, Appendix 2, by permission)

GENETIC-LITHOLOGICAL CLASSIFICATION

Age	Genetic group	Stratigraphical symbol	Rocks and soils: On the surface thickness (m) <2 2–5 5–10; Beneath the surface formations	Engineering formations
Quaternary	Fluvial	${}^{l}_{f}Q_{3-4}$	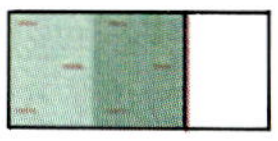	Sandy-loessic to loessic-clayey alluvial loams
		${}^{s}_{f}Q_{3-4}$	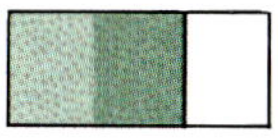	Alluvial sands predominantly medium-grained with loamy intercalations
		${}^{g}_{f}Q_{3-4}$	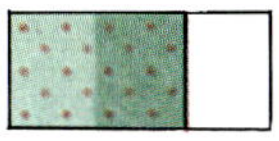	Fine- to medium-grained sandy gravels, loamy in older terraces
		${}_{o}Q_{3-4}$	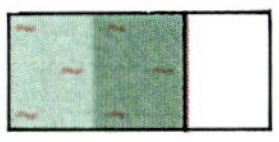	Sapropelic muds in abandoned river channels, sandy-loessic to clayey
	Slopewash (Deluvial)	${}^{l}_{d}Q_{3-4}$	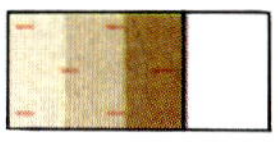	Sandy-loessic to clayey slope loams
	Chemical	${}^{tr}N_2-Q_4$	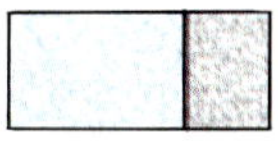	Travertines, mounds and slope sheets
Pliocene	Lacustrine	${}^{l}_{lf}N_2$	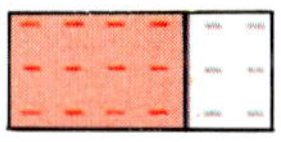	Lacustrine and alluvial clayey-loessic, sandy-loessic, rarely clayey loams
	Lacustrine fluvial	${}^{g}_{lf}N_2$	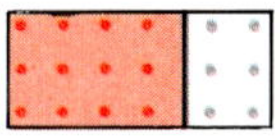	Lacustrine and alluvial sandy to loamy gravels with sand beds
Miocene	Volcanic lacustrine	${}^{tu}N_1$	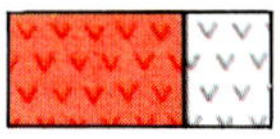	Stratified andesitic tuffs and tuffites

Plate 4 Legend for the map of engineering geological conditions in the Zvolen Basin, Czechoslovakia

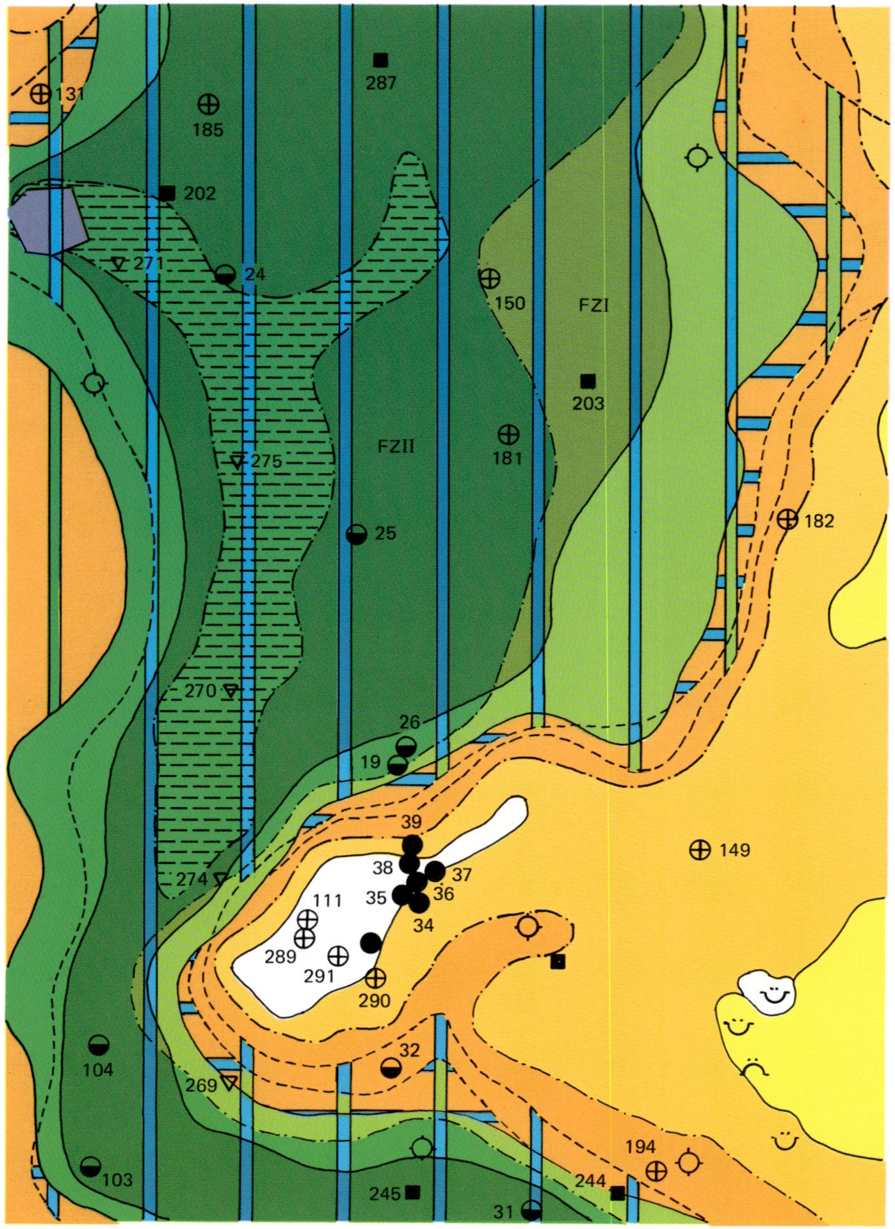

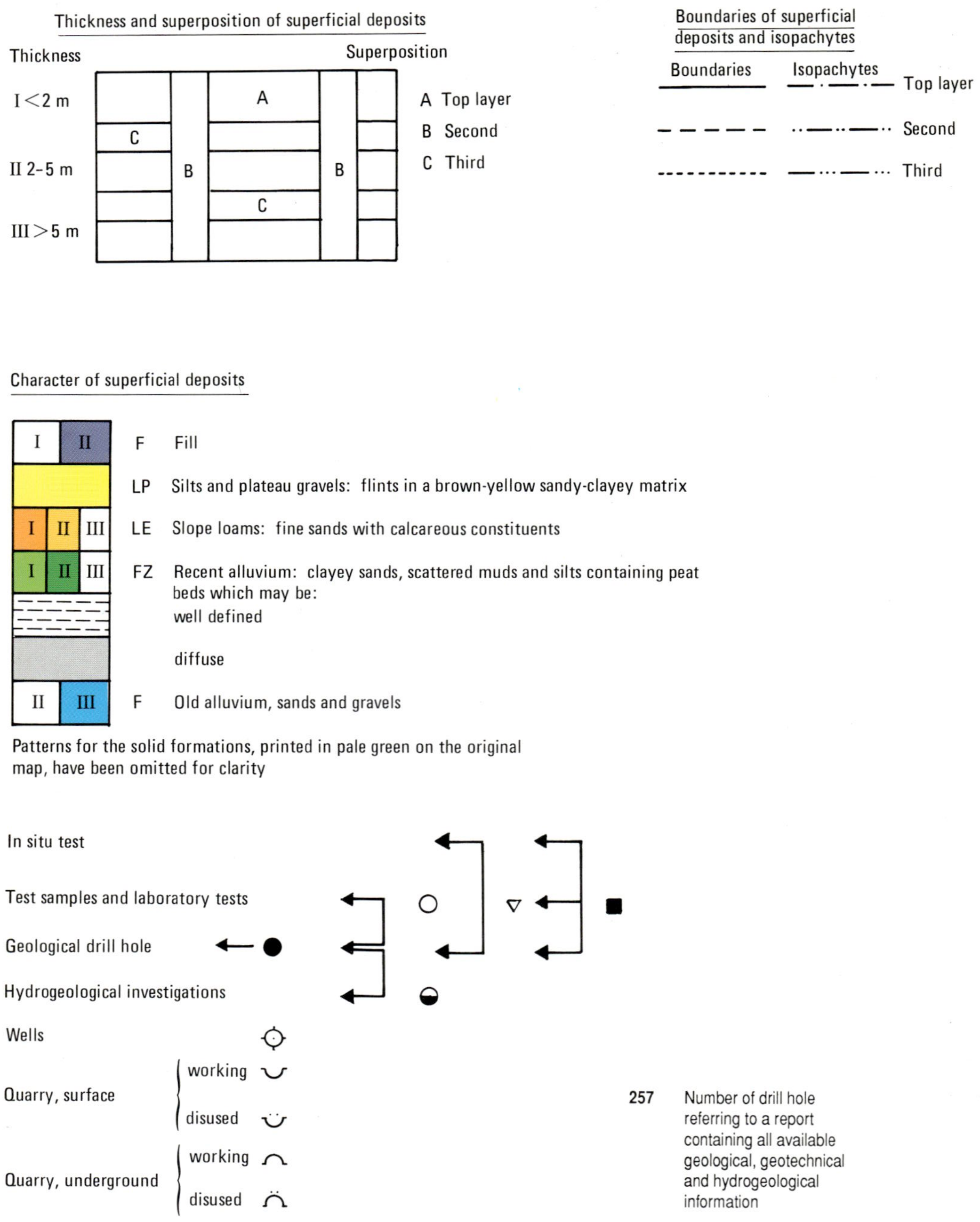

Plate 5 Geological map with documentation details. Part of the geological map of the region of Creil (Oise), France, produced at a scale of 1:20 000 by the BRGM. Reproduced at 1:25 000 (After Sanejouand, 1972, Plate VI, Carte Géologique et de Documentation)

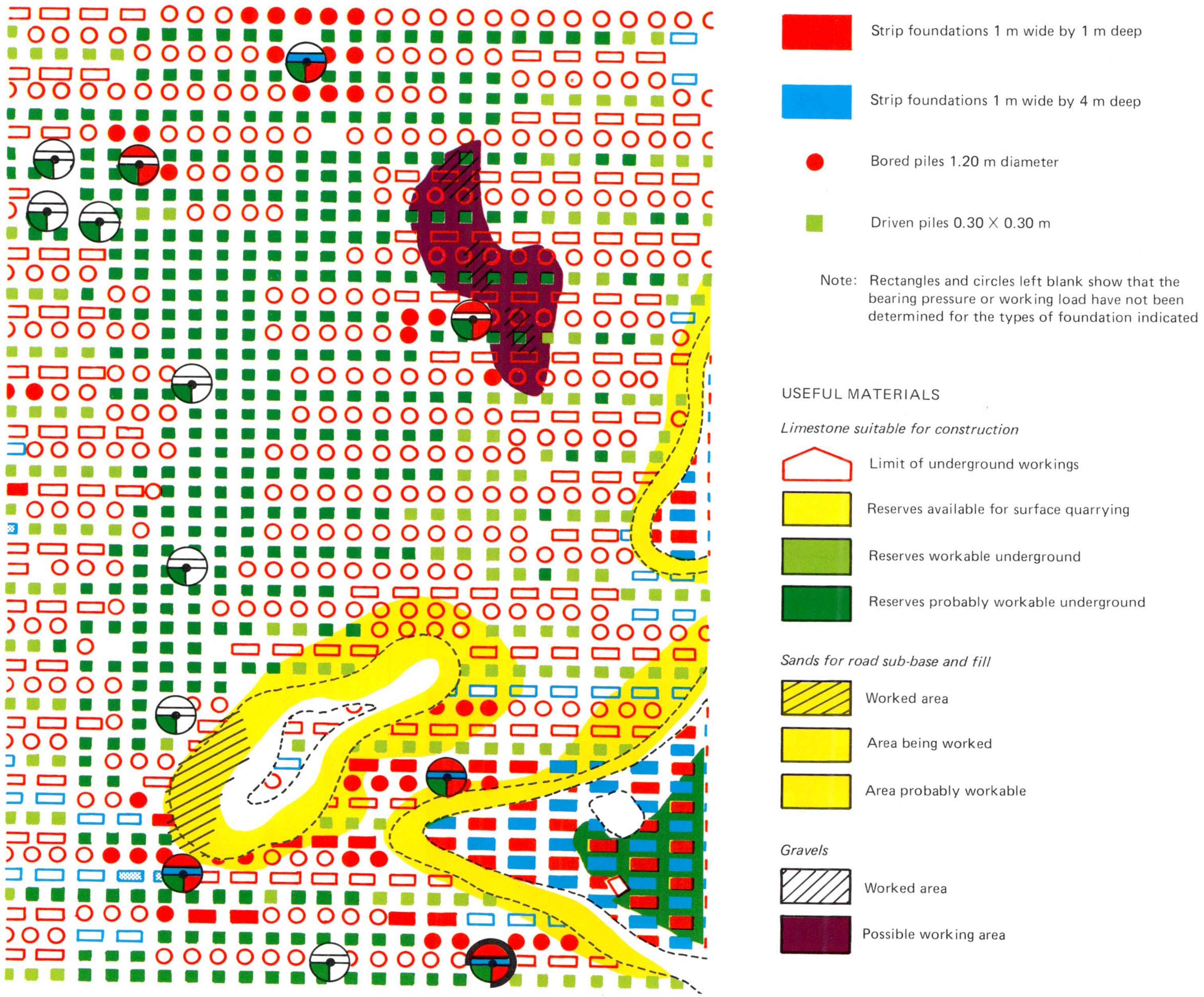

Plate 6 Map of foundation types and construction materials for the region of Creil (Oise), France, produced at a scale of 1:20 000 by the BRGM. Reproduced at 1:25 000 (After Sanejouand, 1972, Plate VII, Carte des fondations types et de substances utiles)

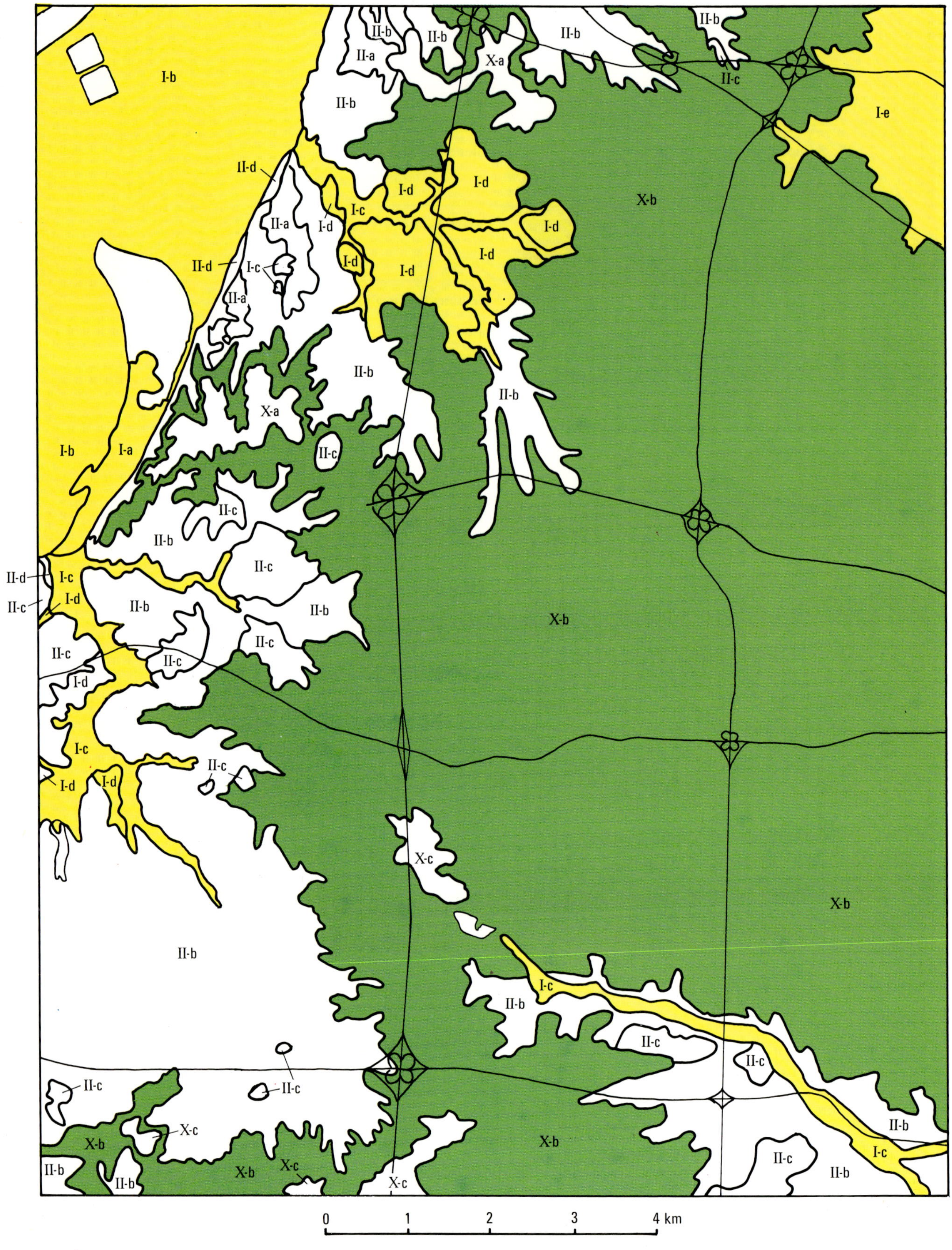

Plate 10 The 1 : 24 000 engineering geological map of the Creve Coeur Quadrangle, St. Louis County, Missouri, USA. Descriptive matter on the legend for each subunit is given in the text; reproduced here at a scale of 1:32 000 (Redrawn from Rockaway and Lutzen, 1970, Platel)

EXPLANATION OF MAP UNITS

Alluvium

I-a

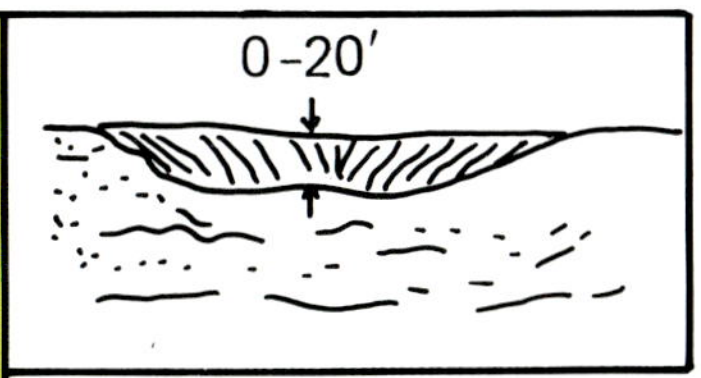

Swamp – Silts and clays with a high percentage of organic material; compressible; high water content; water table generally near surface; subject to flooding

I-b

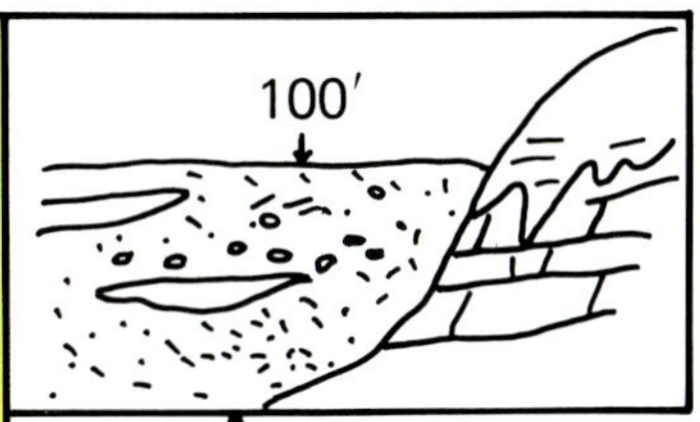

Thick Alluvium – Stratified sands, silts and clays with gravel and/or clay lenses; high water table; subject to flooding

I-c

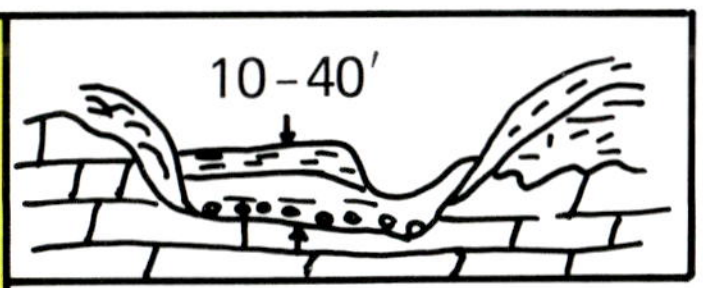

Thin Alluvium – Stratified sands, silts and clays deposited in tributary stream valley; varying thickness; generally silt or clay over sand and/or gravels; subject to flooding

I-d

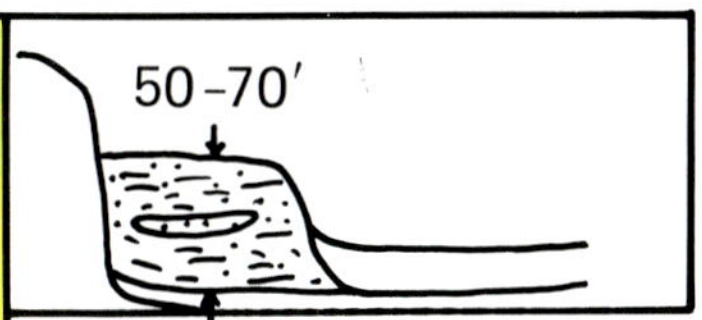

Alluvium Terraces – Stratifield sands, silts and clays; above level of present floodplain

I-e

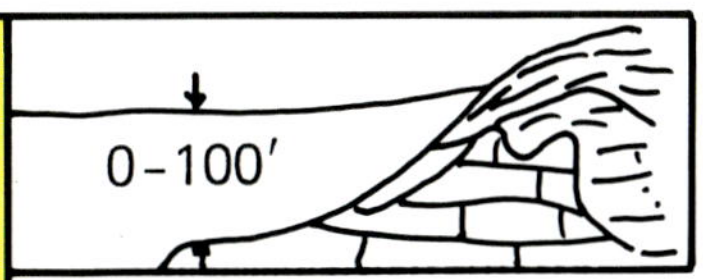

Lake Deposits (lacustrine) – Stratified silts and clays; compressible; high water content; some organic material

Carbonate bedrock (extensively weathered)

II-a

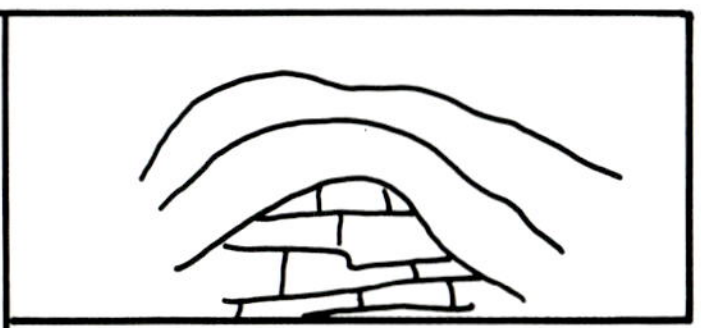

Loess – Loess deposited over carbonate bedrock; two layers; upper layer (silt-rich) will stand on 90° slopes and lower layer (clay-rich); natural moisture near liquid limit at interface of loess layers

II-b

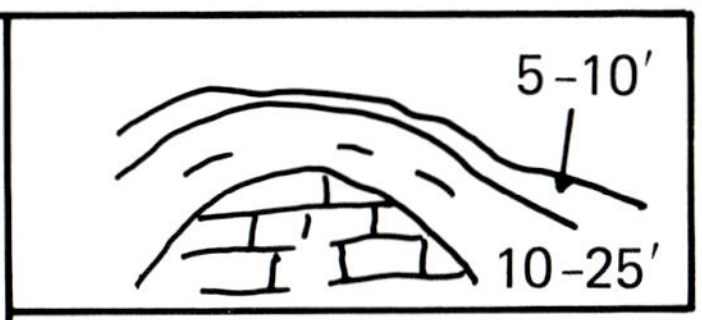

Loess – Loess deposited over carbonate bedrock; two layers; shallow silt-rich layer over thick clay-rich loess; high water content at interface of loess layers

II-c

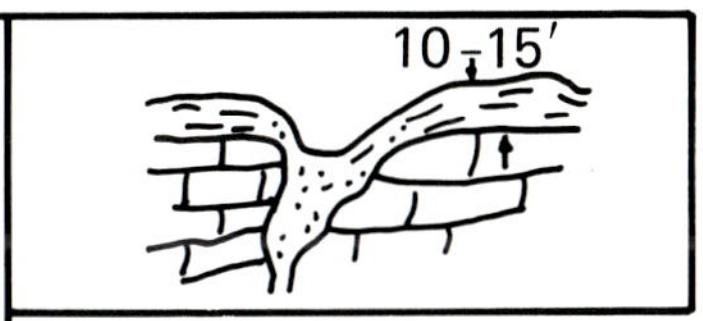

Bedrock – Steep slopes (greater than 60° above horizontal); bedrock at or near surface; thin soil zone has tendency to creep down slope

II-d

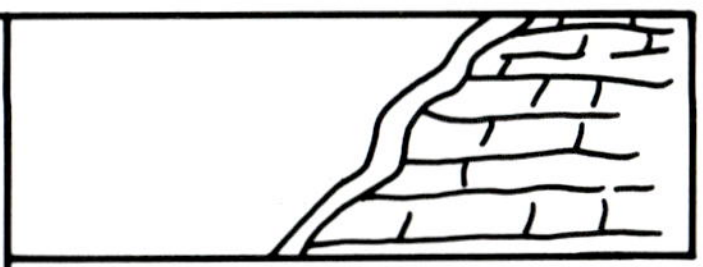

Bedrock – Steep slopes (greater than 60° above horizontal); bedrock at or near surface; thin soil zone has tendency to creep down slope

Cyclic deposits (predominantly shale)

X-a

Loess – Loess deposited over bedrock (predominantly shale); two layers; upper layer (silt-rich) and lower layer (clay-rich); problems similar to subunit II-a, potential slide plane at interface of loess and shale

X-b

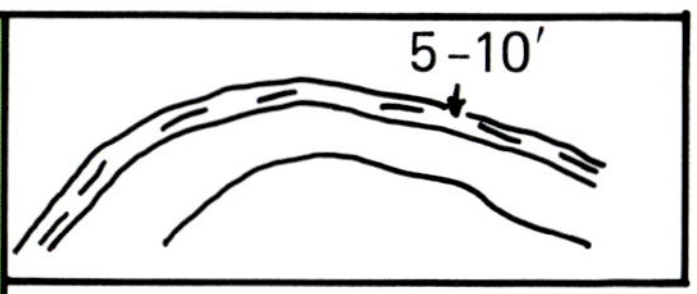

Loess – Loess deposited over bedrock (predominantly shale); two layers; shallow silt-rich layer over thick clay-rich layer; problems similar to subunit II-b, potential slide plane at interface of loess and shale

X-c

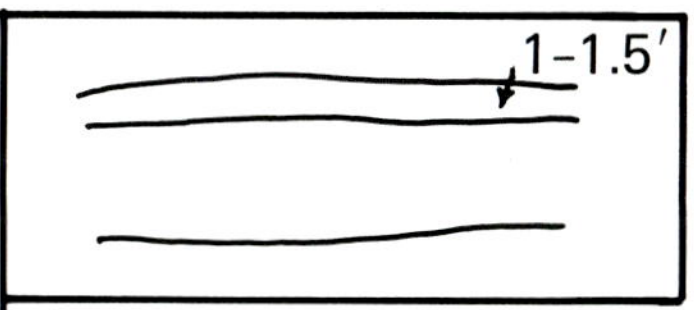

Loess – Loess deposited over bedrock (predominantly shale); swelling clays encountered at shallow depth

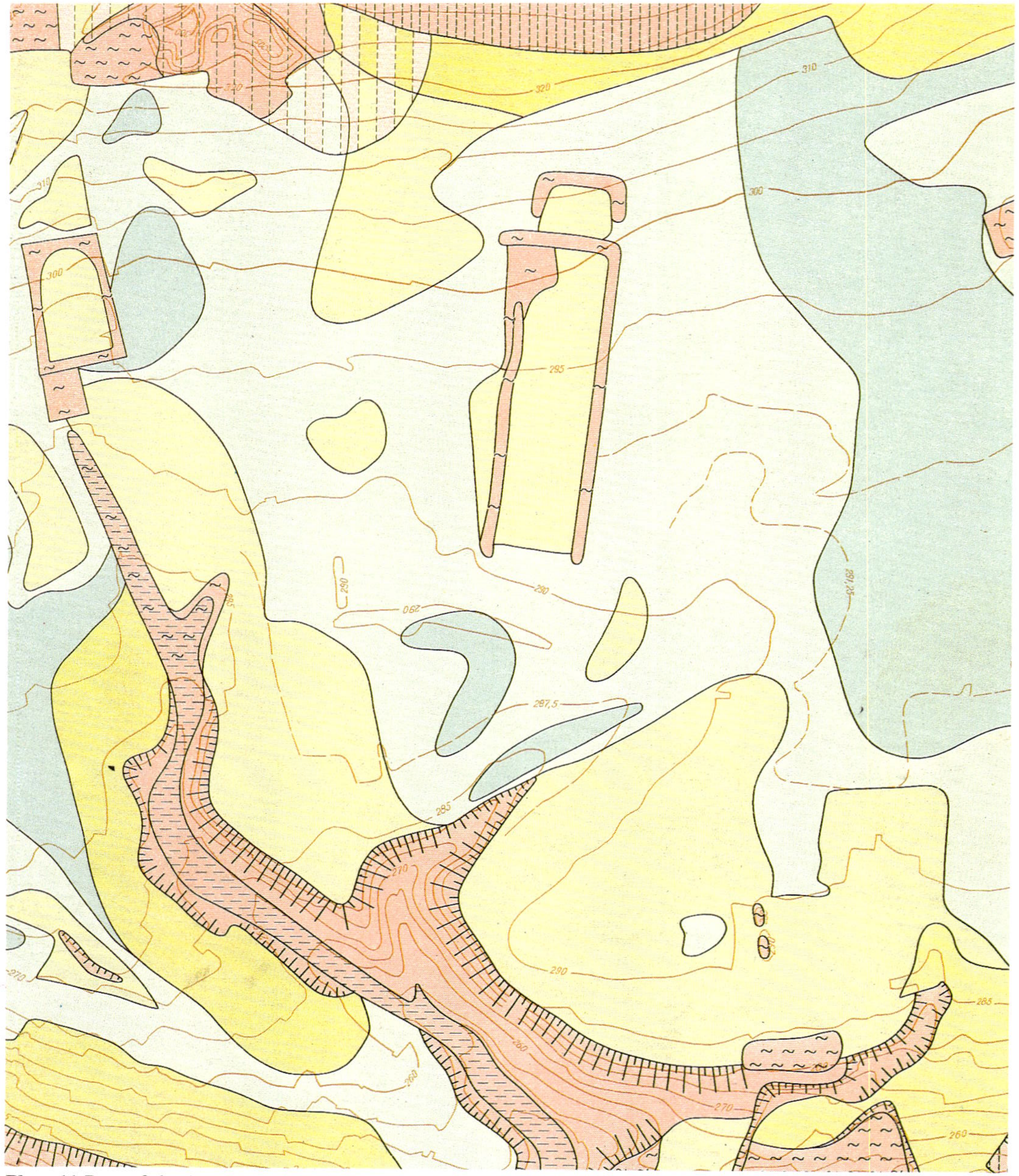

Plate 11 Part of the 1:5000 zoning plan of Prague showing five classes of foundation conditions. Reproduced at 1:10 000

Description and classification of foundation conditions				For construction of standardized multi-storey buildings	For construction of individual houses
I		Depth to suitable stratum up to 1 m	Strata including all rocks which, for the width of foundations of 1 m, have bearing capacity of 2 kg/cm^2 or more, with the exception of rocks which are difficult to excavate (because of high strength and spaced discontinuities). Groundwater level is not within the foundations; maximum inclination of the area is 10–12°	Very suitable	Suitable
II		up to 2 m		Suitable	
III		up to 3 m		Not very suitable	
IV		Bearing capacity is locally lower (from 1.2 kg/cm^2) or it is necessary to deepen the foundations to exclude the effects of weather conditions; inclination of the area is 10–12°		Conditionally acceptable	Acceptable
V		a. Water-bearing valley fills		Unsuitable	Unsuitable
		b. Areas susceptible to landslides and rockfalls			
		c. Steep slopes of more than 20°, rock cliffs and their immediate vicinity			
		d. Man-made fills, thicker than 3 m			
		e. Presently mined areas; colour of stripe shows suitability of the site for construction after excavation			
		f. Areas allocated for mining; where the deposit is not to be mined the colour of the stripe shows the suitability of foundation conditions			
		g. Rocks which are very difficult to excavate			

*As criteria are taken the mean values of various standardized multistorey buildings (founded on strip footing of width 1 m, at the depth of 1 m and allowable bearing pressure of 2 kg/cm^2).

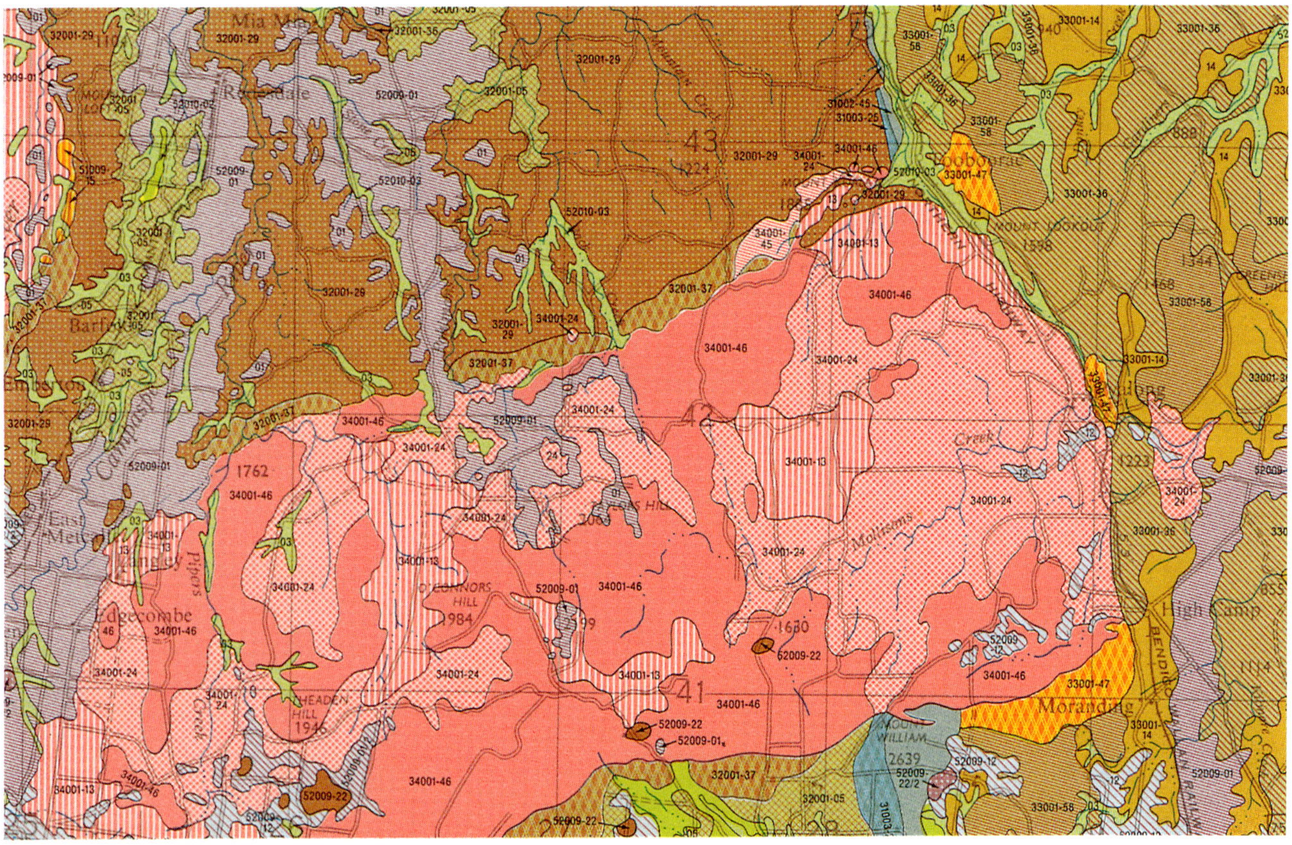

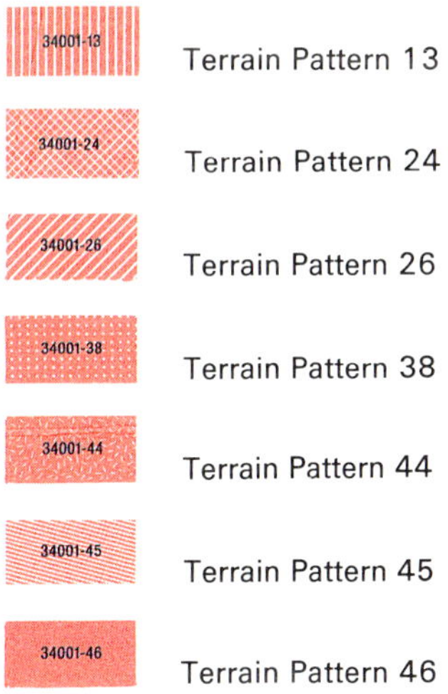

Plate 12 (top) Terrain classification map of part of the Melbourne area, Victoria, Australia, with the legend for seven terrain patterns comprising Province 34.001 on Devonian granite (From Grant, 1972; Map of Terrain Patterns, Melbourne, Victoria, by permission)

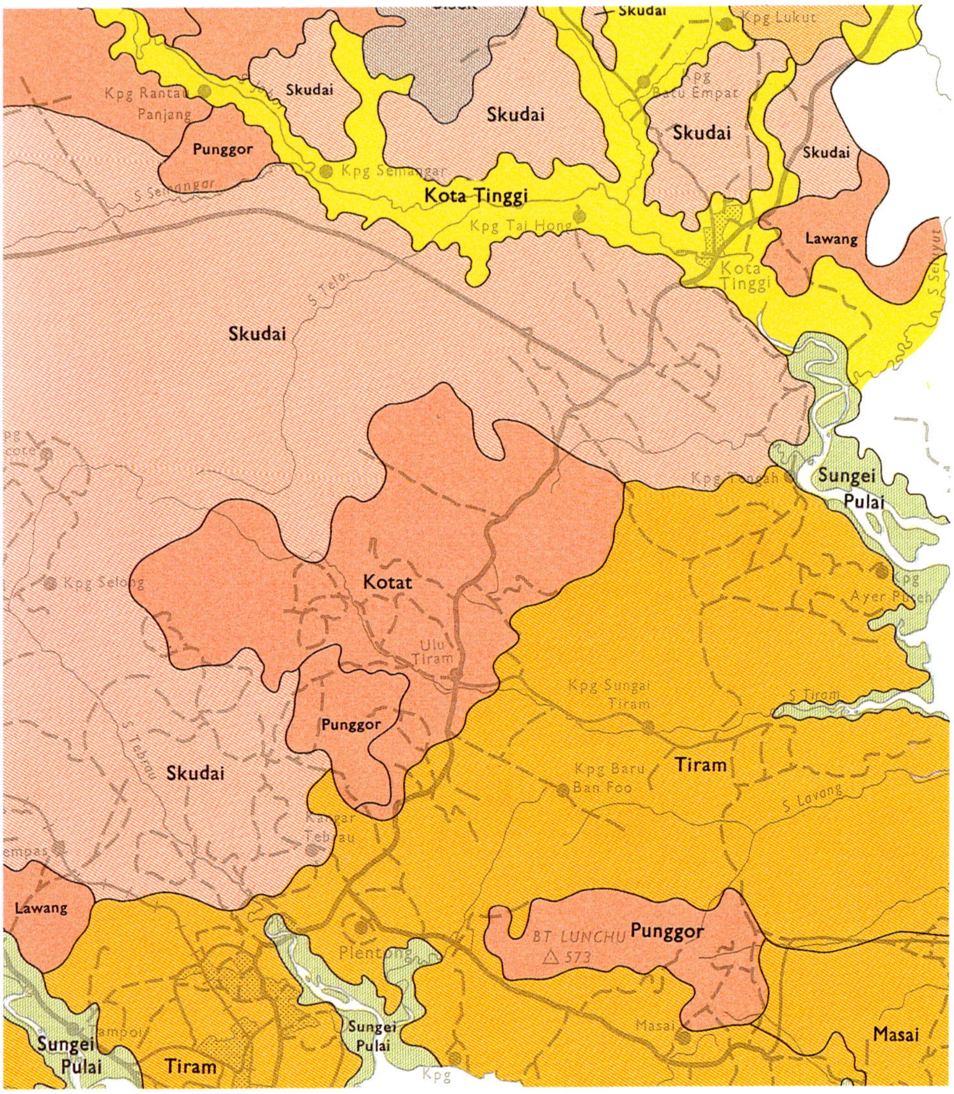

Plate 13 (left) Part of the land system map of South West Malaysia (From Lawrence, 1978: East Sheet, by permission)

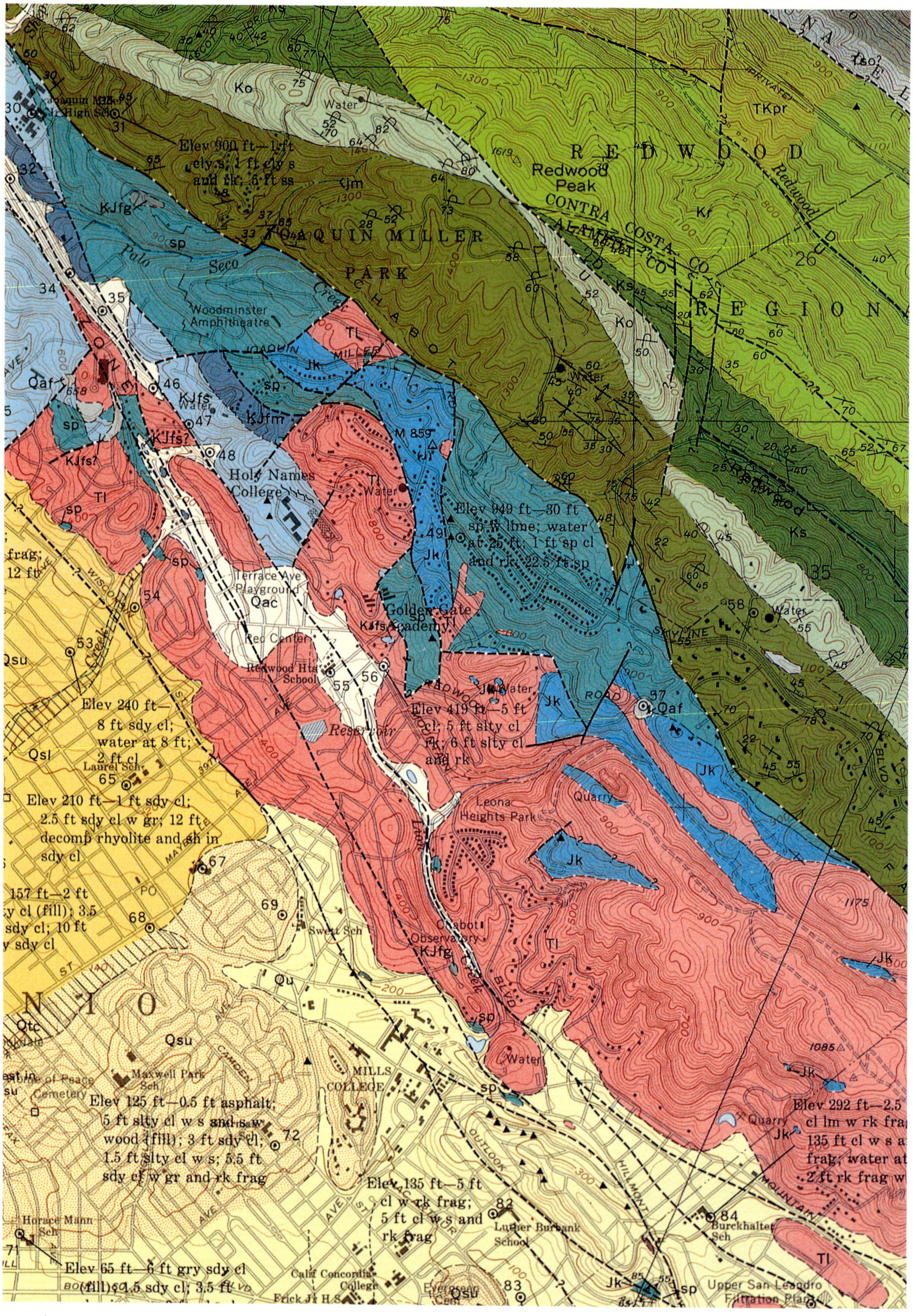

Plate 14 Part of Map GQ-769, Areal and Engineering Geology of the Oakland East Quadrangle, California, USA (From Radbruch, 1969, by permission)

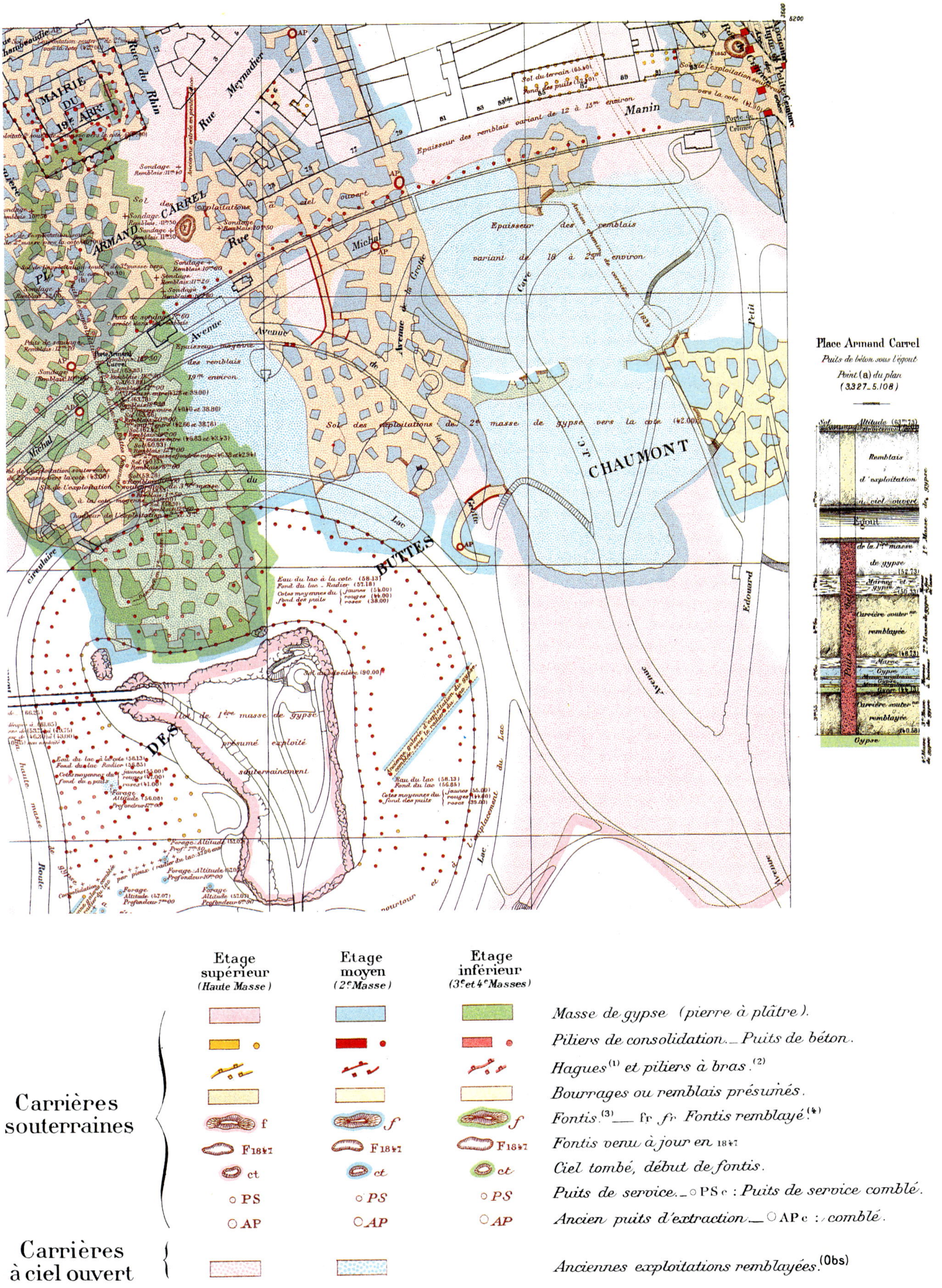

Plate 15 Part of the 1:1000 scale sheet No. 58, Rue Manin, Parc des Buttes Chaumont (Nord), Atlas des Carrières Souterraines de Paris. Published in 1922, revised in 1936 and 1964 (reproduced at 1:2000 scale)

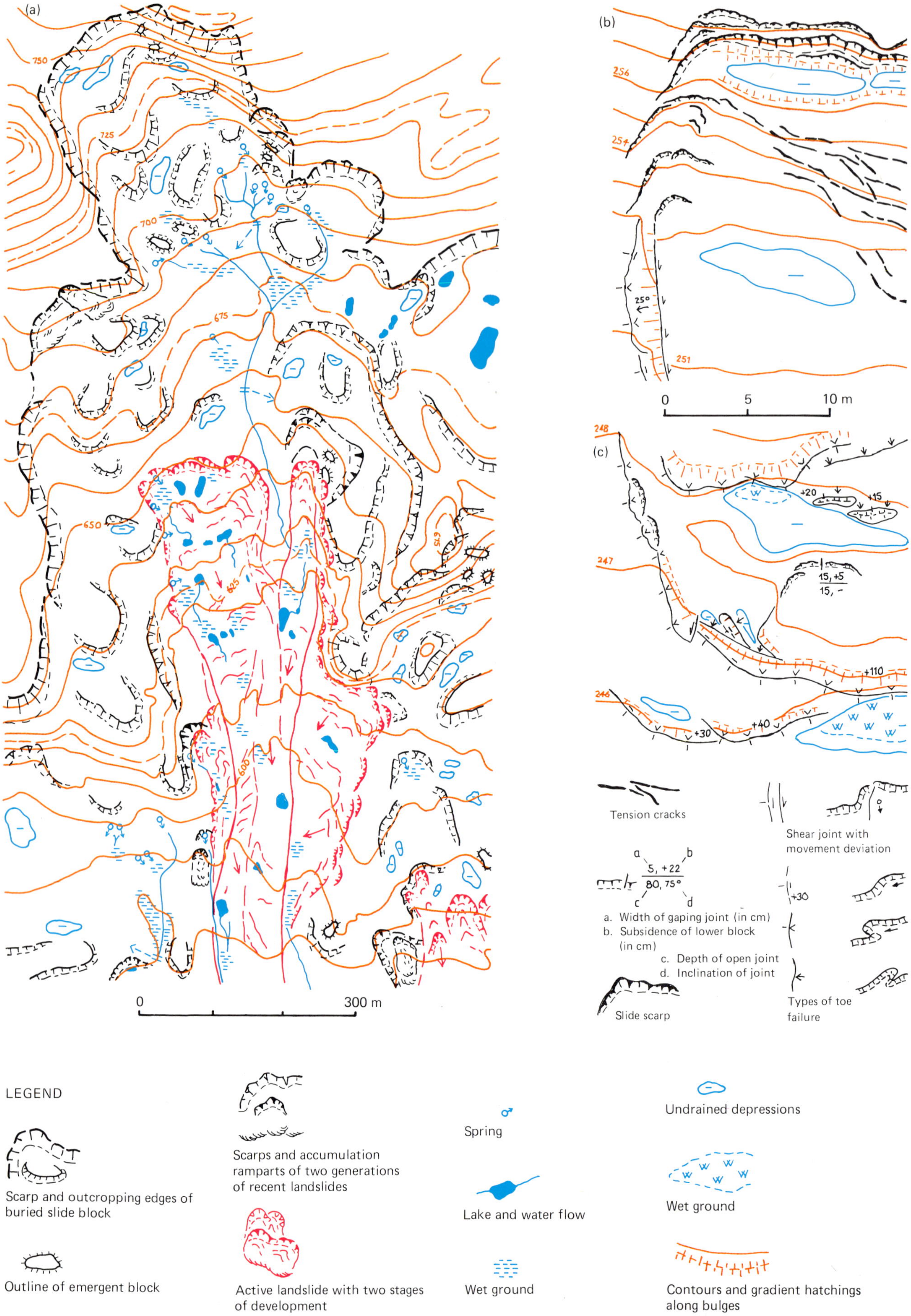

Plate 16 Large-scale plans of landslides: (a) upper part of Landslide No. 1 at Handlova, Slovakia; (b) upper part of Necranice Landslide, Slovakia; (c) toe of Necranice Landslide, Slovakia (After Rybar, 1968)

DEGREES OF HAZARD

- Zones apparently hazard free under present conditions, without prejudging future developments

HAZARD ZONES

- Very slight with regard to engineering construction; practically nil to the population
- Slight with regard to engineering construction; very slight to the population
- Not unimportant to engineering construction; slight to the population
- Important to engineering construction; not unimportant to the population
- Very important to engineering construction; very important to the population

LIMITS OF HAZARD ZONES

Limits of zones with equal hazard

Limits of zones with different degrees of hazard

Hazard zone of higher degree — Hazard zone of lower degree

ENGINEERING CONSTRUCTIONS

- New roads
- Zones of completed construction

SYMBOLS

All hazard zones are defined by:

— An initial letter indicating the dynamic phenomenon responsible

— A figure giving the state of the phenomenon

All deposits resulting from dynamic phenomena have the diagnostic letter suffixed 'Z'

When there is more than one hazard in a zone the colour used is related to the greater hazard and the hazards are indicated by symbols

DYNAMIC PHENOMENA

E	Collapses: very localized block falls (z–ɀ: deposits)
B	Sporadic block falls: ravelling of rocks (z–ɀ: deposits)
r	Gullying
e	Erosion
C	Debris flow (z–ɀ: deposits)
g	Rock slides, bed on bed
G	Slides with back scarps
F	Flow
S	Solifluxion, creep
En	Natural subsidences
Em	Mining subsidences
A	Subsidences — settlements

STATE OF THE PHENOMENON

ACTUAL

High frequency:	high intensity	1
	low intensity	2
Low frequency:	high intensity	3
	low intensity	4

POTENTIAL

High probability of developing	5
Low probability of developing	6

OLD — Actual state doubtful — 7

x x Upper limit of the clayey moraines of the Isère glacier

0 — 1 — 2 km

Plate 17 ZERMOS map of the Région d'Aime, La Plagne (Savoie), France, published at 1:20 000 scale (From Olivier and Bernard, 1976, by permission)

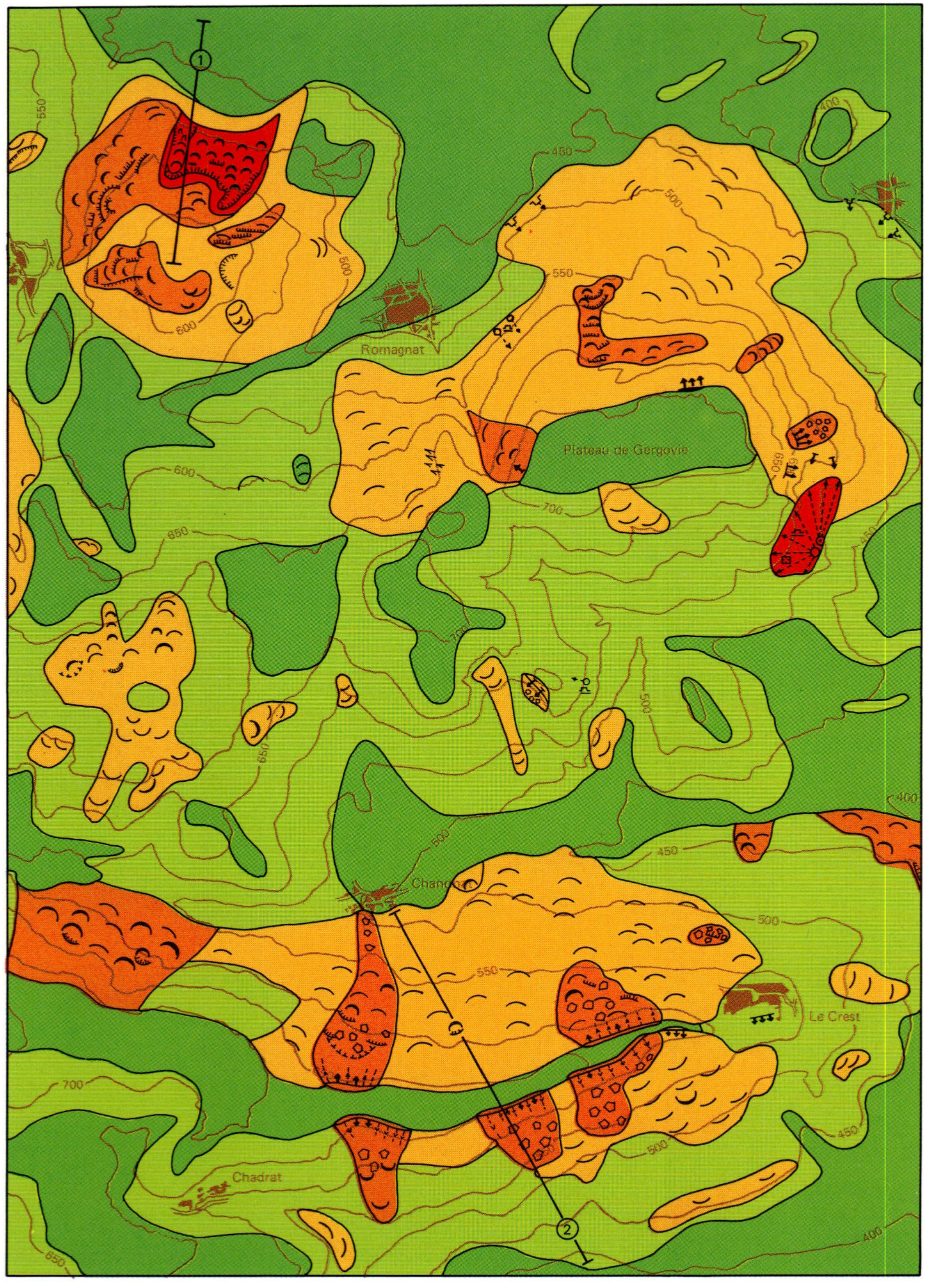

ZONES

Hazard nil or slight	V1	Zone actually free from ground movements
	V2	Zone apparently or practically free from ground movements
Hazard moderate	O1	Zone affected in a diffused way by scattered occurrences of poorly defined ground movements
	O2	Zone affected by known movements of moderate extent or displaying well-defined instability
Hazard high	R	Zone affected by large-scale movements or presenting well-defined aspects of instability

0 1 2 km

DYNAMIC CHARACTERISTICS OF THE GROUND

- Block falls with origin marked
- Block fall zone with origin marked
- Zone of accumulated blocks
- Old collapse of a rock mass
- Old collapse with an active cliff edge
- Subsidence caused by underground quarries
- Underground quarry: disused
- Accessible entrance to a gallery, with direction
- Inaccessible gallery entrance, with direction
- Caves, important, or a cave zone with general direction
- Landslide, single, well located but not represented in detail
- Landslide, recent
- Landslide, recent, localized
- Debris flow (that part below a clay slide)
- Diffuse movements, solifluxion, creep, the most common slope movement
- Topographic irregularity, small fissure
- Back scarp, relatively fresh; crevasse or fissure
- Back scarp, well defined
- Back scarp, old, more or less effaced
- Gullying, concentrated

Plate 18 ZERMOS map of the Région de Chanonat-Sud de Clermont-Ferrand (Puy-de-Dome), France, at a scale of 1:25 000 (From Greffier, 1979, by permission)

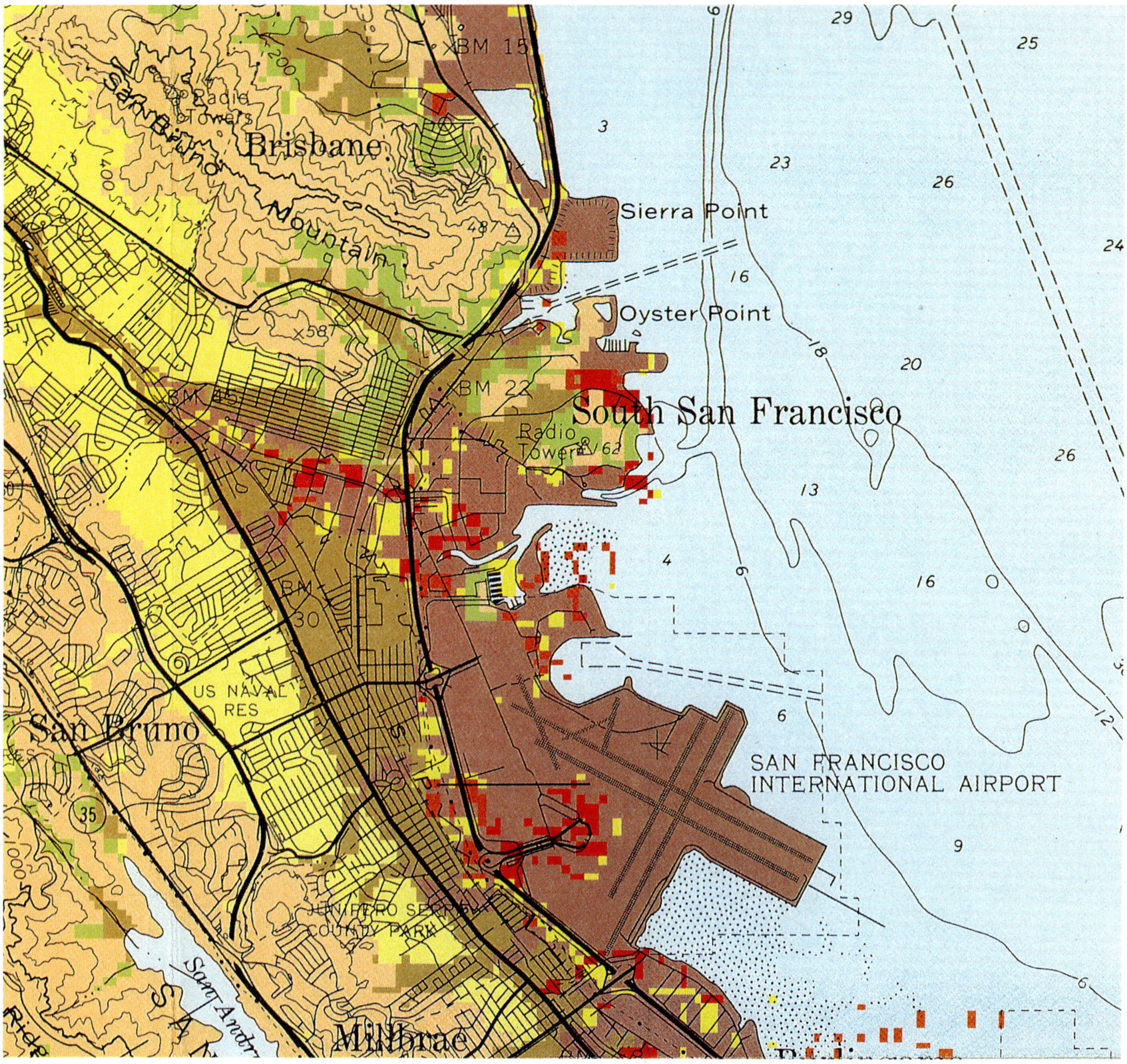

EXPLANATION

Relative liquefaction susceptibility and probability of susceptible sediment in subsurface materials

- LOW TO HIGH——Generally low to moderate, locally high near active and abandoned stream channels. Many of the latter channels may be buried and undetectible without subsurface investigations. Overall, there is a 1 to 10 percent probability of liquefiable sediment in these areas based on the average for the map unit
- LOCALLY HIGH——Possibly liquefiable sediment within or above bay mud confirmed by data from borehole drilled within 1-hectare grid cell
- MODERATE TO LOW——0.1 to 1 percent probability of liquefiable sediment based on average for map unit. Most of the average probabilities for these map units, however, are less than 0.3 percent (table 1), the low end of the range
- LOCALLY MODERATE——Possibly liquefiable sediment immediately below bay mud but not within or above; confirmed by data from borehole drilled within 1-hectare grid cell
- LOW——0.01 to 0.1 percent probability of liquefiable sediment based on average for map unit
- LOCALLY LOW——Liquefiable sediment not found in borehole drilled within 1-hectare grid cell
- VERY LOW——Less than 0.01 percent probability of liquefiable sediment based on average for map unit
- BEDROCK

Plate 19 Part of the 1:62 500 map of San Mateo County, California, USA, showing liquefaction susceptibility (From Youd and Perkins, 1987, by permission)

C. GEOLOGY GEOLOGIE GEOLOGIA ГЕОЛОГИЯ

1a Contact between different units, lines broken where approximate
Contact entre des unités différentes, en traits interrompus si hypothétique
Contacto entre unidades distintas, en trazos discontinuos si es hipotética
Контакт между различными единицами, если гипотетический пунктиром

2 Depth of top of the limestone
Profondeur du toit de la calcaire
Profundidad del techo de la caliza
Абсолютная отметка кровли известняка

Depth of top of the sandstone
Profondeur du tout du grès
Profundidad del techo de la arenisca
Абсолютная отметка кровли песчаника

5 Strike and dip
Direction et pendage
Dirección y buzamiento
Простирание и падение

6 Axis of anticline, with direction of axial plunge
Axe anticlinal, avec direction de plongement axial
Eje anticlinal, indicando su dirección de buzamiento
Ось антиклинали и направление ее погружения

7 Axis of syncline, with direction of axial plunge
Axe synclinal, avec direction de plongement axial
Eje sinclinal, indicando su dirección de buzamiento
Ось синклинали и направление ее погружения

10 Fault, with direction of downthrow side, line broken where uncertain
Faille, avec direction du côté affaissé, en traits interrompus si hypothétique
Falla con dirección del labio hundido, en trazos discontinuos, si es supuesta
Сброс с указанием опущенного крыла пунктиром сброса, если гипотетический

D. LITHOLOGY LITHOLOGIE LITOLOGIA ЛИТОЛОГИЯ

3 Sands and gravels
Sables et graviers
Arenas y gravas
Пески, галечники и гравий.

6 Sandstones with conglomerates
Grès avec conglomérats
Areniscas con comglomerados
Песчаники с конгломератами

7 Limestones
Calcaires
Calizas
Известняки

13 Extrusive rocks, fissured
Roches extrusives, fracturées
Rocas efusivas, fisuradas
Экструзивные породы, трещиноватые

E. HYDROGRAPHY HYDROGRAPHIE HIDROGRAFIA ГИДРОГРАФИЯ

1 Perennial stream with direction of flow
Cours d'eau pérenne, avec sens d'écoulement
Curso de agua perenne con dirección de la circulación
Постоянный водоток с направлением течения

4 Seasonal stream with direction of flow
Cours d'eau saisonnier avec sens d'écoulement
Curso de agua estacional con dirección de la circulación
Сезонный водоток с направлением течения

E.6 Disappearance point of stream
Point de disparition d'un cours d'eau
Punta de desparición de un curso de agua
Пункт исчезновения водотока

9 Marsh
Marais
Marisma
Болото

10 Area inundated during floods
Terrain inondé pendant les crues
Terreno inundado en las crecidas
Участок, затопленный паводковыми водами

11 Surface-water divide
Ligne de partage des eaux superficielles
Línea divisoria de las aguas superficiales
Водораздел поверхностных вод

12 Spring
Source
Fuente
Источник

13 Group of springs
Groupe de sources
Grupo de fuentes
Группа источников

14 Thermal or thermomineral spring (see G.4. and G.6.)
Source thermale ou thermominérale (voir G.4. et G.6.)
Fuente termal o termomineral (ver G.4. y G.6.)
Термальный или термоминеральный источник (см.G 4 и G6)

15 Natural pond or waterhole with no outlet
Etang ou point d'eau naturel sans écoulement
Laguna o punto de agua natural sin escorrentia
Бессточний естественный водоем или водохранилище (пруд)

16 Lake
Lac
Lago
Озеро

F. GROUNDWATER HYDROLOGY HYDROLOGIE DES EAUX SOUTERRAINS HYDROLOGIA DE LAS AGUAS SUBTERRANEAS ДИНАМИКА ПОДЗЕМНЫХ ВОД

1 Groundwater contours (piezometric surface on the section) for the limestone
Courbes isopièzes (niveau piézométrique de la coupe) pour la calcaire
Curvas isopiezas (nivel piezométrico en la sección) por la caliza
Изогипсы — (пьезометрическая поверхность разреза) для известняка

Groundwater contours (piezometric surface on the section) for the sandstone
Courbes isopièzes (niveau piézométrique de la coupe) pour le grès
Curvas isopiezas (nivel piezométrico en la sección) por la arenisca
Изогипсы (пьезометрическая поверхность разреза) для песчанника

2 Direction of groundwater flow with velocity in m per day
Sens d'écoulement de la nappe, avec la vitesse en m par jour
Dirección de la escorrentia de la capa acuífera, con la velocidad en m por día
Направление движений подземных вод с скоростью в м/сут

3 Groundwater divide
Ligne de partage des eaux souterraines
Línea divisoria de las aguas subterráneas
Граница бассейна подземных вод

5 Boundary of area of artesian flow, with date
Limite d'extension de la zone d'artésianisme, avec date
Línea de la superficie de la zona artesiana, con la fecha
Граница артезианского бассейна с датой

Plate 20 Model hydrogeological map and legend (From UNESCO, 1970, by permission)

F.11

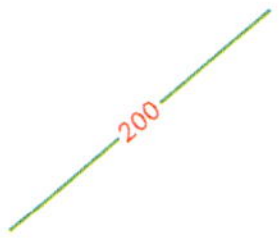

Lines of equal thickness of saturated zone for the limestone
Courbes dégale puissance de la couche saturée pour la calcaire
Líneas de igual potencia de la capa saturada por la caliza
Линии равной мощности известняка

18

750

Natural replenishment of the groundwater in mm per year
Alimentation naturelle de la nappe, en mm par an
Alimentación natural del acuífero, en mm por año
Естественное питание водоносного горизонта в мм/год

G. HYDROCHEMISTRY HYDROCHIMIE HIDROQUIMICA ГИДРОХИМИЯ

2

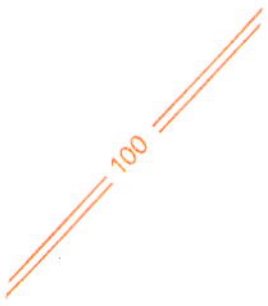

Depth of interface between fresh and saline groundwater
Profondeur de la surface de séparation entre l'eau douce et l'eau salée
Profundidad de la superficie de separación entre el agua dulce y la salada
Глубина залегания поверхности раздела пресных и соленых вод.

5

10°

Water temperature in degrees Celsius
Température de l'eau en degrés Celsius
Temperatura del agua en grados Celsius
Температура воды по Цельсию.

4 and 6

Dug well with sodium chloride water, not fit for use
Puits creusé avec eaux chlorurées sodiques, non utilisables
Pozo excavado con aguas clorurados sódicas, no utilizables
Копаный колодец с хлоридной натриевой водою, неиспользуемой.

Dug well with sodium chloride water, 2 - 4 g/l
Puits creusé avec eaux chlorurées sodiques, 2 - 4 g/l
Pozo excavado con aguas cloruradas sódicas, 2 - 4 g/l
Копаный колодец с хлоридной натриевой водою 2—4 г/л

Dug well with sodium chloride water, 4 - 8 g/l
Puits creusé avec eaux chlorurées sodiques, 4 - 8 g/l
Pozo excavado con aguas cloruradas sódicas, 4 - 8 g/l
Копаный колодец с хлоридной натриевой водою 4—8 г/л

Dug well with sodium chloride water >8 g/l
Puits creusé avec eaux chlorurées sodiques, >8 g/l
Pozo excavado con aguas cloruradas sódicas, >8 g/l
Копаный колодец с хлоридной натриевой водою >8 г/л.

Group of springs with magnesium sulphate water, >8 g/l
Groupe de sources avec eaux sulfatées magnésiennes, >8 g/l
Grupo de fuentes con aguas sulfatadas magnesianas, >8 g/l
Группа источников с сульфатной магниевой водою > 8 г/л.

Group of springs with sodium chloride water, 4 - 8 g/l
Groupe de sources avec eaux chlorurées sodiques, 4 - 8 g/l
Grupo de fuentes con aguas cloruradas sódicas, 4 - 8 g/l
Группа источников с хлоридной натриевой водою 4—8 г/л

8

Highly polluted stream (organic pollution)
Cours d'eau fortement pollué (pollution organique)
Curso de agua muy polucionada (polución orgánica)
Сильно загрязненный водоток (органическое загрязнение).

9

Stream with high salinity
Cours d'eau salé
Curso de agua salada
Соленый водоток

10

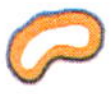

Salt lake
Lac salé
Lago salado
Соленое озеро

H. BOREHOLES, WELLS AND OTHER WORKS FORAGES, PUITS ET AUTRES OUVRAGES SONDEOS, POZOS Y OTRAS OBRAS СКВАЖИНЫ, КОЛОДЦЫ И ДРУГИЕ ВЫРАБОТКИ

H.3

Dug well, non-artesian
Puits creusé, dans une nappe libre
Pozo excavado, en una capa acuífera libre
Копаный колодец в безнапорном водоносном горизонте

Dug well, artesian, not overflowing
Puits creusé, dans une nappe captive, mais non-jaillissant
Pozo excavado, artesiano no surgente
Копаный колодец в напорных водоносных горизонта, но не самоизливающийся

Dug well, artesian, overflowing
Puits creusé artésien (jaillissant)
Pozo excavado, artesiano (surgente)
Копаный колодец артезианский (самоизливающийся)

Recharge well
Puits creusé, d'injection
Pozo excavado de recarga
Поглощающий колодец

Dry well
Puits creusé, sec
Pozo excavado, seco
Сухой колодец

4

Drilled well (see also H.3.)
Puits foré (voir aussi H.3.)
Pozo perforado (ver también H.3.)
Буровой колодец (см. также H.3)

8

Spring used for supply
Captage de source
Captación en una fuente
Каптированный источник

9

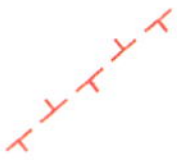

Drainage gallery
Galerie drainante
Galería de drenaje
Дренажная галлерея

10

Groundwater pumping station
Captage d'eau souterraine
Estación de bombeo de agua subterranea
Насосная станция из подземых вод

11

Groundwater observation station, with recorder
Station de jaugeage de l'eau souterraine, avec enregistreur
Estación de aforo de aguas subterráneas, con registrador
Наблюдательная станция за режимом подземных вод

Decade station
Station de la Décennie hydrologique
Estación del Decenio hidrológico
Станция гидрологического десятилетия

12

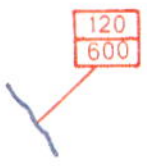

Stream gauging station, with yearly average flow and area of catchment
Station de jaugeage d'un cours d'eau, avec débit moyen annuel et surface du bassin de drainage
Estación de aforo de un curso de agua indicando el caudal medio anual y la superficie de la cuenca de drenaje
Наблюдательная станция на водотоке, указывается среднегодовой расход и площадь дренируемого бассейна

17

Dam (with capacity of reservoir in million m^3)
Barrage de retenue ou de dérivation (avec capacité de réservoir en millions de m^3)
Presa de embalse o derivación (con capacidad en milliones de m^3)
Плотина водохранилища с емкостью в млн. $м^3$

19

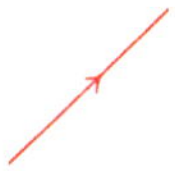

Canal, irrigation canal (perennial waters)
Canal, canal d'irrigation (eau pérenne)
Canal, canal de irrigación (agua perenne)
Канал, ирригационный канал (постоянный водоток)

20

Canal, flood waters
Canal, eaux de crue
Canal, (aguas de crecida)
Канал, (паводковые воды)

21

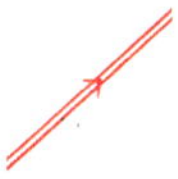

Drainage canal
Canal de drainage
Canal de drenaje
Дренажный канал

HYDROGEOLOGICAL MAP OF SOUTHERN AQUATANIA
CARTE HYDROGÉOLOGIQUE D'AQUATANIA DU SUD
MAPA HIDROGEOLÓGICO D'AQUATANIA DEL SUR
ГИДРОГЕОЛОГИЧСКАЯ КАРТА ЮЖНОЙ АКВАТАНИИ

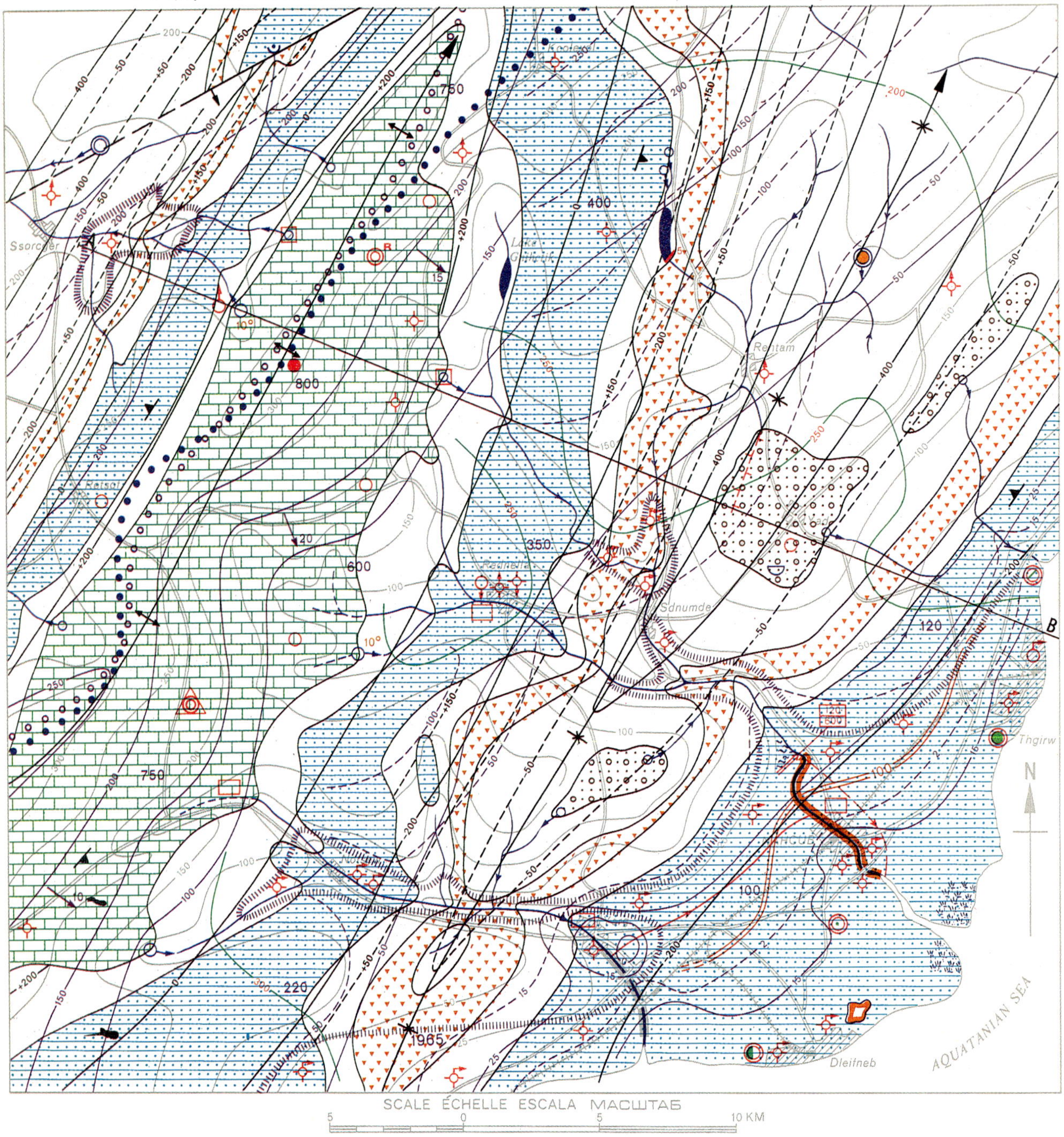

Group of springs
Groupe de sources
Grupo de fuentes
Группа источников

SECTION ALONG A–B
COUPE SUIVANT A–B
SECCIÓN SIGUIENTE A–B
РАЗРЕЗ ВДОЛЬ A–B

Stream
Cours d'eau
Curso de agua
Водоток

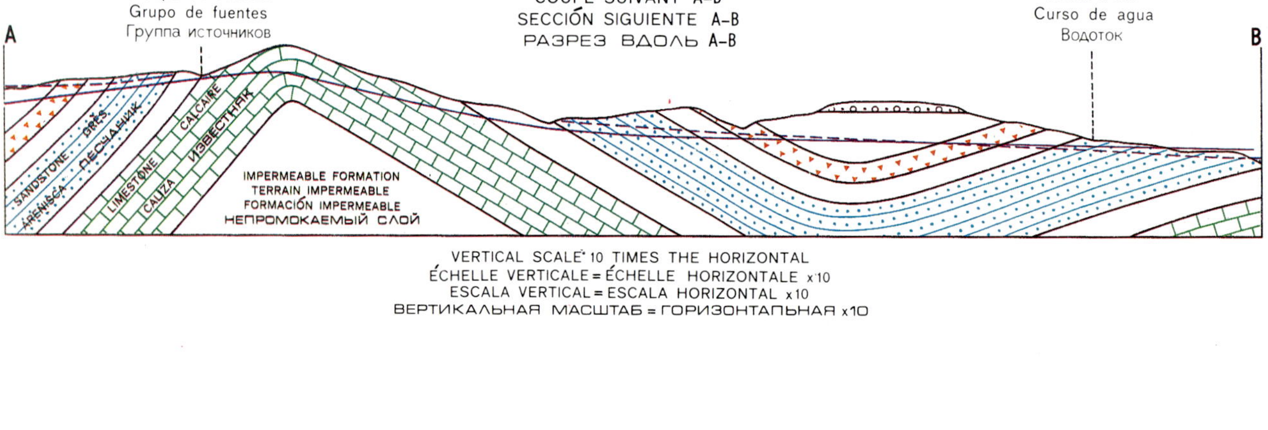

VERTICAL SCALE 10 TIMES THE HORIZONTAL
ÉCHELLE VERTICALE = ÉCHELLE HORIZONTALE x10
ESCALA VERTICAL = ESCALA HORIZONTAL x10
ВЕРТИКАЛЬНАЯ МАСШТАБ = ГОРИЗОНТАПЬНАЯ x10

9

Terrain evaluation: cost-effective mapping

9.1 Introduction

Engineering and environmental geological maps at small scales are generally prepared by a combination of compilation and reduction from existing larger scale geological, topographical and other maps, and interpretation of aerial photographs, with the requisite amount of field reconnaissance or spot checking on the ground. Only rarely are they prepared by any one of these techniques alone. It is possible to prepare them from field reconnaissance, but it is often easier and may be even more accurate to compile much of the map by interpretation of aerial photographs. Many geomorphological features are then very clear. Patterns of soil and rock distribution are sometimes more easily seen from the air than on the ground, and tracing outcrop patterns is much faster than on foot, especially in rough terrain. Air photographs are of doubtful use in heavily forested country, where the only exposures may be in stream valleys, road cuts, or cliffs and quarries. In such areas extensive field work may be required even for small-scale mapping.

It is from small-scale geological maps that the zoning map (see Plate 9) at a scale of 1:200 000 was derived, with the added implication that engineering geological maps are derived from existing geological maps and not directly produced as such. Mapping on the actual land surface may be supplemented, or supplanted, by aerial photograph interpretation and other forms of remote sensing, leading to some form of land classification involving the identification of landscape patterns. There is generally little need to develop the terrain evaluation approach to define morphological units of possibly similar engineering behaviour in Britain, as the country is so well mapped geologically. The value of this method will probably remain confined to developing countries in which little basic mapping has been done, and then only at small scale.

Three aspects of small-scale mapping need to be considered:

- classification of terrain for engineering purposes;
- use of remote sensing to prepare maps;
- derivation of engineering geological maps from existing geological maps.

In considering terrain classification it has been necessary to be flexible by including some medium- to large-scale aspects in order to present a reasonably complete picture of what is involved.

9.2 PUCE system for terrain analysis

A scheme for the classification of terrain for engineering purposes in Australia was discussed by Aitchison and Grant (1967) and described in detail by Grant (1968). The scheme is known as the PUCE (Pattern–Unit–Component–Evaluation) Programme, early versions of which were described and illustrated, e.g. in Grant (1972), with more recent descriptions and definitions of terrain classes given fully by Grant and Finlayson (1978a,b) and reviewed by Finlayson (1984). The system has been metricated.

The basis of the programme is that any area of land can be uniquely defined in terms of its topography, i.e. the slope characteristics; underlying lithological and structural characteristics determined by the geology; and soil and vegetation characteristics. By defining intervals in these terrain characteristics at suitable significant levels, valid natural terrain classes may thus be determined.

The programme operates at four levels of generalization – the province, terrain pattern, terrain unit and terrain components. In order to define these classes, it is necessary to define the limits of variation appropriate to each class allowed in each of the terrain characteristics. The terrain component, the largest scale unit, can be defined uniquely; the terrain unit can then be defined as a limited association of terrain components, and the terrain pattern as a limited association of terrain units.

9.2.1 Definition of terrain classes

Province
A province can be defined as an area of constant geology at the Group level; for example by:

- rocks of sedimentary or volcanic origin, uniform at Group level;

- rocks of plutonic origin with uniform lithology and age;
- alluvium and colluvium occurring in a single drainage division;
- aeolian material with uniform lithology on a continental basis.

A province can usually be determined from a geological map at a scale of 1:250 000.

Members of each class are numerically coded as in Table 9.1. Provinces are classified first on a rock age basis (2 digits) and secondly serially in order of recognition (3 digits).

Table 9.1 The numerical system for nomenclature of provinces in the PUCE terrain classification system*

Eonothem	*Erathem*		*System*	
ARCHAEOZOIC	Archaean	1	General	10
	Proterozoic	2	General	20
			Lower	21
			Middle	22
			Upper	23
PHANEROZOIC	Palaeozoic	3	General	30
			Cambrian	31
			Ordovician	32
			Silurian	33
			Devonian	34
			Carboniferous	35
			Permian	36
	Mesozoic	4	General	40
			Triassic	41
			Jurassic	42
			Cretaceous	43
	Cainozoic	5	General	50
			Tertiary	51
			Quaternary	52

*Provinces representing rocks of sedimentary, volcanic or plutonic origin within each of the above systems are then numbered serially as encountered, using a three-digit system which allows 999 provinces per system, e.g. the Brighton group which occupies an area near Melbourne, Victoria, is Tertiary in age and is the ninth Tertiary province recognized in Australia; therefore it is numbered as Province 51.009. Provinces representing alluvium and colluvium of Quaternary age in Australia are numbered in accordance with the recognized drainage divisions.

Provinces representing material of aeolian origin are numbered as follows:

52.002 = siliceous sand
52.011 = calcareous sand (calcarenite)
52.021 = silt (loess)
52.022 = clay (parna)

Terrain pattern

A terrain pattern can be identified as an area comprising a single landform having recurring topography, soil associations and natural vegetation associations. A terrain pattern has a consistent maximum local relief amplitude, characteristic drainage pattern and uniform drainage density. The numerical system of classification is given in Table 9.2, with the first digit for local relief and the second for degree of dissection.

It is essentially a uniform landscape, that should be coincident with an area represented by a distinctive pattern on an aerial photograph of suitable scale.

Terrain patterns are mapped at a scale of 1:250 000.

Terrain unit

A terrain unit can be defined as the area covered by a single landform having characteristic soil and vegetation associations.

Table 9.2 Numerical system for nomenclature of terrain patterns

Greatest local relief amplitude	*Mean drainage density*
0 to 15 m	0 No surface drainage
1 to 30 m	1 One stream-line per 1.6 km
2 to 75 m	2 Two stream-lines per 1.6 km
3 to 150 m	3 Three stream-lines per 1.6 km
4 to 300 m	4 Four stream-lines per 1.6 km
5 to 600 m	5 Five stream-lines per 1.6 km
6 to 1200 m	6 Six stream-lines per 1.6 km
7 to 2400 m	7 Seven stream-lines per 1.6 km
8 to 3600 m	8 Eight stream-lines per 1.6 km
9 to >3600 m	9 Nine or more stream-lines per 1.6 km

The terrain unit so defined can be regarded as being composed of a limited number of terrain components which always recur in the same spatial relationships within the terrain unit. The characteristic slopes and soil and vegetation associations of the terrain unit will be those produced by a synthesis of those of the terrain components associated to form the terrain unit.

Topographically, the terrain unit will fall into classes, and within each class will have a characteristic association of slopes and a consistent local relief amplitude. The classes so far recognized are listed as follows, but the list is not exhaustive:

1. Surfaces, flat or with varying degrees of undulation, i.e. undissected, dissected and/or eroded. These surfaces are not necessarily tied to specific erosional or depositional surfaces, i.e. the same terrain unit may occur on more than one erosional or depositional surface; such occurrences will be separated by a terrain unit classified under (2).
2. Slopes between surfaces as in (1); gentle, steep, or escarpment-like.
3. Isolated hills, ridges, etc., except those with flat tops; in the latter case the flat tops and surrounding slopes have been given independent terrain unit status in accordance with (1) and (2), respectively.
4. Drainage lines, lakes, etc.

The characteristic soil association of the terrain unit will be dominated by one of the following textural types:

Rock outcrop, pockets of shallow soil or gravel	Sand over clay soils (G)
Clay soils (Ug)	Sand over clay soils (D)
Clay soils (U or G)	Sandy soils (U or G)
Clay soils (D)	Stratified soils
Silty soils	Organic soils

The symbols refer to the classification system of Northcote (1965), given in abridged form in Table 9.3.

The characteristic vegetation formation will be dominated by one of the following classes:

Bare, sparse grass, occasional tree or shrub	Woodland
Grassland	Forest
Shrubland	Rain forest
Open woodland	Fresh-water swamp forest
Savannah woodland	Salt-water swamp forest

Terrain units are classified first in terms of physiography (1 digit) and surface configuration (1 digit), and secondly in terms of texture and subdivided primary profile form

Table 9.3 Abridged soil classification system (After Northcote, 1974)

Primary profile form	*Soil type*	*Subdivided primary profile form*	*Criteria*
U	Uniform—the texture of the soil is uniform throughout the entire depth of the profile	Uc Um Uf Ug	Coarse textured soils (the sands) Medium textured soils (the silts) Non-cracking fine textured soils (the clays) Cracking fine textured soils (the expansive clays)
G	Gradational—the clay content of the soil increases gradually with depth throughout the profile	Gc Gn	Profile is calcareous throughout Profile is not calcareous throughout
D	Duplex—the clay content increases suddenly at some depth in the profile (the minimum amount of increase of clay content is defined in Northcote, 1974)	Dr Db Dy Dd Dg	Clay B horizon is red in colour Clay B horizon is brown in colour Clay B horizon is yellow in colour Clay B horizon is dark in colour Clay B horizon is grey in colour
O	Organic—organic soils	No further subdivision	

Further subdivisions can be made on the basis of soil pH and colour relationships, but these properties are not usually important for geotechnical investigations and are not considered here

of the dominant soil (1 digit) and vegetation formation (1 digit), as shown in Table 9.4.

Terrain parameters relevant to terrain patterns and terrain units are not given in absolute figures. To avoid a multitude of different numbers, they have been classified using class intervals that are considered to be significant, as shown in Table 9.5.

Terrain component

A terrain component can be defined as an area which has:

1. Apart from microtopography, along each of a pair of lines parallel to the major and minor axes of a slope, a constant rate of change of curvature always in the same sense, i.e. either planar, convex or concave but not concavo-convex.
2. Uniform underlying lithology in a uniform structural environment.
3. A consistent association of soils such that each layer of the soil can be expressed within one class of the Unified Soil Classification (USC) and the whole soil profile within one class of a subdivision of the primary profile form (PPF) (Northcote, 1965, 1974). This is except for stratified alluvial aeolian or aeolian soils in which the topmost layer only can be expressed in terms of a subdivision of the primary profile form.
4. A characteristic vegetation association, i.e. although more than one species or genus may be present, the species or genera always occur in the same spatial relationship to each other and there is no discontinuity in their occurrence.

The numerical system for terrain components is set out in Table 9.6.

A hierarchical system for terrain classification can thus be devised based upon natural, recognizable criteria. In this system, each class is composed of a repetitive association of members of the next class in the hierarchy. That is:

- a province consists of a repetitive association of terrain patterns;
- a terrain pattern consists of a repetitive association of terrain units;
- a terrain unit consists of a repetitive association of terrain components.

Terrain provinces and terrain patterns are mapped at small scale, typically 1:250 000, using aerial photographs.

Mapping terrain units, usually at 1:250 000 or larger scales, and terrain components, at 1:2500 or larger, is only carried out in selected areas defined by terrain patterns and terrain units.

9.2.2 Classification nomenclature

The nomenclature used in the PUCE system is numerical:

(a) *Terrain components*

These are allocated eight digits:

1 the types of slopes along the orthogonal axis, i.e. combinations of convex, concave and planar slopes;
2,3 the maximum angles of slope along each of the orthogonal axes;
4,5 the soil horizons and profile (serial);
6 land use, in developed areas, or the surface cover in undeveloped areas – standardized by listing;
7,8 the vegetation association (serial).

(b) *Terrain units*

These are allocated four digits:

1,2 the land form;
3 the soil profile;
4 the vegetation formation.

These are all standardized by listing (see Table 9.4).

Table 9.4 Numerical system for nomenclature of terrain units in the PUCE terrain classification system

Topography	*Soils dominantly*	*Vegetation dominantly*
1. Flat to undulating or sloping smooth surfaces, usually with relatively deep soil	0. Rock outcrop, pockets shallow soil and gravel	0. Bare, sparse grass, occasional tree or shrub
1.1 Flat surface		
1.2 Gently undulating surface (to 2°)	1. Clay soils (Ug)	1. Grassland, occasional tree or shrub
1.3 Moderately undulating surface (to 5°)	2. Clay soils (U or G)	
1.4 Undulating surface (to 10°)		2. Shrubland, occasional tree
1.5 Sloping surface (to 2°)	3. Clay soils (D)	
1.6 Sloping surface (to >2°)	4. Silty soils	3. Open woodland
1.7 Undulating sloping surface		4. Woodland
1.8 Strongly undulating surface (to >10°)	5. Sand over clay soils (G)	
2. Irregular sub-horizontal to undulating eroded surfaces, usually with rock outcrop or shallow soil	6. Sand over clay soils (D)	5. Open forest
	7. Sandy soils (U or G)	6. Closed forest
2.1 Flat surface		7. Rain forest
2.2 Eroded surface	8. Stratified soils	
2.3 Benched surface	9. Organic soils	8. Fresh water swamp forest
2.4 Undulating eroded surface		9. Salt water swamp forest
2.5 Strongly undulating eroded surface		
2.6 Moderately dissected surface		
2.7 Moderately dissected eroded surface		
2.8 Strongly dissected surface		
2.9 Strongly dissected strongly eroded surface		
3. Slopes, including escarpments, between surfaces		
3.1 Smooth steep slope		
3.2 Rough rocky steep slope		
3.3 Gentle slope		
3.4 Gentle dissected slope (to 5°)		
3.5 Dissected slope (to 10°)		
3.6 Benched slope		
3.7 Steep dissected slope (to >10°)		
4. Isolated hills		
4.1 Conical hill including pap and butte		
4.2 Rounded hill		
4.3 Complex hill		
4.4 Razor-backed hill		
4.6 Steep knoll, mamelon or volcanic plug		
4.7 Small volcanic vent		
4.8 Elongated rounded hill		
4.9 Dissected hill		
5. Isolated ridges, etc		
5.1 Low ridge (to 5° slope)		
5.2 Dissected ridge		
5.3 Strongly dissected ridge		
5.4 Rough irregular ridge		
5.5 Linear strike ridge		
5.6 Razor-backed ridge		
5.7 Linear ridge (other than strike ridge)		
5.9 Sand ridge		
6. Irregular eroded sloping surfaces, usually with rock outcrop or shallow soil		
6.2 Eroded sloping surface		
6.3 Undulating eroded sloping surface		
6.5 Undulating dissected eroded sloping surface		
7. Mountains (volcanic)		
7.3 Moderately dissected mountain		
7.8 Heavily dissected mountain		
8. Not allocated (reserved for fold mountains)		
9. Drainage systems		
9.1 Major stream channel		
9.2 Minor stream channel		
9.3 Braided stream channel system		
9.4 Incised gully, ravine, etc.		
9.5 Lake, lagoon, billabong (oxbow lake)		
9.6 Reticulated stream channel system (flood-out)		
9.7 Tidal channel		
9.8 Swamp or marsh		
9.9 Sink hole		

Example. An area of gently undulating surface to 2° slope with clay soils (Ug) and grassland is terrain unit 1.2.11

Table 9.5 Terrain parameters—class intervals in the PUCE system

Classes of lengths and widths of terrain units

1	to 10 m	6	to 2000 m
2	to 50 m	7	to 4000 m
3	to 100 m	8	to 8000 m
4	to 500 m	9	Extensive
5	to 1000 m		

Classes of lengths and widths of terrain components

1	to 2 m	6	to 100 m
2	to 5 m	7	to 500 m
3	to 10 m	8	to 1000 m
4	to 20 m	9	Extensive
5	to 50 m		

Classes of relief

1	to 1 m (microtopography)	6	to 75 m
2	to 3 m	7	to 150 m
3	to 6 m	8	to 300 m
4	to 15 m	9	to > 300 m
5	to 30 m		

Classes of tree spacings, heights, and girth diameters

1	to 2 m	1	to 2 m	1	to 0.05 m
2	to 3 m	2	to 3 m	2	to 0.1 m
3	to 6 m	3	to 6 m	3	to 0.2 m
4	to 30 m	4	to 12 m	4	to 0.3 m
5	Scattered	5	to 18 m	5	to 0.5 m
6	Occasional	6	to > 18 m	6	to 0.6 m
				7	to > 0.6 m

Classes of stream densities

1	< 0.5 total streams/1.6 km
2	0.5 total streams/1.6 km
3	1.0 total streams/1.6 km
4	1.5 total streams/1.6 km
5	2.0 total streams/1.6 km
6–9	increasing at 0.5 intervals

Classes of slopes

0	Level	5	to 20°
1	to 1°	6	to 40°
2	to 2°	7	to 60°
3	to 5°	8	to > 60°
4	to 10°	9	to vertical

Classes of rubble sizes

1	to 0.025 m	5	to 0.3 m
2	to 0.05 m	6	to 0.6 m
3	to 0.1 m	7	to 2 m
4	to 0.2 m	8	to > 2 m

(c) *Terrain patterns*

These are allocated three digits:

1 the maximum local relief amplitude, classified using significant class intervals;
2 the stream density;
3 serial, if required.

Digits 1 and 2 are classified using significant class intervals (see Table 9.5).

(d) *Provinces*

These are allocated five digits:

1 the geological erathem;
2 the geological system within each erathem;
3,4,5 serial.

The numerical system is listed in Table 9.1.

An example of the numerical nomenclature used for terrain classification is given in Table 9.7, but for a complete description reference should be made to Grant (1975a,b).

9.2.3 Example of terrain classification description sheet

The example is from the Melbourne area, Victoria (Grant, 1972), Terrain Pattern No. 46 developed on Devonian granite in Province No. 34.001 (see Table 9.1). Characterizing the province are seven terrain patterns each of which is outlined in Table 9.8; pattern numbers are based on relief amplitudes and stream frequencies (see Table 9.2).

Each terrain pattern is illustrated (Figure 9.1) by a characteristic, scaled, cross-section showing the typical location of terrain units, and a diagrammatic block diagram of the topography and the arrangement of terrain units within the terrain pattern. A typical drainage net for the particular terrain pattern is also given. Tabulated are descriptions of the constituent terrain units in the terrain pattern, in the prescribed format.

9.2.4 Example of map and legend for terrain patterns

The terrain classification map of Melbourne (Grant, 1972) is published in colour at a scale of 1:250 000. A part of the map showing Terrain Pattern No. 46 on Devonian granite has been selected as illustration, together with the complete legend for the seven terrain patterns comprising Province 34.001. With the aid of the legend and the descriptive memoir, the map is particularly easy to interpret (Plate 12).

Although the terrain classification is stated to be for engineering purposes in the Melbourne area, nowhere in the report or on the map is there any specific reference to engineering geology. It would appear that the documents provided should be regarded as a basic assessment or evaluation that can be applied for a number of purposes. Finlayson (1984) has enumerated these as including: location of materials and routes for railway and road construction (Grant and Ferguson, 1979); water resources (Grant and Ferguson, 1979); military and off-road mobility (Grant and Ferguson, 1979); engineering geology (Grant, 1976; Finlayson, 1982); urban and road planning (Grant *et al.*, 1981); regional development (Grant *et al.*, 1982).

9.2.5 Using terrain assessments

Grant and Finlayson (1978), having described the PUCE system of terrain evaluation, consider the application of terrain maps in engineering and planning. This involves assessment and evaluation of the geotechnical resource capability of a terrain. In one particular study (their Appendix II), all soils were selectively sampled to a depth of 2 m.

TERRAIN PATTERN No. 46

PROVINCE No. 34.001
Devonian (granite)

LITHOLOGY — Granite, Granodiorite

OCCURRENCE — Restricted to higher rocky parts of the Harcourt, Cobaw, Kerrisdale Granitic areas (Mt. Alexander excluded)

TOPOGRAPHY — Strongly undulating rough, rocky, eroded, granitic terrain

INCLUSIONS — Terrain pattern 24, province 34.001

CHARACTERISTIC CROSS-SECTION SHOWING TYPICAL LOCATION OF TERRAIN UNITS

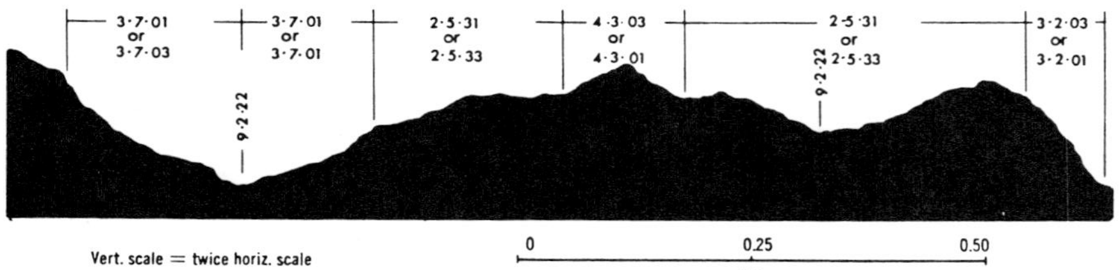

TYPICAL DRAINAGE NET OF TERRAIN PATTERN

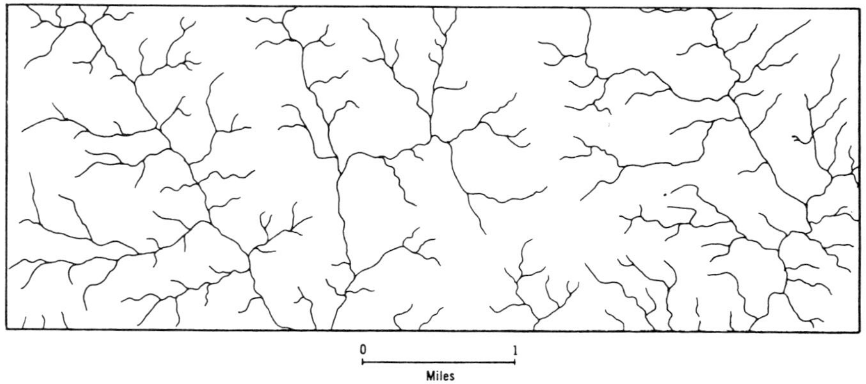

Order	Stream Frequency per Mile 1st	2nd	3rd	>3rd	Total	Drainage Type
N-S	2.0	1.5	0.5	<0.5	4.0	Angulate to sub-dendritic
E-W	2.0	1.5	0.5	<0.5	4.0	

TERRAIN UNITS

Number	Terrain Pattern Area (%)	Occurrence	Description of Dominant Topography	Soil	Surface Cover	Vegetation	Inclusions
1.4.66	<1	Local: Mt. Tallarook only, higher surface	Undulating surface slopes: irregular to 10°	Gradational shallow yellow-red sandy clay (SC-SW;Gc) over decomposed rock; areas of rock outcrop	Forestry	Forest, mountain grey gum, blue gum, manna gum, blackwood, silver wattle, fern	Terrain pattern 24, province 34.001
2.5.31 or 2.5.33	78	Continuous; extensive	Strongly undulating surface, crests: undulating, convex to 20° slopes: undulating to 20°	Crests: uniform shallow red to red-brown coarse granitic sand (S[illegible];Uc) with tor outcrop and associated pockets of red-brown clayey sand (SC) slopes: duplex dark grey clayey sand to 10 in. over yellow-red sandy clay to 3 ft (SC/CL;Dr)	Pasture or unused	Mostly cleared, grassland, mid-height grass, scattered trees, areas of open woodland, peppermint, grey box, yellow box, acacia, Local, Cobaw Range and Barringo area forest, messmate, red stringybark, manna gum (lower slopes only), silver wattle, fern	-
3.2.03 or 3.2.01	2	Mostly marginal to terrain pattern	Steep rough rocky slopes slopes: irregular, planar to convex to 40°	Rock outcrop, piled tors, some pockets of shallow granitic sand	Unused, some pasture	Open woodland, some areas cleared, grassland, mid-height grass, fern, scattered trees, peppermint, grey box, yellow box, acacia, fern	-

Terrain Parameters Terrain Unit No.	Max. Local Relief Amplitude (ft)	Length of Terrain Unit	Width of Terrain Unit
1.4.66	200	5 miles	1 mile
2.5.31 or 2.5.33	500	Extensive	Extensive
3.2.03 or 3.2.01	1000	Extensive	1000 yards

DIAGRAMMATIC REPRESENTATION OF TOPOGRAPHY AND ARRANGEMENT OF TERRAIN UNITS WITHIN TERRAIN PATTERN

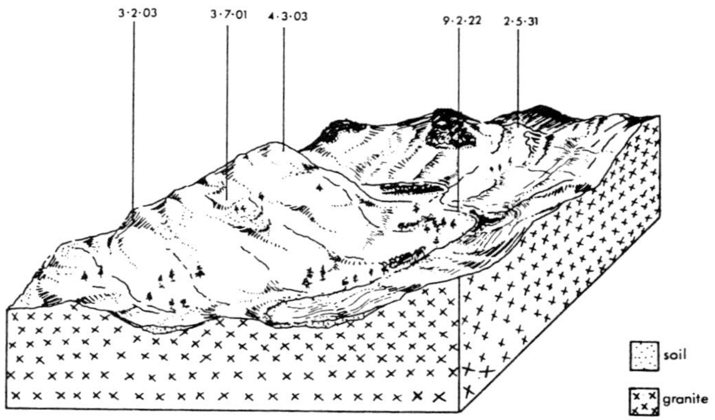

Figure 9.1 Terrain Pattern No. 46 on Devonian granite in the Melbourne Area, Victoria, Australia. Only three of the six terrain units are tabulated (From Grant, 1972, by permission)

Table 9.6 Numerical system for nomenclature of terrain components

	A	B	C	D	E	F	G	H
Slope profile	A							
Maximum magnitude parallel to major axis		B						
Maximum magnitude parallel to minor axis			C					
Soil profile				D	E			
Land-use or surface cover						F		
Vegetation association							G	H

A Slope profile

	Parallel to	
	Major axis	*Minor axis*
1	Planar	Planar
2	Planar	Concave
3	Planar	Convex
4	Concave	Planar
5	Concave	Concave
6	Concave	Convex
7	Convex	Planar
8	Convex	Concave
9	Convex	Convex

B C Slope—maximum magnitude parallel to major axis or minor axis

0	Flat
1	1°
2	2°
3	5°
4	10°
5	20°
6	40°
7	60°
8	>60°
9	Vertical

D E Soil profile—consistent within one class of the Unified Soil Classification and one subdivision of the primary profile form (Northcote, 1974)—numbered serially within each province

F Land-use or surface cover where applicable

	Land-use		*Surface cover*
0	Unused	0	Not present
1	Forestry	1	Silcrete, rounded
2	Pasture	2	Silcrete, angular
3	Agriculture	3	Ironstone, rounded
4	Recreation	4	Ironstone, platy
5	Urban development	5	Porcellanite
6	Not allocated	6	Quartzite
7	Not allocated	7	Calcrete
8	Not allocated	8	Salt
9	Not allocated	9	Rock outcrop and rubble of appropriate lithology

G H Vegetation association—numbered serially within each province

Table 9.7 Example of the numerical nomenclature used for terrain classification in the PUCE system

Province	*Terrain pattern*	*Terrain unit*	*Terrain component*
52.009	*22/2*	*1.7.11*	*431 032 01*

means:

Province	52	Quaternary
	.009	Ninth recognized province of Quaternary age
Terrain pattern	2	Relief amplitude to 75 m
	2	Drainage density 2 stream-lines per 1.6 km
	/2	Second recognized terrain pattern with the above parameters in the particular province
Terrain unit	1.7	Undulating sloping surface
	.1	Clay soils (Ug)
	1	Grassland
Terrain component	4	Slopes major axis concave, minor axis planar
	3	—major axis to 5°
	1	—minor axis to 1°
	03	Soil profile (serial within province)
	2	Land use—pasture
	01	Vegetation (serial within province)

The assessments sought were:

- suitability for road sub-base;
- suitability for road fill and/or untreated wearing surface;
- suitability for foundations – low buildings and high buildings (shallow and/or deep foundations);
- degree of limitation for use as sewerage dispersal field or septic use;
- source of sand and gravel;
- landslide occurrence;
- slope stability;
- suitability for trenching or tunnelling for underground facilities;
- suitability for embankments (water and non-water retaining);
- suitability for grassed waterways (comments on wind and/or water erosion).

Assessments of these matters would be based upon the following tests and observations:

1. Characterization:
   (a) depth of soil profile and depth of water table (where shallow);
   (b) depth to bedrock and type of rock (residual areas only);

   (c) grain-size distribution, Atterberg limits, linear shrinkage;
   (d) moisture content, density;
   (e) pH, dispersivity, soil suction;
   (f) unified soil classification;
   (g) on-site assessment of the drainage characteristics.
2. Engineering (where relevant):
   (a) simple consolidation;
   (b) simple shear strength (triaxial test);
   (c) repeated loading modulus.

Assessments for building foundations were made for small buildings only (footing width 300 mm). It was considered that large buildings would require foundation investigations that would be beyond the scope of soil sampling (to 2 m) planned to be used for the investigation.

Assessments for road construction (specification – sub-base) were made for fill material that could be used as sub-base and as sub-grade. The latter assessment was not called for in the original specification, but in view of the natural materials in the area, it was considered important that this assessment should be made.

Although the assessment and evaluation of the geotechnical resource capability of terrain may or may not be conducted concurrently with the terrain classification stage, it is essentially a separate procedure. Terrain classification is the necessary tool used for assessment and evaluation, but it is the geotechnical assessment and evaluation that supplies the data necessary for engineering and urban and regional development feasibility studies, planning and construction.

9.3 Oxford–MEXE system of terrain classification

The Oxford–MEXE (Military Engineering Experimental Establishment) system of classification, adopted at the Transport and Road Research Laboratory (TRRL) for location of highways in the humid tropics (Lawrence, 1972, 1978), was originally developed in the savanna lands of Africa, where the ground surface is visible from the air and consequently terrain evaluation is relatively easy.

Research into methods of land classification has been summarized by Webster and Beckett (1970). The Oxford–MEXE system recognizes seven categories of terrain, each defined by a combination of climate, geology and landform (Brink *et al.*, 1966). In decreasing order of size, the units are:

Table 9.8 The seven terrain patterns comprising Province No. 34.001 (Age Group: Devonian; Lithology: Granite, Granodiorite)

Terrain pattern no.	*Topography*	*Rock condition*	*Earthen materials (profile)*	*Land-use*	*Vegetation*
1. TOPOGRAPHY: Moderately undulating to undulating terrain					
13	Moderately undulating terrain containing some isolated rocky hills	Crystalline	Duplex grey-yellow sandy clay over medium to heavy-textured clay	Pasture, some agriculture	Most cleared, grassland, or woodland
2. TOPOGRAPHY: Undulating terrain					
24	Undulating terrain containing isolated rocky hills	Crystalline	Uniform grey clayey sand or duplex grey-brown clayey sand over sandy clay, some rock outcrop	Pasture, some unused	Mostly cleared, grassland, or woodland
26	Undulating dissected terrain	Crystalline	Gradational red friable clay	Forestry, unused, or farmland	Forest, or cultivated
3. TOPOGRAPHY: Undulating to strongly undulating terrain					
38	Strongly undulating dissected terrain	Crystalline	Duplex grey-yellow clayey sand over heavy-textured clay, some rock outcrop	Pasture, or unused	Cleared, grassland, or woodland, or forest
4. TOPOGRAPHY: Strongly undulating terrain					
44	Dissected mountain	Crystalline	Rock outcrop, pockets of clayey sand or sandy clay	Unused, some pasture, recreation	Woodland, or cleared, grassland
45	Linear ridge	Crystalline	Rock outcrop, pockets of clayey sandy gravel	Pasture	Mostly cleared, grassland
46	Strongly undulating rough rocky eroded terrain	Crystalline	Rock outcrop with uniform red sand or clayey sand, or duplex grey clayey sand over sandy clay	Pasture or unused	Cleared, grassland, or open woodland, or forest

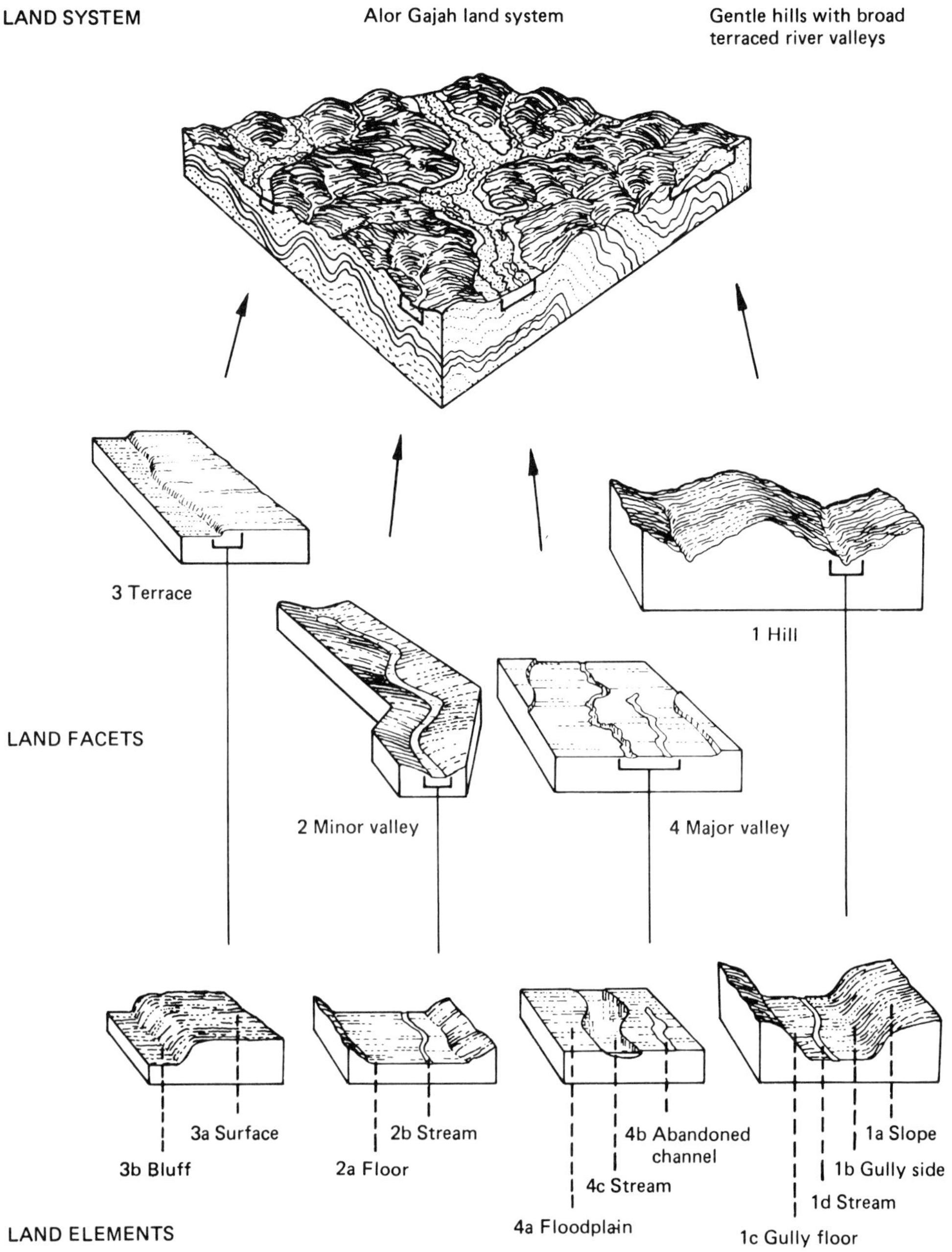

Figure 9.2 Relation between land system, land facet and land element in the Oxford–MEXE system (From Beavan and Lawrence, 1982, Fig. 1, by permission)

- Terrain category – approximate mapping scale;
- Land zone, Land division – greater than 1:15 million;
- Land province – 1:15 million to 1:5 million;
- Land region – 1:5 million to 1:1 million;
- Land system – 1:1 million to 1:250 000;
- Land facet – 1:50 000 to 1:10 000;
- Land element – not usually mapped.

For completeness the whole classification is outlined here, although for most purposes, including the purpose of this chapter, users are only concerned with the lowest three members of the hierarchy. These units provide the basis of the classification from which the larger units are derived.

9.3.1 Land system, land facet and land element

Topography and soils both develop as a response, of the rocks and soils at the Earth's surface, to the same environmental factors. A given set of environmental controls will tend to give rise to specific land and soil types.

A terrain weathering under fairly uniform geological and climatic conditions evolves into a small number of slope types and topographic units. The units are grouped into small associations, recurring over a wide area in a distinctive pattern – the land system. It is defined by geology and climate and by its small topographic units

LAND REGION: Lowland granite.

CLIMATE (Rainfall): 2000 – 2250 mm p.a.

GEOLOGY: Granite.

LANDSCAPE: Low, very convex hills with short deep gullies; very extensive broad valleys consisting of alluvial terraces in several levels, many of which are tin-bearing.

SOILS: Stiff pinkish clays and sandy clays of MSS Rengam Series on hills: sandy alluvium in valleys. BSCS: probably (FS) and (SF).

VEGETATION: Mostly rubber plantations; urban vegetation in Seremban. Scrubland or minor cultivation in valleys.

RELIEF: 30 – 50 m in low hills; up to 130 m on ridges.

ALTITUDE: 50 – 220 m.

AREA: 76 km^2.

N.B. The natural land form (particularly of the valleys) and the vegetation have been very severely disturbed by the urban development of Seremban and its associated alluvial tin workings. For security reasons no aerial photographs of this land system are available.

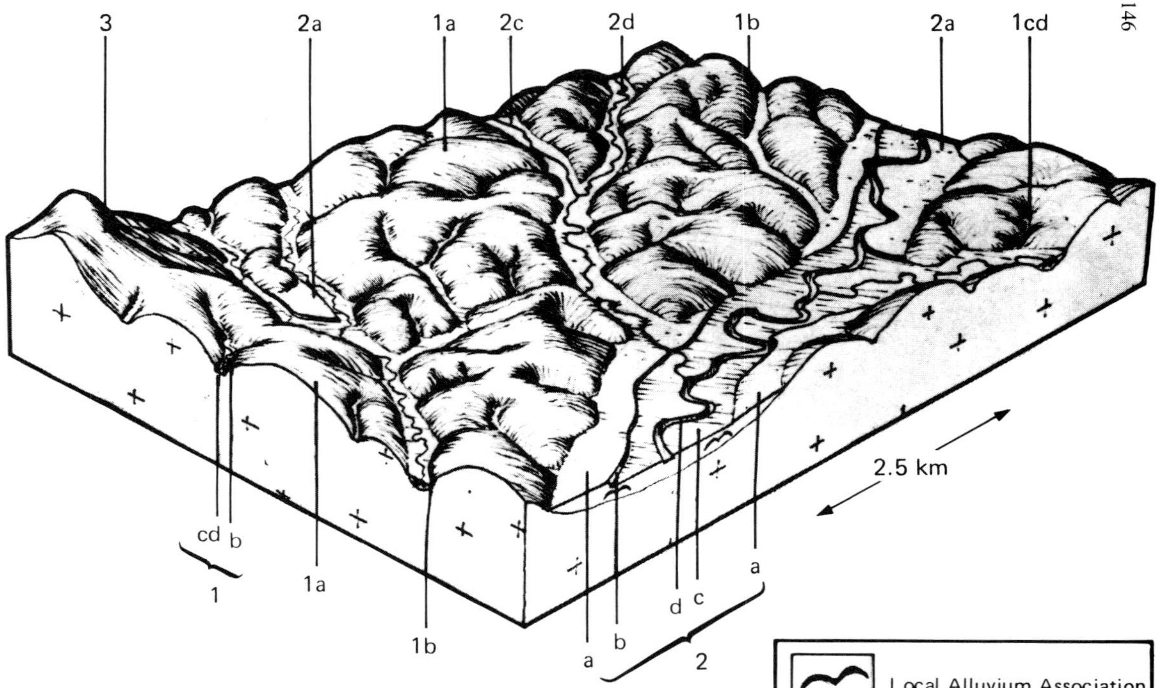

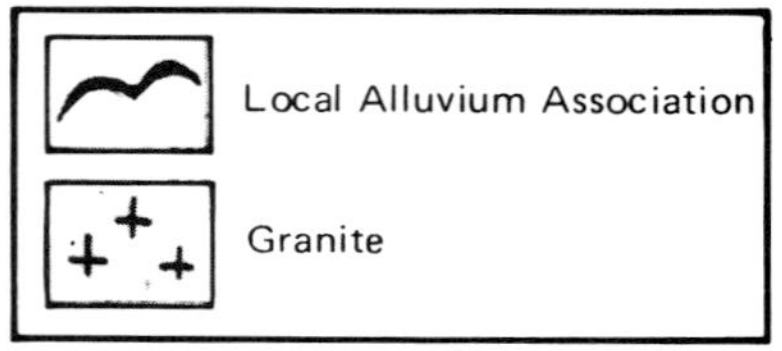

14 RASAH

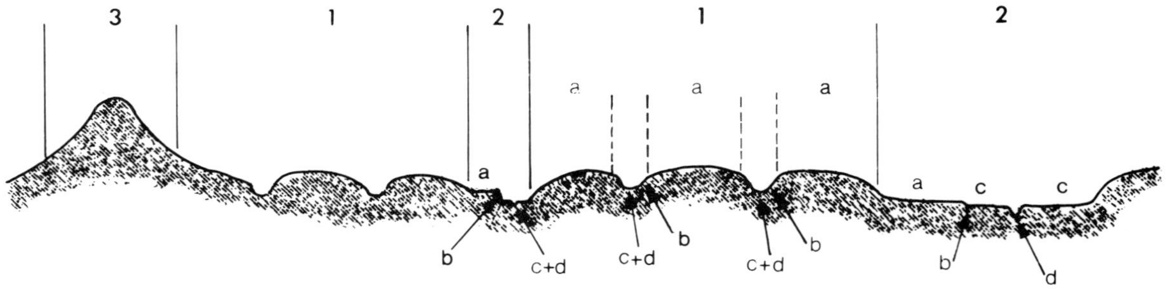

14 RASAH LAND SYSTEM

LAND FACET	LAND FORM	SOILS, MATERIALS, HYDROLOGY AND VEGETATION	ENGINEERING PROPERTIES
1	**Low Hill** (a) Slope 3-5(-10)°, usually gently or markedly convex, sometimes with a slightly flattened crest. 30-50m high and about 200m wide between gullies.	Brownish yellow sandy clay with angular quartz grains, becoming redder below about 2m. Granite rock encountered at depths ranging from 2-20m (usually nearer the latter). Depth to water table not known; soil moderately well drained. (MSS Rengam Series). Mostly rubber plantations but also has extensive urban development (Seremban town).	BSCS: (FS) and (SF). Fine fraction probably of very high plasticity. Moderate earthworks required because of the relatively large number of deep gullies. Soil easily excavated by digger (safe angle of cut about 40-45°) and suitable for fill. Small culvert required in gully floor for stream. Soil test data: LL / PI / LS n: 2 / 2 / 2 $\bar{x}$: 90 / 52 / 18 s: 11 / 5 / 3
	(b) Gully Side and Steep Hill Slope 2-6m high (locally up to about 20m where gully side merges into unusually steep hillside); fairly straight slope, often indented. Head usually abrupt, but may fade into hillside.	Materials and vegetation as 1a.	
	(c) Gully Floor Few - 50m wide; very gently concave, with concave margin to 1b. Curved or fairly straight in plan, few branches, usually 250-400m long.	As 1a, but may be damper. Water table probably within 1-2m. Rubber trees or grassland.	
	(d) Stream Very narrow, often an artificial drainage channel.	Permanent flow.	
2	**Valley** (50-)100-200m wide in minor valleys and upper part of major valleys. Major valleys 300-1200m wide. Floodplains are not well developed as elsewhere in the South West Malay Peninsula. Minor valley "floors" open into terraces on major valley flanks, and several levels of terrace may be found within the major valley. Streams are usually incised 2-3m and are confined to a narrow channel.	MSS "Disturbed Land" and Local Alluvium Association. Sandy soils derived from granite wash. Variable texture. Many valleys given over to tin extraction and remain as flat, poorly drained land or lakes of standing water surrounded by tin tailings.	BSCS not known, but probably (S-F) to (SF) and (FS).
	(a) Terrace Dominant element, forming floor of most smaller valleys and flights of levels in main valleys. Now much disturbed by tin mining. Very flat and level, 100-600m wide and up to several hundred m long, (usually filling entire minor valley floor).	Local Alluvial soils – clayey sands and sandy clays; variable. Mostly freely drained, but locally poorly or very poorly drained. Depth to water table probably 2-3m. Minor cultivation. Urban development around Seremban. Open scrubland or no vegetation in tin mining areas.	Soils well drained and probably generally strong as subgrade. Probably not good as fill because of lack of cohesion. Small amount of earthworks required as terraces change level frequently.
	(b) Terrace Bluff Up to several m high and several hundred m long. Often fades into lower floodplain level.		
	(c) Floodplain Small, found in upper part of smaller valleys. Up to 200m wide and a few hundred (-2000)m long. Occasional very small abandoned channels.	Local Alluvial soils. Organic sandy clays, waterlogged to within 1m of surface; liable to flood. Padi cultivation.	Low embankment (1-1½m) required, but element should be avoided if possible. Poor subgrades, weak and wet.
	(d) Stream Few - 15m wide, winding. Usually incised 2-3m into terrace.	Permanent flow.	Small bridge required across larger streams. Foundation conditions not known.
3	**High Hill and Ridge** Steep or very steep slopes (20-25°) meeting in a pointed or more usually a narrow ridge crest. 50-130m high, occasional; peripheral situation near surrounding granite mountains (Beremban land system).	Yellow sandy clays becoming redder with depth (as facet 1) over granite rock at depths varying probably from a few m to more than 10m. May or may not be a zone of partially weathered rock between the soil and solid granite. N.B. Core of ridge sometimes of quartzite. Jungle, or rubber trees on less steep slopes.	BSCS: not known, probably (FS) and (SF). Steep slopes and possibility of rock near surface indicate necessity of extensive earthworks. Should be avoided altogether. Granite ridges provide good quarry sites where overburden is not too great. Stone for concrete aggregate, crushed stone bases and surfacing material.

Figure 9.3 Description of the Rasah Land System in South West Malaysia

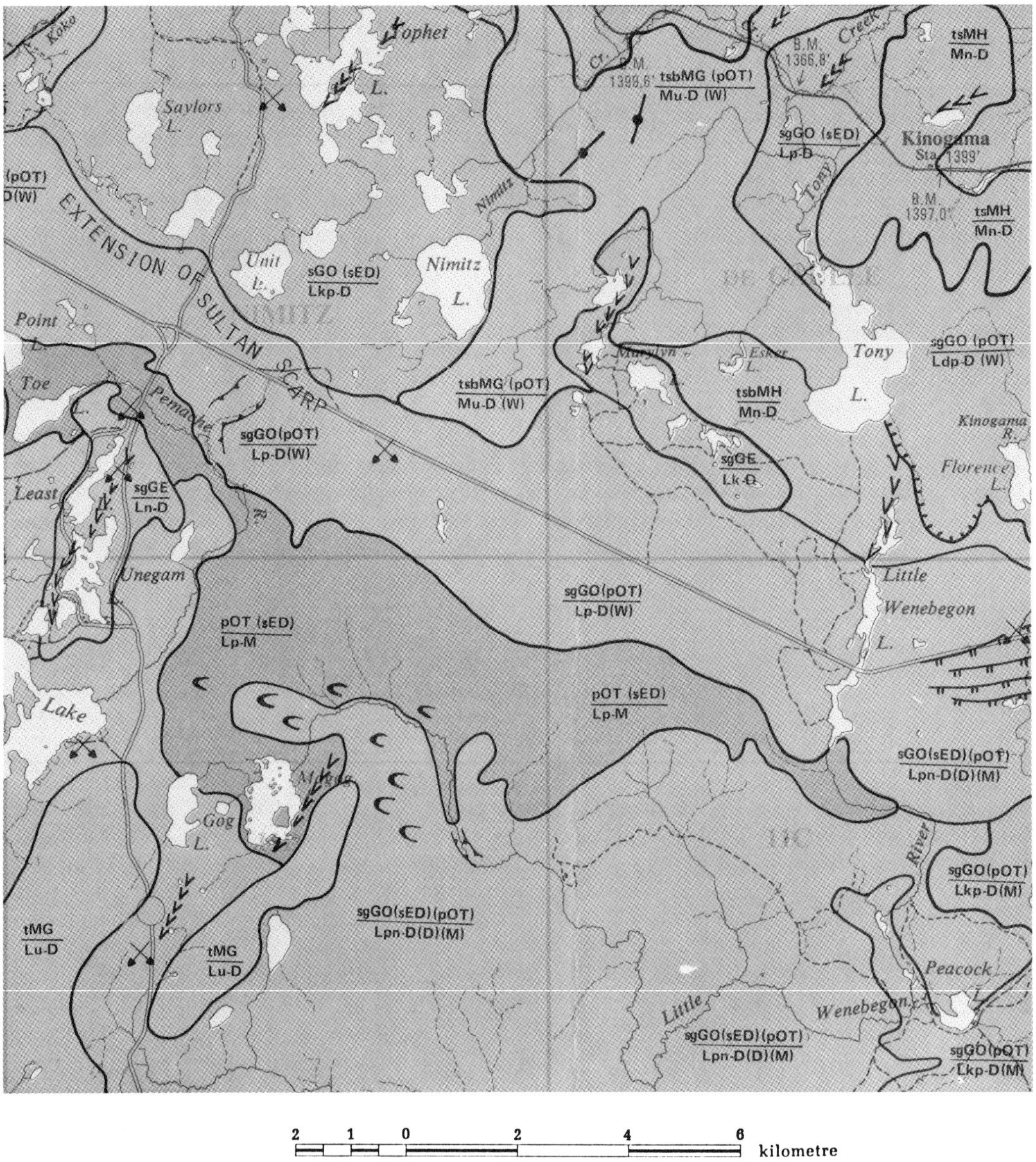

LETTER SYMBOLS

MATERIAL

b boulders, bouldery
c clay, clayey
g gravel, gravelly
p peat, muck
r rubble
s sand, sandy
m silt, silty
t till

LANDFORMS

MORAINAL

ME End moraine
MG Ground moraine
MH Hummocky moraine

GLACIOFLUVIAL

GD Ice contact delta, esker delta, kame delta, delta moraine
GE Esker, esker complex, crevasse filling
GK Kame, kame field, kame terrace, kame moraine
GO Outwash plain, valley train

GLACIOLACUSTRINE

LB Raised (abandoned) beach ridge
LD Glaciolacustrine delta
LP Glaciolacustrine plain

ALLUVIAL

AP Alluvial plain

COLLUVIAL

CS Slope failure
CT Talus pile
CW Slopewash and debris creep sheet; minor talus

EOLIAN

ED Sand dunes

ORGANIC

OT Organic terrain

BEDROCK

RL Bedrock plateau
RN Bedrock knob
RP Bedrock plain
RR Bedrock ridge
/R Bedrock below a drift veneer

TOPOGRAPHY

LOCAL RELIEF

H Mainly high local relief
M Mainly moderate local relief
L Mainly low local relief

VARIETY

c channeled
d dissected, gullied
j jagged, rugged, cliffed
j' cliffy volcanic rock signature
k kettled, pitted
n knobby, hummocky
p plain
r ridged
s sloping
t terraced
u undulating to rolling
w washed, reworked

SURFACE CONDITION

W Wet
D Dry
M Mixed wet and dry

DRAINAGE

h Suspected high water table

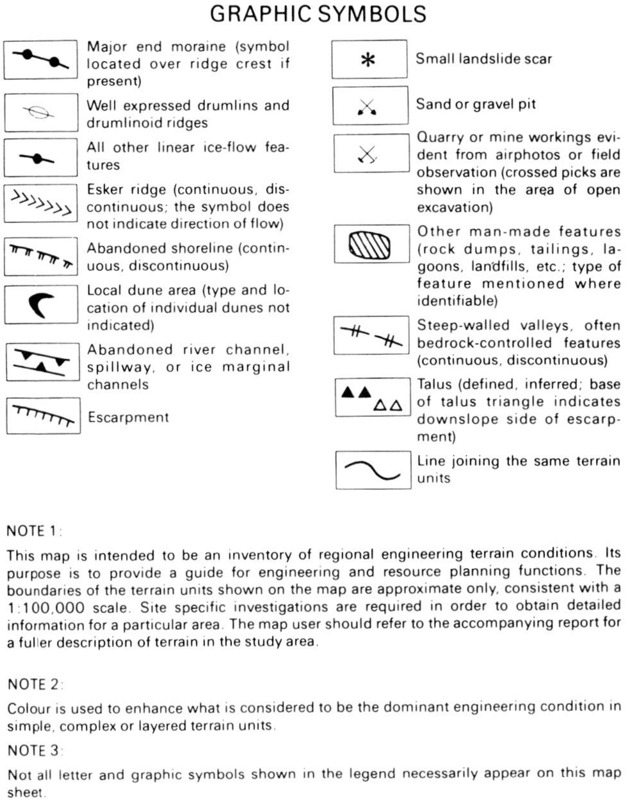

Figure 9.4 Data Base Map prepared at 1:100 000 for the Northern Ontario Engineering Geology Terrain Study, Canada (From Roed and Hallet, 1979, by permission)

called land facets. The pattern persists to the limit of the geological formation upon which it is developed or until the climatic control changes; a new land system is then developed.

Land facets are the basic mapping units. They are defined by geology, water regime and topography. A land facet has a simple form, fitting into one class on a slope map, a fairly homogeneous parent material and a single water regime, both surface and groundwater. Materials developed on it are fairly uniform, such that an engineer would accept a single design specification, for example for a road built on it.

Very small features of a landscape, for example a hill slope in which the upper part is steep and the lower part gentle, may be called land elements. They are too small to be mapped, but are important to an engineer when considering slope stability and amounts of cut and fill.

The relation between the land system, land facet and land element is illustrated in Figure 9.2.

9.3.2 Larger land units

Of these, only the *land region* and smaller landform units are likely to be used in engineering feasibility studies. A land region is made up of a group of land systems having the same basic geological composition and an overall similarity of landform units.

The *land province* is a group of land regions belonging to a major structural, geomorphological or sometimes lithological unit. The *land division* is a major structural unit such as a fold mountain belt or continental shield area, mapped at continental or world scale.

In contrast, a *land zone* is a major climatic region of the world. Its boundaries are diffuse and it may extend across the natural morphological boundaries of the other land units. Incidentally, there appears to be a natural affinity of land divisions and land zones to the divisions used by Ershova and Sergeev in compiling their engineering geological map of the world (see Figure 8.3).

9.3.3 Land system map

A fragment of a land system map prepared for South West Malaysia at a scale of 1:250 000 is illustrated in Plate 13 (Lawrence, 1978).

The colours on the map were chosen to emphasize the major differences in geology and terrain; in fact, they represent the land regions of the area. Land regions are groups of similar land systems developed on the same type of geology. The map fragment has been selected to show the distribution of the Rosah Land System. Lawrence, describing the map, points out that reddish colours are associated with the granite rocks, the blues with consolidated sedimentary rocks, and the yellows and green with various alluvial deposits. The intense weathering of the rocks in Malysia has tended to mask minor geological differences and produce the same type of soil on all rocks of a similar nature. It was found in most cases that differences in the engineering properties of the soils became significant only when a major change of geology occurred, i.e. at land region level. The most obvious differences between terrain and materials should therefore be expected across a colour boundary on the map.

The main purpose of the map is to provide a key to the land facets. The map scale of 1:250 000 is too small to show the distribution of the land facets within each land system, but the identification of a facet at a particular locality, or of several facets along the line of a road, can be quickly carried out. A surveyor can identify a facet on the ground by finding his location on the land system map, and then selecting the facet from the four or five described in the text for that land system. A land facet map can be prepared for an area of special interest, such as the corridor of a proposed road line or the layout of a new plantation. This is done using aerial photographs to identify and delineate the land facets, with the aid of the description in the text. Experience has shown that this can be done without the need of specialist training, provided that the land facet key and air photographs are available. Land facet mapping would normally be the responsibility of the project manager.

9.3.4 Land system descriptions

The map can be used to make a rapid assessment of an area with regard to its suitability for a particular land use. A brief description of the landscape introduces each land system in the report, and this can be used to compare one area with another quickly. An overall view is important at the planning stage, when potentially desirable areas are selected for further study. Having made the choice, the field survey work can proceed with the assistance of the land facet breakdown for semi-detailed soil mapping.

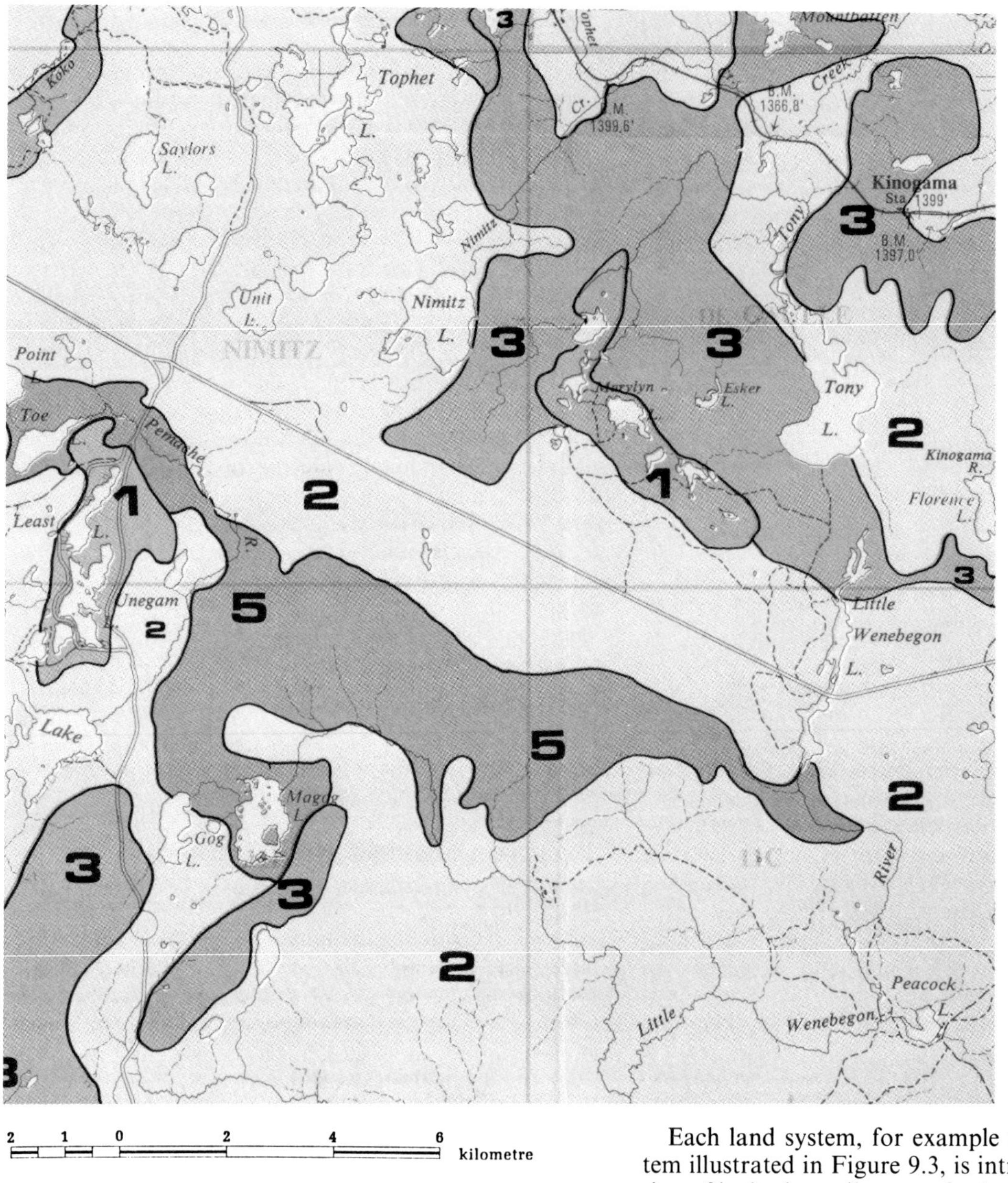

ENGINEERING CAPABILITY MAP LEGEND*

Map Unit	Description	Most Suitable Uses
1	Eskers Kames	aggregate extraction groundwater supply access route location
2	Outwash plains sometimes over- lain by eolian sand sand	aggregate extraction groundwater supply access route location major transportation routes landing strips
3	Hummocky moraine Thick ground moraine	solid waste disposal septic system installation
4	Rock terrain Thin ground moraine	mineral exploration
5	Organic terrain in alluvium	not suitable for above uses

*See Table 1 in accompanying report (Chapleau Area) for soil properties and suitability for various engineering uses.

Figure 9.5 Engineering Capability Map, Chapleau, prepared at 1:100 000 for the Northern Ontario Engineering Geology Terrain Study, Canada (From Roed and Hallett, 1979, by permission)

Each land system, for example the Rasah Land System illustrated in Figure 9.3, is introduced by a description of its basic attributes, referring to the landscape as a whole. This indicates the physical environment in which it occurs: its climate, geology, relief and altitude. A block diagram is used to illustrate the overall appearance of the scene and to show the relationships between the individual land facets. It is considered to be very important that the user associates each land facet with a particular position in the landscape.

After the introduction to the land system there follows a detailed description of the land facets. Their characteristics of slope, size, shape and frequency, by which they may be recognized, appear in the first column. The second column describes the soil profile and its water relations and depth of weathering. Also included in this column is the type of vegetation. In south West Malaysia, where almost the entire land surface is covered with either plantation crops or jungle, this attribute is considered to be of minor importance in the recognition and definition of land facets. The third column describes those aspects of the soils and rocks that are of interest to the engineer. The soils are classified according to the provisional British Soil Classification system (see Section 4.8).

9.4 Other studies of land classification

Terrain evaluation systems, all very similar in their approach, have been developed in many countries for both military and engineering purposes. Highway engineering is a prime example where design is directed towards optimal route alignment and accessibility of construction materials. Geotechnical mapping, so-called, has been developed by the National Institute of Road Research in South Africa, based on the interpretation of aerial photographs and ground sampling. This is the approach adopted in Canada, in the example following, to produce engineering capability maps.

9.4.1 Northern Ontario Engineering Geology Terrain Study, Canada

A final example of the use of air photograph interpretation for small-scale terrain evaluation is that published by Roed and Hallett (1979). It forms part of a series of publications providing regional engineering terrain data for some 370 000 km^2 of Northern Ontario, Canada.

The purpose of the mapping was to provide a guide for engineering and resource planning functions at a level of detail consistent with a scale of 1:100 000. Two maps were prepared – a Data Base Map (Figure 9.4) and a derived Engineering Capability Map (Figure 9.5) – illustrating the suitability of the terrain units in the Chapleau area for aggregate extraction, alignment of transportation routes, groundwater supply, solid waste disposal and septic system installation.

Terrain information was obtained by interpretation of black and white aerial photographs at a scale of 1:50 000. Main roads in the area were traversed to provide spot checks of the desk studies. The map, therefore, represents a reconnaissance overview of the engineering conditions of the glaciated terrain. An engineering terrain legend was developed for the mapping to provide a common information base for the whole map series.

Engineering terrain units

Engineering terrain units comprise a combination of various materials, unconsolidated and/or bedrock, which form recognizable landform with certain engineering characteristics. Major terrain unit groups include bedrock terrain (RN), moraine (MG,MH), glaciofluvial outwash (GO), eskers (GE) and kames (GK), and organic terrain (OT). Less significant terrain units include alluvium (AP) and eolian deposits (ED). A diagrammatic sketch of typical terrain settings is given in Figure 9.6. Table 9.9 summarizes the characteristics and engineering significance of all major terrain units.

Terrain units are coded using a numerator–denominator system in which the letter symbols above the line deal with materials and landforms, whereas below the line drainage and topography are dealt with. For example:

$$\frac{\text{MATERIAL LANDFORM}}{\text{TOPOGRAPHY DRAINAGE}} \quad \frac{\text{RN(tMG)}}{\text{Mjn-D}}$$

indicates bedrock terrain (RN) with very little overburden, although patches of ground moraine (MG), till (t) occur. Most of the rock terrain is of moderate relief (M), rugged (j) to knobby (n) in character, and dry (D). The complete scheme is given in the legend to Figure 9.4.

The principal engineering significance of bedrock terrain is that it is difficult and expensive to excavate. Blasting is required for all subsurface excavations. Extensive cut-and-fill operations are required in rugged terrain for transporation route construction, in order to eliminate extreme horizontal and vertical alignments. Many small sand, gravel and till deposits scattered throughout the terrain provide adequate material for minor access road construction. However, for major transportation routes, aggregate often has to be obtained from considerable distances at high cost.

Bedrock terrain provides excellent foundation conditions in areas that are unaffected by faults and where

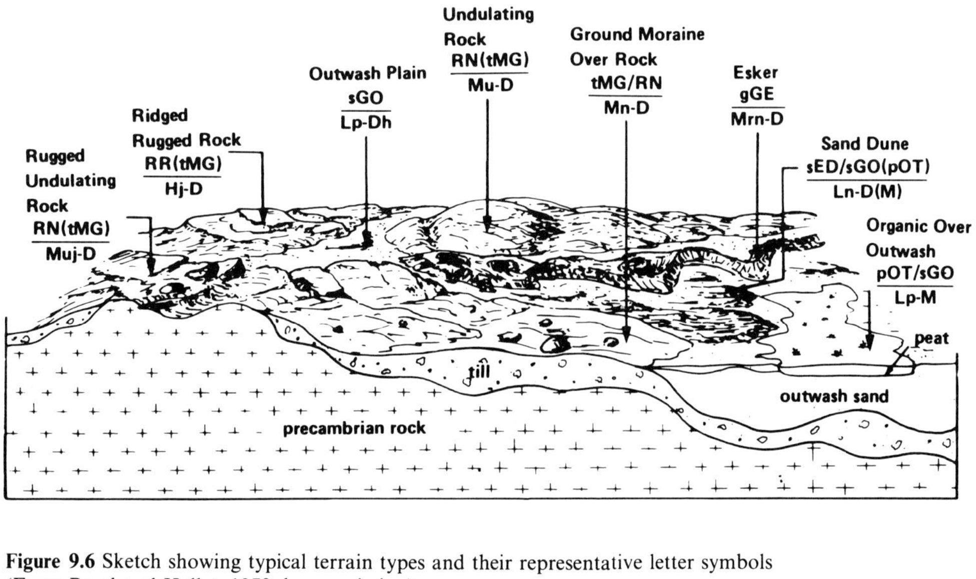

Figure 9.6 Sketch showing typical terrain types and their representative letter symbols (From Roed and Hallet, 1979, by permission)

Table 9.9 Summary of terrain unit characteristics and engineering significance (From Roed and Hallett, 1979, by permission)

Terrain unit	*Principal materials*	*Topography* / *Drainage*	*Workability as construction material* / *Permeability*	*Shearing strength* / *Bearing value*	*Compaction* / *Compressibility*
Rock terrain (R)	Rock	Irregular / Dry	Not applicable / Medium (fracture)	Excellent / Excellent	Not applicable / Not applicable
Hummocky moraine (MH)	Till, sand boulders	Hummocky / Dry	Good / Medium	Good / Good	Good / Low
Ground moraine (MG)	Till, sand boulders	Knobby / Dry	Good / Medium	Good / Good	Good / Low
Eskers (GE) Kame (GK)	Sand, gravel	Kettled and ridged / Dry	Excellent / High	Excellent / Excellent	Excellent / Low
Outwash (GO) Delta (GD)	Sand, gravel	Hummocky-planar / Dry	Excellent / High	Excellent / Excellent	Excellent / Low
Glaciolacustrine plain (LP)	Clay, silt, find sand	Planar / Dry	Poor / Low	Fair / Fair	Poor / High
Glaciolacustrine beach (LB)	Sand	Planar / Dry	Good / High	Good / Good	Good / Low
Alluvium (AP)	Sand, silt, clay	Channelled / Mixed	Fair / Medium	Good / Poor	Fair / Medium
Eolian (ED)	Sand	Ridged, hummocky / Dry	Fair / High	Good / Good	Good / Low
Colluvium soil (CW,CS)	Clay, silt	Sloped / Mixed	Poor / Low	Poor / Poor	Poor / High
Colluvium talus (CT)	Rubble	Sloped / Dry	Poor / High	Good / Poor	Poor / Low
Organic (OT)	Peat	Planar / Wet	Poor / Medium	Poor / Poor	Poor / High

fracture or joint density is low. The unit also possesses considerable potential for aggregate obtained in rock quarry operations.

Groundwater potential is generally poor in bedrock terrain, but is usually sufficient for scattered domestic demand. Considerable groundwater production can be expected in fractured localities or in fault zones. Bedrock is unsuitable for waste disposal except where a thick layer (at least 1.5 m) of soil material is present.

In the report, all terrain units are dealt with in the same way, thus providing a comprehensive *interpretation* of the map units on the Data Base Map (Figure 9.4) which, in the original, is published in colour.

Engineering Capability Map

The Data Base Map forms a basic framework of resource characteristics that allows development of a wide variety of engineering and planning derivative maps. A derivative map can be defined as a graphic representation of the terrain in terms of its potential with respect to a specific engineering or planning use, or a number of uses. Derivative maps are needed because many people who have special interests in the map area do not have the geological or engineering expertise to fully interpret the Data Base Map for their particular application. The object of presenting a derivative map, therefore, was to demonstrate how an easily understood illustration can be produced from the Data Base Map.

Many different types of derivative map can be produced. For example, a 'Granular Aggregate Probability Map' would classify terrain units into high, moderate and low rating categories with respect to potential for finding aggregate. The analysis could be taken a step further through preparation of a Sand and Gravel Resources Map (Roed, 1979). This type of derivative map would incorporate not only the geological conditions of the terrain unit, but also important criteria such as access and size of deposit, in order to outline areas which are best suited for aggregate extraction.

Similarly, derivative maps can be developed which rate the terrain units for other important uses, such as groundwater potential, septic system suitability, solid

Slope stability / *Frost susceptibility*	*Aggregate supply* / *Foundations*	*Highway subgrade* / *Route location*	*Hydro easements* / *Pipeline routes*	*Septic tank suitability* / *Solid waste landfill*	*Domestic groundwater potential* / *Large groundwater supplies*
Good (some talus) / Not applicable	Not applicable / Excellent	Poor / Fair	Fair / Poor	Poor / Poor	Good / Poor
Good / Low	Fair / Good	Good / Good	Good / Good	Excellent / Excellent	Excellent / Good
Good / Low	Fair / Good	Good / Good	Fair / Fair	Excellent / Good	Good / Fair
Excellent / Low	Excellent / Excellent	Excellent / Excellent	Good / Good	Poor / Poor	Excellent / Excellent
Excellent / Low	Excellent / Excellent	Excellent / Excellent	Excellent / Excellent	Poor / Poor	Excellent / Excellent
Poor / High	Poor / Fair	Poor / Good	Good / Good	Fair / Fair	Fair / Poor
Excellent / Low	Good / Good	Good / Good	Good / Good	Poor / Poor	Good / Poor
Fair / Medium	Poor / Poor	Poor / Poor	Poor / Poor	Poor / Poor	Good / Fair
Good / Low	Poor / Good	Fair / Good	Good / Good	Fair / Poor	Good / Fair
Poor / High	Poor / Poor	Poor / Poor	Poor / Poor	Poor / Poor	Fair / Poor
Good / Low	Poor / Poor	Poor / Poor	Poor / Poor	Poor / Poor	Poor / Poor
Poor / High	Poor / Poor	Poor / Poor	Poor / Poor	Poor / Poor	Poor / Poor

Notes:
1. The rating above or below the bars refers to the characteristic or use above or below the bar, respectively, in the column heading.
2. Permeability, compressibility and frost susceptibility are rated as low, medium or high; all other characteristics and uses are rated as poor, fair, good or excellent.
3. This is a generalized table intended for use in a number of reports; therefore, all terrain units dealt with in the table will not necessarily appear in each report.

waste disposal, general construction suitability, route assessment and mineral exploration. It is possible to prepare a derivative map which would outline areas of extremely poor suitability for a specific use such as housing. For example, organic terrain and flood plain units along rivers would be shown as hazard areas with respect to housing.

The derived map (see Figure 9.5) included in the report outlines major terrain units in the Chapleau area which are well suited for the following uses: aggregate extraction, transportation route alignment, groundwater supply, solid waste disposal, and septic system installations. Hazard lands are also shown. It is referred to, therefore, as an Engineering Capability Map and is considered to be a general type of derivative map rather than a specific-use type. Capability maps are considered in more detail in Chapter 13, although it is appropriate to consider this example here as an illustration of the wider application of terrain evaluation using air photograph interpretation.

9.5 Remote sensing for land surface evaluation

9.5.1 Aerial photography

The examples of land surface evaluation that have been quoted all depend on the interpretation of aerial photographs. It is not necessary here to go into detail on the methods available, as they are adequately dealt with elsewhere (Anon., 1982; Beavan and Lawrence, 1982). Rather, it is only necessary to indicate what can be achieved using photography and other methods of remote sensing.

Photographs can be taken horizontally (see sub-section 11.5.1), as low or high obliques, or near vertical or vertically. Horizontal and vertical photographs are of most use as they are nearly true to scale and can be quantified if surveyed ground control is available.

The smallest scales, 1:200 000 to 1:20 000, are used

for reconnaissance photography; preliminary site investigations may be carried out at 1:20 000 to 1:5000 (see Chapter 10); for detailed investigations, or monitoring, scales range from 1:10 000 to 1:500 (see Chapter 11). Air photographs at these scales may be used as field maps, or by combining several photographs as a mosaic of larger areas of ground. Air photographs should always be viewed in pairs under a stereoscope for the identification of landform units, photoanalysis, site analysis, mapping and so on.

The techniques of interpreting air photographs are well established, and an extensive documentation is given in Anon. (1982, Appendices E1 and E3).

9.5.2 *Satellite imagery*

Since 1972, photography can be supplemented by Landsat satellite imagery as a small-scale mapping tool, ideal for reconnaissance studies over large areas. Scales range from 1:3 500 000 to 1:250 000. By sophisticated processing techniques, enlargements to 1:50 000 scale may be achieved, but since the optimal ground resolution is 79 m, enlargement provides no information not already available in the original. However, the larger format may be valuable for compilation purposes. In general, a scale of 1:250 000 is most suitable for engineering purposes. Imagery from Skylab and the Return Beam Vidicon (RBV) cameras from Landsat 3 can usefully be enlarged to 1:100 000.

Most of the information has been collected by the multi-spectral scanner (MSS) which builds up an image of the ground by recording successive west–east scan lines as the satellite moves south. Each scan line is 185 km × 79 m, but standard scenes are made up by compiling scans to give an image 185 × 178 km. The information is issued to customers either as a computer compatible tape (CCT) or as photographs at a scale of 1:1 million.

Within each scan line, reflectance is measured at four separate wavelengths equivalent to green and red light and two bands of infrared. The scan line is sampled at a frequency which is equivalent to a movement of 57 m along the ground and so conventionally the data are reproduced as picture elements (pixels) 79 × 57 m in size. Within a pixel, small features such as buildings or roads cannot be resolved individually, but they contribute to the total reflectance of that pixel and may bias it.

A higher resolution is achieved with the RBV camera on Landsat 3, using two cameras each covering one-quarter of the area of the MSS scene with a resolution of 30–40 m. At present, the data are usually made available in photographic form at a scale of 1:250 000, but CCTs are now being produced. The high resolution of this system generates such a large amount of data that it has been necessary to restrict it to one channel, and so it is not possible to produce an RBV colour image. However, experiments have been made to add the MSS data to the higher resolution RBV data in order to obtain the benefits of both systems and to simulate the imagery that will be obtained by the thematic mapper (TM) on Landsat.

9.5.3 *Interpretation of Landsat images*

The interpretation of Landsat data consists of two phases – data preparation and extraction of information. The latter phase is a similar process to that used to interpret small-scale aerial photographic mosaics.

Data preparation techniques are dependent on the source, i.e. photograph or CCT, although ways exist of obtaining a similar effect on either material.

Interpretation of Landsat photographs

The basic Landsat data are held on four photographic negatives which may be enlarged to a scale of 1:250 000. Larger prints may be useful to correlate with maps but give no extra information. These images may be interpreted individually, but to take full advantage of the system it is necessary to combine data from more than one band. One of the simplest techniques is to view two bands of the same area with a stereoscope which combines the details seen in each separate image. A more complex way is to superimpose a positive transparency from one image onto a negative transparency of the second. If the two images are identical they cancel each other, leaving a neutral tone. Any differences between the bands will show as either darker or lighter areas, to give a simple form of ratio image. More complex ratios are produced by digital processing in a computer.

The more effective way to study three different images together is to generate a colour composite using either original or ratio images or a combination of them. A convenient way to produce a colour composite is to use a colour additive viewer where images are projected through colour filters and accurately superimposed on a screen. The image can be studied as projected, but it is usual to make a colour photograph which can be studied more conveniently.

Digital methods of processing Landsat data

The original Landsat data are recorded on a tape which contains four bands of reflectance data for the 8 million pixels making up a scene, totalling over 30 million reflectance values in all. These data may be mathematically analysed in a computer, for example to count the percentage of pixels of a certain value, to be correlated with some known ground feature. Alternatively, the pixels from three bands can be displayed on a colour television screen to generate a full colour image that may be interpreted in the same way as any other image. However, the data may also be manipulated mathematically before output to the screen, thereby enhancing the picture for interpretation and analysis. Television viewing systems, when connected to a computer to enable the effect of data transforms to be seen immediately, are known as interactive processors.

The use of an interactive processor for the study of Landsat data offers two major advantages over photographic techniques. First, the range of data transformations available in the processor is much greater than those achieved by photographic pre-processing, and the results are available almost instantaneously. There is thus a greater opportunity to find the enhancement best suited to the user's needs. Secondly, the intensity range of the data can be stretched over the full range of the display system, allowing very subtle colour variations to be produced. A computer-generated image, therefore, is a very high-quality product making the most appropriate display of the data in terms of the user's requirements.

The image on the television display may be enlarged to any desired scale, although with present Landsat data at scales larger than 1:50 000 the picture begins to break up into its individual picture elements and to lose its integrity as a whole. The screen can be split into as many as four segments, with different parts or scales of the image placed in any segment. Different processes may then be carried out in each for purposes of comparison. The processing facilities are of two types: geometric, in which the size or shape of the whole image is modified, and spectral, in which the reflectance values of the pixels are enhanced.

The CCT is normally supplied without geometric correction and data are presented as a rectangular array of pixels. Simple processing can skew the rectangle into an image that takes account of earth rotation and which is geometrically fairly accurate, even at a scale of 1:250 000. More elaborate processing, in conjunction with control points, can be used to re-sample the pixel array so that it conforms with a map projection. Experiments have been made to combine topographic maps and Landsat data to produce an image that has height information superimposed onto a detailed topographic base.

Pixel intensity values are generally concentrated within a relatively narrow range of brightness values, occupying perhaps only one-third of the range of the measuring system. If the range of the measuring system is equated to that of the display, a large part of the display range is not used, resulting in a very dull image. A process known as contrast stretching is applied to 'stretch' the data over the whole range and increase the intensity difference between adjacent pixel values; any part of the data range may be stretched in this way to produce the optimum image for interpretation.

In the process known as density slicing, data between certain levels of a single band are displayed to the exclusion of the remainder. The automatic classifier uses a form of density slicing to select pixels having a range of intensity defined for all bands and hence having similar reflectance characteristics on the ground. Ratio images, generated by dividing the pixel values of one spectral band with those of another, emphasize the differences between bands. Ratio images may be combined with normal bands to produce new images, or used in classifications to refine groups of pixels into tightly defined colour populations.

A more general transformation is to calculate the principal components of the data; this is a mathematical technique that takes any number of data sets and calculates an equivalent number of components with the more significant information contained in the first few sets. Thus the four bands of Landsat data yield four principal components of which most relevant information is contained in the first two. The fourth component is almost completely composed of noise, but significant information may be contained in the third component. This means that all the significant information from the four Landsat data sets is condensed into three sets, which can be displayed on the monitor as a coloured image. Interpretations and classifications can then be made in the usual way.

Beavan and Lawrence (1982), from which the above account has been abstracted, conclude that Landsat photographic imagery provides a good base for erecting a coarse terrain breakdown in preparation for more detailed analysis using aerial photographs, digital processing techniques, or a combination of the two. Their account is illustrated by examples of terrain evaluation in Africa and Asia.

9.6 Small-scale engineering geological maps

On small-scale engineering geological maps the choice of map units is made not on age, as with conventional geological maps, but mainly on the basis of lithology, with additional emphasis on the location of potential geological hazards. Such maps, compiled from existing geological data onto adequate topographic maps, are in effect terrain evaluations. As terrain evaluation is most often directed to the selection of transportation routes, from many examples available a ground-based study of a transportation route has been selected as illustration.

9.6.1 Northeast Corridor, Washington DC to Boston, Massachusetts, USA

This report (U.S. Geological Survey, 1967) was compiled as a desk study from available published and unpublished maps, reports and information from many sources. No original mapping was done. It was designed to provide a summary of the regional geology of a highly urbanized area on the east coast of the USA for planners, engineers and geologists concerned with early feasibility studies of a transportation corridor between major cities. The report is in three parts: parts A and B provide maps of bedrock and superficial geology, respectively, a table of geological and engineering characteristics of each map unit, a map showing sources of data, and a list of references. Hydrological data are shown on both maps. Part C contains maps and data on earthquake epicentres, geothermal gradients and major excavations.

The solid and superficial maps prepared at an original scale of 1:250 000 are illustrated in Anon. (1976) and need not be repeated here. There are, however, two cartographical aspects of especial interest. The first is provided by Map I-514-B (Sheet 6 of 8): Selected Cross Sections Through Surficial Deposits, which neatly circumvents the problems of three-dimensional representation of complex sequences in the superficial deposits, including decomposed bedrock, by providing 29 cross-sections. The sections are all drawn to scale, with vertical exaggeration, involving 19 distinct lithological units. Most of the information would have been derived from site investigation data, and the complexity of local sequences is illustrated in the cross-section forming Figure 9.7.

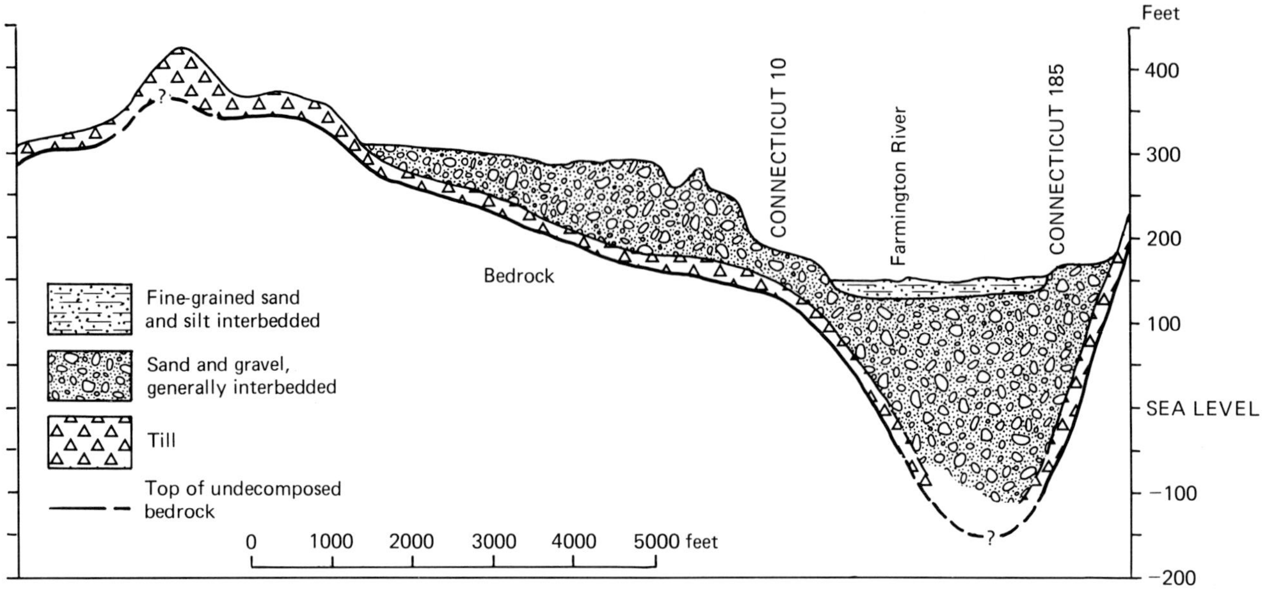

Figure 9.7 Typical cross-section illustrating ground conditions in superficial deposits at a river crossing in the Northeast Corridor, Washington to Boston, USA

One map (Sheet 7) is a table summarizing the geological, hydrological and engineering conditions, from which Tables 9.10 and 9.11 have been assembled to give a flavour of the information provided by the all-important general legend. The table of geological and engineering characteristics for each map unit includes lithology, thickness, alteration, topography, foundation conditions, excavation characteristics, slope stability, significant hydrological features, general hydrological conditions and watertable conditions.

Table 9.10 Tabular description of geological and engineering characteristics of a bedrock map unit in the Northeast Corridor (Washington to Boston), USA, Map I-514-A (Sheet 6 of 7)

BEDROCK MAP UNIT

GEOLOGICAL DESCRIPTION

Map unit. Amphibolite, epidote amphibolite, and metamorphosed gabbro.

Equivalent geological unit.[1] Baltimore gneiss (part); Brimfield formation (part); Glastonbury gneiss (part); Maltby lakes volcanics (part); Marlboro formation (part); Middletown formation (part); Putnam gneiss (part); Tatnic hill formation (part); unnamed units.

Lithology. Massive to banded, tough, strong amphibolite, amphibolite schist, amphibolite gneiss, hornblende gneiss; in part metagabbroic. Forms extensive thick layers, lenses and pods. Commonly epidote-bearing; much epidote-rich amphibolite gneiss, and pods of epidote; includes extensive thick light-red to pink garnet-rich, layers. In places schistose toward margins; locally intruded by quartz diorite dykes.

Structure. Structure variable; rock ranges from massive to gneissic. Large massive bodies rarely show folds, whereas smaller or layered bodies commonly are complexly folded. Joints commonly persistent, but vary greatly in spacing.

Weathering. Weathers to sticky red clay south-west of glaciated area; almost unweathered in glaciated area.

Topography. Hills to low mountains; rugged in places, especially near rivers.

PHYSICAL PROPERTIES[2]

Dry unit weight 3000–3200 kg/m^3.

Compressive strength.[3] High to very high.

ENGINEERING CHARACTERISTICS

Evaluation of rock for construction.[4]

Young's modulus of elasticity.[5] High to very high.

Relative drillability.[6] 2.

Excavation characteristics. Overbreak reported moderate to excessive depending on orientation of excavation to major joint systems. Rock loads mostly slight. Swelling ground probable in wet shear zones. Contains fibrous minerals which may slow excavation by boring machine.

Hydrologic conditions[7]

Significant hydrolic features. Water occurs in weathered zone at shallow depth, probably not much below 45 m. Maximum yields from fault zones.

Permeability. Secondary. Primary fracturing; chemical weathering.

Depth of wells. Range: 10–230 m. Median: 47.5 m.

Yield of wells. Range: 2.2–675 litres/min. Median: 54 litres/min.

Notes:

1. The stratigraphic nomenclature used in this report is that of the authors of the various data sources and does not necessarily conform with usage of the United States Geological Survey.
2. Physical data available only for some rock units; where data are lacking, the physical properties are inferred from comparisons with those of rock elsewhere that possess similar composition, structure and geological histories.
3. Classification is for uniaxial compressive strength of intact rock. Strength is reduced by physical defects and chemical alteration in rock; it may differ with respect to bedding, foliation or direction of principal residual stress.

Strength class	*Range of compressive strength* (kgf/cm^2)
Very high	>2200
High	1100–2200
Medium	550–1100
Low	280–550
Very low	<280

4. Reported construction characteristics are limited in number. Evaluations of rock units, for the most part, are inferred from generalized conditions of structure, alteration, hydrology and state-of-stress. Specific conditions can change within short distance. More refined engineering evaluations must be based on more detailed knowledge of geological conditions.
5. Inferred for intact rock. Modulus is reduced by physical defects and chemical alteration in rock; it differs with respect to bedding, foliation, or direction of principal residual stress.

Modulus class	*Range of static modulus of elasticity* (kgf/cm^2)
Very high	$>8.4 \times 10^5$
High	5.6×10^5–8.4×10^5
Medium	2.8×10^5–5.6×10^5
Low	7×10^4–2.8×10^5
Very low	$<7 \times 10^4$

6. Number 1 indicates rock most difficult to drill. Numbers increase with ease of drilling.
7. The well-yield data used here are based on public-supply and industrial wells in which the maximum potential of the aquifer was being developed.

Table 9.11 Tabular description of geological and engineering characteristics of a surficial deposit map unit in the Northeast Corridor (Washington to Boston), USA, Map I-514-B (Sheet 7 of 8)

SURFICIAL DEPOSIT MAP UNIT

Qt

Younger ground moraine

LITHOLOGY

Till (GM, GC, SM): Chiefly an unsorted mixture of clay, silt, sand, gravel and boulders. Varies from a cohesive, moderately clayey material with embedded pebbles, cobbles and boulders (boulder clay) to a very well graded sand with gravel, cobbles, boulders and minor silt derived mostly from nearby bedrock. Stratification generally lacking, poor to crude where present. Generally very compact, firm and friable; upper few feet may be somewhat loose. Color varies with weathering (oxidation) and with color of dominant source bedrock. Generally gray when unweathered, buff to brown when oxidized; reddish brown in areas of reddish-brown bedrock. Locally contains or underlies thin lenticular sand and gravel. Overlain by many small, thin swamps containing muck and peat, and stream courses with postglacial alluvium that are not shown on map. Locally patchy; numerous bedrock exposures within unit not shown.

THICKNESS

Very irregular, generally less than 25 ft. Unit normally thin to absent on hill crests and along major escarpments, thickens downslope to 50 ft or more on lower slopes. In drumlins may be 150 ft thick or more.

ALTERATION

Upper 3–60 ft oxidized; most stones unweathered or nearly so.

TOPOGRAPHY

A nearly ubiquitous mantle overlying the bedrock surface. Usually thick till underlies streamlined hills, called drumlins, which may be as much as 150 ft high; some drumlins have a core of bedrock.

ENGINEERING CHARACTERISTICS

FOUNDATION CONDITIONS[1]

Bearing capacity generally good because of high density and poor sorting. Expansion negligible.

EXCAVATION CHARACTERISTICS

Generally easy to moderately difficult to excavate with power equipment. Highly compacted till ('hardpan'), and very strong and bouldery till can be troublesome to excavate without special equipment.

SLOPE STABILITY[2]

Cuts higher than 40 ft generally require individual stability analysis. For lower slopes, $1\frac{1}{2}$ on 1 to 2 on 1 generally considered safe. Vertical slopes up to 15 ft common, particularly in more cohesive clayey till.

SIGNIFICANT HYDROLOGIC FEATURES[3]

Small domestic supplies obtained from dug wells. Unit functions as a confining layer over some bedrock aquifers.

GENERAL HYDROLOGIC CONDITIONS[3,4]

Yields of wells (gal/min)		*Coefficient of permeability* (gal/day per ft^2)		*Specific capacity*[5] (gal/min per ft)	
Range	*Median*	*Range*	*Median*	*Range*	*Median*
1–15	6	0.02–4	—	—	—

WATER-TABLE CONDITIONS[3]

Specific yield[6] (%)	*Free water content*[7] (gal/yd^3)	*Excavation permeability*[8] (gal/day per yd^2)
5–17	10–35	1–35

Notes:

1. Bearing capacity—numerical values (tons/ft^2) applied to qualifying adjectives:

Very poor	less than 1
Poor	1–4
Fair	4–8
Good	8–32
Excellent	greater than 32

Compressibility—volume decrease in a soil mass in response to an external load.

Expansion—volume increase that is a function of load, time, density, water content and type of clay minerals.

2. Cut slopes—numerical values (degrees) applied to qualifying adjectives:

Vertical	90
Near vertical	80–89
Steep	45–80
Moderate	30–45
Gentle	0–30

3. The well-yield data used in the preparation of this table are based on public-supply and industrial wells in which the maximum potential of the aquifer was being developed.
4. Data given under general conditions should be used for calculations pertaining to artesian aquifers in tunneling or deep excavations where it may not be possible to dissipate the hydrostatic head.
5. Specific capacity is the discharge expressed as a rate of yield per unit of drawdown. The data used in this column were selected to represent conditions of optimum well development and therefore reflect aquifer characteristics.
6. The ratio of the volume of water which a saturated rock or soil will yield by gravity to its own volume, stated as a percentage. Values used in this table are laboratory determinations or estimates based on field experience.
7. Gallons per cubic yard yielded by gravity drainage.
8. Rate of flow of water in gallons per day through a cross-section of 1 yd^2 under a unit hydraulic gradient at prevailing water temperatures.

10

Urban engineering geological maps

10.1 Introduction

Urban geological mapping is usually undertaken at the medium map scale within the range 1:100 000 to 1:10 000 although, where there is either intensive constructional activity in difficult terrain conditions or an abundance of site investigation information, an appropriate scale may be 1:2500 or even 1:1000. The most generally effective map scales are 1:50 000 and 1:25 000 (1:24 000 to accommodate maps of the U.S. Geological Survey), a range for which quite detailed geological maps of both bedrock and superficial deposits may be available. Such maps provide the basic information for deriving engineering geological maps. Also, it is within this range that two important aspects of engineering geological mapping may be undertaken. The first of these is the direct interpretation, in a general way, of a geological map for engineering purposes; secondly, additional information may be added to the map sheet during compilation to facilitate, for instance, interpretation and understanding of the all-important superficial deposits. Beyond these two aspects, urban engineering geological cartography usually reflects either rather specific ground conditions or the perceived needs of engineers and planners working in the urban environment.

10.2 Interpreting geological maps for engineering purposes

10.2.1 Introduction

A great deal of information can be derived from a geological map, even though it would not have been made specifically for engineering or planning purposes. Valuable interpretations can be made by an engineering geologist provided that he has, or can acquire rapidly, a good background knowledge of the geology of the area, such as can be gained by a limited desk study. A necessary stipulation is that the geologist should appreciate the facts that the engineer needs to know, and also how he would use them. In the same way, the engineer could make the interpretation, but he would need to have somewhat more than a general knowledge of the fundamentals of geology. The same comment applies, perhaps with an even greater force, to an interpretation made for a planner.

Eckel (1951), following up a publication by the U.S. Geological Survey (1949) on interpreting ground conditions from geological maps, and admittedly referring to conditions in the USA at that time, deplored the sparse coverage of the country by any type of geological map at medium scale: 1 inch to 1 mile (1:50 000) or larger.

In addition, available maps rarely gave information on the overburden covering the bedrock, although he said that steps were being taken to rectify this shortcoming. Finally, there was the general lack of quantitative facts that the engineer requires for his work. Care was needed because nearly everything in the field of geology is gradational, and gradations do not lend themselves to exact measurement. He showed, nevertheless, how published geological maps could be interpreted in terms of construction materials, foundation and excavation conditions, surface and groundwater, and soil conditions.

10.2.2 Interpretation of the geological map of the Alpine Quadrangle, Utah, USA

The first example is the anonymous 'how to do it' account published by the U.S. Geological Survey in 1949, mainly in the form of a series of maps with extended legends. The area analysed is in the Alpine Quadrangle on the northern margin of the Bonneville Salt Flats in Utah, USA. After giving the background topographical map (Figure 10.1), the geological version is repeated with a series of different legends.

The introduction to the series of maps makes it clear that intelligent planning for heavy construction, water supply or other land utilization requires advance knowledge of ground conditions in the area. It is essential to know and be able to interpret:

- the topography;
- the geology, including both rocks and engineering

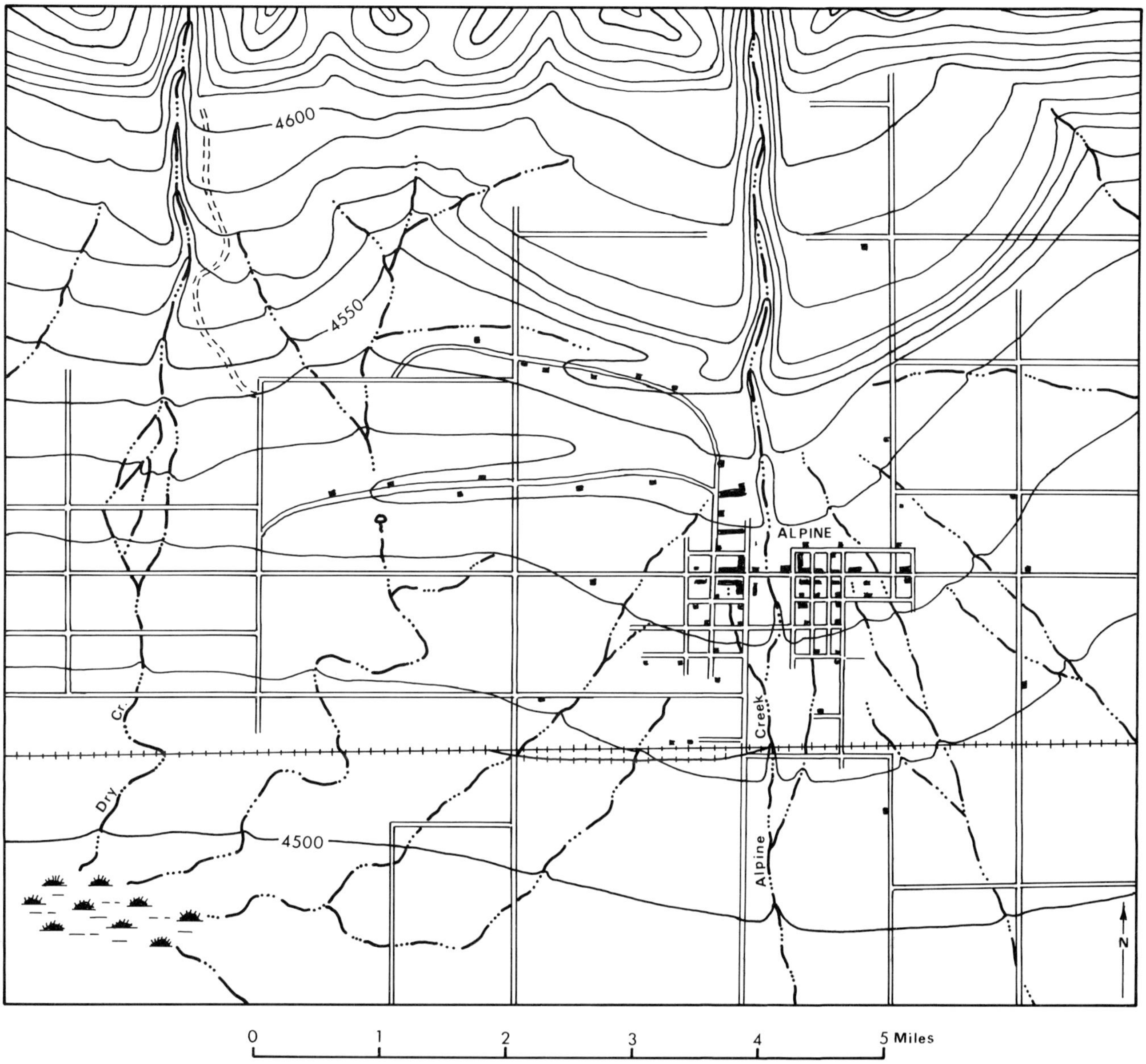

Figure 10.1 Topographic map of the Alpine Quadrangle, Utah, USA

soils, i.e. the deposits that comprise the land and its weathered surface;
- the hydrology, including the occurrence of surface and subsurface water.

These three elements should usually be considered in the planning of land developments that involve much financial investment. At the site selected for development, detailed surveys are usually made, but equally essential and frequently overloooked (at that time) is the need for general surveys before site selection.

Only when the general survey has been made does it become possible to know that a particular site is the most suitable for the purpose, and that no aspects in the tributary areas have been overlooked that might affect the project. Information on both general and specific aspects are needed in order to avoid costly mistakes either during or after development.

The maps accompanying, and forming the main body of the report, illustrate how a geological map can be used for interpreting ground conditions during the planning stage before site election. Not all the maps are reproduced here. The basic topographical and geological maps have been simplified from existing examples. The interpretative sheets of the original report, presented here as additional legends to the basic geological map, were intended to provide some examples of the kinds of information that trained persons can derive from such basic maps.

The topographic map (Figure 10.1) shows quantitatively the configuration of the land surface. Irregularities in the contour lines reflect the ground plan shape of the land forms. South of Alpine the ground rises to the north only 10 ft in one mile – practically flat – and is only slightly incised by ephemeral dry washes, which show up as nicks in the contour lines. North of Alpine, at the north-west corner of the map, curved contours mark the steeper slopes of broadly conical alluvial fans, deeply incised by valleys debouching from the high and steep mountains at the northern edge of the map.

In addition to the form on the ground, topographic maps show the works of man, such as roads, railroads and buildings, and drainage features such as perennial streams, intermittent streams, springs and marshes.

Geological history

Briefly, the geological history of the Alpine area comprises five events. Along the northern margin of the map, ancient folded rocks have been faulted upwards to form mountains flanking a basin to the south. The mountains underwent rapid denudation to form gravel fans (Qbg) as streams debouched onto the flat lands of the basin. Faulting remained active and offset the fan deposits by 5–10 ft before the low-lying land was flooded by glacial Lake Bonneville. A delta of gravel (Qbg) built up at the mouth of Alpine canyon, with shore currents moving sand (Qbs) westward on the delta and in bars in front of the delta. Silt (Qbsi) was deposited nearshore, whereas clay (Qbc) accumulated offshore. As the fifth event recorded on the map, the fault bordering the mountain range again became active with some 20 ft of movement displacing both the old fan deposits and the delta deposits. The final sixth event, still continuing, led to the building of new fans (QR) on top of the Lake Bonneville deposits after the lake had disappeared.

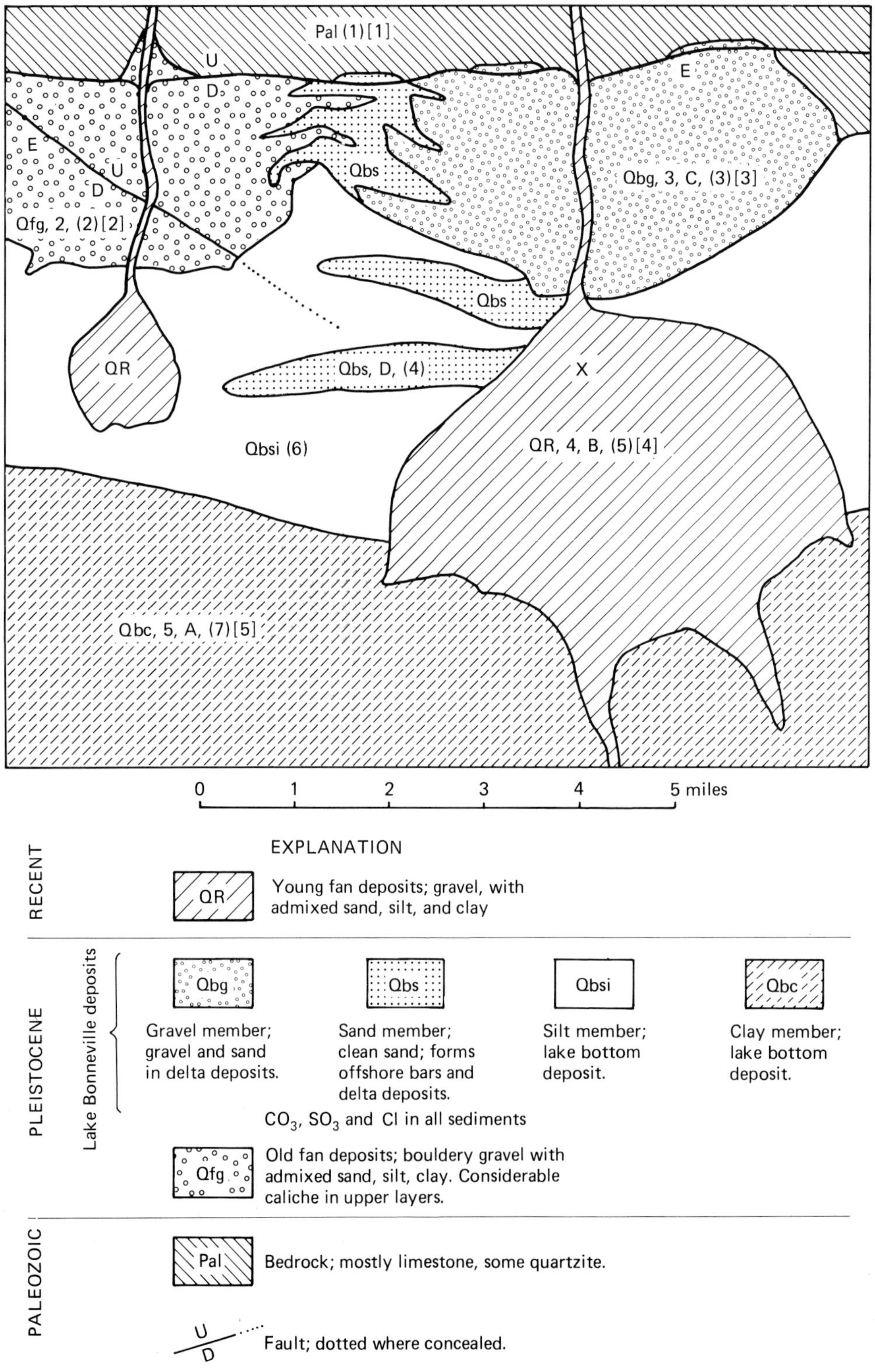

(a) Geological map of the Alpine Quadrangle

(b) Geological map as a guide to construction materials

1 Hard rocks. Good source for limestone or quartzite for building stone, riprap. Quarry operations would require drilling and blasting. Limestone suitable for cement.
2 This gravel is angular but silty; poorly graded and contains considerable secondary lime; not suitable for concrete aggregate; poor source for road metal.
3 This gravel well rounded and well graded but contains considerable lime; not suitable for concrete aggregate; excellent road metal.
4 This gravel is poorly graded; fragments in part angular and in part well rounded; deposit is free of secondary lime; best source for concrete aggregate.
5 Clay deposit contains lime and other water-soluble salts; fair source for structural clay; good source for seal clay; not suitable for high-grade ceramic purposes.

(c) Geological map as a guide to foundation and excavation conditions

A Clay ground poor surface drainage; no subsurface drainage; road metal and fill for subgrades must be hauled from area B or C. Ground easily excavated by power shovel or dozer operation. Will require subdrains.
B Gravelly and silty ground; fairly adequate surface and subsurface drainage. Good foundation for roads or buildings. Basement excavations must be shallow to avoid intersecting the groundwater perched on the underlying impermeable Lake Bonneville beds. This area lies across the projection of one of the recently active faults (see E).
C Gravelly ground with excellent subsurface drainage; ground easily excavated by power shovel or dozer. Excellent road foundation; the deposit rests against one of the recently active faults so buildings should be constructed to withstand shocks of intensity 8, R-F scale (see E).
D Sandy ground underlain by silt at depths less than 8 feet; good surface drainage down to the silt. Easily excavated by hand tools. Basement excavations must be shallow to avoid intersecting ground water perched on the silt. Good foundation for roads but clay is needed for binding sand.
E Two recently active faults. Movement on either one may be renewed at any time causing earthquakes. Buildings within a mile or so of the faults or their projection should be constructed to withstand shock of intensity 8, R-F scale. Five miles from the faults the shocks would not be expected to exceed intensity 5.

(d) Geological map as a guide to soils and land utilization problems

(1) Mountainous area; soil generally thin and stony. Locally there s a fossil soil having 10 feet of leached clay (an excellent source of structural clay). Principal watershed supplying valley area. Forested.
(2) Stony and in part bouldery ground. In places covered by fossil soil (see 1); locally the leached clay has been eroded exposing strongly lime-enriched gravel and silt.
(3) Stony ground. Top foot is brown windblown silt containing well rounded gravel; common large size 1–2 inches diameter. Five to ten feet of lime-enriched gravel beginning a foot below the surface.
(4) Clean quartz sand; grains $\frac{1}{2}$–1 mm diameter, well rounded, stained by iron oxide. No silt matrix, some lime carbonate cement. Locally blown into low dunes.
(5) Stony ground, silt matrix. Slightly lime-enriched zone less than a foot thick under surface layer of leached silt and gravel 6 inches thick.
(6) Silt ground. Contains about 3% of water-soluble salts – 1.5% of calcium carbonate and 1.5% of sulphates and chlorides of sodium and potassium. These salts leached from top 6 inches, and redeposited in next foot.
(7) Clay ground. Contains about 4% of water-soluble salts – 1% of calcium carbonate and 3% of sulphates and chlorides of sodium and potassium. These salts locally form surface crusts around moist depressions.

(e) Geological map as a guide to surface water problems such as flood control, drainage, canal construction, etc.

Area	*Permeability and slope of ground*	*Runoff conditions*	*Flood control, drainage and canal problems*
[1]	Impermeable; slopes steep.	Maximum coefficient of runoff.	Principal source of floods that would be hazardous invalley.
[2]	Moderately permeable; slopes moderate.	Little runoff during moderate storms; considerable runoff during severe storms.	Subject to floods from canyon; during severe storms may discharge floods from surface runoff. Moderate seepage losses can be expected from canals.
[3]	Highly permeable; gentle slopes.	Practically no runoff even following most severe storms.	No flood control problem. Reservoirs and canals require sealing to avoid excessive seepage losses.
[4]	Moderately permeable; low slopes.	Moderate runoff during severe storms.	Subject to flash floods from canyon. Moderate seepage losses can be expected from reservoirs and canals.
[5]	Impermeable and ground nearly flat.	Water stands in pools for long periods after rains.	Ground readily flooded and difficult to drain. No seepage losses from reservoirs or canals.

X Lower limit of perennial flow in streams draining mountain.

Figure 10.2 Four interpretations of ground conditions from the geological map of the Alpine Quadrangle, Utah, USA. (From U.S. Geological Survey, 1949, by permission)

Engineering geological assessments

Descriptions of the lithologies of the various deposits given on the map legend are very abbreviated, mentioning the main rock or soil type and the lesser constituents. Four other versions of the map differ from the geological map only in the legend, and as reproduced here accompany the geological map (Figure 10.2a) as additional legends (Figure 10.2b–e). The legends are imaginative, quite informative and appear to presuppose a good deal of background knowledge. The legends deal with the geological map as a guide to:

- construction materials;
- foundation and excavation conditions;
- soils and land utilization problems;
- surface water problems such as flood control, drainage, canal construction.

In Figure 10.2(a) the map units are letter and number coded to refer to appropriate aspects of the different legends.

Engineering soils, with the obvious exception of the Palaeozoic rocks in the north, dominate the Alpine map.

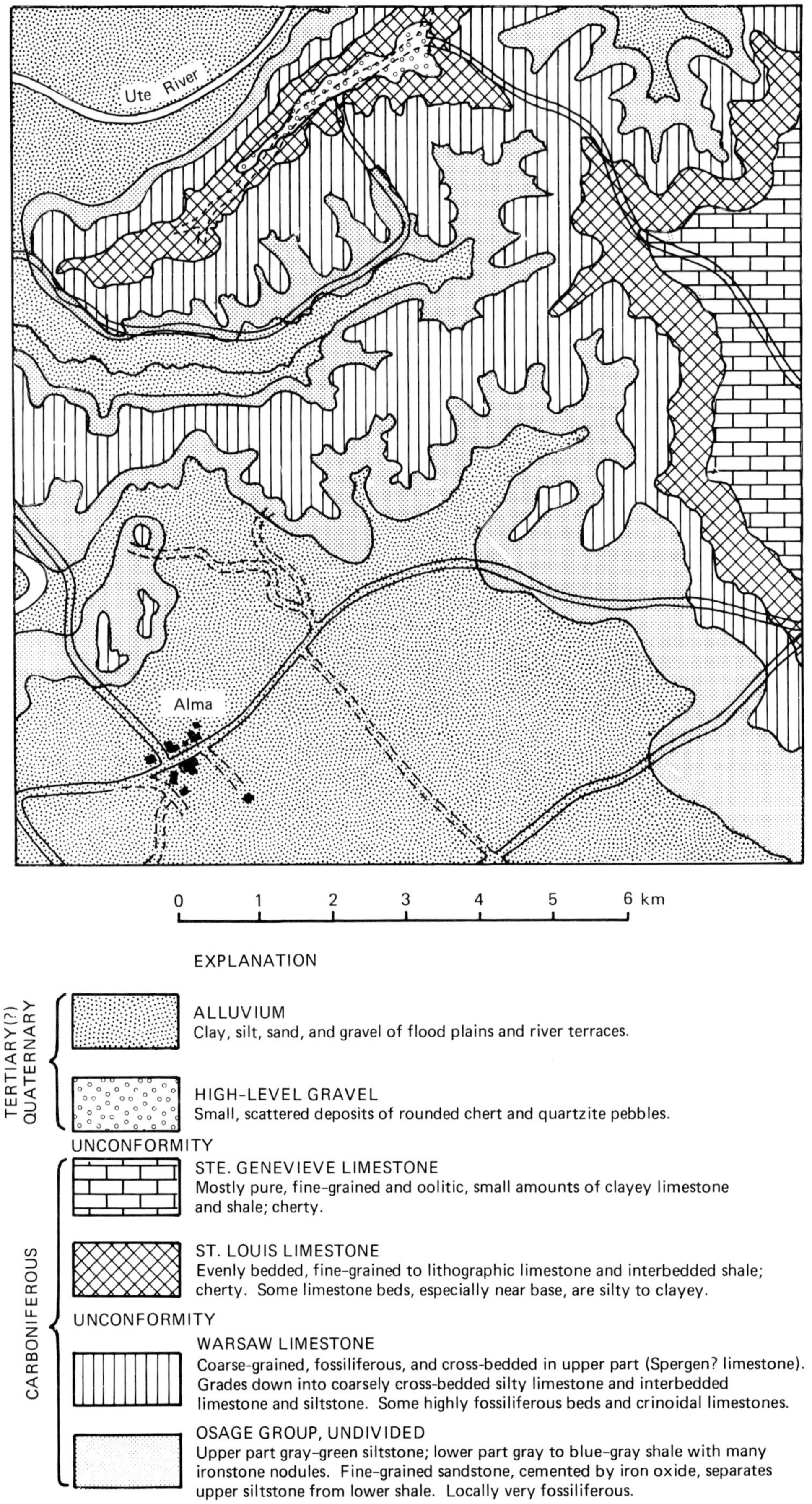

Figure 10.3 Geological map (After Eckel, 1951, Fig. 2)

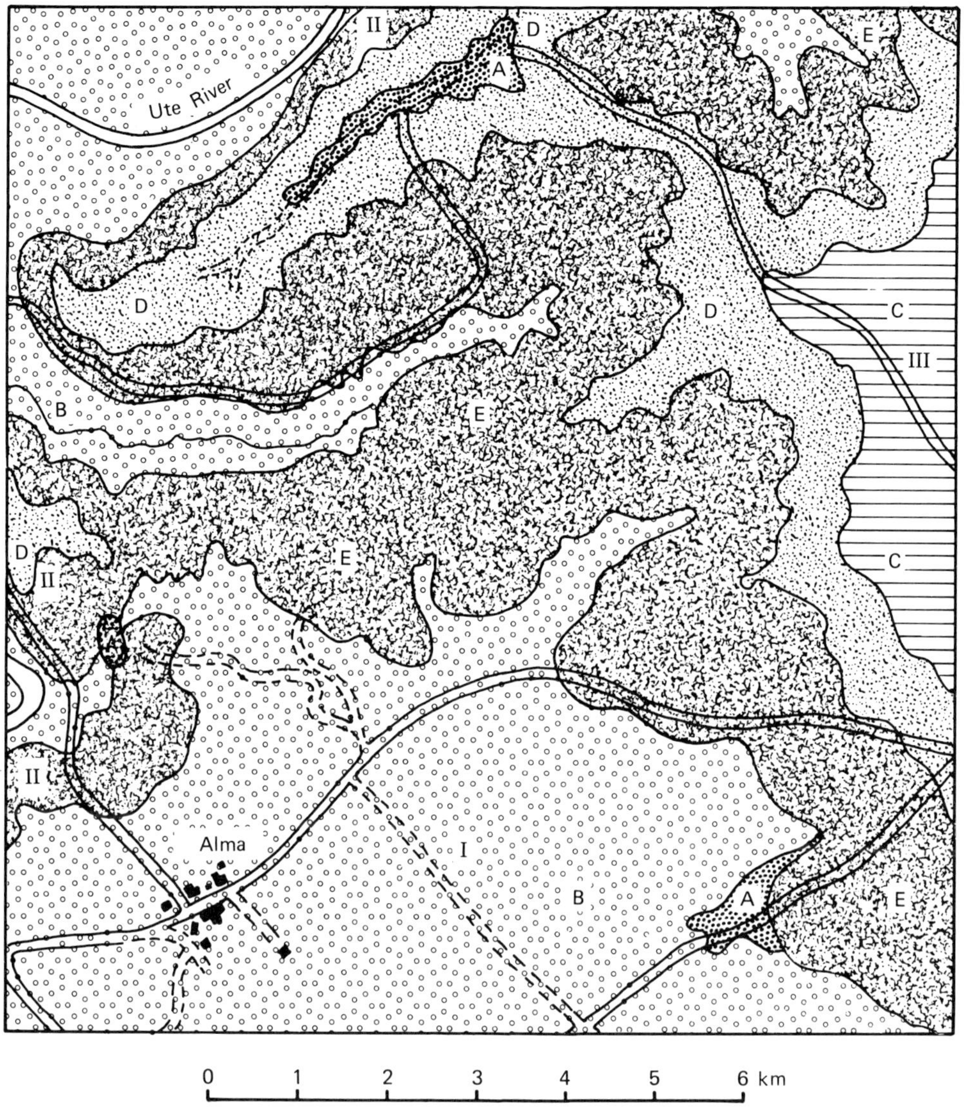

EXPLANATION

A — GRAVEL
Chert and flint, probably suitable for aggregate, base course, surfacing and fill.

B — SILT AND CLAY
Some silty and clayey gravel deposits generally unsuitable.

CLASS I — Flat bottom lands, silt and clay; depth to bedrock more than 100 feet; poorly drained; possible flooding.

CLASS II — Narrow ridgetops, steep wooded slopes, and narrow valleys; relief 100 to 250 feet; slopes covered with thin stony silt and clay soils; ridgetops mantled with 10 to 20 feet of well drained silt and clay soils; slopes, 15% to 45%.

CLASS III — Rolling to hilly uplands with some sinkhole depressions; relief 20 to 100 feet; limestone at depth of 5 to 20 feet; cherty clay soil; local deposits of fine sand. Slopes, 15% to 45%.

C — LIMESTONE
Mostly pure, some layers of shaly limestone and shale, locally cherty; probably suitable for surfacing, aggregate, base course, fill, mineral filler, riprap, and building stone.

D — LIMESTONE
Fine- to coarse-grained, dense, cherty, interbedded shale and siltstone; probably suitable for surfacing, base course, fill, mineral filler, light riprap and rubble masonry.

E — SILTSTONE, AND SHALE
Fine-grained, moderately hard, irregular bedding thin layers of limestone, chert and sandstone; possibly suitable for fill material.

Figure 10.4 Interpretation of the geological map in Figure 10.3 in terms of ground conditions for road construction, excavation and foundations (After Eckel, 1951, Fig. 3)

Interpretation is based on discrete, mappable units, often with distinctive landforms, and no attempt is made to group, for example, the three gravel units. However, such grouping would not appear to be justified, as can be gathered from the detailed descriptions of the gravels as construction materials and so on.

10.2.3 Interpretation of an area of horizontal sediments

In the later paper, Eckel (1951) further develops – for surely he must have played a significant role in the Alpine Quadrangle interpretation – his philosophy behind the interpretation of a general-purpose geological map, again in terms of road construction, excavations and foundations. He gives three maps: topographical, geological and interpretative.

Comparison of contour lines and the outcrops of solid formations shows that the strata are essentially horizontal (Figure 10.3). A sequence of four geological, that is lithostratigraphical, units comprising Carboniferous limestones and siltstones is partially masked by high-level Tertiary gravels and by the general spread of alluvium over low-lying ground. A typical geological map 'Explanation' provides an abbreviated description of each unit.

Alluvium and gravel remain unchanged on the interpretative map (Figure 10.4), whereas there has been an important regrouping of the solid formations with significant lithological types arranged into three separate units. Descriptions of the five engineering rock and soil units on the map are lithological, with comments on possible uses. Topography is grouped into three classes, with comments on such aspects as depth to bedrock, slopes, relief and the presence of karstic features.

Interpreting a geological map for construction materials can be done by separating each rock unit that has distinct physical properties or by grouping rock units whose physical properties are similar. If these map units are described in terms of highlighting their possible uses in construction, rather than their geological characteristics, we have a map that can be used in assessing in general terms the materials situation anywhere in the map area.

Eckel does not use the map he has illustrated (Figure 10.3) to develop his ideas on how map units should be described for engineering use, preferring rather to describe hypothetical map units. Two examples are given here. The essential flavour of an 'applied' description as opposed to a 'pure' geological description can be savoured by comparing the following descriptions of a shale map unit, not necessarily the same in both cases.

The shale unit on a geological map is described as:

REEDSVILLE SHALE

Chiefly olive-green shale, weathering to small slivers, upper part containing argillaceous and ferruginous limestone layers; thick bedded sandstone with abundant brachiopods of the Orthorhyncula zone at top, and a few feet of black shale containing graptolites at base.

Obviously, Eckel comments, there are few facts in this description that would be of direct help or interest to the engineer. Compare that description with that for another shale in which construction aspects are highlighted:

Map Unit H

SHALE

Thinly layered shale of varying hardness with very little sandy material either as grains in the shale or as separate beds of sandstone. Plastic or slippery when wet, especially in weathered parts; makes weak, poorly drained fill and very poor surfacing unless mixed with sandier and tougher material from other units. Many shales included here are excellent material for common brick; some of the more carbonaceous may be useful for lightweight aggregate. Locally soft enough for hand tools, but most requires power equipment.

Such a description applied to all the map units in turn tells the engineer at a glance the kinds of materials that are available in the map area and whether these are likely to meet his needs. The areal distribution of each kind of material, its topographic position and its accessibility to existing means of transportation would be shown on the map. Other aspects of possible concern would be shown on cross-sections, abbreviated logs from which thicknesses, whether or not the unit is close to the surface or is too deeply buried to be of interest, or whether it is horizontal or folded and faulted, could be judged.

One further example gives a description aimed at foundation and excavation conditions in alluvium:

Map Unit G

POOR FOUNDATIONS, EASILY EXCAVATED

FLOOD PLAIN ALLUVIUM

Irregular layers and lenses of gravel, cobbles, sand, and silt, natural bank surfaces generally give exaggerated impression as to quantity of cobbles. Excellent drainage where above water table but low topographic position makes materials subject to floods. Irregular distribution of materials and lack of compaction make it a poor foundation; surfaced roads require excessive maintenance because of differential settling; heavy base course, well-laid and compacted, would overcome this difficulty. Cuts on slopes of $1:1\frac{1}{2}$ appear to be stable.

Of course the description and the map do not give all the information required for the design and construction of a specific structure, but commonly provide sufficient information to guide preliminary planning and tentative choice of sites.

Eckel hastens to point out that not all the engineer's geological problems can be solved by interpretation of the usual geological map. The general-purpose map does, however, answer many if not most of the questions that arise during the early planning stages of any construction programme.

10.3 Areal and engineering geology urban maps

Geological mapping in an urban area has both advantages and disadvantages. The topographic base map is usually kept up to date and it is easy to locate points of interest. Over a period of years, new construction and reconstruction provides the opportunity for many foundation excavations, trenches and site investigations, all of which can be recorded to supplement any existing geological map and borehole data bank. On the negative side, natural exposures may be minimal, and large parts of the area will be covered with buildings and roads. However, it is still possible to prepare geological maps using a combination of conventional geological mapping techniques and the data bank. One such map has already received brief mention: the Engineering Map of Belfast (see Figure 10.22).

10.3.1 Oakland East Quadrangle, California, USA

Between 1957 and 1965, Radbruch (1969), with assistance, mapped this quadrangle with engineering aspects firmly in mind. The trace of the active Hayward fault was indicated for the first time, and the approximate locations of landslides were marked by point symbols. On the face of the map, locations of selected boreholes are marked with abbreviated logs (Plate 14).

Usefulness of the map is enhanced by an explanatory pamphlet giving a generalized tabular description of the engineering properties of the 31 map units (Table 10.1) and a detailed comment on faulting.

Part of the map, redrawn and slightly simplified, has been published as Fig. 5.4.1 in Anon. (1976). The fragment used here (Plate 14) is part of the map as published. Only some of the geological units on the part of the map reproduced are described in Table 10.1.

In the absence of a full descriptive memoir, the summary borehole logs on the map, with additional located borehole logs in the map legend, provide details of successions in the Quaternary deposits in urbanized areas.

Although the Hayward fault does not appear in Plate 14, the table of engineering properties reminds the map user that severe earthquakes were caused by movement along faults within the Hayward fault zone in 1836 and 1868. Therefore the entire length of the Hayward fault zone can be assumed to be active, an observation of particular relevance to the urbanized areas on superficial deposits shown on the map.

10.4 Maps of surface and underground quarries in Paris

10.4.1 Introduction

There is available a twin series of plans of Paris at scales of 1:5000 and 1:1000. Arnould *et al.* (1979) discussed the maps of three arrondissements, compiled since 1977, from the points of view of the presence of open but infilled, and underground, quarries, jointing in Lutetian limestones, and superficial deposits on valley sides. Four factor maps, compiled at 1:5000 but issued at 1:10 000, dealt with such issues as thicknesses of fill; extent, erosion and the top of the limestone; and groundwater conditions. A map of engineering geological zoning was accompanied by a very informative legend on ground conditions.

An essential background to these compilations has been the published large-scale plans, which have been kept up to date, and of course a great deal of additional site investigation information. A part of one 1:1000 plan has been selected for illustration, in colour.

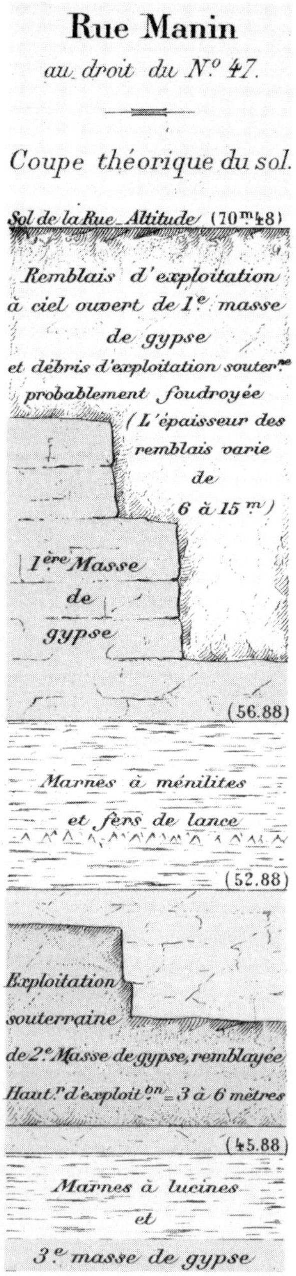

Figure 10.5 Schematic cross-section of the ground on Rue Manin, Paris, where gypsum has been worked in open quarries and underground. Inset on Sheet No. 58

10.4.2 Sheet No. 58, Rue Manin, Parc des Buttes Chaumont (Nord)

Over the area of the map (Plate 15) up to four beds of gypsum have been worked in open quarries or underground. There is a great deal of information, although intensive analysis is necessary to understand the map. Only a few aspects can be dealt with here.

The inset vertical section in Plate 15 illustrates the three beds of gypsum that have been worked in the general area of the Place Armand Carrel. At the surface, the first mass of gypsum has been worked in open quarries which in the south are now an artificial lake with an island of gypsum in the middle that may have been worked underground. Level of water in the lake is

Table 10.1 Generalized description of engineering properties of selected map units, Oakland East Quadrangle, California, USA (From Radbruch, 1969, by permission)

Map unit	*General lithologic description*	*Topographic form*
Artificial fill (Qaf)	Composition varies from place to place. Consists largely of Merritt Sand dredged from bay along shore of Alameda. Most highway fills composed of mixtures of rock and soil derived from nearby cuts or borrow areas. Some fills contain concrete, bricks and other miscellaneous refuse. Only large fill areas shown on map; innumerable fills too small to show on map have been made for highways, streets, and building pads	Flat, level areas a few feet above sea level along the edge of the bay, or flat fills in valley bottoms; also terraces and ramps for highways
Alluvium and colluvium (Qac)	Composition varies from place to place. In small, swift-flowing streams, recent alluvium—largely sand, pebbles and boulders; alluvium in flat valleys and colluvium on hillsides generally finer material, usually dark in color. Alluvium and colluvium underlain by rocks of Contra Costa Group commonly contain much swelling clay. Colluvium as much as 60 ft in thickness on west side of Moraga Valley; alluvium a few inches to more than 75 ft in thickness	Recent alluvium fills stream valleys and forms flat valley bottoms; alluvium fills and obscures many old hillside ravines too small or indistinct to show on map; colluvium mantles sides of hills
Undivided Quaternary deposits (Qu)	Composition and physical properties vary. Consists predominantly of Temescal Formation. Probably includes covered or unrecognized San Antonio Formation and gravel, sand, and clay (Qg), as well as Recent alluvium and colluvium, and artificial fill. Symbols for Qtc, Qts and Qtb shown in parentheses where these units can be positively identified (see Temescal Formation)	Primarily in valleys and on gentle slopes between San Francisco Bay and the Berkeley Hills
Temescal Formation (Qtc, Qts, Qtb)	As used by Lawson (1914) comprises several presumably contemporaneous alluvial units of different origin, lithology and physical properties. Qtc, dark alluvium filling stream channels in the eroded San Antonio Formation; consists of irregularly bedded clay, silt, sand and gravel with organic material and some Claremont chert; dark-yellowish-brown to olive-gray. Poorly consolidated; one consolidation test on Peralta Creek showed compression of 16%; 3–18 ft thick. Qts, material apparently derived from erosion of the San Antonio Formation; consists of lenses of clay, silt, sand and gravel with Claremont chert; yellowish brown; in places lithologically indistinguishable from San Antonio Formation; origin assumed from topographic form and lower compressive strength as indicated by one consolidation test on Sausal Creek (compression 12%). Qtb, alluvium derived from the Berkeley Hills, probably deposited simultaneously with Qts, consists of well-consolidated gravel-sand-silt-clay mixtures with firm pebbles and little or no Claremont chert; moderate yellowish brown; well exposed on Arroyo Viejo where a consolidation test showed a compression of 6%, and on 73rd Street, where it has been tilted to an angle of 72°	In flat valley bottoms and gentle slopes between San Francisco Bay and the Berkeley Hills. Qtc fills meandering channels cut in underlying material

* Underlining indicates geologic conditions that may be critical to planning, design and construction of engineering works.

maintained by a drain seated on the unworked part of the first gypsum layer; the concrete drain is supported by a line of bored *in situ* concrete piles drilled down to the base of the third gypsum that has been worked out and infilled underground.

Underground workings in the gypsum beds have a very irregular pillar-and-stall pattern, and often it will be noted that the pillars in workings at two levels are superimposed. On the map margin, schematic cross-sections (Figure 10.5) are given of the open quarry workings in the first gypsum mass and underground workings in the second mass up to 6 m high. The plan legend illustrates the following:

- Hague: dry stone walls to support infilling;
- Pilier à Bras: pillars of blocks of stone;
- Fontis: bell-shaped opening formed by the collapse of the roof of an underground quarry to the surface;
- fr: infilled collapse;
  PS: access shafts to workings, which may be infilled (comble).

10.4.3 Engineering geological zoning maps of the 19th and 20th arrondissements

The 19th and 20th arrondissements are in the north-east and east of Paris, centred on the hill of Belleville capped by Fontainbleau sandstone (Figure 10.6). Arnould *et al.* (1979) have established three types of phenomena in the subsoil which give constructors particular problems:

1. The presence on the hill of old quarries in gypsum, either worked open to the sky or underground, now masked by man-made fill or packed with broken rock.

Weathering and soil development	*Workability*	*Slope stability and foundation conditions*	*Dry density moisture content and Unified Soil Classification*	*Remarks (includes use and earthquake stability)*
None in most places	Depends on composition; varies from place to place	Depend on composition. Properly engineered fill placed over bay mud generally suitable for light structures. Fill over bay mud may settle differentially as bay mud compacts under load	Vary with composition. 106 (hydraulic sand fill); 23% (14: 92–119; 18–37%)	Properly engineered fill generally suitable for use as foundation for light structures; fill overlying bay mud is susceptible to movement under earthquake stress
Thickness of soil varies from a few inches to several feet. In flat valleys soil as much as 3 ft in thickness has developed on alluvium. In places soil clayey, shrinks and swells	Can be moved with hand tools. Where material is clayey, may be very heavy and sticky when wet, sticking to tools and miring heavy equipment	Depend on composition. Alluvium and colluvium derived from rocks of Contra Costa Group generally contain expansive clay. May cause heaving and cracking of structures in flat areas; susceptible to sliding on hillsides. Alluvium in old ravines may slide in cuts	100; 23%; (66: 86–122; 8–34%) GW–OH	Dry density, moisture content, and Unified Soil Classification determined on sandy and silty clay derived from Moraga Formation and Contra Costa Group
Soil may be as much as 3 ft thick. In places soil clayey, shrinks and swells; may cause damage to buildings	Can be moved with hand or power tools	Depend on composition; generally good. Slides have formed where colluvium apparently derived from gabbro	Varies	Mapped with Temescal Formation in Oakland West Quadrangle (Radbruch, 1957)
In places a soil as much as 3 ft in thickness has formed on the Temescal Formation; soil clayey, shrinks and swells with seasonal moisture changes	May be moved with hand tools. Clay beds may be sticky when wet	Slope stability and foundation conditions fair to poor. Some minor slumping in steep cuts. Qtc generally softer than underlying and adjacent material; buildings founded partly on Temescal channel material and partly on older material may be damaged by differential settlement	86 (Qtc); 42%; (3: 73–104; 25–71%); 100 (Qts); 23%; 110 (Qtb); 14%; (39: 92–133; 3–22%); Qtc, GC–OH; Qts, GC–CH; Qtb, GM–CH	

2. The presence of significant thicknesses of slope deposits, heterogeneous and poorly known, on the hillside.
3. The presence of old and contemporaneous solution features affecting gypsum, giving voids and loose zones, with surface collapses and settlements.

One outcome of the investigation has been the plan of engineering geological zoning with a very detailed legend (Figure 10.7). Zones are not based on stratigraphy of the Tertiary and superficial deposits, but rather on a combination of locality: plateau, plateau edge and valley slopes and valley bottoms; thickness of various deposits and whether quarried; whether the Calcaire Grossier has been worked; thicknesses; depths; investigation requirements.

So far, the study has produced:

- a 1:5000 scale atlas of the extent and stratum contours for the top of various Tertiary formations;
- serial cross-sections at 1:500;
- a collection of estimated lithostratigraphical successions at each nodal point on a 100 m grid over the map area – the intention is to provide an estimate of the likely depth of site investigation boreholes at any location;
- a report on each formation – its lithology, hydrogeology, engineering geological characteristics;
- a table for geotechnicians listing the general characteristics of each formation, and local problems;
- maps directed particularly to the hazards likely to be associated with gypsum solution.

Problems associated with the solution of gypsum, leading to actual cavities and decompressed zones, have

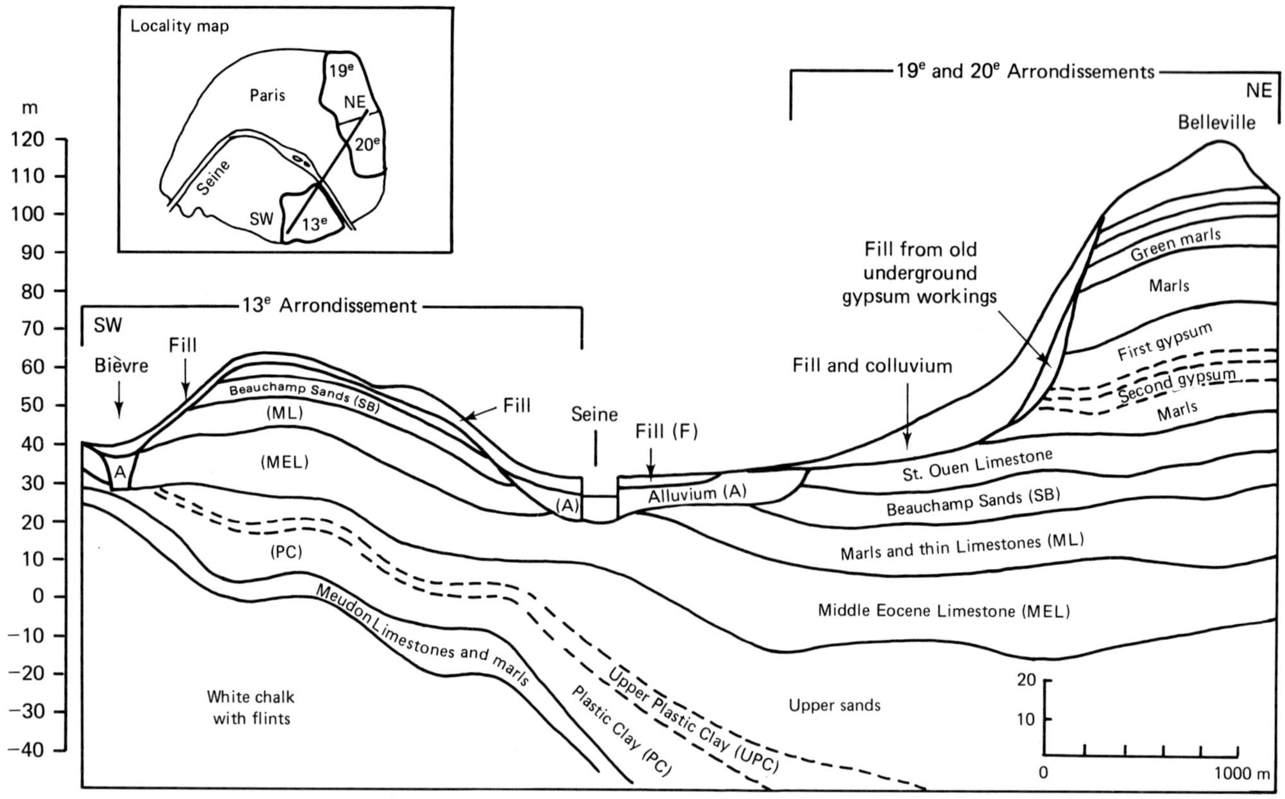

Figure 10.6 Schematic cross-section SW–NE across the east of Paris (From Arnould 1979, *et al.*, Fig. 2, by permission)

been considered by Deveughele and Usseglio Polatera (1979). A map of the zones exposed to the hazards linked to gypsum solution has been prepared.

10.5 Uncovered geological and geotechnical plans of Warsaw

10.5.1 Introduction

Foundation levels for houses and multi-storey buildings are determined by a combination of the character of the soils at and immediately below the ground surface, the allowable bearing capacity of soils and rocks, and climatic conditions. Taking European conditions, for instance, depth of freezing in winter is important, pedological soil layers are always stripped off the site to prepare the foundations which will be pad or strip footings for light structures, or piles of various types for larger structures. Engineering interest may then start at one or two metres below original ground level, and of course the ground is stressed below these levels by the loads imparted by the structure.

In addition, there is always inherent uncertainty on the true position of mapped geological boundaries at the surface, and it may be that the lines on the map really represent the position of boundaries below and not at the surface. Such uncertainties, which become very apparent if a trial pit is dug, led to the proposal that maps should be prepared for engineering purposes representing the geology, and if necessary geotechnical properties, at specified depths below the ground surface. Such maps are called uncovered maps or plans.

10.5.2 Uncovered maps of geological conditions

Specific depth maps are of particular use at large scales in showing the distribution of rocks and soils below ground level. An excellent example, prepared for town planners, architects and the construction industry, is the geological atlas of Warsaw by Stamatello (1965). In most urban areas almost completely destroyed during World War II, it was necessary to move into adjacent rural areas to achieve rapid redevelopment. In these cases, ground conditions had to be determined by drilling. For Warsaw, on the other hand, the first geological map of the city was compiled in 1936 by Rozycki and Sujkowski, at a scale of 1:20 000, from about 3000 borehole records, and observations made on trenches and foundation excavations.

Geological conditions in Warsaw are dominated by Tertiary clays nearly everywhere more than 100 m thick. In drawing up the 1936 map it was assumed that these clays were homogeneous and stiff, thus providing excellent foundation conditions. On the clays, which in fact are very heterogeneous, are deposits of Pleistocene and Holocene. Because of the assumed homogeneity of the Tertiary clays, apart from drawing contours on bedrock, only the superficial deposits were examined in any detail. Immediately after World War II, the map was particularly useful for redevelopment of the central areas, even though it had the grave disadvantage that it could not be used directly by the planners who needed the assistance of geologists, often not very good, to interpret it.

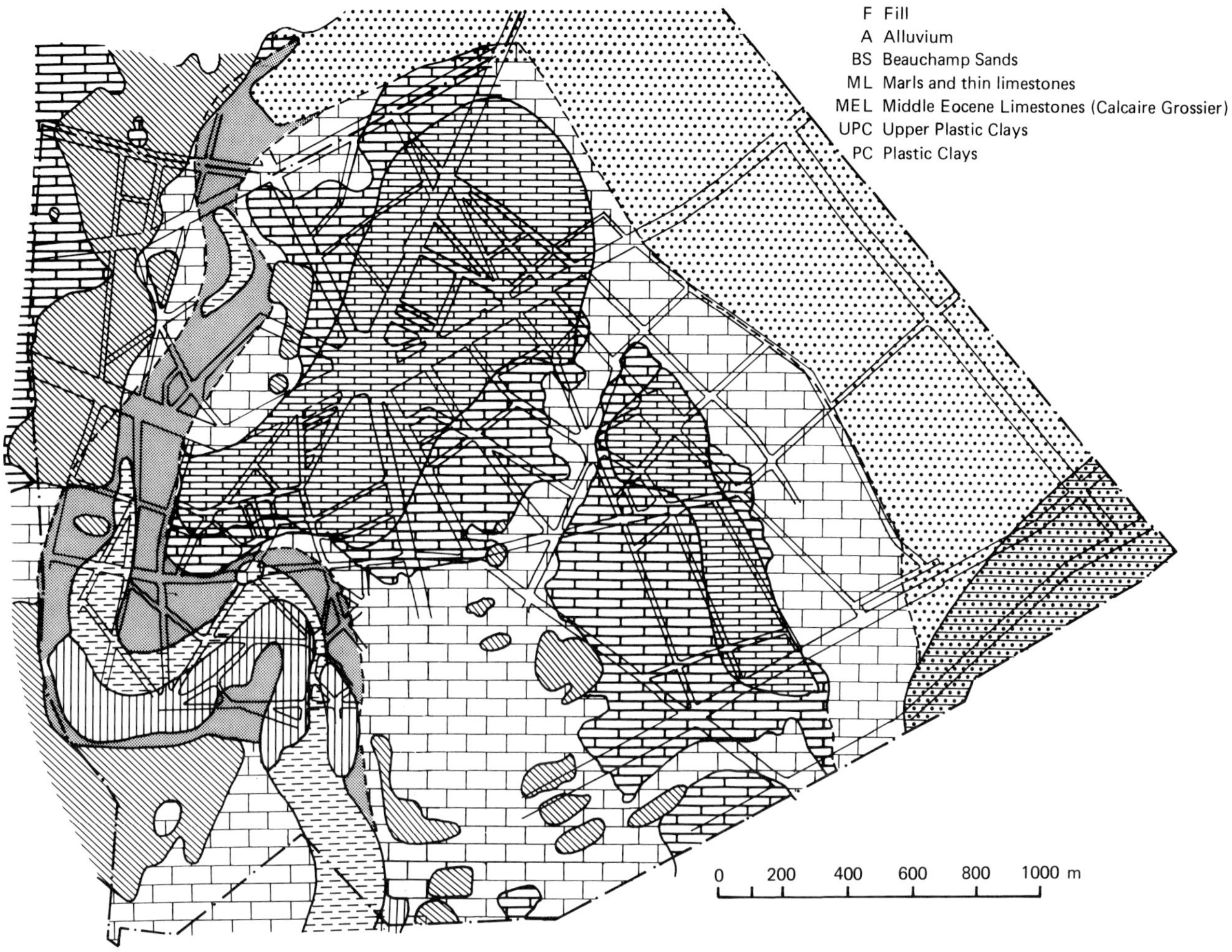

Zone	Pattern		Location	Definition criteria	Workings in Middle Eocene Limestone	Lithology and thickness	Depth	Site investigation, matters to be considered
P1		PLATEAU	Plateau	ML > 10 m underground quarries	Yes	(BS) ML > 10 m MEL ≃ 17 m UPC-PC = 12-20 m	> 15 m	- two foundation levels possible: top of the Middle Eocene Limestone - presence of underground quarries - presence of talus in ML and SB
P2		PLATEAU	Plateau	F < 12 m ML < 10 m no quarrying	No but possibly	(ML) < 10 m (F) < 12 m MEL ≃ 17 m UPC-PC = 12-20 m	> 15 m	- no underground quarries known, but possibility of their occurrence - depth to the top of MEL
P3		PLATEAU	Plateau	F < 12 m ML < 10 m underground quarries	Yes	(ML) < 10 m (F) < 12 m MEL ≃ 17 m UPC-PC = 12-20 m	> 15 m	- underground quarries present - top of MEL
P4		PLATEAU	Edge of plateau	F > 12 m surface quarries	Yes in surface quarries	F > 12 m MEL < 17 m UPC-PC = 12-20 m	> 10 m	- thickness of fill (F) and top of MEL - properties of fill (F) with regard to shallow foundations
V1		VALLEY	Slopes	MEL landslipped	No	F MEL landslipped UPC-PC = 15 m	< 10 m	- slope stability - depth and thickness of UPC-PC - compressibility of UPC-PC
V2		VALLEY	Seine valley	A ML	No	A = 9 m MEL = 17 m	< 5 m	- Old alluvium, depth and properties - problems with earthworks because of groundwater
V3		VALLEY	Seine valley	A UPC or PC	No	A = 9 m UPC-PC = 12 m	< 5 m	- Old alluvium, depth and properties - top of UPC and PC - problems with earthworks because of groundwater
V4		VALLEY	Bièvre valley	MEL upper MEL middle eroded	No	(F-A) MEL < 5 m UPC-PC = 12 m	< 5 m	- thickness and condition of MEL - compressibility of UPC and PC - problems with earthworks because of groundwater
V5		VALLEY	Bièvre valley	MEL completely eroded	No	F = 10 m (A) UPC-PC = 12 m	< 5 m	– compressibility and thickness of UPC and PC – problems with earthworks because of groundwater – investigation of fill

Figure 10.7 Engineering geological zoning plan and the accompanying legend (After Arnould *et al.*, 1979, Fig. 3)

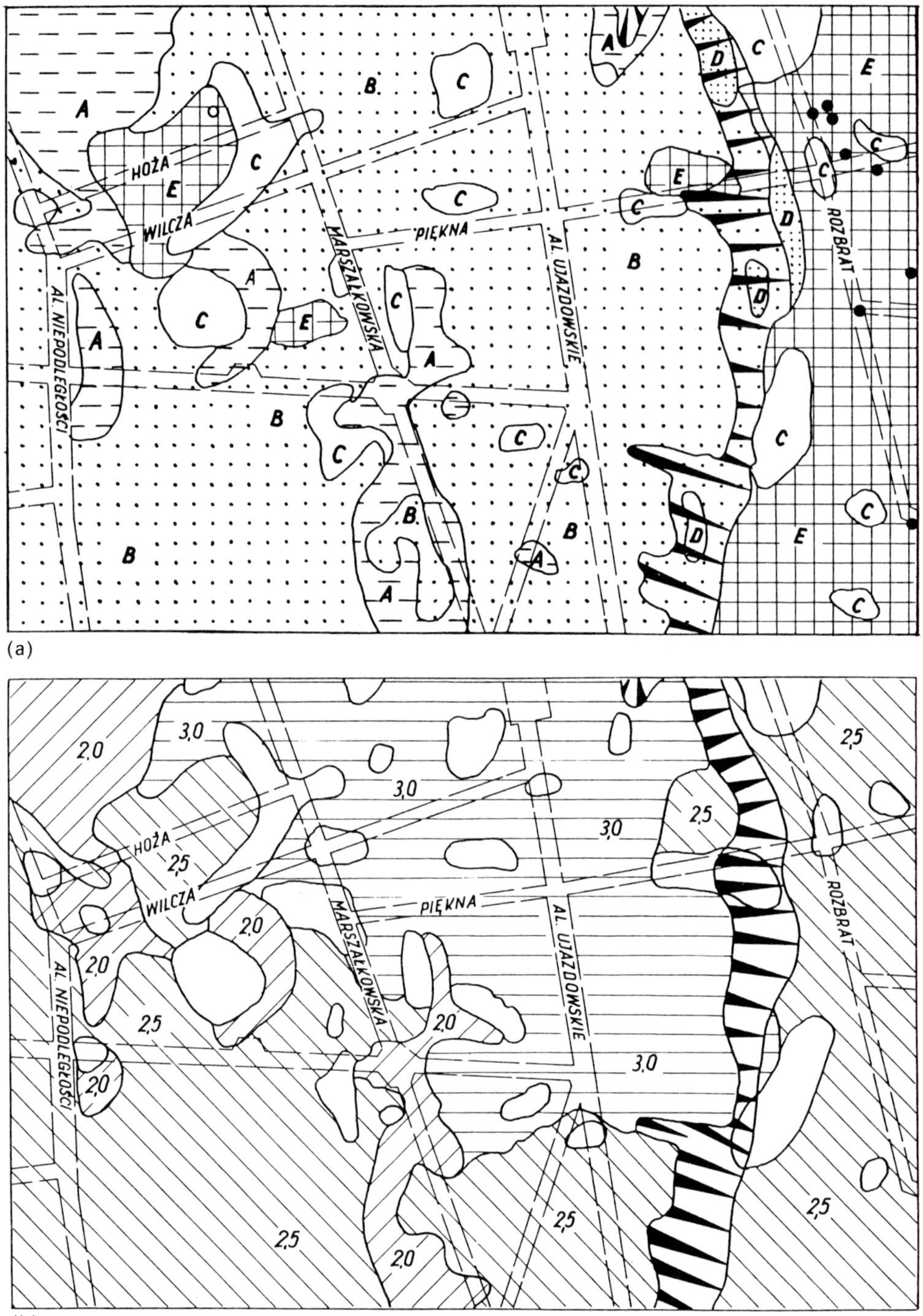

Figure 10.8 Part of the geological map of Warsaw showing the geological units at a depth of 2 m. (a) The geological map – A, varved clay; B, morainic loam; C, fill; D, gravel; E, sand; F, Tertiary and Pliocene clay. (b) The geotechnical map of allowable bearing pressures in kg cm^{-2} (From Stamatello, 1965, Figs 1 and 4, by permission)

These problems were later overcome when engineering geologists became involved in preparing new maps for planners and engineers. Specific depth maps were produced for depths of 2, 4 and 6 m below ground level (Figure 10.8a). Such maps are not easy to interpret in three dimensions unless the surface geological map is available and the specific depth maps are drawn on transparent overlays.

Uncovered maps were also prepared in terms of allowable bearing capacity at the three depths. If the bearing capacity was less than 1.0 kg cm^{-2}, the ground was unsuitable for the footings of buildings. As the original map shows, the highest bearing capacity of 3.5 kg cm^{-2} was suitable for very high buildings, and between 1.0 and 3.5 kg cm^{-2} the ground was mapped in 0.5 kg cm^{-2} steps (Figure 10.8b).

As in most old urban areas, Warsaw had many deposits of fill of different thicknesses, densities and composition. Sufficiently dense fill that had been in place for more than 60 years was treated as a normal soil. Claypits, pools and ditches filled with a great variety of uncompacted fill were generally considered unsuitable for foundations, except when the area was small and the fill could be removed and replaced by compacted soil. In Figure 10.8(a), areas of fill are left plain.

Uncovered maps when superimposed illustrate some of the difficulties inherent in both the preparation and interpretation of unitized maps (see Chapter 7).

The vertical N–S slope on the right of the map between Al. Ujazdowski and Rozbrat is the 25 m high escarpment bordering the floodplain of the River Vistula with slopes of 25–45°. Most of the escarpment is subject to landsliding and it is difficult, and probably inadvisable, to give any indication of bearing capacity.

10.6 Maps of Prague: 1960–1978

In 1979, to mark the eightieth birthday of Academician Quido Zaruba, an account of the engineering geology of the city of Prague was published which included a set of engineering geological maps and details of engineering works carried out in the city (Pichal, 1979).

One of the first engineering geological maps, which sets the geological scene for 36 km^2 of the centre of Prague, is the 'Geological Map of the Inner Prague Area' on a scale of 1:12 500 (Zaruba, 1948). Since that time, various geological and zoning maps have been prepared. Since 1969, engineering geological maps on a scale of 1:5000 have been compiled according to the method devised by Simek (1962). By 1979, 80 map sheets covering 400 km^2 had been completed. The 1:5000 maps are intended for detailed urban studies and plans of housing schemes, but they are less suitable for larger areas for which the 1:25 000 scale is more appropriate. The 1:25 000 mapping was started in 1975, and the nine sheets for the Prague area were completed in 1981. It is appropriate here to give some details of the methodology adopted for the mapping programmes.

10.6.1 Methodology for 1:25 000 maps

Each engineering geological map at this scale consists of three sheets:

Sheet A. The map of engineering geological conditions. It provides an outline of geological and hydrogeological conditions and the location of active geomorphological conditions (geodynamic phenomena). It also contains data on the type and thickness of superficial deposits, type of bedrock, depth to the first groundwater horizon and an indication of the corrosiveness of groundwaters.

Sheet B. The map of engineering geological regions within the mapped area. It defines the regions with similar engineering geological conditions and also the location of mineral deposits, nature reserves, protected zones of water sources, transportation routes, etc.

Sheet C. The map of documentary points.

10.6.2 Methodology for 1:5000 maps

For each map area, four separate sheets, a final report and a list of documentation points are prepared:

Sheet A. The basic and most important map. It shows the bedrock, superficial cover (overburden) and made ground. The map is constructed as exposed at the depth of 2 m, i.e. superficial deposits and artificial fills less than 2 m deep are not shown. In addition, the map includes all features of engineering geological importance, such as quarries, cuttings, sandpits, loam pits, etc. Special attention is paid to fossil, dormant and active landslides. The map also shows location of mineral and nature reserves, protected zones of water sources, transportation routes, etc.

Sheet B. The map of the overburden thickness. These data are presented on a separate sheet in order to ensure legibility by not overfilling Sheet A. Sheet B is useful especially for such areas where there is a great thickness of superficial deposits and their structure is complicated.

Sheet C. The map of hydrogeological conditions. It summarizes the data on surface and groundwaters from the viewpoint of engineering geology. It shows the highest ascertained height of the groundwater table – its depth under the surface. In addition, the map contains basic data on rock environment, water aggressivity toward construction materials, direction of groundwater flow, etc.

Sheet D. The map of documentation points.

The report accompanying the map contains a general description of geological, geomorphological and geographical conditions, hydrogeology and a tabulated outline of geotechnical properties of the rocks. A separate volume contains all the documentary points and surveyed wells in the sheet. An important part of the report is the geological sections which provide an instructive picture of the structure of the bedrock and superficial deposits.

10.7 Hong Kong: terrain classification for geotechnical purposes in deeply weathered rocks

10.7.1 Introduction

The territory of Hong Kong is situated at latitude 25°N on the south-east tip of China. It has an area of only 1050 km^2 and a population of nearly six million. The terrain is mostly very hilly, and building development is concentrated mainly on the reclaimed coastal strips of land and on the footslopes of steep hillsides. Very high seasonal rainfall is often responsible for a large number of landslides, some of which cause human casualties and major damage to property.

The main rock types are granite and rhyolitic volcanic rocks, both of which are deeply weathered to give profiles varying from residual soil at the surface to fresh rock at some depth. Considerable thicknesses of colluvium overlie the tropically weathered rocks in many places, particularly on footslopes.

Scarcity of readily usable land, combined with the geological and climatic conditions, underlines the importance of effective land-use planning in which terrain stability considerations are paramount. Since 1979, the Geotechnical Control Office of the Hong Kong government has devoted considerable resources to a Geotechnical Area Studies Programme (GASP), which also enables a systematic landslide 'inventory' to be maintained. The aims of the programme are: (a) to provide a systematic terrain evaluation for the whole Territory at a scale of 1:20 000, and (b) to enable chosen areas of importance to be evaluated at the more detailed, plan scale of 1:2500.

'Regional' studies at 1:20 000 provide geotechnical input for strategic land-use planning purposes; 'District' studies at 1:2500 give sufficient information for the layout planning and conceptual design of individual development areas. The programme and the methodology have been described by Burnett and Styles (1982), Styles *et al.* (1984) and Burnett, Brand and Styles (1985).

Studies are based entirely on aerial photograph interpretation, site reconnaissance, and existing geotechnical information from borehole logs and technical reports. The land area of Hong Kong is covered by 11 Regional Studies each covering an area of about 80 km^2. Each District Study covers between 2 and 4 km^2. Some of the detailed engineering studies of particular sites, all at large scale, are illustrated in Chapter 11.

10.7.2 Terrain classification

For District Studies, six main characteristics, or attributes, are used for dividing the terrain into land units:

- slope gradient (6 categories);
- terrain component (16 categories);
- terrain morphology (9 categories);
- erosion (17 categories, including 'instability');
- slope condition (8 categories);
- hydrology (3 categories).

The complete terrain classification schedule is presented in Table 10.2. Note that the 'erosion classification' attribute includes the category 'instability' and is subdivided into six subcategories.

10.7.3 Terrain classification mapping at the 1:2500 District scale

The geology of the Mid-levels District of Western Hong Kong Island is shown as an idealized cross-section of a hill-slope in Figure 10.9(a). In some places it comprises up to 28 m of colluvium overlying decomposed Hong Kong granite. The granite intrudes the Repulse Bay Formation exposed in the upper slopes, metamorphosing the constituent tuffs at the contact. The thickness of decomposed rock below the colluvium ranges from 2 to 3 m in the upper slopes, to more than 90 m in the granites on the lower slopes.

Large-scale district studies are undertaken in two stages: Stage 1, an initial geotechnical assessment using existing data, and Stage II, using specially derived data. Stage I, according to Burnett and Styles (1982), combines systematic terrain classification and field engineering geology.

As part of the final report on an area, a series of complementary overlay maps is produced, each dealing with a specific purpose, building up to provide a full understanding of the area. The map series comprises six maps:

- terrain classification map;
- geotechnical land-use map;
- surface hydrology;
- vegetation map;

and two which are compiled from field mapping and a review of existing literature:

- engineering geology map;
- engineering data sheet.

Examples of these maps, and their accompanying legends, are given for the area selected (Figure 10.10), in Figures 10.11–10.15, and two engineering data sheets (Figures 10.16 and 10.17). Techniques used for data acquisition and derivation of the maps is shown systematically in Figure 10.18.

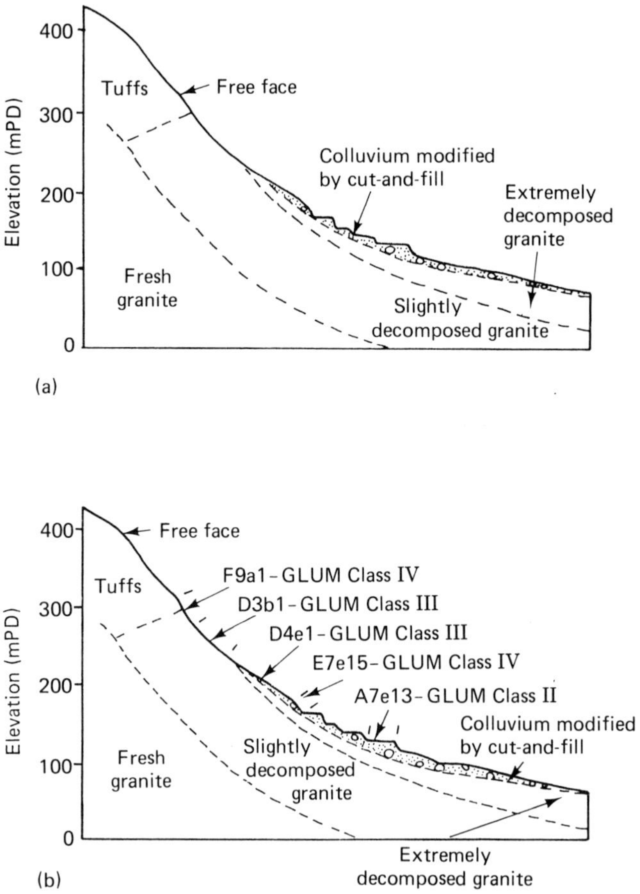

Figure 10.9 (a) An idealized geological cross-section of the Mid-levels District of Western Hong Kong Island. (b) Typical terrain classification units. The legend for the classification system is given in Table 10.2; GLUM classes are outlined in Table 10.3

10.7.4 Terrain classification map

The terrain classification map (Figure 10.11) can be interpreted by reference to the map legend (Table 10.2). Each terrain unit is consecutively assessed in terms of slope gradient, terrain component, terrain morphology and erosion classification. Each aspect is coded, and each terrain unit can be given an alpha numeric code, for

Table 10.2 Legend for the terrain classification map

SLOPE GRADIENT

Code	*Description*
A	0–5°
B	5–15°
C	15–30°
D	30–40°
E	40–60°
F	>60°

TERRAIN COMPONENT*

Code	*Description*
1.	Hillcrest
2.	Ridge
3.	Sideslope
4.	Footslope
5.	Drainage plain
6.	Incised drainage channel
7.	Disturbed terrain
8.	Disturbed terrain/drainage
9.	Rock exposure
10.	Alluvial plain
11.	Floodplain
12.	Coastal plain
13.	Stream course (perennial)
14.	Stream course with rock exposure in bed (perennial)
15.	Stream course with rock exposure in bed (ephemeral)
16.	Man-made channels (including catchwaters)
17.	Water storage
18.	Swamp/marsh
19.	Reclamation
20.	Beach
21.	Dune

TERRAIN MORPHOLOGY

Code	*Description*
a	Straight—in situ terrain
b	Concave—in situ terrain
c	Convex—in situ terrain
d	Straight—colluvial terrain (Relict d)
e	Concave—colluvial terrain (Relict e)
f	Convex—colluvial terrain (Relict f)
g	Straight—<2 m thick colluvium
h	Concave—<2 m thick colluvium
i	Convex—<2 m thick colluvium
j	Straight—corestones on in situ terrain
k	Concave—corestones on in situ terrain
l	Convex—corestones on in situ terrain
m	Straight—alluvial terrain (relict m)
n	Concave—alluvial terrain (relict n)
o	Convex—alluvial terrain (relict o)

* Minor disturbance of terrain is indicated by underlining the terrain attribute code. This is used in three situations:
(i) Squatter development on either natural terrain or terrain subject to major disturbance.
(ii) Development of minor roads, cemeteries or small lots, where the constituent components of cut and fill slopes and platforms are too small to be individually delineated.
(iii) Apparently natural terrain which may have been subject to a minor degree of disturbance during landscaping.

EROSION CLASSIFICATION (TERRAIN DENUDATION)

Code	*Description*	*Code*	*Description*
1.	No appreciable erosion		
2.	Landslip		
	Well-defined landslip	a	Integral (integral should only be used for rotational failures with minimal displacement)
		b	Scar
		c	Debris
	General evidence of instability on	d	Colluvial terrain (hummocky terrain, tension cracks or compression ridges, etc.)
		e	In situ terrain (terracetting, tension cracks or compression ridges without well-defined failures)
3.	Sheet erosion (assessed from ground cover)	a	Minor (1–10% bare ground)
		b	Moderate-severe (11–40% bare ground)
		c	Very severe (>40% bare ground)
4.	Rill erosion	a	Minor (incipient parallel rivulets—spacing generally >2 m apart)
		b	Moderate-severe (deeply incised parallel rivulets—generally <20 cm in depth; spacing generally <2 m apart)
		c	Very severe (deeply incised parallel rivulets—generally between 20 and 50 cm in depth; spacing generally <2 m apart)
5.	Gully erosion	a	Minor (a single gully generally <0.5 m in depth)
		b	Moderate-severe (a single incised gully which is generally 0.5 to 5 m in depth or a multiple system of gullies)
		c	Very severe (a single deeply incised gully which is generally >5 m in depth and/or a multiple system of incised gullies)
6.	Highly jointed boulders or rock exposure, including cut slopes, with adverse jointing.		

SLOPE CONDITION

Description	*Code*	*Height* (m)
Cut	1	0–5
	3	5–10
	5	10–20
	7	20–30
	9	>30
Fill	2	0–5
	4	5–15
	6	15–30
	8	>30

HYDROLOGY

Description	*Code*
Inlet zone	a
Outlet zone	b
Water ponding area	c
Surface groundwater movement	d

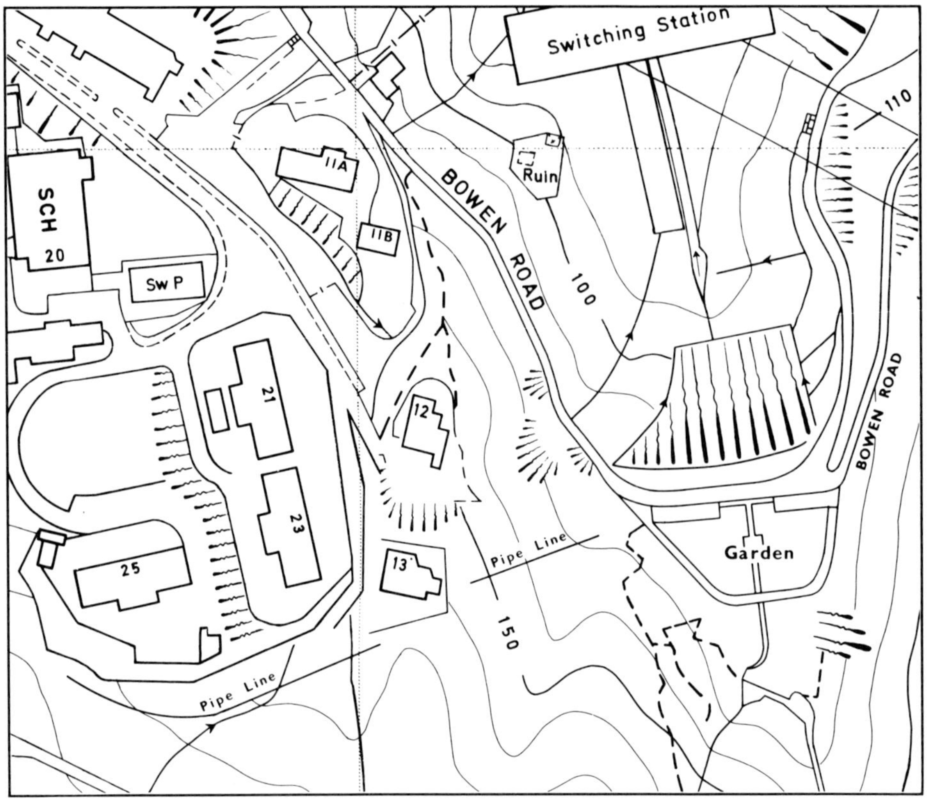

Figure 10.10 Ordnance Survey 1:2500 map of the study area in the Mid-levels District of Western Hong Kong Island

Figure 10.11 Terrain classification map

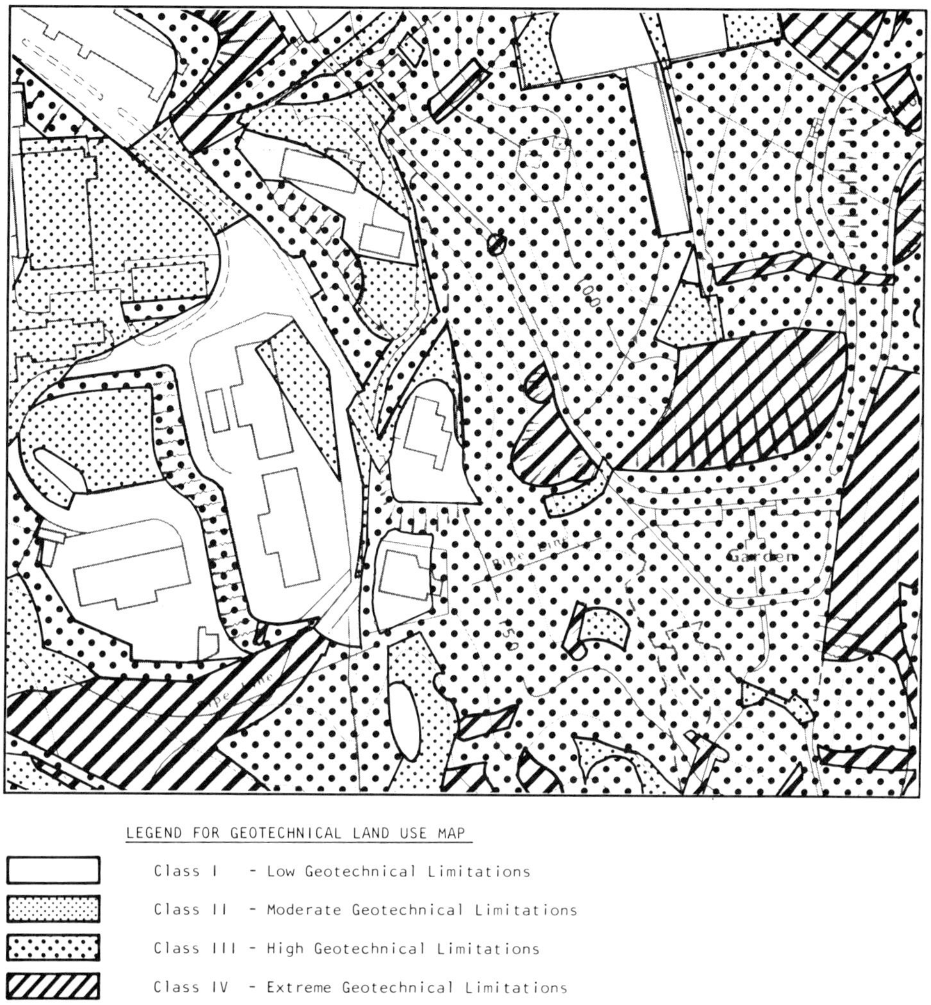

Figure 10.12 Geotechnical land-use map (GLUM)

example D3i1 which can be transcribed by reference to the table into:

D Slope gradient 30–40°
3 Sideslope
i Convex – less than 2 m thick colluvium
1 No appreciable erosion

and the unit can be located on the map. The whole range of coded descriptive terms can be used in order, ending with hydrology.

10.7.5 Geotechnical land-use map

The geotechnical land-use map (GLUM) (Figure 10.12) is prepared from the terrain classification map. Each terrain unit is characterized by a code which defines a set of terrain- or landform-related features. Typical terrain characteristics for each of the four GLUM classes are listed in Table 10.3. The cross-section in Figure 10.9(b) presents a generalized profile of terrain units and their corresponding GLUM classes. The original criteria for the classes of limitation were established as a result of discussion with engineers and engineering geologists. Allocation of a particular combination of terrain attributes to a particular GLUM class was based on sound engineering judgement.

The mapping process is a purely mechanical derivation of GLUM from the terrain classification data base obtained by assigning a GLUM class to each terrain unit using predetermined combinations of the terrain attributes. The purpose of the GLUM is thus to divide the entire study area into relatively small parcels of land which fall into one of four graded classes, each representing a different degree of likely geotechnical difficulty for development.

As the maps are designed for use in planning situations they do not provide reasons or explanations why the classes were designated, but merely present a geotechnical assessment of the terrain.

10.7.6 Surface hydrology map

The surface hydrology map (Figure 10.13) shows catchment boundaries, main drainage lines, springs and man-made channels.

10.7.7 Vegetation map

The vegetation map (Figure 10.14) has an importance in aerial surveying, on which the general terrain analysis is based because it gives some indication of the reliability

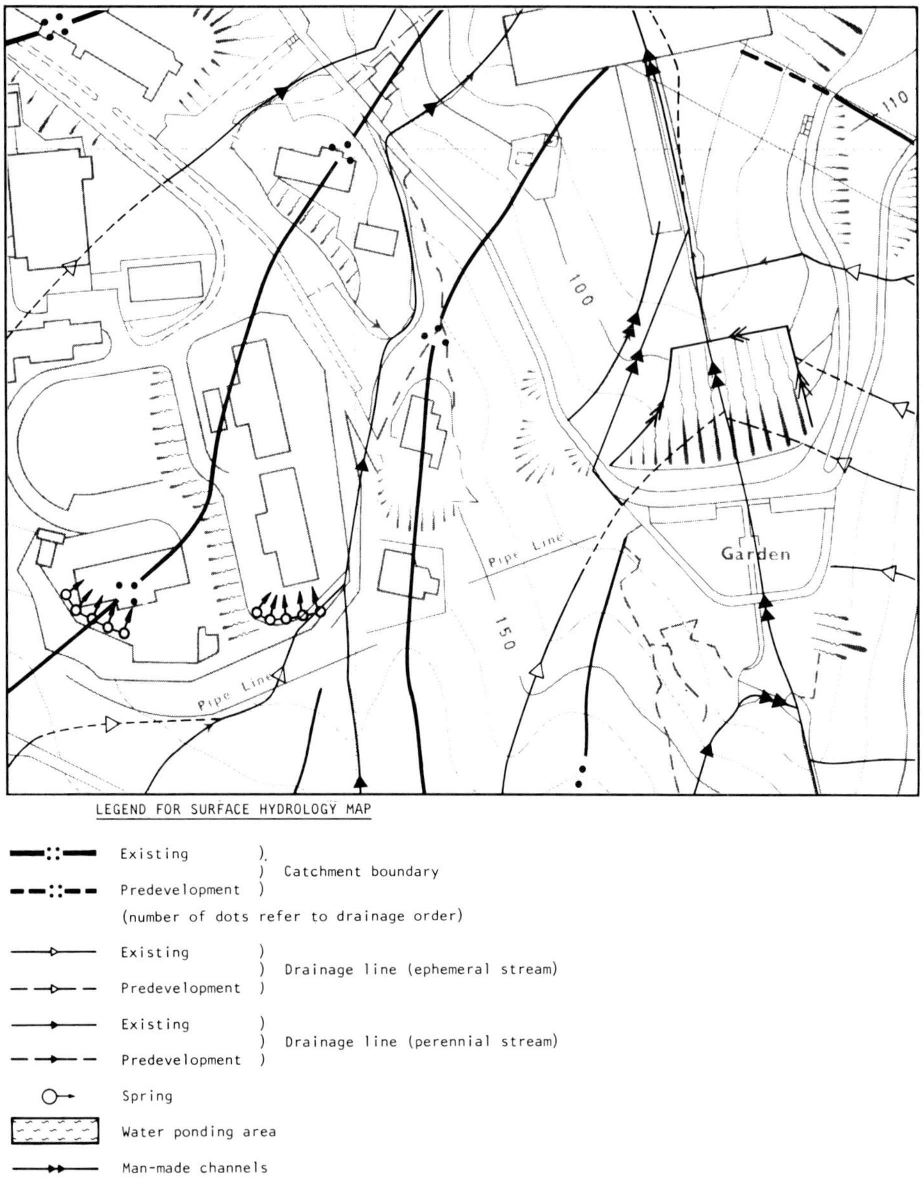

Figure 10.13 Surface hydrology map

Table 10.3 Geotechnical land-use map (GLUM) classification system

Characteristics of GLUM classes	CLASS I	CLASS II	CLASS III	CLASS IV
Geotechnical limitations	Low	Moderate	High	Extreme
Suitability for development	High	Moderate	Low	Probably unsuitable
Engineering cost for development	Low	Normal	High	Very high
Intensity of site investigation required	Normal	Normal	Intensive	Very intensive
Typical terrain characteristics (Some, but not necessarily all, of the stated characteristics will occur in the respective Class)	In situ terrain with gentle slopes (0–15°) without severe erosion or instability Cut platforms in in situ terrain	In situ terrain with slopes between 15 and 30° without severe erosion or instability In situ terrain of gentle slopes associated with drainage but no instability Colluvial terrain with gentle slopes (0–15°) without severe erosion or instability	In situ terrain with slopes between 30 and 60° without severe erosion or instability In situ terrain less than 15° with history of landslips Colluvial terrain less than 15° with evidence of instability High to moderate fill slopes	Very steep in situ slopes (>60°) and cliffs Steep to very steep in situ and colluvial slopes with history of instability Colluvial terrain with gentle slopes, but associated with instability and drainage

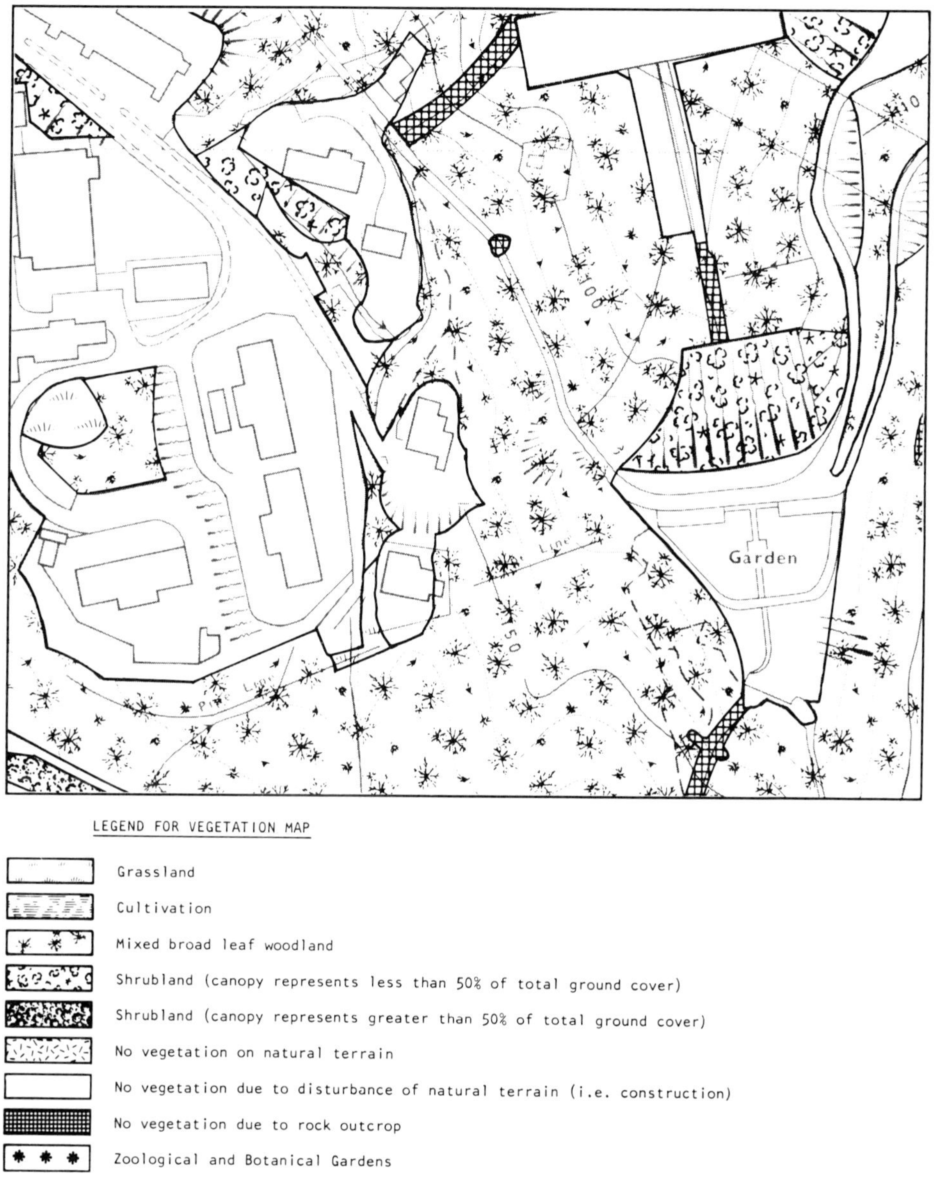

Figure 10.14 Vegetation map

of the surveying where ground cover is high. Also, and this is in a general sense, the distribution of types of vegetation can give an indication of soil and underlying rock type.

10.7.8 Engineering geology map

The engineering geology map (EGM) is included in the map series for a specific purpose which is quite different from that of the GLUM. The purpose of the map is stated to be twofold. First, it presents a basic geological interpretation of the terrain. Secondly, it presents the geomorphological and geological constraints which influence the allocation of the GLUM classes. The map (see Figure 10.15) also provides information that will be of use in the planning of further investigations of the site or even for preliminary design.

Burnett and Styles (1982) provided the following account of how the EGM is compiled. Using a 60% screened topographic map (Figure 10.15) as a base, the solid bedrock geology within the study area boundaries is drafted onto the map using 75% screened symbolic notation contained within dotted or solid interface lines depending on mapping certainty. The superficial geological deposits including alluvium, colluvium and even reclaimed land are superimposed upon the map, in the process often covering up large tracts of bedrock. The disposition of both solid and superficial geology is established using aerial photograph interpretation, field traversing, existing geological mapping and plotting of existing borehole information.

Structural geological information is also presented on the map and this takes the form of aerial photograph interpreted major lineations, field determined bedding and contact measurements, and localized and regional field measured discontinuity (mainly joint) orientation and rose diagrams. This information, when taken in conjunction with the bedrock geology and the proposed layout of new works, can provide useful preliminary guidance as to likely stability problems in hardrock cuttings, tunnelling difficulties, etc.

A series of geological and morphological features is

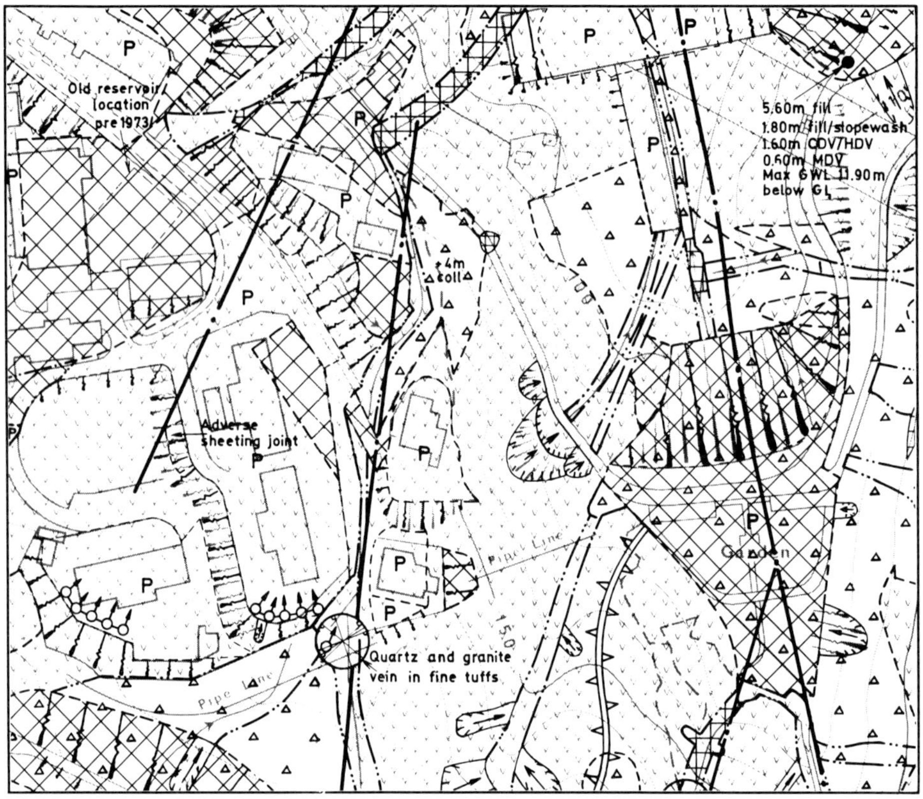

LEGEND FOR ENGINEERING GEOLOGY MAP

Fill (made ground)

Fill over alluvium

Fill over colluvium

Colluvium (< 2 metres thick)

Hong Kong Granite

Repulse Bay Formation

Geological boundary (approximate)

Boundary of superficial deposits (approximate)

Photolineation

Rose diagram giving regional trends of joint planes

Slope failure

Zone of general evidence of instability

Boulders

Spring

Seepage line or front

Major areas of exposed rock

Man-made slope

Man-made platform

Borehole location with simplified borehole log

Severe gully erosion

Major joint orientation pattern at individual exposures, dip direction indicated.

Major drainage channel (natural and man-made), subject to periodic flooding

Convex break in slope)
Centreline of spur line) unaffected by development
Centreline of ridge crest)

Dv Decomposed volcanics) prefixed by weathering grades (RS Residual soil
Dg Decomposed granite) (C Completely
(H Highly
(M Moderately
(S Slightly

e.g. CDv is completely decomposed volcanics

All Alluvium (undifferentiated) GL Ground level

Coll Colluvium (undifferentiated) m metre

GWL Groundwater level max maximum

Figure 10.15 Engineering geology map (EGM) and legend

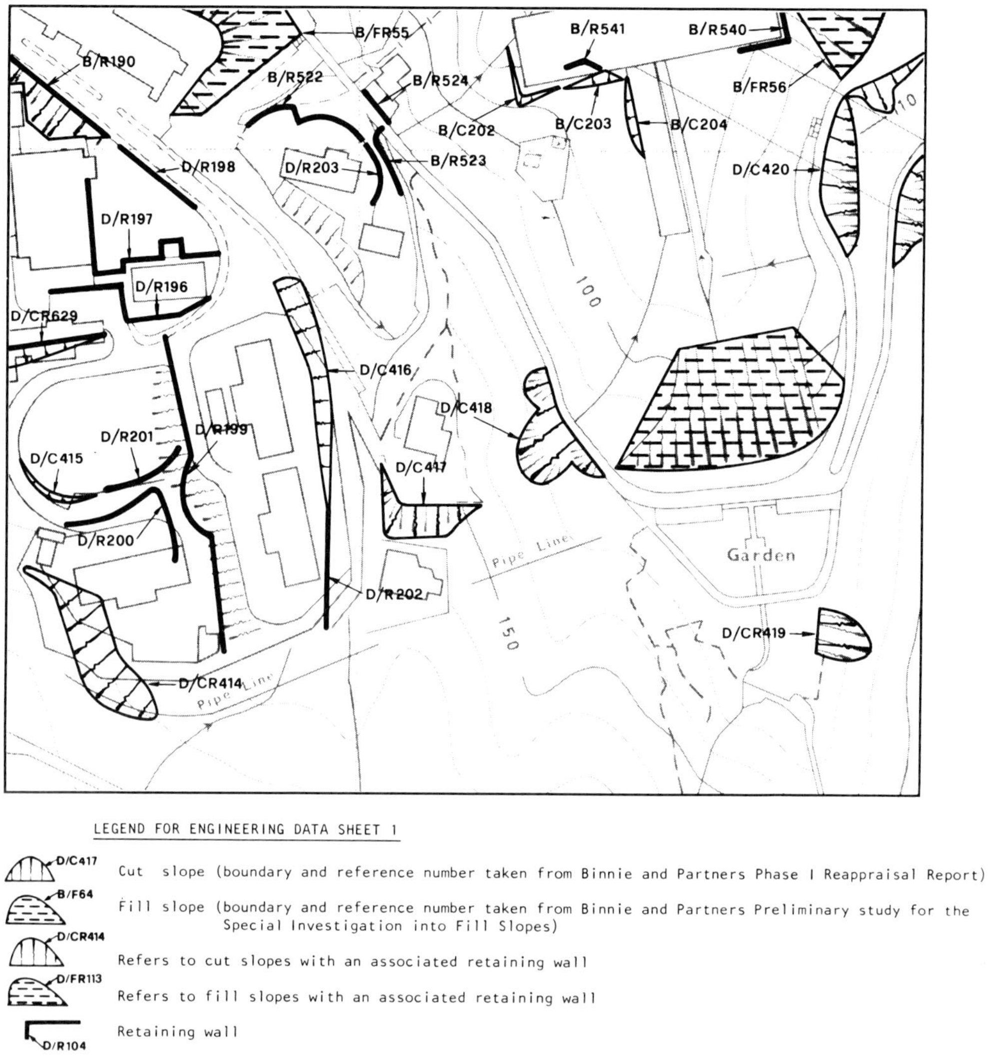

Figure 10.16 Engineering data sheet 1 (EDS1)

also shown on the map to illustrate the type of limitation associated with the terrain. These features, generally detected on the aerial photographs, include relict and recent landslips, zones with general evidence of instability, boulder fields and major areas of surface rock exposed at the surface.

10.7.9 Engineering data sheets

Due to the intensive development in a relatively small area, Hong Kong has a great deal of existing factual and interpretative site investigation data available for each study area. Even for areas as small as 200 ha there may be over 500 existing borehole records, with testing data, available. This information is carefully summarized and systematically tabulated in the report appendices, as well as being plotted on maps. On the 1 : 2500 scale maps, the location of every borehole is plotted and referenced on the separate engineering data sheet (Figure 10.17).

Engineering data sheet (EDS) maps, best presented as transparent overlays, give an immediate impression of:

- the number and location of all field observations and measurements made;
- the number and location of soil and rock samples;
- the number, location and file cross-reference of all existing boreholes, related laboratory testing data, and piezometers;
- the location and report reference number of all previously investigated cut slopes, fill slopes and retaining walls (Figure 10.16).

The EDS is designed for engineering geologists and engineers who are using the reports as a summary of existing information and site specific investigations. As an indication of the location of raw data, the map aids in the evaluation of the GLUM and the EGM, but saves users considerable time and effort at the desk study review stage for other studies.

10.7.10 Technique of data acquisition

The range of maps and the steps needed to acquire information are set out in Figure 10.18. Preparation of engineering geological maps needs the combined input from terrain classification, existing information from geological maps and engineering data sheets, all backed up by geological fieldwork (Table 10.4). The series of factual and interpretative thematic maps are aimed at urban planners and design engineers.

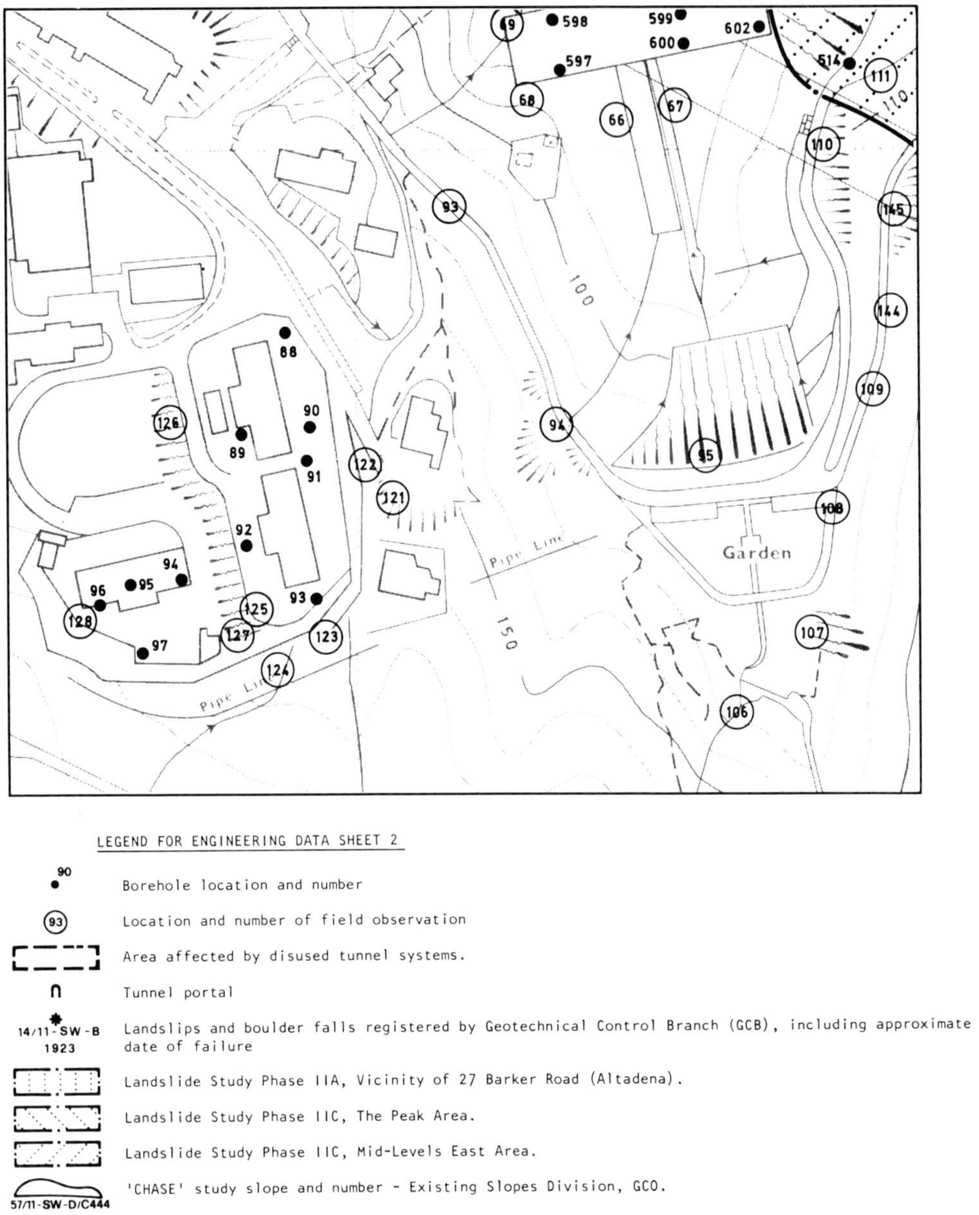

Figure 10.17 Engineering data sheet 2 (EDS2)

10.7.11 Terrain classification mapping at the 1:20 000 Regional scale

A series of maps is produced for each study area at a scale of 1:20 000:

1. Ordnance Survey base map.
2. Terrain classification map superimposed on base map. The legend includes three coded aspects – slope, terrain component and erosion – and each map unit is given a double code where erosion is not significant, e.g. 1Z, and three where it is, e.g. 4Cn.
3. Terrain classification maps without the superimposed base map.
4. Landform map, with an informative legend.
5. Erosion map, with a legend.
6. GLUM, with five classes of geotechnical limitations identified.
7. EGM, with legend.
8. Physical constraints map.
9. Generalized limitations and engineering appraisal map. This is a fragment of a larger map sheet, and can be understood by study of the map legend.

Two maps are reproduced here as examples: Figure 10.19 illustrates the GLUM, whereas the EGM forms Figure 10.20.

10.8 United Kingdom: mapping in the urban environment

10.8.1 Introduction

The urban environment comprises existing towns and cities in which construction may take place in older areas after demolition, or spread into the peripheral rural areas. Alternatively, new towns may be established, again in rural areas, some distance from established towns.

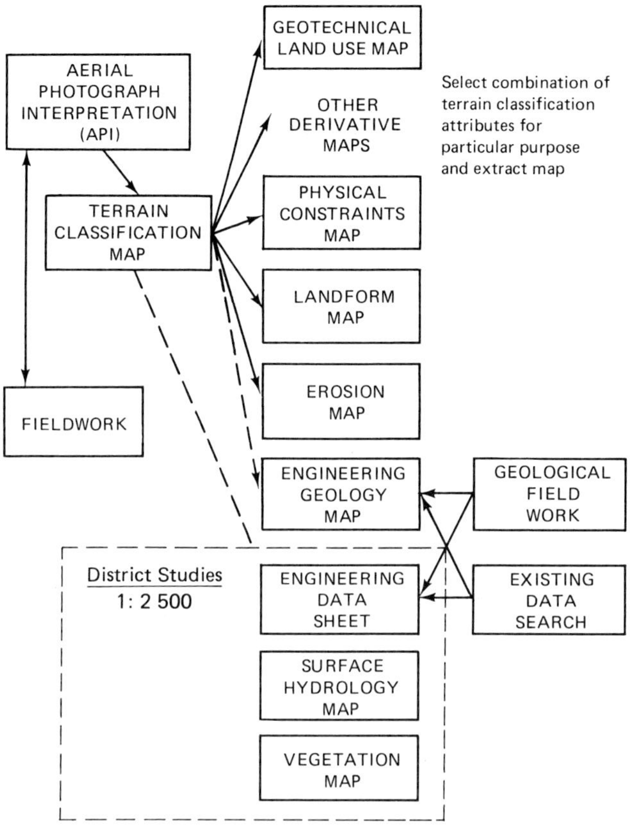

Figure 10.18 Techniques of data acquisition and derivation of maps

Table 10.4 Stages in the preparation of an engineering geological map (EGM)

Stage I. Information derived from terrain classification map
1. Colluvium—thin
   Colluvium—thick
2. Drainage plains and stream channels—including man-made channels where applicable
3. General instability on colluvium
4. General instability on in situ terrain
5. Well-defined landslip on colluvium
6. Well-defined landslip on in situ terrain
7. Fill slopes and platforms—identify material underlying, e.g. in situ, colluvium or alluvium
8. Rock outcrop or cliff
9. Alluvium
10. Unusual groundwater—nature of constraint to be determined by cartographer, e.g. spring, seepage front or inlet zone
11. Severe gully erosion

Stage II. Information derived from engineering data sheet (EDS)
12. Borehole and piezometer locations

Stage III. Geology—derived from existing geological survey information and field mapping
13. Individual rock units—prepared on separate base map; cartography to transfer onto EGM
14. Photolineations, faults, structural discontinuities, dykes
15. Rose diagrams and joint/bedding plane information
16. Summary comments regarding lithology, degree of weathering, joint pattern and stability as required

Stage IV Morphological information derived from terrain classification map
17. Breaks in slope—concave/convex
18. Centre lines of ridge crest and spurs
19. Spot slope angles
20. Centre line of ridges coincident with catchment boundaries

Most existing cities were established in relation to available water supply, communication routes and defensive positions; consequently many are sited on major rivers at the first easily fordable position upstream from the river mouth. Alternatively, a good harbour promoted settlement. In Britain, urban growth followed industrialization which, apart from the south-east of the country, was based on mining coal and in many instances the associated iron ores. This may readily be appreciated by comparing the locations of the present main centres of population with the coal fields (Figure 10.21). Although mining is now concentrated in fewer, larger mines, some in green-field sites away from the old population centres, past mining in urban centres has left a legacy of problematic ground conditions.

Despite this, the first published map of an urban area in Britain was of an estuarial region around the city of Belfast in Northern Ireland. The other example, unpublished although a summary account is available, was the green-field development remote from coal mining and industrial development of the new town of Milton Keynes in Southern England.

10.8.2 Engineering geology map of Belfast

The earliest published example of engineering geological mapping in Britain is the 'Geology of Belfast and District – Special Engineering Geology Sheet, Solid and Drift' produced by the Institute of Geological Sciences (1971). A portion of the map is reproduced as Figure 10.22. The choice of scale (1:21 120) was limited to that of the only suitable topographic base map available, but as this was prepared from the 1:10 560 maps it is possible to locate sites precisely (Bazley, 1971). The topographic map is overprinted with a conventional coloured geological solid-and-drift map with isopachytes drawn for the Estuarine Clay (Sleech). On the front of the sheet there is also a simple geological legend and horizontal cross-section. On the reverse side is a summary of the geology and separate map showing the general form of rockhead contours for the central part of the city. Some 2000 borehole records were utilized for the preparation of the map; the sites of 60 selected boreholes with condensed records of the sequence of superficial deposits are located on the map. A table of rock and soil characteristics summarizes the various soil and rock types with data on their thicknesses, main rock types, structure and occurrence, and their hydrogeological characteristics. A typical entry is given in Table 10.5 (Wilson, 1972).

The map covers an area of 174 km^2 (108 square miles). Compilation began in 1967 from published and unpublished work of the Geological Survey, and from the records of site investigations accumulated over about a decade. Since publication in April 1971, the map has been well received by engineers and may be regarded as having met a local working need (Wilson, 1972, p. 85).

10.8.3 Milton Keynes

Another example that is unlikely to be published is from the new town of Milton Keynes, situated 50 miles north

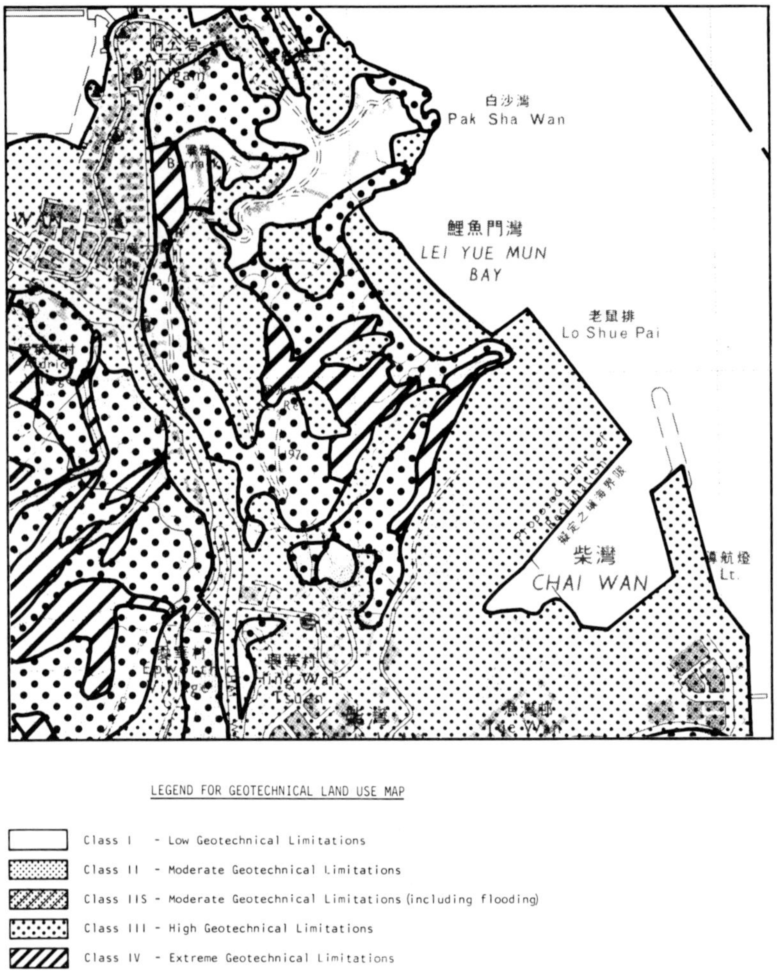

Figure 10.19 Geotechnical land-use map (GLUM)

Table 10.5 Typical group entry from table of 'Soil and Rock Characteristics' on Special Engineering Geological Map of Belfast, UK

SYSTEM	Post-glacial and Recent
ROCK GROUPS	Estuarine clay (Sleech)
THICKNESS	0–15 m
ROCK TYPES	Unconsolidated grey silt or sand; variable organic content
STRUCTURE AND OCCURRENCE	Shallow water deposits in estuarine regions
CHIEF GEOTECHNICAL PROPERTIES	Typically soft or very soft. Highly compressible, with very low bearing strength. Undrained shear strength 5–25 kN/m^2. Coefficient of volume decrease (m_v) 0.001–0.015 m^2/kN at a superimposed load of 25 kN/m^2. Water sulphate content may be high, requiring sulphate-resisting cement
ECONOMIC MATERIALS	Used in the past for brick clay. In an adjacent area is used for the manufacture of cement
UNDERGROUND WATER	Poor quality and generally contaminated. Watertable often near surface
SOILS	Slobland. Largely built-over

of London, where fairly extensive pilot studies were carried out, albeit started at a late stage, perhaps too late to influence planning (Cratchley and Denness, 1972).

Information from the whole investigation has been presented on an engineering geological base map accompanied by a series of overlays, a table of measured geotechnical properties and interpreted behavioural characteristics, and a detailed explanatory report which includes results of the other special studies. Base maps on a 1:25 000 scale were available for the complete area and at 1:10 000 for presentation of more detailed data on the superficial deposits of the Ouzel Valley.

Figure 10.23 shows a simplification of the base map using a system of stippling instead of the original colours. The original colours refer to engineering divisions and regroupings of the geological units and parts of units, largely on a lithological basis, supported by qualifying index data from tests on borehole samples. Symbols refer to stratigraphical units.

Some of the special study results have been drawn on transparent overlays, for example a thickness-of-drift map for the whole area, contoured at 3 and 10 m depth, for the superficial deposits. Another example is the special 1:10 000 scale map for the Ouzel Valley (Lukey, 1974) which shows the limits and thickness of workable sand and gravel and the thickness of overburden, both contoured at 1 m intervals.

The table of properties consists of two basic parts – an 'input' section of ranges of measured properties, and a 'readout' section of interpretations for various engineering conditions based on these results. This amounts

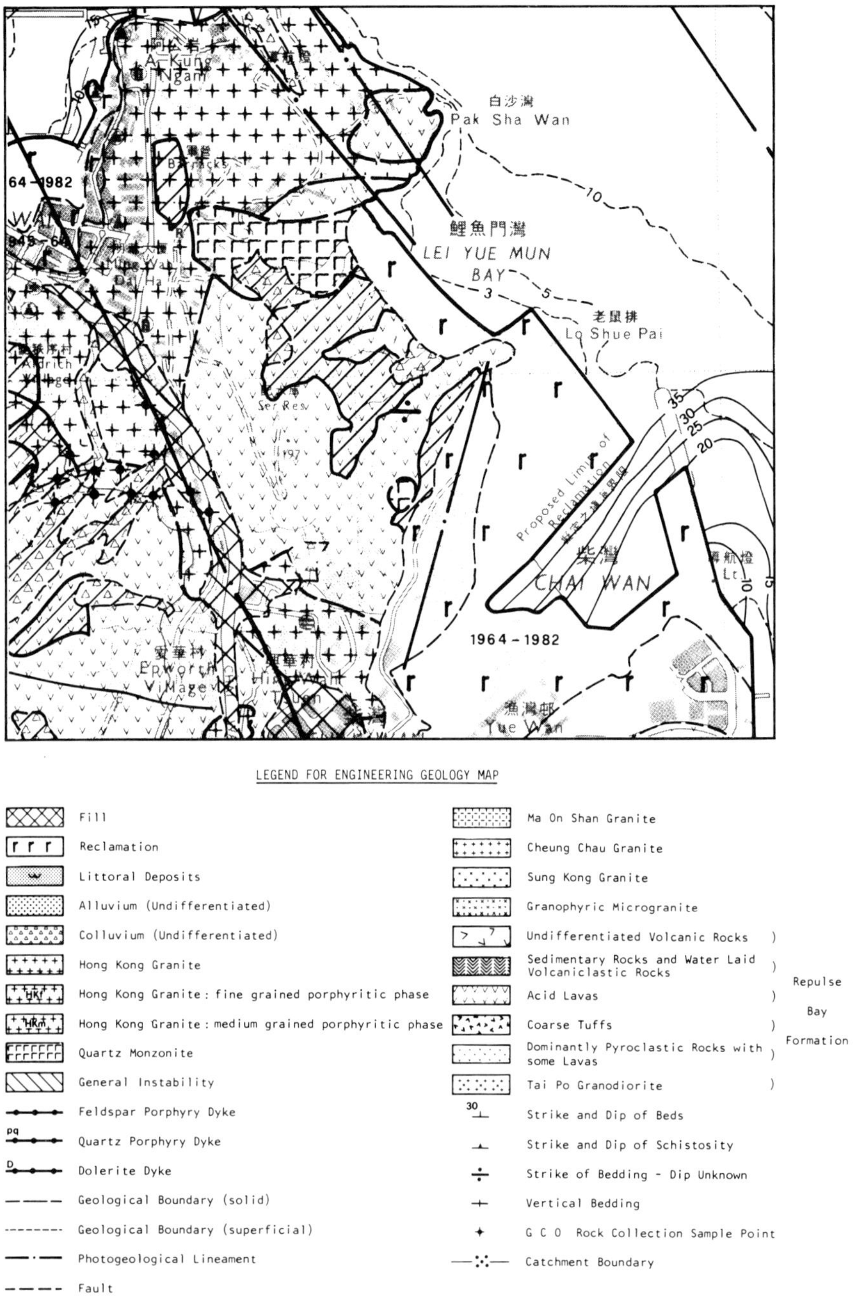

Figure 10.20 Engineering geology map (EGM) and legend

to seven input and seven output sections for each of the 16 defined geotechnical units. The Institute of Geological Sciences point out the possible limitations of this system, which involves considerable interpretation, and informed readers are referred particularly to the input system so that they may make their own judgement.

The report accompanying the map contains the new data from which the tables were prepared and an explanatory section running to some 10 000 words. The degree of confidence in interpreting the results for each geotechnical unit is also explained.

Table 10.6 shows two of the 14 input and output sections for two of the 16 geotechnical units – one solid (Oxford Clay) and the other superficial (Clayey Till).

The 1:25 000 Special Geological Sheet of Milton Keynes has been published (British Geological Survey, 1970) with a descriptive booklet (Horton, Shephard-Thorn and Thurrell, 1974).

10.9 A case study of engineering geological mapping at various scales in North-East England

10.9.1 General setting of Tyne and Wear County

Urban geology is concerned with the complex interplay of natural features, including the solid and superficial deposits, topographic features and general geological history, with the changes and modifications to these brought about by man's activities. Also to be considered is the combined influence, often hazardous, of all these features on existing man-made structures and those under construction. Various aspects have been dealt with in some detail in Price (1971), Dearman and Matula (1976), Dearman *et al.* (1977) Dearman, Baynes and Pearson (1977) and Dearman *et al.* (1979).

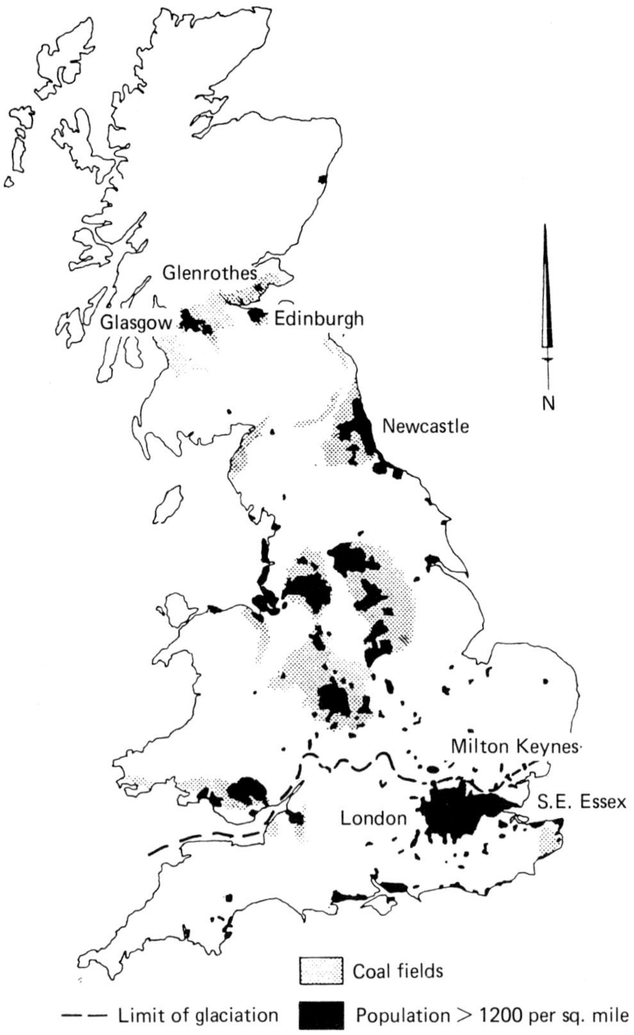

Figure 10.21 Coal fields of Britain in relation to the main centres of population (After Price, 1971, Fig. 3)

The Tyne and Wear Metropolitan County in North-East England includes the urban areas of Newcastle, Gateshead and Sunderland (Figure 10.24). Much of the outcrop of the solid rocks is masked by a variable cover of drift deposits which may be of considerable thickness. A major part of the county is underlain by productive Coal Measures, but in the south-eastern part of the county the coal field is concealed by Permian sands and limestones. Early coal mining took place around the medieval settlements of Newcastle and Gateshead, but as the towns expanded, principally in the nineteenth century, the urban areas spread over the mined areas. Price (1971) has illustrated the extension of urban development over old mining areas in Newcastle and indicated that by 1968 the city had embraced many mines with shafts sunk before 1778 (Figure 10.25). The city is therefore spread over areas of hidden mineshafts and surface subsidence resulting from long-abandoned mine workings. The precise location of shafts and the state of their present condition may not be known; the same doubts arise over the nature, depth and extent of old workings, including bell pits in the shallowest seams. Age gives some indication of the most likely method of shaft construction (Hair, 1839; Dearman, Baynes and Pearson, 1977).

Coal mining started very much later where the Coal Measures are concealed beneath the highly permeable Permian deposits. The Permian limestones have been extensively exploited for lime burning and now many of the disused quarries have been infilled with industrial waste and household refuse.

Expansion of the city also involved the early complete or localized infilling of many of the steep-sided valleys, locally known as denes, which run down to the Tyne. One of the case histories presented later deals with the history of infilling of Pandon Dene, a major barrier to easterly development and a present cause of many foundation difficulties, especially near the edges of the dene. The denes cut down through the glacial drift deposits into the Coal Measures beneath, and the valley sides are unstable. Glacial deposits generally obscure the Coal Measures, except for parts of the higher sandstone ridges. Exposed sandstones were quarried for building stone and grindstones, whereas pits in the drift provided sand and clay for the manufacture of bricks, tiles and pottery. Most of the old quarries and pits have been infilled and are now built over.

Ballast brought in by sailing colliers and the residues of various coal-based industries, including lead, iron, glass and a variety of chemical works, cause additional problems when developing the made ground. Periodic redevelopment over many centuries has left a general cover of fill some 2–3 m in thickness over the older parts of the urban area.

With the general decline of mining inland and concentration on large mines exploiting undersea reserves from the coast, some of the general areas of dereliction have been rehabilitated and the old mine tips either removed or landscaped and planted. Back-filling and reclamation after large-scale open-cast mining of shallow seams has contributed locally to the variety of ground conditions in the conurbation.

Against this background of complex geological conditions and intense industrial activity spanning at least two centuries, it is possible to review the likely subsurface conditions which have to be taken into account both in planning a site investigation and in the preparation of engineering geological plans of the urban areas and the surviving open spaces within the limits of the county.

Some of the problems that the Tyne and Wear urban environment presents to the engineer and the engineering geologist involved in site investigation are presented as case histories of three small areas. An area in central Gateshead (Figure 10.24) illustrates the importance of old workings in near surface coal seams; at Preston Grange, near the mouth of the River Tyne, open-cast mine workings and variable superficial deposits have been mapped. In central Newcastle, Pandon Dene illustrates the problems associated with the infilling of a natural dene.

10.9.2 Geological setting

Coal Measures

Because of the correlations made possible by the presence of extensive fossiliferous marine bands, the stratigraphy of the Upper Carboniferous Coal Measures is now well understood. Divisions generally accepted are: Lower Coal Measures 215 m thick, Middle Coal

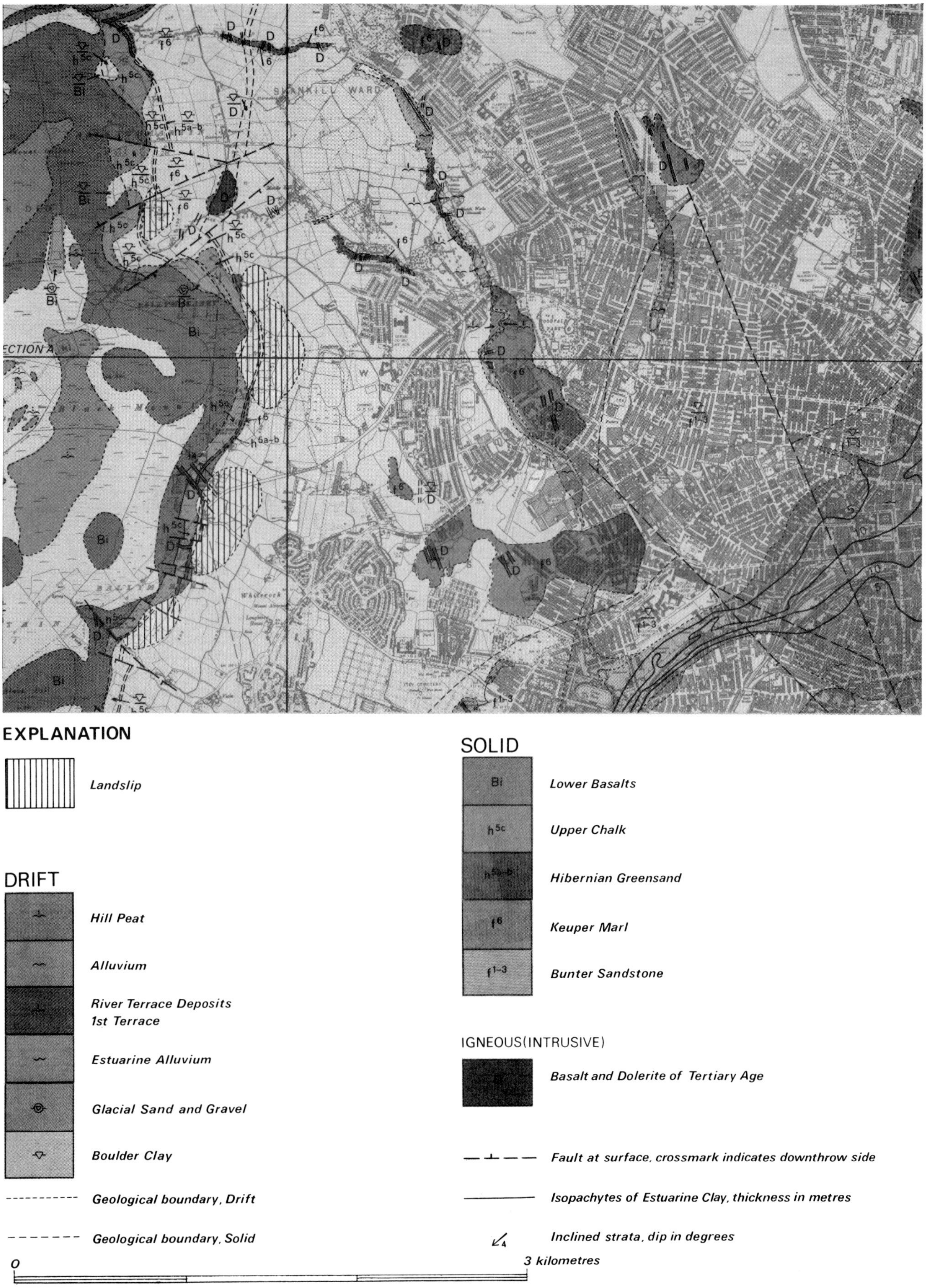

Figure 10.22 Part of the Special Engineering Geology Sheet, Solid and Drift, of the Belfast district, 1971. Scale of the original map is 1:21 120; reproduced here at 1:31 000 approximately (Courtesy of the Director of Geological Survey in Northern Ireland)

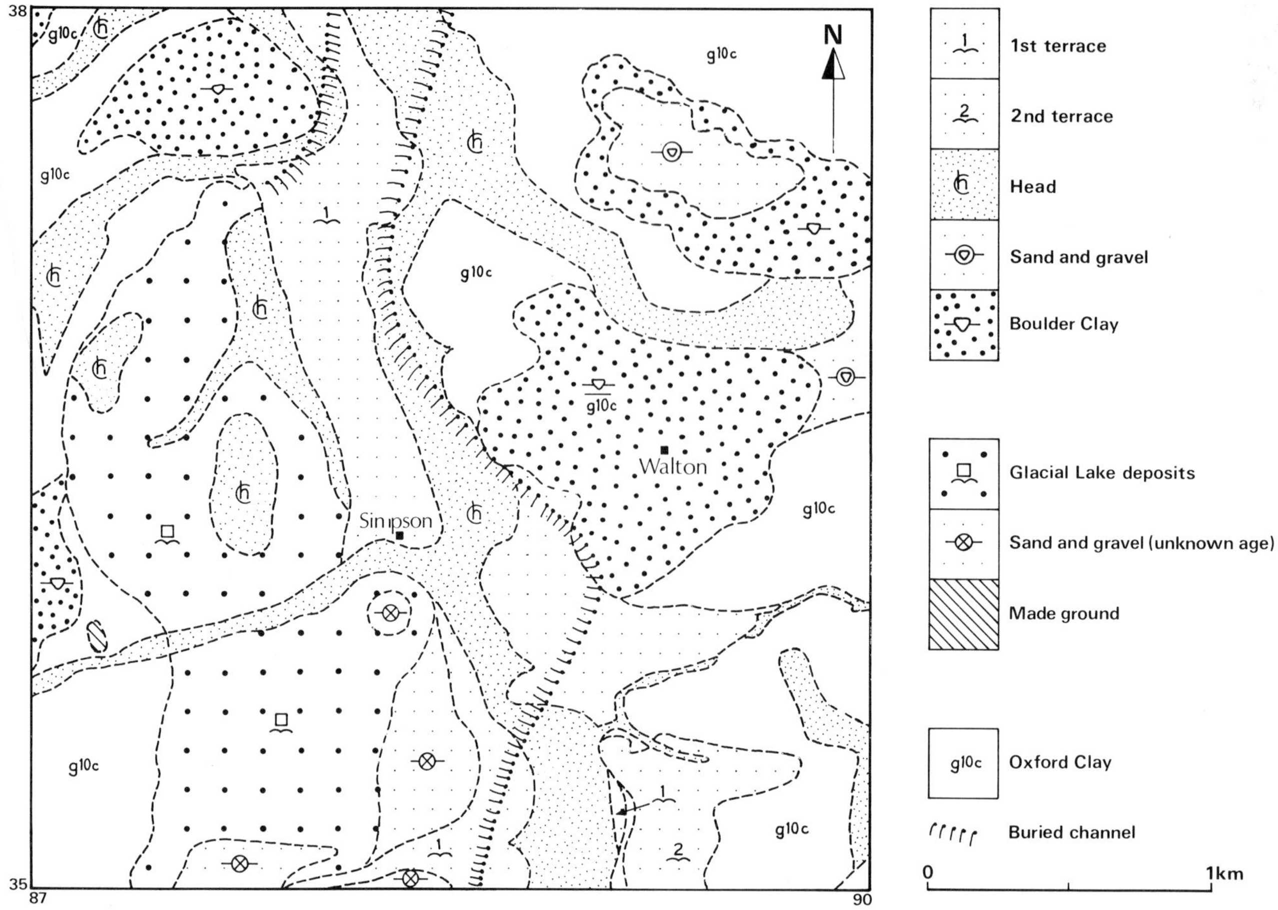

Figure 10.23 Part of the 1:25 000 scale geological map of Milton Keynes, UK, produced by the Engineering Geology Unit of the Institute of Geological Sciences. Simplified for presentation at a smaller scale in black and white (After Cratchley and Denness, 1972, Fig. 4)

Table 10.6 Extract from 'Table of Engineering Geology Characteristics' to accompany 1:25 000 engineering geology map of Milton Keynes

	Bedrock–Oxford Clay	*Superficial–Clayey Till*
4. Other features:		
(a) Range of geotechnical		
parameters: LL%	50–75	40–60
PL%	20–30	20–30
PI%	30–45	20–30
LI	(−0.3)–(−0.1)	(−0.1)–(0.2)
SG	2.6	2.45–2.6
γ(mg/m^3)	1.9–2.1	1.8–2.0
Intact c_u (kN/m^2)	100–250	50–150
pH	6.5–8.5	
Sulphate (% init. dry wt)	0–15	Probably 5
(b) Collective implications of 4(a)	Good foundation and fill	Good foundation and fill
(c) Variability and openness of fissuring	Variable with lithology: usually closed	Variable and closed
(d) Implications of 4(c)	—	—
(e) Bedding orientation	Gentle SE dip	—
11. Foundation stability:		
(a) Light loads (< 150 kN/m^2) on		
thick deposit: (i) Loading rate	Fairly rapid	Rapid
(ii) Loading intensity	High	High
(iii) Foundation type	Any	Any except point or strip
(b) Heavy loads (⩾ 150 kN/m^2) on		
thick deposit: (i) Loading rate	Slow	Slow
(ii) Loading intensity	High	Moderate
(iii) Foundation type	Strip or spread footings	Raft
(c) Susceptibility to moisture content change during and after construction	Low	Moderate
(d) Metastability	None	None
(e) Swelling pressure	Fairly low (occasionally moderate)	Low
(f) Pre-existence of slip surfaces	None	Possibly abundant
(g) Strength variation	Low	High laterally and vertically
(h) Compressibility variation	Low	High laterally and vertically
(i) Special conditions	None	Differential settlement near surface slip zones

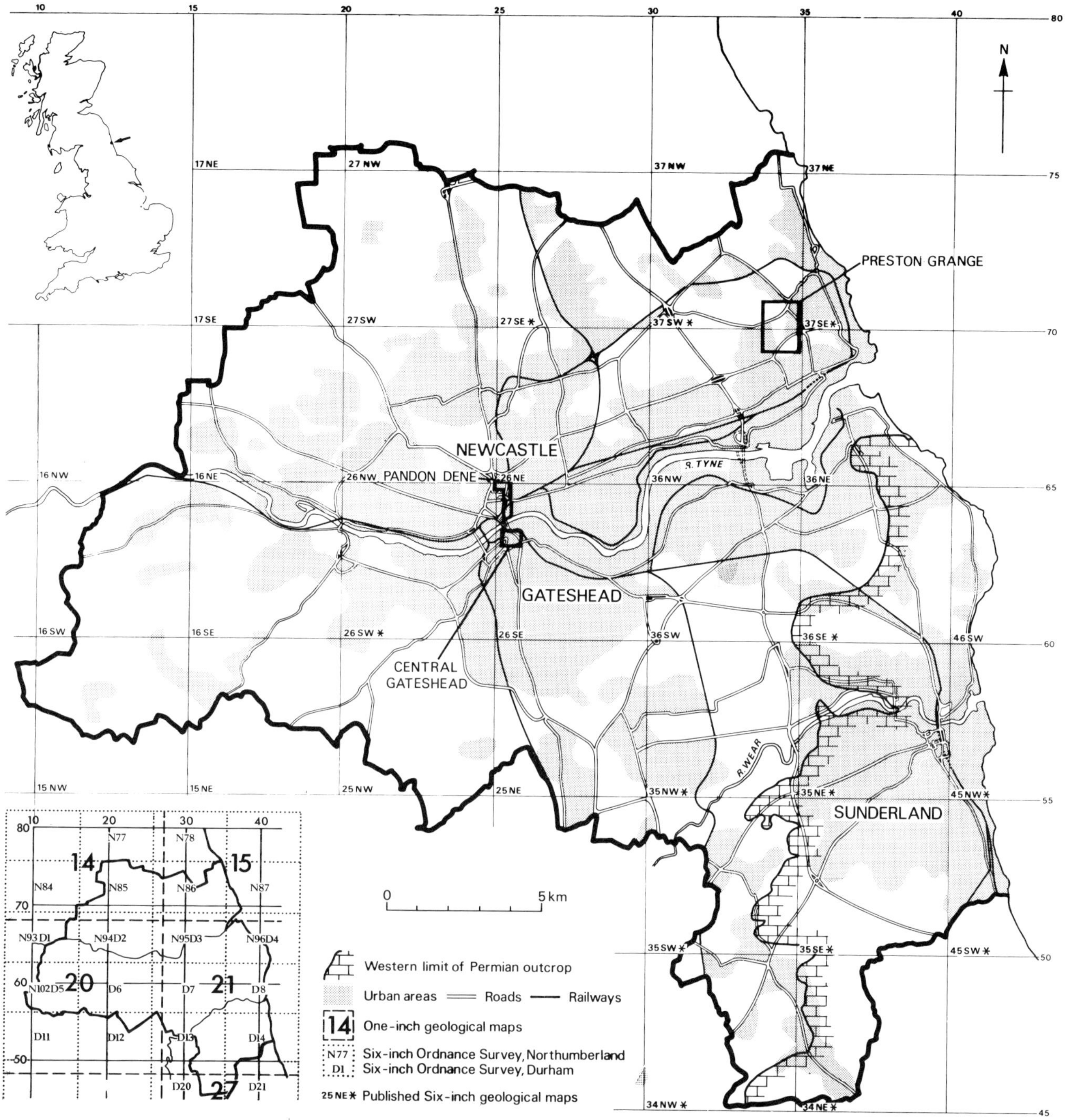

Figure 10.24 Map of the Tyne and Wear Metropolitan County, UK, showing the location of three areas discussed in the text, and the extent of the 1:10 000 revision of the basic geological maps carried out by the British Geological Survey

Measures 510 m thick, and the incomplete Upper Coal Measures is represented by 150 m of strata.

Shales, mudstones and sandstones dominate the succession comprising ill-defined cyclothems each consisting of a general sequence – shell bed, or rarely, marine bed; shale; siltstone; sandstone; seatearth and coal – in ascending order. Repetition and exclusion of certain beds are common. Thick, local developments of coarse sandstone occur at the base of the Lower and at the top of the Middle Coal Measures (Taylor *et al.*, 1971; Johnson, 1972); they are associated with shales and a few thin coals. The main productive group, with 15–20 of the best seams from the Brockwell seam in the Lower Coal Measures to the High Main Coal in the Middle Coal Measures, was formerly regarded as the Middle Group of the Coal Measures in North-East England (Eastwood, 1935). The Upper Coal Measures are mainly mudstones and siltstones with subordinate sandstones and a few thin coals.

The Coal Measures dip gently eastwards, but towards the coast the strata rise in a broad syncline. In the south-west coastal strip the coal field is concealed beneath Permian rocks, but the gentle easterly inclination of the basal Permian unconformity has facilitated exploration and exploitation of the coal field both on land and beneath the North Sea.

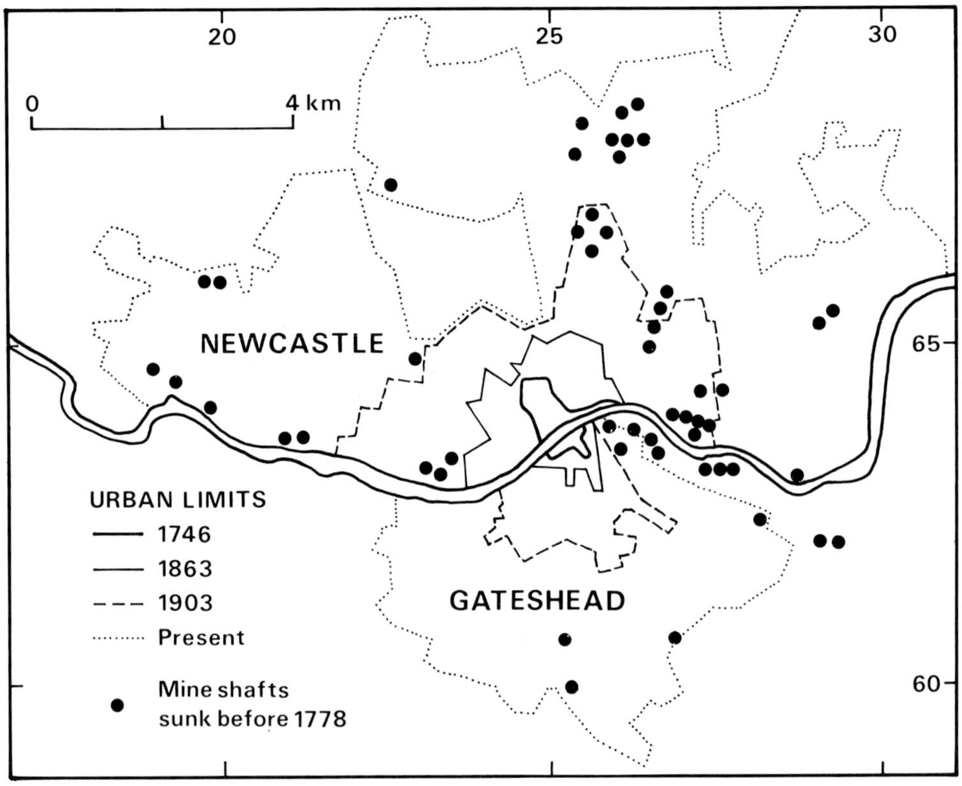

Figure 10.25 Spread of urban development over old mining areas in Newcastle and Gateshead, UK (Redrawn from Price, 1971, Fig. 2)

The Permian

Following the Hercynian earth movements, uplift and extensive erosion during the Late Carboniferous and the early Lower Permian left the area as a broad rolling peneplain inclined gently to the east (Smith, 1972). Late in the Lower Permian, the aeolian yellow sands accumulated in discrete WSW–ENE masses separated by areas in which the sands are thin or entirely absent.

Continental conditions were brought to an end by a marine transgression. The Upper Permian sequence, incompletely represented in Tyne and Wear County, begins with the Marl Slate, some 3 m of grey to black laminated bituminous calcitic or dolomitic siltstone or silty dolomite. At outcrop it is generally weathered to a yellow or buff colour. Succeeding Magnesian Limestone sequences are thick accumulations of bedded limestone and evaporites. In the present area, the evaporites have been removed by solution, with consequent collapse and brecciation of overlying limestones.

The Lower Magnesian Limestone, 76 m thick, is an evenly bedded, compact, silt-grade calcitic dolomite. Relatively deep-water deposits of the Lower Magnesian Limestone were followed by accumulation of a shallow-water organic reef, with associated lagoonal oolitic dolomites, and contrasting reef-front talus deposits which thin rapidly basinwards into a few metres of silt-grade carbonate. In contrast, reef deposits may be over 200 m thick. Former thick basinal anhydrite deposits are now represented on the present land area by a very thin silty-clay solution residue: the Hartlepool Residue.

Only two of the three carbonate formations of the Upper Magnesium Limestone are present, a Lower Concretionary Limestone some 116 m thick and 90 m of the overlying Hartlepool and Roker Dolomite. The Concretionary Limestone has long been famous for the bewildering variety of its calcitic concretions, but large parts with few or no concretions are thin-bedded buff slightly calcitic dolomites. Solution of the underlying Hartlepool Anhydrite brought about varying degrees of brecciation in the Concretionary Limestone. Brecciation decreases irregularly upwards and the higher beds may be unaffected. Hartlepool and Roker Dolomite is almost entirely cream or buff soft dolomite which is often oolitic.

Bedding in the limestones is gently inclined eastwards, except where flexed adjacent to faults. Encrusting algal stromatolites on the Middle Magnesian Limestone reef-front, up to 90 m high, may be steeply inclined and even vertical.

Faulting

Many of the faults in the Coal Measures have been rejuvenated during the Tertiary and thus offset the Permian as well. Two sets of normal faults are present: the main set trends within 25° of east and west; less important are those slightly west of north.

Igneous dykes

Dolerite dykes of two ages, late Carboniferous and Tertiary, cut the Upper Carboniferous rocks of the area. The dykes are impersistent and are arranged in a WNW/ESE trend. Coal seams cut by the dykes are extensively altered to coke.

Superficial deposits

Superficial deposits include alluvium and occasionally peat, preserved in the bottom of preglacial valleys, the often complex glacial sequences infilling preglacial valleys together with the general cover of glacial deposits, postglacial river alluvium, beach deposits and associated wind-blown sands, general fill and reclaimed land, infilled quarries and reclaimed open-cast coal sites.

The preglacial valleys were, in many cases, graded to a glacial sea level at or about 46 m below Ordnance Datum (BOD).

Glacial deposits

Glacial deposits are of three types: (a) boulder clay; (b) laminated clays, silts and sands; (c) outwash sands and gravels. Only the general characteristics of each type of deposit need to be discussed.

The boulder clays are typically overconsolidated, silty clays or clayey silts with varying amounts of boulders ranging up to at least 1 m in size. There are compositional differences between different boulder clay sheets, mainly in the size, type and amount of boulder materials present. Where exposed to the land surface the boulder clays are weathered from dark grey to a light brown colour and are extensively fissured both horizontally and vertically.

Laminated clay sequences, separating upper and lower boulder clay sheets, are of particular importance when encountered in excavations. They are usually normally consolidated with high pore water pressures, markedly anisotropic, with their geotechnical properties dominated by the horizontal laminations.

Outwash sands and gravels may occupy channels in the boulder clays, or may occur as surface spreads with marked topography.

In the very thick sequences plugging preglacial valleys, the boulder clays may be underlain by alluvial sands and gravels. More than one horizon of laminated clays may be present; they represent accumulations of sediments in standing bodies of ice melt-water in boulder clays beyond local ice margins. There is, however, a general lack of lateral continuity in these deposits over a large area, and the deeper deposits are confined to the major river valleys.

Alluvium

Alluvial sands and gravels, with occasional interbedded peat, are found along the present drainage courses.

Beach sands in the boulder clay-plugged coastal bays are locally underlain by 'submerged forests' – postglacial peat deposits which may extend up to 15 m below sea level.

Sandy bays are usually backed by sand dunes.

10.9.3 Variations in ground conditions

Although from the point of view of civil engineering construction significant variations in ground conditions are related more to the superficial deposits than to the solid formations, problems presented by areas underlain by Coal Measures are, in the main, different from those encountered where Permian strata occur. Superficial deposits, particularly till (boulder clay) sequences, are similar over the whole area and very deep, preglacial channel infills are present on all types of solid rock.

Of particular value to the planner or engineer are schematic cross-sections, and this is a general point to make, that are accurate in detail but not drawn to scale, which generally illustrate the topography and geology and some of the likely effects of man's activities on the environment. Two examples have been prepared, one for areas, including the River Tyne, underlain by Coal Measures (Figure 10.26), with the other the Permian outcrop crossed by the River Wear (Figure 10.27).

Areas underlain by Coal Measures

Central Newcastle may be taken as a typical example of ground conditions likely to be encountered on the Coal Measures (Figure 10.26). The present River Tyne, deeply entrenched, follows its overdeepened preglacial channel, now partly infilled with a variable complex of glacial deposits, generally comprising at least three sheets of till separated by two horizons of fluvio-glacial laminated clays, sands and gravels. Alluvium is present above bedrock, in the deepest part of the channel. In places, the present river valley is eroded down to bedrock beneath the drift.

The river is flanked by steep slopes, and most of the level riverside areas are reclaimed land. Beyond the river gorge, an extensive drift-covered 'plateau' with charac-

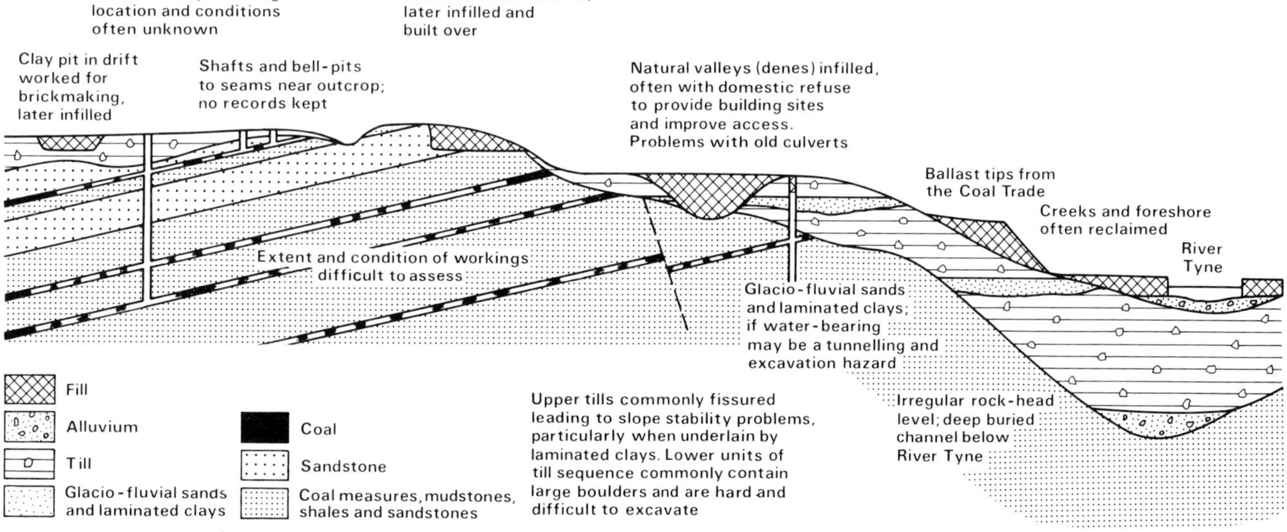

Figure 10.26 Variations in ground conditions in areas, including the valley of the Tyne, UK, underlain by Coal Measures (From Dearman and Matula, 1976, by permission)

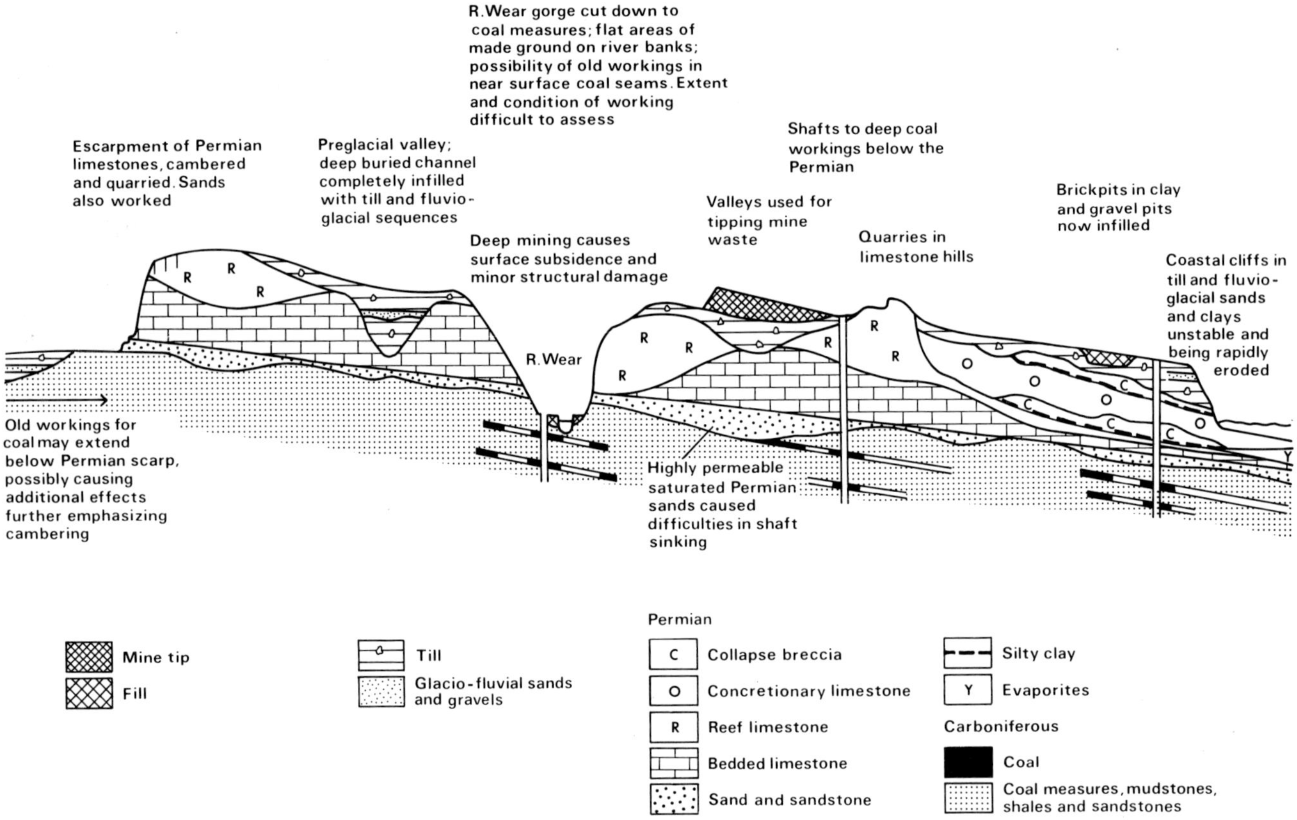

Figure 10.27 Variations in ground conditions in areas underlain by Permian strata (From Dearman *et al.*, 1977, by permission)

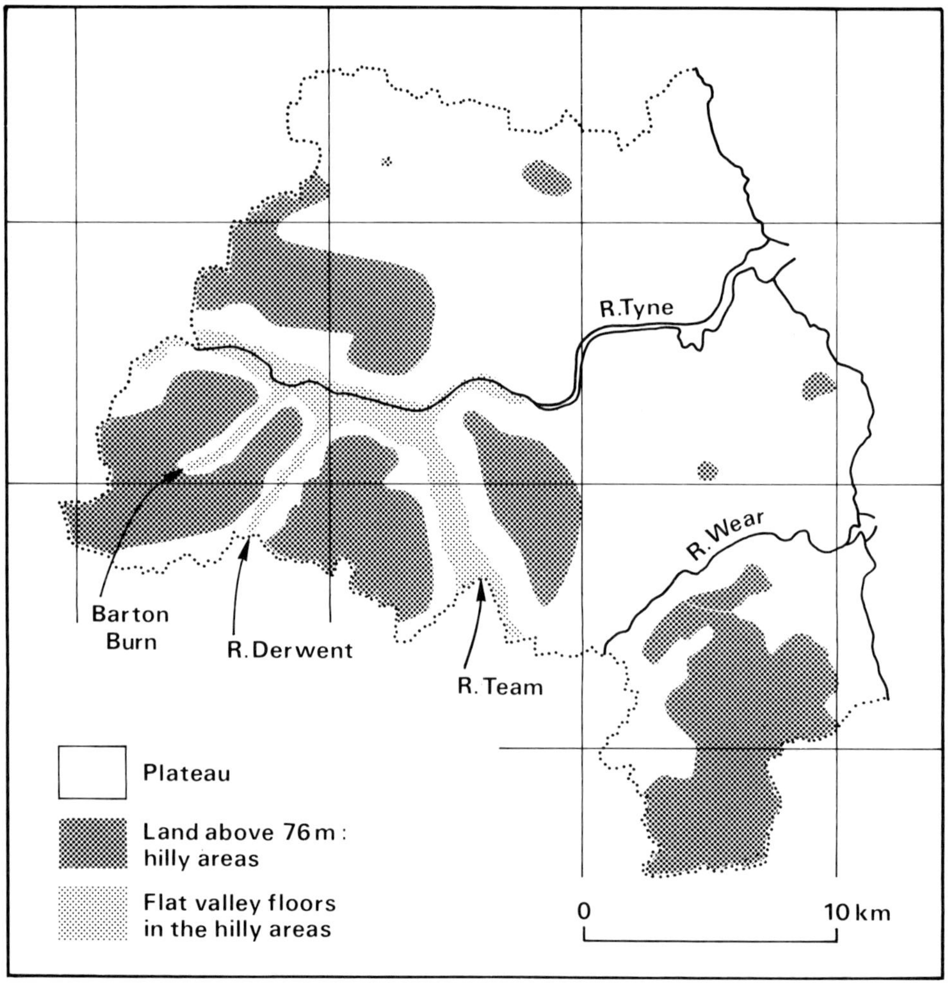

Figure 10.28 General morphology of Tyne and Wear County, UK

teristic gentle slopes lies mainly above 30 m OD (Figure 10.28). Prominent denes – the steep-sided tributary valleys – are incised in the plateau, and Coal Measures strata may be exposed in their sides.

Hills rising from the drift-covered plateau mark the outcrops of massive sandstones in the Middle Coal Measures. Formerly quarried for such diverse uses as building stone and grindstones, many of the sandstone quarries are now infilled and built over. Where redevelopment was necessary, as for the Metro system, uncertainty arises from the final extent of the quarry faces, the distribution of quarry waste, and the nature of the fill material.

Near surface coal seams, for example, the High Main Coal at the base of the High Main Post (sandstone), were frequently worked by bell pits, shallow shafts and where topography was appropriate by adits in from the valley sides. No records were kept of such early workings, and their physical state and the possible presence of gas are matters for conjecture. Working in the deeper seams has been carried on for several centuries: the extent and condition of older workings are generally not known, but the main geotechnical problem is the occurrence of abandoned mine shafts. Back-filling using various expedients (Dearman *et al.*, 1977) may have effectively obscured any surface indications of a former shaft, while available archival maps may give more than one locality for the same shaft. Surface subsidence, due to the progressive collapse of old pillar and stall workings or to relatively recent, controlled, long-wall working, affects surface levels, drainage and may damage existing buildings.

Areas underlain by Permian limestones

Engineering geological conditions are different where the magnesium limestones crop out. The limestone escarpment locally rises above the general drift-covered plateau of the Coal Measures; within the limestone outcrop the land falls gently towards the coast, with small rounded hills, marking the line of the reef complex, emerging from beneath the general drift cover (see Figure 10.27). Minor denes, draining to the coast and to the River Wear, are cut down into the glacial cover and in places to the bedrock beneath. These are postglacial features which do not necessarily follow the courses of preglacial valleys.

A general picture of other ground conditions may be gained from Figure 10.27 and two aspects are worthy of comment. One is the great variability of the limestones themselves, which makes it extremely difficult to predict the local development and distribution of various rock types listed in Figure 10.27. Typical karstic features are largely absent, but the limestones have undergone complex changes involving dolomitization and dedolomitization resulting in the formation of the concretionary limestones. In the concretionary limestones, hard calcitic concretions may be set in a loose silt-grade dolomite matrix; elsewhere bedded dolomites, which stand unsupported in vertical cliff faces, are in reality hard silt-grade soils which slake in water. There is also the brecciation at some levels considered to have been caused by removal of general support on solution of anhydrite in the Tertiary.

The other aspect is the response of the limestones to subsidence caused by deep mining. In Figure 10.27, cambering in the limestones on the escarpment may be further emphasized by the effects of mining subsidence. Elsewhere, surface subsidence may have associated with it deep fissuring, also controlled by the major joints in the limestones.

Areas underlain by glacial deposits

For both glacial and periglacial ground conditions it is possible to recognize distinct terrain types, each having a characteristic topography and distinctive subsurface conditions. Tyne and Wear is a typical glaciated lowland with a mainly subdued rockhead surface almost entirely obscured by glacial soils which were deposited by large ice lobes. This is the subglacial terrain type where the surface of the till deposited by the glacier at its base has been modified by later erosive agencies.

Subsurface and surface features associated with the subglacial terrain model are illustrated in the block diagram forming Figure 8.16. This conceptual model from Eyles and Dearman (1981), combining features at rockhead, in the lodgement till itself, and outwash features at the surface, is presented as an aid to the interpretation and understanding of the results of site investigation boreholes.

10.9.4 Examination of selected ground conditions at the plan scale

Engineering geological plans of old coal workings

One of the areas for which a complete set of 1:10 000 engineering geological plans has been prepared is that part of central Gateshead (see Figure 10.24) on Ordnance Survey Sheet NZ 2560 (Scott, 1976).

This is an area for which no recent geological map was then available and one had to be compiled at a scale of 1:2500 from available archival information. Glacial deposits are absent from the centre of Gateshead, where Coal Measures sandstone is overlain by man-made fill generally to a depth of 2–3 m, but exceptionally up to 6 m. The fill is very varied and includes spoil tips from the earliest mining periods; clay, stone, wood and bricks from residential development and demolition spanning the past 200 years; domestic refuse, chemical slag and ash, tarmacadam and concrete.

Where present, the underlying till is a stiff, sandy, silty clay with cobbles and boulders and occasional lenses of sand and silt. A maximum thickness of 7 m is found in small buried channels in rockhead but, as already noted, it may be entirely absent on higher ground.

In the Coal Measures there are two worked coal seams close to the surface, the High Main and the Main, both nearly 2 m thick and 7 m apart. Above the High Main seam a thick, medium-grained micaceous sandstone forms a distinctive marker horizon: above and below the Main seam there are alternating shales, sandstones, siltstones and mudstones. The succession dips gently eastwards at about 15°, but is dislocated by a normal fault trending NE–SW with a downthrow of 10 m to the north, although the precise location and attitude of this fault is not known.

A major geotechnical problem in Gateshead is the close proximity to the surface of cavities resulting from pillar and stall extraction of coal. None of the workings in these seams is recorded accurately, but it is generally

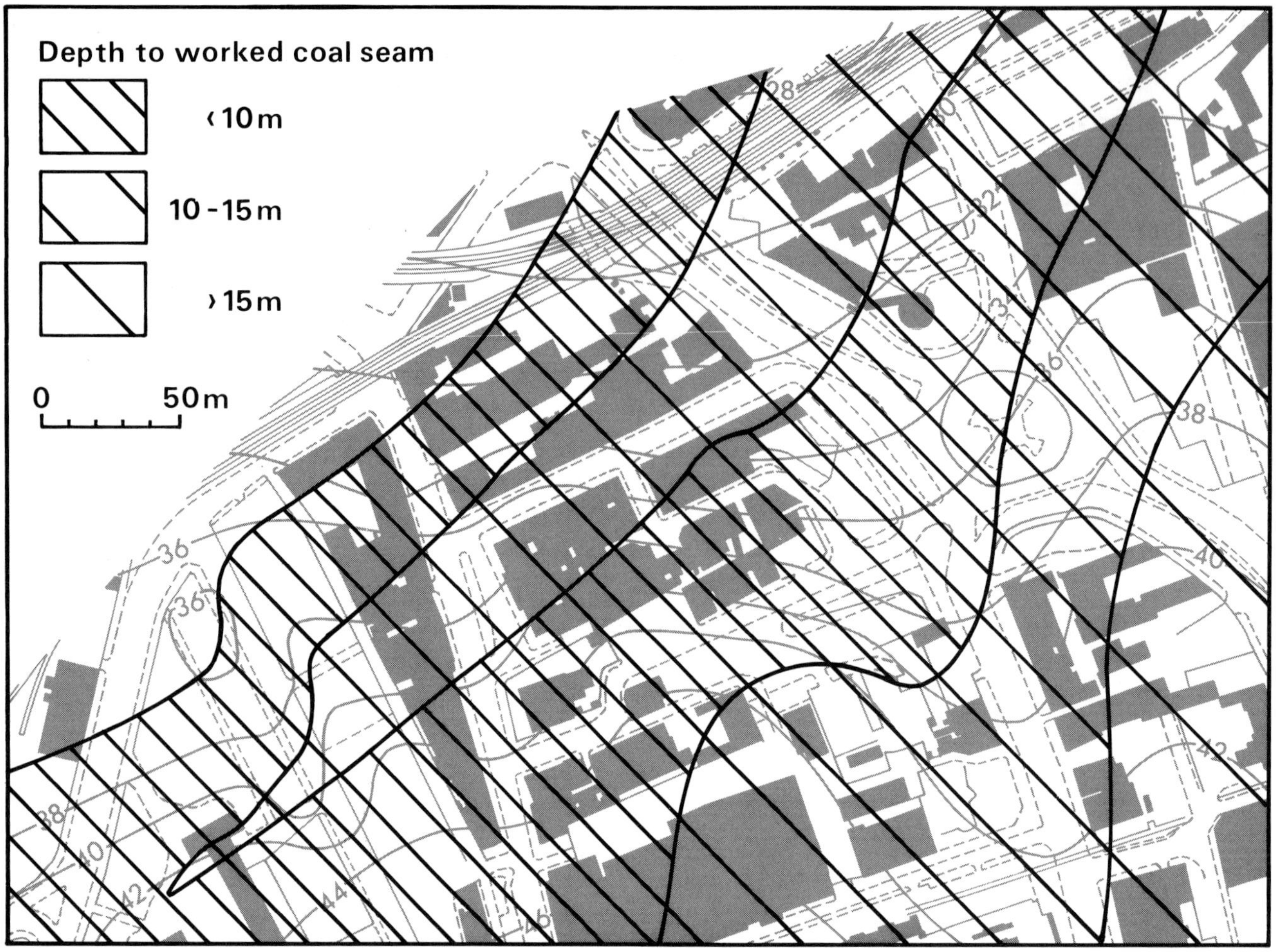

Figure 10.29 General-purpose plan of central Gateshead, UK, showing zones within which worked coal seams are believed to occur less than 10 m and 15 m from the ground surface. Produced at a scale of 1:2500 as reproduced here.

accepted that the coal seams have been fully exploited in the past. Work ceased well over 150 years ago, with the result that no plans exist.

The precise position of cavities and old workings cannot be predicted, but a knowledge of the depths of the seams makes it possible to define existing and potential zones of collapse. A general-purpose large-scale map (Figure 10.29) has been derived from the geological map to show zones within which a worked seam is believed to occur within 10 and 15 m of the ground surface. The depth ranges have been selected only to illustrate the type of map that can be prepared.

Engineering geological mapping of superficial deposits

A group of plans has been prepared for an area of Tynemouth where the Middle Coal Measures with workable coals are completely obscured by glacial deposits. Housing covers only parts of the area and in open country selected coal seams have been worked by open-cast mining. Some parts of the restored workings, infilled with excavated Coal Measures strata and the original superficial deposits, are now being developed for industrial use and site investigations have been carried out in the fill deposits.

Open-cast prospecting over a large part of the map area, some site investigation boreholes and other investigations have provided a wealth of subsurface information. This has been assembled as a documentation plan (Figure 10.30) with a legend detailing, for example, methods of drilling, types of sample taken, wells and mineshafts. Intepretation of the data provides first a large-scale plan showing contours on the natural rockhead and the base of the open-cast workings (Figure 10.31). Slopes on the walls of the excavations were not recorded and have therefore been shown as vertical.

From all the borehole evidence, supplemented by a few specially augered holes, a complementary geological plan has been compiled of both the solid and the superficial deposits (Figure 10.32). Outcrops of coal seams, sandstones and associated Coal Measures strata are as they would appear at rockhead if the superficial deposits, including the infill of the worked opencast coal sites, were removed. The High Main seam is the highest in the succession and the strata dip gently to the west-south-west.

Outcrops of superficial deposits include fill, till, sands and gravel. Sand and gravel also occur locally as part of the till sequence, but this is not apparent from the outcrop plan of the superficial deposits.

A general-purpose engineering geological plan of the superficial deposits (Figure 10.33) evaluates both type of deposit and thickness. Four mapping units, comprising

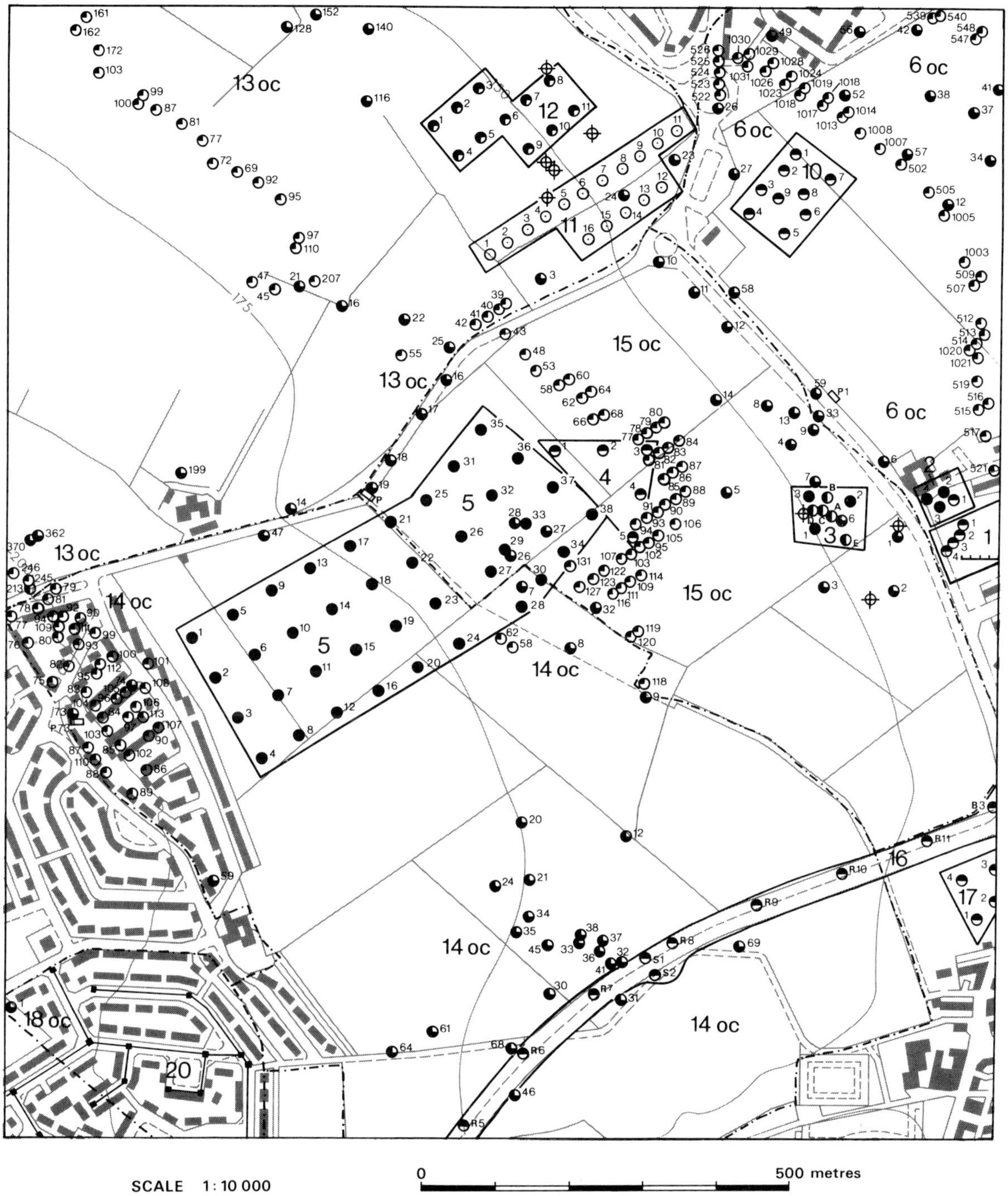

LEGEND

Opencast Prospecting Boreholes

- Diamond, drill rock cores taken
- Water flush – chip samples
- Hand auger in soil

Engineering Site Investigation Boreholes

- Shell and auger
- Shell and auger with rotary core in rock
- Shell and auger with rotary in rock
- Rotary – rock roller

Research Boreholes

- Power auger – disturbed samples

- ⊗w Well, backfilled or inaccessible
- Mine shaft, abandoned
- Trial pit
- Sewer Trench
- Geophysics – constant separation resistivity traverse
- 12 Area of site investigation, with reference number
- 6 oc Opencast prospecting area, with reference number

Figure 10.30 Documentation plan recording the location and nature of archival information for a small area of North Tyneside, UK. The original was prepared at the scale of 1:10 000; reproduced here at 1:8800 (From Anon., 1976, Fig. 5.5, by permission)

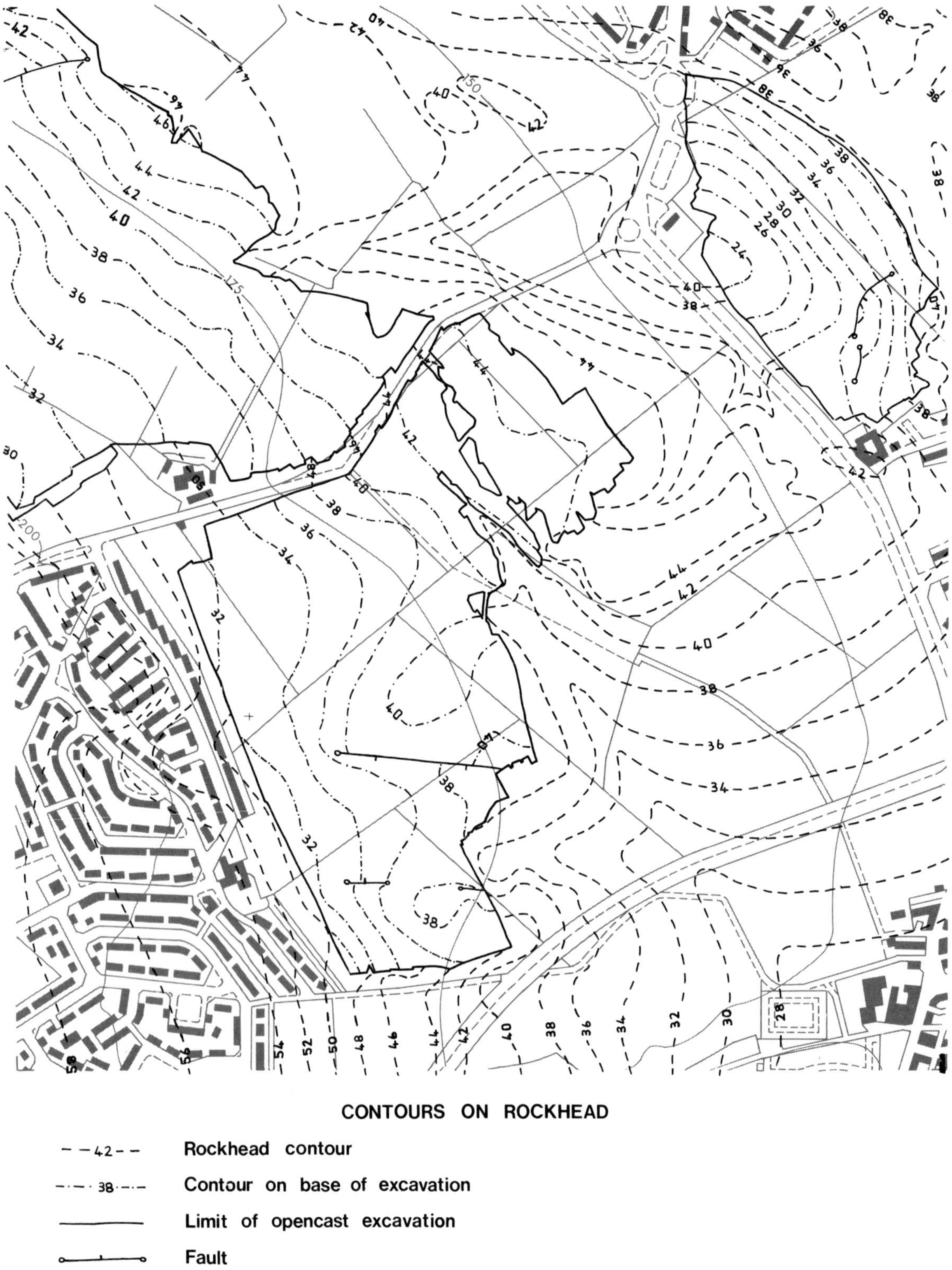

Figure 10.31 Rockhead contour plan of the area of Figure 10.30, with contours on natural rockhead and the base of open-cast coal workings

till, sand and gravel, till with sand and gravel and open-cast backfill, are each divided into three subunits on the basis of thickness. Thicknesses, which are depths of deposit to rockhead of 0–3, 0–10 and more than 10 m have been selected to illustrate how the plan is derived from the geological data. Twelve units result from these combinations, but no indication is given of the nature of the solid rocks beneath each unit.

Comparison of Figure 10.33 with combined Figures 10.31 and 10.32 illustrates the essential difference between an engineering geological plan and a basic geological map. Differences lie in type of mapping unit

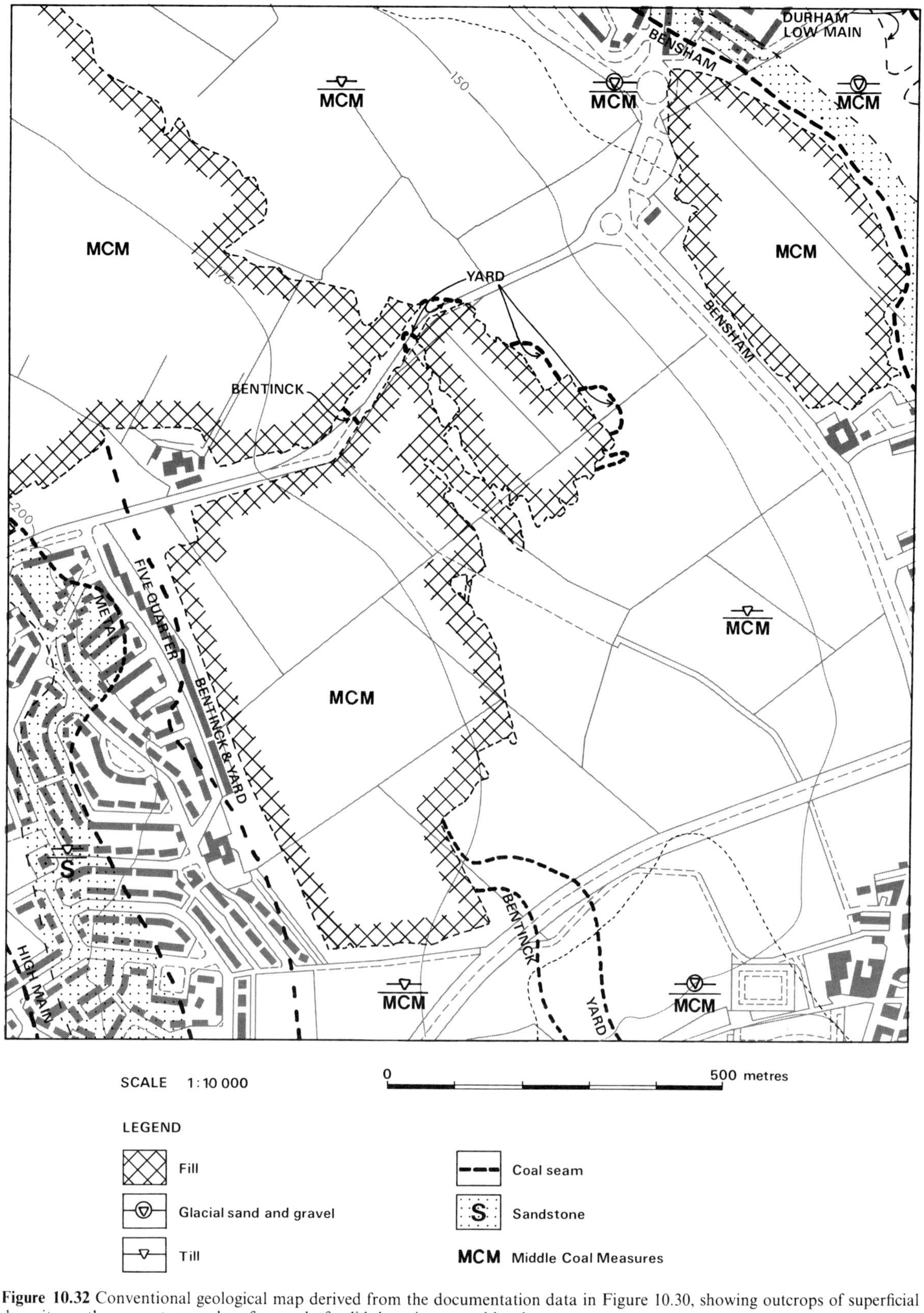

Figure 10.32 Conventional geological map derived from the documentation data in Figure 10.30, showing outcrops of superficial deposits on the present ground surface and of solid deposits on rockhead

and the direct utility of the engineering geological plan to the planner and engineer unversed in conventional geological interpretation.

Pandon Dene: an exercise in the use of existing archival information

Study of the history of a site is very desirable in urban site investigation and forms an essential component of engineering geological mapping.

The use of archives in site investigation is not new; the British Standard Code of Practice on Site Investigations (CP 2001 1957, 1981) recommends the use of old editions of Ordnance Survey maps, and Holmes (1972) has illustrated the geological uses of old maps. In England,

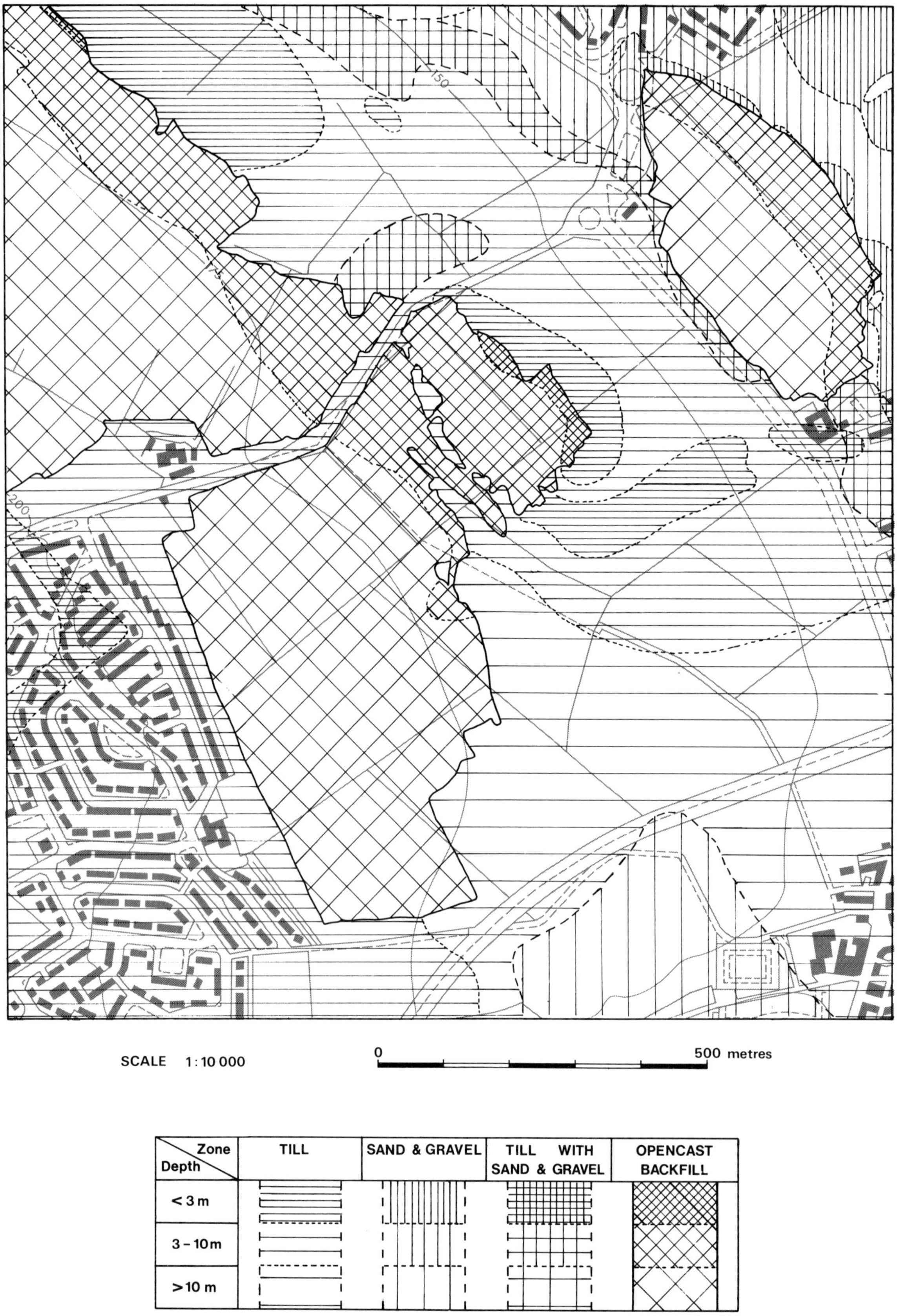

Figure 10.33 General-purpose engineering geological plan evaluating both type and thickness of superficial deposits on the area of Figure 10.30

Local Record Offices, Public Libraries and local history societies hold a wealth of archival information. Although much of this material consists of written and printed documents whose study is extremely time consuming, there are also numerous maps and plans that can provide information of direct interest to the engineer. Such maps include the various editions of the Ordnance Survey 1:500 and 1:2500 plans and the 1:10 560 maps together with mining and estate records, plans of Tithes, Enclosures and Public Works, and plans produced by local surveyors and land agents. Useful plans may even be attached to the deeds of the site itself.

The abundance and varied nature of this information inevitably produces problems for the engineer attempting to track down data relating to this site. The information has first to be found and it may be distributed among several libraries; at least nine separate libraries in Newcastle upon Tyne hold copies of Ordnance Survey plans of the city, but none holds a complete set of all scales and editions. Access to some libraries may be restricted, loans are often impossible and photocopying, especially of large maps, may be difficult. Once discovered, the information must be interpreted and here further problems arise; old plans may be at awkward scales such as 4 chains to the inch (1:3168) or oriented with respect to the magnetic north of the time and difficult to relate to present-day topography. Faced with these difficulties it is probable that exhaustive investigations of site history are rarely carried out and then mainly by engineers who already have detailed local knowledge. In preparing engineering geological plans, however, systematic searches of archives can be made, and the information plotted and indexed on a documentation plan. In an urban area this approach will be both more effective and economical than random searches for individual sites, and a properly prepared series of engineering geological maps could largely eliminate the preliminary stages of site investigations.

This aspect of mapping may be illustrated by a study of Pandon Dene. Pandon Dene is a natural valley in central Newcastle which originally ran down the eastern side of the city and is clearly shown in the view of the city from the south drawn about 1590 (Figure 10.34).

Other prints of eighteenth to early nineteenth century hachured maps show the Dene in its original form and a succession of watermills (Figure 10.35). There is some evidence of a small tidal inlet at the confluence with the Tyne, but this was probably filled in or silted up by the time the city wall was built along the original quay about 1280. The exact course of the old stream in this lower portion is uncertain, but the position of the culvert is known from recent engineering surveys. This culvert extended upstream to the point where the city wall again crossed the stream. The date of the culvert is unknown, but is thought to be before about 1600.

Culverting and filling of the remainder of the Dene started at the beginning of the nineteenth century, but Ordnance Survey plans were not surveyed until 1858 by which time much of the Dene had been filled. Other sources had to be used to reconstruct the form of the valley.

Plans, sketch elevations and a few dimensions of the 'New Bridge' built across the Dene in 1812 enable a rough cross-section to be plotted. A detailed plan of Newcastle and Gateshead surveyed in 1830 by Thomas Oliver, a local architect, shows the stream bed and the sharp break of slope along the edges of the valley (Figure 10.35).

Downstream of the New Bridge the Newcastle and South Shields railway crossed the Dene on a 24 m high embankment; this was widened in later years as the railway expanded. These developments can be traced from the Parliamentary Plans and Sections, some of which are well surveyed whereas others were clearly produced in haste and are of doubtful reliability. Upstream of the bridge, filling was at first by tipping of

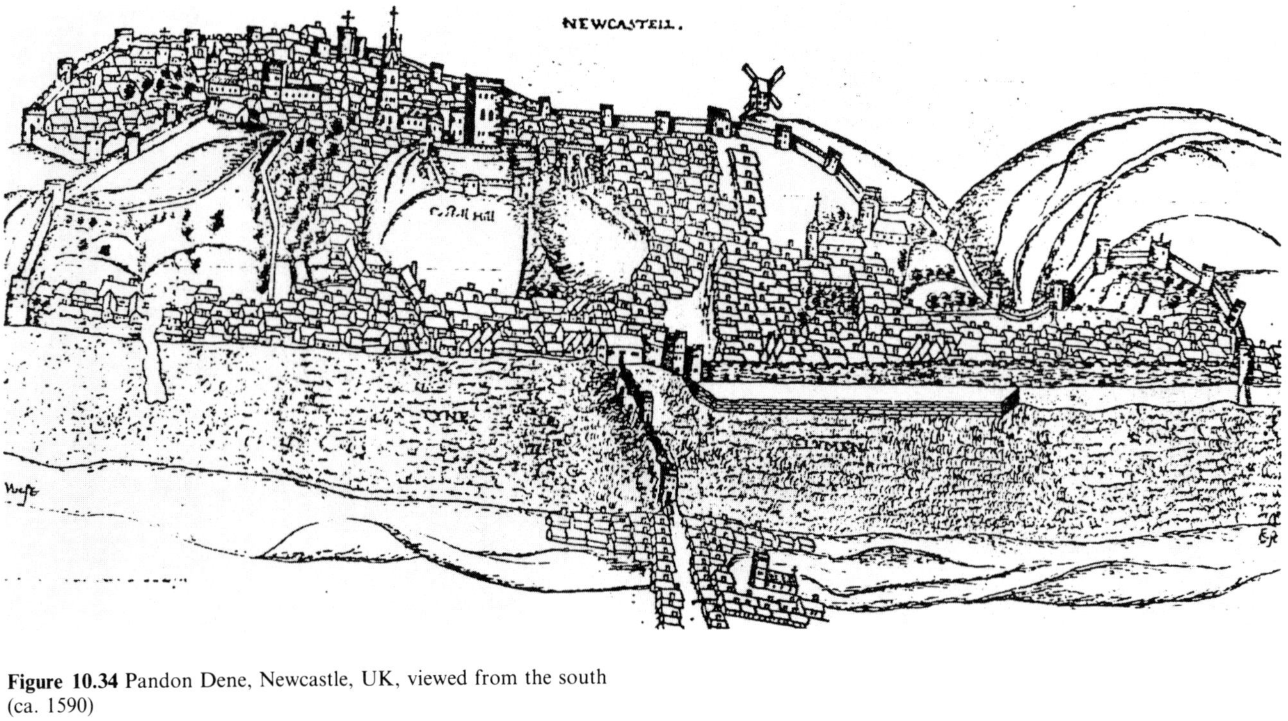

Figure 10.34 Pandon Dene, Newcastle, UK, viewed from the south (ca. 1590)

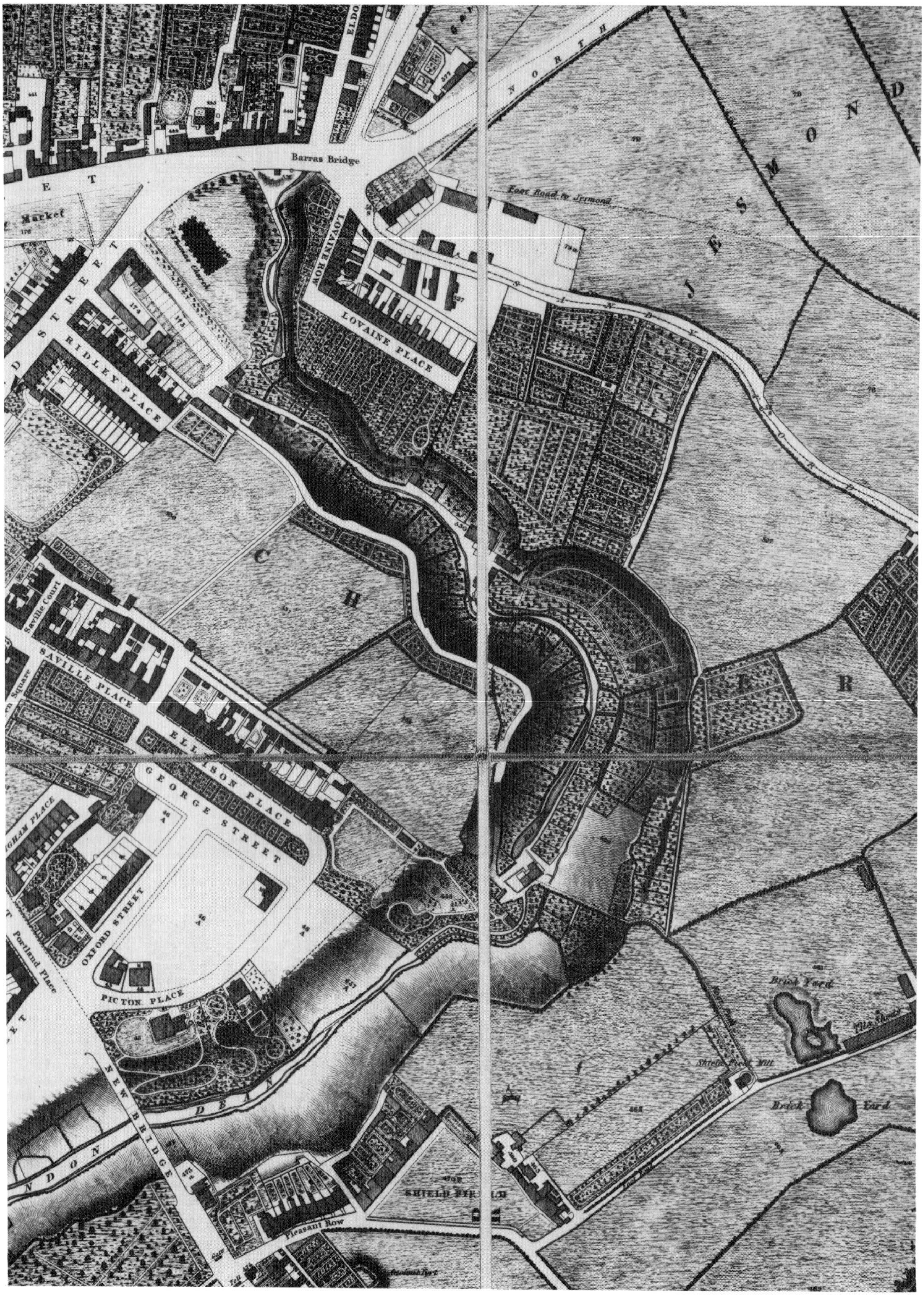

Figure 10.35 The Thomas Oliver 1830 map of Pandon Dene, Newcastle, UK. The part reproduced here is at a scale of 1:6336

Figure 10.36 The present topographic expression of Pandon Dene, Newcastle, UK, at its downstream end

Figure 10.37 St. George's Hall, College Street, Newcastle, UK, sited astride the edge of Pandon Dene, showing structural defects caused by differential settlement. The hall has since been demolished

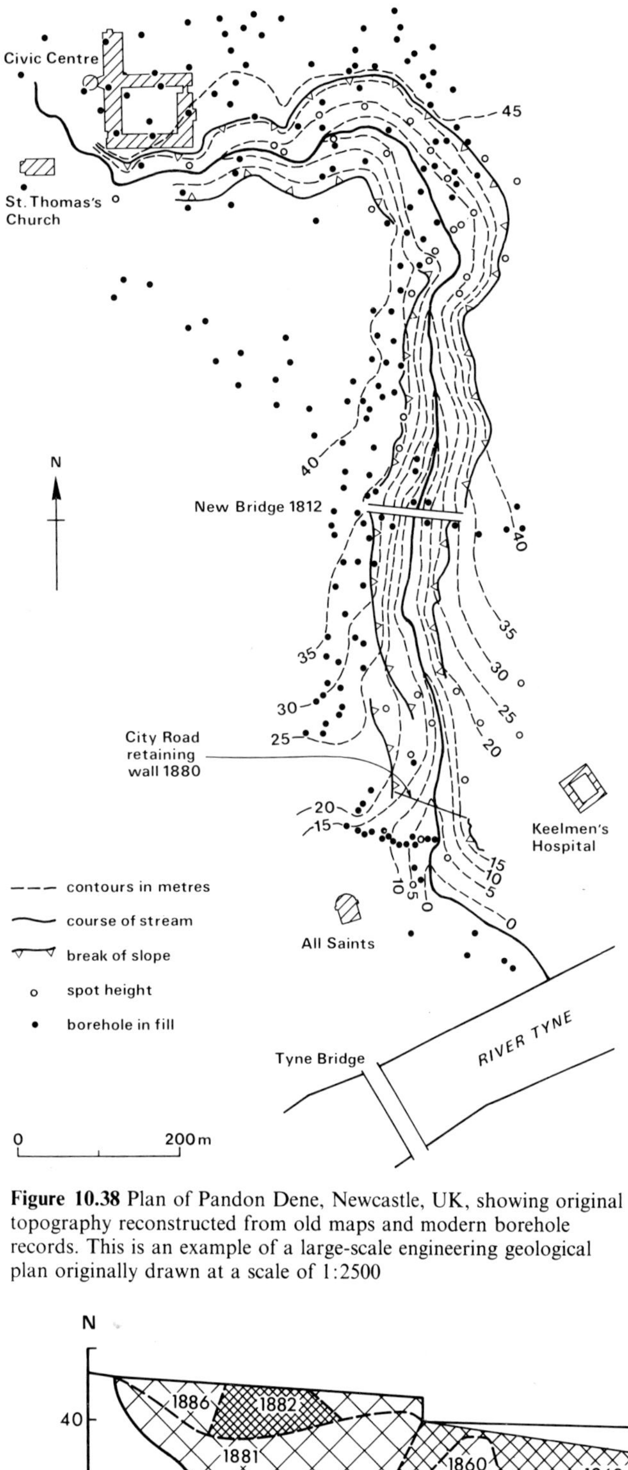

Figure 10.38 Plan of Pandon Dene, Newcastle, UK, showing original topography reconstructed from old maps and modern borehole records. This is an example of a large-scale engineering geological plan originally drawn at a scale of 1:2500

domestic and demolition refuse, but in the early 1860s the Blyth & Tyne Railway filled in part of the valley to form the site for a new terminus.

Upstream of this site the first edition of the 1:2500 Ordnance Survey Plan surveyed in 1858 provides the first accurate plan of the unfilled part of the Dene together with a few surface levels. Additional levels are shown on the contemporary 1:500 plan. The 1:2500 plans are particularly useful because they can be overlaid directly with transparent copies of the modern map and changes of topography and vanished features may be plotted accurately.

The remainder of the Dene was filled with refuse to complete road extensions across the old valley site, but this stage is not well documented.

Urban development then spread across the site of the Dene and subsequently well beyond. The Dene ceased to exist as a topographic feature, except for a short length at the downstream end (Figure 10.36) and in the upper reaches west and north of the A1 trunk road. Careful geomorphological mapping shows, however, that the position of the sharp break of slope at the top of the banks of the Dene can still be identified as changes in present street levels.

The fill, which is extremely varied in its engineering properties, is in places up to 24 m thick and redevelopment involving multi-storey or urban motorway construction requires a careful analysis of this man-made substrate. Some older buildings within the fill zone have suffered structural damage due to settlement, notably St. George's Drill Hall in College Street (Figure 10.37) and a range of shops and offices in New Bridge Street where the buildings straddle the junction between natural ground and the fill.

Results of the new survey are illustrated in plan in Figure 10.38 which, in the manuscript version, is an example of a special-purpose engineering geological plan. Figure 10.39 shows in longitudinal section the historical stages of infilling the Dene.

10.9.5 Tyne and Wear data bank

Introduction

The aims of the Tyne and Wear Engineering Geological Survey (TWEGS) were:

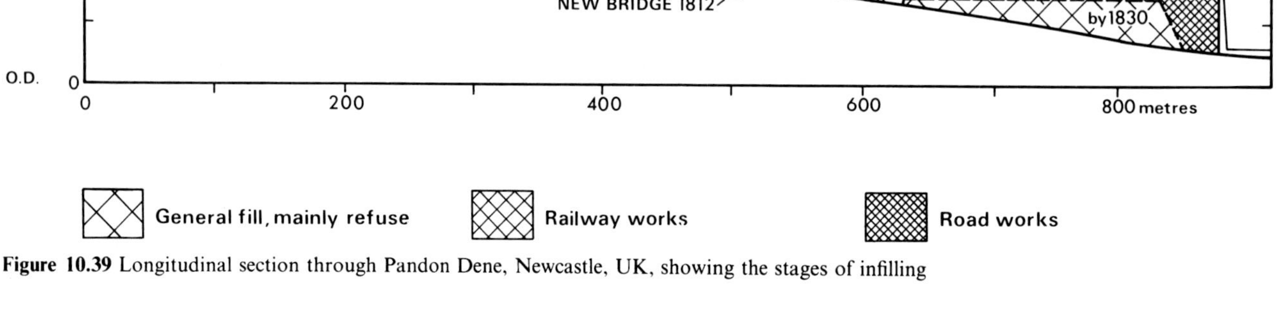

Figure 10.39 Longitudinal section through Pandon Dene, Newcastle, UK, showing the stages of infilling

- to compile a data bank of engineering geological information, and complementary informative or document plans;
- to compile derivative maps and interpretative memoirs;
- to investigate specific engineering geological problems relating to Tyne and Wear.

Presented here is an account of the data bank and the manual information retrieval system that has been developed. Documentation plans (UNESCO/IAEG, 1976) form an essential part of the data recovery system. They have also been used as the basis for the compilation of engineering geological maps and plans of Tyne and Wear County (Dearman *et al.*, 1977, 1979).

The data system

The data system had to be designed to cope with the anticipated large amount of information that was to be collected in the initial stages of the mapping project, and as a result had to allow for the following:

- ease of storage and retrieval of information;
- low cost, as only limited finance was available;
- ease of use of the data system by inexperienced operators.

The scheme for classifying and storing data was originally set up by Scott (1976). Data collected by him were stored as paper copies. The present system closely follows that of Scott, except that data are now stored on microfilm. The data system is essentially a geographically based filing system with the 1 : 10 000 scale Ordnance Survey quarter sheet numbers forming the basic file titles under which each item of information, located either totally or mainly within that area, is numbered consecutively. Site investigations are not necessarily confined within the boundaries of the Ordnance Survey maps, and therefore it will be appreciated that boreholes, put down for say a new road or sewerage system, could be located on one or more maps. It was not considered to be good policy to separate boreholes either from the parent report (especially if it contained geotechnical data) or from the group to which they belong. Therefore all boreholes relating to one particular site investigation, either with or without accompanying text and data, were kept together. This proved to be good practice, especially when trying to avoid duplicating material or completing missing parts of reports, when another source of the data was discovered. It follows that an adequate cross-indexing system was also needed for data that extended beyond the limits of the area under which they are filed.

The data system comprises two parts:

- the data bank, within which the raw data are stored, now wholly on microfilm;
- the associated informative or documentation plans, which show spatially where the information is and also indicate where it is filed.

These two parts will be considered in more detail.

The data bank

The data bank is used to store existing ground information. Most of the information is in the form of site investigation reports carried out for civil engineering purposes within Tyne and Wear. The major sources of such information include local government offices, commercial site investigation firms, consulting and contracting engineers, the British Geological Survey and British Coal. Thus the sources are both local and national.

A general flow sheet from the collection stage to the completion of storage in the data bank is illustrated in Figure 10.40. Three types of information may be assembled, and of these existing site investigation reports are of prime importance. An important step, once the Intro-sheet is completed, is to show on an index map (Figure 10.41) the 1 km squares within which site investigation reports are available.

Each report, or item of information, is allocated its own index sheet (referred to as an Intro-sheet by Scott) which gives specific information relating to it, e.g. title, index number, source, client, etc. These sheets (Figure 10.42) are kept together for each file and provide a good

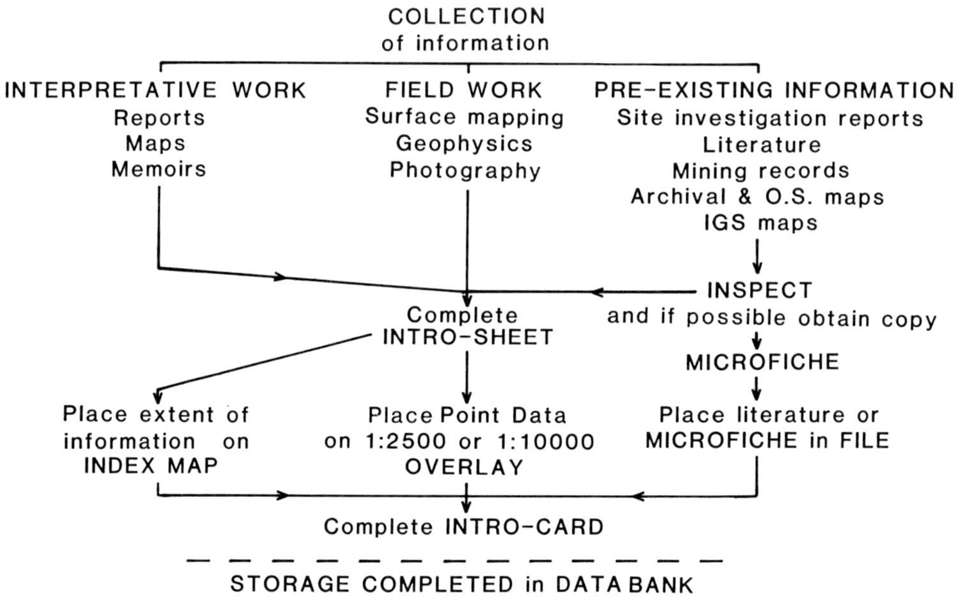

Figure 10.40 TWEGS system for storage of information in the data bank

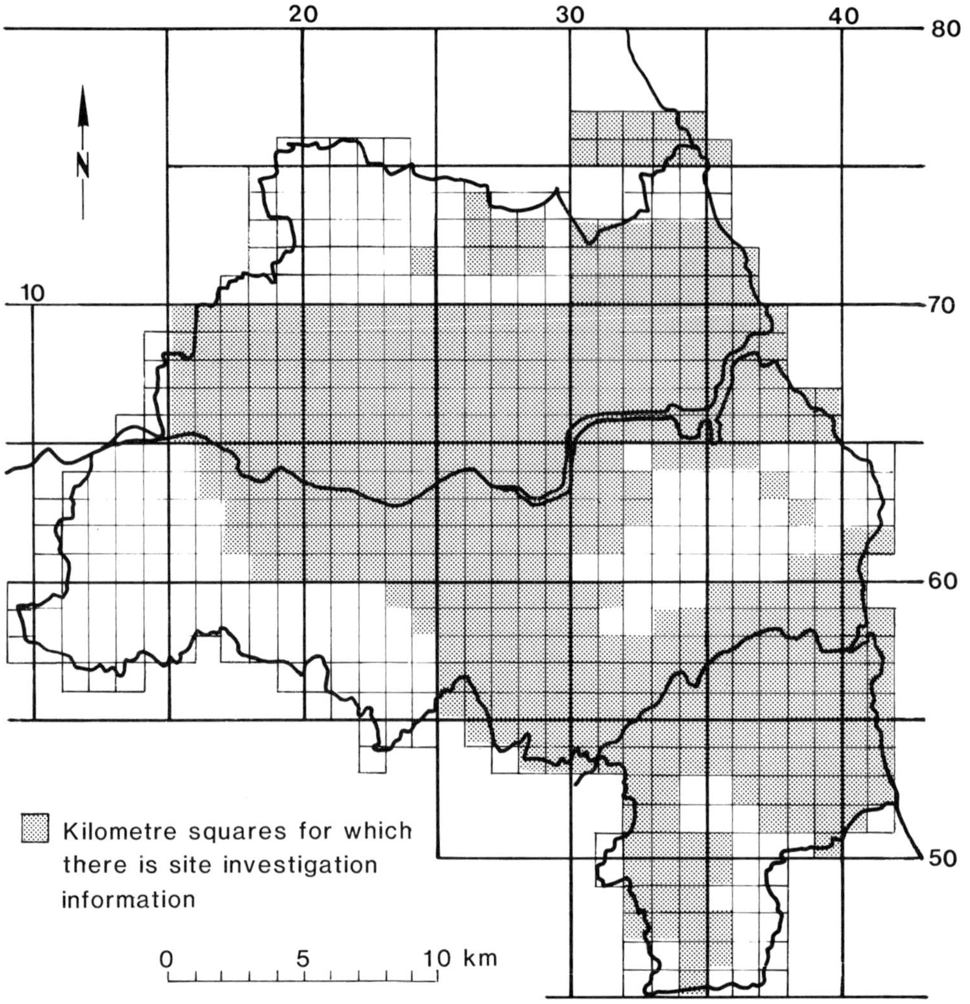

Figure 10.41 Index map showing the 1 km squares for which there is site investigation information

summary of the data held without any need for searching through the individual reports. At present these sheets are held as paper copies in lever arch files, but could easily be transferred to microfilm (one microfilm jacket holds 60 pages of original A4 size sheets). These index sheets are also used for cross-indexing. In addition, there are collective index sheets for the whole file which summarize its contents (Figure 10.43). These are useful to avoid duplication of material when other potential sources of data are contacted.

At an early stage of the survey, a punched card system, the Intro-card, was used as a readily accessible permanent file of the information on the Intro-sheets (Figure 10.40). It is now considered that this is an unnecessary step in data processing, and has been discontinued.

All data are stored on microfilm. Initially three forms of storage are considered: computer storage, microfilm storage and storage of photocopies of the original data. Microfilm storage was considered to be the most suitable of these for a number of reasons.

The storage of photocopies was only briefly considered, as it soon became apparent that a very large amount of shelf space would be needed to cope with only a modest quantity of data. Thus, due to the lack of available space, the idea was dismissed. This form of bulk paper storage was also not desirable from the point of view of security and confidentiality and the fact that large quantities of paper give rise to handling problems. This left a choice of either microfilm or computer storage.

Under ideal conditions computer storage would have been best, with only a terminal being needed. However, the most important point concerning storage was the desire to restore the raw data in its original format. This is necessary because no matter how the data are processed, it is essential to be able to get back to the original form. In storing only processed data, any attempt at comparison of both the raw data and the resulting interpretation for different areas would not be possible. All interpretation is subjective and carried out for a particular purpose, and again, storage of processed data would not be very useful if the data were required for an alternative purpose.

Microfilm storage was felt to be the only possibility. This method does have one great advantage over a computer. In the event of a company being willing to contribute data to the new project, but not wishing to have the information removed even temporarily from its premises due to reasons of security or confidentiality, the microfilming equipment can be taken to the premises as it is reasonably portable and would take up little space. The task of microfilming a large quantity of data, often including tables and diagrams, could be accomplished

TITLE	Coast Road Extension CRADLEWELL BRIDGE JESMOND, NEWCASTLE	1:10 000 NZ 26 NE	05
CONTRACTOR AUTHOR	Cementation Ground Engineering		C
CLIENT PUBLISHER	Newcastle City Council		CL
CONSULTANT			CS
SOURCE	Newcastle Civic Centre		S
Report Ref No.	430/72/MD McQ	Report Date	May 1972
6" Sheet	NZ 26 NE		
1 Km	2565 and 2665		
INFORMn TYPE	Site investigation	TWEGS CENTRE	S
RESTRICTION	None	RESEARCHER	JCR
		INSPECTn DATE	Dec 74
LITERATURE	P ✓ Borehole Logs	COPY	✓
BOREHOLE ✓	METHOD: RCD; No.: 2; Depth (16?) metres	Fill; Drift ✓; Solid ✓; Mn./works; Water	✓
LABORATORY & IN SITU TESTING ×	Ll & Pl; Moisture Content; Densities; Chemical; Gradings; Ground Water	Undrained Triax; Rock Field; Rock Lab.; Consolidation; Geophysical; S.P.T.	Drained Triaxial; Vane Shear; Shear Box; Organic
	Holes drilled by rock roller bit to 'prove' bedrock. No Laboratory tests undertaken. Only chippings recovered.		
MAP-PLAN ✓	Site location 1:500		
REMARKS	Holes drilled to prove bedrock.	RESEARCH USE	Plotted and logged.

Figure 10.42 Typical completed index sheet, or Intro-sheet – the first stage in processing information

more quickly than by putting it onto a computer storage disk. The basic equipment to read the microfilmed documents is also portable, and a data bank of 20 000 boreholes does not fill a single filing cabinet drawer. As such, it is relatively easy to keep secure. In case of any accidents which would destroy the microfilmed documents, copy negatives are kept in reel form in a fireproof safe. In addition, paper copies can be supplied from the microfiches by using commercially available equipment.

Informative plans: the borehole documentation plans
Most of the subsurface data are in the form of borehole logs and where possible these boreholes are located on 1:2500 scale plans (Figure 10.44). This scale was chosen as being the largest available that was common to both rural and urban areas. Larger scales are used in urban areas by planners, surveyors and engineers of the local authorities. But for the purposes of the project it was felt that this scale would be adequate for the amount of data likely to be retrieved for any one particular area. Also, there was a desire for consistency in using only one scale which could cover the whole county.

Transparent grids were printed for use as overlays to the Ordnance Survey plans, and after the boreholes have been put onto them they could easily be reproduced using the dyeline process. The boreholes are represented by Letraset symbols which reflect their type and quality, i.e. whether the borehole penetrates superficial deposits

SHEET NUMBER NZ 26 NW									1:2500		1:10000		
No.	Title	Ref.No.	Date	BH:TP	Data	Plan	km.Sq(s)	M	P	L	P	L	A
01	S.I. Harewood Road, Gosforth	C1018	1973	10 4	✓	✓	2468	✓	✓				
02	Ext. J. Marley School, Wellfield Lane	C981	1972	7	✓	✓	2067	✓	✓				
03	39 Montagu Avenue, Gosforth	C1182	1974	2		E	2366	✓	✓				
04	Housing Lambert Sq. & Riddell Terrace	C1213	1974	6	✓	✓	2368	✓	✓				
05	Trial Boreholes - Fenham Brickworks	C1266	1975	2				✓					
06	North Kenton Housing - test boreholes	C265	1967	2				✓					
07	The Fairway, Gosforth	C368	1968	4				✓					
08 p	Stamfordham Road	Job 369	1967	5		6 No G.R	2066	✓	✓				
09 p	Proposed Flats, Newbiggin Hall	S/3660	1963	9	✓	✓	2068	✓	✓				
10 p	Proposed baths, Montagu Estate	S/2235	1960	6	✓	✓	2167	✓	✓				
11 p	Prop. Sanderson Orthopaedic Hosp. School	S/3480	1963	9	✓	✓	2367	✓	✓				
12 p	Prop. development at the R.V.I.	S/6881	1969	25 3	✓	✓	2465, 2464	✓	✓				
13 p	Swimming Baths, Kenton Secondary School	S/2967	1961	5	✓	✓	2267	✓	✓				
14 p	Proposed Redevelopment at Blakelaw	S/5377	1966	28 8		✓	2066, 2166, 2167	✓	✓				
15 p	Prop. Housing, Kenton Bankfoot	S/8292	1971	20		✓	2067, 2167, 2068 2168	✓	✓				
16 p	Prop. Blakelaw Neighbourhood Unit	S/4630	1965	21	✓	✓	2166	✓	✓				
17 p	Shields Road Motorway Stage I	S/7373	1970	19	✓	✓	2365, 2465, 2366	✓	✓				

Figure 10.43 Collective index sheet for the Intro-sheet file for a single Ordnance Survey 1:10 000 quarter sheet

and rock, or only superficial deposits, and also whether testing was carried out (Figure 10.45).

In order to identify any particular borehole, its reference number is given adjacent to the symbols. A borehole reference has three components:

- the basic file title (1:10 000 scale Ordnance Survey sheet number, e.g. NZ26NW);
- the report index number (e.g. 12);
- the borehole number (e.g. 3).

Normally, only the second two numbers are used – in this example, 12/3. This is because the majority of boreholes are located entirely within the area under which number they are filed, and this number can be obtained from the Ordnance Survey Plan. However, if data are filed under a different area number, then this number always prefixes the other two, and clearly indicates that reference must be made to a different file for the data. (Note: the 'NZ' prefix is not normally used, as it is common to all maps and plans within Tyne and Wear.) Figure 10.44 illustrates this.

If any research staff wished to look at one particular area they would begin by locating the appropriate geographical plan; then by having these borehole documentation plans available, any boreholes of interest can quickly and easily be obtained from the data bank by noting the report and borehole number. Alternatively a rapid appraisal of data can be made by referring to the collective file index. This indicates the 1:2500 scale plan on which the data are located and, in the case of borehole data, whether there are any accompanying geotechnical data. More detailed information is given in the index or Intro-sheets for each report, which are kept separately.

Use of the data bank

Despite restrictions both on finance and time, within a relatively short period (3–4 years) a large data system has been initiated from which data can easily be retrieved by relatively inexperienced operators. A knowledge of the National Grid is virtually all that is required.

One important aspect of the project, however, is that all data are treated as confidential, but if permission is sought and obtained from the owners of specific data, then it can be released. The normal use of the data is in producing interpretative maps and plans to show the underlying geological conditions.

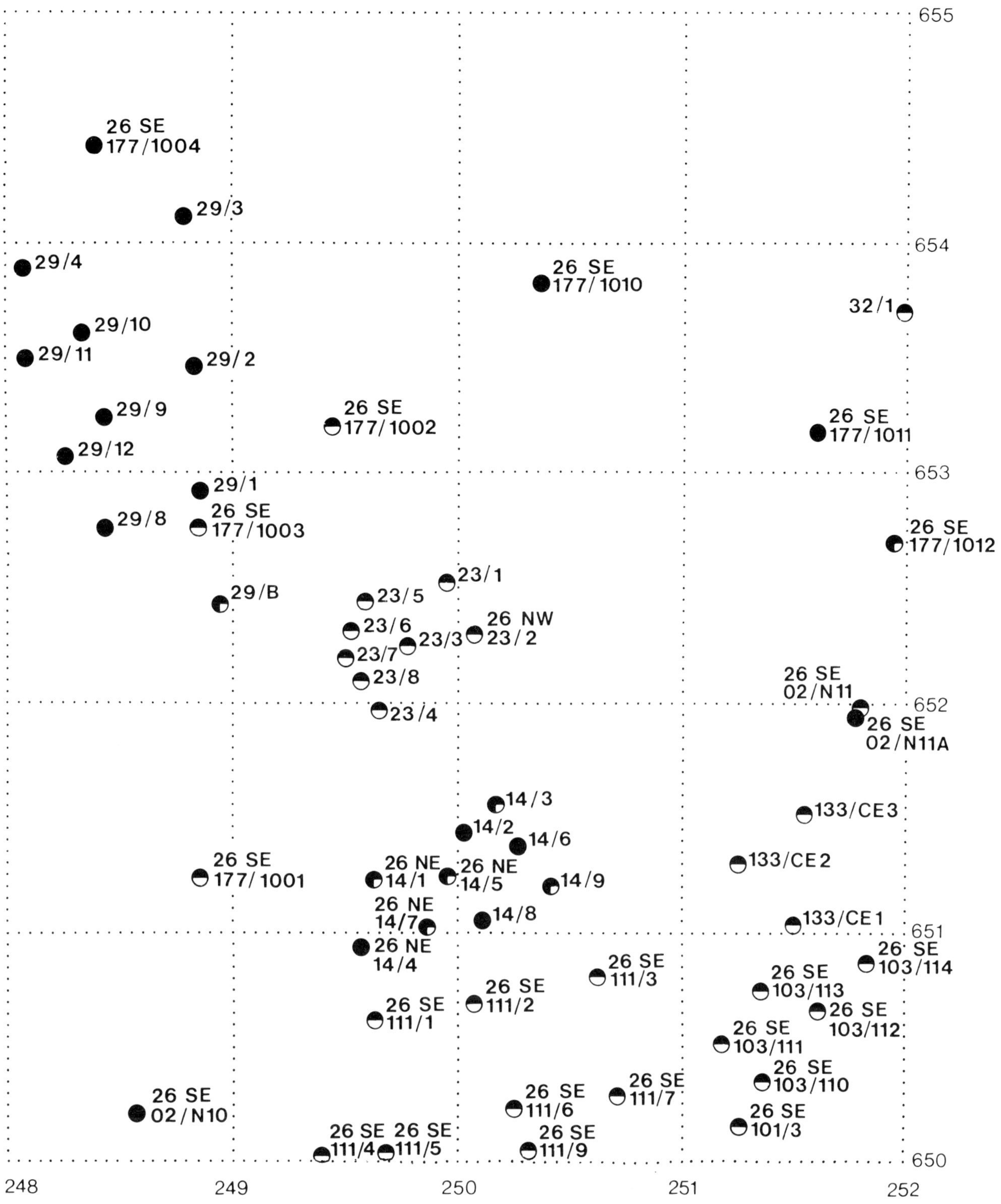

Figure 10.44 Part of a typical borehole documentation plan of Ordnance Survey Plans NZ 2465 and 2565. Original plan as reproduced here is 1:2500 scale

It is also considered that the formation of this type of data system in any region would be of considerable value to local authorities. Within Tyne and Wear the situation has occurred where a particular site has been investigated several times over a period of years. Sometimes this has been deliberate, but at other times it has occurred because all previous data have been lost. A regional data system, to which all relevant personnel had access, would be of immense benefit and save time and money. In many cases it would also help if a geological interpretation was given (Dearman and Strachan, 1982). Time would be saved in unnecessary searching through personal and departmental files on the offchance of there being any information. Money could be saved if under the circumstances a modified site investigation was all that was required. However, such a situation of reasonably easy access to a data system could not be seen to apply to professional organizations because of the commercial aspect. Nevertheless it is possible that arrangements could be made for exchange of information.

Access to the site investigation data bank would not dispense with site investigations as it would not be prudent to base recommendations solely on data from outside sources. Moreover, if pre-existing data were available, it could help in the design of site investigation with regard to specific features. One other feature of regional data systems, apart from concentrating all the information at one source, would be to preserve data that might otherwise be lost as well as increasing the knowledge of subsurface conditions. Apart from indicating ground conditions in site investigations, the data would also be of interest to both geology and geography departments of universities as a research tool.

BOREHOLE SYMBOL

SUPERFICIAL DEPOSITS								
Described	X	X	X	X	X	X		
Sampled and tested			X	X		X		
Thickness only							X	
SOLID DEPOSITS								
No details		X						
Described				X		X	X	
Core details					X	X		
Chippings described								X

Figure 10.45 Borehole symbols used on borehole documentation plans

A case has been presented elsewhere (Tuckwell and Sadgrove, 1977), following a pilot scheme in the south of England, for the establishment of a National Registry of Site Investigation Reports. The system recommended, with a retrieval system based on copy microfiches or paper copies from microfiches, is very similar to that already in operation in Tyne and Wear.

10.9.6 Analysis of the site investigation data bank

An account has already been given of the first stage of the Tyne and Wear Engineering Geological Survey, namely a data bank of easily recovered original site investigation borehole logs (Strachan and Dearman, 1982). Each coded borehole is located on a 1:2500 scale documentation plan based on the appropriate Ordnance Survey sheet.

Completed documentation plans show quite clearly the distribution of borehole information within the 100 m square grid of each 1:2500 plan sheet. Before undertaking the drafting of engineering geological plans at the same 1:2500 scale as the documentation plans, a decision was taken that an attempt should be made to characterize the geology of each 100 m square in terms of one average succession. These unit squares would then provide, essentially, point data from which the 1:2500 plans, together with plans and maps at smaller scales, could be derived. It was felt that with such small units, each clearly defined on the Ordnance Survey Plans, there was likely to be a good chance of uniform ground conditions being present. In practice this was unfortunately rarely the case, as will be demonstrated later.

For this initial analytical purpose a data summary sheet was devised on which summary logs could be drawn up for all site investigation boreholes and pits in one square. A critical step in this evaluation was assessment of the quality and reliability of each borehole log, a process inevitably involving subjective judgements.

Before completing the data summary sheet for any 100 m square, each borehole log had to be metricated, scanned for inconsistencies, and the soil and rock descriptions translated as far as possible into the standard lithological types.

Standard lithological types

A considerable background knowledge was already available of the stratigraphy of both the superficial deposits and the bedrock. With this in mind, and following inspection of the available borehole logs, eight lithological types were sufficient to deal with the superficial deposits, and six with the solid deposits beneath (Figure 10.46). These were also colour coded and given standard national (British Standards Institution, 1981; BS 5930) and international (IAEG/ISO) symbols (Anon., 1981c). Some explanation of these is necessary.

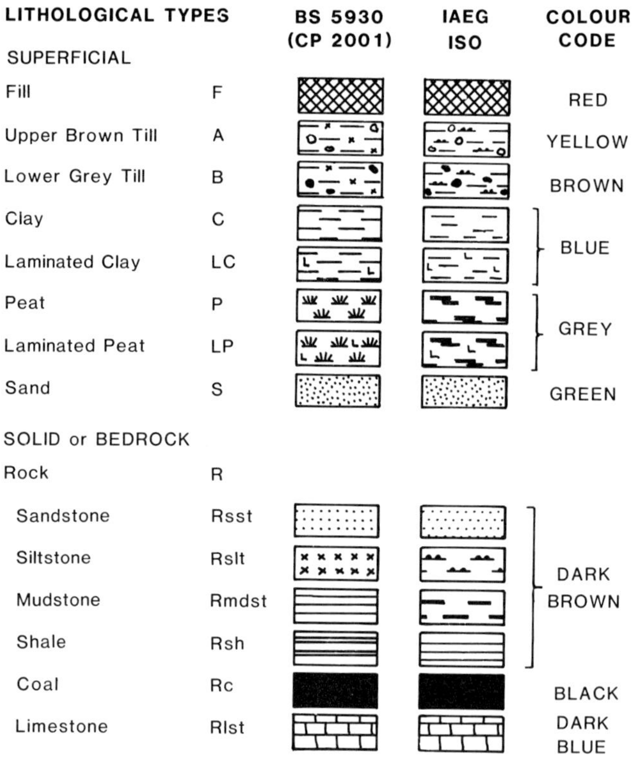

Figure 10.46 Standard lithological types, symbols and colour code

A single symbol used for sand is also intended to cover gravel, gravelly sand, and so on. Composite symbols are used for the upper and lower tills: the Brown Till A is considered to be a weathering derivative from Grey Till B. The symbol combines clay with silt and an otherwise unused symbol for boulders.

For fill and other man-made deposits, different spacings of the diagonal lines on the symbol are used to indicate different depths of fill when the symbol is used on a plan.

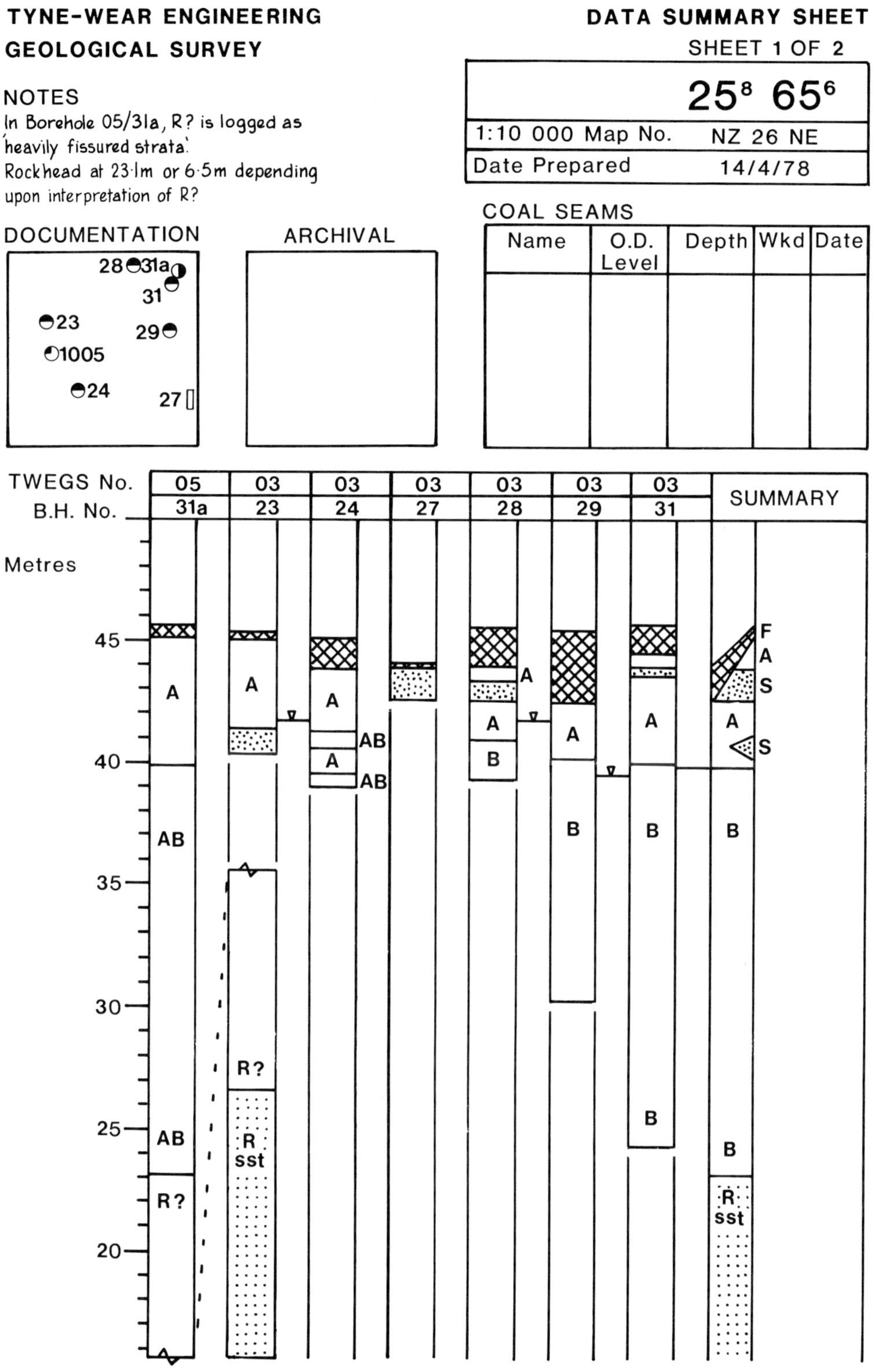

Figure 10.47 Completed data summary for 100 m square 258656

Example of summary sheet

Seven boreholes are set out in Figure 10.47. The two lines of the parallel column are broken to indicate the base of each borehole, which it must be remembered is not necessarily a boundary between lithological units. Borehole 31A is continued in the vacant space on the column for B.H.23, but an extension sheet could equally well have been used. Water levels in the bores are also shown.

Problems in characterizing each 100 m square

There is obviously no problem if only one borehole occurs in a square: that borehole record is held to characterize the whole of the area.

It was in the hope that a general succession could be derived from a rapid inspection of all boreholes in a particular square that the summary sheet was devised. Set out side by side, and using the colour code (Figure 10.46), seven or more borehole records can be appraised.

In the example cited (Figure 10.47), surface fill partly obliterates the original ground contours. Below is the till sequence with the Upper Brown Till A, extending down to a sensibly even level over the whole area. Within the brown till, however, sand and gravel is sporadically present but at different levels; the sands are thus obviously lenticular, as has been shown on the summary log.

Weathering of Grey Till B to give the overlying Brown Till A has penetrated to the same general level, and in some borehole records there is a variable transition zone.

Clearly, the distribution of sand and gravel lenses in the upper till can only be satisfactorily determined from the distribution of boreholes in which sand has been recorded. Each individual square cannot be considered in isolation, thus disregarding evidence from the surrounding eight squares. This is an aspect of interpretation that is even more important at the mapping stage.

Only one borehole 05/31a, has penetrated rock. Once a decision has been taken on the interpretation of the record (see Notes on Figure 10.47), this level has to be taken as representative of rockhead level over the whole square.

10.9.7 The 1:2500 plans

It is sensible to start making maps at the largest plan scale because maps at smaller scales can readily be derived by successive reduction and simplification of the necessarily very detailed large-scale plans. The basic methods of map preparation are the same regardless of scale. However, it is only at the larger scales that rocks and soils can be represented as engineering types (see subsection 2.5.2.1), and there is a limit to the scale at which lithological types can be shown.

At the 1:2500 scale, suitable cartographic methods have had to be developed and justified. These include the selection of symbols for rock and soil types, boundaries and structures. Special symbols are also needed for hydrogeological conditions, geomorphology and the indication of geodynamic processes, but the most important aspect of large-scale cartography is the scope that the scale gives for the representation of the sequence of layers in the superficial deposits and the solid rocks beneath. It is this aspect that is dealt with here.

Methods of representing multi-layer sequences

The engineering geological plans and maps that are being produced of Tyne and Wear are intended to be readily understood and interpreted by users with only a background knowledge of both geology and geological maps. Interpretation of a two-dimensional map in terms of three-dimensional geology requires training and is often a daunting prospect.

Standard method adopted

A modified single stripe system has been adopted and depending on how many layers have been recognized in a single 100 m square, the stripe may be absent, single or divided into two or three parallel stripes.

When there is no borehole or other information for a square, it is assumed that the surface layer is A: Upper Brown Till. Where the surface has not been penetrated, the stripe is omitted. If only two layers have been proved, an undivided stripe is used. The main part of the square indicates the surface layer and the stripe the second layer. Average thicknesses may also be shown against the code letter for the layer (Figure 10.48), but variations on this method have also been used, as will be shown later. For three or four layers the stripe is subdivided (Figures 10.49 and 10.50), although the three-stripe–four-layer pattern is reserved for use wherever rockhead had been penetrated; the fourth layer is thus always rock.

Applying the trench analogy, on the map the deepest layer is the southernmost one. Consequently the code letters arranged in a vertical column (Figure 10.50) can be read from top to bottom as a lithological sequence down from the surface.

Mapping sequence

A necessary preliminary stage to the preparation of engineering geological plans of the superficial deposits is the completion of both the 1:2500 Borehole Documentation Plans and the associated Data Summary Sheets for each 100 m square. Each 100 m square is analysed separately and the results are combined in a checker pattern of those squares for which data on ground conditions are available. As has already been explained, where no information is available it is assumed that the surface deposit is brown till over the whole square, unless the distribution of lithological types in adjacent squares suggests otherwise.

There are three separate stages to producing the plans which can be completed in any order. These stages are:

- preparation of rockhead contours;
- illustration of the distribution and thickness of man-made fill;
- preparation of stripe maps of the superficial deposits.

Initial versions of each stage are drawn on dyeline prints of the Ordnance Survey sheets; final versions are prepared on special transparent grids that may be used as overlays to each Ordnance Survey sheet.

Rockhead contours

Rockhead levels, determined from relevant borehole logs, are written against the borehole position. Where a borehole has not encountered rock, the maximum level of penetration is shown. Contours on the rockhead surface are then drawn in by inspection.

Wherever there is insufficient information, the analysis may be done at the 1:10 000 scale, and if appropriate the resultant more general contours may then be transferred to the 1:2500 sheets.

Fill thickness

Fill thicknesses are derived from borehole logs and also by comparing present ground contours and spot heights with those on the earliest Ordnance Survey maps. The latter approach is especially useful in the case of the infilled natural denes (see section 10.9.4).

Fill is represented by a standard pattern (Figure 10.46), with different line spacings representing the four thickness ranges: 0–2, 2–6, 6–20 and more than 20 m.

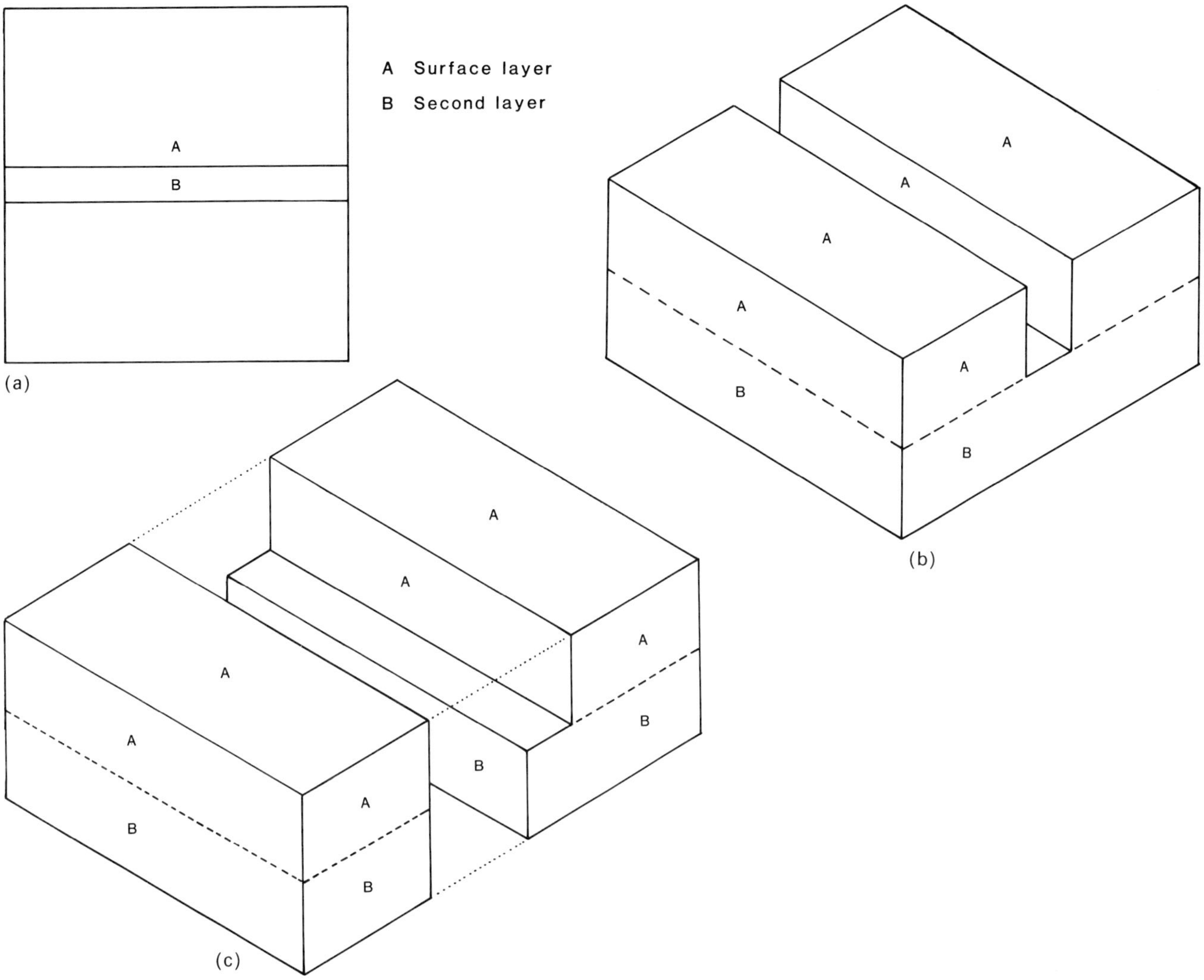

Figure 10.48 Single stripe method as applied to Tyne and Wear, UK: (a) the 100 m square; (b) block diagram showing the interpretation of (a) in three-dimensions; (c) expanded block diagram showing the stripe 'trench' more effectively than in (b)

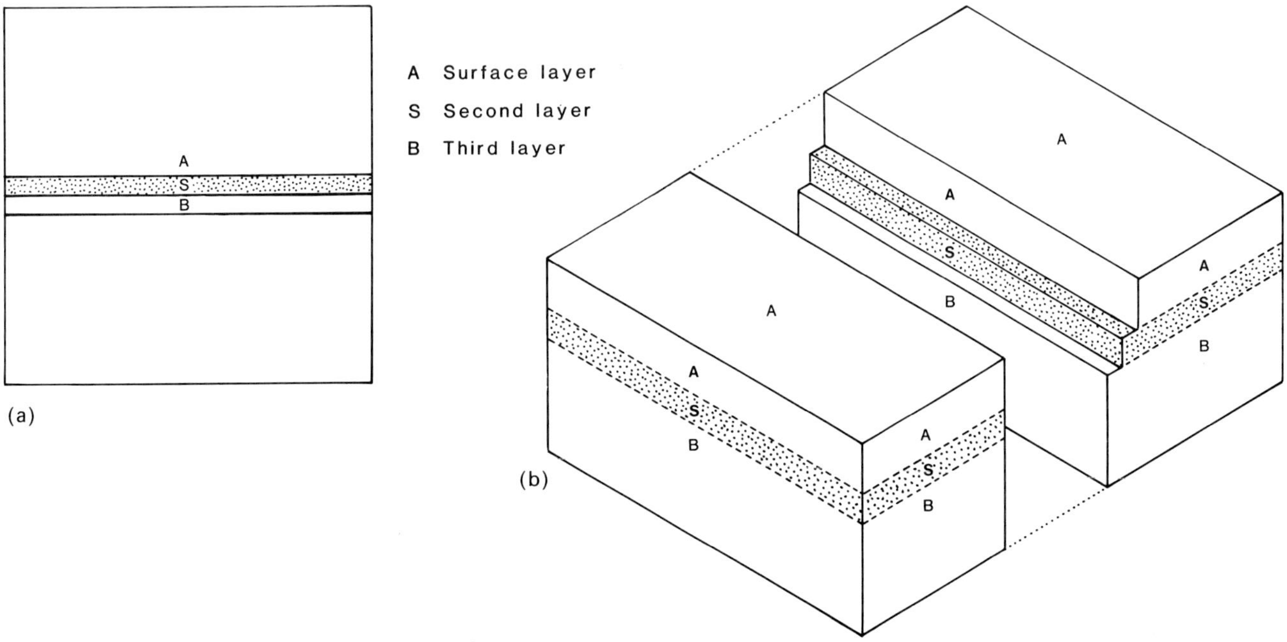

Figure 10.49 Subdivided stripe used to indicate three layers over the 100 m square: (a) the map square; (b) expanded block diagram

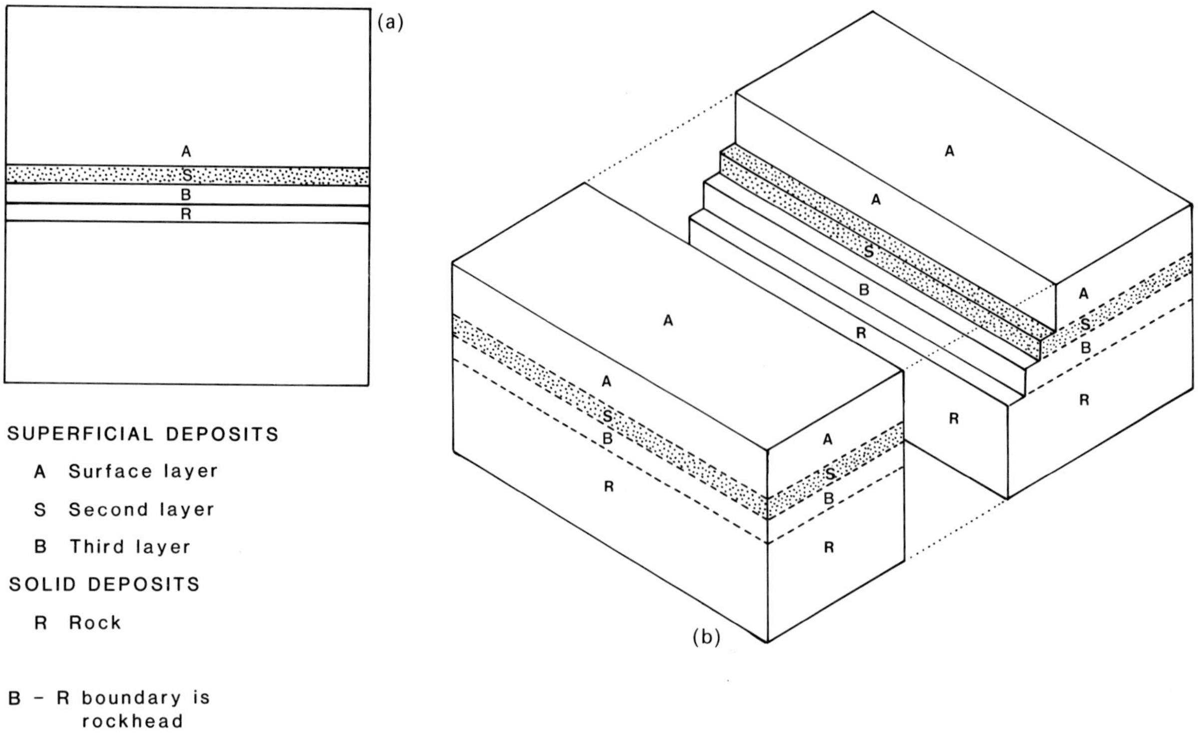

Figure 10.50 Three-stripe–four-layer pattern applied to the general conditions in a 100 m square. The lowest stripe is used for rock, so that the level of the stripe represents rockhead. (a) The map square; (b) expanded block diagram

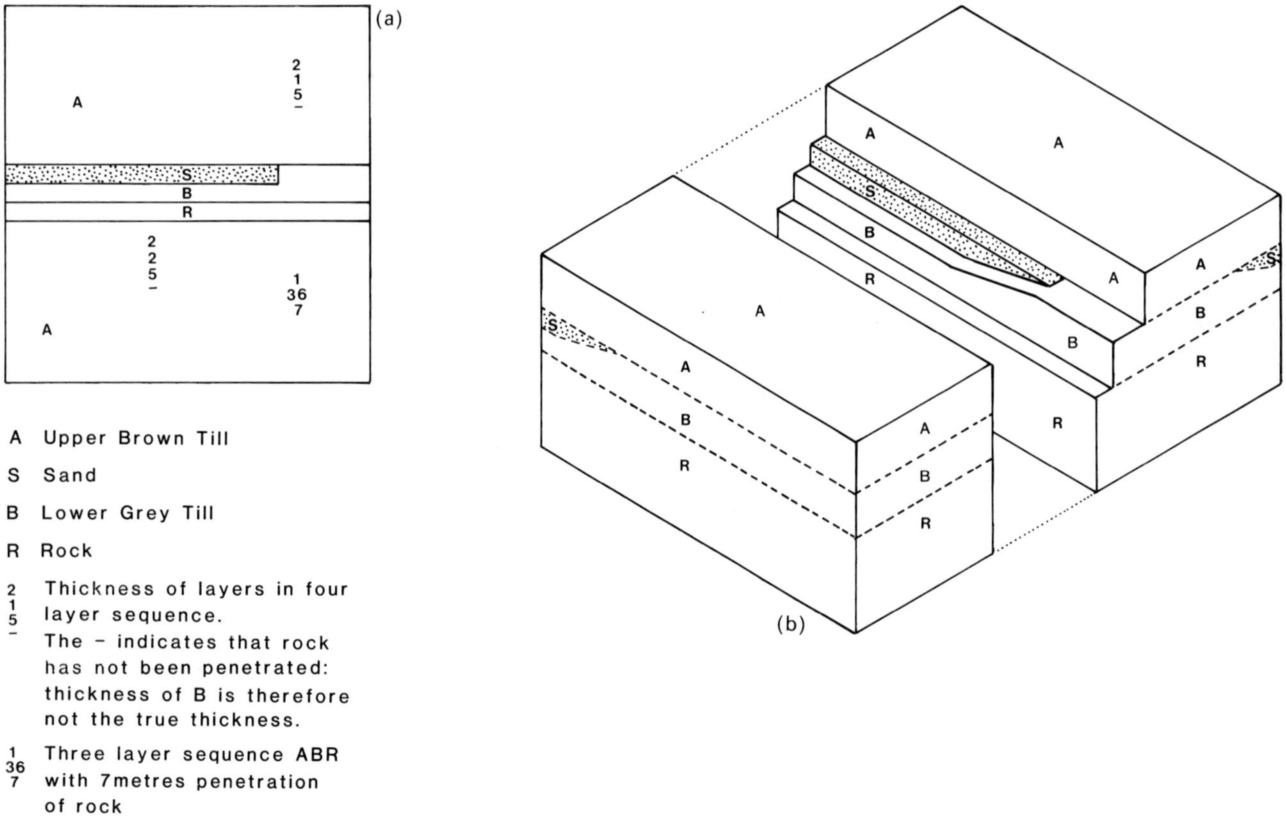

Figure 10.51 Three-stripe system with a discontinuous second layer of sand: (a) the map square with thicknesses shown; (b) interpretative expanded block diagram

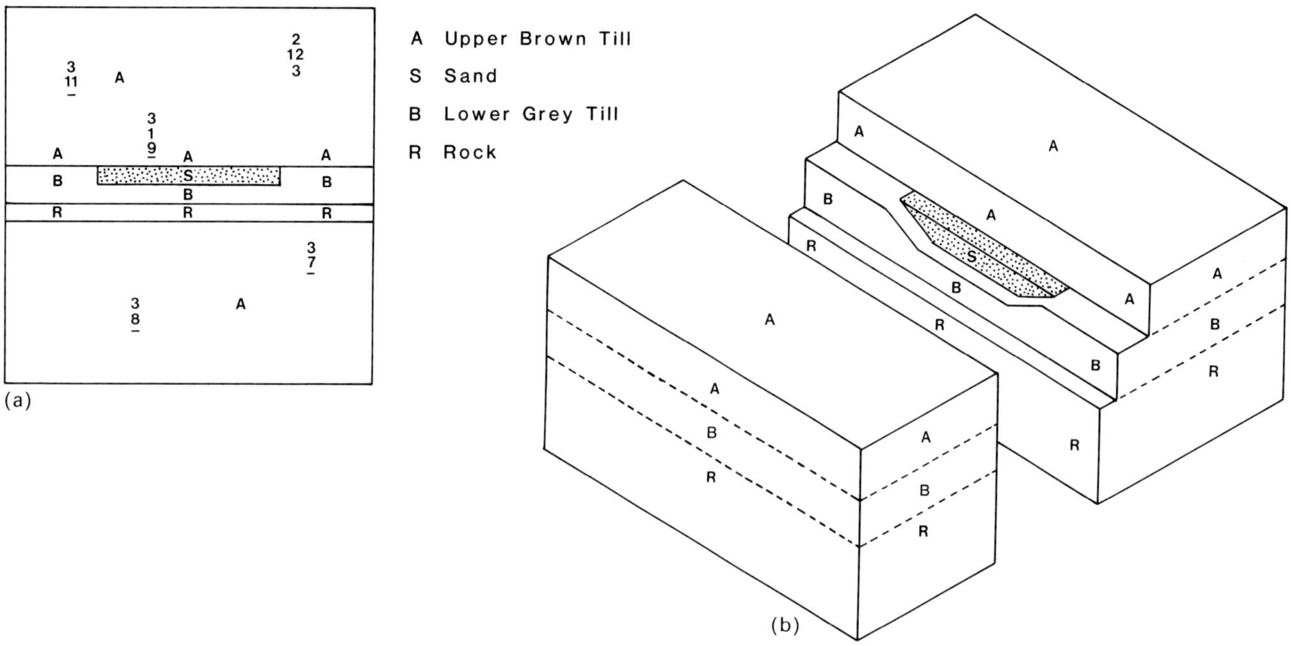

Figure 10.52 Three-stripe system with a central discontinuous layer of sand: (a) the square with variations in sequence from three to four layers and variations in thickness; (b) interpretative expanded block diagram

10.9.8 Engineering geological plans of the superficial deposits

If the sequence of superficial deposits can be summarized for a 100 m square, the lithological types in the layers are letter coded (Figures 10.46, 10.48, 10.49 and 10.50). Thicknesses of the different units, taken to the nearest metre, are numbered in descending order of layers *over the borehole location*. It should be remembered that the last number only indicates the depth of penetration into the lowest layer, and not the thickness of the layer.

Only four layers, including rock, can be dealt with satisfactorily and consequently there are difficulties if the summary log contains more layers. For example, in Figure 10.47 the Upper Brown Till apparently contains sand lenses at two distinct levels within a depth range of 3.5 m. It is therefore necessary to devise variations in the standard stripe method in order to accommodate these more complex successions.

Some of the cases that had to be considered follow. In practice the standard method of representation of the the layered sequence (Figure 10.48) can be used with certainty only when there is one borehole in a square. It is then assumed that brown till is present at the surface and in the example illustrated is underlain by grey till over the whole square. The same criterion should be applied regardless of what lithological type occurs at the surface and as the second layer.

With two or more borehole records there may be different lithological types at the surface, and indeed different second and third layers beneath. Where there is ample coverage by boreholes, boundaries between two or more lithological types at the surface can be drawn in by inspection. Since the surface layer is generally brown till, other lithological types can be expected to occur as closed areas within one square or a number of adjacent squares (e.g. map square 248652, see Figure 10.54). An arbitrary decision has been taken that a single lithological type extends for a radius of 20 m around a single borehole (e.g. map square 251653, Figure 10.54).

Variations in lithology in the second layer can also be shown. Two cases with a discontinuous sand layer beneath the surface layer are illustrated in Figures 10.51 and 10.52. In the first, sand is missing from the south-east corner of the map square. Sand, as a second layer, therefore does not extend over the whole length of the stripe, and from west to east a four-layer sequence passes into a three-layer one. On the block diagram (Figure 10.51b), the sand layer is shown wedging out in the centre of the diagram. From additional localized borehole information on the map square (Figure 10.51a), the distribution of sand in the second layer can be inferred for the southern and eastern faces of the block diagram (Figure 10.51b).

The second example (Figure 10.52) shows a central four-layer sequence, with sand as the second layer, surrounded by a three-layer sequence without sand. It is not possible to determine the precise distribution of the sand lens, except that it is circumscribed by four boreholes in which brown till is directly underlain by grey till. Bearing in mind the large scale of the plan, a suitable degree of accuracy is attained if a strong indication is given of sand as a second layer in the centre of the map square; the short central sand strip achieves this.

Figure 10.53 is given as a final example of the complexity that may be present in a single square: fortunately it is not common. Thickness sequences are placed over the relevant borehole locations and each succession is different. The block diagram (Figure 10.53b) is drawn to scale as an aid to understanding the map square (Figure 10.53a). There are three boreholes: a central one

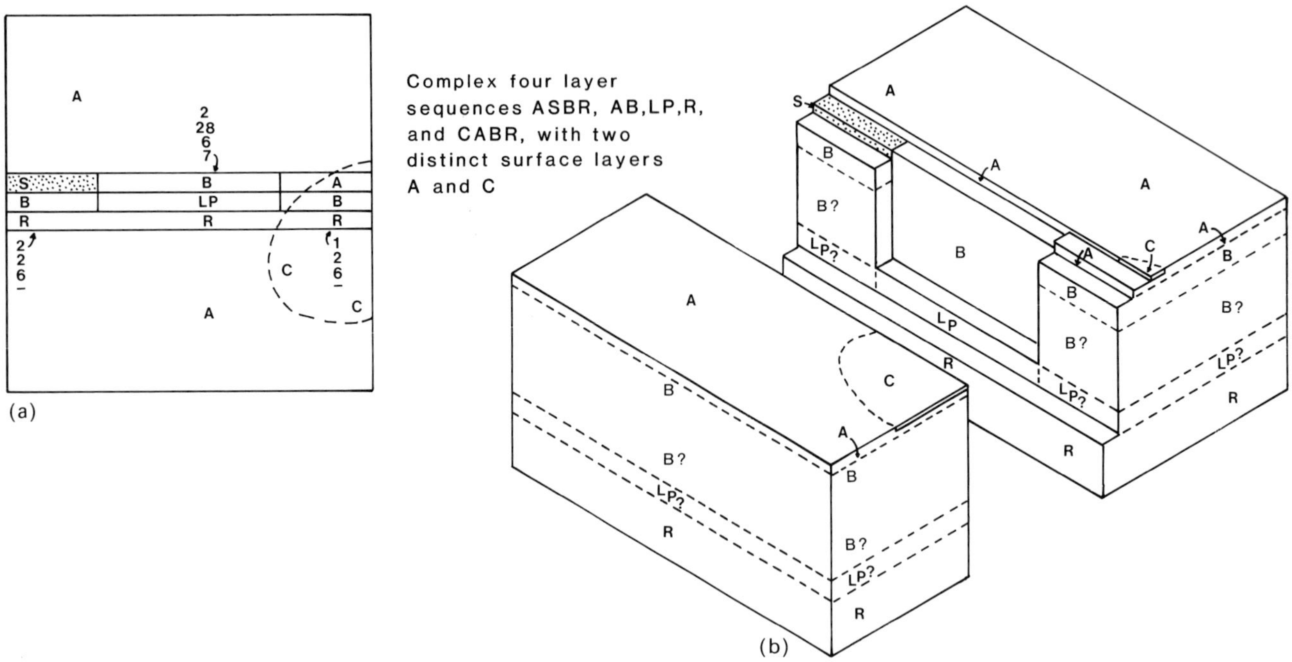

Figure 10.53 Complex four-layer sequences with two distinct surface layers: (a) the map square showing the distribution of soils at the surface together with local successions and thicknesses; (b) interpretative expanded block diagram

very deep and penetrating rock; and two relatively shallow ones towards the margins of the square. Three different complex four-layer sequences have to be related if possible. A sequence CABR, with R not penetrated, on the eastern side of the map square indicates a local occurrence of clay at the surface underlain by a normal sequence of brown till resting on grey till (normal so far) resting on 'laminated' peat on rock. Another type of normal sequence (cf. Figure 10.47) was revealed by the western borehole: brown till, sand, underlain by grey till.

In interpreting the likely subsurface conditions over the whole square, it has to be assumed that the single depth recorded to rockhead applies to the whole area. Whether or not the laminated peat is of local occurrence, it is probably safer to assume that the distribution is that recorded by L in Figure 10.53(b). Grey till is most likely to be very thick over the whole area. At or about the interface between the brown and grey tills, sands can be expected and the single record should be sufficient to warrant caution in interpreting near surface conditions over the area.

Specimen plan

Twenty 100 m squares are shown in Figure 10.54 which is presented as an example of an engineering geological plan at a scale of 1:2500. For three squares there is no information and it is therefore assumed that brown till crops out. For 11 squares there is some evidence for depth to bedrock, recorded in the appropriate squares by the southernmost R-stripe. Based on this evidence and that from adjacent areas, contours have been drawn on rockhead at 30 and 40 m AOD.

The distribution of three distinct lithological types at the surface is clearly shown. There are two isolated areas of clay that have been dealt with in the arbitrary way already discussed. Larger areas of clay also occur, and an anomalous area of grey till which is usually present as the second or third layer down. The discontinuous nature of sand lenses beneath the brown till, and occasionally in the grey till, can also be appreciated without difficulty.

Over the map area, the stripe method of conveying an impression of variations in lithological types with depth has coped quite adequately with the soil types recorded in the site investigation boreholes. Continuous single, double and triple stripes are easily interpreted by reference to the block diagrams in Figures 10.48–10.50. Lateral changes from single to double stripe, or double to triple stripe, have also been illustrated (Figures 10.51 and 10.52).

Vertical columns of numbers give examples of local thicknesses of two-, three- and four-layer sequences, read in the same vertical order as the adjacent letter codes for lithological types. It will be recalled that the amount of penetration, and not thickness, is given for the lowest layer. Where rock has been recorded in a particular square, but has not been penetrated in a particular borehole for which thicknesses are given, this is indicated by a bar at the lower end of the column of numbers. This ensures that the appropriate lithological sequence can be determined by inspection; the thickness of the lowest layer recorded is then only the depth penetrated and not the true thickness of that layer.

General comment on the 1:2500 plans

Preparation of large-scale plans of Tyne and Wear County was undertaken as an experiment in engineering geological cartography. Although the plans are available in manuscript form, most use has been made of the associated documentation plans which provide ready access to the established data bank of site investigation records, providing the raw geological and geotechnical data. The direct route to original records is most likely to be taken by the engineer, who will then make his own

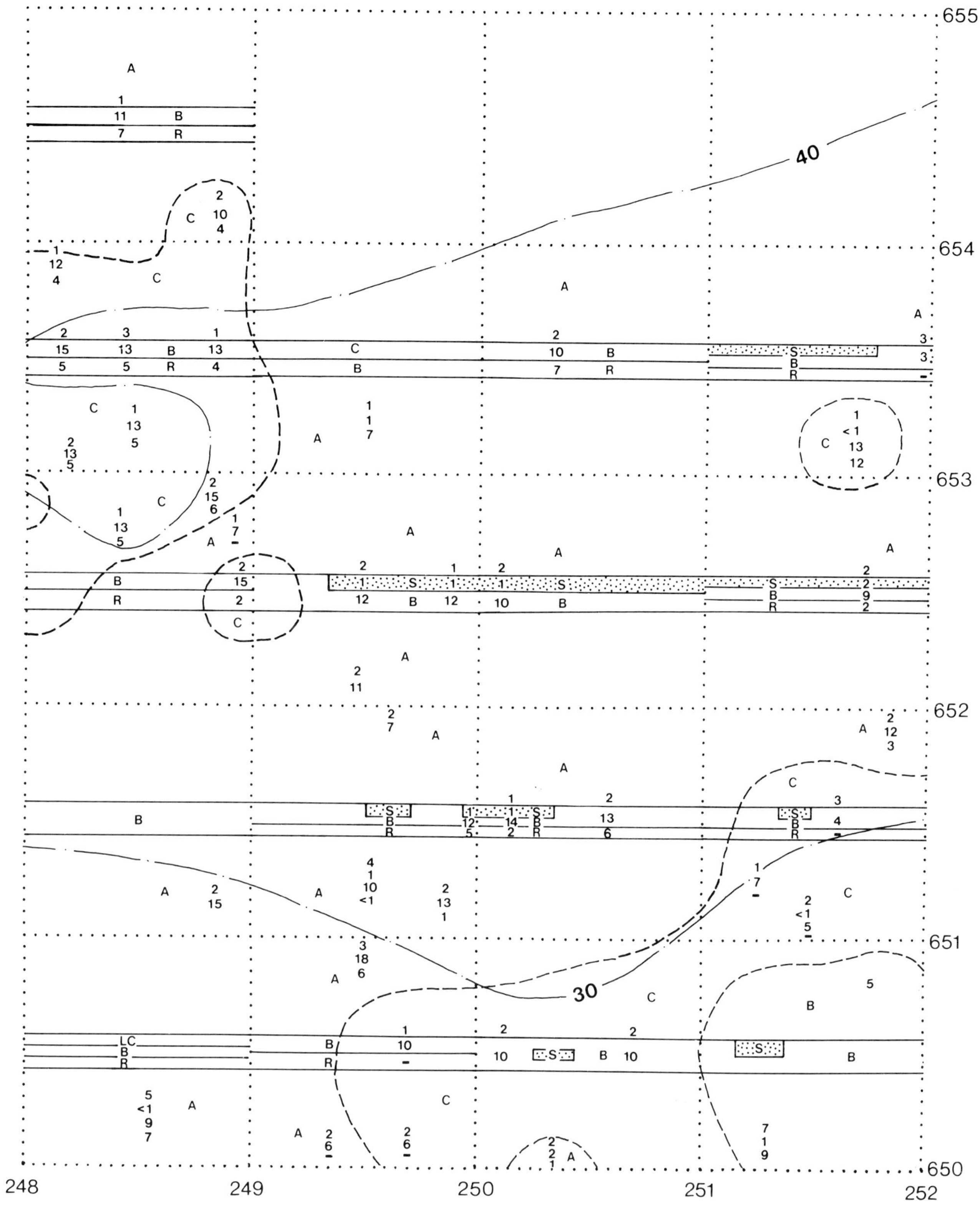

Figure 10.54 Part of the 1:2500 geological plan on Ordnance Survey Plans NZ 2465 and 2565 covering the same area as the documentation plan (Figure 10.44) published in Strachan and Dearman (1982). Contours are drawn on rockhead at 30 and 40 m AOD

initial appraisal of ground conditions. For the planner, usually concerned with areas larger than a specific construction site, the scale of the plans is too large and the consequent impression of ground conditions much too detailed and difficult to interpret.

The flow sheet (Figure 10.55) for compiling the engineering geological plans shows that the next stage in producing, eventually, documents at the map scale, i.e. 1:10 000 and smaller, should be direct reduction from the 1:2500 map plans. In this there is international agreement (UNESCO/IAEG, 1976, p. 15) even though experiments at reduction, some generalization, and the use of the stripe method are considered to have been less than successful (Strachan, 1982).

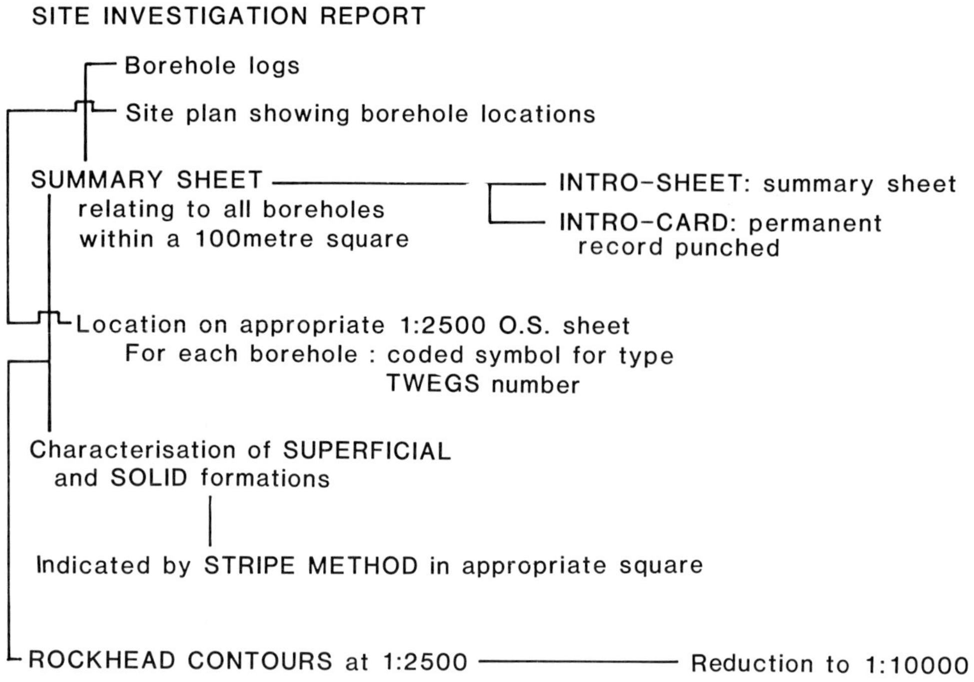

Figure 10.55 Flow sheet for compiling engineering geological mapping from the Tyne and Wear data bank, UK (TWEGS system)

Problems encountered are mainly related to:

- variations in the surface layer, which can be handled by conventional geological cartography;
- the presence of discontinuous sand layers at differing levels in the tills.

These sand layers are of particular importance in assessing engineering conditions, as usually sand is water-bearing and is associated with normally consolidated laminated clays and silts. Sand layers, and the laminated clays and silts, are geotechnically the most significant aspect of the till succession, leading in conjunction with vertical fissuring in the tills to a marked propensity to instability of the walls of excavations.

Obviously a simpler approach is necessary to prepare satisfactory 1:10 000 plans still using, if appropriate, the stripe method.

10.9.9 The 1:10 000 geological plans

Plans at the 1:2500 scale were designed to complement the 1:10 000 geological plans being produced at the same time by the British Geological Survey as part of their basic 1:10 000 remapping for the Newcastle 1:50 000 sheet. As such the data bank, more extensive than that held by the survey, was made available to them. Reduction of the 1:2500 plans (Figure 10.54) to 1:10 000 by Strachan (1982) appeared in the same year as manuscript geological maps of central Newcastle, Sheet 26 NW (Figure 10.24). The geological approach can therefore be compared with a type of engineering geological map.

Geological survey 1:10 000 maps

As the final output of the British Geological Survey revision or primary geological mapping, now carried out at a scale of 1:10 000 (1:10 560 formerly), it is customary to prepare maps on transparent film from which dyeline copies can be produced on request. For Sheet 26 NW the following maps were prepared:

- Map of general information (mainly site investigation boreholes);
- Rockhead contour map;
- Drift map (details of the superficial deposits);
- Solid and drift map

The 1 : 10 000 stripe map

After some experimentation, it became clear that the reduction from 1:2500 to 1:10 000 scale could not be achieved by using the single, divided, horizontal stripe. Accordingly, two stripes, horizontal for the second layer and interrupted vertical for the third layer, were used. The resultant plan (Figure 10.56) was drawn. On the plan are shown rockhead contours, and the boundaries of the surface and second layers. Thickness ranges are coded I, II and III.

It has to be admitted that although qualitative appreciation of geological conditions can readily be worked out at a single point on the plan, a full appreciation of the geology demands the drawing of cross-sections along N–S and E–W directions.

10.9.10 The 1:50 000 geological maps

The first maps of the whole of Tyne and Wear County produced by engineering geologists with engineers and planners in mind were drawn at a scale of 1:50 000 and are reproduced here at 1:100 000. They are, therefore, medium-scale geological maps, or equally effective small-scale maps as reproduced, showing only a broad distribution of the main lithological formations in the solid rocks (Figure 10.57) and the general distribution of the superficial deposits, alluvium, till and made-ground

(Figure 10.58). Despite lacking most of the detail of the modern geological maps prepared by the British Geological Survey, it was nevertheless considered that they would be of use as a broad indication of variations in ground conditions for site investigations and also for general planning purposes. The latter comment is particularly applicable to the map of superficial deposits which departs from convention in simply indicating the distribution of four thickness ranges in the general till cover.

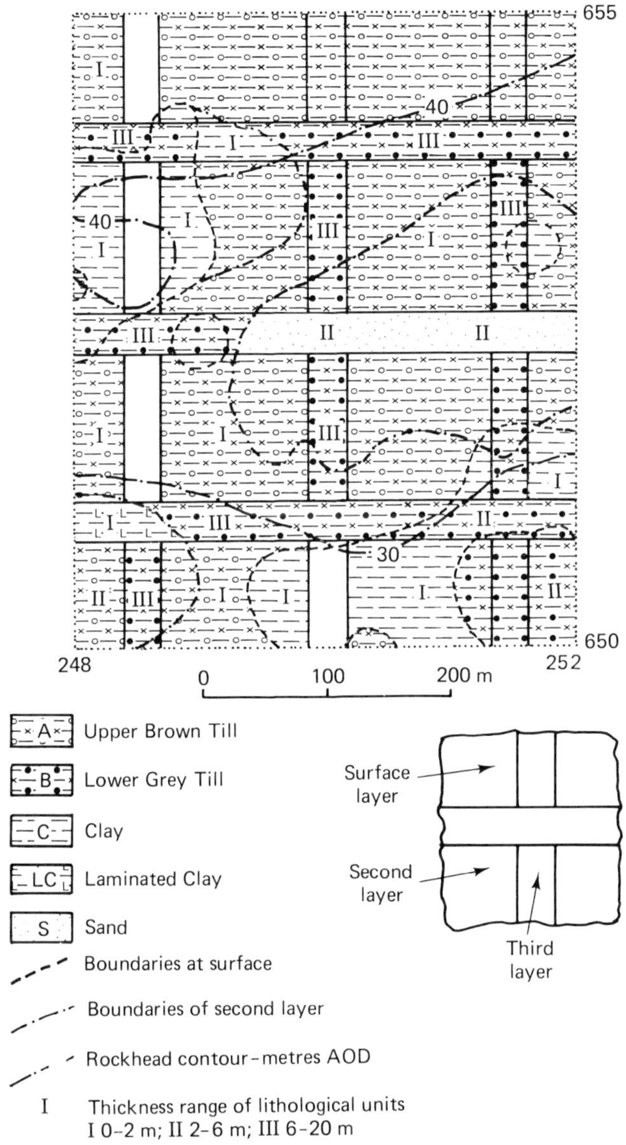

Figure 10.56 A 1:10 000 scale version (reproduced at 1:6000) of the 1:2500 stripe map illustrated in Figure 10.54. The double stripe method is used rather than the divided single horizontal stripe. Vertical stripes without ornament are rock

Geological maps are best and most easily understood when they are reproduced in colour. Single coloured copies of the 1:50 000 maps of the superficial deposits and the solid deposits have been prepared. On these the mapping units are more easily distinguishable than on the black and white versions. Both have all the boundaries for the superficial deposits, with the four thickness ranges in the tills, and also the solid formations. A combined single map would have been possible using colour, but in black and white the complexity would have been bewildering.

Map of solid formations

The solid formations comprise four lithological units – three involving Coal Measures and the fourth Magnesian Limestone. Of course, the map (Figure 10.57) represents the distributions of the different rock units at rockhead, i.e. as though the superficial cover had been stripped off.

On account of the general easterly dip of the strata, the Coal Measures below the main worked unit crop out only in the west of the conurbation, mainly flooring two deep river valleys.

It follows that the outcrop of the main worked unit of the Coal Measures is also in the west, beyond the main urban areas centred on Newcastle and Gateshead, except for the coastal area north of the River Tyne where the rocks are brought to the surface again by a broad synclinal structure.

Coal Measures above the main worked unit have the largest outcrop, flanked to the south of the Tyne by the outcrop of the Magnesian Limestone.

Map of thickness units in superficial deposits

The map (Figure 10.58) of thickness ranges in the superficial deposits is an isopachyte map derived from a combination of surface and rockhead topographic contours. A pattern of deep valleys at rockhead, more or less coincident with the present valley systems, emerges with some unexpected features.

One of the most intriguing is the general impression of the interception of the upstream part of the preglacial River Tyne valley by a combination of the Ouseburn and the Team valleys discharging southwards, and picking up the preglacial River Wear and other tributaries further south. However, the valleys at rockhead could have been moulded by subglacial streams which, because of their confinement beneath the ice, could erode a very irregular bed. The general course of the Tyne valley complex could then have been continuous to the sea beyond Tynemouth with the change of valley depth in the mid-reaches of the Tyne of little significance.

Bases for discrimination of mapping units

In the solid formations, the primary distinction is between the Upper Carboniferous Coal Measures and the Permian strata. No distinction is made between the Yellow Sands and the Magnesian Limestone comprising the Permian strata and neither are the various limestone types separated into lithological types (Dearman and Coffey, 1981). Although the Yellow Sands have important engineering geological properties, the outcrop is small in relation both to the extent of the rest of the Permian and the scale of the map.

For the distinction of mapping units in the Coal Measures, it has been expedient to revert to the earlier (Eastwood, 1935) practical subdivisions rather than to use the more formally correct, modern stratigraphical divisions (Taylor *et al.*, 1971). Separation of a middle unit containing the main worked coal seams, the older Productive Coal Measures formation, leaves an overlying unit affected indirectly by coal mining and an underlying unit that is largely unaffected by mining.

Local areas of fill, including both landfill and restored open-cast coal mines, and alluvium, are shown on the maps, but they are included in the separation of the glacial deposits into four subunits discriminated on the

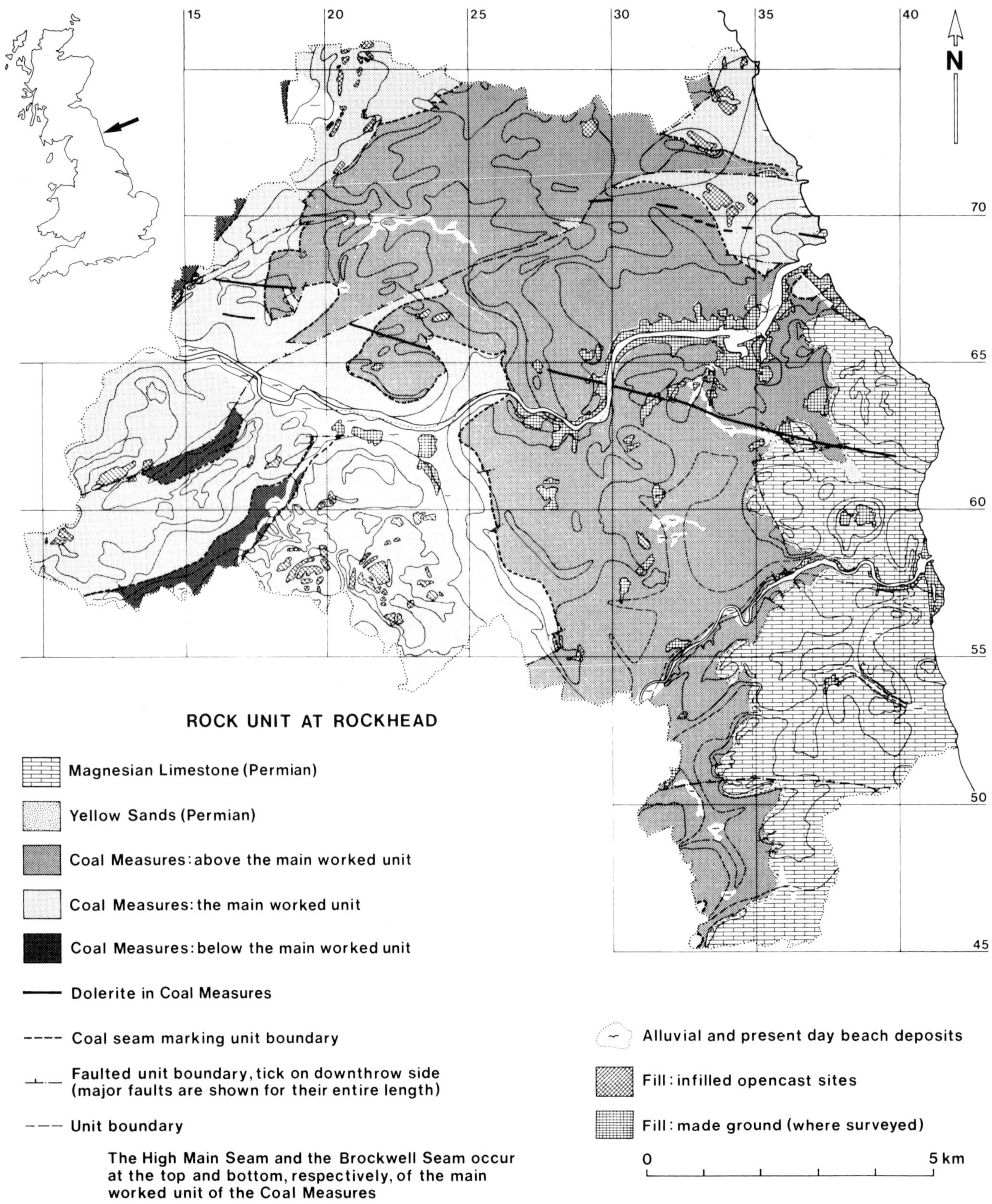

Figure 10.57 The 1:50 000 geological map of Tyne and Wear, UK, reproduced here at approximately 1:100 000, showing the distribution of the main lithological units in the solid rocks

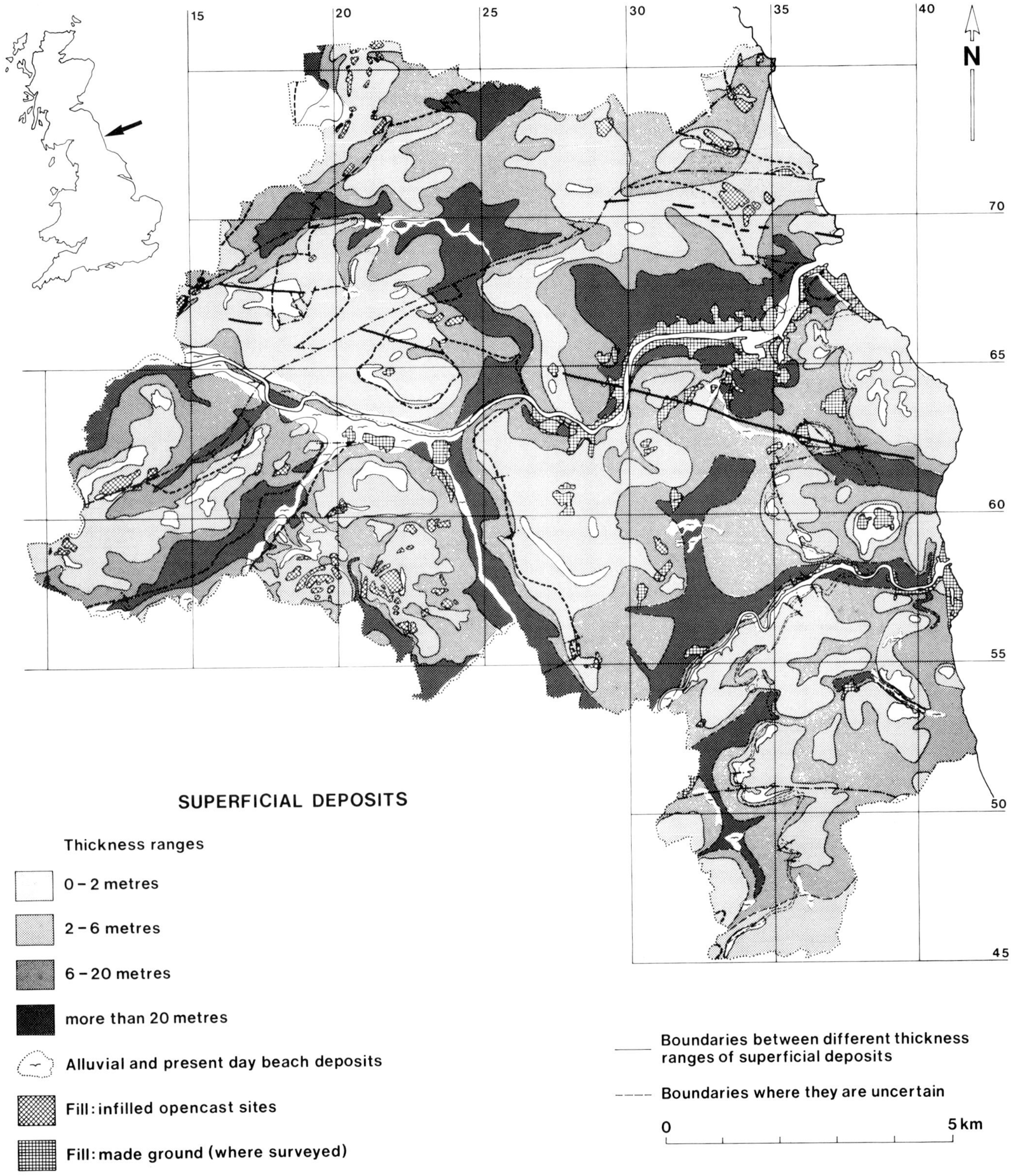

Figure 10.58 The 1:50 000 engineering geological map of Tyne and Wear, UK, reproduced here at approximately 1:100 000, showing the distribution of the superficial deposits in four thickness range categories

basis of thickness ranges. Further subdivision is not considered practicable for the present scale of map, although as has been mentioned earlier, there are horizons of normally consolidated laminated clays, silts and sands in the boulder clays that are of particular engineering interest (Figure 10.26).

Mapping units

Distinctive mapping units are based on combinations of classes of thickness of superficial deposits and the character of the solid formation at rockhead, to give 16 mapping units if the alluvium and fill are grouped with the glacial deposits. Categories of thickness ranges adopted for the superficial deposits are given in Table 10.7. The lithological divisions within the solid formations are listed in Table 10.8. The 16 possible combinations of superficial deposits and the underlying lithological complexes are shown in Table 10.9.

Table 10.7 Classes of thickness ranges adopted for mapping units in superficial deposits

Class	*Thickness range* (m)	*Term*
1	0–2	Thin
2	2–6	Medium
3	6–20	Thick
4	20	Very thick

Table 10.8 Classes of lithological divisions in solid Coal Measures and Permian deposits

Class	*Lithological divisions*
a	Coal Measures below the Main Worked Unit
b	Coal Measures: Main Worked Unit
c	Coal Measures above the Main Worked Unit
d	Magnesian Limestone including basal sands

Table 10.9 The 16 mapping units resulting from combining Tables 10.7 and 10.8

1a	1b	1c	1d
2a	2b	2c	2d
3a	3b	3c	3d
4a	4b	4c	4d

Each of the 16 possible combinations of superficial and solid deposits presents distinct engineering geological conditions based on such factors as economic exploitation of superficial and solid deposits and consequential changes related to such exploitation. These are generally indicated in Figures 10.26 and 10.28 and may be listed as:

1. Quarrying of limestones, sands, sandstones, shales and mudstones.
2. Open-pit (open-cast) working of coal seams and ironstones in Coal Measures.
3. Bell pits to near surface coal seams, ironstones.
4. Adits to near surface coal seams, ironstones, fireclays.
5. Shafts to deep coal seams, ironstones, fireclays.
6. Clay pits in glacial deposits.
7. Sand and gravel workings in glacial deposits.
8. Sand and gravel workings in alluvial deposits.
9. Reclamation of open-pit mines by mining waste.
10. Partial infill of quarries with overburden and quarry waste.
11. Selective infill of sand and gravel workings with fines produced by washing, settled in lagoons.
12. Partial or complete infill and reclamation of quarries, clay pits, sand and gravel working with industrial and household waste.
13. General fill from urban demolition and renewal.
14. Pillar-and-stall workings in coal seams.
15. Longwall working in coal seams.
16. Collapse of workings in (3), (4), (14) and (15) and associated subsidence features which may affect the surface.
17. Settlement of fill.

Justification of selected thickness ranges in superficial deposits

In terms of foundation conditions for civil engineering structures, thin superficial deposits may be ignored, as they would be removed before construction began, and may thus be accepted for practical purposes as rock outcrop. The older, central, urban areas are likely to have a general thin cover of fill, the accumulation of repeated urban demolition and redevelopment. Allowable bearing capacities of standard foundations will be those related to lithological type, and, where various rock types are present at rockhead, assessments will have to be made in terms of individual lithological types and associations of more than one type.

Medium thickness range deposits imply that it may be feasible and economic to excavate to rockhead as a foundation. Thick deposits could, in the main, be dealt with by end-bearing piles socketed in underlying rock, whereas in very thick deposits friction piles would provide an economic foundation design. In the latter case, the presence of rock at depth would be of little practical significance, except for example in deep tunnelling operations.

Areas of fill

Former open-cast coal-mining sites

All of the larger sites shown in Figure 10.58 are, with a few exceptions that will be discussed later, on outcrops of the Main Worked Unit of the Coal Measures, where the superficial deposits have thickness ranges in Classes 1 and 2. A few sites either overlap onto, or are within, the Class 3 thickness range. Typical of exceptions are two areas of fill shown on the Town Moor in the centre of Newcastle upon Tyne (GR 2466), located on Coal Measures above the Main Worked Unit at sites where the High Main seam was extracted.

These areas of fill obviously always rest on bedrock. A general sequence of working an open-cast coal site will give some idea of the likely distribution and physical conditions of the fill materials. In order to make the initial excavation, superficial materials will have been stripped from at least part of the site and dumped in the vicinity. Coal Measures will have been treated in the same way for the initial cut down to the coal seams. Once this has been done, the next and succeeding panels of rock are drilled and blasted if necessary, and then removed, usually with a dragline, and dumped in the worked-out area. The fragmented material accumulates

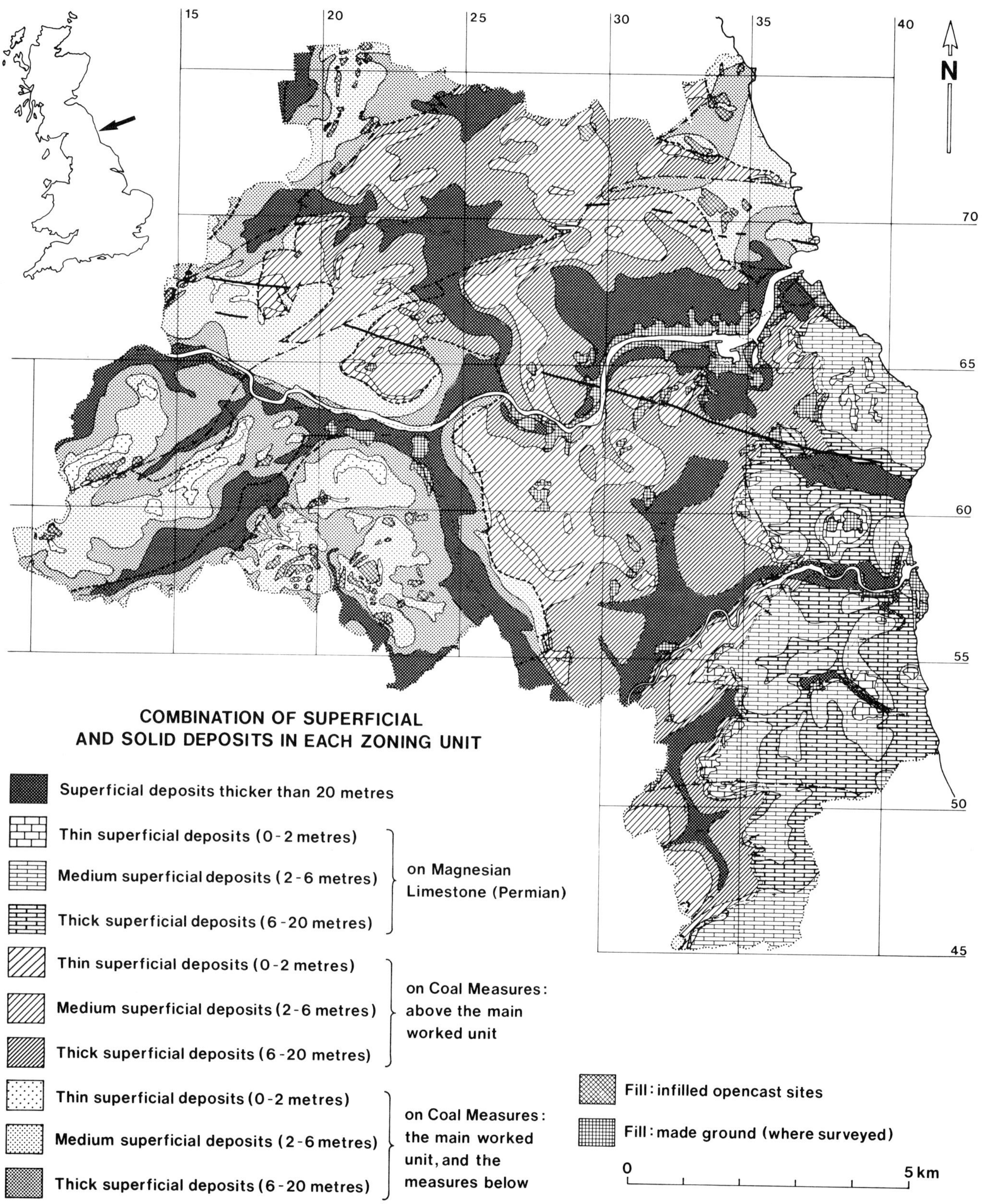

Figure 10.59 Zoning map of Tyne and Wear County, UK, showing the distribution of the 10 zoning units

THIN SUPERFICIAL DEPOSITS MEDIUM SUPERFICIAL DEPOSITS THICK SUPERFICIAL DEPOSITS ON COAL MEASURES ABOVE THE MAIN WORKED UNIT

Thin superficial deposits

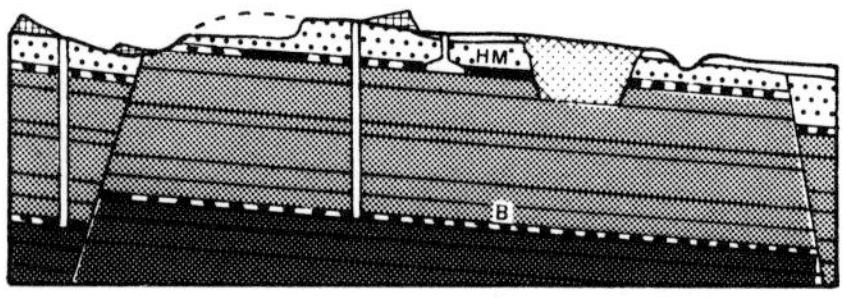

SUPERFICIAL DEPOSITS

Up to 2 m of weathered and fissured boulder clay, general fill in old urban areas; alluvium.

Sandstone and other rocks at outcrop may have been fragmented by freeze-thaw and/or weathered, grading down into rock in place.

SOLID DEPOSITS

Massive sandstones, thinly bedded siltstones, shales, mudstones and possible thin coals with seatearths.

Rock may crop out.

MAN-MADE FEATURES

Massive sandstones quarried for building stone; shales worked for brick and tile manufacture.

Quarries partly filled with quarry waste during working life, subsequently used as landfill sites.

Shallow coals in Main Worked Unit beneath may have been exploited by bell-pits, adits and shafts for which there are inadequate records.

High Main Seam has been worked opencast.

Deeper coals in Main Worked Unit beneath worked from shafts from the surface; attendant surface tips and capped shafts.

Solid deposits may be affected by subsidence due to mining; effects may reach the surface.

Collapse of old workings and consequent subsidence may still be taking place.

Danger of gas in old workings and in ground above fractured by subsidence.

DISTRIBUTION

Isolated hilltop areas where massive sandstones crop out, for example the High Main Sandstone overlying the High Main Seam.

Medium superficial deposits

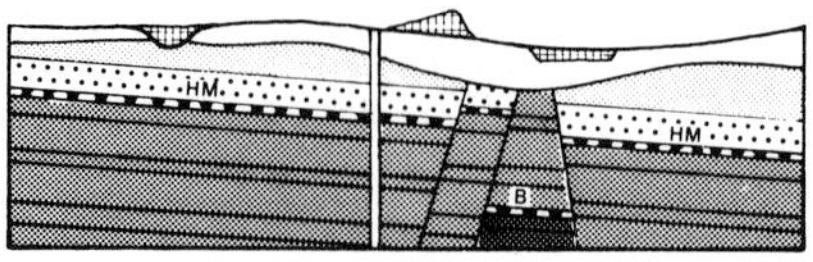

SUPERFICIAL DEPOSITS

From 2 to 6 m of boulder clay weathered and fissured at the surface; outwash sands and gravels; alluvium.

SOLID DEPOSITS

Massive sandstones, thinly bedded siltstones, shales, mudstones and possible thin coals with seatearths.

At rockhead, transition to superficial deposits may be abrupt, or gradual through a disturbed zone which is incorporated into the boulder clay above.

MAN-MADE FEATURES

Clays in superficial deposits may have been worked for brick making; sands and gravels for aggregates; workings may have been infilled.

Coals in the Main Worked Unit below have been worked from shafts, with attendant surface tips and capped shafts.

Solid and superficial deposits may be affected by subsidence due to mining; effects may reach the surface.

DISTRIBUTION

Largely on the hills east of River Team south of the River Tyne, and on the hills north of the River Tyne. Also on the interfluves on rockhead in the plateau area north and south of the River Tyne.

Thick superficial deposits

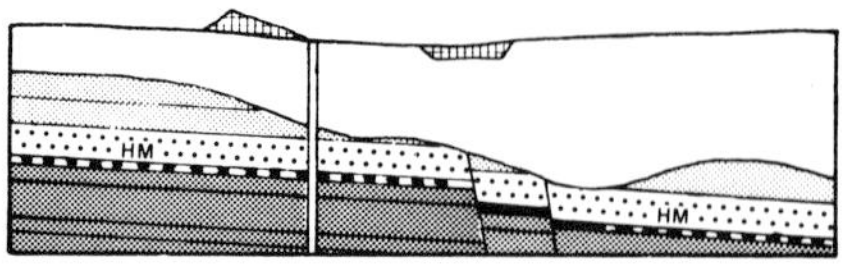

SUPERFICIAL DEPOSITS

From 6 to 20 m of boulder clay; outwash sands and gravels; alluvium.

SOLID DEPOSITS

Massive sandstones, thinly bedded siltstones, shales, mudstones, and possibly thin coals with seatearths.

At rockhead, transition to superficial deposits may be abrupt, or gradual through a disturbed zone which is incorporated into the boulder clay above.

MAN-MADE FEATURES

Clays in superficial deposits may have been worked for brick making; sands and gravels for aggregates; workings may have been infilled.

Coals in the Main Worked Unit below have been worked from shafts, with attendant surface tips and capped shafts.

Solid and superficial deposits may be affected by subsidence due to mining, but because of the thickness of solid deposits above the worked seams, and the thickness of the superficial deposits, subsidence is unlikely to affect the surface.

DISTRIBUTION

On the north-eastern slopes of the hilly areas north of the River Tyne and east of the River Team south of the R. Tyne. Elsewhere, in those regions underlain by slopes in rockhead topography leading down into the very deep preglacial valleys.

THIN SUPERFICIAL DEPOSITS MEDIUM SUPERFICIAL DEPOSITS THICK SUPERFICIAL DEPOSITS ON THE MAIN WORKED UNIT OF THE COAL MEASURES

Thin superficial deposits

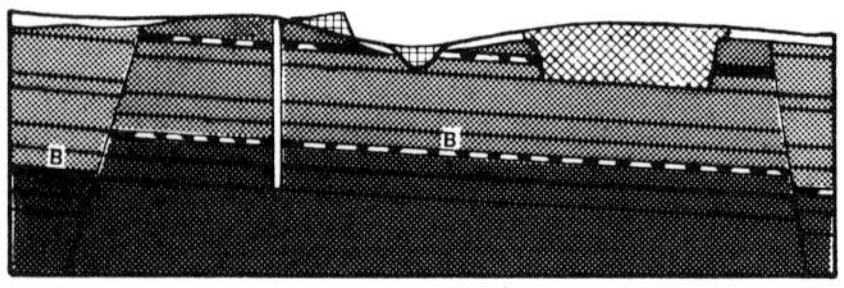

SUPERFICIAL DEPOSITS

Up to 2 m of weathered boulder clay; general fill; alluvium.

Rocks at outcrop may have been fragmented by freeze-thaw and weathered, grading down to rock in place.

SOLID DEPOSITS

Sandstones, siltstones, shales, mudstones, coal seams with seatearths, ironstone nodules in shales.

Rock may crop out.

MAN-MADE FEATURES

Shallow coal seams may have been exploited by bell-pits, adits, and shafts, for which there are inadequate records.

Near surface seams have been worked opencast, and the workings restored.

Deeper seams have been worked from deep shafts from the surface; with attendant surface tips and capped shafts.

Solid deposits may be affected by subsidence due to mining; effects may reach the surface.

Collapse of old workings and consequent subsidence may still be taking place.

Danger of gas in old workings and in ground above fractured by subsidence.

DISTRIBUTION

Small areas on the crests of the hilly region south of the River Tyne and west of the River Team.

Medium superficial deposits

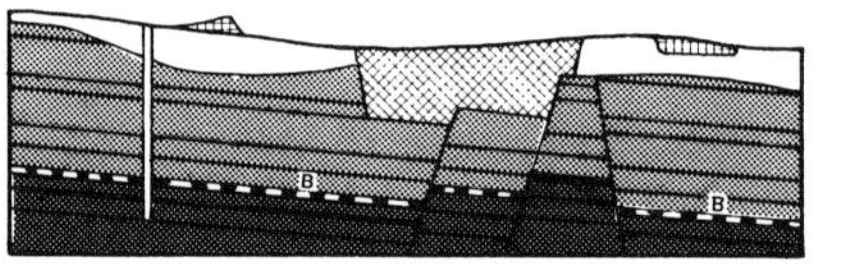

SUPERFICIAL DEPOSITS

From 2 to 6 m of boulder clay weathered and fissured at the surface; outwash sands and gravels; alluvium; blown sand on coast.

SOLID DEPOSITS

Sandstones, siltstones, shales, mudstones, coal seams with seatearths, ironstone nodules in shales.

MAN-MADE FEATURES

Shallow coal seams may have been exploited by bell-pits, adits and shafts, for which there are inadequate records.

Near surface seams have been worked opencast, and the workings restored.

Deeper seams have been worked from deep shafts from the surface; with attendant surface tips and capped shafts.

Solid deposits may be affected by subsidence due to mining; effects may reach the surface.

Collapse of old workings and consequent subsidence may still be taking place.

Danger of gas in old workings and in ground above fractured by subsidence.

DISTRIBUTION

In the west of the county, mainly underlying the higher parts of the hilly areas, and the southward facing slopes of the Tyne valley down to river level. Local distribution north of the River Tyne in the coastal area.

Thick superficial deposits

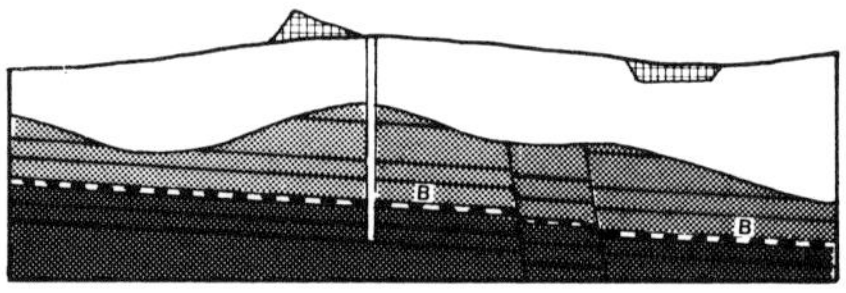

SUPERFICIAL DEPOSITS

From 6 to 20 m of boulder clay weathered and fissured at the surface; outwash sands and gravels; alluvium; blown sand on the coast.

SOLID DEPOSITS

Sandstones, siltstones, shales, mudstones, coal seams with seatearths, ironstone nodules in shales.

MAN-MADE FEATURES

Near surface seams have been worked opencast, and the workings restored.

Deeper seams have been worked from deep shafts from the surface; with attendant surface tips and capped shafts.

Solid deposits may be affected by subsidence due to mining.

Danger of gas in old workings and in ground above fractured by subsidence.

DISTRIBUTION

In the west of the county on the slopes in rockhead leading down into the deep preglacial valleys. Two local areas on the coast north of the River Tyne.

VERY THICK SUPERFICIAL DEPOSITS

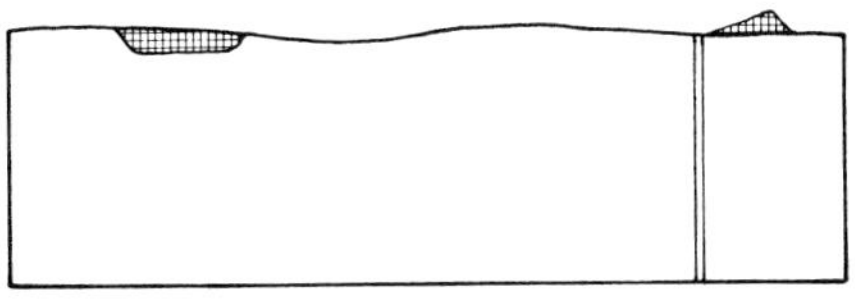

SUPERFICIAL DEPOSITS

More than 20 m of boulder clay weathered and fissured at the surface; outwash sands and gravels; alluvium.

MAN-MADE FEATURES

Clays in superficial deposits may have been worked for brick making; sands and gravels for aggregates; workings may have been infilled.

Coals beneath may have been worked from shafts, with attendant surface tips and capped shafts.

DISTRIBUTION

Along the preglacial valleys of the R. Tyne and the tributary valleys on the south bank now occupied by the Stanley Burn, Barlow Burn, R. Derwent and R. Team, and by the Ouseburn on the N. bank. Along the preglacial valley of the R. Wear and tributary valleys. A local east-west area reaching the coast north of Sunderland at Whitburn Bay.

EXPLANATION

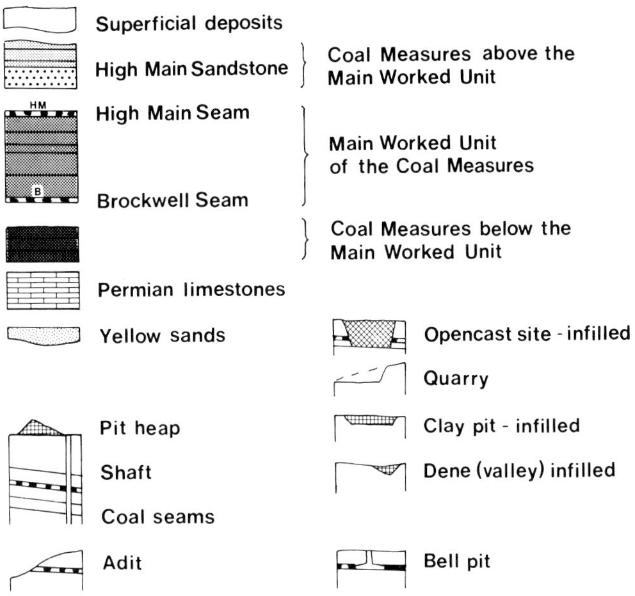

THIN SUPERFICIAL DEPOSITS ON PERMIAN LIMESTONES

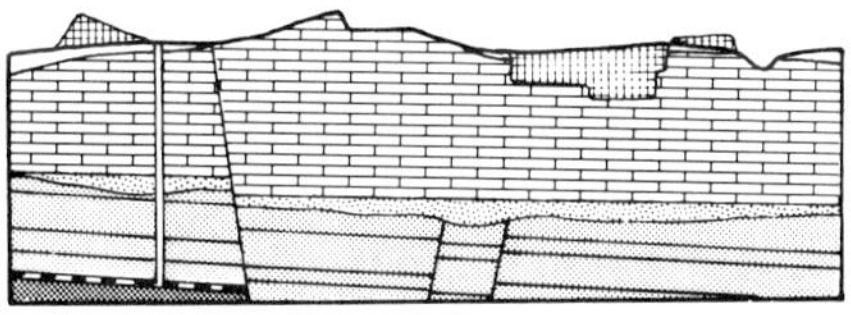

SUPERFICIAL DEPOSITS

Up to 2 m of weathered and fissured boulder clay; general fill in old urban areas; alluvium.

Limestone at outcrop may have been fragmented by freeze-thaw and/or weathered, grading down to rock in place.

SOLID DEPOSITS

A variety of types of limestone overlying Yellow Sands.

Rock may crop out.

MAN-MADE FEATURES

Limestones quarried for aggregates and for lime burning.

Quarries partly filled with quarry waste during working life; subsequently left or used as landfill sites.

Coals underlying the limestones worked from deep shafts; attendant surface tips and capped shafts.

Limestones may be affected by subsidence effects due to mining coal seams beneath; effects may reach the surface.

DISTRIBUTION

In isolated areas scattered along the western escarpment of the Permian outcrop; hills on concretionary limestones north of Sunderland, and mainly reef limestones to the south.

MEDIUM SUPERFICIAL DEPOSITS ON PERMIAN LIMESTONES

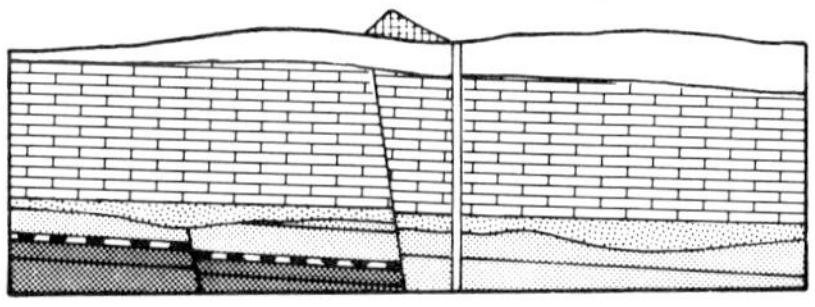

SUPERFICIAL DEPOSITS

From 2 to 6 m of boulder clay weathered and fissured at the surface; outwash sands and gravels; alluvium.

SOLID DEPOSITS

A variety of types of limestone overlying Yellow Sands.

MAN-MADE FEATURES

Clays in superficial deposits may have been worked for brick making; sands and gravels for aggregates; workings may have been infilled.

Coals beneath the limestones have been worked from shafts, with attendant surface tips and capped shafts.

Solid and superficial deposits may be affected by subsidence due to mining; effects may reach the surface.

DISTRIBUTION

North of Whitburn Bay. Between Whitburn Bay and the R. Wear from western scarp to the coast, but dissected by two N-S preglacial valleys.
South of the R. Wear, very local small area on coast, and extensive area extending east from the scarp but dissected by preglacial valleys.

THICK SUPERFICIAL DEPOSITS ON PERMIAN LIMESTONES

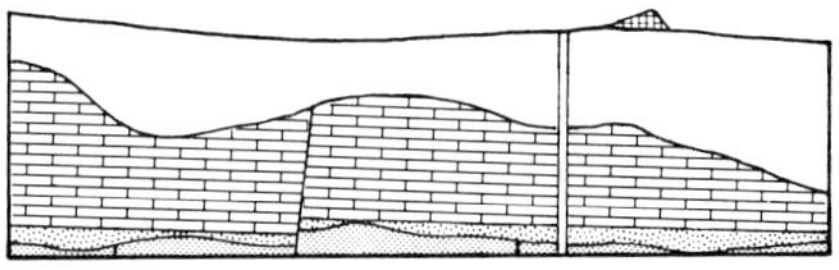

SUPERFICIAL DEPOSITS

From 6 to 20 m of boulder clay weathered and fissured at the surface; outwash sands and gravels; alluvium.

SOLID DEPOSITS

A variety of types of limestone overlying Yellow Sands

MAN-MADE FEATURES

Clays in superficial deposits may have been worked for brick making; sands and gravels for aggregates; workings may have been infilled.

Coals beneath the limestones have been worked from shafts, with attendant surface tips and capped shafts.

Subsidence effects unlikely to reach the surface.

DISTRIBUTION

North of the R. Wear, restricted to the flanks of the Whitburn Bay deep preglacial valley, and N-S connections with the N. bank of the R. Wear.
South of the R. Wear, dominating the coastal plain, and extending inland along preglacial valleys which breach the limestone escarpment in two places.

Figure 10.60 Description of the 10 units for the zoning map of Tyne and Wear County, UK loosely at the angle of repose, and remains in an uncompacted state. Progressive reclamation of the site involves replacement of glacial deposits, subsoil and topsoil. Not in every case, however, is the original sequence maintained in the fill: the fill then becomes more heterogeneous and unpredictable.

Infilled quarries
Two types of fill may be present on former quarry sites: locally derived quarry waste, often loosely tipped, and imported material which may be household and industrial wastes or soil and rock from construction sites.

Perhaps of more importance is the likelihood that the final extent of the quarry including depth may not be that shown on the map. Towards the end of the useful life of a quarry, partly worked faces may have been abandoned early, so that volumes of apparently solid rock might have been disturbed by quarrying; for example, discontinuities may be more open than in adjacent *in situ* strata.

Fill associated with land reclamation
Areas of land reclamation are associated with the valleys of the rivers Wear and Tyne, particularly in the lower reaches at their mouths. Some of the denes, particularly those in Newcastle, have been infilled (Figure 10.58).

On sloping sites, cut-and-fill with a retaining wall on the downhill side may have been used to provide a level building site.

10.9.11 Engineering geological map of the area

It is a general principle, as discussed in Chapter 2, that the map units on small-scale engineering geological maps should be firmly based on lithology and have no stratigraphical implications. If this had been adhered to in compiling Figure 10.57, some important aspects of the engineering environment would have lacked very necessary emphasis. By far the most obvious example is the presence of worked Coal Measures beneath the Magnesian Limestone outcrop. It might be said, with an element of truth, that engineering works in the limestones are all relatively shallow; foundation excavations, road and other cuttings are examples. Even so, some of the unpredictable near surface and engineering problems are caused by deep mining (see section 10.9.3), whereas others are due to natural geological causes.

Division of the Coal Measures into three lithostratigraphical units is equally expedient because the broad stratigraphical sequence with a middle sandwich of worked coal seams has important engineering implications for the areas where each unit crops out. Here, it is the man-made, or at least man-induced features that are important in assessing likely engineering geological conditions.

Clearly a different approach to mapping is needed if the result is, uniquely, an engineering geological map. A version of the zoning principle (see section 7.1) has to be applied, in which the map units are not lithostratigraphical.

Map of engineering geological zones
Excluding alluvium and the various kinds of fill, there are 16 distinct combinations of ground conditions involving the superficial cover and the solid rocks immediately beneath (see Table 10.9). For practical purposes this number can be reduced to 13 if the total thickness range being considered is kept to 20 m. Superficial deposits greater than 20 m thickness then constitute a single unit regardless of the underlying lithological type in the solid formation.

Again, because the outcrop area is so restricted, it is tempting to group the main worked unit in the Coal Measures and the unit below, thus reducing the mapping combinations to 10: three solid formations covered by three thickness ranges of superficial deposits plus the superficial deposits thicker than 20 m. In justification of this, it can be pointed out that there are unimportant seams beneath the Brockwell Seam that have been worked. In addition, large parts of the two outcrops of the unit below the main worked measures at the WSW margin of the area, to the south of the Tyne, are covered by more than 20 m of superficial deposits and are thus excluded from individual consideration on the zoning map.

The zoning map, with 10 zoning units and two types of fill, is reproduced in black and white as Figure 10.59. As a map it is easy to understand, if it is remembered that each pattern represents a specific set of ground conditions down to 20 m depth, involving superficial cover and where appropriate the solid rocks beneath. Schematic cross-sections show typical geological relations for each zoning unit (Figure 10.60), and do not need further comment.

Distribution of zones
Notes on the distribution of the zones are given in Figure 10.60, so that here only general comments need be made. The most obvious feature of the area is the very limited outcrops of the solid deposits. There are only sporadic outcrops along the discontinuous western escarpment of the Magnesian Limestone. Between the escarpment and the sea, the reef limestones are exposed in conspicuous conical hills (Maiden's Paps). On the Coal Measures, isolated hills rise from the coastal plain (Figure 10.27) and on these the drift cover is very thin or absent. Similar conditions are found on the high hills in the west and south-west of the area.

Preglacial topography has determined the distribution of superficial deposits of medium thickness on flat-topped interfluves, with progressive increase in thickness into the deep preglacial valleys.

General comment
Ground conditions in the Tyne and Wear conurbation are determined by:

- variation of thickness of the general cover of glacial deposits, dominated by a preglacial valley system infilled with more than 20 m of superficial deposits;
- the outcrop of the Permian Magnesian Limestones which have been extensively quarried, and undermined for coal;
- the preponderant outcrop of Coal Measures above the main worked unit, with attendant difficulties associated with surface subsidence due to coal workings and with the uncertain location of old coal shafts;
- the western outcrop of the main worked unit with extensive open-cast mining sites and deep mining.

General geological maps have been prepared to show the distribution of the solid deposits, and the variations of thickness of the superficial deposits. A map has also been prepared showing 10 zones in which various thickness ranges of the superficial deposits are associated with the four lithological units in the soil formations. The latter can lay claim to being an engineering geological map, whereas the others are geological maps.

10.9.12 Future of urban engineering geological mapping

Geological mapping in an urban environment is usually undertaken at two quite specific levels. There is, on the one hand, mapping at the plan scale which has two facets:

1. Recording of ground conditions at an individual engineering site that may be either part of a site investigation, or an investigation aimed at problem solving.
2. Preparation, mainly from available archival information, of a standard set of plans including geological and documentation, backed up by a data bank, mainly to provide engineers with a detailed background into ground conditions as an aid to site investigation.

Both aspects tend to be complex, demanding a high level of combined geological and engineering expertise for their effective understanding and practical use. Generally, and even if the geological map is a conventional 'Geological Survey' type of compilation, the significance of variations in ground conditions has to be established by detailed study.

On the other hand, maps at the medium scales of 1:25 000 and 1:50 000 may be prepared either by direct reduction from detailed 1:10 000 plans or *de novo* from all available information, with a particular output in mind. With the latter approach it is possible to produce maps of direct application to planning problems. Such maps are variations of zoning maps, for example those discussed in Chapter 7 and illustrated in this chapter in Figure 10.59, and for a variety of reasons are often referred to as environmental geological maps rather than geological maps.

Recording at the plan scale involves combinations of techniques that are well established, cost effective, of direct use to the engineer in geotechnical engineering, and recognizably geological in character (Chapter 11). With added geotechnical data the plans can justifiably be called geotechnical plans, but even so the geological basis of the plans remains evident.

Transition to map scales can lead to successive derivations from basic geological data, and eventually the maps cannot easily be recognized as geological. This is particularly the case when financial considerations of development are involved in the assessments.

11

Construction sites

11.1 Introduction

Engineering geological plans are usually site specific or, in the case of roads and other linear constructions, specific to the site corridor. The plans may be geological, dealing with rocks, soils and structures. Depending upon terrain and amount of available exposure, the geomorphological approach may be particularly effective at the plan scale in highlighting local problem areas. The plans may also be geotechnical, presenting a single engineering design parameter or a combination of a small number of parameters.

Frequently, the final plan will be a combination of geological detail with the remedial measures applied; in other words it will be one type of geotechnical plan.

In the UK, a significant suite of engineering geological plans has been developed by site experience, and these range from quick reconnaissance mapping to detailed in-construction recording of data.

Dearman and Fookes (1974), reviewing engineering geological mapping for civil engineering practice in the UK, stated that the art of preparing such maps and plans was then in its formative period. This is surely no longer true, although there still remains the problem of why and what to map; 'why' reflects the common geological dictum: 'if you don't understand it, map it', although in the present-day climate of black-box geology some form of remote sensing is more likely to be used than conventional field mapping methods (Barnes, 1981). Particular skills and techniques are needed to produce and interpret maps and plans. Even more so is the individual development of an appreciation of situations in which they are required to solve an engineering problem, and of their value in the assessment and design of an engineering project.

The aim of this chapter is, therefore, to highlight these two aspects by illustrating and reviewing current practice in the UK and elsewhere in the world. Reference will also be made to methods of recording the detail necessary at the site investigation and construction stages of a project. At the construction stage the plans may be made to provide a record of as-found conditions for record or maintenance purposes. They may also be made as a predictive tool, for example of the stability of a rock face as excavation proceeds, or to solve particular engineering problems.

Plans are also made for research purposes, although the number of published examples is small with most of them related to the distribution of weathering grades.

Mapping specifically for civil engineering purposes is usually carried out at large scales, which may range from 1:10 000 to 1:100 and exceptionally to even larger scales. These maps, formerly referred to as geotechnical plans (Fookes, 1969), are now called engineering geological plans (Anon., 1972), the information presented on them usually being supported by the results of field and laboratory testing. They are made by an engineering geologist, and on larger projects several geologists may be required in order to produce them, such is the amount of recording needed to be done.

Plans are prepared at all stages of a civil engineering project, for example: (a) at the feasibility of reconnaissance stage; (b) at the site investigation stage; and (c) at the construction stage.

Mapping at the feasibility or reconnaissance stage is mainly but not exclusively at small scale, and extensive use may be made of aerial photographs and the imagery produced by other remote sensing systems. These maps have already been discussed in Chapter 9 and attention is focused in this chapter on the particular requirements for large-scale mapping. Pre-construction mapping is undertaken at the site investigation stage and its aim is to produce a plan or plans on which features of engineering significance are shown, together with an indication of the expected engineering behaviour and engineering classification of the ground.

Mapping at the construction stage serves to record the geological conditions found on site which may then be checked against the expected conditions. Comparison of expected ground conditions with those actually found may lead to additional remedial measures, or local redesign of engineering aspects of the project, both of which inevitably lead to additional mapping for record purposes. The final plan serves as a permanent record of what was found and what in consequence was done, and is therefore invaluable for maintenance purposes.

Engineering geology plans are therefore of two

types – the pre-construction or site investigation plan and the construction or foundation plan.

Pre-construction or site investigation plan
This is prepared in the early part of the investigation to permit efficient planning and the anticipation of engineering problems. The scale varies with the size of the site, the engineering requirements and the time available, for example:

- 1:5000 may be adopted as a general scale for many purposes, including extended or compact sites, small reservoirs, large dam sites, borrow areas, tunnels, airfields, parts of road systems or the land installations for harbours;
- 1:1250 is used for more detailed representation of important compact sites including bridges, tunnels, dam sites, parts of road systems or large buildings.
- 1:500 to 1:100 or even larger is necessary for recording the details of trenches, open pits or sides and parts of excavations.

Construction or foundation plan
This is made during construction when foundations are exposed. It provides a record of the actual ground conditions found, which may be of use later if additions or modifications are made to the engineering works or if there are claims based on actual foundation conditions. The scale used may be either that of the pre-construction plan, or, because it may be more convenient, that of the construction drawings.

The engineering geology plan may be based on an existing large-scale topographic map, but with large scales or in complex areas it may be necessary to produce a special base map using photogrammetric, plane table or other techniques. A recent useful innovation is the orthophoto map which is a dimensionally corrected air photograph with contours overdrawn.

One other aspect of mapping is the preparation of as-built plans when there has been a gross departure from ground conditions predicted at the site investigation stage. Such plans may be prepared for litigation concerned with extra costs incurred during construction. The plans may be a true-to-scale record or an interpretative cross-section or elevation used in illustration of the as-found ground conditions.

Large engineering undertakings, for example hydroelectric schemes or integrated water supply schemes involving dams and possibly distribution tunnels, necessitate engineering geological mapping, much of it at plan scales, as part of the site investigation and later during construction. The engineering setting provides detailed survey grids so that, although record photographs will be taken as a matter of course, recording will mainly be by theodolite survey supplemented by chainage and offset methods from a surveyed grid of stations. Many examples can be found in the literature, where variations in ground conditions and how to record them are the main points of interest.

11.2 Methods of mapping at the plan scale

Plans are made using a small number of highly trained staff and simple survey techniques (Anon., 1972, Appendix I.10). The mapping method consists essentially of recording observations and points of measured data, e.g. the log of a hand-augered hole in soil. Boundaries are inserted by observation of field changes in geomorphological form, in rock or soil characteristics, or by interpretation based on observed features and judgement. Mapping teams commonly consist of one or two engineering geologists occasionally accompanied by soil or rock engineers or geomorphologists. Commonly used simple surveying and recording techniques include the following principal methods.

- by drawing directly on a photograph or onto a transparent overlay on a photograph;
- by adding to an existing or specially prepared topographic plan;
- by sketching onto squared paper using the chainage-and-offset method, or some modification of it;
- by instrumental means using surveying equipment such as the theodolite or a plane table.

More sophisticated methods may be applied under special circumstances, such as the need to record rock faces that are only accessible with difficulty or after scaffolding has been erected. These methods include:

- terrestrial photogrammetry.

11.2.1 Use of photographs

Photographs are taken as a matter of routine on engineering construction sites to record progress of the work, but more importantly to record the as-found ground conditions that may depart markedly from the expected conditions predicted from the results of a site investigation. The position from which a photograph has been taken and the direction of the camera axis should, ideally, be recorded on a site plan of suitable scale. A special symbol is used for the purpose. Each photograph should contain a scale and should be dated.

Photographs are ideal for recording accessible rock faces in foundations and road excavations, and also in quarries and open-pit mines, and natural cliff faces. It is assumed that normally the camera will be used at ground level at the foot of the vertical or steeply inclined rock face, with the inevitable result that vertical lines on the face will converge towards the top of the picture.

Details on horizontal or vertical planes in a simple record photograph are distorted by convergence of parallels with distance from the camera, unless the camera axis is normal to the plane being photographed. This applies to fixed lens types of camera, but by using a technical camera with independent movements to the lens panel and the camera back, limited corrections can be made in order to achieve a more nearly plan view. Use of such equipment requires considerable technical expertise, and is best left to the professional photographer.

One problem with photography is that, although it provides a permanent and very detailed record that can be re-examined or even reinterpreted at a later date, some time usually elapses between the photographs being taken and the photographic prints being available for mapping. If certain limitations are acceptable, chiefly related to the size of photographic print, this shortcoming can be overcome by the use of instant film, for example of the Polaroid type. With one type of film, Polaroid 55, a large format negative 4 × 5 in (10 × 12.5 cm) is also produced that, with a little care over processing in the field to stabilize the negative image, can be used later to produce high-quality, enlarged record prints.

Colour photographs, although more expensive than black and white, have considerable advantages as a recording medium. The different hues and tones recorded on a colour photograph are much more expressive of the lithological variations, in an exposure, and of the localization of weathering effects.

With a minimum of apparatus it is possible to take stereoscopic pairs of photographs, in either black and white or colour. Using either prints or transparencies, the three-dimensional view afforded by this method makes the interpretation of conditions on a rock face very much easier than using a conventional photograph.

Photographs may be used in the following ways:

- as a straightforward record;
- as the base for a line drawing made directly on the photographic print;
- as the base for a transparent overlay on which significant geological and geotechnical details are recorded.

Record photograph

A photograph taken for record purposes should ideally contain an easily read, unequivocal scale, an indication of magnetic orientation or a reference to plan grid points, and an identifying code number.

Line-drawing directly onto a photograph

Normal black drawing-inks can be used directly on the gelatine surface of a photograph. Once dry, the ink lines are permanent and waterproof. This latter characteristic makes it possible to bleach out the black and white photographic image, using a suitable photographic reducer, leaving a black and white line drawing. If the bleaching is carried out without touching or rubbing the ink lines and the bleached print is carefully washed, the drawing will be permanent when dried. The three stages of producing a line drawing in this way are shown in Figure 11.1.

The method was used extensively before 1960 for recording the geological details of rock faces in quarries (Dearman, 1959). Other illustrations of the use of the method are given later in Figures 11.5–11.10 for weathering studies carried out by Fookes, Dearman and Franklin (1971).

Base for a drawing on a transparent overlay

A transparent or translucent tracing medium is taped along one edge to the photograph so that reference can be made readily to the details on the picture. Significant lines are marked on the overlay, which may also be annotated. Use of this method is described in the next section, which also gives some additional practical hints on how to make the best use of this recording method.

11.2.2 Mapping by chainage-and-offset measurements

On construction sites where there is ample survey control for setting out and measurement purposes, mapping of the as-found or as-built foundation geology can easily be carried out by chainage-and-offset measurements from survey pegs. The details are then plotted directly by a resident geologist onto a hand-carried sheet of squared paper on which the construction layout has been faintly outlined. This method is fast, accurate and convenient, and reliable plans can be produced at large scales, e.g. 1:1000 up to a largest scale of say 1:100. Indeed, in the chaos of excavating and other operating machinery it may be the only possible method unless the work is halted to allow the plan to be made. As foundation surfaces are cleaned, it is often imperative that a blinding layer of concrete or the appropriate earth fill be placed immediately to avoid natural degradation of the exposed rock or soil surfaces. A map-making geologist is probably the last person a contractor wants cluttering up his operations in a confined area. In round-the-clock tunnelling operations, the interval between mucking-out operations and guniting the exposed walls, for example, may

Figure 11.1 Drawing on a photograph in Indian ink and bleaching out the photographic image to produce a line drawing: left, line drawing; middle, photograph with ink lines; right, the original photograph. Scale given by the ranging poles calibrated in feet

be as little as 20 min (McCrae, Powell and Yu, 1979), so that special recording techniques may have to be used.

Mapping during construction is carried out as the foundations are being excavated and its aim is to produce an as-built plan of the geological conditions in the foundations, as they are revealed. The plan is used as a record to enable comparison to be made of the ground conditions predicted at the site investigation stage and the as-found conditions. It can be used as the basis for design changes if the as-found conditions make this necessary, and later to assess performance, such as leakage through dam foundations.

As has been mentioned earlier, this type of mapping has been referred to very appropriately as 'geotechnical mapping' because details of geotechnical treatment to foundations or the walls of excavations may be recorded on the geological base plan. However, it is often more convenient to produce a series of observational and record plans of each area mapped.

11.2.3 Mapping by theodolite, plane table and compass methods

Direct surveying by instrumental means is more accurate than the sketching methods already described. One important aspect of plane tabling is that the lines of the plan can be logically developed on site, with obvious errors and omissions corrected at the time. Both methods are easiest to use where the surface to be mapped is reasonably plane and horizontal, with easy access to instrument stations and recording points.

Where a high degree of accuracy is not needed, simple compass and measured baseline traverses may be used. Detail is mapped from a survey tape laid along the line by offsetting or recording where features intersect the baselines. If the network of traverses is plotted first before recording offset detail, the closure error may be acceptably small or may be corrected.

11.2.4 Terrestrial photogrammetry

Vertical or steeply inclined rock faces, that may be partly obscured by vegetation or locally unstable, can be mapped by terrestrial photogrammetry. Powell and Mercer (1979) have assessed the cost effectiveness of the method of producing engineering geological plans of inaccessible rock faces during a rock slope stability investigation at Tremadoc in North Wales (see section 11.5.1). They pointed out that the procedure has mainly been applied to open excavations (Wickens and Barton, 1971; Ross-Brown *et al*, 1973; Moore, 1974), although there are examples of its use on natural slopes (Cheffins and Rushton, 1970).

There are a number of ways in which the information required can be obtained, including locating discontinuities on a plan of a pit by determining co-ordinates at specific points, by vertical contouring, or by plotting directly from an orthophotomap. Vertical contouring shows up any sharp breaks in profile, corresponding in most cases to particular discontinuities. A contour interval as close as 5 cm has been used in profiling, thus enabling very fine details to be identified on the elevations. This type of plan requires an engineering geologist to identify and assess all the geological detail, involving a great deal of work. Location of major discontinuities in an open pit is useful, in that complete exposure enables those parts of the face in between to be mapped easily.

Optimum recording conditions should be provided by an orthophotomap, as all the information is reproduced photographically to scale, leaving the geologist to select and record significant information on an overlay.

11.3 Recording at plan scale using descriptive rock and soil characteristics

In the engineering geological description of rocks and soils, as set out in detail in Chapter 4, there is an implicit, clear distinction between those properties that are descriptive of the rock or soil in the conventional geological sense, and those properties such as strength, deformability and permeability that have a more engineering cast. Engineering geological mapping is based on the recognition of uniform, homogeneous lithological units, differing from classical geological mapping only in its lack of insistence on chronostratigraphy.

It is also possible to map in terms of individual descriptive aspects – for example weathering grade, discontinuity spacings and strength – particularly at large scale on exposed faces. The individual maps or plans are still engineering geological in character. It is only when an assessment is made of the combined effect of a number of parameters, for example on tunnel support requirements, that the plan becomes geotechnical both in character and intent.

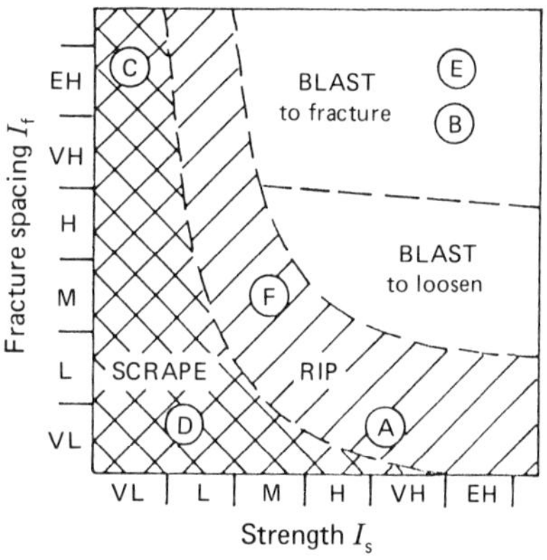

Figure 11.2 Classification diagram in terms of fracture spacing I_f and strength I_s subdivided according to ease of excavation (After Franklin, 1970, Fig. 2, and Fookes, Dearman and Franklin, 1971, Fig. 9)

Franklin (1970) showed how two parameters – strength and fracture spacing – could be combined, somewhat subjectively, to assess ease of excavation in rock (Figure 11.2). He first illustrated this as an interpretation of site investigation boreholes, demonstrating the difference between a geological cross-section based on lithological similarities and an engineering geological cross-section in which rock units with similar properties are correlated (Figure 11.3). Critical to the interpretation is the relation between any pair of rock properties. It is self-evident that broken or weak rocks should be easy to excavate, but complete appraisal of an exposure

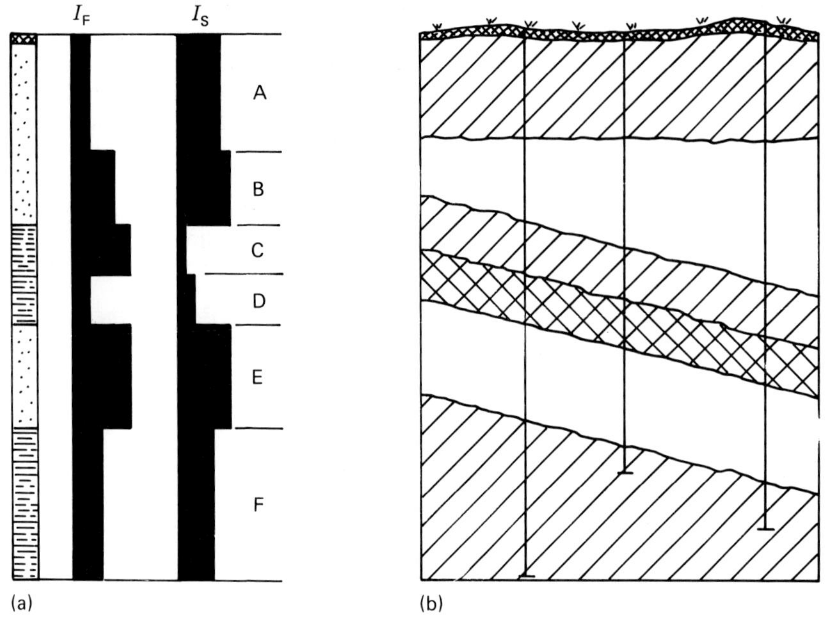

Figure 11.3 Fracture–strength classification applied to rock mapping: (a) sandstone and shale core logged in terms of fracture spacing and strength indices; (b) interpretation of three boreholes in terms of rock quality based on Figure 11.2

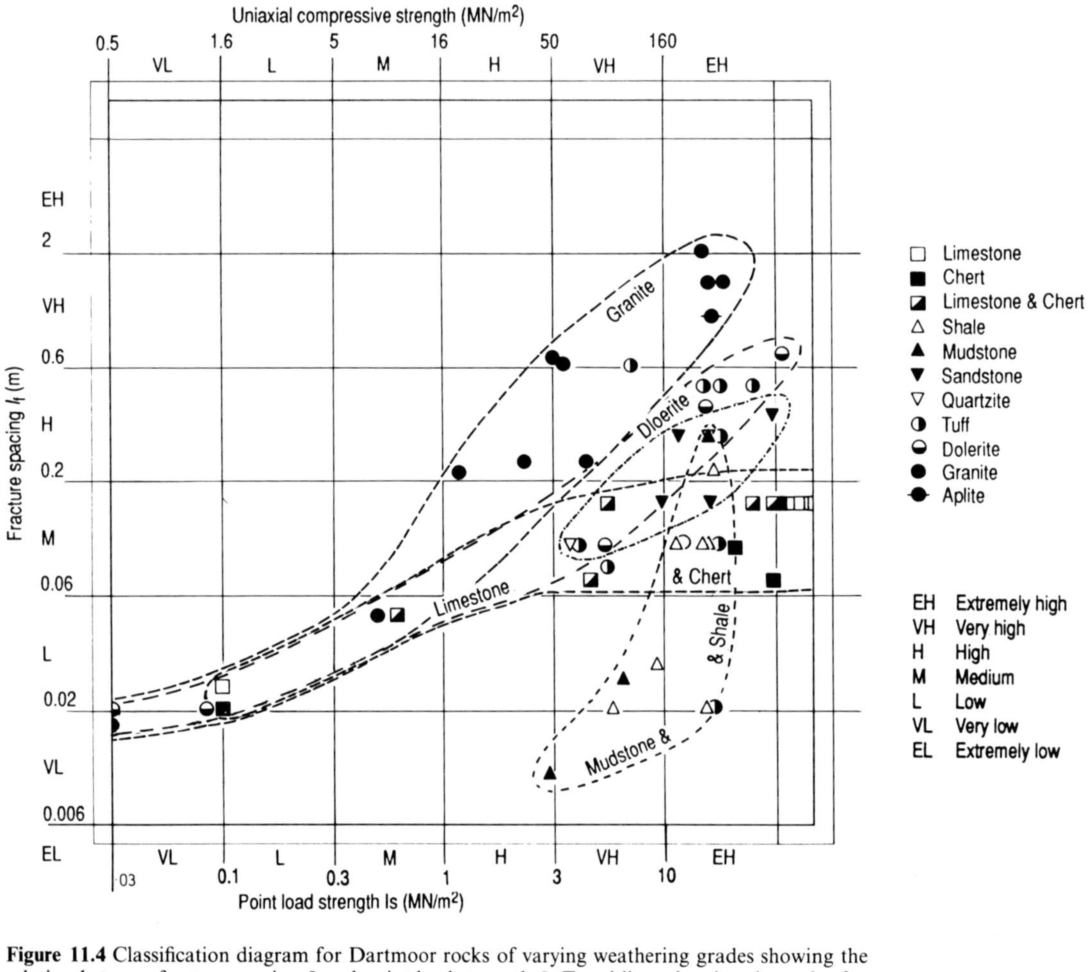

Figure 11.4 Classification diagram for Dartmoor rocks of varying weathering grades showing the relation between fracture spacing I_f and point load strength I_s. Trend lines showing the path of weathering relative to fracture spacing and strength index have been drawn for each rock type except tuff, for which there was not sufficient field data

would also have to take into account other factors such as weathering grade, openness of fractures, etc.

Precise location of the blast, dig and scrape or rip areas on the graph would depend upon the influence of other rock properties, on local conditions and on the mechanical capability of the excavating equipment (Franklin, Broch and Walton, 1971). It is possible to prepare plans showing the boundaries between rocks having different engineering properties by using as parameters the divisions of such a classification diagram.

As part of an early research programme into the engineering geology of weathered rocks, Fookes, Dearman and Franklin (1971) applied a recording method previously used extensively by Dearman (1959) to record the lithologies and structures in successive working faces in Meldon Quarries, near Okehampton, on the northern margin of Dartmoor in South-West England.

The area was chosen because it had escaped glaciation, and large remnants of deep tropical weathering, possibly Tertiary in age, were preserved on high level

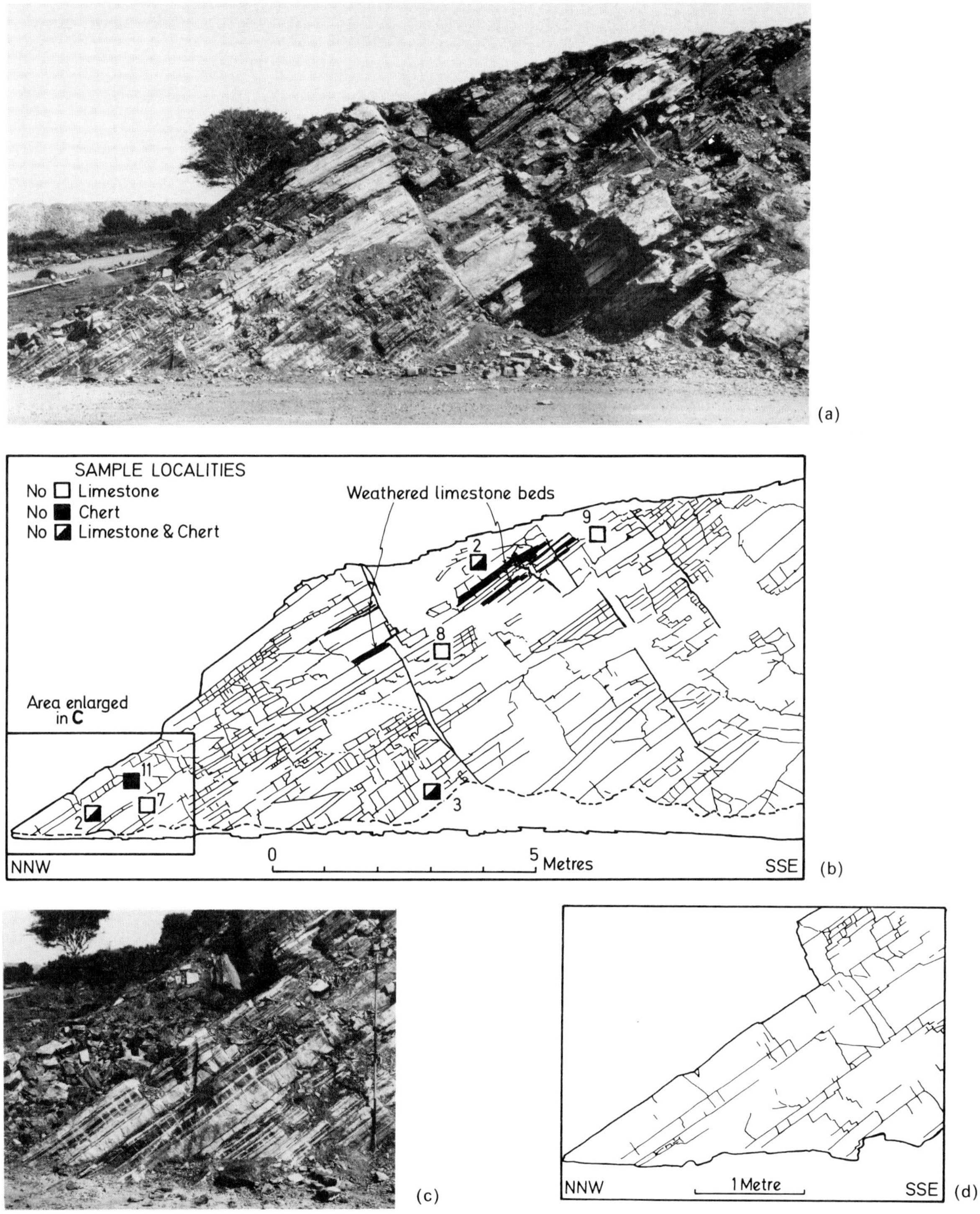

Figure 11.5 Quarry face in evenly dipping cherts and limestones, 1010 level, Meldon Quarries, Okehampton, UK: (a) quarry face at present; (b) drawing of the face showing the main discontinuity patterns, the location of extremely weathered limestone beds and locations of specimens tested in the field and laboratory; (c) part of the face, located on (b); (d) drawing of discontinuities for comparison with joint, bedding and lamination frequency apparent in (c)

erosion surfaces. A wide range of rock types – dolerite, limestone chert, sandstones, shales and mudstones – were involved in intense chevron folding on a variety of scales, the whole broken by faults along the bedding. The Dartmoor granite later intruded the folded complex of Carboniferous rocks. Much later, strike-slip brecciated faults, related to the Alpine orogeny, jostled and offset the early Tertiary landscape, breaking up the continuity of the folds and permitting localized deep penetration of the weathering agencies.

Fracture spacing and point load strength were determined for all rock types (Figure 11.4), and the graph was put to a practical use as a means of assessing ease of excavation (Figure 11.2).

11.3.1 A weathered face in chert and limestone

In a disused quarry face, cherts interbedded with thin limestones dip evenly at 33° NNW (Figure 11.5a), and in the outcrop the discontinuity pattern has four elements, namely bedding, vertical dip joints, strike joints normal to bedding, and a subordinate set of near horizontal joints. Photographs (Figure 11.5a,c) show that the rocks are both bedded and laminated and that the sedimentary banding has a higher frequency than bedding planes which are present as open or potentially openable discontinuities. Eight open bedding planes are present in 1.6 m of sediments; in the same thickness over 80 changes in lithology are discernible on the photograph, and each of these smaller units is seen to be finely laminated when examined in hand specimen. In the same way, joint planes may be picked out by white reaction selvedges in otherwise black cherts, but are not necessarily discontinuities, as becomes apparent when Figures 11.5(c) and 11.5(d) are compared.

By visual inspection, the face has been divided into three zones on the basis of the presence of Grades II, III and IV on the mass weathering scale (see section 4.7.3). These estimates of the weathered state of the rocks are supported by detailed measurements made along five traverses on the quarry face, at right angles to the bedding. The thickness of extremely weathered limestones, present as dark brown soil, was measured to determine rock–soil ratio as an assessment of mass weathering grade. These determinations, together with additional information from Figure 11.5, and point load strength of selected samples, provide the basis for the measurements presented in Figure 11.6.

The engineering appraisal (Figure 11.6d) represents an assessment of the relative influence of discontinuity spacing and point load strength (Figure 11.2), backed up by weathering grade, on likely excavation conditions. There are three zones on the face; the limited area which would have to be broken by blasting coincides with very high discontinuity spacing, extremely high point load strength, but Grade III weathering. The two other areas on the quarry face with the same characteristics are surrounded and penetrated by rocks with the same weathering and strength characteristics but with a high discontinuity spacing. Consequently, blasting only would have to be used to loosen the rock fractures. In contrast the area of weathering Grade IV, with mainly medium fracture spacing and of very high strength in the unweathered rock blocks, could be removed by ripping.

Mapping was undertaken as an experimental exercise to test the application of the method of geotechnical

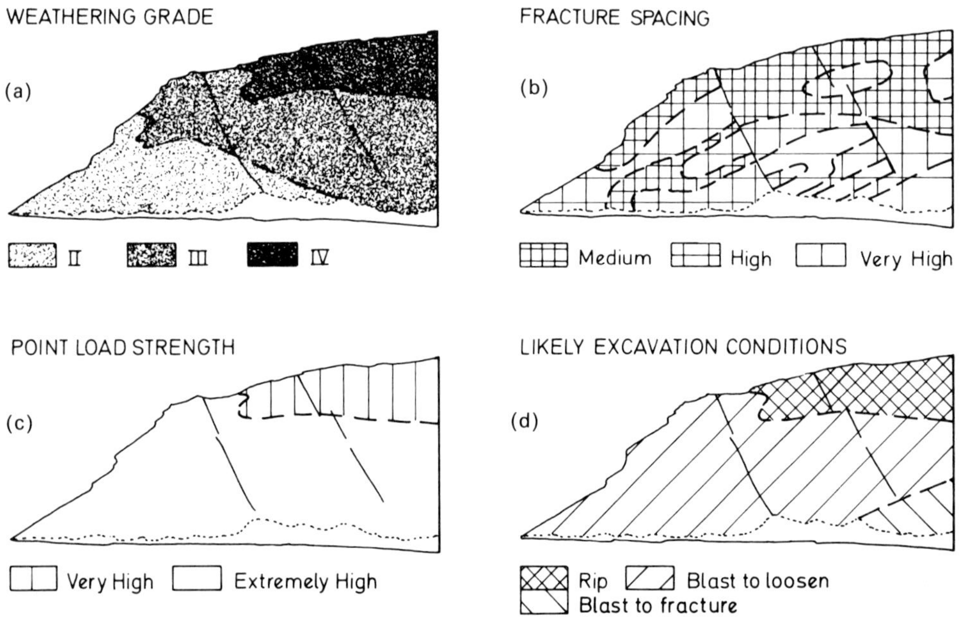

Figure 11.6 Geotechnical properties and engineering appraisal of the quarry face in the cherts and limestones, 1010 level, Meldon Quarries, Okehampton, UK, illustrated in Figure 11.5: (a) weathering zones defined in terms of grade; (b) discontinuity spacing and variation of fracture spacing values are those set out in Figure 11.4; (c) strength zones determined by point load strength tests – values refer to those set out in Figure 11.4; (d) engineering appraisal in terms of likely excavation conditions determined partly from Figure 11.2

assessment. In quarrying practice, the rock face would be dealt with as a single unit, using drilling and blasting with vertical holes.

Elsewhere in the quarries both cherts and limestones have been completely weathered. On the strike extension of the exposure described above, at a lower quarry level, the face reveals 3 m of completely weathered beds intercalated between less weathered cherts. This exposure has been used to illustrate how line drawings can be made on photographs (Figure 11.1) to give the type of permanent record illustrated in Figure 11.5.

Details, for example, of the rock–soil ratio determined on a measured traverse of an exposure can also be recorded directly on a photograph. Elsewhere in the Meldon Quarries limestones are more abundant, weathering at the surface to dark brown soils. One exposure has been measured and recorded (Figure 11.7). Here, weathering in the interbedded limestones and cherts increases in grade down dip. This is convincingly demonstrated by measurements made along lines 2 m apart; on traverse 2 (Figure 11.7) there is only 20 cm of soil in 213 cm of beds, whereas in the adjacent traverse 3 further down dip, there is 152 cm of soil, that is extremely weathered rock, in the same group of strata. Visually there is a change in weathering from Grade III to IV down dip, and finally into Grade V.

11.3.2 Weathering in folded sandstones and shales

The combined influence of structure and lithology on weathering sandstones and shales was determined in temporary exposures in Knowle Quarry, north of Okehampton, Devon. As part of the quarrying activity, overburden has been removed and a vertical quarry face provides exposures to depths of about 5 m below original ground level (Figure 11.8a). Sharp chevron folds bring the same beds repeatedly to the surface, and a normal fault parallel to steep beds breaks the transition from one limb to the other in the syncline occupying the northern part of the face. Interbedded shales are strongly cleaved at an acute angle to bedding in the fold limb.

In passing, it is worth mentioning that the plan of the quarry face in Figure 11.8(a) was accurately plotted from a taped baseline with 2 m long ranging poles set vertically at 1 m intervals along the baseline to provide a rectangular sketching grid. This method, complete with taped measurements on the face of, for example, bed thicknesses and spacings, provides a satisfactory, rapid and reasonably accurate method of plotting to scale both structural and lithological detail. Indeed, an ability to sketch detail within a broad grid is a very useful aid to recording detail at the plan scale, as has been and will be repeatedly demonstrated (see Figures 11.36 and 11.88); the requisite skills can readily be acquired with a little practice.

At a later stage in the recording programme it became necessary to survey the layout of the whole quarry, to prepare an elevation of the faces on the three working levels, and to plot in the main fold and fault structures (Figure 11.9). This was achieved using the simplest equipment: ranging poles (sticks would have done), measuring tapes, a clinometer, a prismatic compass and two engineering geologists – one at each end of a tape!

Drawing on photographs was used to record the discontinuity patterns and frequency (Figure 11.8b,c). As the photographs incorporate a scale, it becomes possible to map the quarry face in discontinuity spacing classes (see Table 4.7) once the basic drawings have been done. By this means, parts of an exposure that are difficult or impossible to reach can still be mapped.

Each lithological type has a distinctive discontinuity pattern. In the sandstones, bedding surfaces tend to be smooth and plane, but, for example in bed a (Figure 11.8a), three beds form an apparently single unit; this amalgamation is only revealed by splitting induced by weathering or the strains imparted by blasting. As usual in this type of structure, jointing comprises a steeply inclined dip set which is common to both limbs in the fold. Strike joints are present normal to the bedding

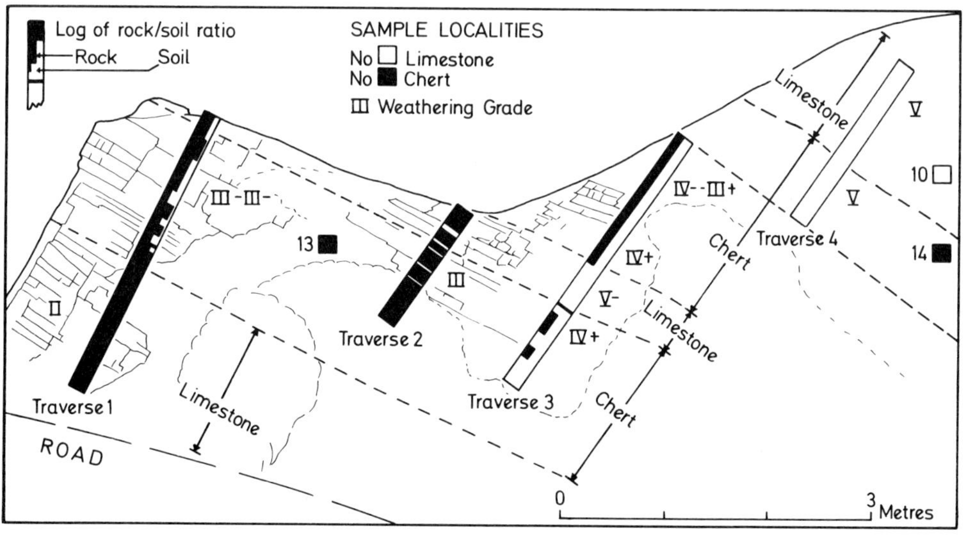

Figure 11.7 Sketched elevation of the south-western wall of the 950 South Quarry Bay cutting, Meldon Quarries, Okehampton, UK, where limestones and cherts of the Meldon Chert Formation are extremely weathered. Note the positions of the two traverses mentioned in the text

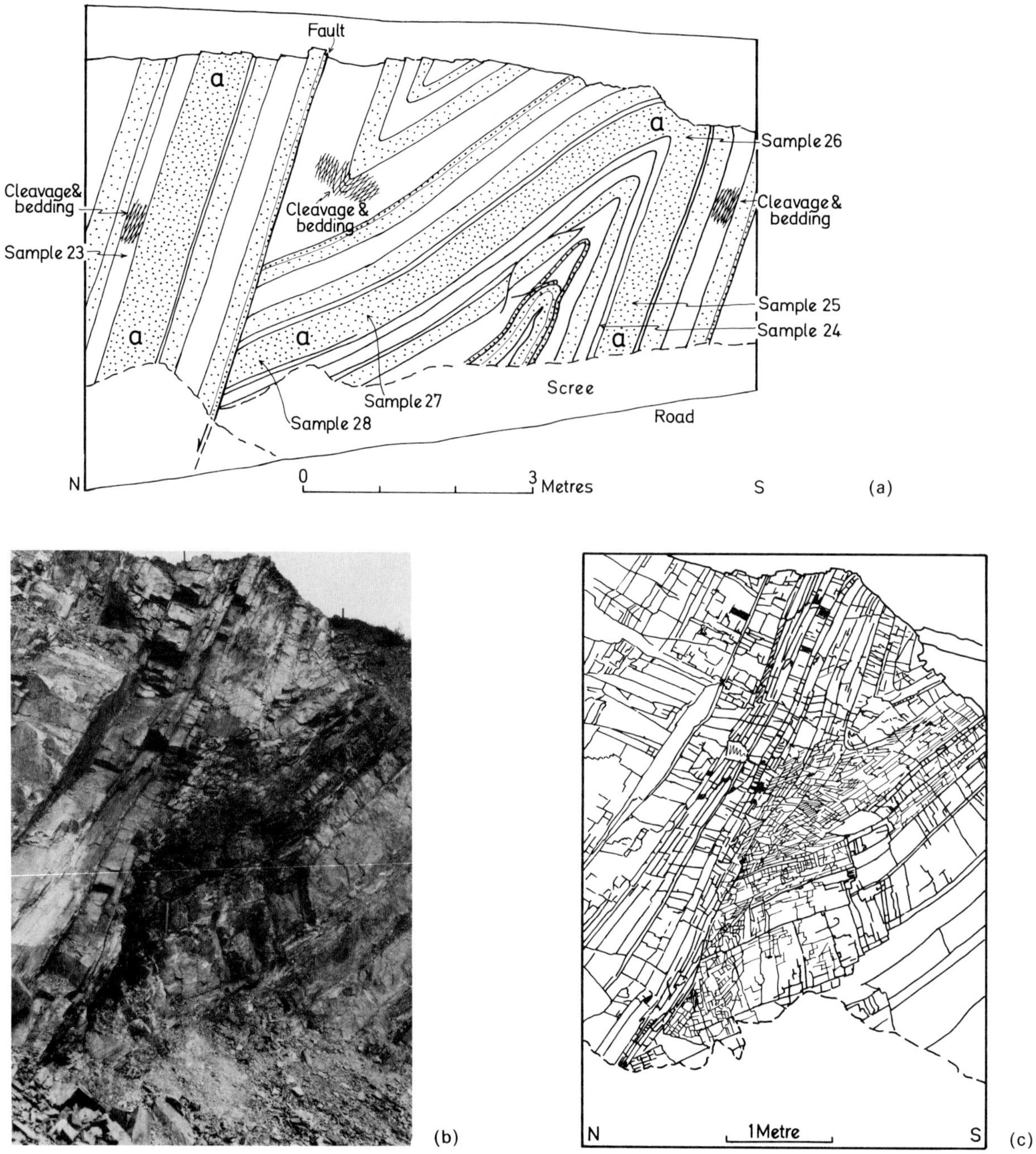

Figure 11.8 Quarry exposure in Upper Carboniferous sandstones and shales, Knowle Quarry, Okehampton, UK: (a) quarry face showing the distribution of sandstones (dotted) and shales, and the sample localities; (b) part of the syncline; (c) drawing showing discontinuity patterns in (b). Note the increase in fracturing in the core of the syncline and in gently inclined beds adjacent to the fault

planes, so that there is change in attitude in passing from one fold limb to the next. Major dip and strike joints may cut indifferently across whole groups of individual beds, otherwise planar continuity is confined within beds. The same pattern of joints is present in the shales, and there is a high degree of fissility along both the laminations and the cleavage (Figure 11.8b,c).

The weathering pattern in this quarry exposure (Figure 11.10a) shows the expected decrease in weathering grade with increasing depth from the surface without any detectable differences between the weathering of the sandstones and the cleaved mudstones. There are, however, some interesting departures from this simple pattern which are related to the tectonic structure. Along the region of maximum curvature in the anticline, weathering Grade III penetrates to quarry floor level. There is a similar deep penetration of Grade III weathering in the adjacent syncline, but here, because of the combination of faulting and intense cleavage in the mudstones, there is a localized development of Grade IV weathering underlying the zone of Grade III weathering. Such an inversion of weathering zones is unexpected and unless the structure is understood it is unpredictable; hence the possibility of its occurrence should not be overlooked in any investigation of highly folded terrain.

Discontinuity spacing (Figure 11.10b) and strength

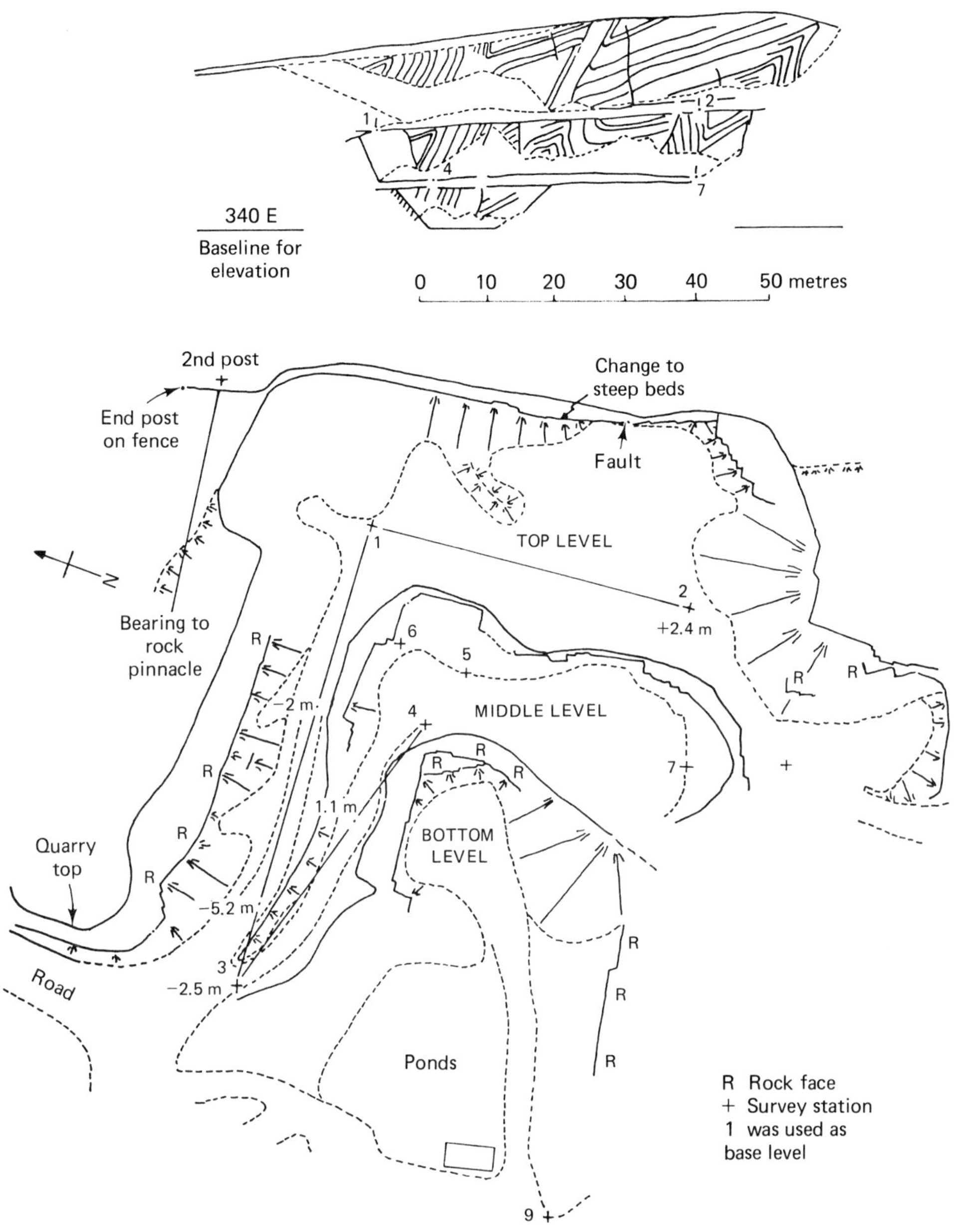

Figure 11.9 Plan and elevation of the face in Knowle Quarry, Okehampton, UK, plotted using simple surveying and recording techniques

values (Figure 11.10c) show a clear relationship to the weathering pattern and the distribution of sandstones and mudstones in the face (Figure 11.8a). The engineering appraisal (Figure 11.10d) reflects the combined effects of these influences on the excavatability of the rocks.

11.3.3 Geotechnical mapping on the projected Edinburgh Ring Road scheme

The rock in the area of the projected Edinburgh Ring Road scheme (Cottiss, Dowell and Franklin, 1971) is quite variable and includes basalts and welded tuffs and a variety of sandstones and siltstones. Most of the site is overlain by glacial or fluvioglacial materials, recent river, lake and beach deposits, and fill which may reach thicknesses of up to 50 ft. Investigations of rock conditions, therefore, depended on core samples taken from some 400 boreholes ranging in depth from 20 ft to 150 ft.

The variety of geotechnical problems and rock types 'suggested a broad coverage of the site using classification testing and mapping rather than sophisticated testing at one or two locations' (Franklin, 1970). Tests and observations included petrography, fracture spacing in the cores, uniaxial and Brazilian strength classification tests, Schmidt hardness tests and 'wet' tests for durability, porosity, density and sound velocity. In order to simplify the data for engineering application, a rock quality score was computed by combining the assessed results of tests by means of a standard formula.

On the basis of the test results the whole site was divided into five zones (Figure 11.11) corresponding to percentage composition of five principal rock types, each rock type being characterized by the classification tests

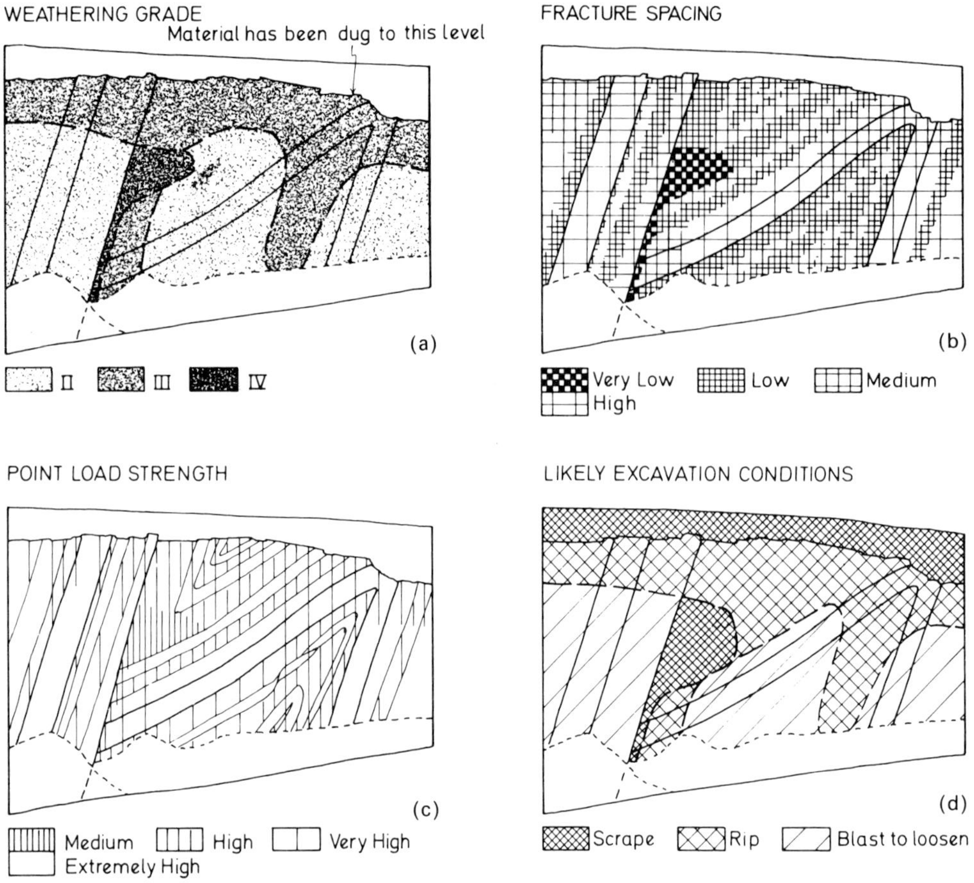

Figure 11.10 Geotechnical assessment of the quarry face exposed in Knowle Quarry, Okehampton, UK, illustrated in Figure 11.8(a): (a) weathering zones defined in terms of grade; (b) discontinuity spacing variation – fracture spacing classes are those set out in Figure 11.4; (c) strength zones determined by point load strength tests – classes refer to those set out in Figure 11.4; (d) engineering appraisal in terms of likely excavation conditions determined partly from Figure 11.2

and observations. Each zone was shown to differ significantly from the next in mechanical characteristics and it was considered that each zone offered information on engineering behaviour that would not have been available from conventional geological description. Conditions in Zone 2 along the line of the Calton Hill tunnel are shown in Figure 11.12. Sections along the line of tunnels show the distribution of joint frequency (Figure 11.12a) and the rock quality scores (Figure 11.12b) as determined from the borehole records and tests on core. The sections and the plan provide an example of a multivariate approach to rock characterization in geotechnical mapping carried out at the site investigation stage of an engineering project.

Basic differences in the geological and geotechnical approach to mapping are brought out in the comparison between the classical geological map of the area (Figure 11.11a) and the same area mapped in terms of five different engineering zones (Figure 11.11b).

11.4 Plans produced at feasibility and site investigation stages

Mapping to determine the feasibility of an engineering undertaking entails reconnaissance mapping in terms of descriptive rock or soil classification, engineering geomorphology, and geomorphological processes. It is not uncommon for mapping at the actual site investigation stage to overlap the initial feasibility studies. There is little practical point in considering the two separately – air photograph interpretation and walk-over survey are adequate at the feasibility stage; more precise methods of location, sampling and testing are usual at the site investigation stage.

11.4.1 Logging pits and trenches

When producing plans of the walls and floors of pits and trenches, care must be taken to support the walls against collapse. Support can be installed after mechanical excavation is completed, which has the disadvantage of obscuring part of the section. Offsetting this is the potential advantage of the support as a framework for the mapping grid. Mapping is done from vertical scanlines – usually a tape dropped down the pit wall from a known point on the surface – using, as necessary, offsets to locate significant detail. Basically the method is chainage-and-offset.

Where ground conditions are such that complete support of the walls of a pit is considered to be absolutely essential, then recording can be done as successively lower levels are excavated before the support is advanced. In such circumstances a borehole would probably be a more suitable way of determining ground conditions.

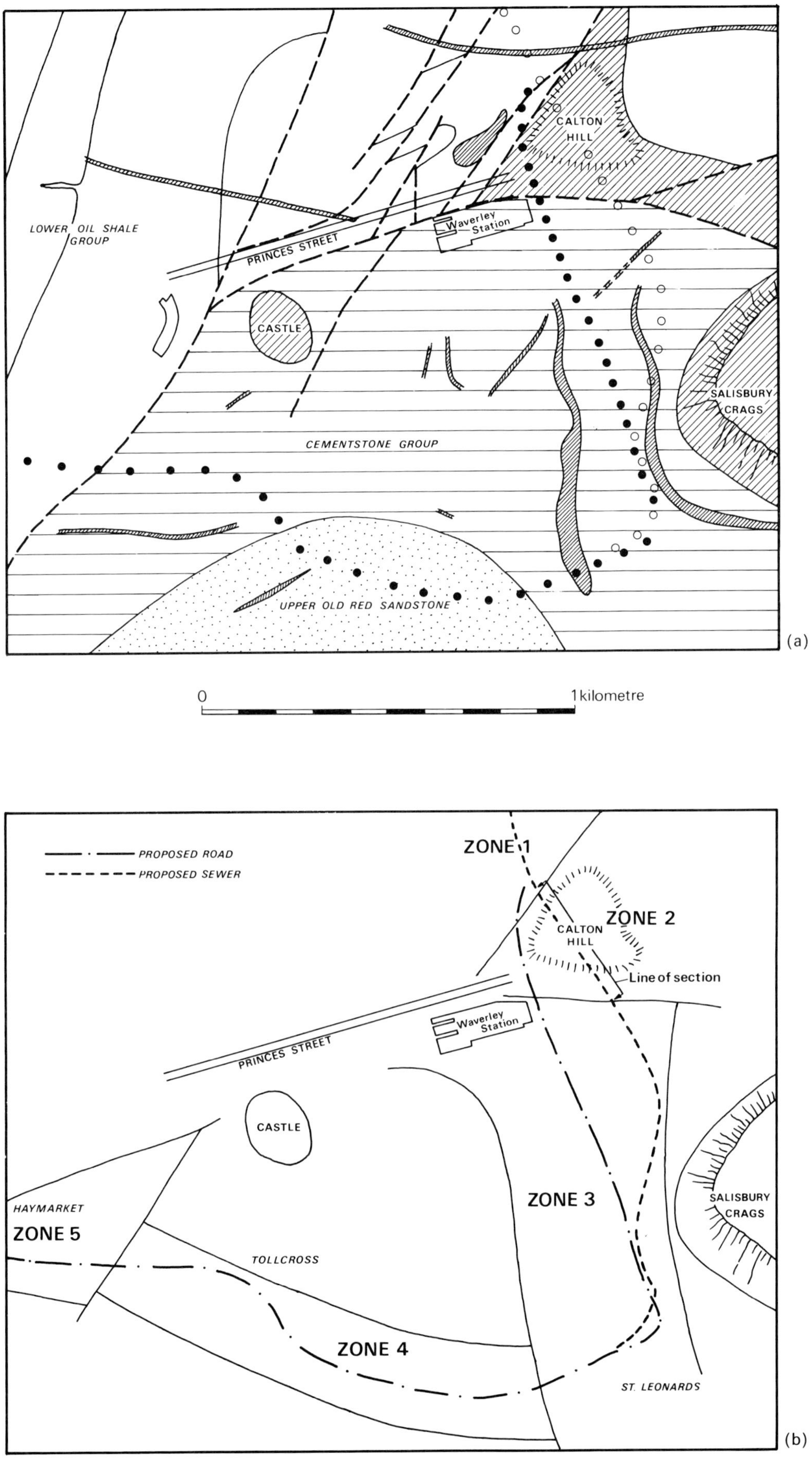

Figure 11.11 Edinburgh, Scotland, UK, inner ring road: (a) conventional geological map; (b) subdivision of the site into zones of characteristic mechanical and engineering nature. The road line is shown by closed circles, the sewer line by open circles (After Cottiss, Dowell and Franklin, 1971)

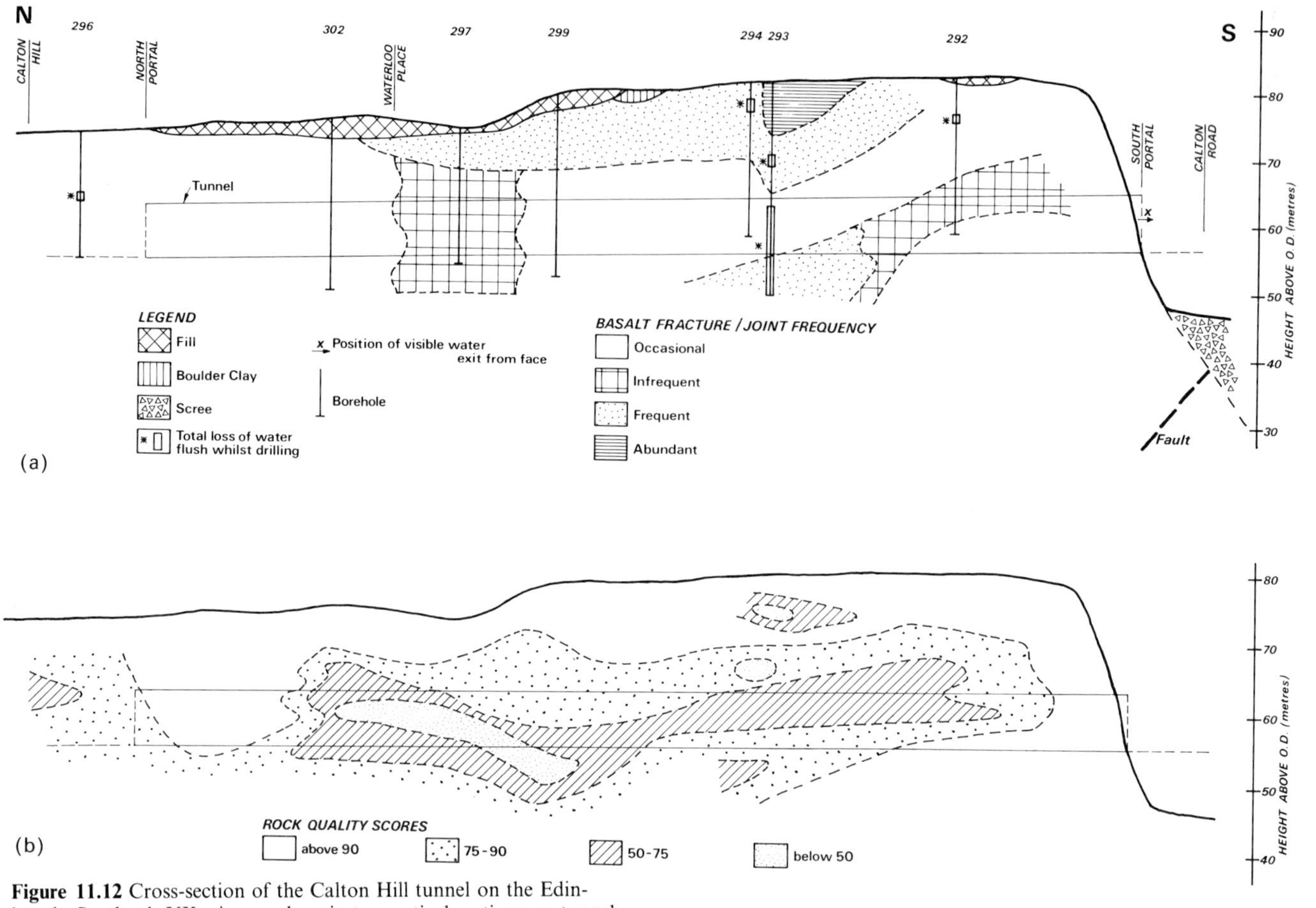

Figure 11.12 Cross-section of the Calton Hill tunnel on the Edinburgh, Scotland, UK, ring road project – vertical sections on tunnel centre line: (a) superficial soils and rock fracture frequency; (b) relative hardness (rock quality scores). The line of section is marked in Figure 11.11(b) (After Cottiss, Dowell and Franklin, 1971)

11.4.2 Field record of a pit face

Pits may be dug economically to a depth of approximately 5 m. Information gained by careful description of the pit faces and bottom and the taking of bulk samples aids considerably in making a correct assessment of superficial deposits above bedrock. This applies particularly where boulders and cobbles are present in the soils, making drilling and taking driven-type undisturbed samples very difficult. Additionally in those ground conditions, the value of any samples taken is questionable. Groundwater may cause difficulties, with pit faces likely to collapse as water flows in.

Sketched plots on squared paper should record the distribution of rock and soil types, described in the recommended manner (see Chapter 4 and Anon., 1972, 1977, 1981a,b).

An example of the amount of detail that may be recorded, in this case in glacial deposits above sandstone bedrock, is given in the field sheet log of a trial pit illustrated in Figure 11.13 (Fookes, Hinch and Dixon, 1972, Fig. 17). The original field log may be all that is required or it may be fair-drawn for the final site investigation report.

Fair-drawn copies

Observations made on all four faces of a trial pit, with a summary graphic log and detailed descriptions of the strata, are illustrated in Figure 11.14 (Dumbleton and West, 1974, Fig. 12.).

11.4.3 Trenches

Site investigation trenches – long pits – are usually dug one bucket width by a backhoe, or bulldozed out as a broad cut, perhaps one-sided, with a vertical or stepped face. An important problem with trenches is the stability of the walls while they are being recorded; thus the dozer cutting, with benches, is usually the safest method.

A backhoe trench dug through topsoil into weathered basalt overlying weathered shale is illustrated in Figure 11.15. The trench has locally been deepened until the groundwater table was reached, marking the limit of practical investigation. Distribution of the rock and soil types, annotated by extended descriptions, and the main structural features are shown.

Careful layout of trenches will save money, and the investigation should be related to the geometry of the structure being investigated. Trenches normal to fault directions or parallel to the dip in bedded sequences provide most information from the shortest length of trench. A typical trench recorded from the Sukian dam site at Mangla, West Pakistan (Fookes, 1967, Fig. 10) illustrates how the local details of superficial and bedrock deposits are recorded, together with details of the

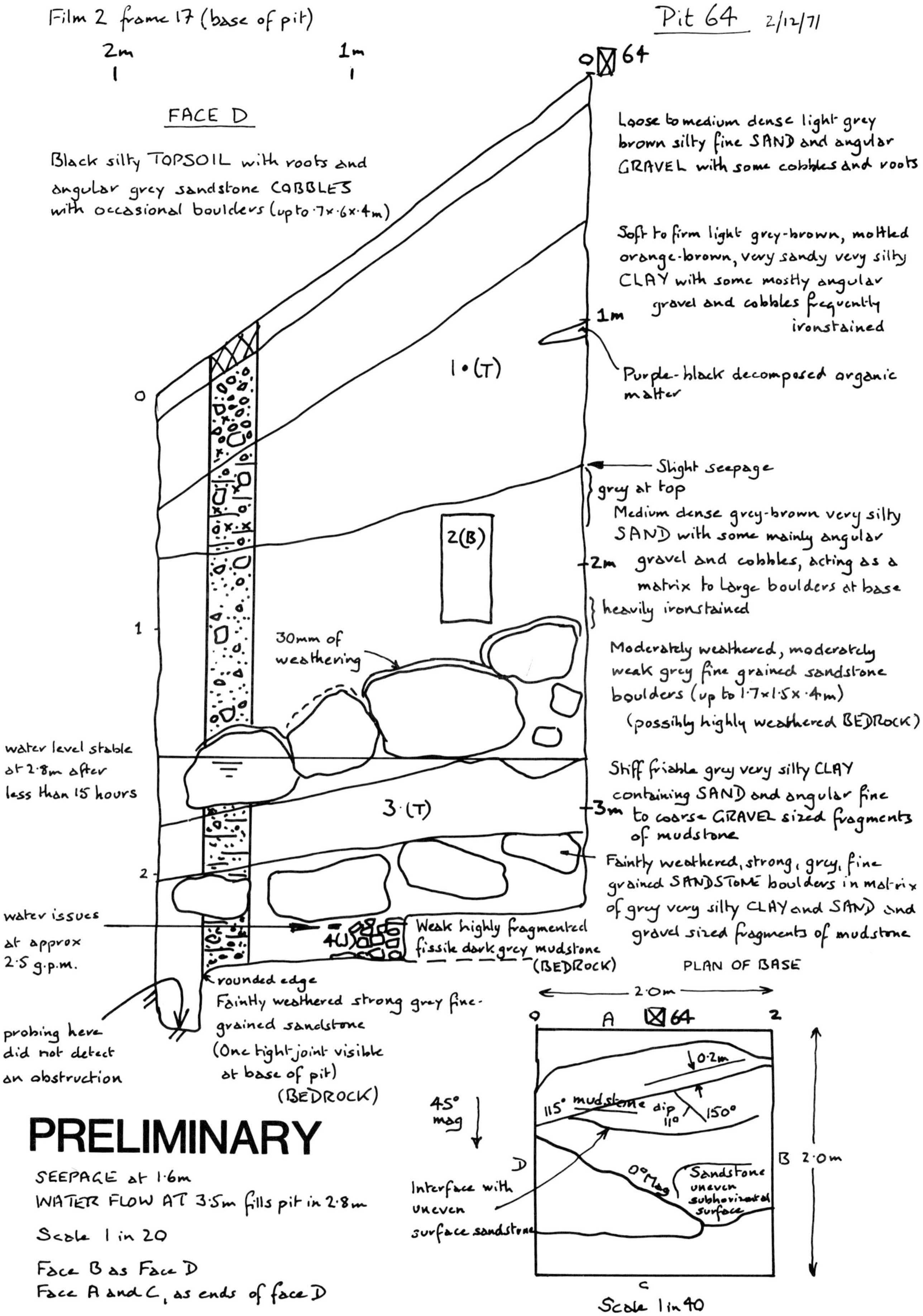

Figure 11.13 Plan of site investigation trial pit. An original field sheet drawn at a scale of 1:20, reproduced here at 1:22 (From Anon., 1972, 13.1.9, by permission)

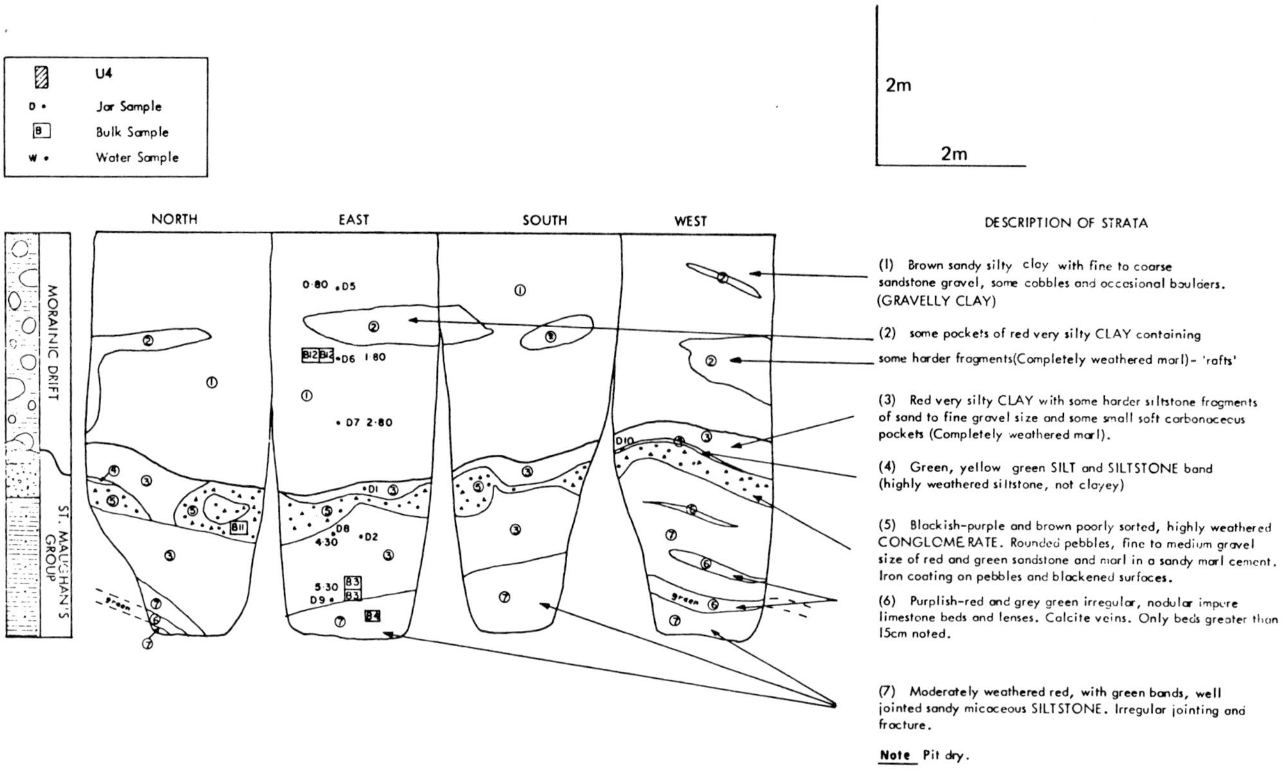

Figure 11.14 Fair-drawn log of the four faces of a trial pit, with summary log, and a description of strata (From Dumbleton and West, 1974, Fig. 12, by permission)

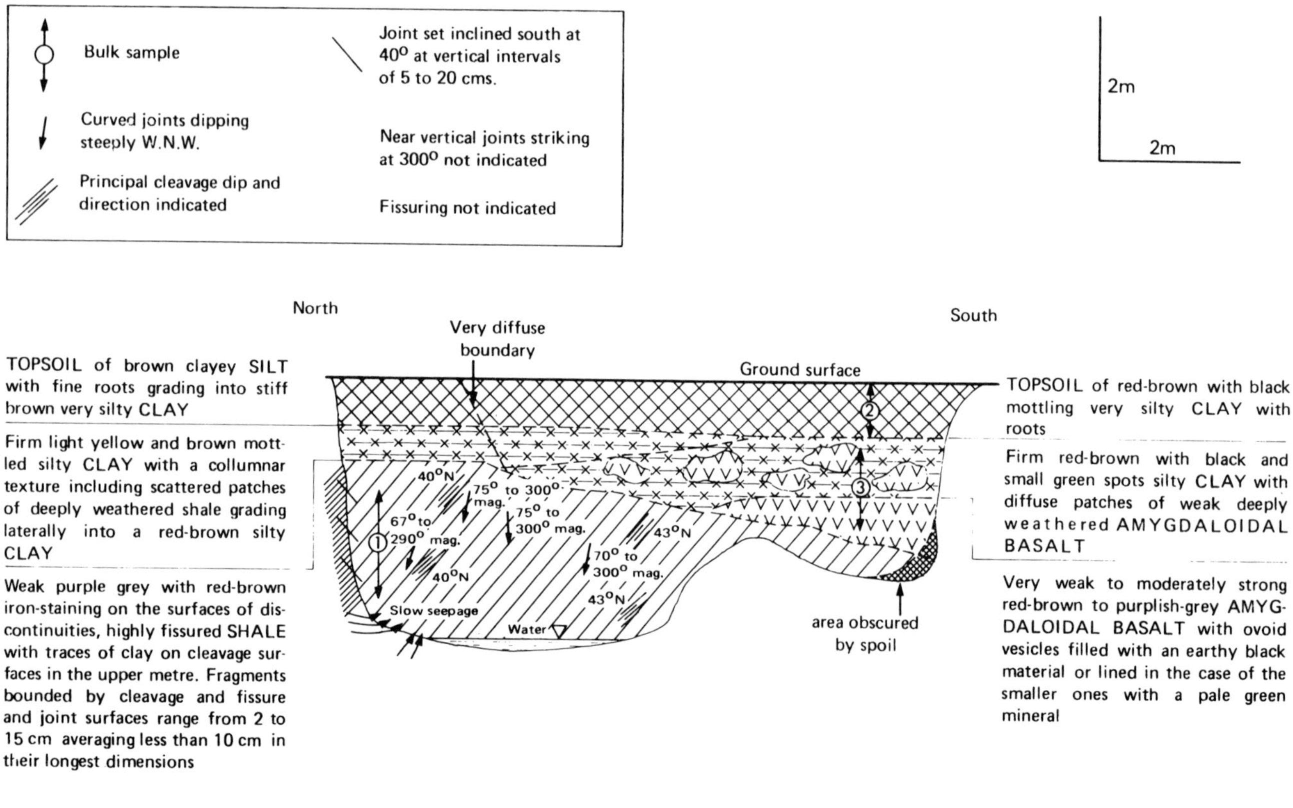

Figure 11.15 Section showing observations made in a trial trench (From Dumbleton and West, 1974, Fig. 11, by permission)

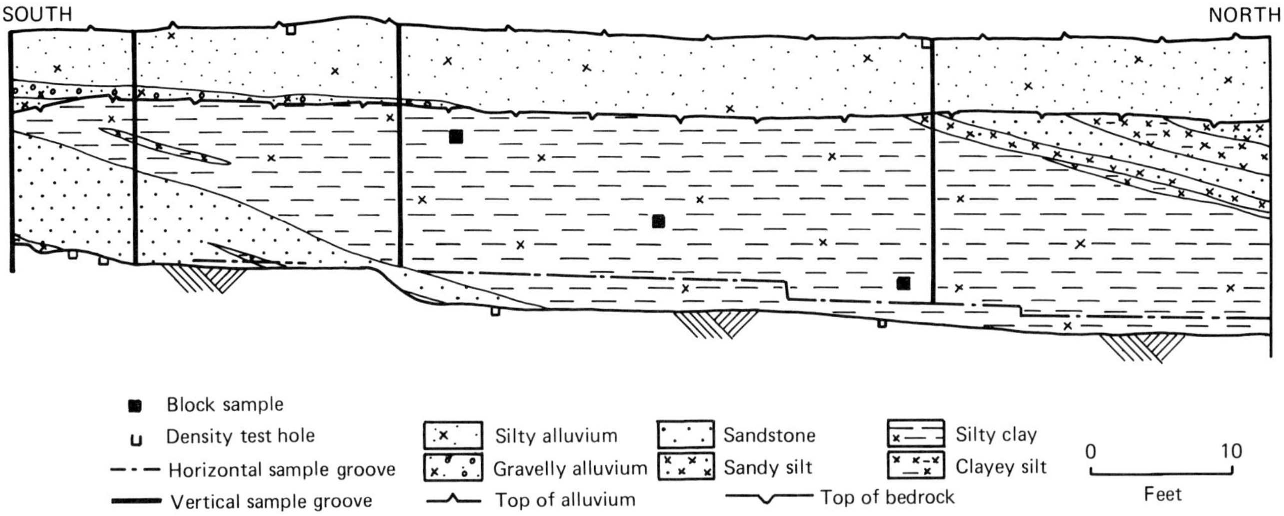

Figure 11.16 Section of an investigation trench, Sukian dam site, Mangla, West Pakistan (After Fookes, 1967, Fig. 10)

sampling programme (Figure 11.16). The floor of the trench, when cleaned off, would effectively complete the three-dimensional picture.

Bulldozer trenches across capable faults

An illuminating example of the results of mapping, in considerable detail, the effects of reverse faulting on a sequence of Quaternary terrace deposits overlying Miocene deposits has been provided by McClay (1985) and Asquith (1985). The site concerns a proposed liquefied natural gas terminal site in California where it was necessary to demonstrate, in an area of known late Quaternary faulting, that the site was seismically and geologically suitable for the proposed purpose.

Fault locations were extensively trenched by bulldozer, and the resultant exposures were mapped in the walls of the vertical cuts. A part of one wall is reproduced in Figure 11.17, from Asquith (1985, Fig. 4). Units designated '1' are modern soils; those marked '3' are alluvial deposits of the terrace sequence; those designated '4' are marine sand deposits of the terrace sequence; and unit 5 is the Miocene Sisquoc Formation. Detailed descriptions of individual units are appended to the original. From the plan, complexities of the fault are apparent and could only have been worked out from a detailed correlation of the units across and through the many minor fault offshoots. The plan is a first-class example of detailed mapping at the plan scale.

By coincidence, the same exposure was used by McClay (1985) to illustrate the use of photographs to record geological data. It is not possible to reproduce the photographs here. Drawings were done on an acetate film overlay. Figure 11.18 is of a detail on the Arroyo fault near the base of the excavations; unit 5, as far as it was then exposed, is shaded. Figure 11.19 is the graphic log of the exposure in the trench wall as presented by McClay (1985, Fig. 4.). It is considerably simplified as compared with Figure 11.17, although manifestly of the same exposure. The difference between the two seems to be that Figure 11.19 is an uncorrected drawing on a photograph, whereas Figure 11.17 has been recorded on a surveyed 1 m grid and is therefore dimensionally true. Whether a dimensionally true drawing of this exposure is really necessary is, of course, open to question.

11.4.4 Cross-sections

Where boreholes are used to determine the local site geology, the core logging results have to be displayed graphically in the form of cross-sections.

A cross-section should show the mapped units into which the foundations have been divided, together with any additional divisions used in engineering calculations. The data used in calculation can be usefully shown at one side of the section. Where relevant, other information as indicated in section 5.3 should be given. The level of groundwater and other piezometric observations and date of recording should be indicated.

Quantitative information on field test results, e.g. standard penetration test, cone resistance, water content of soils, and locations of samples and photographs, should be added on the line of section if available.

A small-scale key plan with north point in the top right-hand corner of the drawing should show the position of the section in relation to the main features of the site. North points should always be marked A.M. (astronomical meridian) or M.M. (magnetic meridian); if M.M. is used, the date should be added. The dimensioned positions of the boreholes should be shown on this or another plan. The position of the cross-section must also be shown on the accompanying engineering geological plan or map.

For horizontal ground, it is often convenient to have a vertical scale of 1:100 and a horizontal scale of 1:200 or 1:500. For sloping ground, the vertical and horizontal scales should usually be the same. Differences in section scale and plan scale should be noted in large lettering.

For any connections drawn between subsurface details, for example between adjoining boreholes, the assumed lines notation should be used. A typical cross-section at the plan scale is illustrated in Figure 11.20.

11.4.5 Geotechnical reconnaissance mapping: a case study from Jamaica

A proposed site for an industrial complex in Jamaica, one of a few under consideration for development as an

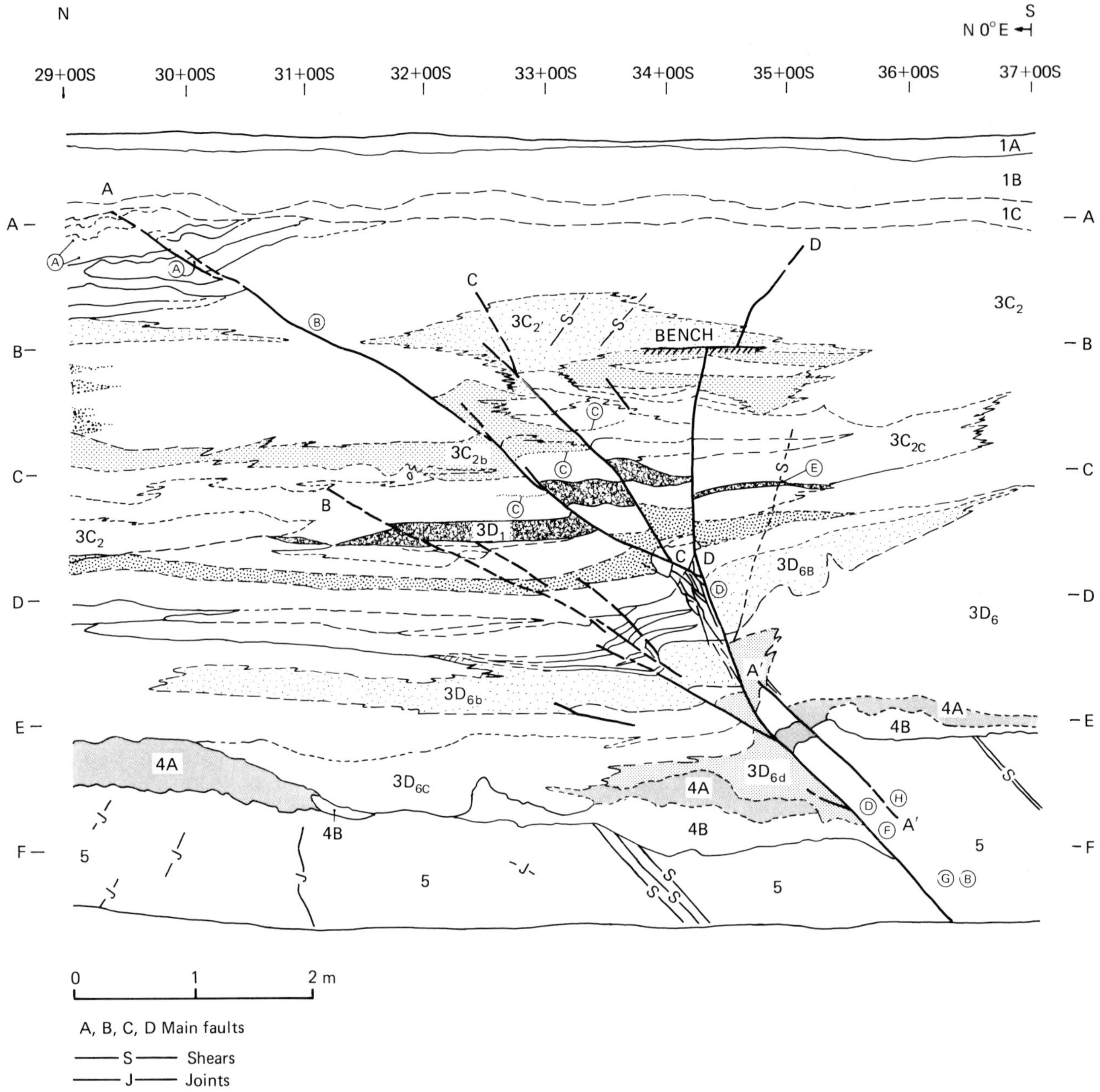

Figure 11.17 Geological plan of part of the east wall of a trench through the Arroyo fault in Quaternary terrace deposits and the uppermost part of the Miocene Sisquoc Formation, California, USA (From Asquith, 1985, Fig. 4, by permission)

oil port, oil refinery, petrochemical works, power station and bauxite smelter, is illustrated by a 1:1250 scale plan in Figure 11.21.

The plan presents some of the results of a second-stage feasibility study which included a limited amount of pitting and boring, and marine and some land geophysical surveys carried out by a specialist firm of site investigation contractors. Mapping, general supervision, and preparation of the geotechnical feasibility report were undertaken by the engineers.

The first-stage reconnaissance consisted of a day visit by an engineering geologist, together with engineers and planners, when the initial feasibility was considered and the second-stage feasibility investigations designed. During the first stage the Geological Survey of Jamaica was consulted, together with inspection of air photography and relevent maps and literature. Because of the lack of any previous detailed geological work and the highly variable nature of the alluvial and marine deposits overlying the bedrock, it was considered that for planning purposes, and for subsequent detailed site investigation, an engineering geological appraisal of the site would be valuable. A plan (Figure 11.21) was therefore prepared while the pitting and boring work was being carried out for the second-phase feasibility investigation. An initial 1:2500 photogeological plan of the site was prepared by the Geological Survey. The field sheets were prepared using the best available topographic surveys (1:2500 enlarged to 1:250) for use as a base plan. The field locations were determined by inspection, supplemented where necessary by compass resection. The mapping took about one month and was carried out by a young engineering geologist as a part of his duties as resident engineer for the boring and pitting contract. He was assisted for part of the time by staff of the Geological Survey. Some additional pitting and seismic geophysical traverses were also carried out for the mapping.

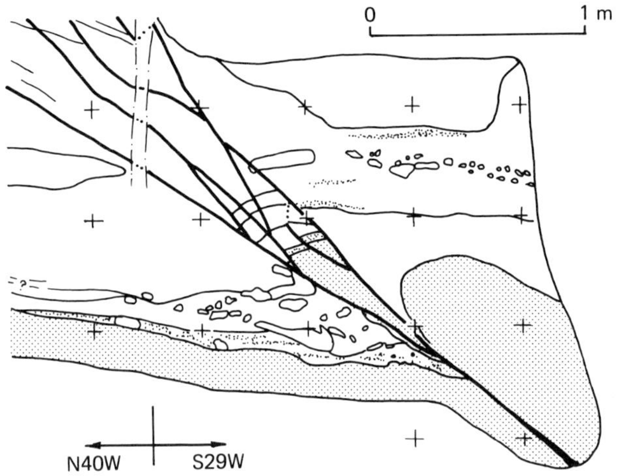

Figure 11.18 Geological plan of the initial Arroyo fault exposure, California, USA, recorded from a survey grid (From McClay, 1985, Fig. 2, by permission)

Mapping units were based on deposit type to which a field estimate of bearing capacity had been added. This enabled further subdivision in terms of estimated foundation performance when watertable and depth to bedrock were considered. The field estimates were subsequently improved on the basis of laboratory and field test data. The completed geotechnical plan was originally produced on a 1:7500 scale.

11.4.6 Conventional mapping on a plan scale topographic base

The Taff Valley trunk road in South Wales runs from the northern outskirts of Cardiff through Pontypridd to Merthyr Tydfil as the southern part of the main South–North Wales highway.

Geotechnical considerations of the site investigations for Stage IV of the road (Fookes, Hinch and Dixon, 1972) had taken into account the difficult alignment conditions imposed in part by the terrain and by considerable stability and foundation problems that were likely to occur. The valley runs north–south and is generally U-shaped and deeply incised to over 1000 ft below the general upland plateau surface formed by the Upper Carboniferous Pennant Sandstone. Boulder Clay (till) occurs along the valley sides, although nearer the river, terrace gravels overlie bedrock. River alluvium occurs locally. Site investigation mapping was carried out in 1971–72.

The engineering geological plan made during the site investigation by an engineering geologist, with assistance as necessary by soils engineers and design engineers, is illustrated in Figure 11.22. The plan was made on 1:1250 scale topographic sheets supplied by the engineer to the site investigation specialist firm. Field work for the portion of the plan shown took about four man-days and consisted of conventional mapping techniques with locations determined by inspection. It was carried out

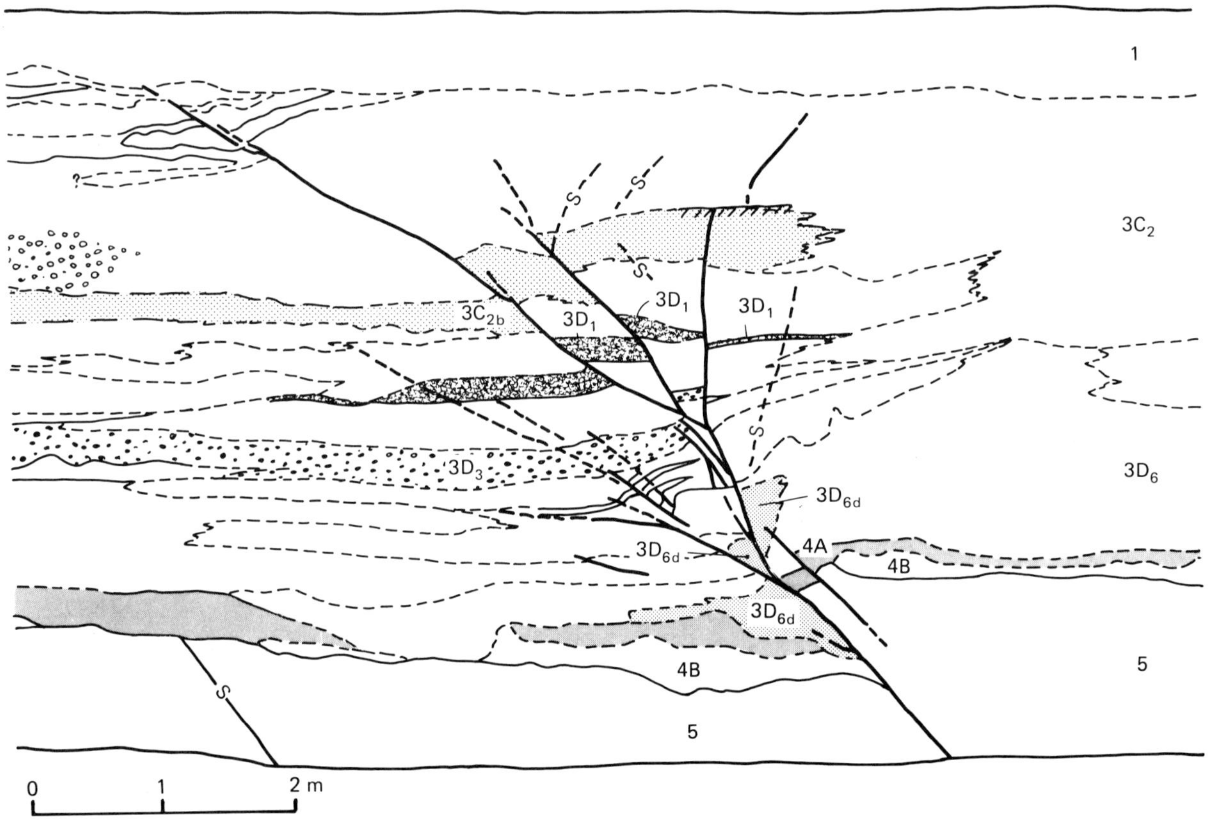

Figure 11.19 Plan based on a photograph of the trench wall illustrated in Figure 11.17 (After McClay, 1985, Fig. 4)

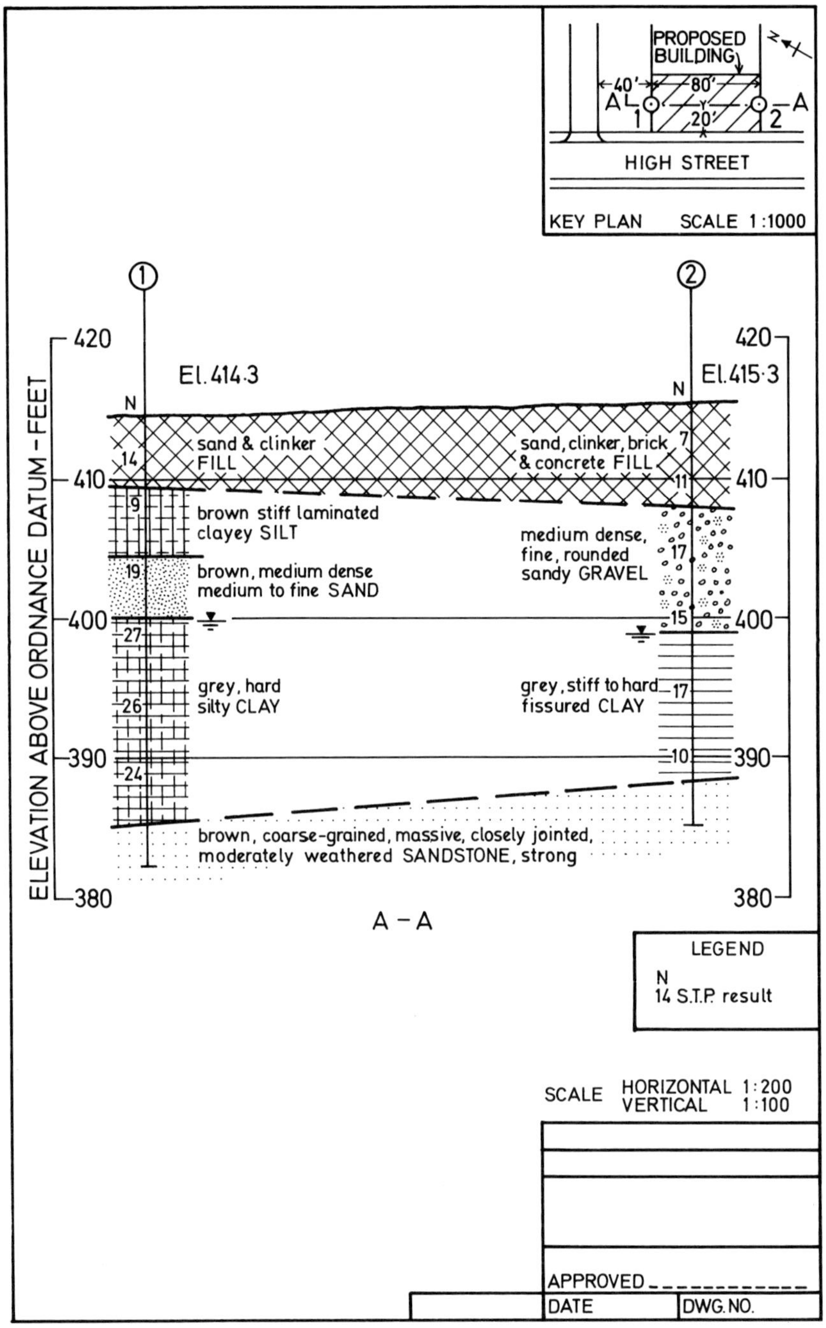

Figure 11.20 Typical plan scale cross-section based on the logs of site investigation boreholes (From Anon., 1972, 13.1.13, by permission)

approximately halfway through the overall field work programme when some of the pits and borings in the area had been completed, but others were still to come after the completion of the field mapping. The tender for the investigation had contained an item based on time for the services of an engineering geologist working on the mapping.

As can be seen the area, unfortunately somewhat typical of this part of the valley, contains many small slips and is generally a wet and geotechnically unattractive location for major roadworks. The plan helped to determine the location of additional pits and boreholes put down in the second half of the investigation and was of great assistance in writing the full investigation report. It was also significant in highlighting to the designers the location of unstable and wet areas which led to small but important changes in the proposed road alignment, and in the development of foundation designs appropriate to the conditions. The same area was mapped using the then newly developing techniques of engineering geomorphological mapping (Figure 11.23).

This work was carried out after completion of the site investigation by an experienced geomorphologist assisted by a research student. Mapping was completed in about three man-days on 1:1250 base topographic plans supplied by the engineer. It was also deliberately done as an experiment carried out by the mappers without prior knowledge of the findings either of the site investigation or the previous mapping. As can be seen, the two plans have been produced in terms suitable for

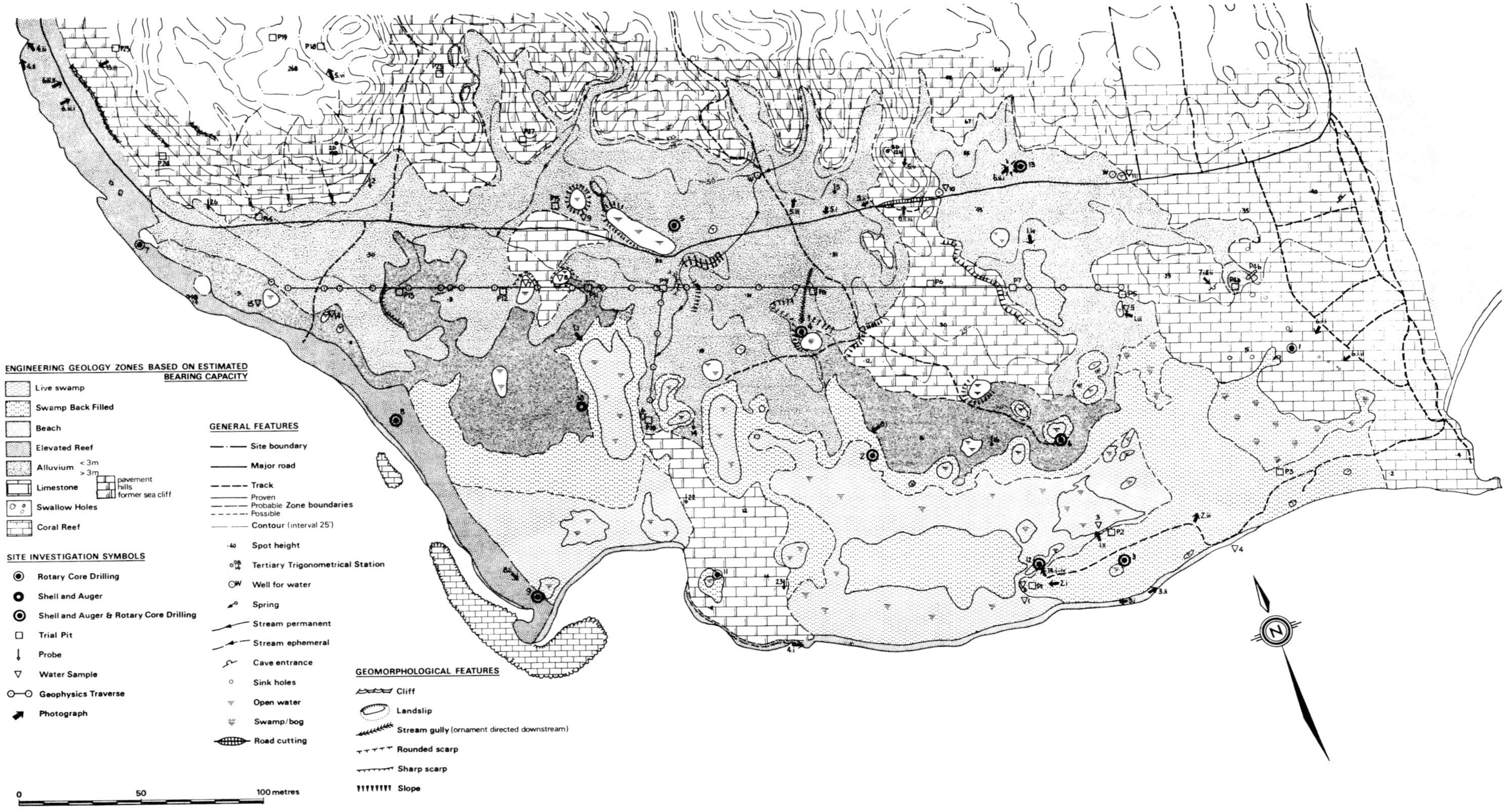

Figure 11.21 Geotechnical mapping, plan scale, at the reconnaissance stage of a proposed site for an industrial complex in Jamaica. The engineering geological zones are characterized in terms of estimated bearing capacity (From Dearman and Fookes, 1974, Fig. 10, by permission)

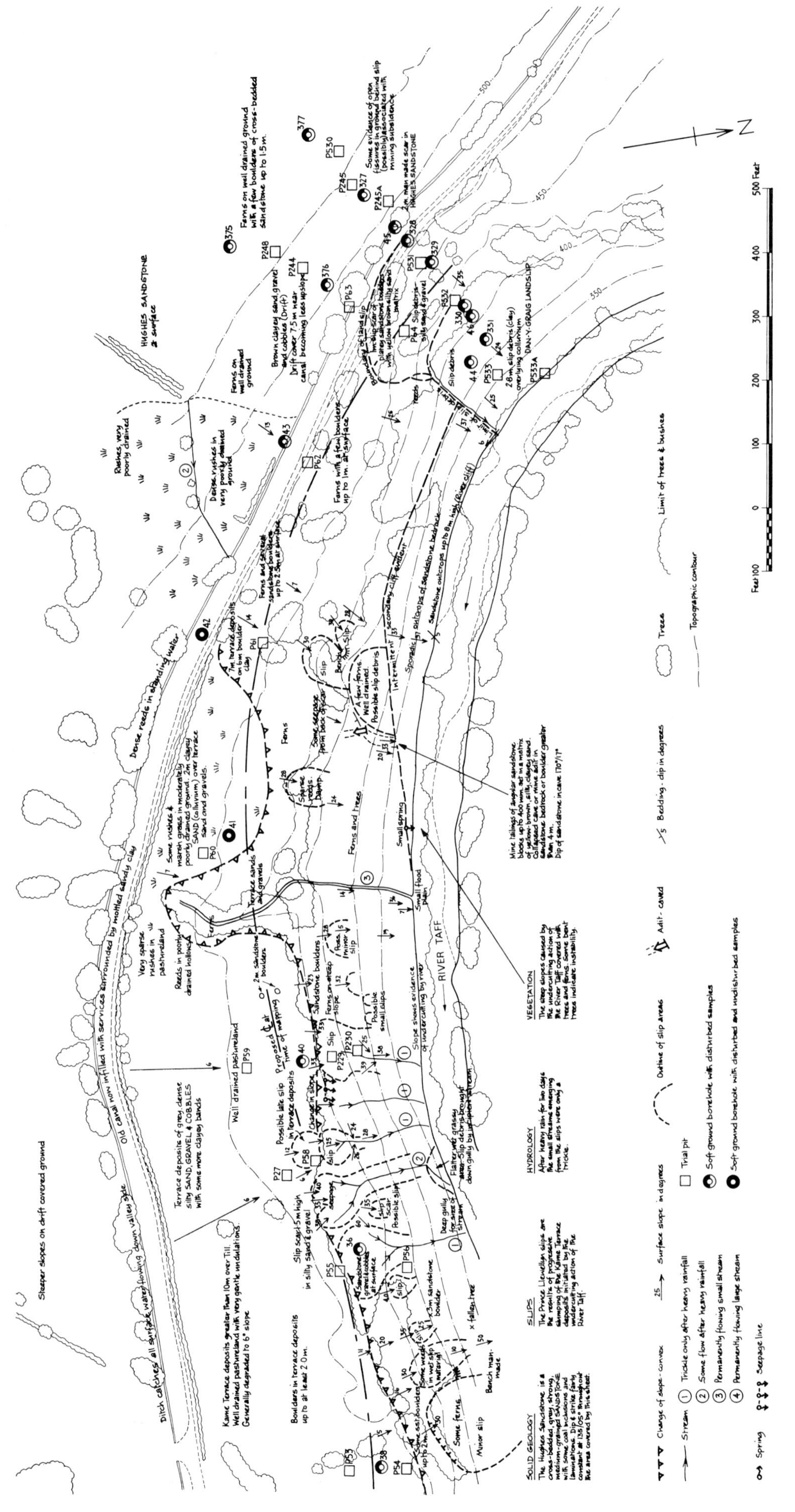

Figure 11.22 Engineering geological plan produced at the site investigation stage, Prince Llewellyn area, Stage IV, Taff Valley trunk road, South Wales, UK. The contours are in feet AOD

the area – one highlighting the ground slope conditions, the other ground stability and similar significant conditions. These plans again helped in foundation designs, in settling the proposed alignment of the road, and in the design of drainage measures.

11.5 Examples of plans produced by terrestrial photography

11.5.1 Recording a relict cliff face in Tremadoc village, North Wales

Introduction

Tremadoc village (Figure 11.24) is one of the first examples of town planning in Britain, its most striking feature being a large market square dominated by a prominent rock face on the north side of the village (Figure 11.25). A relict cliff related to an earlier shoreline its presence is due to a continuous dolerite sill intruded into the softer Ordovician shales of the area. The rock face, varying in height from 20 m to 30 m from east to west, supports an abundant growth of stunted oak trees, ferns, bracken and grasses on the lower slopes of the rock face, except where scree is present (Figures 11.25 and 11.26). The present configuration is of rock buttresses alternating with V-shaped gullies, partly attributed to current weathering.

Engineering geological considerations

The rock face is very close to the row of terraced houses (Figure 11.25), leaving no doubt that a major slope failure would cause extensive damage. A failure involving a 3 m cubed block which slid to within 10 m of Sunnyside Terrace prompted a walk-over survey of the area. It became evident that the rock face was in very poor condition and that ravelling of small to medium sized (0.5–5.0 m^3) joint bounded blocks would be a continuing hazard. Of greater importance were tension cracks in the east section, some associated with substantial overhangs. The potential for a failure in one particular section involving about 600 t led to a strong recommendation that a detailed survey should be made in order that stability analyses could be made. Powell and Mercer (1979) have described the application of terrestrial photogrammetry to solve the problem of access, giving at the same time an estimate of its cost effectiveness.

Engineering geological mapping

A large proportion of the time that any rock slope stability investigation takes is spent on collection and interpretation of geological data – chiefly significant discontinuities within the rock mass that determine potential failure mechanisms.

On natural or existing rock slopes, such as Tremadoc, where exposure is good, the feasibility stage of a site investigation may only involve a walk-over survey. If remedial measures are envisaged, most of the effort can be divided into design stage investigations. Choice of mapping method is influenced by a number of physical factors, mainly the size of the area, the height of the slope, accessibility both from above and below, the amount of exposure and the density of vegetation. The relative importance of each aspect may vary, but at Tremadoc the following points arose:

- a slope height of 20–30 m and an unfavourable geometry;
- the generally unstable nature of the face;
- poor access to slopes below and above the rock face;
- lack of survey drawings on which to map;
- the density of trees on the lower slope which obscured some detail and hindered surface photography.

A number of mapping methods were considered (Table 11.1). Coupled with photography, direct mapping was considered to be just possible. Structural geological features could be located with greater precision at the top and bottom of the face, if the mapping was combined with a theodolite survey, but a number of critical joint surfaces would not be accessible for measurement. For both methods it was considered that the work would be hazardous and would limit the survey to identifying and positioning major discontinuities.

Table 11.1 Summary of mapping methods

Method	*Accuracy*	*Remarks*
Direct mapping	At best, ±1000 mm	Very slow rate of working Surface photography essential + access by rope, where possible, to critical surfaces Overall and local accuracy very poor Only a limited number of joint orientation measurements possible Assessment of discontinuity properties slightly better than achieved using indirect mapping methods Dangerous if parts of the face are highly unstable
Direct mapping and theodolite survey	±500 mm	Same as above but increased local accuracy where access is reasonable Will be more expensive because of the survey involvement
Indirect mapping—terrestrial photogrammetry	±200 mm	Considerably more accurate than direct mapping with a local accuracy of ±100 mm for detail and ±50 mm for co-ordinated points Joint intersections and positions obtained accurately Orientations of critical surfaces obtained provided they are at a suitable angle to the camera axis The overall assessment of discontinuity properties is not quite as good as that from direct mapping (Anon., 1977) Less hazardous than direct mapping Permanent monitoring stations can be established Can also be used for topographic survey purposes

It was apparent after discussion that the same drawbacks would not apply to terrestrial photogrammetry, which would enable base elevations to be drawn up and relevant geotechnical data to be plotted. Also the method offered a possibility of determining the orientation of most joint surfaces, except those striking

SLOPE ANGLE CATEGORIES
0 - 10
10 - 20
20 - 30
30 - 40
>40
PROPOSED REALIGNMENT OF ROAD
RIVER TAFF
Line of maximum stream velocity
N
100 0 100 200 300 400 500 feet

(a)

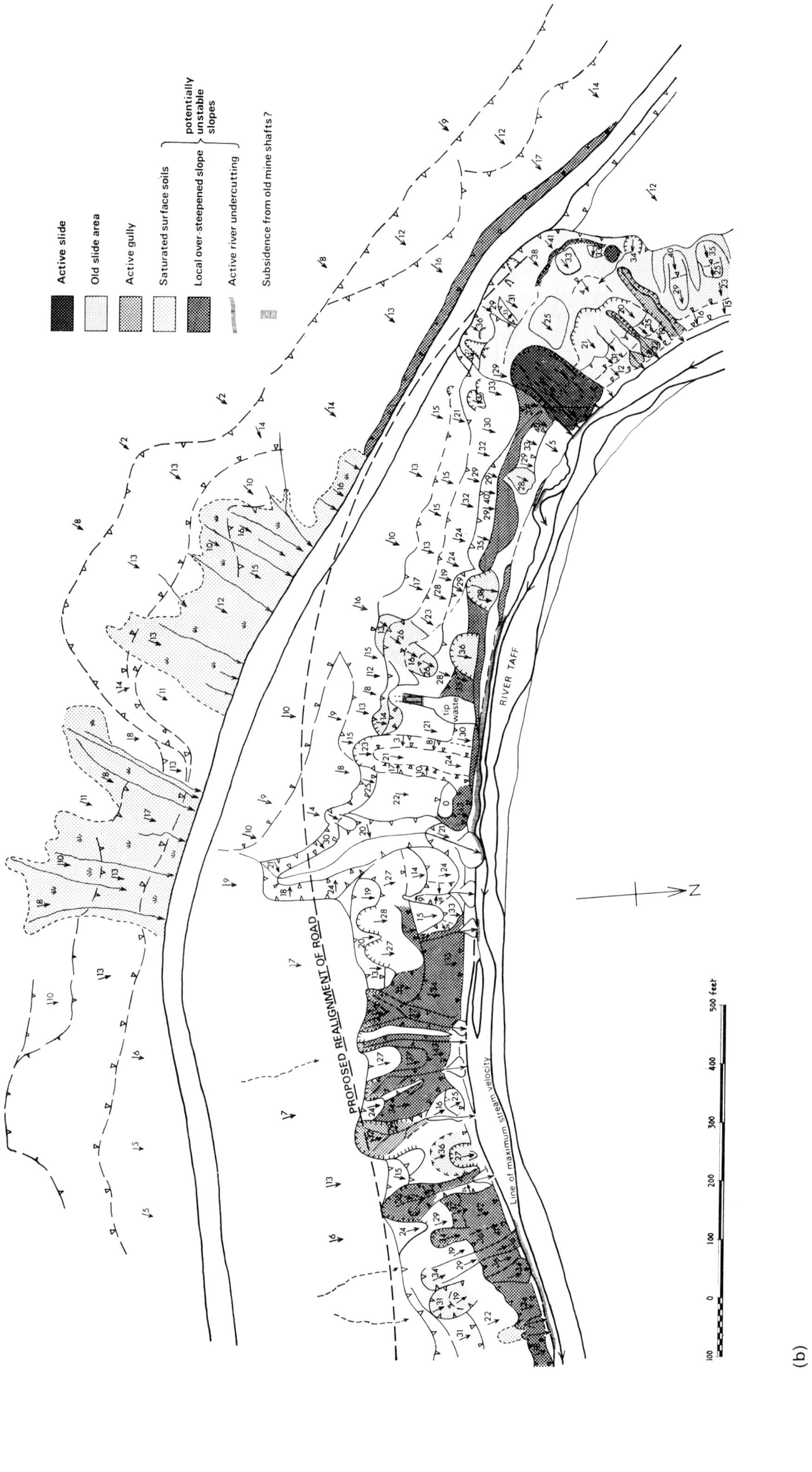

Figure 11.23 Engineering geomorphological plan of the Prince Llewellyn area (Figure 11.22): (a) slope angle categories; (b) active geomorphological processes

parallel to the common axis, producing cross-sections at predetermined intervals, and contributing to general topographic work. After discussions with a specialist firm, photogrammetry was recommended for the survey.

Figure 11.24 Location of Tremadoc village in North Wales, UK

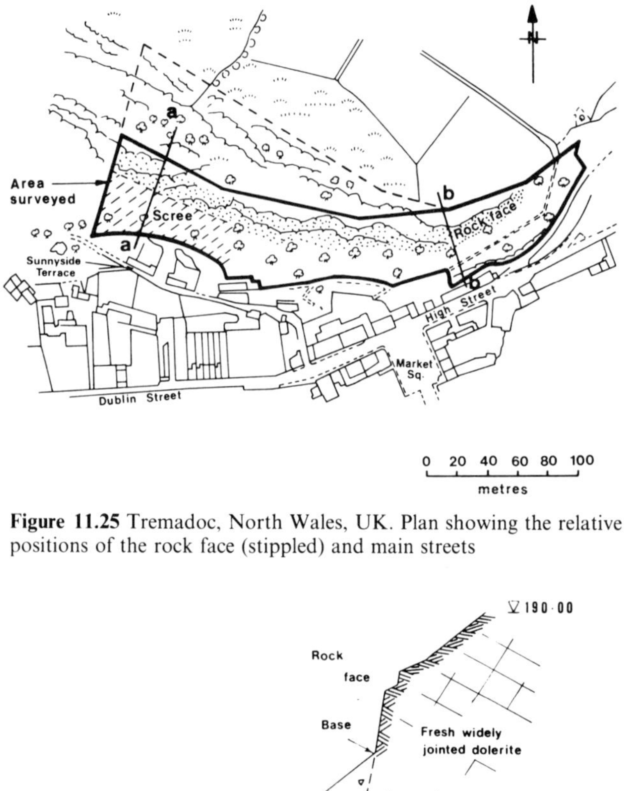

Figure 11.25 Tremadoc, North Wales, UK. Plan showing the relative positions of the rock face (stippled) and main streets

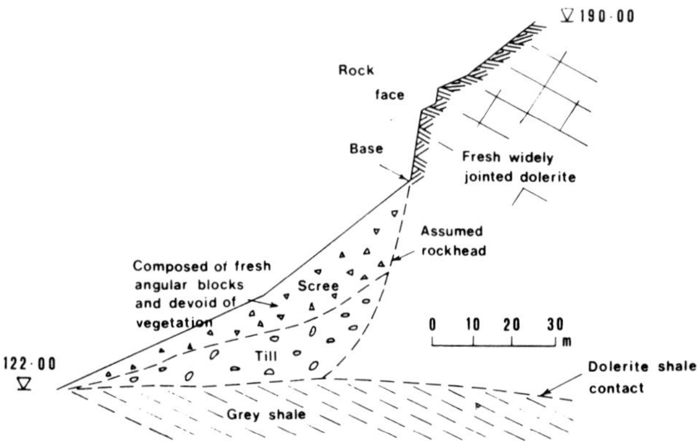

Figure 11.26 Cross-section of west part of the rock face, a–a (Figure 11.25)

Methods used in the survey

Vertical contouring – the method finally adopted – produces profiles like those in Figures 11.26 and 11.27 at predetermined intervals along the face at right angles to the baseline. It has the advantage that any sharp breaks in profile are immediately revealed; it can be assumed that the breaks coincide with discontinuities, although they may not significantly influence the stability of the face.

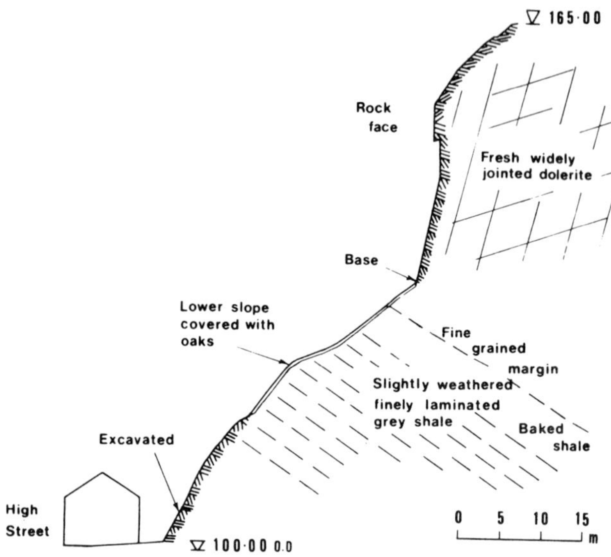

Figure 11.27 Cross-section of east part of the rock face, b–b (Figure 11.25)

It is not necessary here to go into actual survey methods in any detail. Briefly the rock face is 500 m in length, with a slight change at the east end, and varies in height from 60 to 120 m above ground level. Suitable photographic baselines had to be located along or near main roads (Figure 11.25) because of the houses at the foot of the slope. Photographs were taken with a Carl Zeiss Jena UMK1318 Terrestrial Survey Camera: six at intervals of 30 m for the western section, and 13 over the eastern section at intervals varying from 10 to 30 m, depending on the position of obstructions and the closeness of the camera to the rock face. The photographs were taken at a height of approximately 10 m above ground level using a hydraulic platform. Despite the method of positioning, the maximum deviation from line and level was $\pm 2^\circ$ and this was later corrected during the photogrammetric procedure.

Twenty-two control points were established on the rock face and adjoining buildings to control the photographs, and the co-ordinates of these were later transformed to relate to each baseline. Photo-control points were located both at the top and bottom of the rock face, with at least three for each pair of overlapping photographs.

Gross irregularity, combined with the depth of the face, meant that mapping methods adapted to a plane excavated rock face, mainly involving location and plotting of major discontinuity surfaces, could not be used. Accordingly, work proceeded as follows with the provision of:

(a) one copy (contact prints) of all stereo-models;
(b) photographic enlargements at 1:200 scale, reducing the stereo-models to a number of mosaics;
(c) one negative of all the stereo-models;
(d) an elevation survey 1:200 scale parallel to the camera stations on a stable transparent material with all geological and related features plotted as defined by the engineer – accuracy to be ± 50 mm;
(e) three-dimensional co-ordinates for specific points supplied in a form suitable for processing on a desk-top computer;
(f) three-dimensional co-ordinates at nominated sections through the slope.

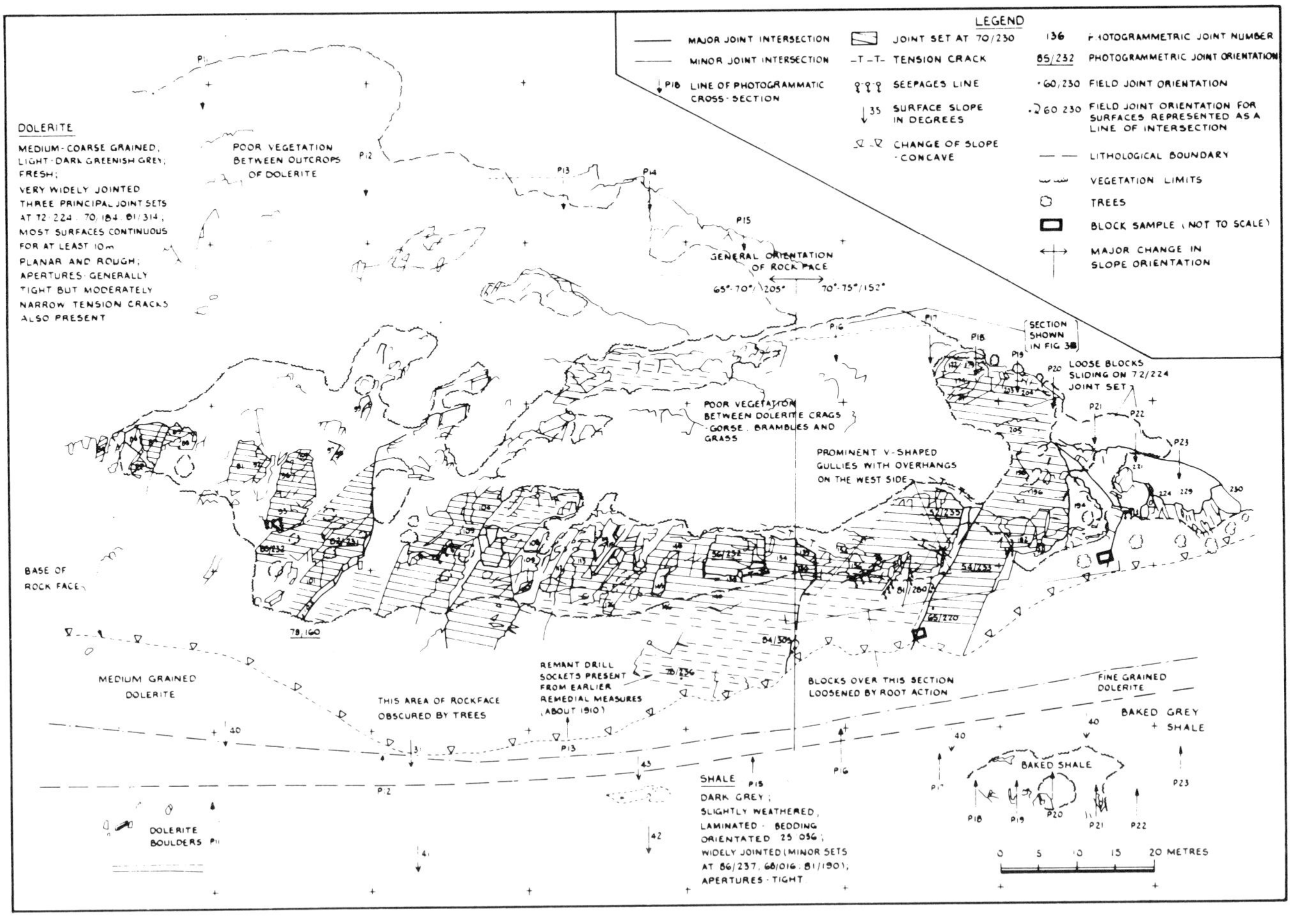

Figure 11.28 Geotechnical elevation of the eastern half of the rock face (Figure 11.25)

Sections (a) and (b) of the specification form the basis for producing the elevations. On completion of the photography (carried out in March when all vegetation had died down), a set of contact prints and enlargements was forwarded to the engineer. Initially, a field check was carried out to determine what information could be obtained from the stereo-models and an assessment made of how much of this would be fundamental to design stage investigations. The required information was then plotted on the enlargements and compiled by assessing the stereo-pairs of each section of the face using a mirror stereoscope.

The main element of the work, apart from providing an accurate grid of the rock face, was to make certain that all significant joint intersections, areas of exposure, vegetation limits, positions of properties, scree limits and changes of slope were plotted on the elevation. In addition, co-ordinated points were required for specific joint surfaces and survey control. Each of the parameters was marked on the enlargements using permanent colour fibre-tip pens.

The process of using stereo-pairs and enlargements is straightforward and enables a large amount of detail to be plotted quickly and accurately. This ensures that the engineering geologist's time is used in assessing each part of the slope in an overall context rather than spending long periods in the field mapping limited amounts of outcrop which, because of poor accessibility and the topography, would not be seen in relation to adjacent sections. This is particularly important in some areas where significant discontinuities need to be identified. The marking up of detail is followed by the positioning of co-ordinated points on the joint surfaces (at least four points on each surface and given as large a spread as possible to minimize errors in computing the dip and dip direction), and the nomination of cross-sections both at set intervals and at selected locations where stability problems were anticipated.

On completion, the enlargements were forwarded to the survey firm for elevations to be produced to scale.

Production of plans of the rock face

A completed engineering geological plan is illustrated in Figure 11.28. Once the main joint sets were established, shading enabled a partial three-dimensional effect to be obtained, thus making the plan (elevation) easier to understand.

The representation of orientation data on the plans can be accomplished either by utilizing the joint reference number allocated during the photogrammetric stage or by replacing this number with the actual reading. A combination of the two is probably to be preferred, with the most significant discontinuities given an orientation and the remainder represented by the number and shading relevant to the joint set. In addition, a number of field readings will have to be recorded anyway, mostly those surfaces at an unsuitable angle to baselines and important in the analysis of particular sections of the face. The plans when finally completed must be reasonably easy to interpret and useful throughout the design stage.

Selection of recording method

Commercial mapping is always subject to financial constraints, and one important outcome of the Tremadoc mapping exercise was an estimate of the cost effectiveness of the method finally adopted. The various stages which should be gone through in deciding which method of mapping should be used for a rock slope stability investigation are summarized in Figure 11.29.

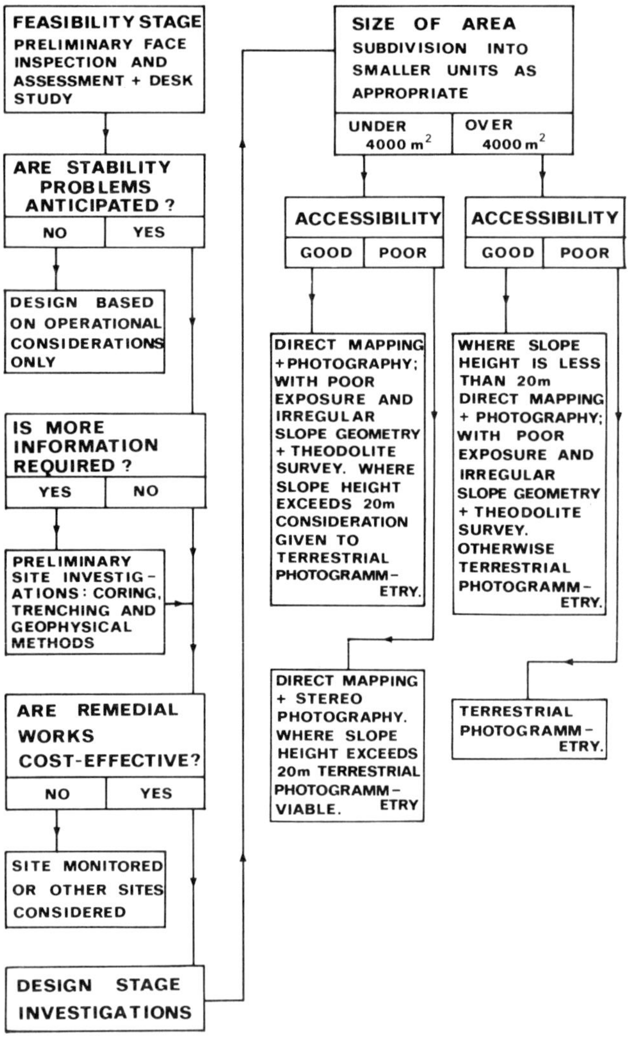

Figure 11.29 Selection of mapping methods for slope stability investigation

Relative costs of different mapping methods (Table 11.2) have been assessed on the following factors:

(a) the initial inspection of the slope;
(b) a desk study;
(c) determining mapping procedures;
(d) the length of mapping stages;
(e) the production of plans;
(f) analysis of survey data, e.g. joint readings;
(g) preparation of reports;
(h) any additional derived benefits, e.g. the use of photogrammetry for topographic purposes.

with major differences being chiefly related to items (d), (e), (f) and (h).

Table 11.2 Relative costs of different mapping methods at Tremadoc, North Wales, UK

Method	*Cost* (in units)	*Remarks*
Direct mapping	5500–6100	
Direct mapping and theodolite survey	7000–8500	Overall cost includes for a surveyor with electronic distance-measuring equipment and an engineering geologist holding the reflector
Indirect mapping—terrestrial photogrammetry	2300–3500	Overall estimate obtained using a budget price. Final cost fell within this range, but it should be noted that if an average value of quoted rates was taken for the photogrammetry content of the work, then the range of values would need to be increased by 30%

As a result of the investigation, Powell and Mercer (1979) concluded that a flexible approach to the mapping of rock slopes was necessary. Slopes with an area greater than 4000 m^2 can be mapped with advantage by terrestrial photogrammetry. At different sites the method can be used to define joint intersections, either between sets or with the rock face; obtain joint orientations; and compile cross-sections from which accurate engineering geological elevations can be produced. Although the work needs to be supplemented and corrected at various stages by field checks, the ease of operation, amount of data which can be processed accurately and the additional survey facility which it offers suggest that there is scope for increasing the use of terrestrial photogrammetry in rock slope stability investigations.

11.6 Application of geomorphological mapping to site investigation

Geomorphological mapping appears to have its main application in the investigation of road alignments, and indeed one example has already been given (see Figure 11.23) of a trial of the method in site investigation. Since that time a considerable body of experience has been built up in mountainous areas for which a combination of aerial photograph interpretation and geomorphological analysis has been particularly cost effective. One example will suffice.

11.6.1 Methodology

The very extensive range of international symbols developed for geomorphological mapping, some of which are illustrated in Chapter 5, are used in a formal context in Figures 11.30–11.33. It should be remembered that the recording of topographical features, as a guide to the disposition of unexposed soils and rocks, has always been an important aspect of conventional geological mapping (see subsection 3.2.2.1).

11.6.2 General comments

To geotechnical engineers the general advantages of geomorphological mapping techniques are:

- the speed at which the work can be carried out;
- the extent to which the form and origin of landforms can be identified;
- the inclusion on the map of features of importance to engineering interpretation, such as ancient and recent slope features, surface and groundwater characteristics, and superficial deposits.

The aims of a geomorphological survey for highway engineering are to guide and complement a site investigation by:

1. Identifying the broad terrain characteristics of an area to be crossed by the route corridor, and hence to provide a basis for the evaluation of possible locations, enabling hazard areas to be avoided wherever possible.
2. Identifying influences upon the corridor from outside its area, such as landslides and rockfalls. Mapping beyond the corridor boundaries may provide evidence that could not have been gained from an investigation of the site alone, and thus would not have been included in the site investigation report.
3. Providing a synopsis of the geomorphological development of the area with reference to:
   (a) the availability of construction materials for fill, sub-base, base and wearing course;
   (b) the presence of active processes that might affect road construction or long-term safety of the road.
4. Defining specific hazards, especially aspects of slope stability and vigorous fluvial activity.
5. Describing the surface and subsurface drainage patterns, thereby enabling early design and costing of drainage measures.
6. Providing a classification of slopes on the basis of steepness, material composition, development and stability to assist in estimating land cut or fill and costing for stability measures.
7. Characterizing the extent, depth and susceptibility of an area to weathering, karst development, erosion or mining subsidence.
8. Clearly defining the shape of geomorphological units, in order to provide a framework within which:
   (a) the detailed site investigation can be designed with optimum location of boreholes and trial pits and decreased costs;
   (b) the findings of the geotechnical survey can be extended from the sample points.

11.6.3 Geomorphological mapping for routes in mountainous regions

Mapping in mountainous regions can only be undertaken by walking the route corridor, recording on available maps and aerial photographs the landforms, sources of material and major hazards to road construction. For the linear engineering structures involved, cuttings, embankments, portals to tunnels and bridge sites are all required. For surface structures, engineering geomorphology is particularly important because of the steepness of the slopes, slope instabilities, localized intense gully erosion and associated fan accumulations, talus, screes and mudflows. In general, the important engineering problems are associated with the nature of near-surface rocks and soils, slope stability and drainage; all of these require plan-scale mapping for the development of engineering solutions.

Mountain environment

Over the surface of the globe, mountains are extremely diverse, varying from the isolated accumulations of volcanic mountains to the uplifted tectonic fold and nappe mountain chains. Mountains are characterized by high relative relief, with highlands and uplands rising 700–1000 m above their surroundings, and mountains exceeding 1000 m.

High relief is associated with very rapid erosion and consequent steep slopes. Particularly for the Alpine-Himalayan and the Circum-Pacific belt of young mountain chains, active neotectonic uplift sustains the erosional activity and hence the high relief. World climatic zones are modified within such tectonically active areas and accumulation of characteristic weathering crusts is often limited by the rapid erosion rates, and development of particular types by altitude.

It is not proposed here to develop a classification of mountain types for mapping purposes. Rather, emphasis will be given to a conceptual model of one mountain environment and its microzonal features. The microzonal features will then be reviewed as they apply to engineering geological mapping for highway construction.

Typical landform associations in mountain areas

Although mountainous terrains are diverse in their topography, geology and climate, there are a number of landforms which occur frequently. A simple model of fold-mountain topography is shown in Figure 11.30(a) (Fookes *et al.*, 1985; Brunsden and Allison, 1986). Five main units are defined, although Unit 1, the high altitude glacial and periglacial unit, will not be considered further. Similarly in Unit 5 most landforms and materials are related to fluvial processes that are common to other natural environments.

The model illustrates a two-cycle development of valley slopes in which an early phase of rapid downward incision by major streams and tributaries produced the very steep rock slopes of Unit 2. As downcutting eventually gave way to lateral erosion, meandering of the rivers widened the valley floor and formed a floodplain with terraces. This is Unit 3, and continued fluvial activity at this stage would lead to the degradation of the steep rock slopes of the Unit 2 valley walls, except where meander bends impinging on the valley slopes maintain active erosion of the steep rock slopes. Rejuvenation of the stream by tectonic uplift with renewed active downcutting would incise the flat valley floors of Unit 3. Steep geodynamically active slopes would be the result, with Unit 2 type topographic features forming in rock and soil slopes in the ancient alluvium of the Unit 3 valley floor. Minor tributaries would actively erode into gullies discharging alluvial cones where the valley slopes flattened onto the new valley floor.

Depending on the scale of the original mapping, the simplified model illustrated in Figure 11.30(a) can be

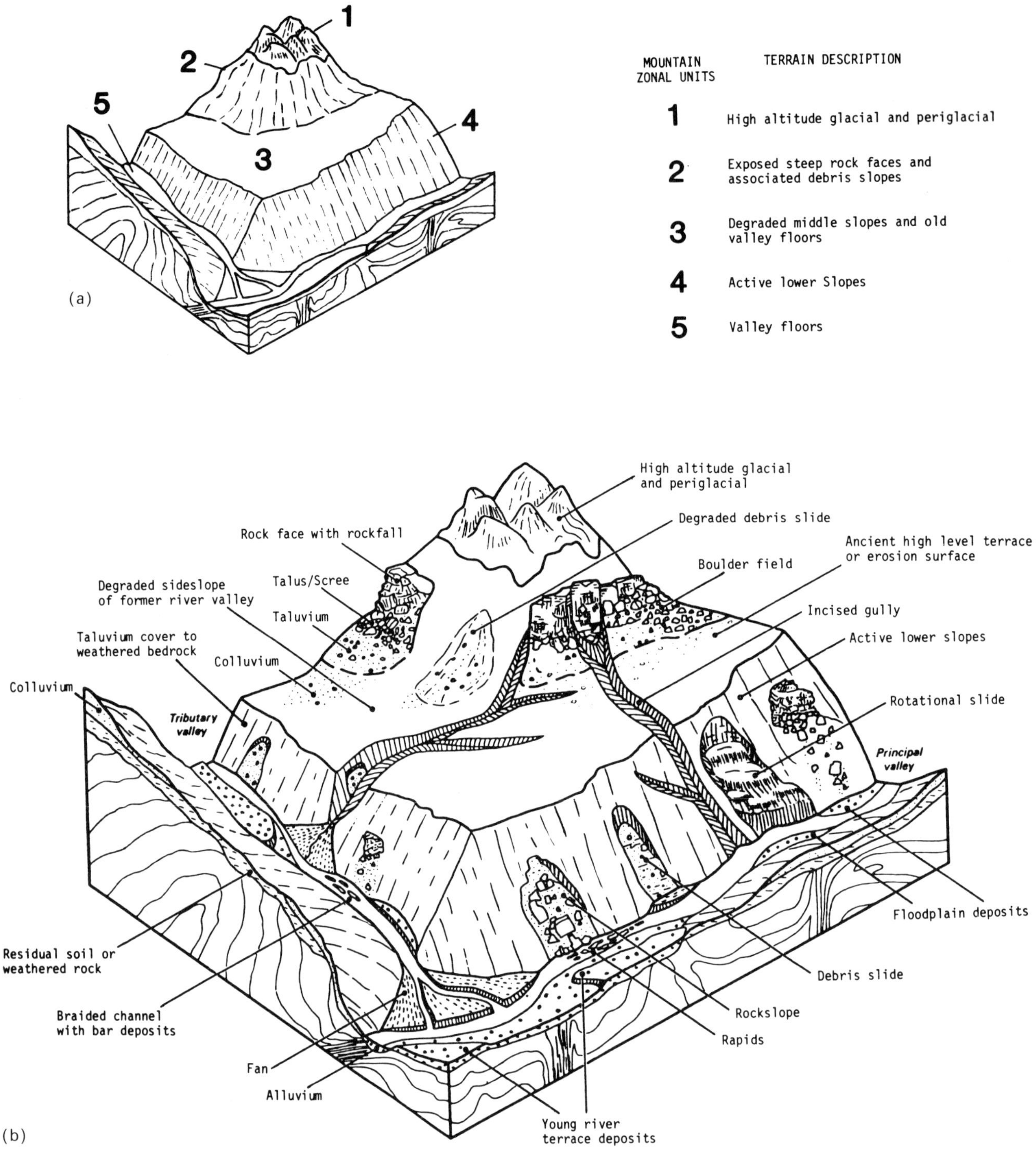

Figure 11.30 Block diagrams illustrating typical landform associations in mountainous areas: (a) the different zonal units; (b) the main zonal units with associated microzonal geomorphological features (After Fookes *et al.*, 1985)

regarded as a uniform natural *unit* from the point of view of general-purpose zoning. The model represents a uniform *region* in terms of geotectonics; a uniform *area* from the broad geomorphological point of view; a uniform *zone* because rocks and soils, although diverse, each have an integral homogeneity and genetic association; a uniform *district* because there is limited variability in both hydrogeological and geodynamic conditions.

On the other hand, from the point of view of a special-purpose engineering geological map, it could be grouped into more than one type of *area* on the basis of geomorphological conditions. Units 2 and 4 and Units 3 and 5 have common genetic backgrounds and could comprise two distinct types of area. Because of the landscape, the youthful steep slopes of Unit 4 can be distinguished from the older, more mature and less geodynamically active steep slopes of Unit 2. There are also sufficient differences between the flat slopes of Units 3 and 5 to consider these as sufficiently distinctive to warrant separation. Units 2, 3, 4 and 5 then each become separate *areas* in the general-purpose zoning scheme, but a similar subdivision on the basis of region, zone and district could not be justified so readily.

On a special-purpose map, for example an engineering geological map prepared at large scale for road alignments and engineering, each of the terrain units in the model, with the possible exception of Unit 1, would be highly distinctive in terms of materials, denudation processes and slope forms. The terrain units comprise distinctive mountain zones with the following characteristic engineering geological conditions:

Unit 2 *Exposed steep rock faces and associated debris slopes.*
MATERIALS: rock exposures; coarse debris from weathering of rock faces, boulder fields and screes; taluvium of mixed sand, gravel and cobbles with some fines.
DENUDATION PROCESSES: dominant mechanical with some chemical weathering; instability in rock and soil masses; subsurface water erosion, talus creep, soil creep; possible frost wedging and freeze-thaw activity.
SLOPE FORMS: rock cliffs; moderate to high angle, course debris mantled slopes; boulder fields, scree slopes; fresh rock scars and debris cones associated with slope failures.

Unit 3 *Degraded middle slopes and old valley floors.*
MATERIALS: old river terrace and fan deposits; *in situ* residual soil mantles; colluvium.
DENUDATION PROCESSES: chemical weathering; soil creep, unconcentrated and concentrated surface water erosion.
SLOPE FORMS: low angle or flat remnants of erosion surfaces; partly degraded low to moderate angle valley side slopes; ancient river terraces; ancient fans; ancient soil and rock failure forms, e.g. degraded rock scars, vegetated subdued accumulations of landslide debris.

Unit 4 *Active lower slopes.*
MATERIALS: taluvium and colluvium overlying *in situ* weathered rock; alluvium of ancient river terraces associated with Unit 3; some bare rock exposures with associated boulder fields and screes.
DENUDATION PROCESSES: instability in rock and soil masses; chemical and mechanical weathering; surface water erosion by sheet flow and gullying; soil creep; subsurface water erosion.
SLOPE FORMS: high-angle, usually straight, debris-mantled slopes; small rock faces and cliffs; steep-sided seasonal and perennial gullies; soil and rock failure forms both active and degraded.

Unit 5 *Valley floors.*
MATERIALS: rock exposure in channel beds; mainly coarse river alluvium, channel floor, terrace and fan materials covering a complete range of grain sizes.
DENUDATION PROCESSES: concentrated surface water erosion in perennial streams and rivers; small-scale instabilities by undercutting alluvial terrace banks.
SLOPE FORMS: low-angle or flat river channel, terrace and slopes separated by small, steep bluffs; landforms of fluvial erosion; meander scars, bedrock cut terraces; and of fluvial deposition – alluvial terraces, fans, gravel bars and sheets.

Most of these features are illustrated on the block diagram in Figure 11.30(b) which shows the detailed erosional and depositional forms likely to be associated with four of the five units of the basic model of Figure 11.30(a).

The mountain terrain model has been progressively developed over a number of years by a team of engineering geomorphologists and engineering geologists working on a highway project in Nepal and Pakistan (Brunsden *et al.*, 1975, 1981; Jones, Brunsden and Goudie, 1983; Fookes *et al.*, 1985; Brunsden and Allison, 1986). Abundantly illustrated by geomorphological maps and plans, these papers present the reality of transferring the general detail of the model to site-specific engineering geomorphological maps and plans.

In the literature there are many similar examples of detailed mapping in mountainous areas. These usually involve slope deformations, many of which are actively moving or may be reactivated by the proposed engineering works (Mahr and Malgot, 1978, 1985; Malgot and Mahr, 1979; Matula *et al.*, 1985).

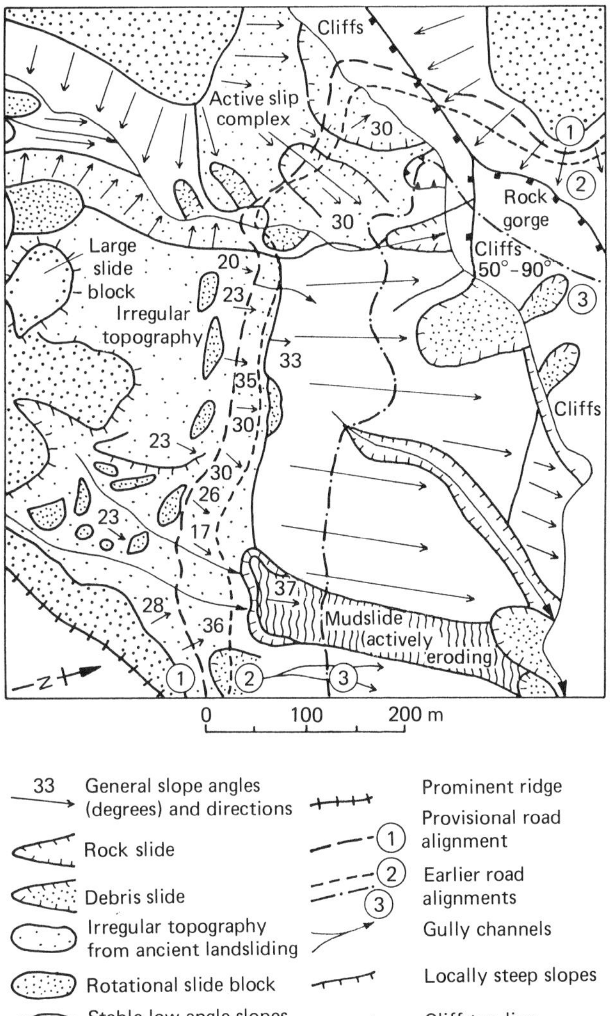

Figure 11.31 Example of low-cost detailed mapping in a mountainous area for a road alignment. This is large-scale mapping, mainly of geomorphological features, done at the reconnaissance stage by field sketching (After Fookes *et al.*, 1985, Fig. 14)

11.6.4 Examples of engineering geological mapping from Nepal

To illustrate the application of mapping techniques at the various stages of a project, Fookes *et al.* (1985, Fig. 14) gave as a case study a small part of the Dharan–Dhankuta road in Nepal. This is an example of mapping in Units 3 and 4 of the mountain terrain model (Figure 11.30a), where the steep valley side slopes are dissected by two major gullies. The relative relief is more than 700 m. The route corridor was forced by other constraints to cross the gullies at about mid-height approximately at the boundary between Zone 3, agricultural land, and the steep slopes of Zone 4. Air photography interpretation at the feasibility stage was made difficult by the generally dense sub-tropical forest, which meant that it was not clear at this stage that the road would have to cross extensive landslides and erosion gullies. Feasibility mapping was at an approximate scale of 1:22 000, followed by field sketch mapping at 1:3300 at the reconnaissance stage. It is a slightly modified version of the latter map that is illustrated in Figure 11.31. Contours have been left off for clarity. The map is clearly geomorphological, geodynamic and hydrogeological, and the only reference to solid geology is the note 'Importance of mica schist dipping steeply north was appreciated'.

Apart from symbols for a slope angle and a variety of form-lines, what mapping units are shown? There are six main units:

- cliffs in bedrock with rock exposed;
- steep slopes;
- stable low-angle slopes;
- irregular topography from ancient landsliding;
- rotational slide blocks;
- landslides of different types.

Within these map units there are, variously, a lithologically homogeneous bedrock unit and five distinct types of 'micro-districts', in each of which hydrogeological conditions and geodynamic phenomena are uniform. The map in Figure 11.31 could therefore be simplified into six distinct types of microzones which, if topographic contours are retained, present an easily assimilable map free from possibly confusing detail. Use colours to represent the six microzones and the simplification process for ease of understanding by the non-engineering geological specialist is complete.

In the mapping process it is obviously essential that each unit be circumscribed by a boundary which of course may have varying degress of certainty. Figure 11.32 is another road example from Nepal (Fookes *et al.*, 1985, Fig. 38A) which illustrates this last point. Rock outcrops are clearly indicated, as is the extent of a possible planar rockslide. But the scree run is not connected to the uphill scarps of the ancient debris slide presumably feeding the scree run. Reviewing the literature, there are many examples of geomorphological mapping for road alignments in mountainous areas.

Another map from Nepal is for the site of a proposed bridge crossing in Mountain Unit 5 with an associated series of hairpins to gain height rapidly on the steep slopes of Mountain Unit 4 rising above the river. The map is very complex, with a legend containing 52 distinct symbols for drainage, sediments, bedrock, terraces, fans, unstable ground and other information including the road alignment. Fifty-two symbols on a map reaches more or less the limit for comprehension.

The map (Brunsden *et al.*, 1975, Fig. 4) is reproduced

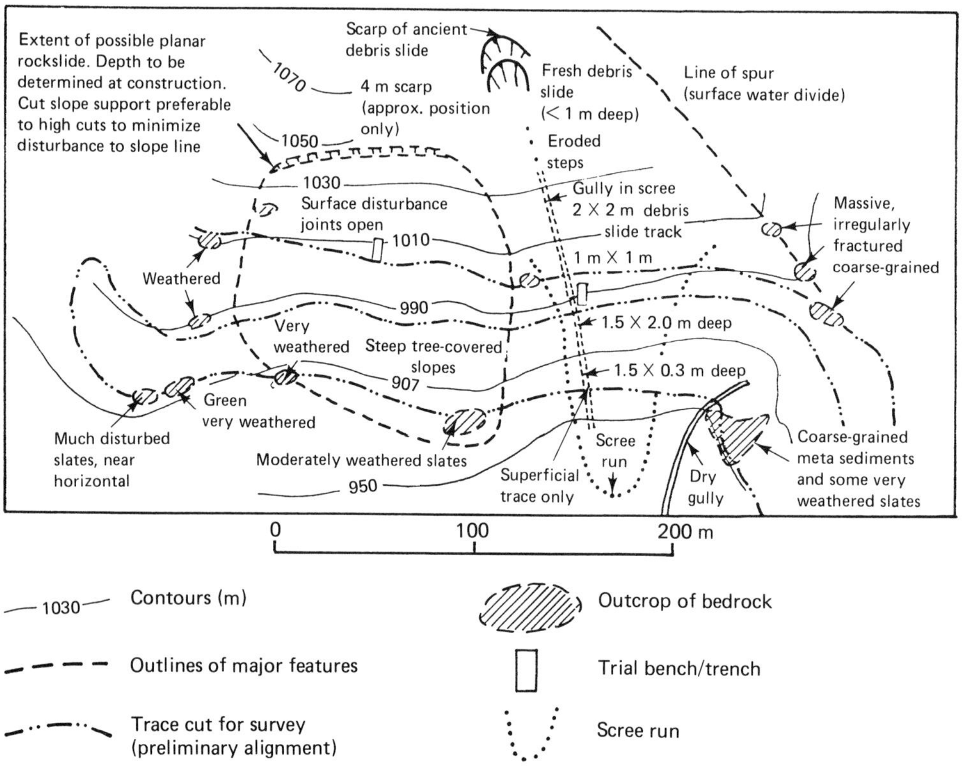

Figure 11.32 Geomorphological and engineering geological assessments at the site investigation stage. Plan of part of the Kumlintare hairpins, East Nepal (After Fookes *et al.*, 1985, Fig. 38A)

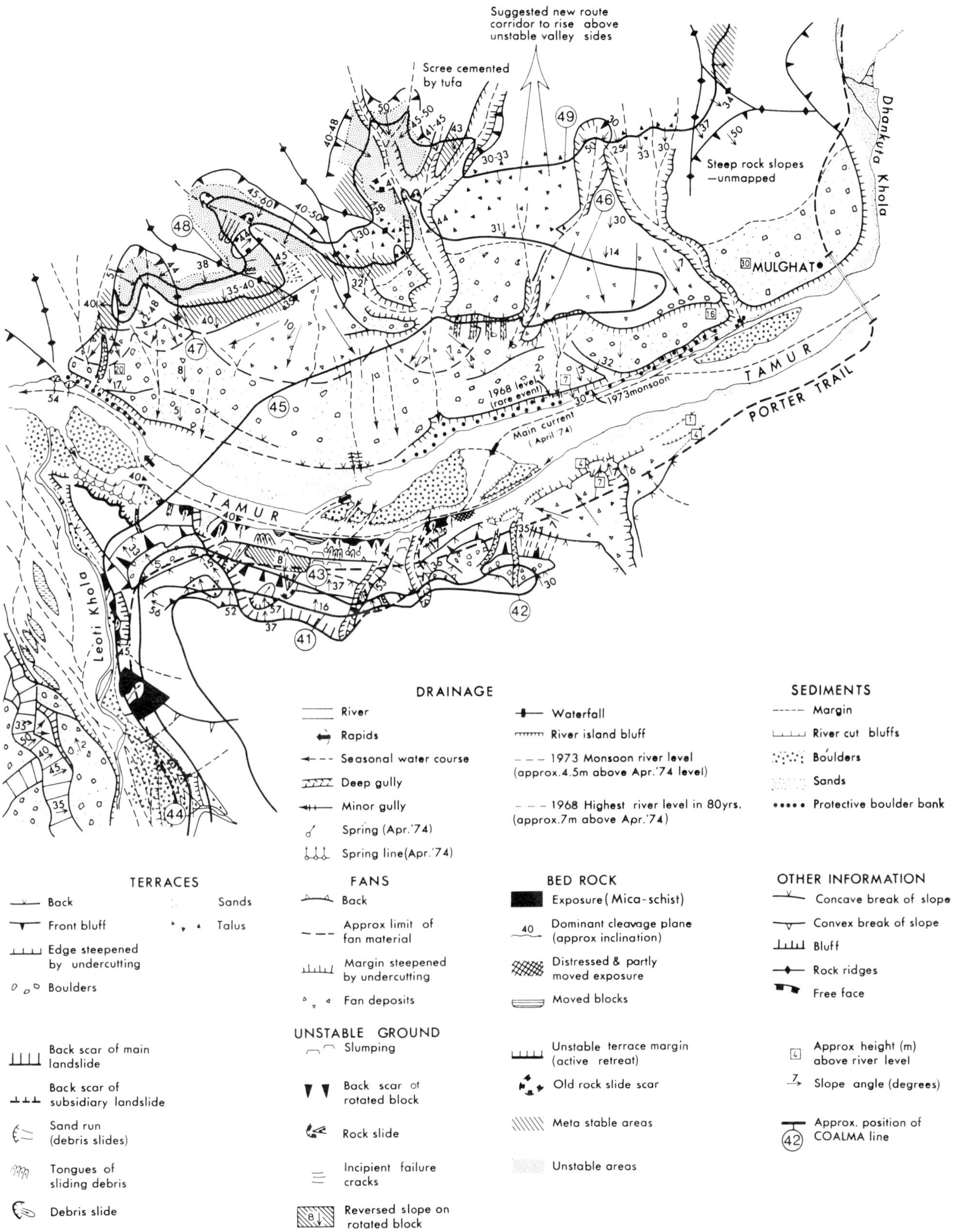

Figure 11.33 Geomorphological map of the site and situation of a proposed bridge crossing on the Tamur River, Eastern Nepal. Scale approximately 1:12 500 (From Brunsden *et al.*, 1975, Fig. 4, by permission)

here as Figure 11.33, and illustrates the importance of understanding the detailed geomorphology as a basis for defining the engineering geological conditions in mountainous areas. It is interesting to note how little detail of the solid geology is given in Figures 11.31 and 11.32, with only the outcrops marked on Figure 11.32 described in terms of lithology, fracturing and weathering grade.

Use of conceptual models in mapping

For geological mapping at medium and large scales, conceptual models of the terrain showing the surface geomorphological features, as in the mountain example in Figure 11.30, or the association of surface and subsurface features in the glacial and periglacial ground conditions illustrated earlier in Figures 8.16–8.19, have two functions. First, they aid in the understanding of engineering geological maps and plans prepared for a specific project. Secondly, they provide an *aide-mémoire* as to what geomorphological features are likely to be present in a natural environment, and therefore should be looked for and assessed.

This aspect of engineering geology is often referred to as engineering geomorphology. Most of the typical environments are discussed in Fookes and Vaughan (1986), a useful source of general geomorphological information.

11.7 Foundation mapping

Depending upon the type of construction, the entire foundation excavation or the area of the core trench is usually mapped at a dam site, and many examples have been published. Some impression of the variety of ground conditions and of the mapping methods adopted can be gained from the following sections, although no claim is made that the treatment is exhaustive.

11.7.1 Examples of mapping by the chainage-and-offset method from the Mangla Dam Project West Pakistan,

One of the earliest accounts of how large-scale engineering geological plans are actually made is that given by Fookes (1969) in his very detailed description of the preparation of 'geotechnical maps', as he then called them, of the Mangla Dam Project in West Pakistan. Site investigation mapping was being carried out during the period 1958–1963, and the examples of foundation mapping that he used are dated 1964 and 1965. It is a great tribute to the work then being done that the methods developed at that time are those that are still used with but little modification, at least in UK practice.

The Mangla Dam Project is a multi-purpose hydroelectric and irrigation scheme that impounds the headwaters of the River Jhelum. Impounding structures are all on a massive scale; the largest, the Mangla Dam closing the gap across the Jhelum River, is 1 mile long and contains 84 million yd^3 of fill. Lower Siwalik deposits in the foundation areas are friable, fine- to medium-grained sandstones, siltstones, and fissured overconsolidated stiff to hard clays. Of particular importance to the design of the structures was the discovery of bedding plane shears in the foundation excavations (Binnie, Clark and Skempton, 1967). These were mapped in great detail at a large scale. Here the account of the project will be largely confined to mapping aspects; fuller geological and geotechnical information will be found in Fookes (1965), Fookes and Wilson (1966) and additional engineering information in Eldridge and Little (1967) and Binnie *et al.* (1967).

In contemplating the organization required for mapping in detail the very large areas of the original site and the exposed foundations, it was necessary first to consider the methods to be used in the various branches of mapping. It was therefore imperative to decide on, at a very early stage, the following:

- how the various rocks and soils should be described;
- the symbols to represent the different rock and soil types on the maps;
- a variety of symbols for boundaries, structures, sample locations, etc;
- the mapping scales.

Description of rocks and soils

The code of practice for site investigations, CP2001 (British Standards Institution, 1957) had been published the year before mapping started on the project. Recommendations for the description of soils and rocks for engineering purposes, although specifically applicable to UK geological conditions, were adopted for field descriptions.

Descriptions were given in a standard order; for example, for clay soils the descriptive terms were: consistency, structure, colour, grading, soil name, additional information. Bedrock clay was described as a stiff fissured brown silty clay, or firm slightly fissured sandy clay with occasional $\frac{1}{4}$ in diameter chalky concretions, or hard red brown clayey silt. All the characteristics were determined by eye or by feel, checked later by laboratory determinations on samples.

Sandstones and sands were similarly described, e.g. friable, yellow brown fine to medium sandstone with some silt.

Symbols for rock and soil types

The legend for rock and soil types was apparently specially developed for the Mangla mapping. It is derived from CP2001. Solid or bedrock formations were shown by a regular pattern or ornament (see Figure 11.16), whereas superficial soils such as alluvium, although geotechnically similar to bedrock lithologies, were shown by a random ornament. By the use of this convention, bedrock material was clearly differentiated from alluvium and both were readily distinguishable on the map. Composite materials such as sandy clay or clayey silt were shown by combining the two appropriate symbols, with the dominant material shown by a greater intensity of ornament.

Other symbols

Various symbols were used to represent geological structures such as faults, shear zones, fissures, geological boundaries and so on. A selection of these, together with geotechnical symbols to indicate foundation treatment, drains, sumps, springs, seepages and sample locations, is given in Figures 11.34 and 11.35.

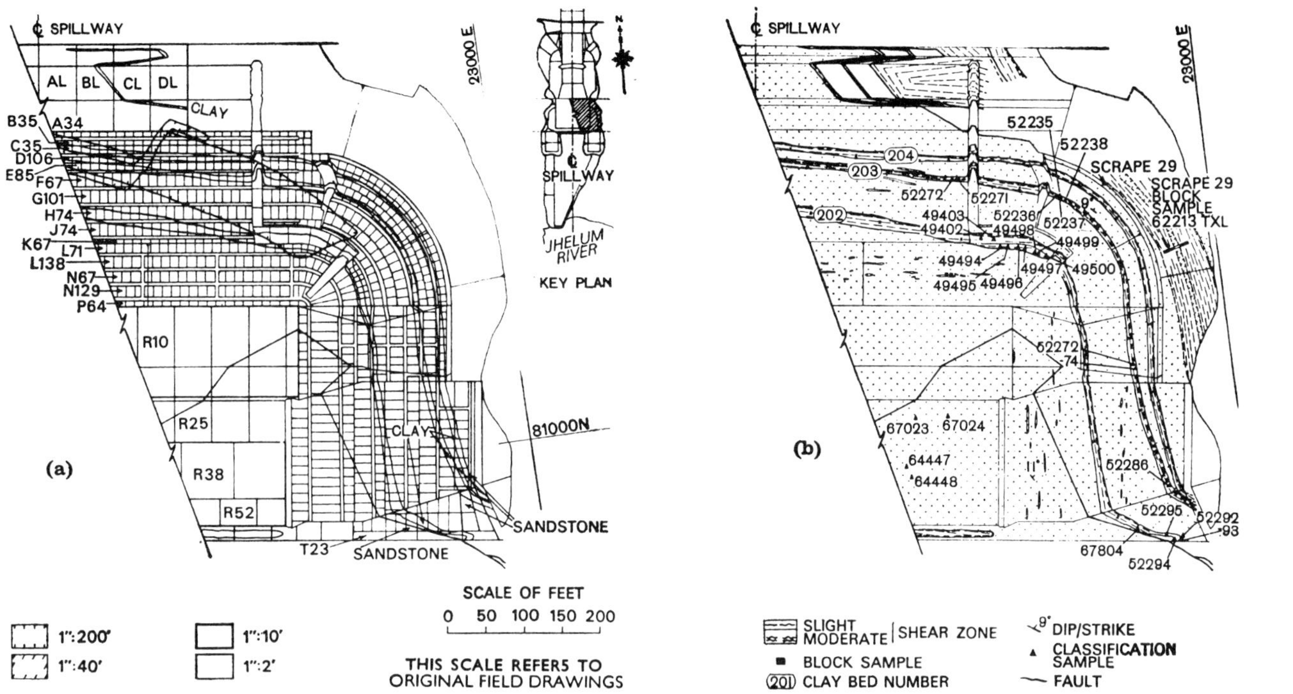

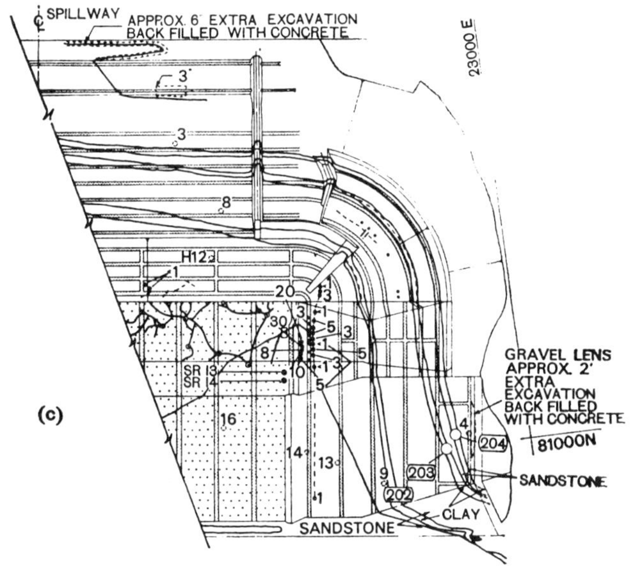

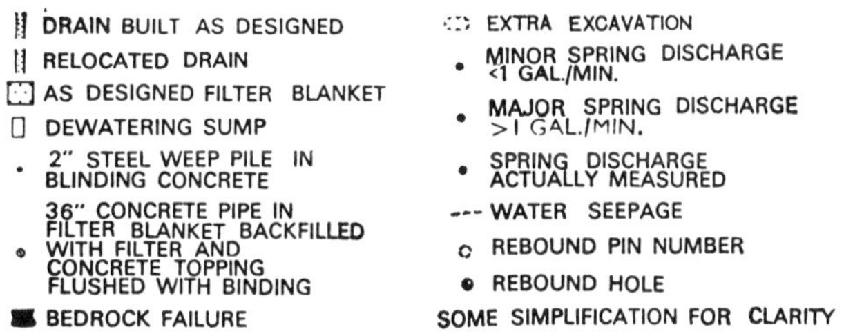

Figure 11.34 Foundation investigation geotechnical mapping, Mangla, West Pakistan, 1964–1965: (a) key map to show location and mapping scales of original field sheets; (b) simplified geotechnical map and sample locations reduced from the original field sheets; (c) simplified 'foundation treatment' map (From Fookes, 1969, Fig. 10, by permission)

Mapping scales

Conventional imperial engineering scales were used: for site investigation mapping, 1 in to 500 ft (1:6000) and 1 in to 200 ft (1:2400); for detailed mapping, 1 in to 40 ft (1:480) and 1 in to 20 ft (1:240); and for logging drill core or the floor of a trench, 1 in to 2 ft (1:24) or 1 in to 4 ft (1:48) or 1 in to 10 ft (1:120).

Mapping method

Mapping was carried out on the foundation after the excavated surfaces had been prepared, by hand cleaning or by using a compressed air blast, to take the concrete or earth fill. The sequence of preparation of the foundation surface, mapping and recording, treatment, and placing fill or concrete, was generally as follows:

(a) bulk excavation by bulldozer and scraper to about 1–2 ft above final grade;
(b) foundation excavation to grade by specially adapted blades on backhoes and by small bulldozers;
(c) hand clean-up by brushing and air blasting. If the exposure was below the watertable, sumps were sunk and water that collected was pumped out;
(d) inspection, quantity measurement and engineering geological mapping, photography and sampling of the exposure for record purposes;
(e) dental treatment of suspect areas with weak cement grout or other treatment as required;
(f) covering with concrete or fill.

Time between operations (a) and (b) varied between a few days and several months depending on the contractor's programme. That between operations (c) and (f) was usually an hour or so.

On foundations that were to have concrete placed against them, e.g. the foundations of the main spillway, plans were produced on squared paper usually at a

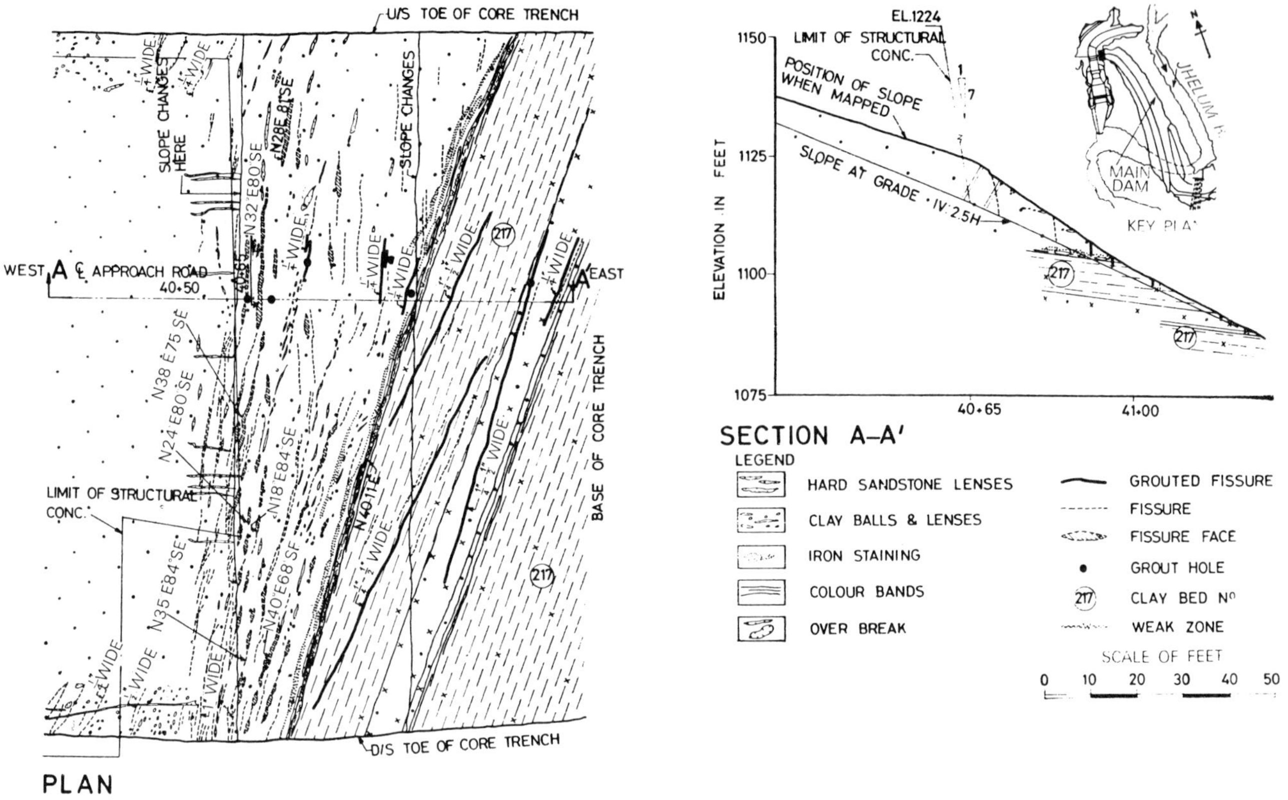

Figure 11.35 Foundation investigation geotechnical mapping. Part of the junction main spillway headworks, Mangla Dam, West Pakistan, mapped in the field at 1 in:10 ft scale, 1964 (From Fookes, 1969, Fig. 11, by permission)

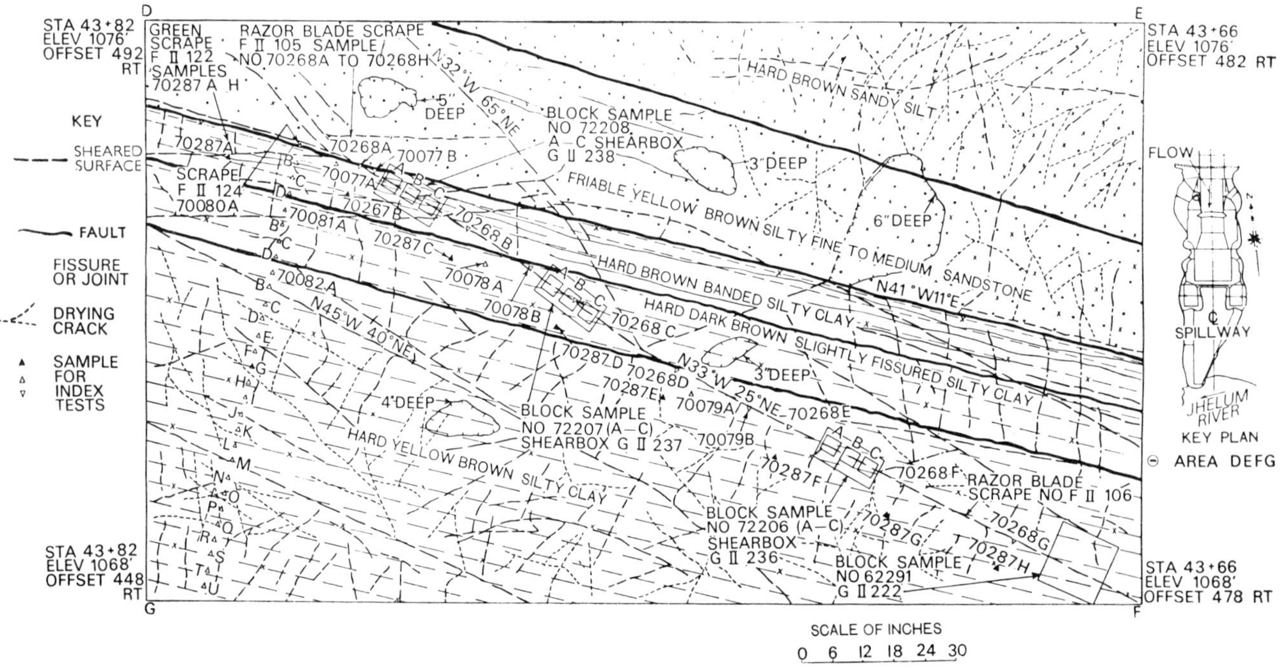

Figure 11.36 Foundation investigation geotechnical mapping. Detail from a thrust shear joint sample location, main spillway, Mangla, West Pakistan, mapped in the field at 1 in:6 in scale, 1964 (From Fookes, 1969, Fig. 13, by permission)

1:120 scale, by measurements made from foundation survey pegs. Then, after copying on a reduced scale, the original field sheets were stored for future reference. From the field sheets, three plans were prepared on the same 1:480 scale as the basic construction drawings for each part of the engineering structure. One plan showed the locations and reference numbers of the original 1:120 scale field sheets; another showed the results of engineering geological mapping and also the sample locations; the third showed the geotechnical foundation treatment, e.g. the grade of filter material placed against the foundations, areas of dental excavation, and groundwater conditions such as lines of seepage, measured flow from springs and damp patches.

To illustrate this procedure, Figure 11.34 is part of a 1:480 plan showing locations of the 1:120 scale field sheets. The simplified engineering geological plan and sample locations from the same area are shown in Figure 11.34(b) and part of the foundation treatment plan is shown in Figure 11.34(c).

Foundations that were to be covered by earth fill, e.g. the core trench of the main dam, were mapped directly on a 1:120 scale: one plan showed the geotechnical mapping and sample locations and the other the details of foundation treatment. There was no need for a map showing locations of field sheets.

In areas where special detail was required, e.g. at the junction of the spillway headworks and the main dam, where bedrock grouting had previously been carried out through a loading embankment before the foundations were cut down to final grade, foundation mapping was carried out on a 1:120 scale and then presented on the same scale; part of this is shown in Figure 11.35.

As the careful mapping of foundations and trenches had revealed geological structural weaknesses in the bedrock, such as faults, joints, fissures and shear zones, detailed field drawings were produced at selected locations in addition to the routine engineering geological mapping. These illustrated various features such as the attitude and frequency of joints and fissures, together with the exact locations of selected samples. Some drawings (Figure 11.36) were on as large a scale as 1 in:6 in.

11.7.2 Brianne Dam site, South Wales

The 300 ft high Brianne Dam is sited in a narrow V-shaped gorge cut in Lower Palaeozoic indurated mudstones with subordinate sandstones (greywackes). Nearly vertical cleavage, opened up at the surface, predominantly controls the excavation pattern, except where occasional dominant more gently inclined bedding planes crop out disadvantageously. There are minor fault planes and shears in the foundations and abutments. It is a rockfill dam with a clay core, and the general location, in particular the base for the clay core, was intensively investigated (Figure 11.37). The site investigation plan shows the outline of the dam, details of outcrops and local geology including faults, trial trenches, pits and boreholes. As such, the plan is a good example at 1:960 scale of a site investigation plan.

At this dam site, the core trench exposures were recorded in detail, and here two illustrations are given as plans recording the geology (Figure 11.38a) and the pre-construction treatment of the foundation (Figure 11.38b). The geology plan illustrates the distribution of different rock types, distribution and types of prominent discontinuities, and the significant areas of very close jointing. Foundation treatment involved grouting, guniting the closely jointed areas, and mass concreting where a dolerite dyke weathered to an engineering soil had been excavated. The record is completed by marking grout infilling discontinuities at the surface of the excavation, a seepage line, locations of samples, and photographs. An impression of the whole site can be gained from Figure 11.39.

11.8 Mapping cuttings

Cuttings in soil or rock may be mapped once excavation is started or complete either to solve a particular engineering problem, usually involving stability, or to provide an as-built record for maintenance purposes or to support litigation. Examples have been selected to illustrate some of the mapping techniques that are available.

11.8.1 Taff Valley trunk road scheme, South Wales

This scheme has already been referred to in section 11.4.6 when discussing plan-scale mapping for site investigation. Cuttings along the road provided more examples of incipient failures that required urgent remedial action during construction.

The Stage II section of the road runs between Nantgarw and Glyntaff. North of Nantgarw Colliery, at the southern end of this section, the road is in deep cutting on the east side for some 800 yd. The cut slope in this area failed during construction on a laminated silt horizon cropping out half-way up the cut and it was considered that strong measures to control surface and groundwater flows were required as a major part of the remedial works (Fookes, Hinch and Dixon, 1972).

A small boring, pitting and testing programme was mounted to investigate the cutting failures in this and other areas of the road. As part of the investigation, an engineering geological plan was commissioned by the engineers on a 1:500 scale, with particular emphasis on water conditions. The work was carried out by two experienced engineering geologists and about 4 man-days were spent on the part area shown on the plan (Figure 11.40). The field work was carried out by conventional mapping onto field slips prepared from enlargements of 1:1250 photogrammetric plans.

Stabilization works were designed to take account of the groundwater regime, and these included capture and drainage of surface water flowing into the slipped area, dewatering of depressions which had no drainage outlets, and the filling of open tension cracks to prevent further water penetration. Part of the resulting drainage design is illustrated in Figure 11.41.

At Taff's Well, on the Stage I section of the road, lack of space required the road to be cut through the toe of a steep hillside in limestone (Figure 11.42). The cut slope in this area had a small rock slide and several other small slides appeared imminent during construction (Fookes, Hinch and Dixon, 1972). The rock cut was composed of

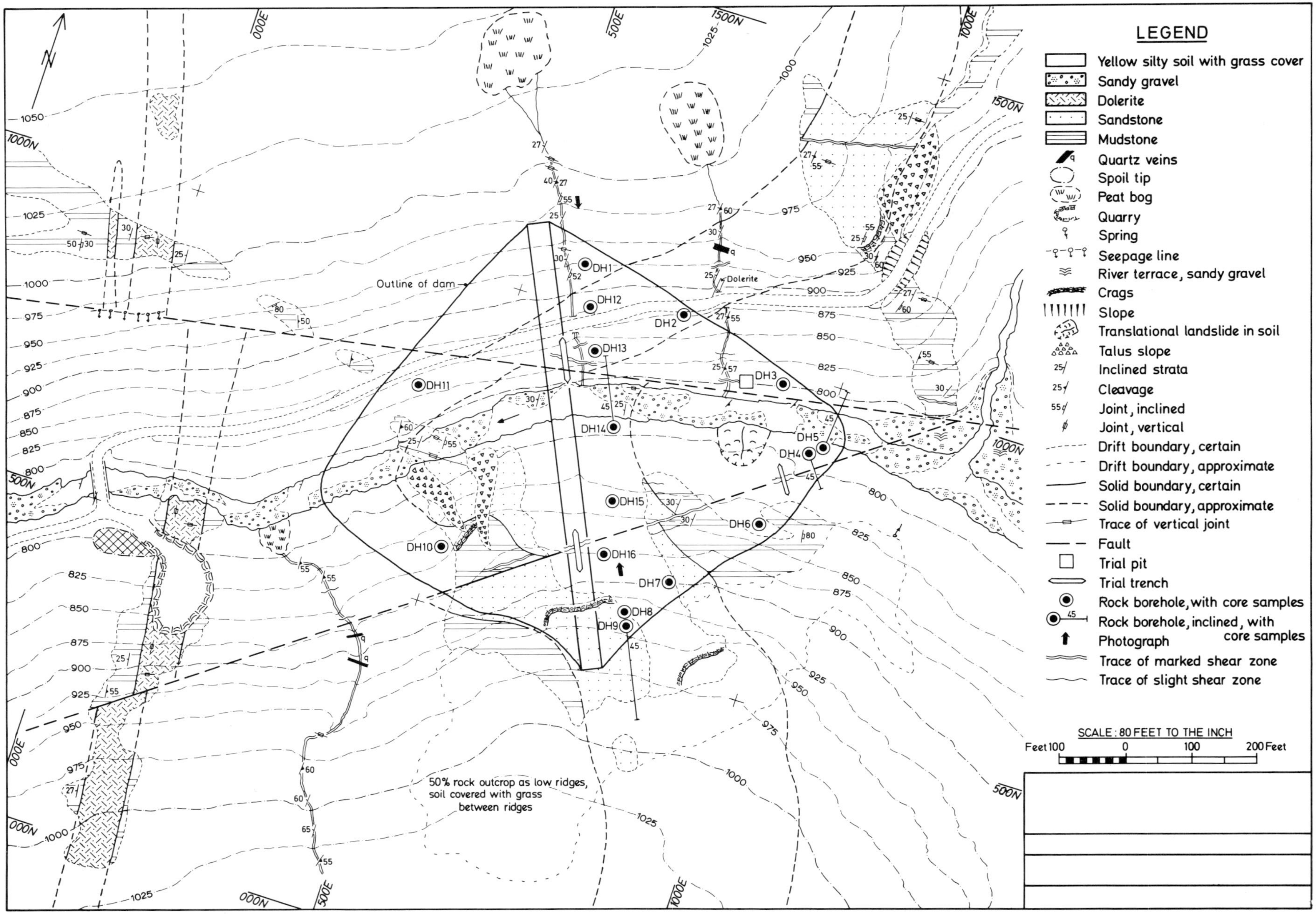

Figure 11.37 Plan of a dam site showing the details derived from the initial walk-over survey and the boreholes, trenches and pits used in the initial site investigation. This drawing has been prepared, using stencilled lettering, as a record drawing from the original site investigation drawing (From Anon., 1972, 13.1.3, by permission)

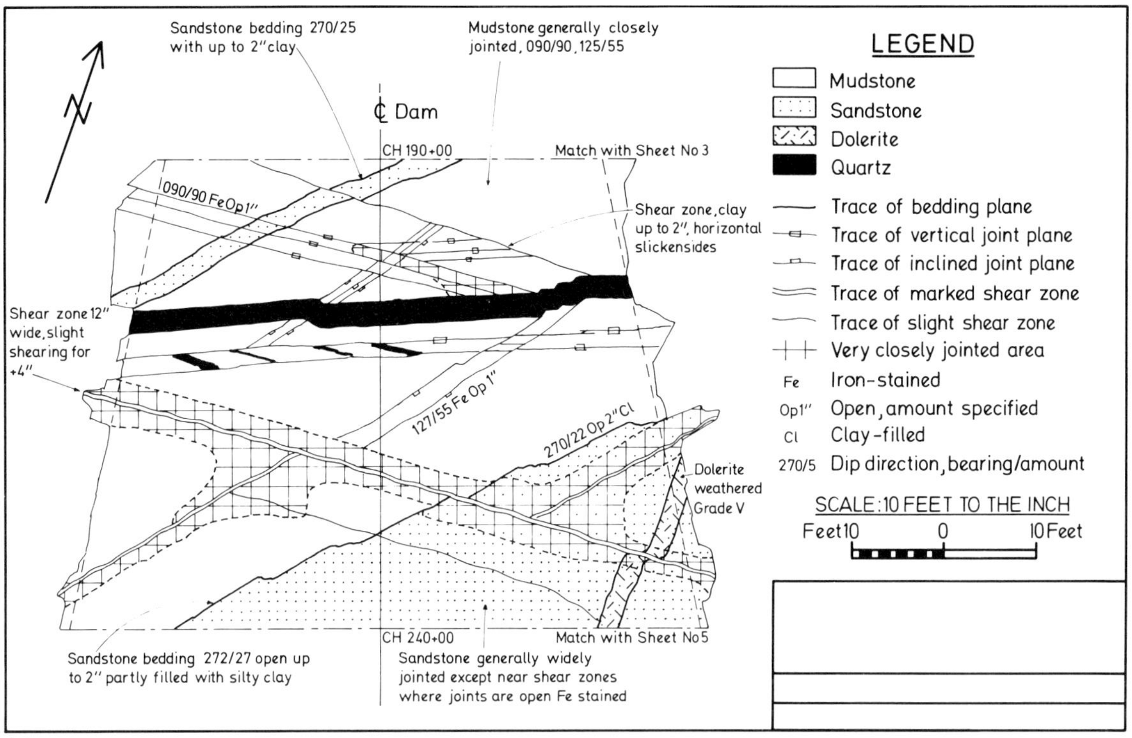

(a)

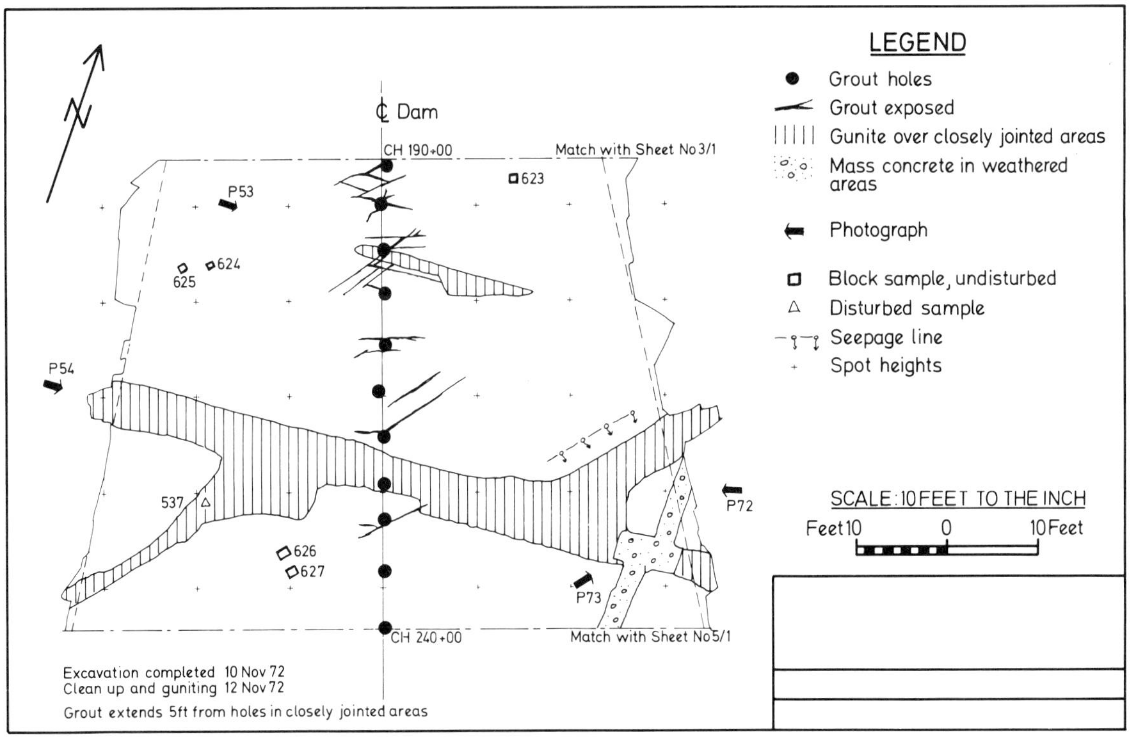

(b)

Figure 11.38 Foundation plan of one section of the core trench of the left abutment of the Brianne Dam site shown in Figure 11.37: (a) engineering geology plan; (b) plan of pre-construction treatment of the foundation

Figure 11.39 Brianne Dam site (Figure 11.37) looking upstream, showing work in progress on the core trench on the left abutment

thin to thick bedded jointed dolomitic limestones with occasional thin clay seams, dipping unfavourably into the cut. Immediately the poor stability of the situation became clear during excavation, the face was cut back about 3 m to give a greater distance between the face and the road, and later a low mass concrete wall was placed adjacent to the road to help prevent any rock-fall debris reaching the road. The height of this wall was largely determined by the sight line of the road which was on a gentle inside curve along the length of the rock cut.

The whole length (nearly 100 yd) was then surveyed, from a mobile platform mounted on a hydraulic arm, by an experienced engineering geologist and rock engineer working together. A dimensioned sketch of the rock face was produced from this survey together with a continuous overlapping set of colour photographs of the face. The photographs and the sketch section were referenced to chainage marks painted on the face. This work took about three days. A working construction drawing was then prepared using the sketch section as a base plan, with significant geological features shown in outline.

The area of the face between chainage marks 11–13, marked on Figure 11.43, presented the most difficult problem. Layers of thinly bedded limestone associated with thin sheared beds of shale showed water seepages and rapid deterioration even during the summer months. It was therefore decided that the whole of the face should be protected. Details of the work are shown in Figure 11.43, whereas Figure 11.44 shows the face before and after treatment.

In order to design the remedial works, and incidentally to provide a base from which the work could be undertaken, a carefully assembled working scaffold was set up parallel to the face. The scaffolding provided the basic rectangular grid for the survey. Offsets were taken from the nodal points of the scaffold to the face, which enabled vertical profiles to be plotted at 5 ft intervals along the face.

Remedial measures consisted of bolting, anchoring and dowelling; dentition and walling; guniting and pitching; draining, scaling and grouting (Figure 11.43). The annotated, detailed plan scale elevations of the face are true geotechnical plans made during the constructional stage of the project to solve perceived, localized small-scale instabilities.

Above the cut is a high steep forested slope with a thin cover of slope debris and soil. This was fenced and drained by ditching.

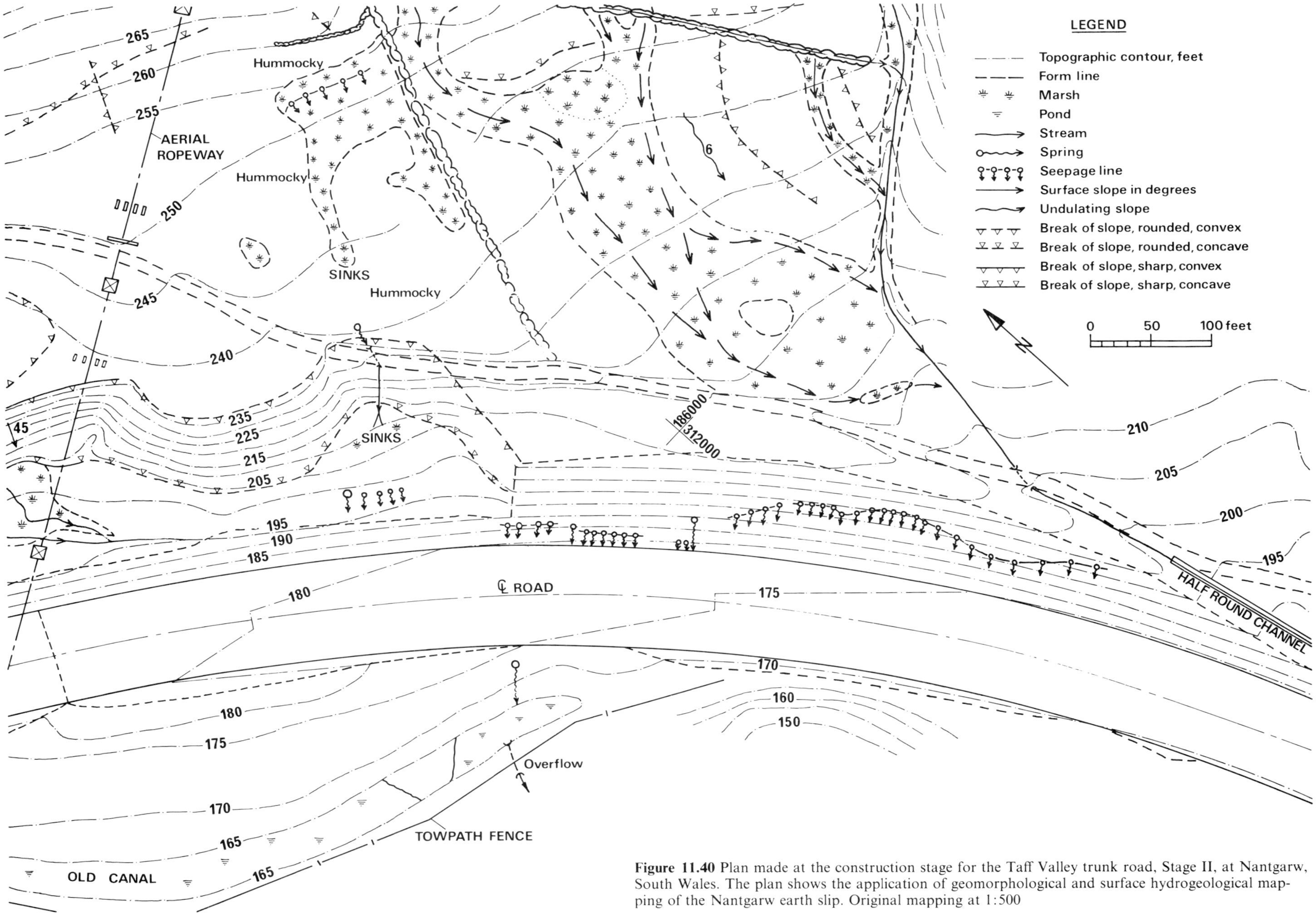

Figure 11.40 Plan made at the construction stage for the Taff Valley trunk road, Stage II, at Nantgarw, South Wales. The plan shows the application of geomorphological and surface hydrogeological mapping of the Nantgarw earth slip. Original mapping at 1:500

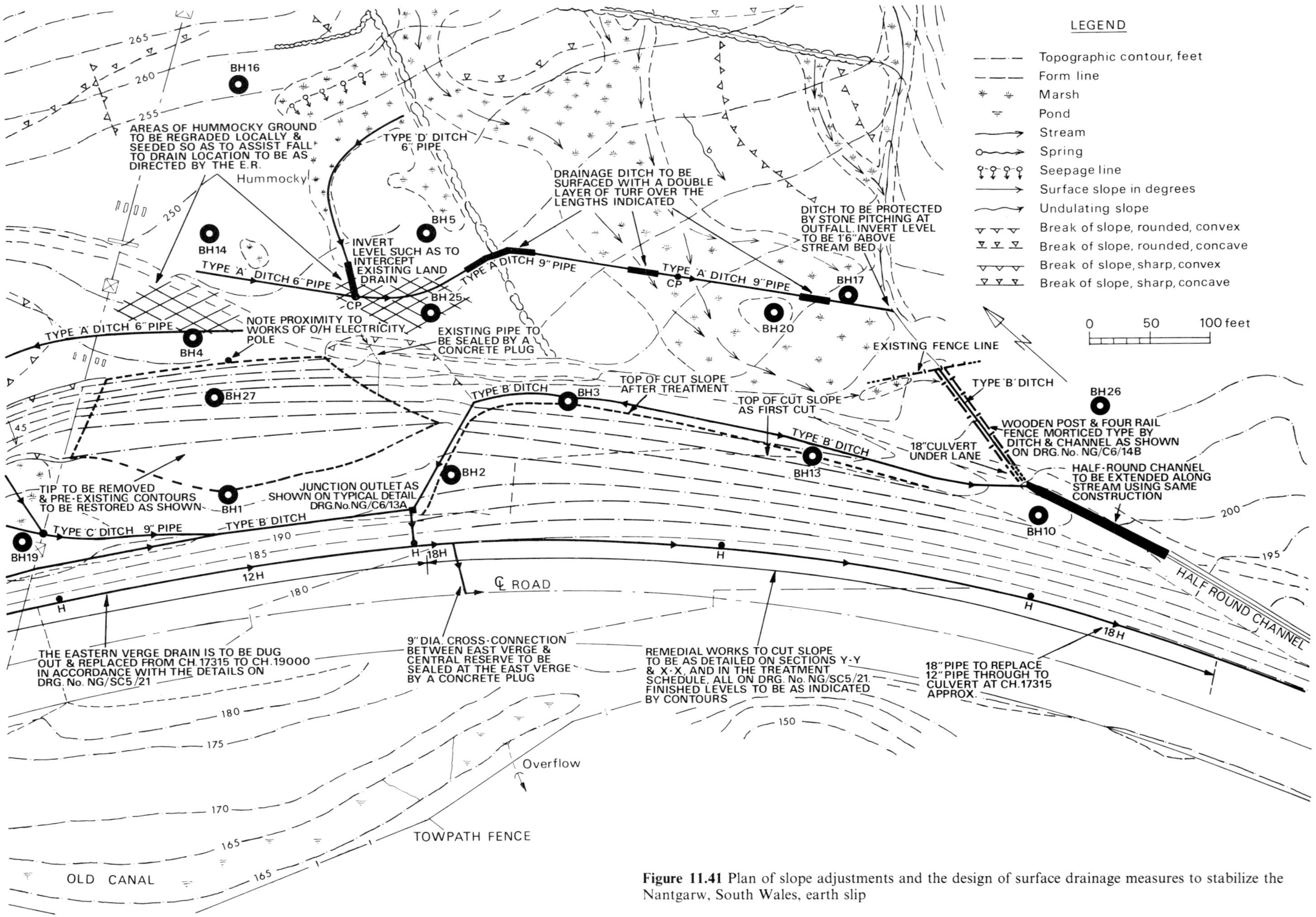

Figure 11.41 Plan of slope adjustments and the design of surface drainage measures to stabilize the Nantgarw, South Wales, earth slip

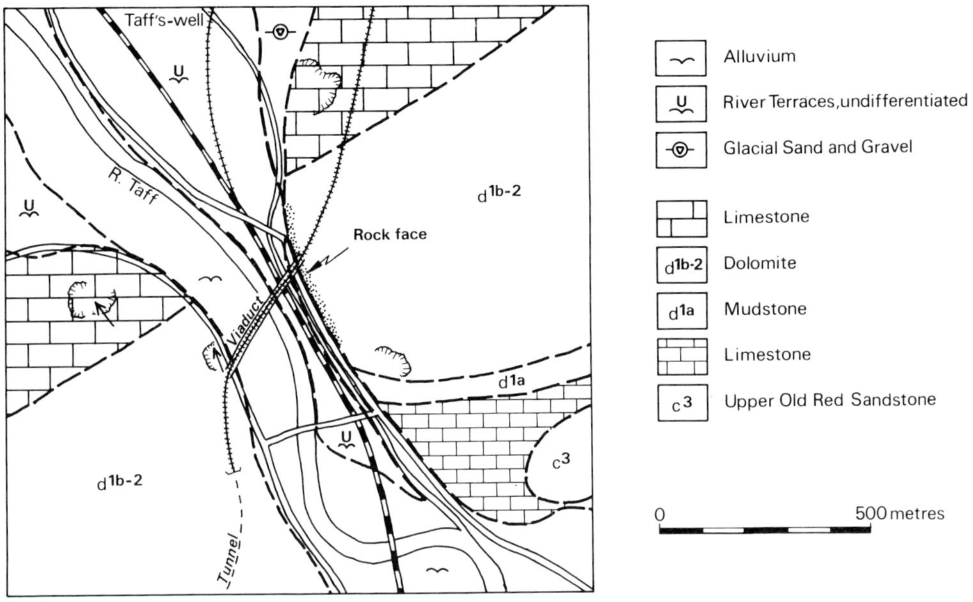

Figure 11.42 Simplified geological map of the site of the Taff's Well road cut, Stage I of the Taff Valley trunk road, South Wales (From Dearman and Fookes, 1974, by permission)

11.8.2 Recording the Clevedon road cutting on the M5 Motorway, Bristol, UK

An example of engineering geological mapping using purpose-taken photographs as base-plans is the maintenance record made of what was in 1973, apart from quarry faces, the largest rock cutting in England. The work was carried out at a late stage of construction when the cutting was completely exposed and in addition to providing a construction record was also used to check the design of some of the stabilization measures (Eyre, 1973; Dearman and Fookes, 1974).

The route of the M5 Motorway west of Bristol crosses the River Avon just upstream of the docks at Avonmouth and gradually ascends the 100–180 m high Tickenham Ridge. Climbing the hillside, the road is in a side-long cutting at two levels in Carboniferous Limestone and Pennant Sandstone, with a maximum height of face of 30 m (Figure 11.45).

At the western end of the ridge the Carboniferous Limestone has been thrust northwards over the Coal Measures sandstones of the Pennant Series. For about 2500 m the cutting followed the general line of the thrust that cropped out for a short length in the upper cutting and more extensively in the lower cut. Generally the limestone dipped into the face of the cutting at between 10 and 40°; consequently the main problems were solution cavities filled with clay and calcite veining, and the possibility of joint bounded blocks sliding out of the face. The Pennant Sandstones are folded (Figure 11.45) and locally beneath the thrust are highly shattered. Mudstones in the fault are weathered to clay, which has migrated into adjacent open discontinuities.

A complete series of photographs was taken from the level of the upper carriageway at 50 ft (15 m) centres. In order that the various features on the face could be located, ropes were hung down the face to mark the end of each 15 m photographic panel. The ropes were marked at 10 ft (3 m) intervals, related to Ordnance Datum, so that a reference grid could be drawn over the photographs (Figure 11.46). Geological data, the location and type of remedial treatment of cavities, together with the position and code number of each rock bolt, were recorded on transparent plastic overlays on the photographs. The overlays were checked and added to directly by field inspection. Each field record was then transferred to a plan of the panel filed with the actual photograph (Figure 11.47).

In the drawing office, the record for each panel was corrected for vertical and horizontal distortion and from these individual drawings (Figure 11.47) a complete record plan of the face was produced. A small part of the completed record plan of the face (Figure 11.48), drawn at a scale of 1:400, also provides analytical data on geological conditions and a record of stabilization works.

11.8.3 Some additional aspects of mapping cuttings in limestones

Crystalline limestones of Devonian and Carboniferous age in Britain give rise to strong topographic features that have to be negotiated by roads. Examples have already been given (Figures 11.42 and 11.45) where mapping was undertaken to design support measures and to provide a construction record. Two examples follow where mapping was used to support litigation arising from unforeseen ground conditions.

Problems that can be anticipated are those associated with the direct solution of limestone by circulating groundwaters leaving open or filled solution cavities along many structural planes in the rock mass. The karstified zone may be of limited depth, having mainly

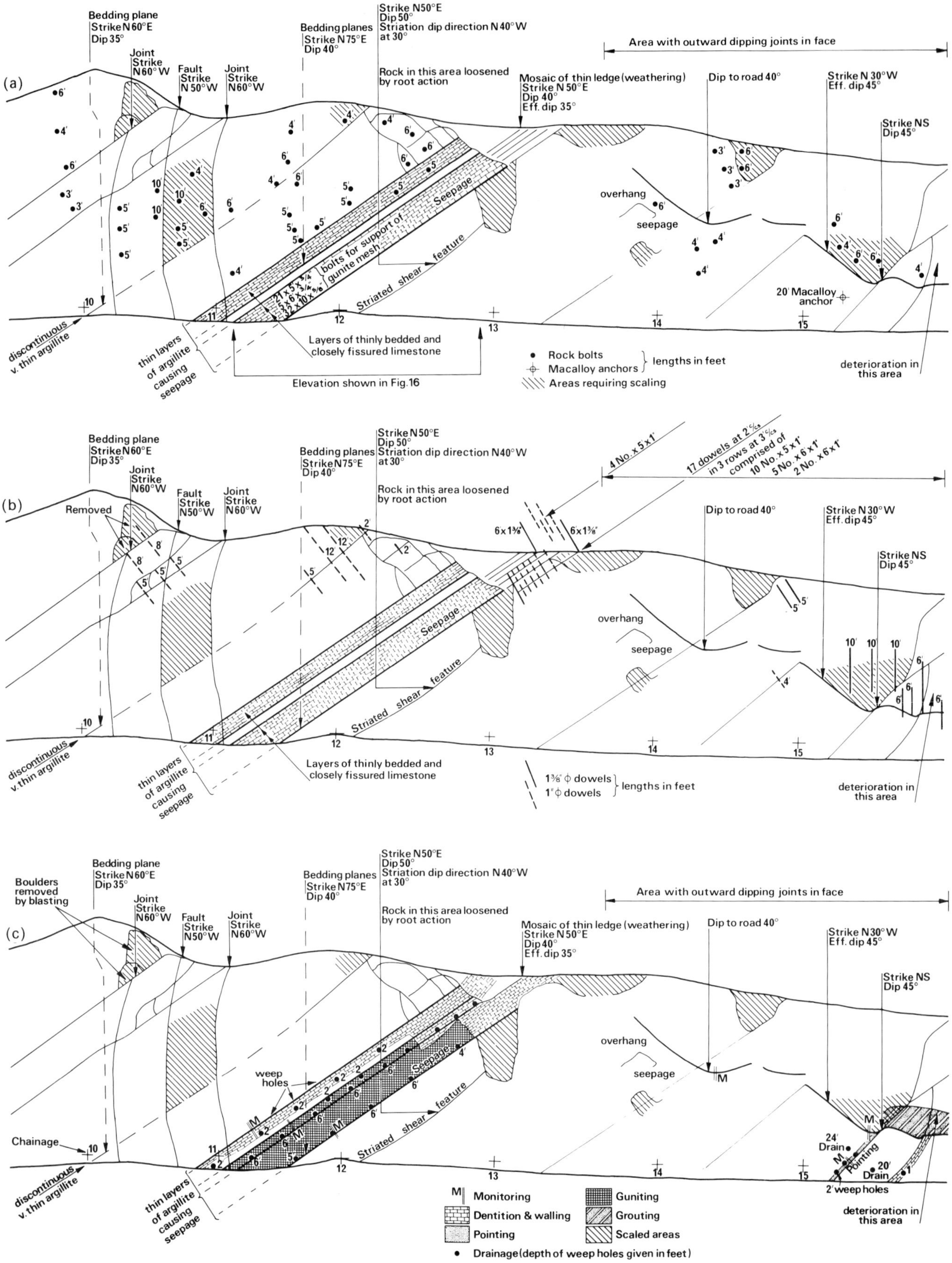

Figure 11.43 Geotechnical plans made at the construction stage of the Taff's Well road cut, Stage I of the Taff Valley trunk road, South Wales: (a) elevation of part of the face showing geological conditions, the location and lengths in feet of rock bolts and Macalloy anchors, and areas requiring scaling; (b) location of dowels; (c) location of monitors, and areas requiring dentition and walling, pointing, guniting, grouting and scaling (From Dearman and Fookes, 1974, by permission)

(a)

(b)

Figure 11.44 Photographs of part of the rock face in Figure 11.43, (a) before, and (b) after, treatment. The location of the photographs is marked in Figure 11.43 between stations 11 and 13, from which details of the geology and the security measures can be ascertained (From Dearman and Fookes, 1974, by permission)

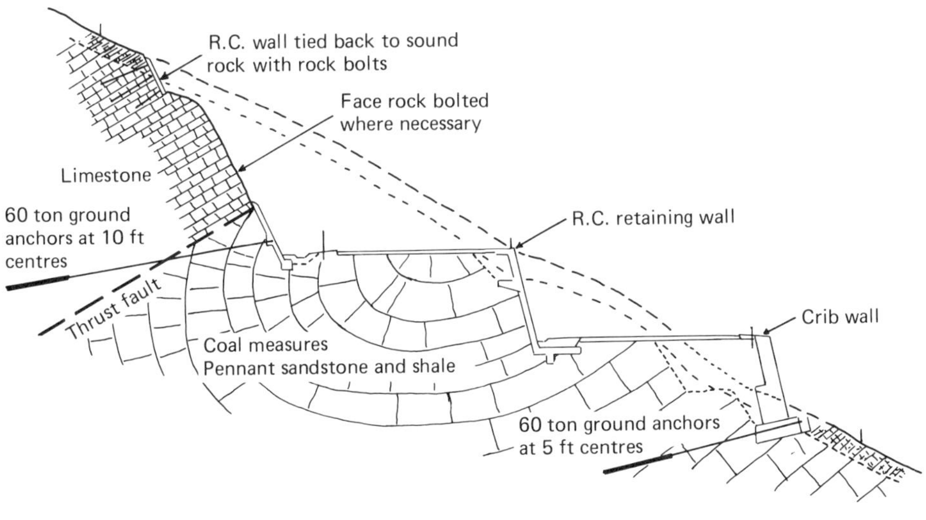

Figure 11.45 Cross-section through the Clevedon, UK, M5 Motorway in rock cutting showing the geological structure and the main types of support measures (From Eyre, 1973, Fig. 1, by permission)

formed within the range of fluctuation of the groundwatertable, and in addition may have very strong surface expression. This is common knowledge, so that it is of interest to consider some practical examples of rock excavation in which minor variations from this simple pattern give rise to difficulties during construction. The examples all refer to road cuttings, the sites of which were subject to a standard site investigation, presumably following a desk study.

Interpretative plans, as sections connecting the sites of logged boreholes, would have been produced as a matter of course. The sections would have shown the base of superficial deposits and any variations in lithology in the solid. What would not necessarily have been shown was

Figure 11.46 Photograph of a single panel in the limestone rock slope from which an individual plan (Figure 11.47) was drawn

the possible implications of juxtaposition of certain rock types in the solid, or the character of the interface between limestones and overlying superficial deposits.

Exeter to Plymouth A38 trunk road in Devon, UK

As part of a general upgrading of the main road between Exeter and Plymouth in Devon, a new dual carriageway was constructed bypassing bottlenecks and incorporating bypasses where they existed. The latter usually involved doubling-up the carriageway and at Buckfastleigh an existing cutting in steeply inclined Devonian limestone had to be widened for this purpose.

For much of the cut there was no difficulty until the faulted junction with weathered slaty tuffs was reached. There, drilling for blast holes revealed limestone blocks set in an earthy matrix of wad – black manganese oxides (Figure 11.49); blasting was ineffective and excavation had to be started at the top of the cut using a backhoe. Mixed soil and rock made excavation of the wad from between massive rounded limestone blocks extremely difficult and undoubtedly increased costs.

Naturally, the question arose as to whether or not these conditions could have been foreseen. Deposits of black earthy wad, usually relatively shallow, have been worked at many places in the general region, although they were all exhausted many years ago. However, the character of the deposits is recorded in the descriptive memoir accompanying the 1 in geological map of the area, and a favoured location is at the faulted junctions between limestone and volcanics where manganese oxides replace the limestone. Alternatively, umber or ochre deposits may be present and these were also exploited.

Examination of the limestone face in the original cutting revealed nearly vertical solution joints, cutting across inclined bedding, partly infilled with dark brown soil. Presence of iron- or manganese-rich soils in the

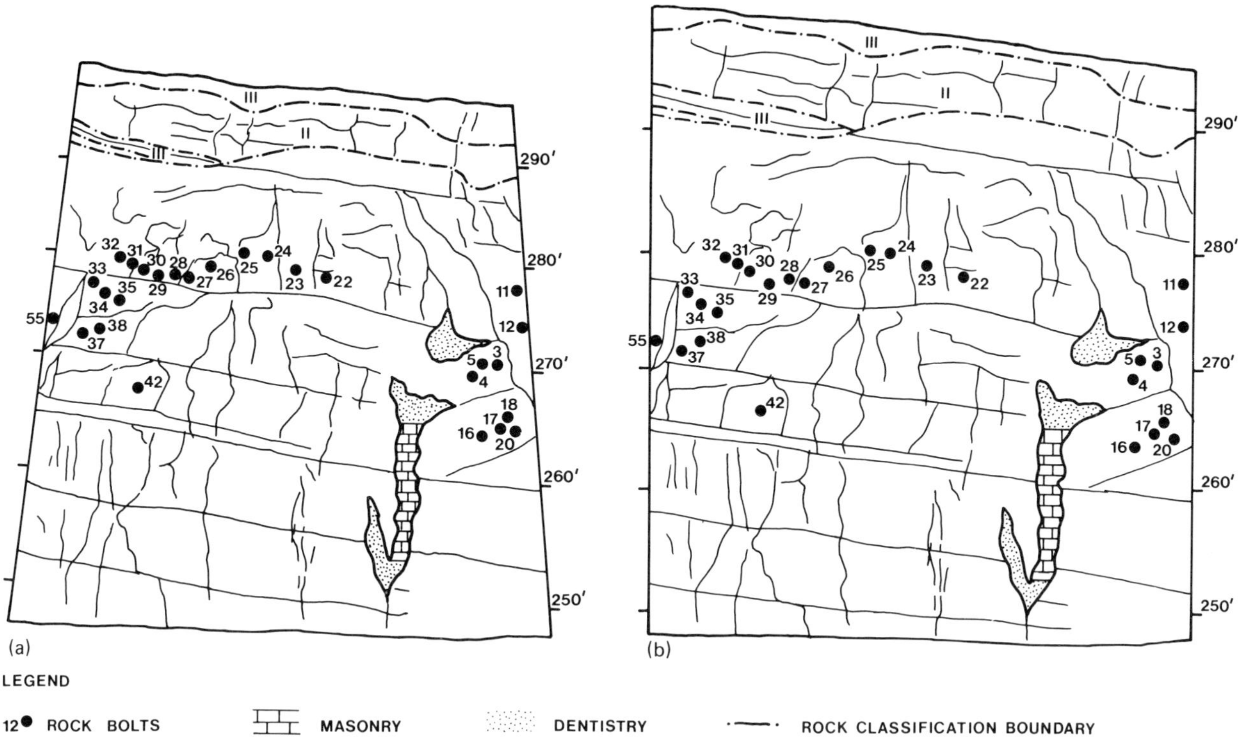

Figure 11.47 Plan of panel (Figure 11.46) in two versions, one uncorrected and the other corrected for vertical convergence, with grade zones and significant discontinuities marked, together with the locations of rock bolts and other surface remedial measures

limestone joints could therefore have been established at an early stage of site investigation. In turn, this discovery could have highlighted the possibility – or even probability – of a wad deposit at the faulted margin of the limestone and steps could have been taken to trench across the boundary. This is of course hindsight, but an engineering geologist carrying out a desk study, or a geologist familiar with the area, could have suggested that such conditions were at least possible in anticipation of possible construction difficulties.

A sketched plan elevation (Figure 11.50), which could have been an actual record plan of the cutting face after excavation, is the type of illustrative engineering geological plan that could have been produced in support of a claim for extra payment because of unexpected ground conditions.

Bypass to Newton Abbot, Devon, UK

Construction of a road cutting for the Kingsteignton–Newton Abbot Easterly Bypass revealed large sediment-filled solution hollows in Middle Devonian limestone. A very irregular rock surface is concealed beneath the relatively smooth terrain through which the cutting passes, with steep-sided masses and pinnacles separated by unconsolidated gravel and clay of probable Cretaceous and early Tertiary age (Figure 11.51). In vertical sections, steep-sided pipes are visible, and these appear as circular outcrops on the cutting slopes. The cutting walls were mapped by Brunsden *et al.* (1976), and part of the elevation is reproduced in Figure 11.52 and illustrated in Figure 11.53.

The larger pipes are several tens of metres across and are known to extend to depths greater than 30 m, the greatest depth of the cutting. Brunsden *et al.* (1976) commented on their likely mode of origin by slow settlement of a sequence of unconsolidated sediments, which can be correlated with others nearby, as the limestone was removed by solution.

A weathering residue next to the limestone, either a slickensided red clay or a deposit of umber and ochre, has also been involved in the subsidence. This, together with relatively undisturbed bedding at places in the sediments, appears to indicate a slow settlement by solution rather than by collapse into a solution hollow. It has been suggested that solution may have been controlled by groundwater circulation in the immediately overlying permeable sands and gravels.

A conventional site investigation using boreholes had not revealed the highly variable depth to rock, and consequently the excavation was conceived as an upper soil cut and a lower rock cut, both with appropriate side-slopes, and involving different excavation methods. As it turned out, excavation was extremely difficult and attended by considerable additional expenditure.

Solution features on this scale in the Devonian limestones have not been recorded in the literature, although extensive cave systems in the neighbourhood are tourist attractions. In the nearby seaside resort of Torquay, a road cutting (Figure 11.54) in the same limestones, without surficial cover, broke to a very irregular profile, with deep but relatively small solution cavities. An irregular profile could perhaps have been foreseen at the Newton Abbot cutting, but the scale of the solution pipes and attendant complexities are, without doubt, surprising.

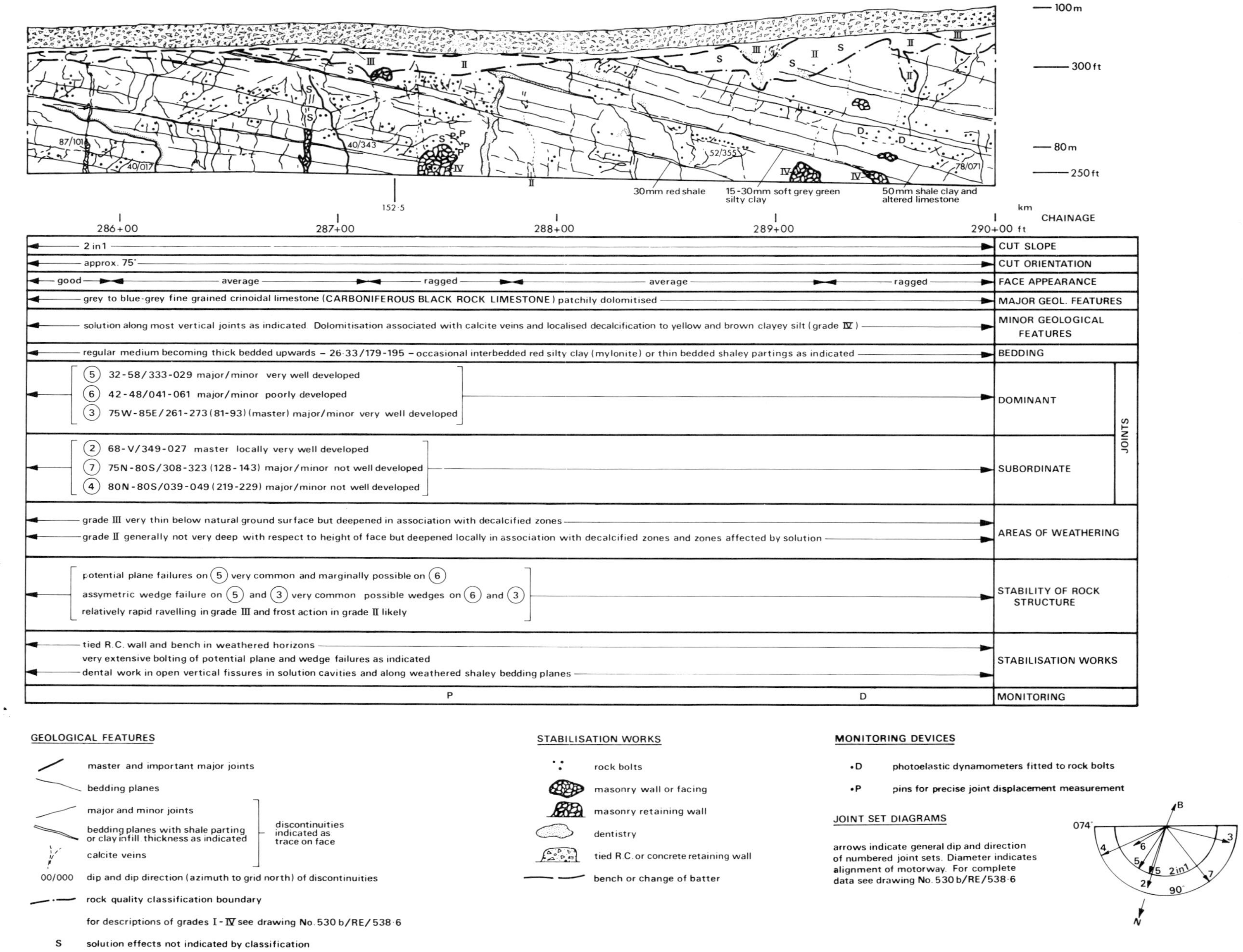

Figure 11.48 Record plan of the Clevedon, UK, rock cut built up from field plans as illustrated in Figure 11.47. Drawn at a scale of 1 : 400, the original drawing of the rock face is 3.2 m long

Figure 11.49 Wad deposits at the weathered junction between limestone and tuff giving rise to difficult excavation conditions (Plymouth, UK)

11.8.4 Sandstone cuttings on the M62 trans-Pennines Motorway, UK

Arrowsmith, Morris and Bradley (1972) have described and illustrated the geological setting of major slips which occurred either during construction or after completion of cutting excavations on the M62. With the need to establish stable slopes rapidly, detailed investigations could not be undertaken, but there is sufficient information to suggest a causal relationship between the slope failures and intraformational shears in faulted sediments.

At the sites described, the strata are either alternating beds of sandstone and shale of the Carboniferous Millstone Grit or are beds at the base of the Lower Coal Measures. Bedding dips gently into the cuttings at 6–20°, and the cut slopes of 2 vertical in 3 horizontal in shales and 4 in 1 in sandstones were expected to be stable, and should have presented no problems.

They failed, however, and remedial measures had to be undertaken using a buttressing technique without affecting the existing road alignment. A substantial part of the toe of each slip was excavated, including the basal shear zone, and the surface of the underlying sandstone bed was roughened by ripping to improve the key between the sandstone fill of the buttress and the sandstone base. The geology and remedial works at the Windy Hill cutting (Figure 11.55) are shown in Figures 11.56 and 11.57. The even slope on the right, where beds dip into the cutting, has been reprofiled, and the lower part of the slope is a sandstone-fill buttress. In contrast, the profile of the southern slope is determined by the geological sequence with steep slopes (4 in 1) in sandstone, access benches at the top of each sandstone bed, and slacker slopes (2 in 3) in the shales. A fault can be seen almost halfway along the south face of the cutting from the abrupt cut-off in the prominent lowest sandstone bed. The depth of the cutting here is about 30 m, and the shales in the toe of the northern slope were cut back for 15 m and backfilled with sandstone bulldozed in from above.

The plan (Figure 11.56) is a typical engineering record plan showing the basic construction forms of the cutting and the broad outline of the geological conditions encountered on the north batter.

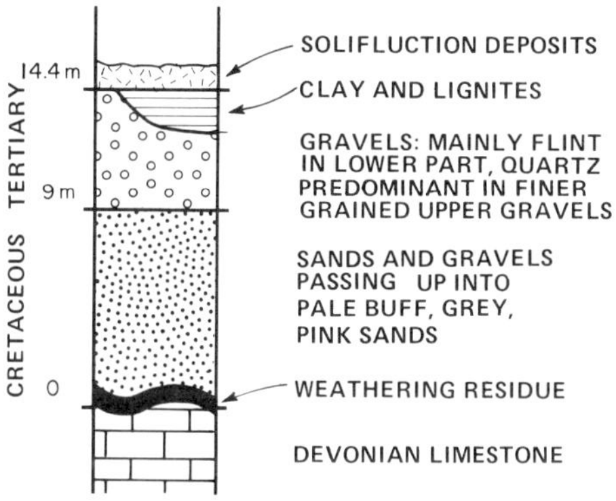

Figure 11.51 Succession exposed in the Newton Abbot bypass, UK

11.8.5 Failure during excavation of a cutting in slates, Cornwall, UK

With the Pennine examples in mind, the stability of any excavation in inclined mudstone and shale is suspect if the rocks are faulted, even if the beds are only gently inclined. Slate derived by low grade regional metamorphism of mudrocks is abundant in the Devonian of south-west England, a region where there is extensive normal faulting, with renewal of movements during the Tertiary when major wrench faulting sliced the peninsula from the present coast to coast. Coupled with this, intense tropical weathering caused slates to revert *in situ* to fine soils, and fault planes provided deeply penetrative channels which were selectively weathered.

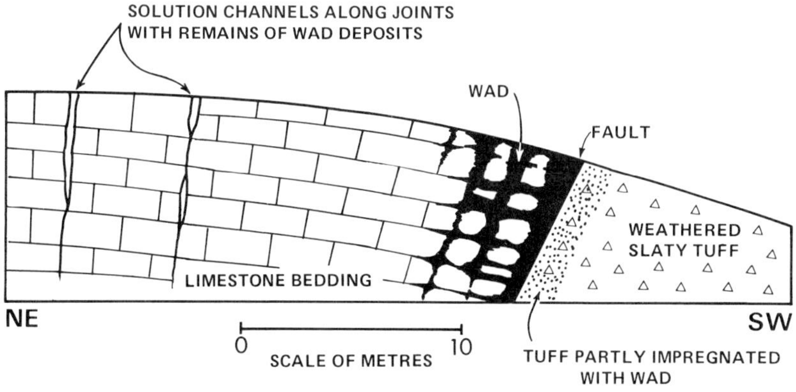

Figure 11.50 Sketch elevation of the cutting wall (Plymouth, UK)

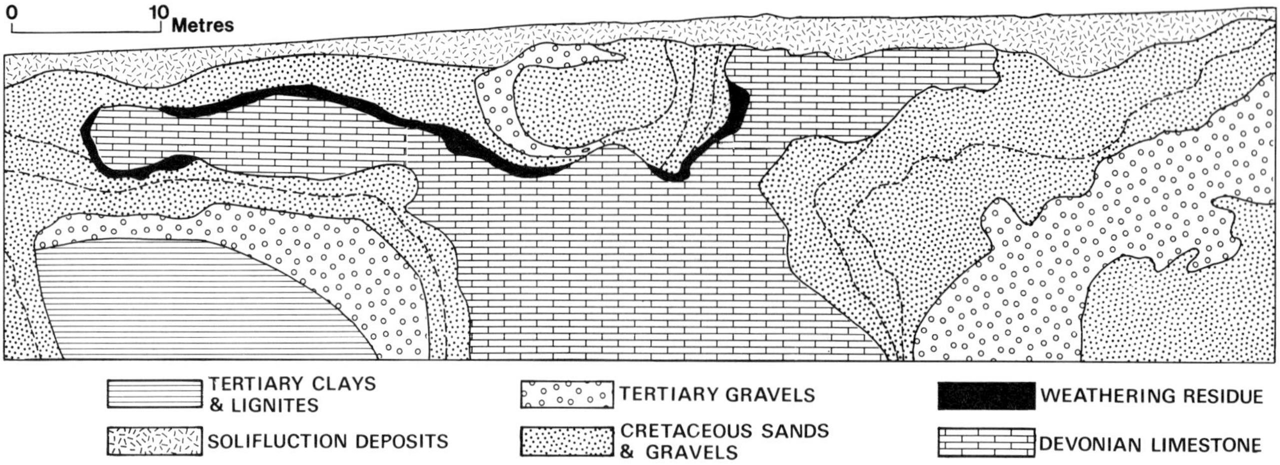

Figure 11.52 Part of the excavation record plan produced for the Newton Abbot bypass, UK

Figure 11.53 View of the actual finished cutting mapped in Figure 11.52 (Newton Abbot, UK)

Figure 11.54 View of a limestone cutting near Torquay, UK, illustrating the prevalence of strong solution features

Most site investigations are done on an *ad hoc* basis, with focus maintained on the immediate area under investigation. Such was the case on the banks of the River Fowey in south Cornwall where a new store for china clay, and additional loading and docking facilities, were to be built as part of a programme of expansion and modernization (Fookes, Reeves and Dearman, 1977).

The leased site was restricted to a narrow strip of land between 20 and 30 m wide in the form of an S, following the curve of the river, at the foot of a steep wooded hillside (Figure 11.58). In 1968, a series of boreholes was sunk within the leased land and in the river, located to suit an early version of the scheme, but not extending up the hillside. Subsequent design changes required an assessment of the amount of slope trimming that would have to be done. It had been hoped that the slate could be trimmed to a 60° slope in order to minimize excavation, and to avoid clay holding up on the rear slope of the store. However, a walkover survey in 1971 led to the conclusion that, since natural slopes along the strike of the slates – the direction of excavation – ranged from 45 to 50°, a slope of 48° was the maximum that could be considered in good slate. In the regolith, natural slopes of about 28° showed marked signs of hillcreep, so that for the upper part of the cut slope in weathered slate an inclination of 25° was estimated. At these slopes, however, the excavation would have extended for a considerable distance up the hill, beyond the property boundary, if there was much weathered material.

It was decided to cut the rock face to the full height of 48°, leaving a margin of 5 m for any flattening of the slope in the expected thin layer of weathered material. Excavation was commenced at the top of the slope and after some small rock falls it became apparent that the 5 m margin would be insufficient.

After the slope failures, excavation was stopped, and engineering geological mapping of the rock face, a site investigation of the rock mass behind, and back-analysis of the small failures were undertaken as a basis for the design of remedial measures. Mapping highlighted faulting in the face (Figure 11.59), and the borehole investigation behind the part of the face that was insufficiently exposed to map effectively revealed zones of highly weathered rock and thin chlorite-rich clay bands. As the chlorite layers were considered to be along the slaty cleavage, i.e. they were intraformational shears of some sort, they could be expected to daylight on the finished rock face, and therefore were an additional factor to be taken into account in the redesign of the excavation.

As a result of the investigations, the layout and shape of the clay store were changed; a maximum slope of 48° was retained in the lower part of the excavation which would form the backwall of the store, and the inclination

Figure 11.55 Windy Hill cutting looking westwards from the Pennine Way footbridge, M62 Motorway in Lancashire, UK. The face is about 35 m high

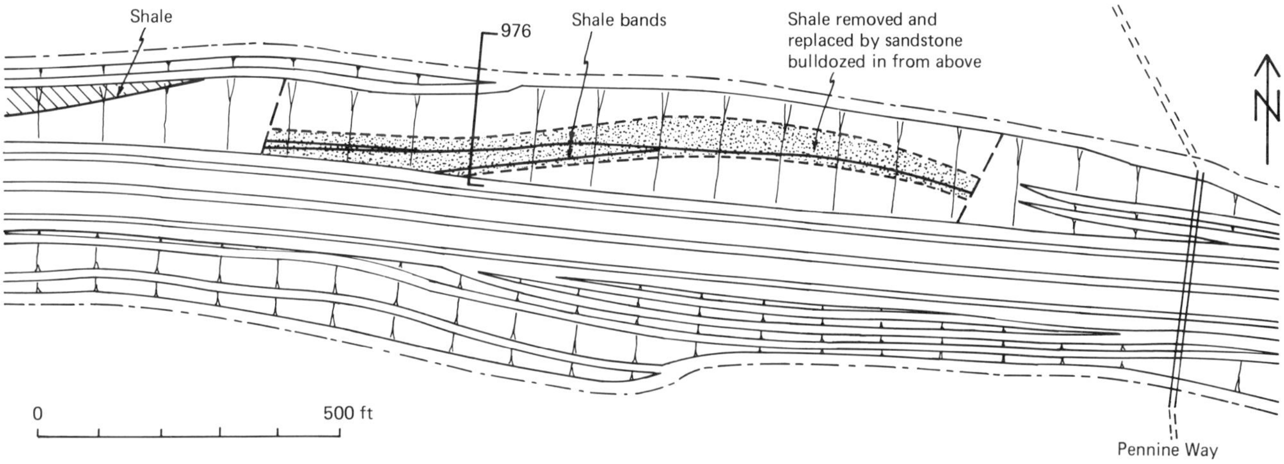

Figure 11.56 Plan of Windy Hill cutting on the M62 Motorway, Lancashire, UK, showing the remedial works on the north batter of Cut 15, and location of cross-section 976 illustrated in Figure 11.57 (After Arrowsmith, 1971, Fig. 4)

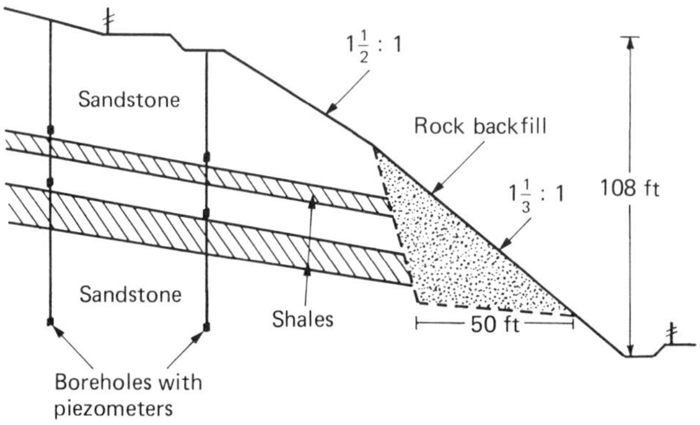

Figure 11.57 M62 Motorway, Windy Hill cutting, Lancashire, UK. Remedial works on north batter of Cut 15, cross-section 976, marked on Figure 11.56

of the higher permanently exposed part of the cut was increased to 32° to avoid excavation beyond the boundary of the site (Figure 11.60). Analysis indicated that the upper and lower parts of the face would need to be supported both during and after excavation. Consequently, as excavation proceeded downwards, an anchored reinforced retaining wall, a rock anchor system and drainage measures were installed.

In this particular case it is of interest to consider what was known of the engineering geological conditions at four stages of the project:

1. At the initial design stage it will be recalled that an assessment was made of natural slope angles, leading to 48° for the angle of cut in slates and 25° in weathered material.
2. Immediately after the initial small failures, the exposed face where the slips had occurred was

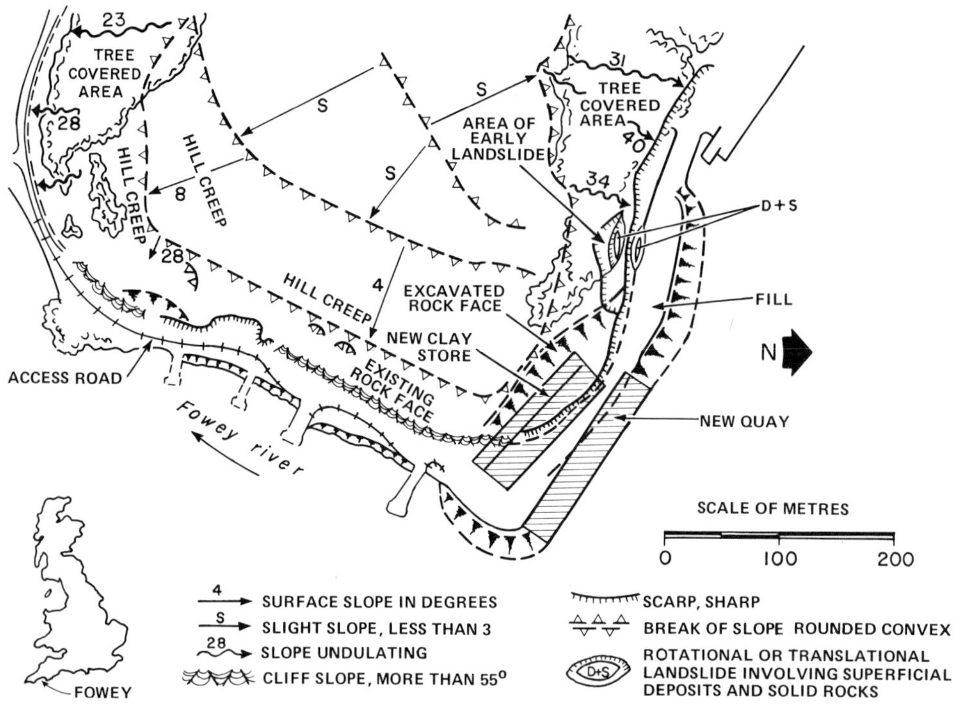

Figure 11.58 Plan of the site of the clay store and new quay, Fowey, UK, with geomorphological details of slope angles and areas of instability. Inset: location of Fowey

mapped into zones (Figure 11.59a) and the engineering geology of each zone was tabulated.

3. After the second site investigation, predictions were made of the likely geological conditions in the final rock cut, including the outcrops of potential failure planes involving weathered slate and chlorite along cleavage, as predicted from the borehole logs (Figures 11.59b and 11.60).
4. During excavation to the final profile, an as-built engineering geological elevation of the 48° rock slope was produced by an independent observer (Figure 11.59c).

The findings in the fourth stage of the investigation present a detailed account of engineering geological conditions in the rock slope.

Early and late descriptions of lithology differ slightly, but the as-built plan confirms to a remarkable degree the earliest predictions in Figure 11.59(b). For example, the fault zone at section 12 and the broader area of faulting between sections 10 and 11 are present. The predicted weathered area at section 8 (Figure 11.59b) is present only in the lower part of the face bounded by faults (Figure 11.59c); the area of fault-shattered zone III material at section 7 (Figure 11.59b) has been mapped as fault-bounded. What have not been revealed by the later mapping are the projected zones I and II (Figures 11.59b and 11.60) parallel to cleavage and containing chlorite and weathered slate. Instead there are many faults, fault zones and clay seams, all generally cutting across the cleavage. The complex of faults cutting the slates would contribute to the general relaxation of the face on excavation, and to the possibility of failure on through-going cleavage planes (Fookes, Reeves and Dearman, 1977).

The first three stages of the investigation highlight an important civil engineering requirement, namely the need to arrive at decisions rapidly so that the complex interplay of design by the consultant and construction by the contractor suffer as little interference as possible. This demands an appraisal, more often than not on very limited information, of the most likely significant engineering geological conditions. In this case it involved interpretation of limited borehole information in terms of the geological controls of the most probable mode of failure of the whole slope. Once this decision had been made, then the whole train of the civil engineering approach to design of remedial measures was set in motion; as a result the slope was constructed without failure and without endangering those involved in the construction.

11.8.6 Soil and rock cut slopes in Hong Kong granite

Introduction

There can be little doubt that one of the most intensively mapped regions, from the point of view of engineering geology, is the island of Hong Kong and the associated Kowloon mainland area. It will readily be recalled that this is classical territory in which the basic weathering grades for granite in the mass were first formulated, although at present weathered rhyolite and dolerite have an increasing significance.

Many of the companies engaged in engineering in Hong Kong have close associations with the UK, and consequently mapping methods first applied at home have been applied to, and modified for, the very different ground conditions in Hong Kong. The physical environment is dominated by intensive tropical weathering, resulting in deeply decomposed granite and rhyolite profiles that have been, and still are, subject to vigorous denudation such that fresh granite can be exposed at the surface adjacent to granite weathered to an *in situ* granite soil.

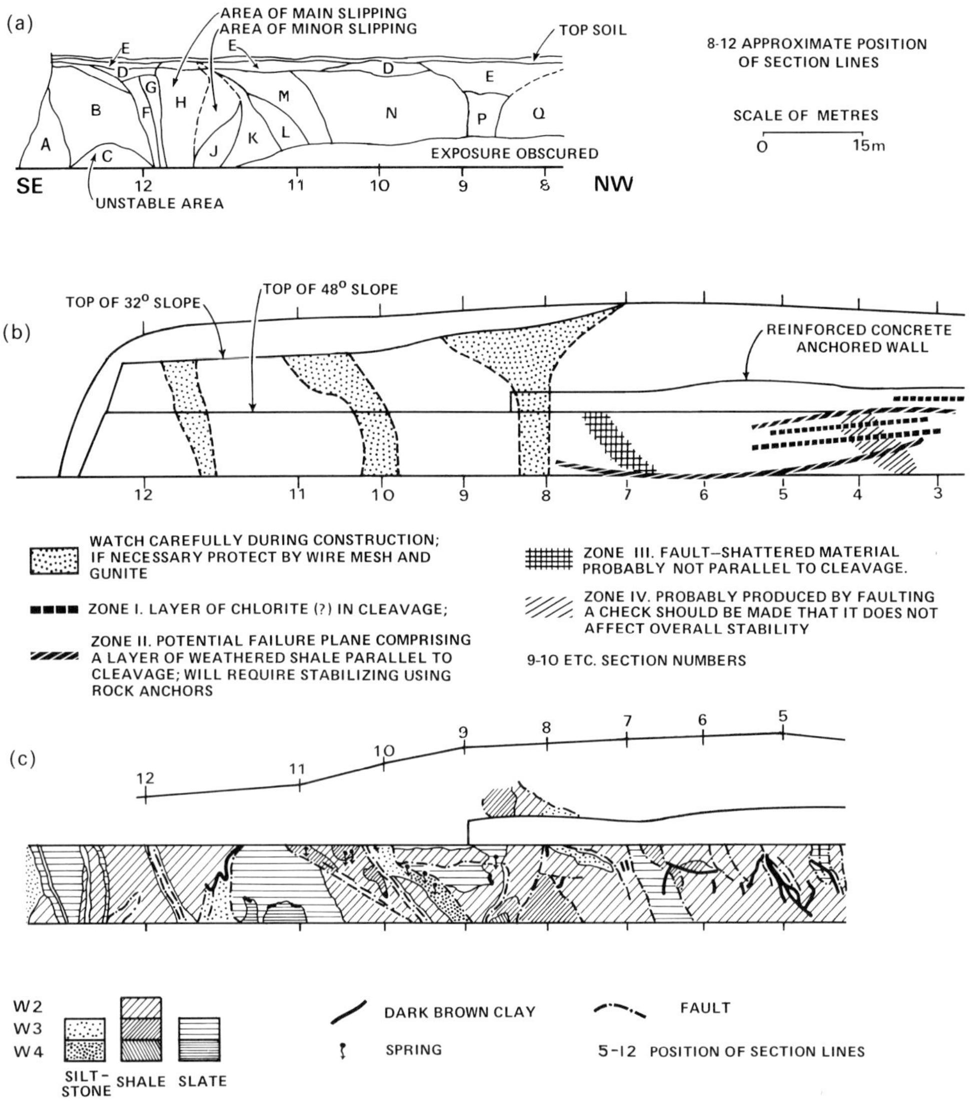

Figure 11.59 Comparison of engineering geological conditions in the 48° slate slope at Fowey, UK, as predicted or mapped at different stages of construction: (a) mapped homogeneous zones determined after minor failures at the south-east end of the face in 1971; (b) important geological features on the finished rock slope predicted from the second site investigation, and suggested treatment of the cut slopes; (c) the as-built engineering geological elevation of the 48° slope exposed between sections 5 and 12 in 1972 (After Fookes, Reeves and Dearman, 1977, Figs 5, 9, 17)

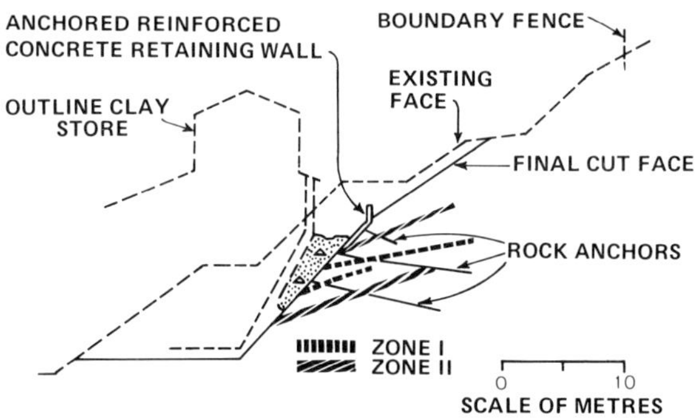

Figure 11.60 Predicted positions of potential failure planes at section 4 showing the clay store, the anchored reinforced concrete retaining wall, and positions and lengths of proposed rock anchors (Fowey, UK) (After Fookes, Reeves and Dearman 1977, Fig. 10).

As elsewhere, mapping is predominantly logging and characterization of discontinuities, identification of the different weathering grades, and plotting the distribution of rock and soil mass zones in exposed excavations. For this purpose it is very necessary to have a standardized approach to ensure consistent mapping.

Classification of weathering grades

The weathering grade classification used in Hong Kong is a modification of the scheme developed by Ruxton and Berry (1957) for the description of the different rock and soil materials found around a typical corestone (Figure 11.61/1). Criteria used for the field identification of six grades from fresh granite to residual soil are:

1. Fresh
   (a) Material: 'rock';
   (b) Colour: light pinkish grey;
   (c) Grain size: fine or medium to coarse;
   (d) Texture: crystalline – fine-grained granite may be porphyritic;
   (e) Structure: no microfractures;
   (f) Weathering grade: fresh granite (FG);

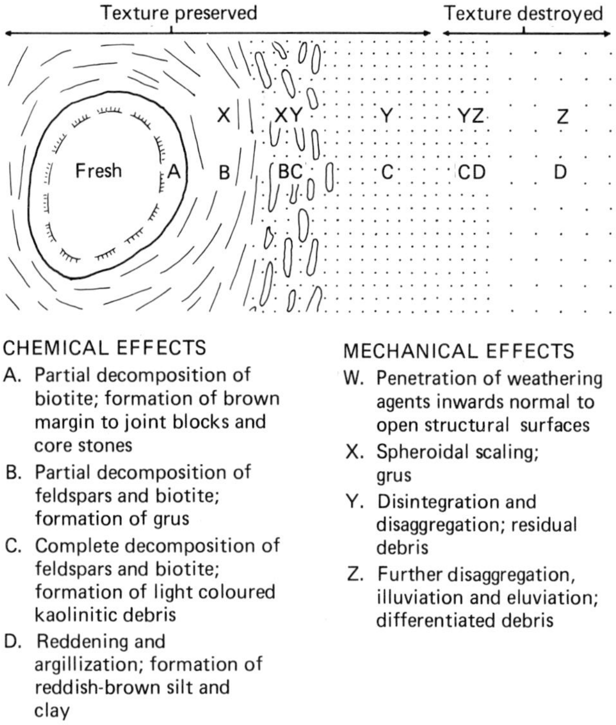

Figure 11.61/1 Rock and soil materials associated with a typical corestone in Hong Kong granite

(g) Strength:
  (i) very strong,
  (ii) Is(50) point load Index = 4.0–12.5 MN m^{-2};
(h) Rock name: GRANITE;
(i) Additional information: discontinuity surfaces are fresh or very slightly stained.

2. Slightly weathered
  (a) Material: 'rock';
  (b) Colour: light pinkish grey;
  (c) Grain size: fine, or medium to coarse;
  (d) Texture: crystalline – fine-grained granite may be porphyritic;
  (e) Structure: unlikely to have microfractures;
  (f) Weathering grade: slightly weathered granite (SWG); biotite and some plagioclase beginning to decompose; feldspars not scratched by a pin;
  (g) *Strength:
    (i) strong to very strong,
    (ii) Is(50) point load index = 4.0–12.5 MN m^{-2},
    (iii) Schmidt hammer '*N*' value greater than 45,
    (iv) not likely to be marked by geological pick;
  (h) Rock name: GRANITE;
  (i) Additional information: discontinuity surfaces have orange-brown discoloration.

3. Moderately weathered
  (a) Material: 'rock';
  (b) Colour: light pinkish grey to grey;
  (c) Grain size: fine, or medium to coarse;
  (d) Texture: crystalline – fine-grained granite may be porphyritic;
  (e) Structure: unlikely to have microfractures;
  (f) Weathering grade: moderately weathered granite (MWG); biotite, plagioclase and orthoclase decomposing; feldspar minerals generally have a 'gritty' consistency and can just be scratched by a pin;
  (g) *Strength:
    (i) strong,
    (ii) Is(50) point load index = 1.0–8.0 MN m^{-2},
    (iii) Schmidt hammer '*N*' value = 25–45,
    (iv) likely to be marked by geological pick only;
  (h) Rock name: GRANITE;
  (i) Additional information: discontinuity surfaces mostly iron stained and pitted.

4. Highly weathered
  (a) *Material: 'soil';
  (b) Colour: generally light pinkish grey but may be light yellow, pinkish yellow, pinkish brown;
  (c) Grain size: medium to coarse SAND to fine GRAVEL;
  (d) Texture: only quartz grains have original shape and hardness;
  (e) Structure:
    (i) likely to have microfractures,
    (ii) relict – original grain structure and joints preserved;
  (f) Weathering grade: highly weathered granite (HWG); feldspar minerals are still distinguishable, but plagioclase is generally decomposed to a clayey consistency and orthoclase is 'gritty'; feldspar can be grooved by a pin;
  (g) Strength:
    (i) very weak rock or hard soil,
    (ii) does not disintegrate in water,
    (iii) Is(50) point load index = 0–0.5 MN m^{-2},
    (iv) Schmidt hammer '*N*' value = 0–25,
    (v) geological pick cannot be pushed into the surface,
    (vi) large pieces can be broken by hand and individual grains can be plucked from the surface.

5. Completely weathered
  (a) Material: 'soil';
  (b) Colour: light brownish yellow, light yellow brown;
  (c) Grain size: silty fine to coarse SAND with some gravel;
  (d) Texture: only quartz grains are intact;
  (e) Structure:
    (i) unlikely to have microfractures,
    (ii) relict – original grain structure and joints preserved;
  (f) Weathering grade: completely weathered granite (CWG); some orthoclase is 'gritty', plagioclase and rest of orthoclase is 'clayey'; all biotite is decomposed;
  (g) *Strength:
    (i) loose,
    (ii) disintegrates in water,
    (iii) Is(50) point load index – not suitable for testing,
    (iv) Schmidt hammer '*N*' value – no rebound,
    (v) geological pick easily indents when pushed into surface.

6. Residual soil
  (a) Material: 'soil;
  (b) Colour: orangish brown or brownish red;
  (c) Grain size: silty clay SAND with some medium to coarse sand;
  (d) Texture: only quartz grains are recognizable by shape, colour and hardness;

(e) *Structure:
 (i) unlikely to have microfractures;
 (ii) no relict grain structure or joints preserved;
(f) Weathering grade: residual soil (RS); virtually all feldspars decomposed to a clayey consistency; biotite completely decomposed;
(g) Strength:
 (i) loose,
 (ii) Is(50) point load index – not suitable for testing,
 (iii) Schmidt hammer '*N*' value – no rebound,
 (iv) disintegrates in water,
 (v) likely to be indented by thumb.

* Diagnostic features.

The order of the descriptive indices (see section 4.6.1) is that used in Hong Kong:

(a) Material type – either an engineering rock or soil;
(b) Colour;
(c) Grain size – where material is an engineering soil, the principal grain size should be written in capitals, e.g. silty coarse SAND;
(d) Texture;
(e) Structure;
(f) Weathering grade represented by the material;
(g) Strength;
(h) Rock name, applicable to rock materials only, and is shown in capitals;
(i) Additional information.

Generally, classification and description of the indices themselves follows standard practice; consequently only some aspects peculiar to Hong Kong need be touched upon.

Microfractures in rock are planes surrounding individual crystals or small aggregates of grains causing the rock to be friable. These may be related to joints, to weathering processes, to unloading by erosion, or to tectonic activity. Microfractures in soil are usually found only in highly and completely weathered grades of granite.

Relict major joints in parent rock are still visible in weathered soil, and their attitudes can still be recorded. Such joints are usually stained by iron and other minerals. Relict joints in highly weathered granite are often of significantly lower strength than the surrounding material. In completely weathered granite their strength is comparable to that of the material.

Additional information should include any factors that are likely to affect long- or short-term stability which cannot be recorded elsewhere. As examples, signs of rapid relaxation after excavation, minor incipient instability, extra loading or vibration near to the slope crest, inefficient or damaged drainage works, and drainage/infiltration characteristics of natural catchments beyond the crest of a cut slope.

Weathering zones in the rock mass

Once again, the weathering zone classification can be traced back to that of Ruxton and Berry (1957) for the Hong Kong granite. The present scheme of five zones beneath the thin topsoil is illustrated in Figure 11.61/2 which is largely self-explanatory. Some points, however, are worth comment. The first is that the jointing becomes closer as the ground surface is reached. Some of the larger rounded corestones in zones 3 and 4 are jointed, with iron staining penetrating inwards. Large corestones of fresh granite with stained selvedges persist throughout the weathering profile, and may remain as surface boulders isolated by erosion.

An important aspect of a weathering profile, and this is applicable to all rock types, is that weathering may have penetrated deeply along major discontinuities, such as faults or shear zones, into otherwise fresh granite (Figure 11.61/2).

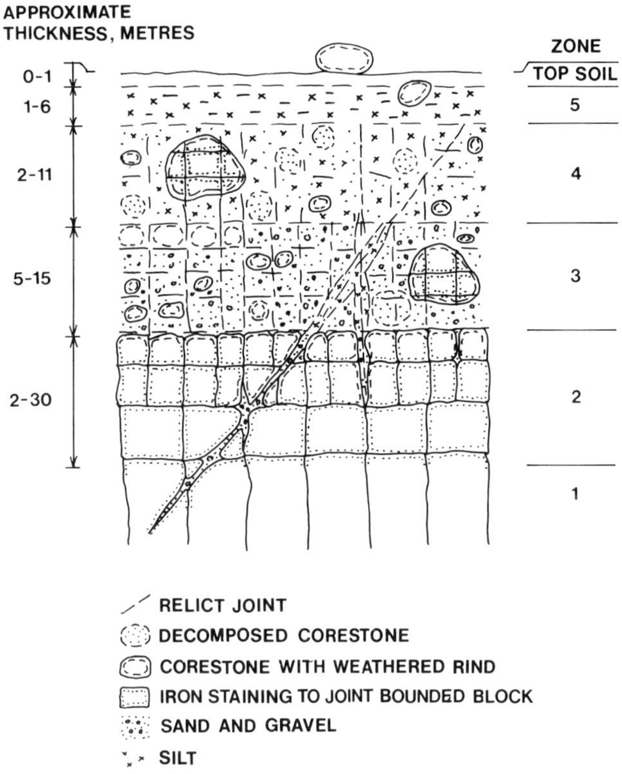

Figure 11.61/2 The five zones in the weathering profile in Hong Kong granite

The diagnostic features of the five zones are set out in Table 11.3, together with those for fill, colluvium and topsoil. Boundaries between the various zones may not be well defined.

Discontinuities in the rock mass

In the rock zones and in soil zones with relict structures, slope stability is controlled by the discontinuities. A standard approach has been adopted in Hong Kong to the description and quantification of all discontinuity types. Details are given on the discontinuity data sheet, discussed later (Figure 11.61/10). Usually all that is mapped is the trace on the rock face, and the dip and dip direction of the discontinuity. Fault zones and major sheeting joints are usually mapped and described in more detail (Figure 11.61/9).

Engineering geological and geotechnical mapping

The general profile for weathered granite has already been well documented (Figure 11.61/2) and it is in this context that ground conditions at engineering sites have to be interpreted and recorded. A standard approach to

Table 11.3 Classification of zones in weathered Hong Kong Granite at Kornhill Development Site, Hong Kong

Zone	*Diagnostic features*	*Design slope angle*	*Dominant failure mode*	*Appropriate remedial measures*
Fill	Heterogeneous, can comprise soil and/or rock of any grade. Rock may show fresh angular fracture surfaces. Extremely variable density and the only structure visible may be layering. Organic-rich layer of topsoil may be present at base of zone where seepage may also occur. May contain foreign material such as tree stumps, building materials, etc., and is commonly associated with old building structures. Can occasionally be identified by sharp changes in topography	30°	1. Erosion 2. Non-circular	(a) Backfilling with approved compacted backfill (b) Regrading to lower design slope angle (c) Lowering groundwater table (d) Retaining wall
Colluvium	Heterogeneous, appears as subangular to rounded irregular arrangement of weathered boulders in a sandy silt or clayey sand matrix. Structure rarely visible. Joints in boulders terminate at the boulders' edge. Organic-rich layer of topsoil may be present at base of zone where seepage may also occur. Generally encountered at the base of steep slopes, and can be identified by sharp changes in topography	40°	1. Erosion 2. Non-circular 3. Translational	(a) Regrading to lower design slope angle (b) Lowering groundwater table (c) Retaining wall
Topsoil	Dark yellowish brown. Upper part sandy, generally depleted and occasionally organic, lower part silty. Structureless	40°	1. Erosion 2. Non-circular	As the thickness of this zone is usually less than 2 m, the material should be removed
5	Dark orangish brown, brownish red, silty clayey sand with no relict structure. Zone comprises 80% RS material; may contain small pockets of CWG material or occasional boulders at the surface. Generally encountered on the tops of low ridges	40°	1. Erosion 2. Circular 3. Non-circular	(a) Regrading to lower design slope angle (b) Lowering groundwater table (c) Retaining wall
4	Zone comprises greater than 50% CWG material and may contain patches of less weathered HWG material (up to 30%) or corestones of MWG to SWG (up to 10%) material. Relict structure and texture commonly preserved, with RS material picking out relict joint planes	40°	1. Circular 2. Non-circular	(a) Regrading to lower design slope angle (b) Lowering groundwater table (c) Retaining wall
3	Zone comprises greater than 50% HWG material and up to 15% subrounded to rounded corestones of MWG or SWG. Relict structure and texture preserved with CWG or RS material picking out relict joint planes. Limit of rippability generally assumed to coincide with Zone 2/3 interface	40–55°	1. Non-circular 2. Planar along relict joints	(a) Regrading to lower design slope angle (b) Lowering groundwater table (c) Retaining wall Higher shear strength parameters may be assumed where the corestone (greater than 60 mm) content is in excess of 30%
2	Zone comprises greater than 50% SWG to FG material. Original rock fabric preserved throughout. Joint surfaces show discoloration which can extend to a depth of 25 mm from joint surface. Joint apertures generally narrow to tight but occasionally wide and infilled with more extensively weathered HWG or CWG	50–75°	Failure mode dependent on joint spacing: 1. Moderately narrow to moderately wide	Potential failure mode Rock fall — Shotcrete with/without mesh
			2. Moderately wide to wide	Rockfall; Planar, Wedge, Toppling } Minor — Shotcrete with one or two layers of mesh, possibly pinned with dowels
			3. Wide to very wide	Planar, Wedge, Toppling } Minor to major — Rock dowels, Rock bolts, Buttressing
			4. Very wide to extremely wide	Planar, Wedge, Toppling } Major only — Rock bolts, Rock anchors, Buttressing
1	Zone comprises 95% FG material, with some slight surface staining. Generally encountered at depth in tunnels, shafts and at the base of deep cuttings. Joints are generally moderately widely spaced with tight apertures, but spacing depends on original grain size		Generally wide to extremely wide	Planar, Wedge, Toppling — Rock bolts, Rock anchors, Buttressing

Notes: 1. The material grades are referred to on pages 275–277.
2. The boundaries between the zones may not be well defined.

mapping has been formalized, leading to a stage-by-stage sequence of recording of ground conditions and assessment of engineering remedial measures. The boundary between engineering geological mapping and geotechnical mapping is blurred, and most of the published examples of mapping, usually at the plan scale, include remedial measures and are therefore strictly geotechnical plans.

Engineering in Hong Kong is extremely varied, ranging from dam sites to shafts and tunnels, cut-and-fill tunnels for the Metro, underground caverns, and extensive site levelling with high rock cuts for housing development. Typical of the latter is the Kornhill Development situated on the northern shore of Hong Kong island (Figure 11.61/3). General details of the scheme have been given by Muir, Smethurst and Finn (1986), from which much of the following account has been taken.

The Kornhill site

The Kornhill Development comprises an S-shaped 22 ha land-site for 44 30-storey residential tower blocks. The approximate location and the original topography, the site and the final site layout are shown in Figures 11.61/4 and 11.61/5. In three years, $3\frac{1}{2}$ million m^3 of rock and $1\frac{1}{2}$ million m^3 of weathered granite soil and superficial deposits had to be removed, lowering the original ground level by up to 80 m. The main perimeter rock

Figure 11.61/3 General view of Kornhill Development, Hong Kong, from the north in mid-construction. The main perimeter rock slopes can be seen in the background and partly formed slopes and bulk excavation in the foreground (From Muir, Smethurst and Finn, 1986, Plate 1, by permission)

slope, 800 m long and up to 65 m high, was mapped in panels as excavation proceeded. The engineering geological plans so produced provided the basis for assessment of stability and the design of remedial measures.

Site investigation

Before construction began, the natural topography comprised three broad subrounded ridges separated by NNE and subsidiary WNW trending steep valleys (Figure 11.61/4). In soils the valley slopes were up to 30°, whereas in rock they were locally vertical. The original slopes of the eastern half of the site had been extensively modified by quarrying, and excavation and backfilling for structures.

Fault lineaments inferred from aerial photographs were checked by seismic refraction surveys. Two types of Hong Kong granite were present, intruded by four deeply weathered dolerite dykes picked out as marked surface features, and cut by a deeply weathered fault zone. The main features are shown in the geological site plan (Figure 11.61/5). At formation level it can be seen that extensive soil zones were still present, principally in the areas adjacent to the deeply weathered dolerite dykes.

On granite, vertical transition from soil to continuous rock occurred over 1 m or less. The seismic refraction interface and ripping limit coincided with the top of the moderately weathered granite. The normal criterion for

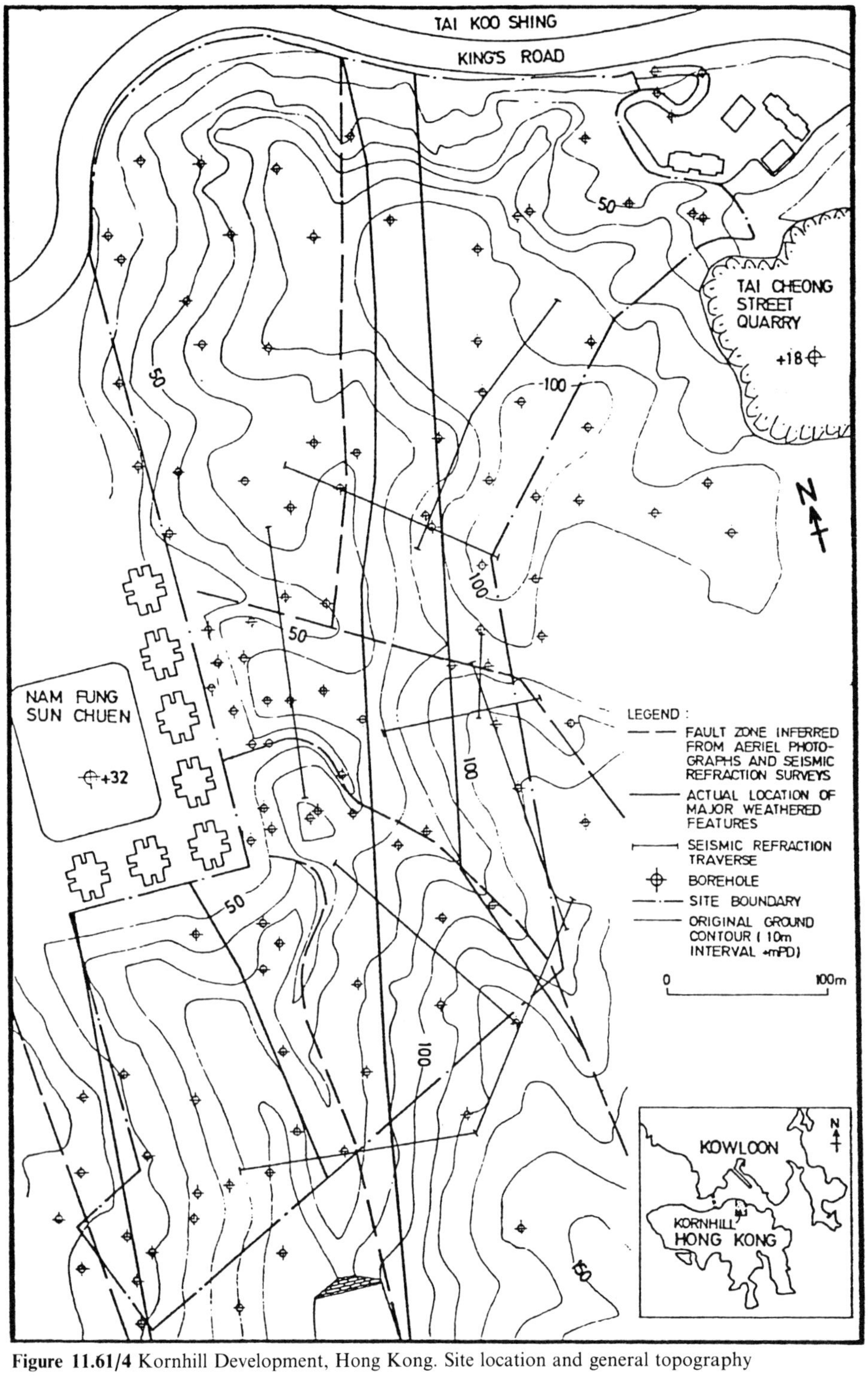

Figure 11.61/4 Kornhill Development, Hong Kong. Site location and general topography

proving bedrock was the recovery of 5 m of continuous sound core. Orientation of discontinuities in the boreholes was determined by the Borehole Impression Device which provided a plan of discontinuities intersecting the borehole wall.

Residual soils were typically 10 m thick on ridge crests, thinning on valley sides to nil in valley bottoms.

Logging cut slope faces

The geotechnical parameters for the design of cut slopes for the Kornhill Development were obtained from detailed site investigation with associated laboratory testing, coupled with comprehensive field mapping. Engineering construction, particularly that related to excavated rock slopes, was subjected to intensive checking of critical design parameters by the government agency. Consequently, in order that this could be done it was necessary to record all salient geological features during excavation on large-scale engineering geological plans showing both these and proposed permanent protection measures. For this purpose a standardized procedure was evolved for the logging of cut slopes and for the design of any necessary remedial measures.

The slopes along the perimeter of the site were up to 65 m high, composed of soil and rock, whereas the international slopes were generally in rock up to a maximum height of 125 m. Rock slopes were generally excavated at 65–70°, with 2 m wide horizontal berms at

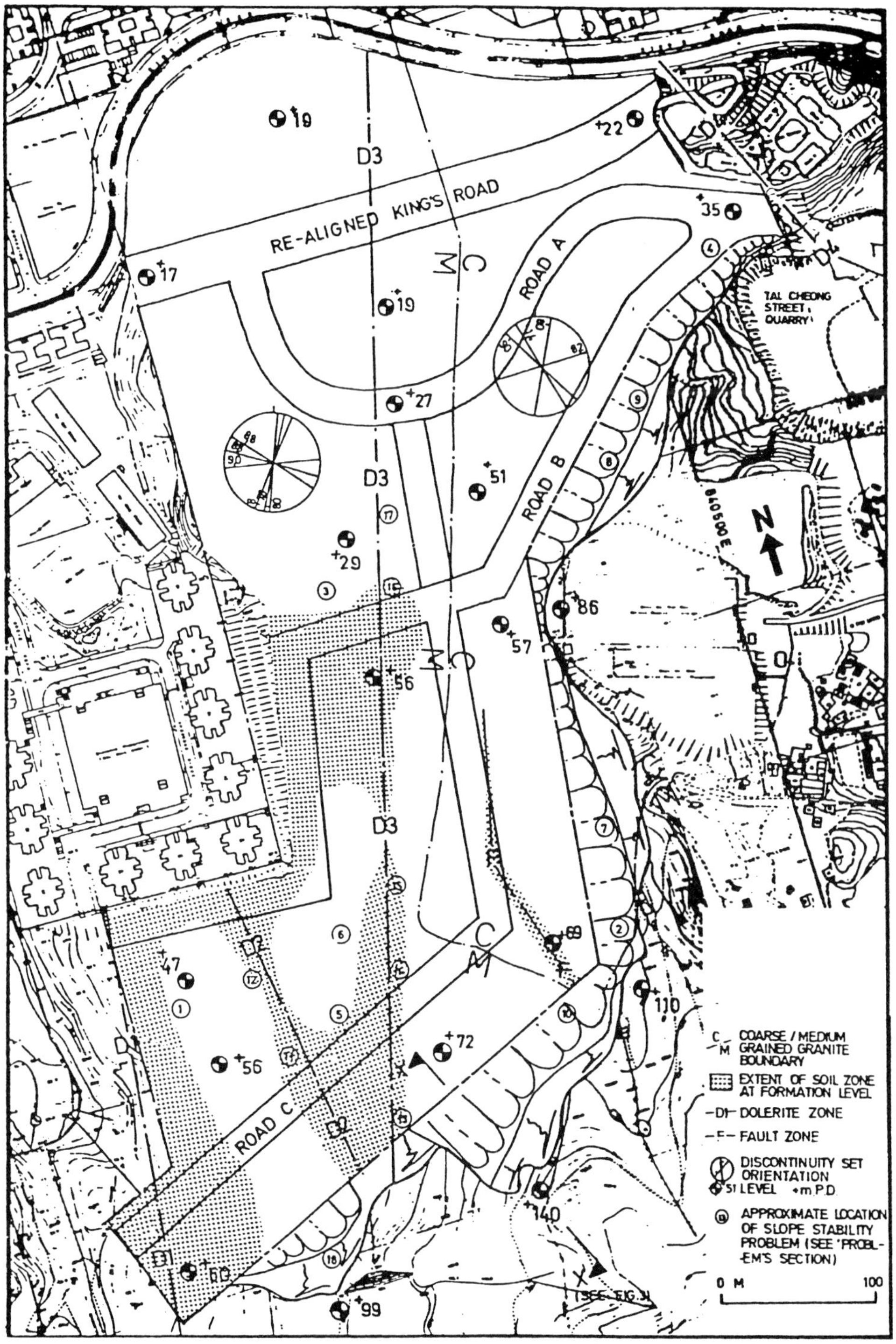

Figure 11.61/5 Kornhill Development, Hong Kong. Geological site plan

12 m vertical intervals, whereas soil slopes were cut at 40°, with 1.5 m wide horizontal berms at 7 m vertical intervals (Figure 11.61/6).

Engineering geological mapping methods

The methodology of mapping was based on past experience by the engineers involved (Eyre, 1973; Dearman and Fookes, 1974), and as it may be regarded as a model procedure, a full description is justified here. Systematic mapping or logging of the face of a cut slope is essential to check original geotechnical design assumptions based on the site investigation and to identify areas of potential instability at an early stage of construction so that necessary remedial measures can be designed.

Accordingly, it was decided that a complete record of the whole of the exposed face should include:

- a description of the materials exposed;
- a log of the discontinuities;
- detailed logging of roughness of joint planes where this was considered important in stability analyses;
- recording on a plan of all information that might possibly affect the stability of the final slope.

All data that had been collected from the final cut slopes were documented on sheets for description, sketch and discontinuity survey, and on photographs. To make retrieval of information easy, the face was divided into panels, and all the information for each panel was kept together. Each panel was given a distinctive number, and all relevant sheets for the same panel were coded with this number. A panel was located accurately by at least two survey stations with co-ordinates and levels, and the plan position of each panel was shown on a site plan (Figure 11.61/7).

The mapping sequence provided the following information:

- Panel description sheet (Figure 11.61/8);
- Panel sketch sheet (Figure 11.61/9) – scale 1:100 except for critical areas, where additional larger scale drawings would have to be prepared;
- Discontinuity data sheet (Figure 11.61/10).

Panel description sheet

The lithologies present in each panel, both rock and soil, and where appropriate chunam, were described

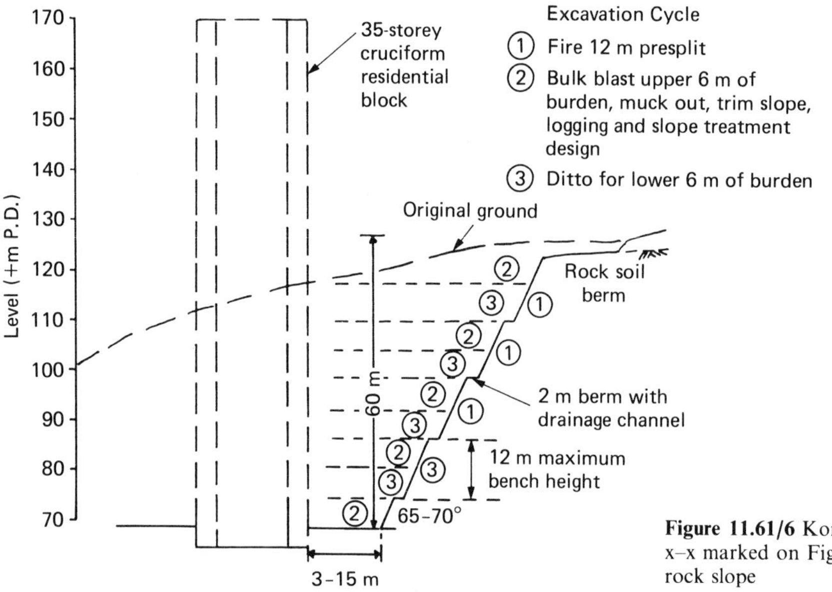

Figure 11.61/6 Kornhill Development, Hong Kong. Typical section x–x marked on Figure 11.61/5 and the excavation sequence for the rock slope

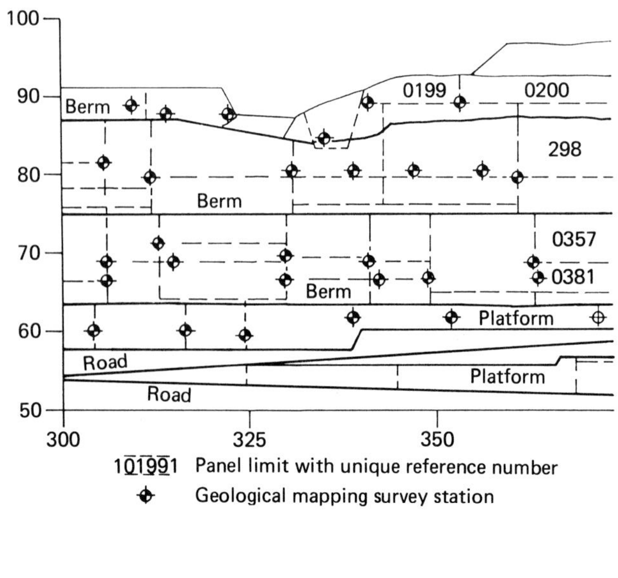

Figure 11.61/7 Kornhill Development, Hong Kong. Positions of mapping panels on the site plan of the rock slope

GENERAL INFORMATION

PANEL No.					DATE	DAY	MONTH	YEAR	LOCATION PLAN No.			
	0	1	0	1		21	09	84		0	0	2

SCANLINE No.						DISCONTINUITY DATA SHEET No.												
0	1	0	1	—	1	0	1	0	1	—	1		0	1	0	1	—	5
0	1	0	1	—	2	0	1	0	1	—	6	TO	0	1	0	1	—	9

LOCATION (IN WORDS) SOUTHERN MAIN SITE
ZONE L9
SECTION 1–23

OPERATOR		PHOTO No.													
	R.F.		0	1	0	1	—	1	TO	0	1	0	1	—	5

MATERIAL	COLOUR			GRAIN SIZE				TEXTURE	STRUCTURE	WEATHERING GRADE OR ORIGIN			STRENGTH	ROCK NAME	ADDITIONAL INFORMATION	% OF FACE
2		Y	B		Si	MS	Gv			C	W	G	De			40
2	L	Y	P		Si	CS	Gv		MF	H	W	G	De			40
1	L	P	G				C	PH		S	W	G	V.St	1		20

SURVEY STATION No.							NORTHING										EASTING										LEVEL						
	0	1	0	1	—	1		8	1	5	8	0	0	·	7	0		8	4	0	3	9	5	·	6	6		1	3	1	·	2	3
	0	1	0	1	—	2		8	1	5	7	7	8	·	9	2		8	4	0	4	0	0	·	2	5		1	3	4	·	6	7
	0	1	0	1	—	3		8	1	5	7	9	4	·	6	3		8	4	0	3	9	0	·	2	7		1	2	8	·	4	2
	0	1	0	1	—	4		8	1	5	7	7	5	·	2	5		8	4	0	3	9	7	·	6	3		1	2	4	·	9	7

MATERIAL	COLOUR					
1. ROCK	Light (L)	Pinkish (P.)	Pink (P.)			
2. SOIL	Dark (D)	Reddish (R.)	Red (R.)			
		Yellowish (Y.)	Yellow (Y.)			
		Brownish (B.)	Brown (B.)			
		Greyish (G.)	White (W.)			
		Greenish (GR.)	Grey (G.)			

STRENGTH	
Very strong (V. St.)	Very hard rock—more than one blow of geological hammer required to break specimen.
Strong (St.)	Hard rock—hand held specimen can be broken with single blow of geological hammer.
Very weak rock (V. Wk.) or hard soil	Brittle or tough, may be broken in the hand with difficulty.
Dense (Gravel and Sand)	Requires pick for excavation, 20 mm steel peg hard to drive.
Loose (Gravel and Sand)	Can be excavated with a spade, 20 mm steel peg can be easily driven.

TEXTURE	STRUCTURE	
ROCK	ROCK	SOIL
PORHYRITIC (PH.) PEGMATITIC (PF.) CRYSTALLINE (CR.)	MICROFRACTURE (MF.)	RELICT (RL.) MICROFRACTURE (MF.)

Very coarse grained (V. C.)	>60 mm	Boulders and Cobbles (Bo. CB.)
Coarse grained (C.)	(20–60 mm)	Coarse Gravel (C. GV.)
	(6–20 mm)	Medium Gravel (G. GV.)
	(2–6 mm)	Fine Gravel (F. GV.)
Medium grained (M.)	(0.6–2 mm)	Coarse Sand (C.S.)
	(0.2–0.6 mm)	Medium Sand (M. S.)
	(0.06–0.2 mm)	Fine Sand (F. S.)
Fine grained (F.)	(0.002–0.06 mm)	Silt (Si.)
Very fine grained (V. F.)	>0.002 mm	Clay (Cl.)

WEATHERING GRADE OR ORIGIN	ROCK NAME
Topsoil (T. S.) Fill (F.)	1. GRANITE
Colluvium (COL.)	2. DOLERITE
Residual Soil (R. S.)	3. PEGMATITE
Completely Weathered Granite (C. W. G.)	
Highly Weathered Granite (H. W. G.)	
Moderately Weathered Granite (M. W. G.)	
Slightly Weathered Granite (S. W. G.)	
Fresh Granite (F. G.)	

Figure 11.61/8 Kornhill Development, Hong Kong. Panel description sheet

in a coded form relating to the standard Hong Kong sequence of descriptive indices.

The indices are shown on the panel description sheet using abbreviations or numbers. Typical descriptions of materials are:

Rock, light pinkish grey, medium- to coarse-grained porphyritic, slightly weathered GRANITE very strong.

In abbreviated form on the panel description sheet this would be: 1, LPG, M to C, PH, SWG, GRANITE, V.St.

Soil, light yellowish brown, silty coarse SAND with some gravel, completely weathered granite, dense.

In abbreviated form: 2, LYB, S : C.SGv, CWG, De.

The panel description sheet relating to Figure 11.61/9 is given in Figure 11.61/8.

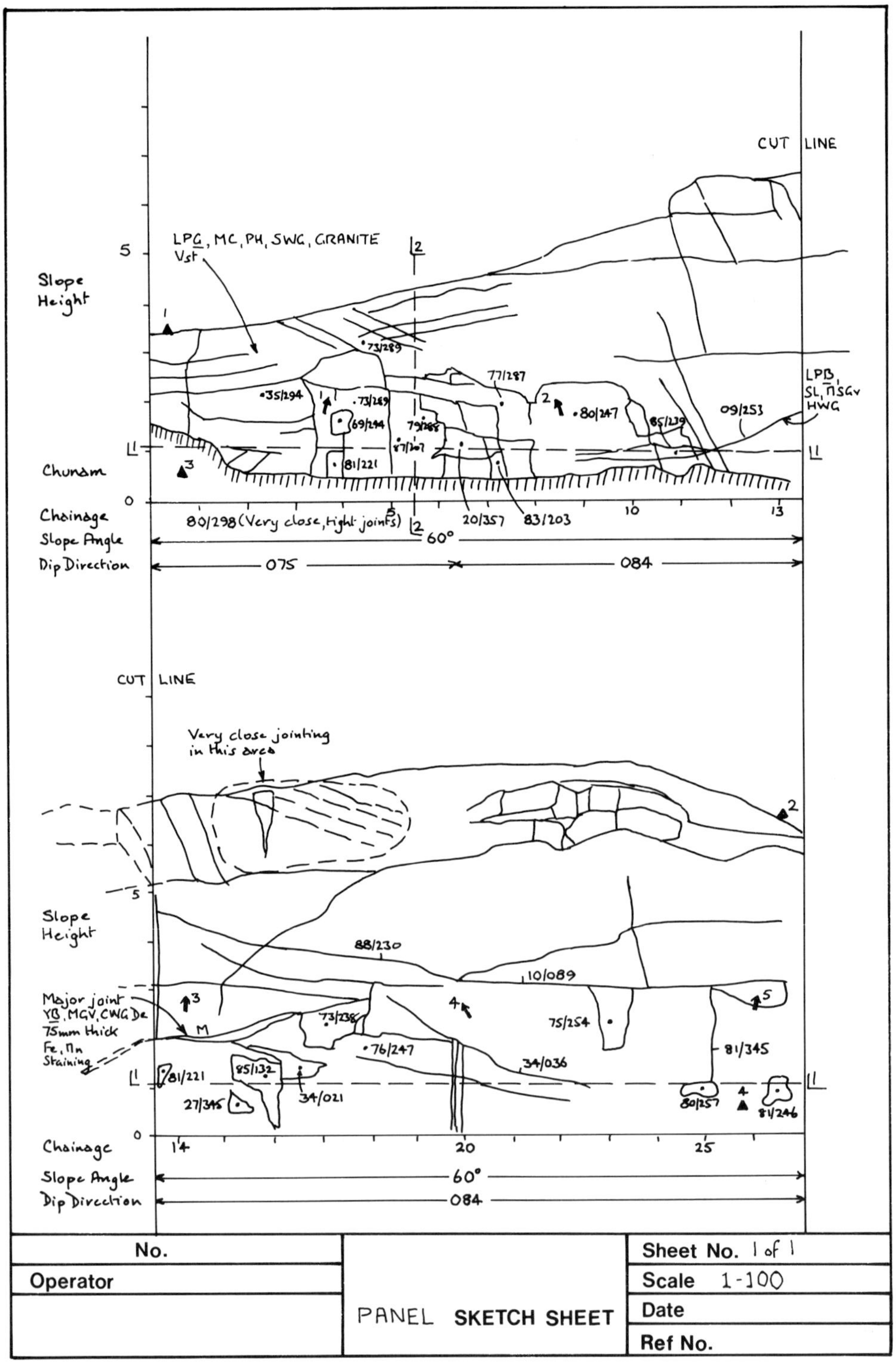

Figure 11.61/9 Kornhill Development, Hong Kong. Panel sketch sheet

Panel sketch sheet

The descriptive scheme is useful in describing individual areas of homogeneous soil and rock materials that can be mapped on an exposed face. In a cut slope through a typical weathered sequence, it is rarely sufficient to describe the individual material types because the general stability of the slope is dependent on the relationship between the various weathered types or weathering grades. In practice it was necessary to use the simple zoning scheme of five zones which could be recognized in the *in situ* weathering profile (Figure 11.61/2), with separate classifications for topsoil, colluvium, residual soil and fill.

As an aid to mapping and subsequent engineering stability assessments, the diagnostic features for each zone were set down in a table (see Table 11.3), with an indication of potential failure modes and appropriate remedial measures.

Two examples of the field sketching will suffice. The first is of Panel 0671, 10 m long and 7.3 m high (Figure 11.61/11a), in uniform rock with a simple joint pattern.

The second example is from the top of the rock cut,

GENERAL INFORMATION SHEET No. 2407-5	SHEET 5 OF 8	TYPE OF SURVEY 2			
PANEL No. 0407	OPERATOR RPF	DATE (DAY MONTH YEAR) 15 09 82	1 SUBJECTIVE 2 OBJECTIVE		
SCANLINE No. 1	PLUNGE 10	TREND 055	LENGTH 29.5	DISCONTINUITY SPACING 3	

Chainage or No.	TYPE	Dip	Dip Direction	Persistence	Spacing	Aperture	Nature	Thickness	Consistency	Planarity	Waviness	Amplitude	Roughness	Water	Fabric	Block Size	Additional Information
20.6	2	12	164	5.0	3	6	5 6	6		1			2	1		3	
21.3	2	10	156	7.5	3	6		6		1			2	2		3	
21.5	2	13	158	6.5	3	6	6	6		1			2	1		3	
21.6	2	47	227	4.2	3	6	5	6		2			2	1		3	
22.1	2	52	220	5.2	3	6		6		1			2	1		3	
22.4	2	50	220	4.2	3	6		6		2			2	1		3	
22.6	2	47	229	6.3	3	6		6		2			2	2		3	
23.0	2	47	226	6.3	3	6		6		2			2	2		3	
23.2	2	35	007	4.7	3	6		6		1			2	1	3	3	
23.5	2	84	015	6.5	3	6		6		3	8.0	0.5	2	1	3	3	
23.9	2	88	006	4.5	3	6		6		1			2	1	3	3	
24.3	2	80	009	3.3	3	6		6		1			2	1	3	3	
24.5	0	87	007	6.8	3	5		5		2			5	3	3	3	WITHIN FAULT ZONE, [illegible]
24.8	2	88	012	3.6	3	6		6		2			2	1	3	3	
25.5	2	86	011	3.6	3	6		6		1			2	2	3	3	
26.2	2	84	013	5.5	3	6		6		1			2	1	3	3	
26.8	2	82	004	5.8	3	6		6		1			2	2	3	3	
27.0	2	87	003	7.5	3	6		6		1			2	1	3	3	

TYPE	SPACING	APERTURE	NATURE	INFILLING THICKNESS	CONSISTENCY	PLANARITY	ROUGHNESS	WATER	FABRIC	BLOCK SIZE
1. BEDDING PLANE 2. NON-CONFORMITY 3. JOINT 4. RELICT JOINT 5. FAULT ZONE 6. FAULT 7. SHEAR 8. TENSION CRACK	1. Extremely Wide (2 m) 2. Very Wide (600 mm - 2 m) 3. Wide (200 - 600 mm) 4. Moderately Wide (60 - 200 mm) 5. Moderately Narrow (20 - 60 mm) 6. Narrow (6 - 20 mm) 7. Very Narrow (< 6 mm)	1. Wide (200 mm) 2. Moderately Wide (60 - 200 mm) 3. Moderately Narrow (20 - 60 mm) 4. Narrow (6 - 20 mm) 5. Very Narrow (2 - 6 mm) 6. Extremely Narrow (< 2 mm) 7. Tight	1. Clean 2. Surface Staining 3. Cohesive 4. Non-cohesive 5. Iron 6. Manganese 7. Cholorite 8. Quartz 9. Kaolinite 10. Other	1. > 200 mm 2. 60 - 200 mm 3. 20 - 60 mm 4. 6 - 20 mm 5. 2 - 6 mm 6. < 2 mm	V.WK. Rock or Hard Soil DR. Dense (Gravel and Sands) Lo. Loose (Gravel and Sands) So. Soft (Cohesive) St. Stiff (Cohesive)	1. Planar 2. Curved 3. Undulating (Express Waviness Wavelength (W) and Amplitude (A) in Metres	1. Smooth 2. Rough 3. Stepped 4. Polished 5. Slickensided	1. Dry 2. Seepage 3. Flow 0.1 L/S 4. 0.1-1 L/S 5. 1 L/S	1. Blocky 2. Tabular 3. Columnar 4. Irregular	1. V. Large > 2 m 2. Large (0.6 - 2 m) 3. Medium (0.2 - 0.6 m) 4. Small (0.05 - 0.2 m) 5. V. Small (0 05m) (Maximum dimension in metres shown in brackets)

FOR DEFINITION OF ALL DESCRIPTIVE INDICES REFER TO LOGGING MANUAL

FREEMAN FOX & PARTNERS (FAR EAST) CONSULTING ENGINEERS HONG KONG	KORNHILL DEVELOPMENT DISCONTINUITY DATA SHEET FIG. 7	DATE OCT 1982 Ref No FE/272/469

Figure 11.61/10 Kornhill Development, Hong Kong. Discontinuity data sheet

with the original ground beyond the slope crest rising at 30° as bare soil on rock (Figure 11.61/9). On the A4 size standard sketch sheet, the recording has been done in two 13.5 m length contiguous panels broken by a cut line across which some connecting detail from both sketches has been shown. Attitudes of significant joints are shown and joints sub-parallel to the face are marked by a dot. Two discontinuity scan lines are indicated, as are the positions of survey points and photographs. Although the face is in slightly weathered granite (SWG), one major joint has been highlighted containing up to 75 mm of highly (HWG) to completely weathered granite (CWG). The lower part of half of the slope had been plastered with chunam before the record was made.

Later, this early record is compared with the final working drawing (Figure 11.61/12), when some major discrepancies between the two are discussed. According to the standard method of working, the field panel sketch had to be checked by the engineering geologist or engineer before being worked up to the final version in the drawing office. The importance of checking at all stages of recording cannot be too greatly emphasized.

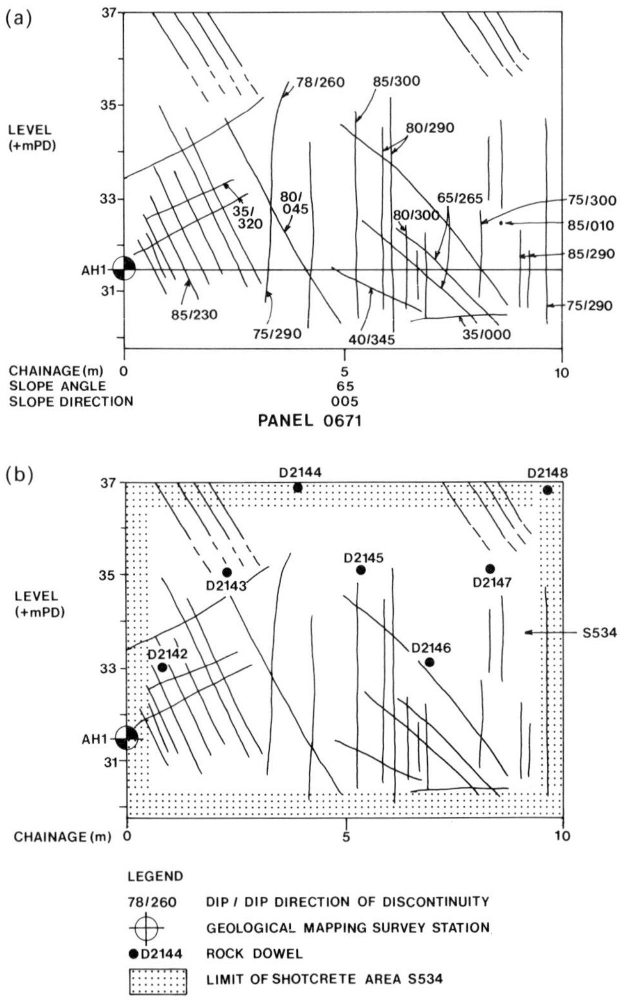

Figure 11.61/11 Kornhill Development, Hong Kong. Typical panel logging evaluation: (a) the geological details; (b) treatment applied

Discontinuity surveys

All the main discontinuities visible on the face were recorded on the panel sketch sheet (Figure 11.61/9), together with areas of very close jointing. It was considered to be very important that the structure of the rock mass should be recorded in a systematic manner, and for this purpose a standard descriptive scheme was evolved and quantified as far as possible (cf. Chapter 4, subsection 4.6.4.2). These descriptive indices were usually recorded on a discontinuity data sheet on which abbreviations or numbers were used (Figure 11.61/10). In some cases, the details are written in abbreviated form on the panel sketch sheet and in full wherever it was necessary to avoid ambiguity.

The discontinuity readings should be plotted on a lower hemisphere Schmitt net (see Figure 4.6) and contoured if there are enough poles. Pole maxima will be checked by subjective examination of the mapped panel in order that possible failure modes may be checked, and potentially unstable blocks or rock masses identified. A subjective assessment would already have been made by the geologist or engineer during the mapping and logging.

Line logging, in which the full details of every discontinuity intersecting a stretched tape are recorded, was used as an objective check at the beginning of site formation work, and whenever there was a change of logging personnel. Line logs should ideally be run in three directions at right angles, although this is rarely possible and horizontal and vertical lines on the face often had to suffice.

Record photographs

Each face was photographed, taking care to ensure reasonable overlap, to provide a complete record of the face. The photographs could be in colour or black and white, and should contain a scale and reference number. The centre of the photograph was marked, with reference number, on the panel sketch sheet.

Presentation of data

The raw data collected in the field, together with the photographs, had to be checked, coded and sequentially numbered. All originals were then photocopied and preserved in the archives; the copies were used for the subsequent design stages.

It was very important that sufficient detail should be recorded in the field so that, using the panel sketch sheets supplemented by photographs, large-scale engineering drawings could be prepared. These were true geotechnical plans because they included not only the geological details, but also the location and types of remedial works. Clearly, the final drawings had to be produced as soon as possible after logging, particularly as in Hong Kong all remedial works had to be checked and approved by the Building Authority.

Final geotechnical plan

The final plans differ in appearance, depending upon whether they are of panels at the top of the cut slope or mid-slope panel, and on the weathering zone, or zones, that have been recorded.

Slope treatment for Panel 0671 (Figure 11.61/11b) was simple; seven rock dowels were inserted and the whole area shotcreted.

Panels 0657 and 0673 (Figure 11.61/12) are more interesting. The plan is in three parts: the elevation of the rock face; tabular data on the geology, stability, and remedial measures; and the legend.

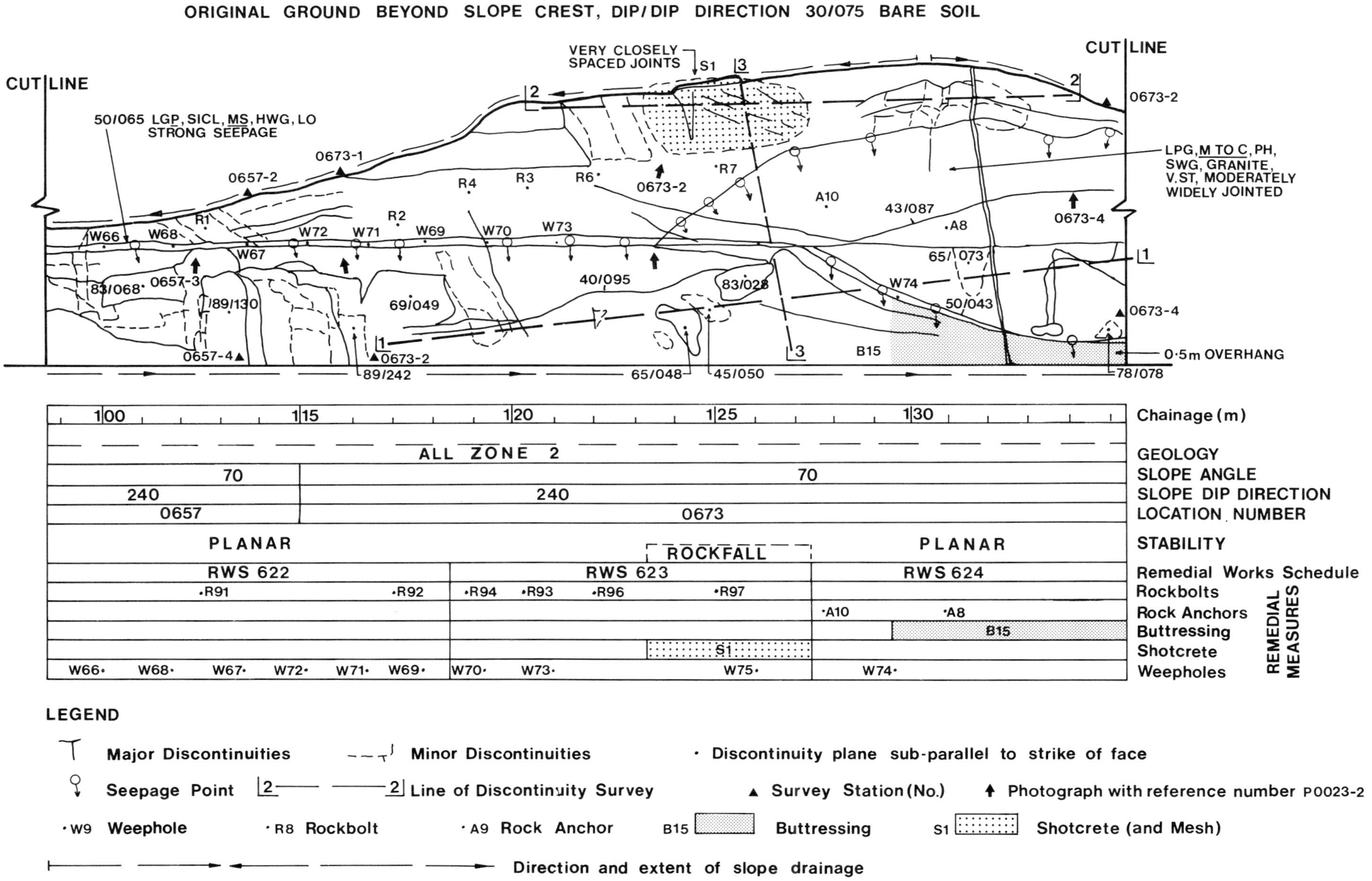

Figure 11.61/12 Kornhill Development, Hong Kong. Final record geotechnical plans for panels in slightly weathered granite (SWG) with prominent layer of highly weathered granite (HWG)

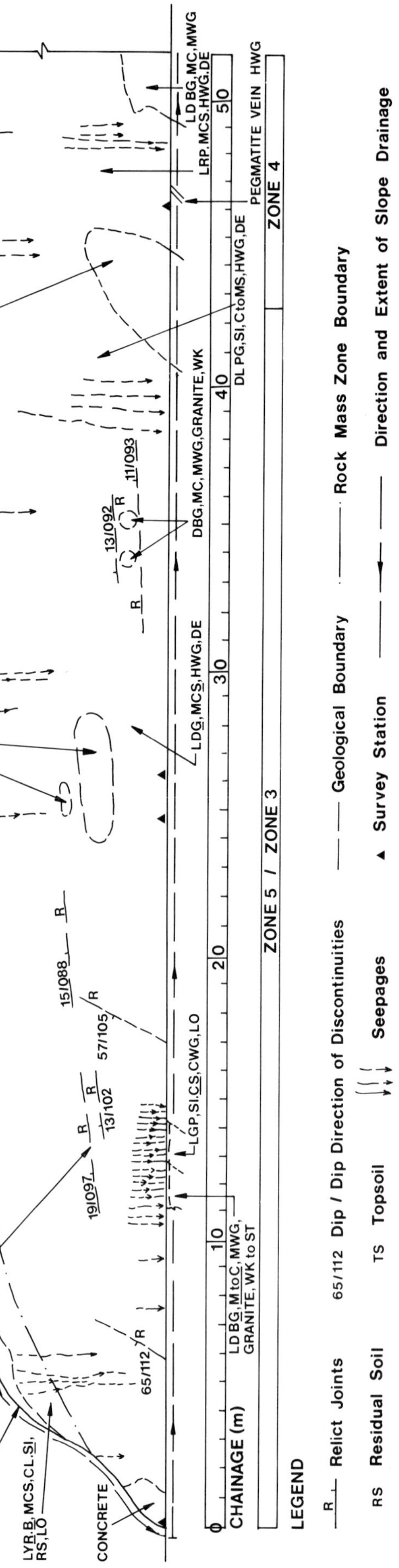

Figure 11.61/13 Kornhill Development, Hong Kong. Final record geotechnical plans for panels in highly to extremely weathered granite

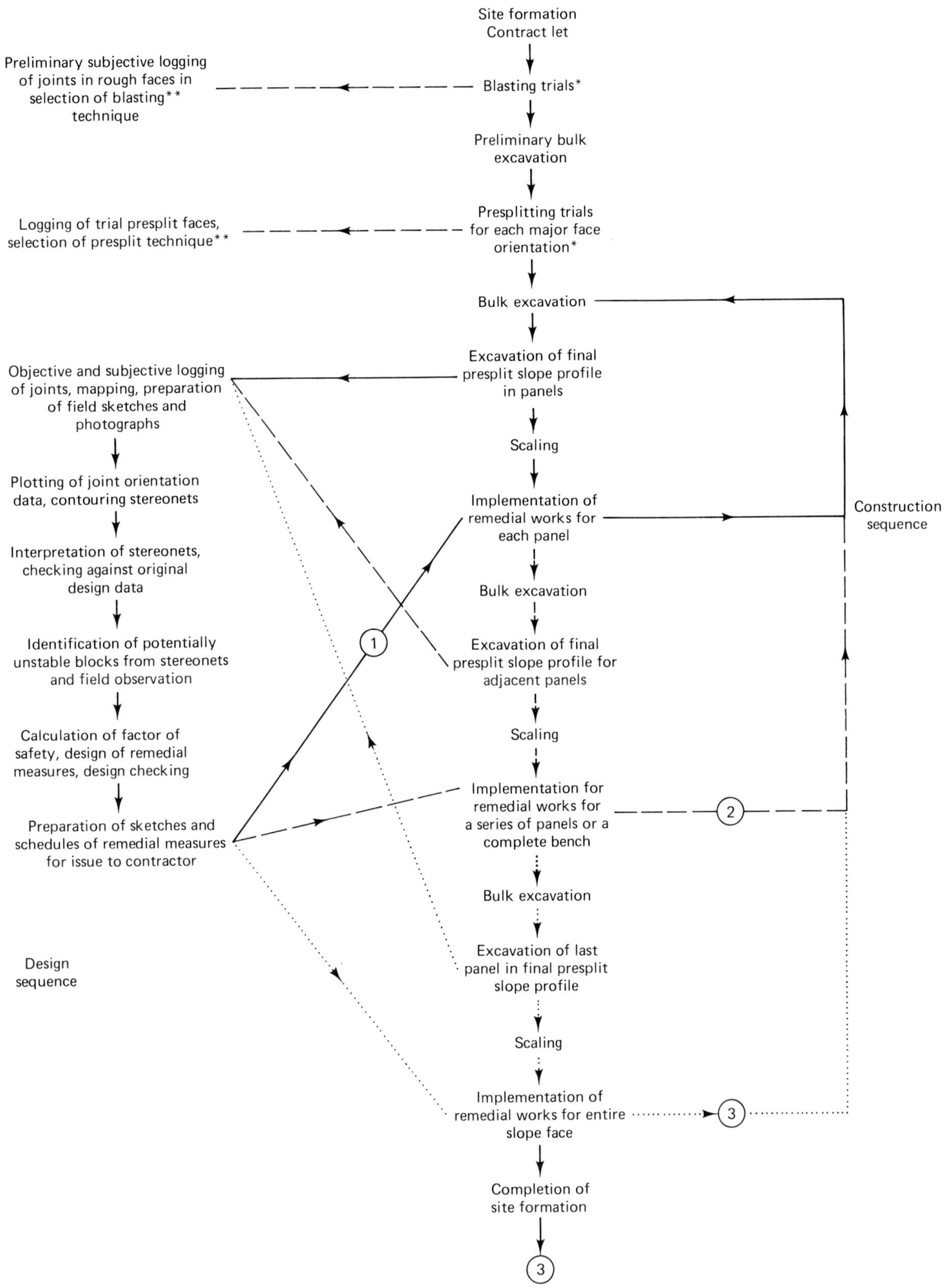

* Subject to inspection by Mines Division and BDD

** Subject to consultation with Mines Division and BDD

(1) Duplicate copy of sketches and schedules issued to BDD, cc GCB for information

(2) Record drawings and calculations submitted to BDD for approval

(3) Large-scale engineering geological drawings and calculations, prepared in accordance with BOO Practice Note 1979.61, submitted to BDD for approval

Figure 11.61/14 Kornhill Development, Hong Kong. Flow chart for the logging of rock faces and the design of remedial measures

The first point to notice on the panel sketch is the very marked differences of emphasis between the information recorded on the panel sketch sheet, the preliminary assessment (Figure 11.61/9), and the final record drawing – but first there are discrepancies between the two. On the panel sketch sheet, structural data and even the inclination and orientation of the face appear to be given in the conventional way used by geologists: dip and strike are used. In Figure 11.61/12, on the other hand, dip and dip direction are used. In addition, the survey stations, photographs and discontinuity line surveys are now placed in their correct positions.

One aspect of the geology that does not appear on the panel sketch is the line of seepage along a thin layer of highly weathered granite, running horizontally about half-way up the face from the left, eventually dipping down nearly to berm level at the right. Above it, the slightly weathered granite is porphyritic, and below crystalline and microfractured.

It is evident that a great deal of additional recording was done on the face to produce the final record. The locations of all remedial measures are shown: weepholes along the highly weathered band which presumably was concreted over; an area of shotcrete and a buttress to support the overhang and the right-hand side of the face; rock bolts and rock anchors. Technical details would have been recorded in the three Remedial Works Schedules.

Apart from the possibility of rockfall from the very closely jointed area at the top of the face, subsequently shotcreted, potential instabilities due to planar failure were countered using mainly rock bolts and two rock anchors.

Plan of a panel in highly to extremely weathered granite
Plans of Zones 4 and 5 granite are, in general, simpler than those of the rock face just described. The chief points of interest are the relict corestones of moderately weathered granite in the highly weathered granite, the relict joints, and individual lines of seepage down the face (Figure 11.61/13).

The face has been mapped as two distinct areas: the larger Zone 5/Zone 3 part on the left, with the remainder to the right as Zone 4. Gently undulating natural ground at the top of the slope has a local, thin covering of fill. Beneath is up to 2 m of residual soil (RS) mapped as two separate units; light yellow brown soft silty sand clay overlies highly weathered granite (HWG) until, as the natural slope angle increases, a layer of light yellow brown loose clayey silt intervenes. The residual soil of Zone 5 dies out on the steep face of the slope, and the outcrop is thus confined to the crest of the hill.

The Zone 3 material is composed of more than 50% of highly weathered granite (HWG) with scattered corestones of moderately weathered granite (MWG). Some traces of relict joints have been mapped as two sets, with one dipping gently out of the face, candidates for planar failures if the shear strength parameters are sufficiently low, although individual joints do not persist along the face.

Zone 4 at the right-hand end of the face is highly weathered granite (HWG), coarse to medium sand with large corestones of moderately weathered granite (MWG). No relict structures have been recorded. From the descriptions in Figure 11.61/13 the zone could equally well have been recorded as a continuation of the Zone 3 of the rest of the panel.

Recording and design team
It was realized at an early stage of the preliminary designs for the site that it would be necessary to set up a design team on site for the cut slopes in order to minimize turn-round time from completion of a panel design to its approval by the Government Geotechnical Control Office. The requirements for approval, it will be recalled, was the submission of 1:100 scale as-built engineering geological drawings for all the major rock slopes (Figures 11.61/11 and 11.61/12), photographs with overlays, stereographic projections demonstrating overall stability, and calculations justifying the proposed slope treatment measures.

As always in Hong Kong, the programme for bulk excavation, slope forming and building work was very tight. It was essential to install slope support measures as the formation was lowered so that problems of access for installation were reduced, and building work could commence as soon as final platform levels were reached. Consequently, a separate team, comprising a geologist and three technicians, was included in the site supervision staff to map slopes immediately after trimming. Approximately 300 m^2/day of completed slope had to be assessed in the first year of the project. A design team of a geotechnical engineer, an engineering geologist, four assistant engineers and two technicians was used at the peak of design work. Once potential failure modes and types of slope treatment had been established, a slope treatment design manual was produced.

A flow chart (Figure 11.61/14) sets out the design and construction sequences for individual panels, a series of panels and the entire slope.

11.9 Large underground openings

Apart from mining, increasing use is being made of specially excavated underground caverns for a variety of purposes, including power stations, sewage works, warehouses and factories, swimming pools, and civil defence shelters. Over the past decade, not only has the variety of uses increased dramatically, but also the size and complexity of the caverns. As just one example, whereas a few years ago the largest caverns being excavated in Precambrian granites in Finland were oil storages 25 m high, 15 m wide and several hundred metres long (Saari and Ylinen, 1979), caverns with a width of 35 m have now been constructed. Increased size brings with it greater problems of locating the cavern in a suitable rock mass, to ensure the stability of the cavern walls and roof. Location is mainly a problem of the site investigation stage of a project, whereas stability depends to a very great extent on assessments made during construction. Making such assessments requires a detailed knowledge of all aspects of the geology of the walls and roof and immediate environment of the cavern such as can only be gained by detailed geological mapping at a sufficiently large scale.

11.9.1 Plans of large underground openings

For engineering structures deep below the ground surface, the initial site investigation, including surface mapping, can only give a broad, generalized view of the geology. Distribution of rock and soil types and the major structures can be determined, although it is notoriously difficult to project surface information downwards to the level of the works with sufficient certainty for engineering design and construction. Reliance has to be placed on recording as the construction proceeds, although trial headings may be used if they are part of the final design or if the project is sufficiently large to justify the cost. Mapping has therefore to be done during the construction cycle as a rapid predictive and recording tool.

11.9.2 Methods employed in mapping underground caverns at Dinorwic, North Wales, UK

Techniques adopted for mapping the large caverns and tunnels associated with the Dinorwic pumped storage scheme in North Wales have been described by McCrae, Powell and Yu (1979).

General description

The scheme comprises nine major caverns and associated hydraulic and access tunnels, all excavated by drilling and blasting in Cambrian slates and grits intruded by dolerite dykes. Extensive quarry workings above the site of the complex provided a substantial amount of information on the geological structure at an early stage of design. Engineering geological investigations at the exploratory stage confirmed that the slate was a highly discontinuous rock mass. These early studies indicated that detailed information would be needed in order to evaluate the support requirements for each section of the caverns. Trial enlargement of part of the machine hall (Figure 11.62) was mapped in detail and selected monitoring instruments were installed to provide information on displacements. The purpose of the trial was to assess support pressures needed to maintain roof stability. Results of the monitoring recorded that variations of movement were related to geological features, and lithological changes, and in particular to dyke margins. This confirmed that during excavation, attention would have to be paid to geological mapping and a thorough appraisal made of all discontinuities.

The dolerite dykes have been metamorphosed, with a well-developed planar fabric parallel to the cleavage in the slates. Near vertical dykes are very competent, except at the margins where up to 1.5 m of strongly sheared rock is present. There had been extensive chloritization, and the joint sets are generally chlorite coated and extremely closely spaced.

It is not necessary here to discuss the geologial structure in detail. Five principal discontinuity sets with smooth and planar surfaces were identified. Of particular importance were minor faults, with displacements of a few centimetres to several metres. They have a vertical and horizontal continuity of several hundreds of metres and are regularly developed throughout the slates. With an average spacing of 2 m, the clay-lined faults contribute to the formation of discrete blocks. Although individual faults could readily be traced across an excavation, it was much more difficult to project fault positions from one excavation to an adjacent one with any degree of accuracy.

Definition of block geometries, necessary for the design of support measures where required, could only be done once each area had been excavated. McCrae and his co-workers developed mapping methods that could

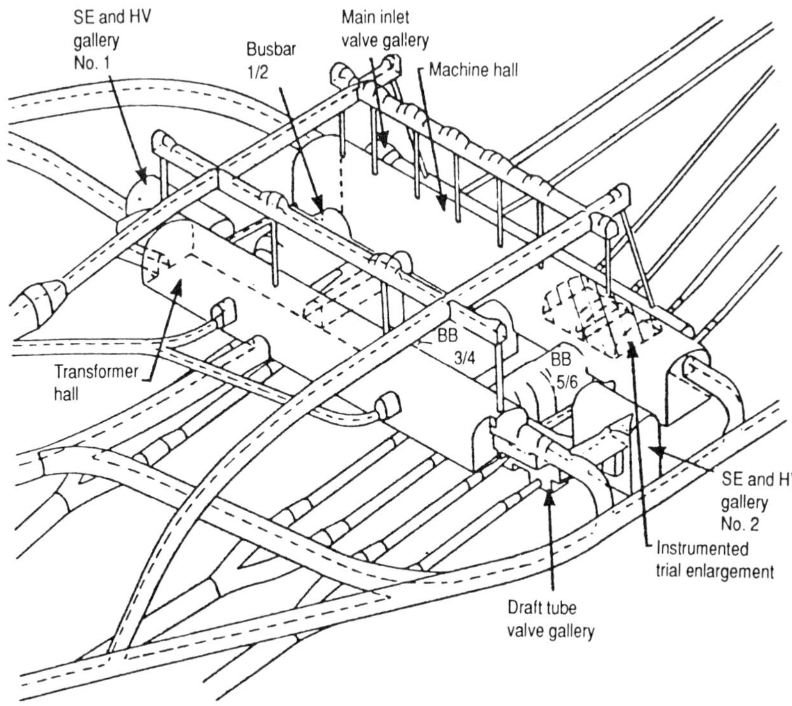

Figure 11.62 Main cavern complex and associated tunnels at Dinorwic, North Wales, UK, and the location of the trial excavation of the machine hall

be slotted into the continuous working cycle that was adopted at Dinorwic. In an excavation sequence involving drilling, blasting, mucking out, shotcreting and installation of support, the only period when the heading is free to be mapped is between mucking out and the application of sprayed concrete for temporary support purposes. The length of time available was usually about twenty minutes, or occasionally longer if the face was considered to be stable. A well-organized simple mapping method was essential when 12 or more headings were being advanced at the same time, and all needed to be recorded.

Mapping methods

Of paramount importance in mapping was the need to hinder, or interfere with, the progress of the engineering works as little as possible. In such circumstances the main requirement is that the mapping technique employed should be quick and reasonably accurate; a simple unsophisticated method was therefore necessary.

As a first approach, the possibility of setting out a grid on the excavated surfaces was appraised. If a hydraulic hoist was available on site, setting out a grid of vertical and horizontal lines spaced at 1 m intervals was possible. A 1 m spacing was considered an optimum spacing as each blasting round pulled 3 m. The time taken to mark the grid on the walls, using a paint spray or similar process, was rarely less than 20 minutes, but varied with the plant available. Marking out a grid on the roof of the excavation took much longer. At a very early stage it was concluded that the necessary preparation took too long, and gridding was abandoned as the basis of a practical mapping method.

Very careful consideration was then given to the possible use of photography. Provided that a scale, such as a ranging rod, was included in the photograph it was expected that important discontinuities could be picked out and recorded. Following many trials, the results were generally poor because of the dust on the walls, the dust in the atmosphere leading to poor definition on the photographs, inadequate lighting levels and hence poor contrast on the photographs, and an inability to identify overbreak and any blasting damage behind the excavated face.

The method finally adopted was related to the installation of rock support, which was closely preceded by the application of sprayed concrete. It was clear that photography, except for instant photography that

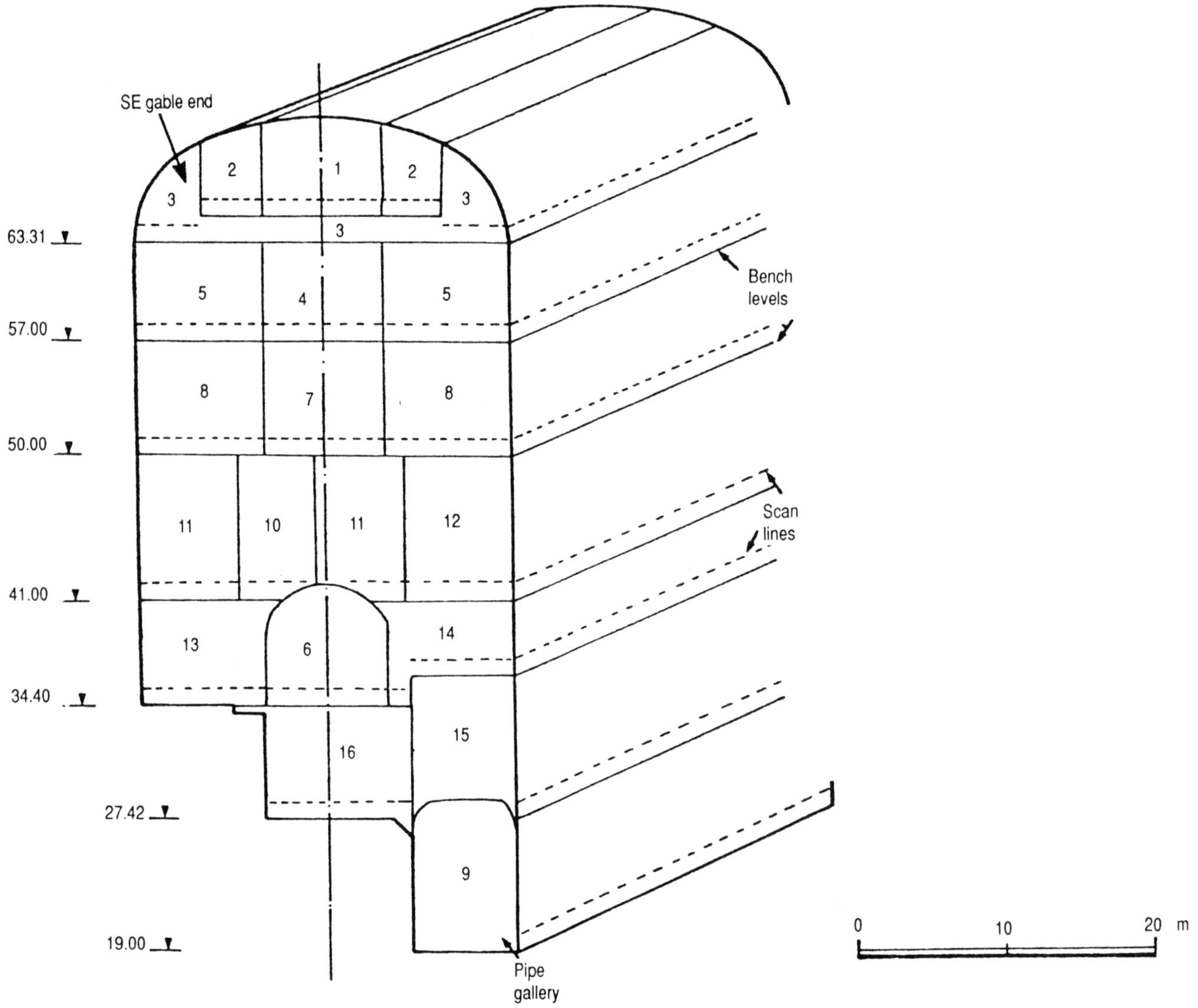

Figure 11.63 Excavation sequence numbered on the gable end of the main machine hall and associated scan lines used for detailed mapping at Dinorwic, North Wales, UK, and illustrated in Figure 11.64

apparently was not considered, would have taken too long, as would recording using the grid system that would rapidly have been overtaken by shotcreting. A quicker, sufficiently accurate method of mapping had to be devised, because, as emphasized by Douglas, Richards and O'Neill (1977), during construction there was a need to identify areas of potential instability for which remedial measures would need to be designed.

For an underground excavation, under ideal conditions, the walls, roof and invert should all be mapped in order to gain as much information as possible, in particular on the three-dimensional aspects of the structural geology. There was little doubt that different mapping methods would have to be used for the different parts of the excavation, if only from the point of view of difficulties of access to the roof and the general cover of broken rock on the invert.

It is the mapping of the enormous excavation for the machine hall that has been dealt with in some detail. The excavation sequence for the machine hall is shown in Figure 11.63; at roof level, six walls were available for mapping from south-east to north-west gable ends, and at most other levels, four or five.

Wall mapping

It should be remembered that the main purpose of mapping during construction was to monitor local geological conditions that might give rise to support problems. In practice the main requirement was the identification of blocks that, because of their geometrical shape, could possibly move outwards into the excavation and would therefore require support. To this end the line method of mapping adopted for tunnels (Anderson, Arthur and Powell, 1977) had to be modified to give a reasonably accurate plan of the face being monitored. The method of mapping that finally evolved was a combination of a more accurate version of the line mapping used in tunnels extended by visual offsets vertically above and below the main logging-line – in other words, a simple adaptation of the chainage-and-offset method of mapping.

A standard field sheet (Figure 11.64) was used for mapping all geological and geotechnical information. The sheet represents an elevation of the rock wall with a single horizontal line and vertical lines spaced at regular 1 m intervals, thus providing a co-ordinate system for plotting purposes. At the beginning of mapping, a measuring tape is set out along the horizontal line from one of the known chainage and level markers installed at 10 m intervals along each wall of the excavation. Using the field sheet, the intersections of all discontinuities with the tape-line are measured off accurately and marked on the sheet. All relevant information on rock types, attitude and character of discontinuities is also recorded on the sheet, before the various structures and other features are sketched in below and above the line.

Because most of the significant discontinuities encountered were steep and planar, the sketch-mapping could achieve a high degree of accuracy. In these circumstances a painted grid was hardly necessary to locate features away from the line-log.

Confidence in the accuracy of the mapping meant that the field sheets could be fair drawn immediately, and used in conjunction with adjacent sheets to assess any stability problems in the face. As such they were readily acceptable to the engineer because, unlike the line-surveys for tunnels, the plan-type presentation was in a conventional and therefore easily understood engineering format.

Field mapping of the cavern walls was completed at a scale of 1:50, a sufficiently large scale to record all relevant detail. The same method of mapping was applied to the advancing faces of each heading in the cavern excavation (Figure 11.63). This provided information on the spacing of discontinuities parallel to the main cavern walls, giving a more reliable estimate of the geometry of rock blocks in three dimensions.

Roof mapping

Mapping the roof in the same detail as the walls of the excavation proved to be practically impossible because of the difficulty of access without special plant being always available. In consequence, recording was confined to sketching in major features such as minor faults, dykes, dyke margins and bedding surfaces. This could be carried out from the floor of the excavation, after wall mapping had been completed.

Invert mapping

Mapping of the invert could only be undertaken when the floor had been completely cleared of debris, and this only occurred when permanent blinding concrete was to be placed. The floor at the 34.40 m level in the machine hall was mapped from a rectangular grid painted on the cleaned-up rock surface. Unhurried surveying meant that the grid nodes could be tied into the main survey and, consequently, the geological details could be mapped to a high degree of accuracy.

Supplementary mapping techniques

Where the assessment of support levels required block geometries to be defined with great precision, increased accuracy was obtained in several ways. For specific surfaces where accurate dip amount and dip direction measurements were required, co-ordinated points were obtained and the dip amount and dip direction determined in a manner similar to that employed in terrestrial photogrammetry (Ross-Brown, Wickens and Markland, 1973). Electronic distance-measuring equipment was employed to obtain at least four points on each surface.

For potential failure surfaces behind the excavation, e.g. dyke margins or major faults, percussion drilling was used with some measure of success to obtain the positions of the surfaces. While results from this type of probing are normally highly suspect, the changes in water colour due either to moving from green dyke margins into purple slate or in passing through a clay-lined fault were quite obvious.

Synoptic engineering geological plans

Individual field mapping sheets were transferred to 1:100 plans of the wall elevation (Figure 11.65). Reduction from the 1:50 field scale involved little loss of accuracy. Positions of important discontinuities are known accurately at the level of each line scan, and only minor adjustments were needed at the edges of individual plans. Printing the 1:100 plan on a negative enabled the geological details to be overlaid on the rock support drawings. McCrae, Powell and Yu (1979) discuss the likely accuracy of the mapping of the large caverns,

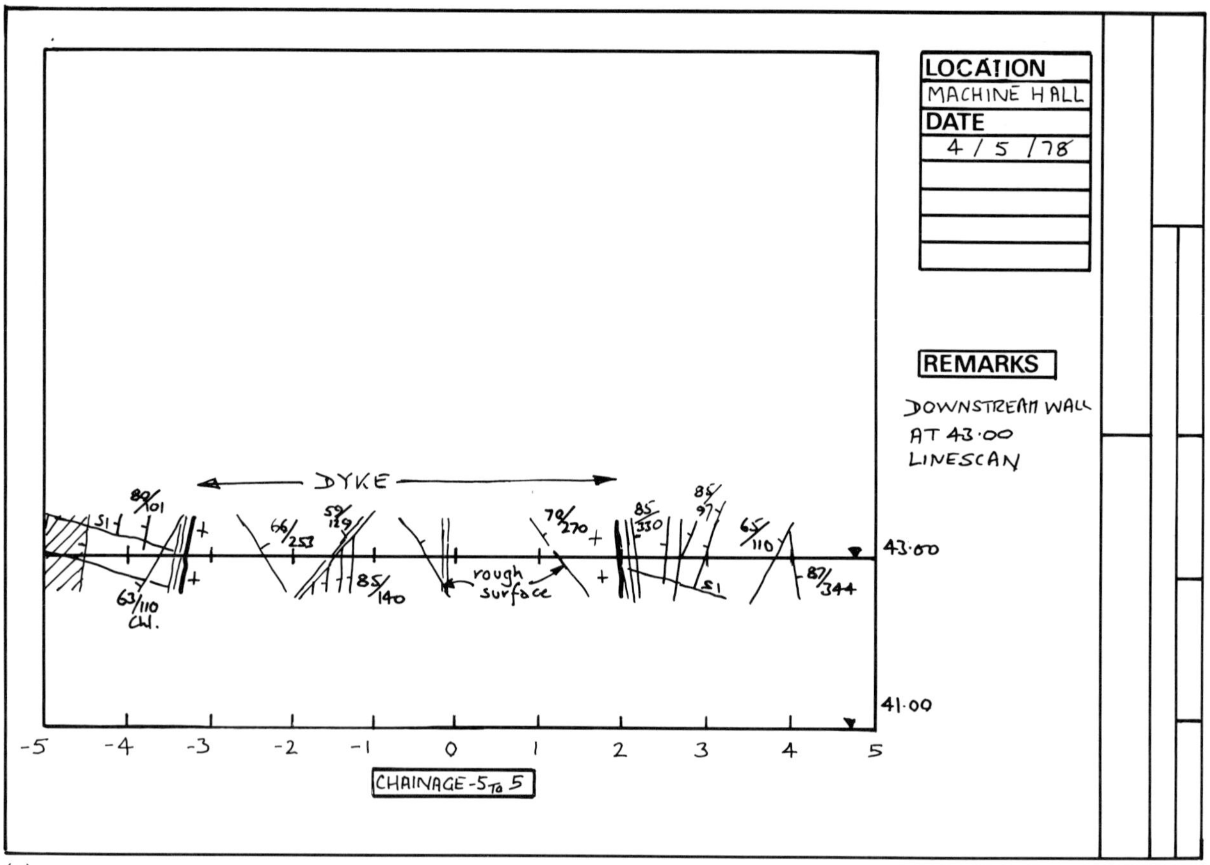

(a)

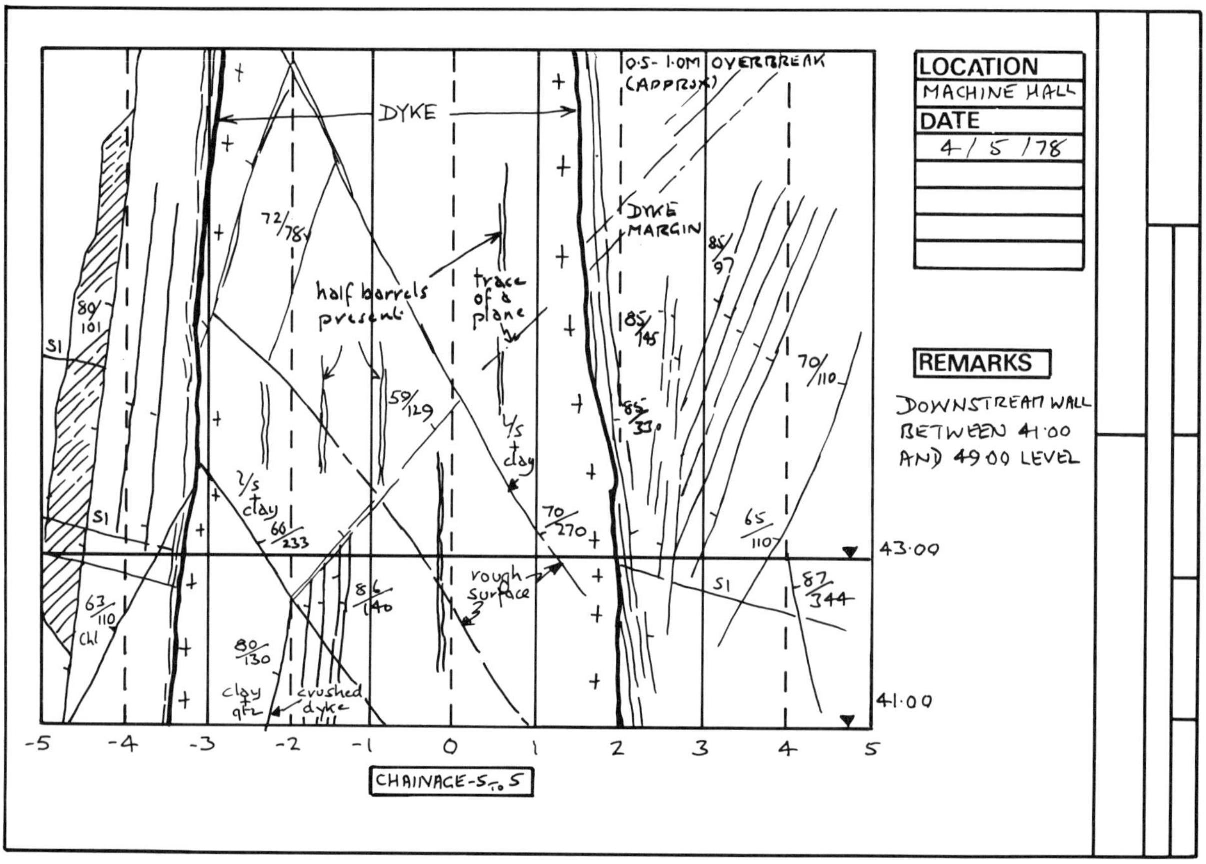

(b)

Figure 11.64 Example of a field mapping sheet: (a) initial data recorded on the scan line; (b) 'eye-ball' extension from the line scan survey over the whole area of the rock face. Original mapping scale 1:50

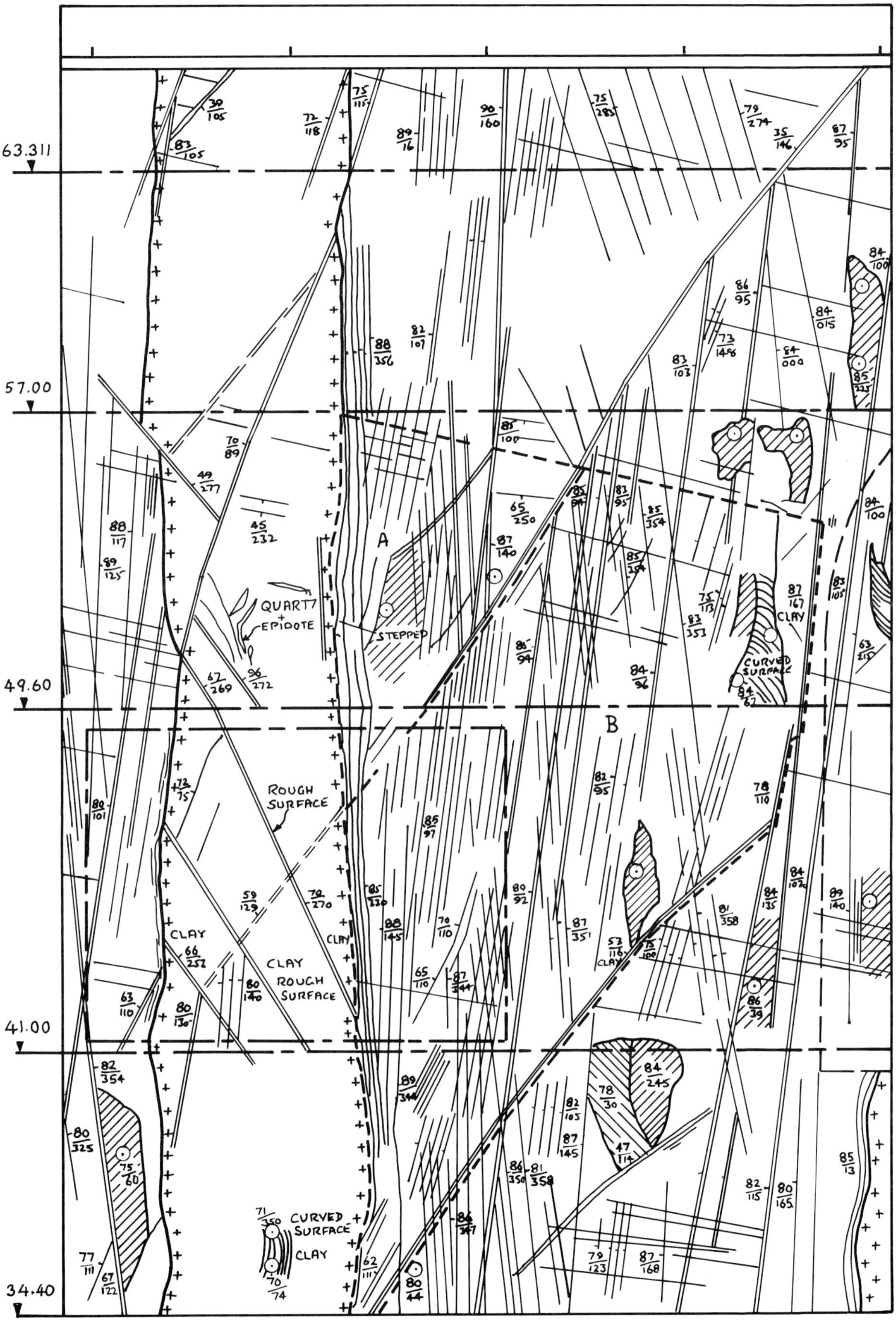

Figure 11.65 Synoptic geotechnical evaluation of part of the machine hall downstream face at Dinorwic, North Wales, UK. The area covered by Figure 11.64 is shown, as are the potential wedge features A and B. Original scale of the plan 1:100

pointing out that the incorrect location of chainage markers is the most likely source of errors. This in fact occurred and a number of field sheets and plans needed revision on this account. It was considered that the general accuracy of location of discontinuities of ± 50 cm was reasonable, but along line surveys may locally be ± 50 mm. In determining the dip and strike of discontinuities, usually only one reading could be taken on the small surface area available. An accuracy of $\pm 5°$ for both readings was probably achieved for most discontinuities, but this could of course lead to very considerable errors of location in projecting a discontinuity across the cavern from wall to wall or, indeed, in projecting structural details from the exploratory heading into the underlying position of the machine hall excavation.

An important application of the synoptic 1:100 plans was the general view afforded of the whole wall as the main excavation proceeded, thereby enabling potential failure wedges to be identified that were very much longer than the areas analysed on individual field mapping sheets (Figure 11.65). Down to the 57.00 m level, mapping of the wall of the top heading revealed no immediate problems apart from the weak wall of the dyke, but identified a possible wedge (A on Figure 11.65) defined by the dyke margin and the 62/129 bedding plane at chainage 10. An approximate analysis was undertaken of the likely factor of safety with and without support, and particular attention was paid to mapping the area as the next bench was removed to the 49.60 m level. Excavation to the 41.00 m level revealed another bedding plane discontinuity that determined wedge B on Figure 11.65.

Cross-sections at the plan scale

Engineering geological mapping at the plan scale is not only concerned with the recording of the as-found conditions already described for the Dinorwic caverns. There is also the question of local interpretation in three dimensions in order, for example, to solve a rock support problem.

An example is provided by the 1:100 wall elevation illustrated in Figure 11.65. Mapping, it will be recalled, had revealed first the small potential wedge failure A defined by the dyke margin and the bedding. As excavation proceeded to lower levels, the larger potential wedge was revealed, controlled by the same combination of structures. The dyke margin passed obliquely behind the slates to the right, as illustrated in the cross-section at chainage 4.5 m (Figure 11.66), constructed from the detailed recordings in Figure 11.65. This is the type of situation in which supplementary techniques, paying attention to the boreholes for the multiple point borehole extensometers (MPBX) and additional support tendons, could provide confirmation of the projected positions of the dyke margins.

11.9.3 Application of tunnel mapping methods to large caverns

Engineering plans of the Dinorwic caverns were presented as simple elevations or cross-sections at a scale of 1:100. With planar vertical walls and a nearly flat roof, there was no difficulty in visualizing geological detail in

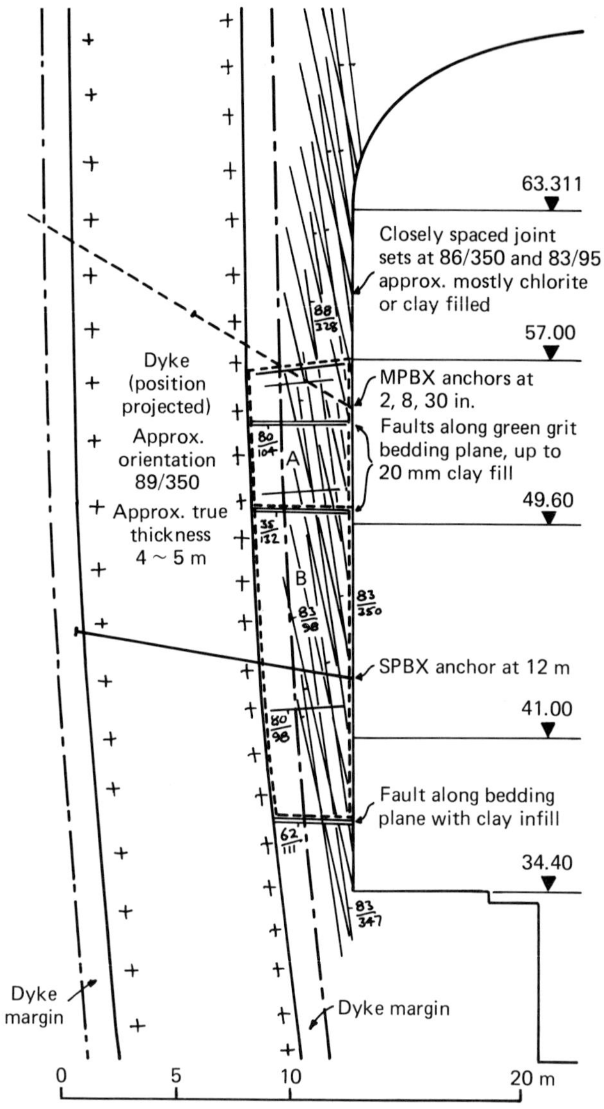

Figure 11.66 Cross-section of the machine hall at chainage 4.5 m, showing the projected position of the dyke margins and the extent of blocks A and B behind the excavated rock face at Dinorwic, North Wales, UK

three dimensions from such plans. Other cavern cross-sections, e.g. egg-shaped or horseshoe with flat invert, demand a different approach. Recording is done on projections of the surface of the excavation opened out into the horizontal plane – a method commonly used in tunnels (see section 11.10).

Cirata Powerhouse Cavern, Western Java

Engineering plans of the geotechnical details of some caverns, especially those with rounded cross-sections, are presented in the same way as plans of tunnels. The Cirata Powerhouse Cavern in Western Java has a width of 35 m, a height of 49.5 m and a length of 253 m. An egg-shaped cross-section with a flat floor was selected, with support provided by shotcrete, rock bolts and prestressed rock anchors applied quickly after partial excavation of each bench. Details of wall geology were recorded before shotcrete was applied to each excavated section of the final cavern wall.

Very complex geological conditions at the site (Reik and Soetomo, 1986), involving highly weathered volcanic breccia, sandstone, mudstone and river alluvium, made it advantageous to build underground rather than at the surface, where large open cuts would have required extensive slope stabilization methods and foundation treatment.

At cavern level, fresh volcanic breccia and andesite, without well-developed bedding planes recorded by site investigation, were negatively favourable. During excavation, continuous geotechnical monitoring, and mapping of the cavern walls, provided the basis for a rock mass classification, using the Rock Mass Rating (RMR) system (Bieniawski, 1974), for the assessment of primary support requirements.

Geological boundaries, although not the distribution of rock types, are shown on the sketch map of the cavern wall (Figure 11.67). The frequency and continuity of discontinuities varied considerably. Apart from some zones of massive rock, small, mostly closed and randomly oriented joints are spaced at 10–30 cm. Larger discontinuities are either master joints or sheared and fractured zones. From the orientation of the larger fractures it became apparent that wedge-type failures were unlikely. However, monitoring revealed significant movements in the south face of the cavern, associated particularly with a major shear zone (K4 in Figures 11.67 and 11.68) and adjacent master joints. Newly developed fractures, orientated nearly parallel to the

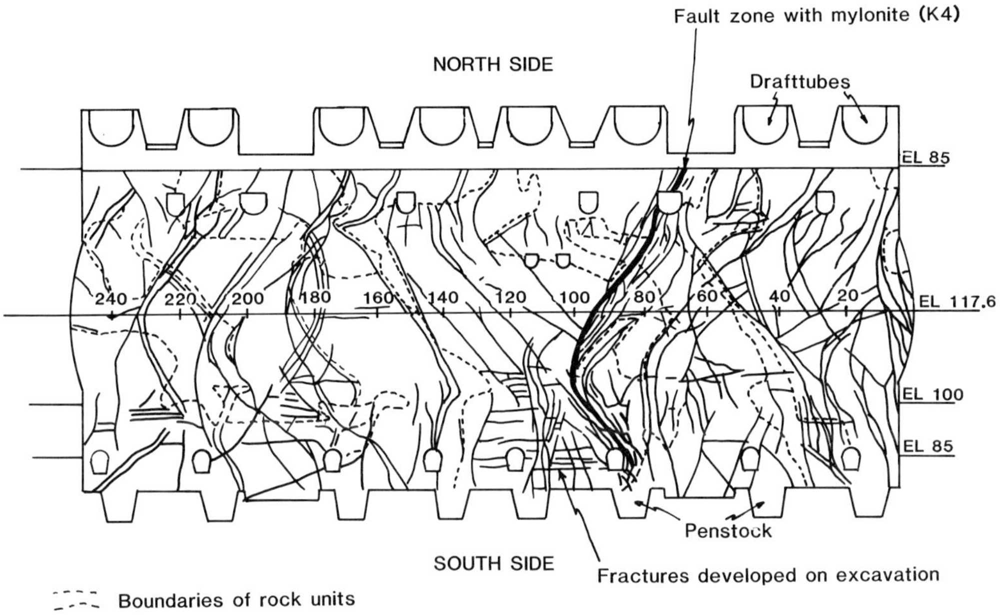

Figure 11.67 Geological plan of the walls of the Cirata Powerhouse Cavern, Western Java, showing the boundaries of lithological units, discontinuities and the main fault zone K4 (From Reik and Soetomo, 1986, Fig. 7, by permission)

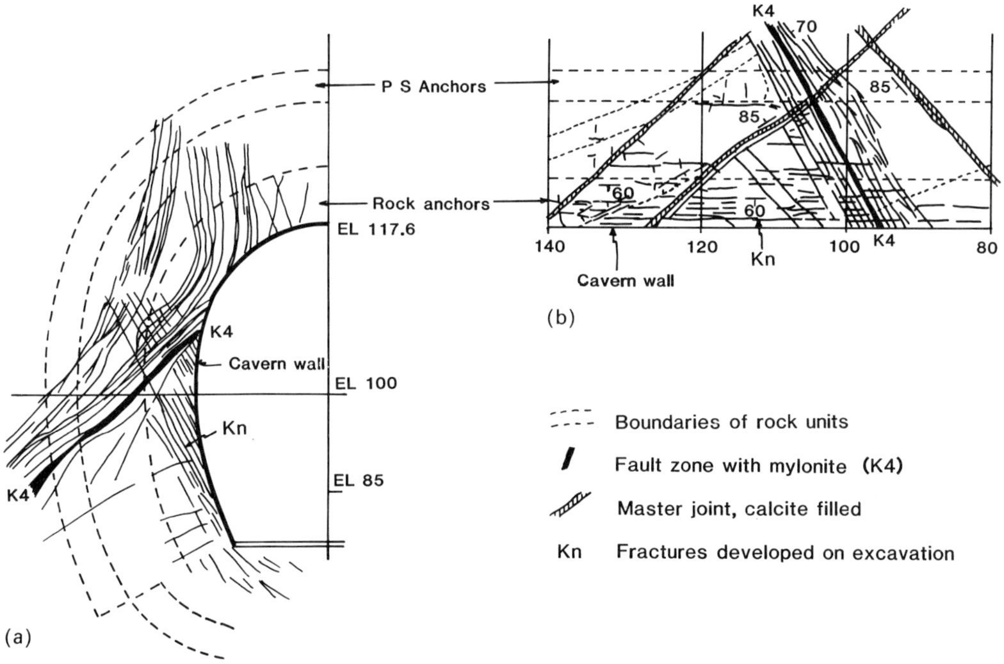

Figure 11.68 Sketch of discontinuities in the south side wall between Stations 80 and 140 m with indication of potential wedge structure (Cirata Powerhouse Cavern, Western Java): (a) the left diagram is an interpretative cross-section and (b) a horizontal plan at elevation 100. Distances and elevations are in metres (From Reik and Soetomo, 1986, by permission)

excavation and inclined steeply into the cavern, formed to a depth of 3–5 m in massive rock between the main fractures. Movement, cracking the shotcrete as excavation continued, was controlled by increasing the number of rock bolts and anchors.

For engineering purposes, plans were prepared to illustrate what was going on in the south wall of the cavern between Stations 80 and 140 m (Figure 11.67). One was an interpretative vertical cross-section at Station 100 (Figure 11.68a) showing the main fault (K4), associated smaller fractures and the fracturing induced by wall displacement. To complete the three-dimensional picture, the other plan is a more detailed representation of the fracture patterns within the rock on a horizontal plane at elevation 100 (Figure 11.68b).

The plans, on the originals, locate extensometers and recorded displacements on them in the main movement zones. An additional geotechnical plan of the whole of the critical area of the cavern wall (Figure 11.69) locates the fractures that developed in the shotcrete and the area of heavy additional support applied to counteract the effects of discontinuity K4.

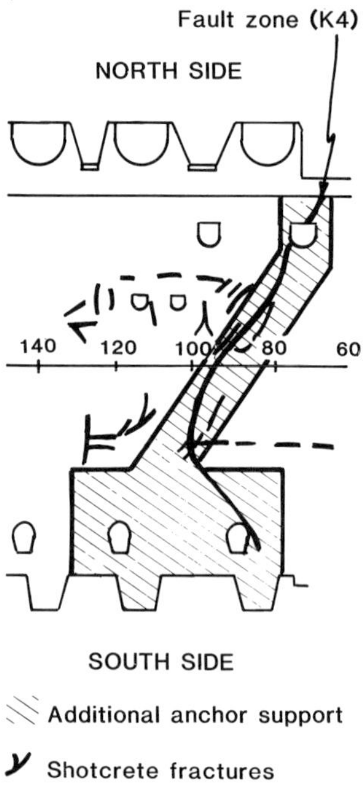

Figure 11.69 Area of heavy additional support as a countermeasure to the effects of discontinuity K4 and the calcite-filled master joints (After Reik and Soetomo, 1986, Fig. 12)

11.9.4 A mass transit concourse in Hong Kong

An interesting case history, involving geological prediction, initial geotechnical design assessment and final as-built engineering construction requirements, is that provided by Matson, Choy and Gibson (1986) for a concourse for the mass transit railway in Hong Kong. A 35 m deep basement for the concourse was to be built 'top down' through soft alluvial soils overlying decomposed granite resting on granite bedrock. Because of the location in a very densely populated urban area, stability of the vertical walls of the excavation had to be guaranteed.

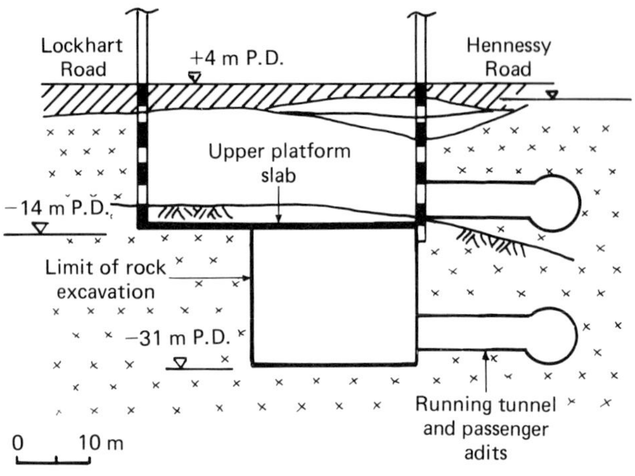

Figure 11.70 Typical geological cross-section through the mass transit concourse, Hong Kong (After Matson, Choy and Gibson, 1986, Fig. 1)

The running tunnels and the platforms are at two levels – the upper just below rockhead in the granite, the lower deep in the granite (Figure 11.70). It was originally intended to excavate down to the level of the upper platform slab, and up from the lower platform slab, in order to speed up construction. A shoulder of rock was to be left supporting the upper diaphragm wall (Figure 11.71) and thus, unless jointing was unfavourable, no additional support would be required.

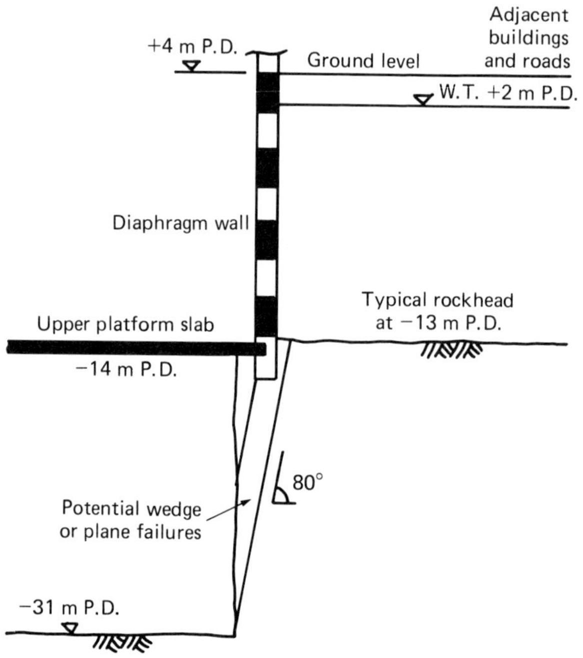

Figure 11.71 Cross-section to illustrate possible failure mode in rock side wall of the lower concourse, Hong Kong (From Matson, Choy and Gibson, 1986, Fig. 2, by permission)

Site investigation drill cores had revealed that the slightly weathered granite was closely to moderately widely jointed, with joints inclined at 30°, 60° and occasionally vertical, crossed by nearly laminated joints. Below 5 m in the solid rock, rock quality designation (RQD) was often greater than 75%.

Because of delays in excavating down to the upper platform slab, the contractor proposed that the lower

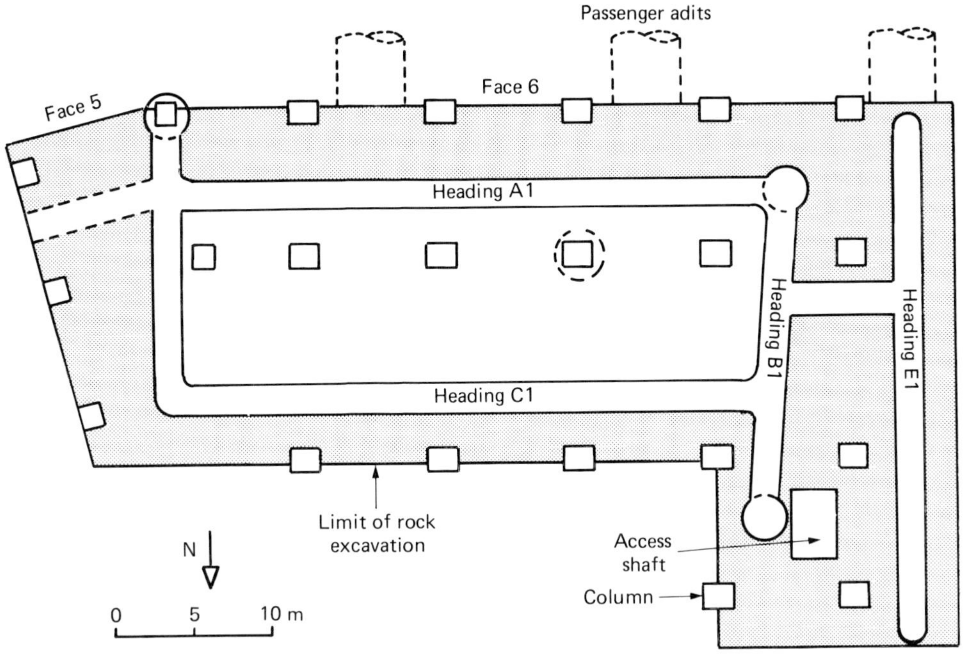

Figure 11.72 Plan of the upper headings, pillar locations and the limits of rock excavation – mass transit concourse, Hong Kong

levels of rock should be removed by underground mining while the upper soil excavation was going on. This meant that the upper platform slab would have to be supported by columns constructed in specially sunk shafts. The intention was to excavate the lower half of the rock mass as two caverns, leaving rib pillars and a crown pillar for support. Two metre wide headings were driven at the top and bottom of the initial mass to be excavated (Figures 11.72 and 11.73). Following completion of the caverns, the crown and rib pillars were to be removed by top-down excavation.

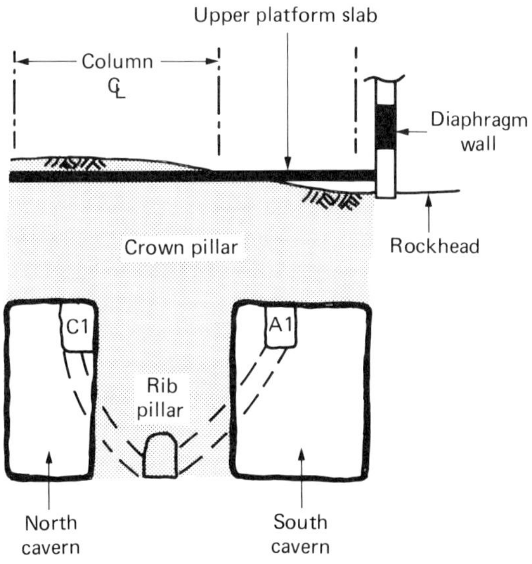

Figure 11.73 Cross-section illustrating the sequence of underground mining for the lower concourse, Hong Kong

Predictive plans

Construction of the headings and the column shafts gave much needed data on rock quality *in situ* and the discontinuity pattern and spacings. Preliminary mapping (Figure 11.74) along headings A1 and C1 (Figure 11.72) revealed that most of the discontinuities were joints with only few shears. The joint patterns, a subvertical conjugate set, occurred in repetitive sets with one continuous joint following behind another. Stability of the southern faces, Faces 5 and 6, gave cause for concern with the possibility of planar failure along one set of joints in Face 5, and potential wedges elsewhere.

Stability analyses were based on assumptions of shear strength parameters along the joints, groundwater conditions, and the prospect that the joints would be continuous from top to bottom of the excavation. These analyses led to the production of the first predictive plans of vertical Faces 5 and 6 (Figure 11.75a). It was recognized that total support of the faces would not be required. Four probable major wedges were predicted on Face 6, one of which, Wedge 9, would not need support. The potential plane failures on Face 5 could not be verified, and it was recommended that the top half of the face should be supported, with associated mapping of the major joints.

As-built plans

Detailed mapping of major discontinuities, undertaken as the final excavated face was achieved, confirmed the predicted joint system. Only two wedges were exposed at their predicted locations (Figure 11.75b) on Face 6. Moreover, no plane failures were identified on Face 5 because of the lack of joint persistence on this part of the face. The final support installed on the face comprised spot bolting, shotcrete and mesh reinforcement.

Conclusions

The authors concluded that if time had been available for a more detailed site investigation, including core orientation, at the design and documentation stage, it is likely

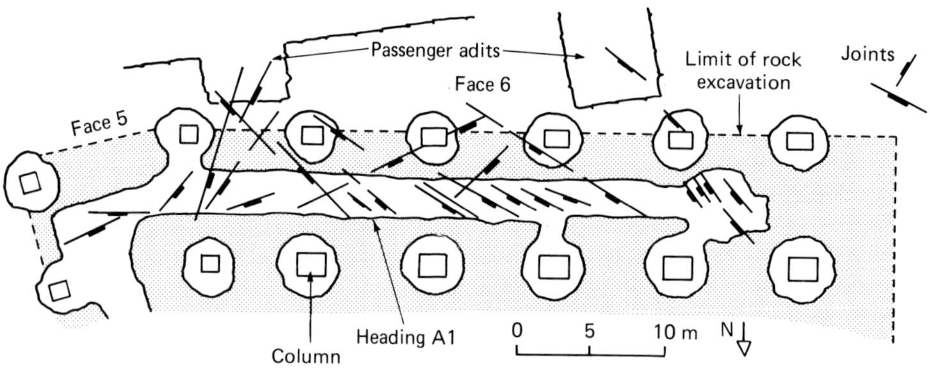

Figure 11.74 Joint pattern in the lower concourse, Hong Kong, as revealed by mining heading A1 and shafts for the columns, projected onto the rock wall in Faces 5 and 6

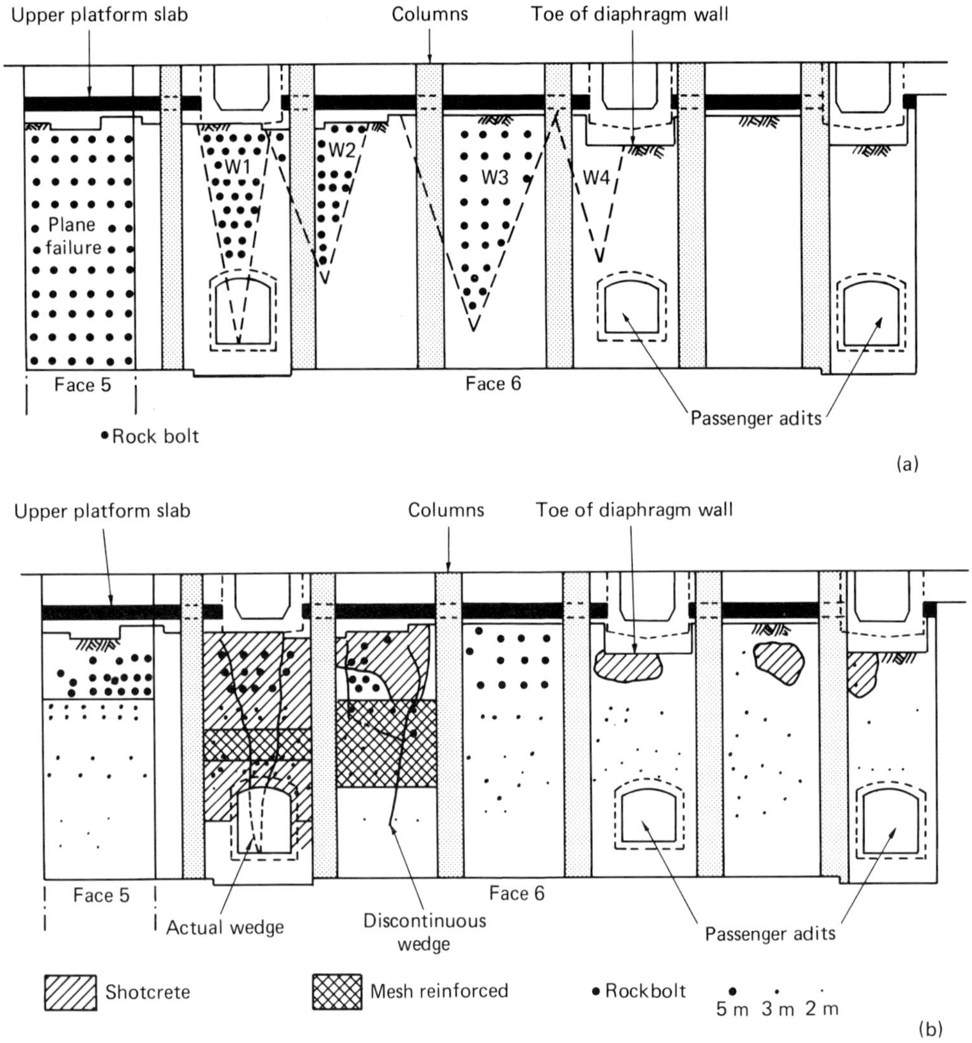

Figure 11.75 Lower concourse, Hong Kong – elevation of excavated rock wall of Faces 5 and 6: (a) predicted unstable areas and proposed support; (b) as-found conditions and the support installed (After Matson, Choy and Gibson, 1986, Figs 10 and 11)

that the joint sets which affected Faces 5 and 6 would have been identified. However, predicting the persistence of individual joints within a set cannot be done from drillholes. What is clear from the record of the various stages of construction is that the predicted rock bolt support varied significantly at each of the three stages, very largely as a result of the different availability of geological information and its interpretation. Geological information from mapping the headings, providing a three-dimensional view of the structures in the granite (see Figure 11.74), was of particular importance in the predictive second stage (Figure 11.75a), although the final outcome (Figure 11.75b) was markedly different from predicted conditions.

11.10 Tunnels

During the construction stage of a project, tunnels are logged or mapped geologically in order to assess and record rock mass condition and weathering grade, geological structures including faults and other discontinuities, different rock types and their interrelationships, and water inflows. The engineering use of the geological information is in the design and control of temporary and permanent support works. Temporary support measures would be recorded, together with details of overbreak, as the mapping proceeds.

Regardless of the shape of the tunnel in cross-section, geological details may be recorded as either: (a) a line-log, or (b) a plan of the tunnel walls. Additionally, the geological details may be presented as: (c) a plan section along the centreline from crown to invert.

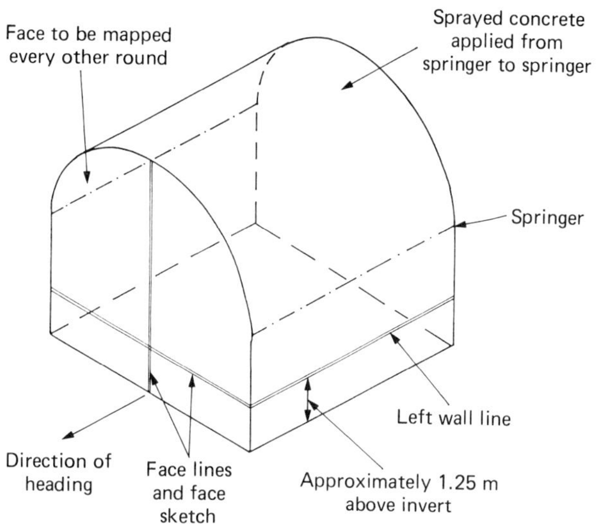

Figure 11.76 Scanline logging system used for the Dinorwic tunnels, North Wales, UK

11.10.1 Line-logging

All tunnel faces may be mapped on a regular basis, although the time available to the geologist in the excavation cycle usually restricts the amount of information that can be gathered. For this, a rapid scanline system is used (Figure 11.76). At least the horizontal and vertical face lines were logged as the tunnels for the Dinorwic pumped storage scheme in North Wales were driven but, if time did not permit, only the left wall line was logged later as the tunnels were left unlined.

West and Ewan (1981) outlined the methods adopted. Face sections, as shown in Figure 11.77, were prepared periodically at points of particular interest, on average one every 7.5 m along the tunnel. These face sections consisted of a sketch of the face, together with annotation of the different lithologies present, cleavage, bedding and fracture orientations.

In areas which the engineering geologist thought critical, scanline joint surveys were carried out. These consisted of 10 m lines, 1.25 m above invert level on the left wall of the tunnel (Figure 11.76). Measurements included dip and dip direction of the joints, with a note of the infilling if necessary, and the rock type. This continuous monitoring, with data plotted as stereograms of joint orientation, facilitated estimation of support measures and changes in tunnel design if necessary.

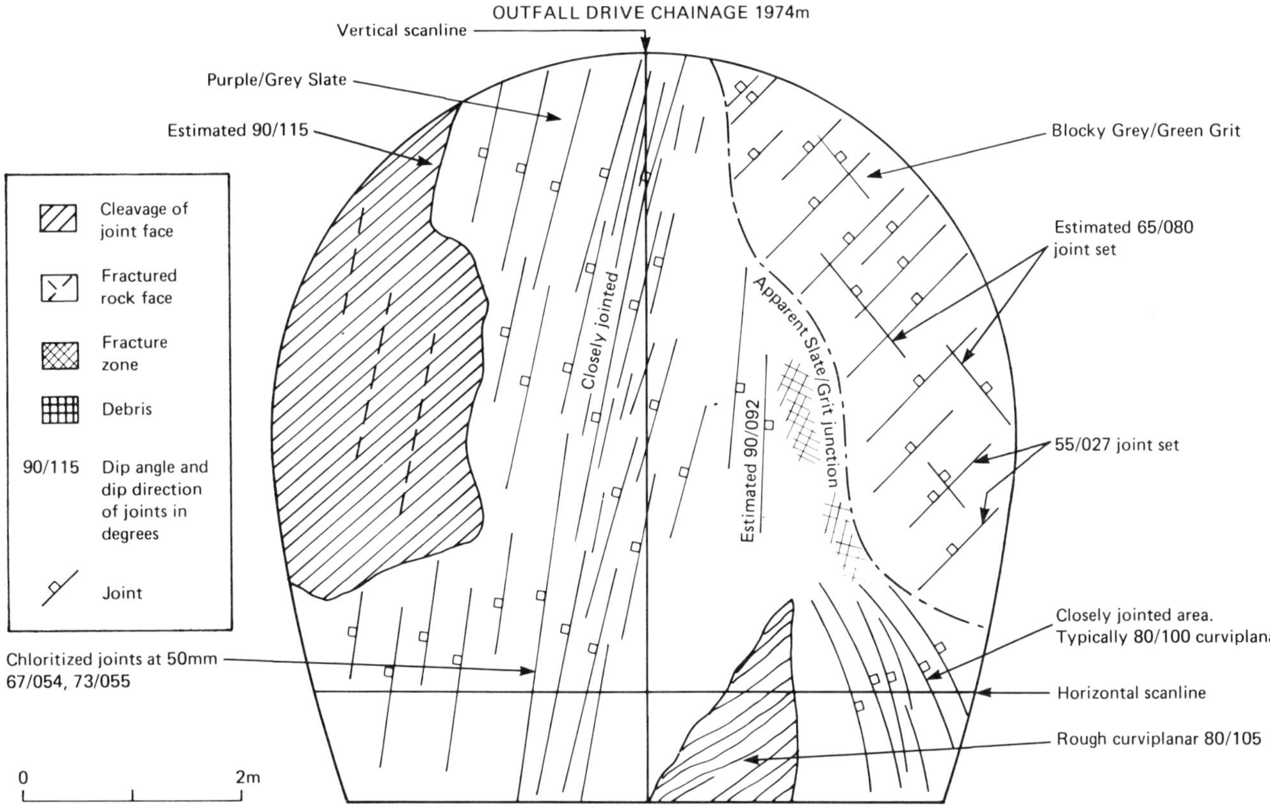

Figure 11.77 Face section prepared during the construction of the diversion tunnel at Dinorwic, North Wales, UK (From West and Ewan, 1981, Fig. 6, by permission)

11.10.2 Plans of tunnel walls

As a general rule it is difficult to map the invert of a tunnel, regardless of cross-sectional shape, because of the usual need to install track and drainage channels. In any case the floor of a tunnel is difficult to clean up until the lining is to be placed, and mapping can be delayed until that phase of construction if there are geological or other details that might need recording. The walls and crown of a tunnel are a different matter, however, because of the need to install temporary support. Design of temporary support measures, and permanent support for that matter, is very dependent upon geological conditions, and by far the best way to understand the geology is to map the walls and crown of the tunnel.

If a tunnel is to be mapped on the crown and the walls, the plan is laid out so that if it is wrapped round a scale model of the tunnel the geological features are shown from the outside looking into the tunnel (Figure 11.78). There are many examples of tunnel mapping in the literature and for a particular tunnelling project a standard plan sheet, on which all relevant information can be noted, is usually devised. Layout of the plan sheet usually has the cachet of the consulting or contracting firm involved with the project, and it is instructive to compare several of the published versions.

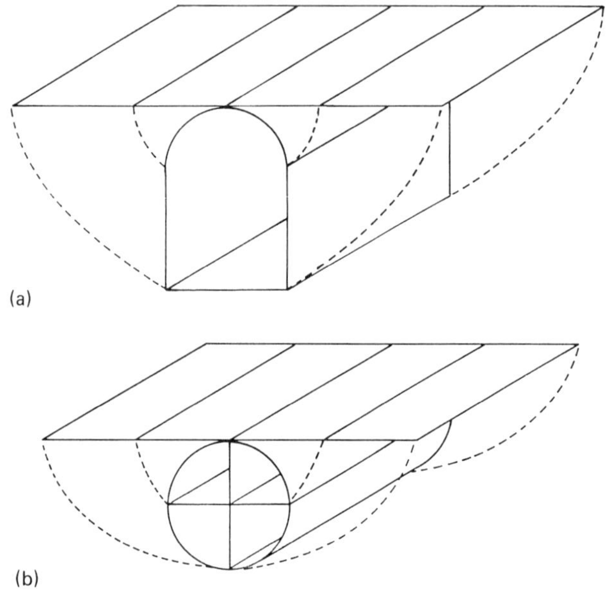

Figure 11.78 Layout of drawings for tunnel walls and crown for two typical cross-sections: (a) horseshoe cross-section; (b) circular cross-section

Vertical sections

When the geology is simple, e.g. a tunnel in gently inclined strata, a final record drawing may comprise a vertical section along the centreline. Information for the construction of the section is usually provided by full-face sections at intervals along the tunnel line, with additional details projected from localized mapping of the tunnel walls (Figure 11.79).

The method is particularly appropriate for tunnels with circular cross-section, as direct mapping of the curved walls is difficult. A three-dimensional impression of the geology is given by the vertical centreline section and cross-sections at intervals.

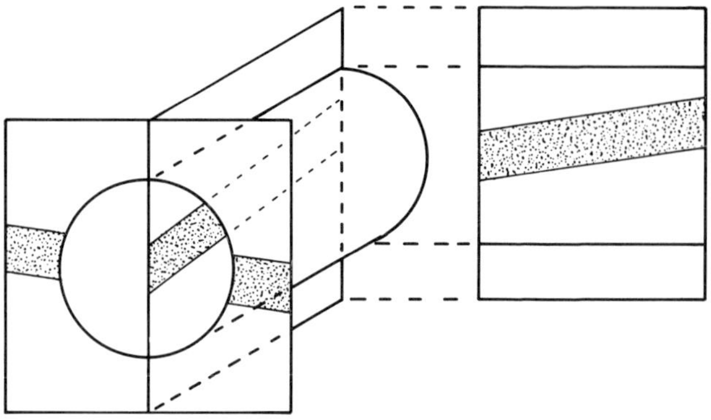

Figure 11.79 Layout of vertical sections for recording tunnel geology

Mapping tunnels at Mangla, West Pakistan

Two illustrations of tunnel mapping (Fookes, 1969, Fig. 12) at 1:120 scale refer to small drainage tunnels within the Mangla project. The object was to record rock types and other features of engineering significance such as seepage, joint or fault patterns. Some tunnel construction information and sampling detail were also included. The geology was projected onto the centreline of the horseshoe-shaped tunnels, with occasional cross-sections.

Record drawings of a 14 ft diameter, circular section, bored tunnel (Figure 11.80) were presented as a vertical section along the centreline, cross-sections and an unrolled plan of the walls from invert to invert. The geological section shows the alternations of thick beds of gently inclined sandstone and hard brown clays, with a record of overbreak in the tunnel crown. On the wall plan, in addition to the geology and overbreak, support measures are indicated for every 15 ft of tunnel length and there is a description of the strata, samples taken, overbreak and water seepage.

Record plan for a horseshoe-shaped tunnel

The example chosen (Figure 11.81) looks deceptively simple, but the drawing in fact conveys a great deal of engineering geological information. Lithology, major discontinuities and the presence of water are supplemented by weathering and jointing indexes. On a grid pattern, the number of fractures per square metre indicates the variation in brokenness of the ground.

The Kielder aqueduct, North-East England

The Keilder tunnels are part of a major scheme distributing water from the Kielder reservoir by aqueducts linking the rivers Tyne, Wear and Tees (Davies *et al.*, 1981, Fig. 1). A hard rock tunnel, 32 km long, crosses a varied sequence of gently dipping Carboniferous sediments with some intrusive igneous rocks. Full-face tunnel boring machines were used for most of the tunnel, with shorter sections driven by drill-and-blast methods and a roadheader (Figure 11.82). Geological investigations have been described by Carter and Mills (1976).

As one of a series of case history studies of tunnel projects, the report (Davies *et al.*, 1981) describes the planning and execution of the investigations, and illustrates the data collected during construction. A continuous record was kept of lithologies, discontinuities (joint pattern, frequency, filling), and structural features such as faults and variations in bedding dip. Inspectors kept shift by shift records on specially designed log sheets (Figure 11.83). In machine-driven sections the tunnel face was usually inaccessible, and information was obtained by observation of the roof and tunnel walls, with sketches being made at regular intervals, mainly showing strata levels (Figure 11.83).

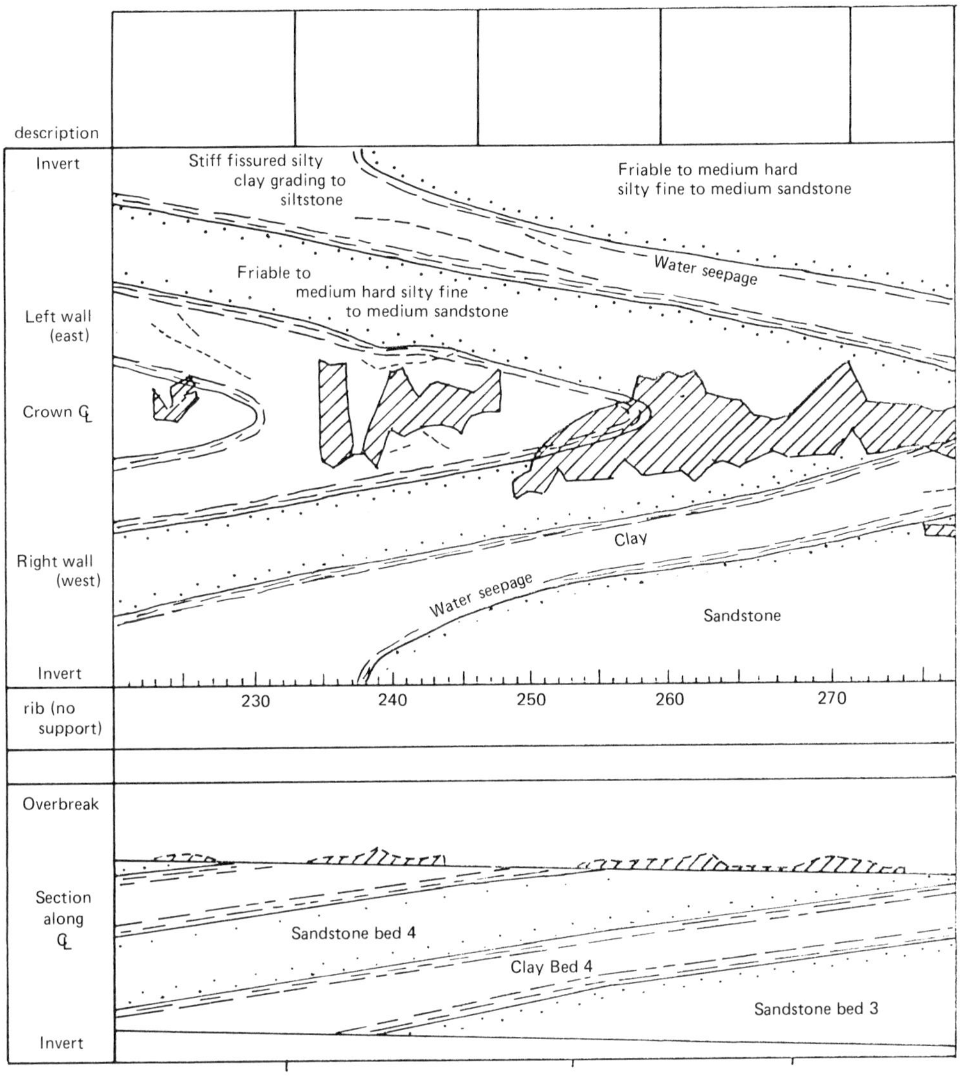

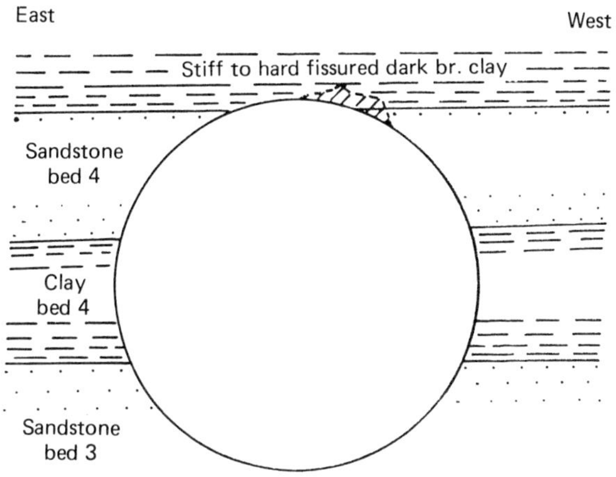

Figure 11.80 Example from the Mangla, West Pakistan, tunnels of recording on a vertical section along the tunnel centreline; and an unrolled plan of the walls of the circular tunnel

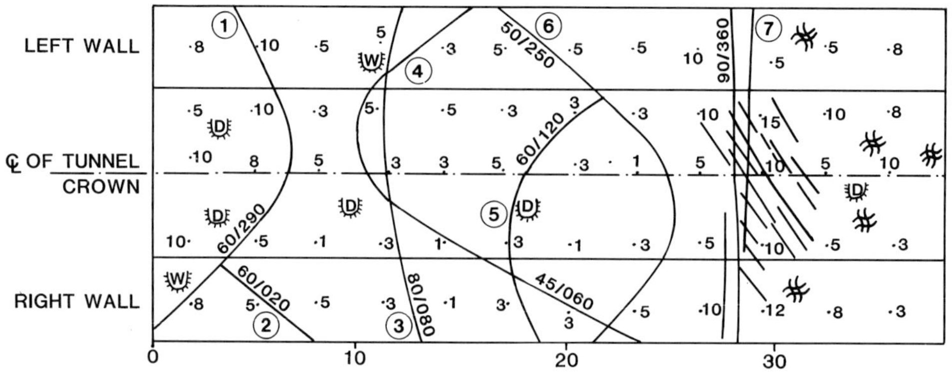

<table>
<tr><td>LITHOLOGY- L</td><td colspan="4">GREY GNEISS</td><td colspan="3">PINK GNEISS</td></tr>
<tr><td>WEATHERING INDEX - W</td><td>W2</td><td colspan="2">W2/W1</td><td colspan="2">W1</td><td>W2/W3</td><td>W1</td></tr>
<tr><td>JOINTING INDEX - J</td><td colspan="2">J3</td><td>J2</td><td colspan="2">J1/J2</td><td>J3</td><td>J1/J2</td></tr>
</table>

Presence of water: W Wet; D Dripping; J Jet

60/290 Major joint, dip and dip direction degrees

·5 Number of fractures per sq metre

(7) Joint number

Chaotic joints

Scale 1:200 0 5 10m

Figure 11.81 Typical engineering geological mapping of a horseshoe-shaped tunnel using the layout in Figure 11.78(a) (After Pupo, Vêncio and Trinidade, 1979, Fig. 2)

In drill-and-blast tunnels with lagged steel arch support, sketches were made of tunnel faces, but most information was gained from the walls and roof. For the Letch House–Derwent tunnel (Figure 11.82), detailed 1:50 scale wall maps were produced to record details of weathering and discontinuities (Figure 11.83). An important outcome of the recording was an appreciation that the cored, site investigation boreholes had not revealed the poor condition of joints at tunnel level, and the extent of weathering was also underestimated from the borehole information.

11.10.3 Mapping tunnels in terms of rock condition

Estimation of the stand-up time of a tunnel excavation, or the support requirements for stability and safety of working, is an important aspect of tunnel engineering in rock. Various systems have been devised, usually based on back-analysis of geological conditions, among the best known being the CSIR Geomechanics Rock Mass Rating (RMR) (Bieniawski, 1973, 1976) and the NGI Rock Mass Quality (*Q*) Classification (Barton, Lien and Lunde, 1974). These and other systems have been reviewed by Dearman (1983), including the early rock condition assessment developed by Terzaghi (1946) for use with steel arches as support.

Examples in the literature of the practical use of the Terzaghi method are not common. The one reproduced (Figure 11.84) referring to the Sepulveda tunnel in Southern California (Proctor, 1971, Fig. 1) is a good example of an as-built record plan. Only selected parts of the plan are given, excluding the specification of support recommendations based on the usual compromise between estimated rock condition and the actual installed support.

11.10.4 Tunnel log for the High Island Scheme, Hong Kong

The tunnel log, dated 1973 (Watkins, 1979), provides a clear picture of what geological and other aspects were mapped and recorded in the 40 km of tunnels and shafts associated with the scheme.

The scheme is in an area underlain by rhyolite and granite, where the geological variables of engineering significance are the effects of weathering and the discontinuity systems. Rock material weathering grade was the main basis for site investigation mapping, and rock mass condition for mapping during construction. Mapping carried out, and its general purpose, are summarized in Table 11.4.

The classification used for the weathering of rock material in site mapping was the conventional grade scale described by Little (1967) and Newbery (1970). In their fresh or slightly weathered states (Grades I and II), the igneous rocks, both rhyolite and granite, have an average strength of 160 MPa. In comparison, uniaxial compressive strength of moderately weathered (Grade III) rock is an order of magnitude less than fresh rock, and highly weathered (Grade IV) another order of magnitude less. Weathering therefore produces a far more marked variation in rock strength than does rock type.

During the construction stage it is asserted (Watkins, 1979) that the classification of the rock mass is much more important than during site investigation mapping, because in a section of dam core or a tunnel wall there may be combinations of materials varying from Grade I to Grade V. The proportions and actual distribution of these grades, coupled with their structural control, are particularly relevant to tunnel support, and incidentally to foundation treatment.

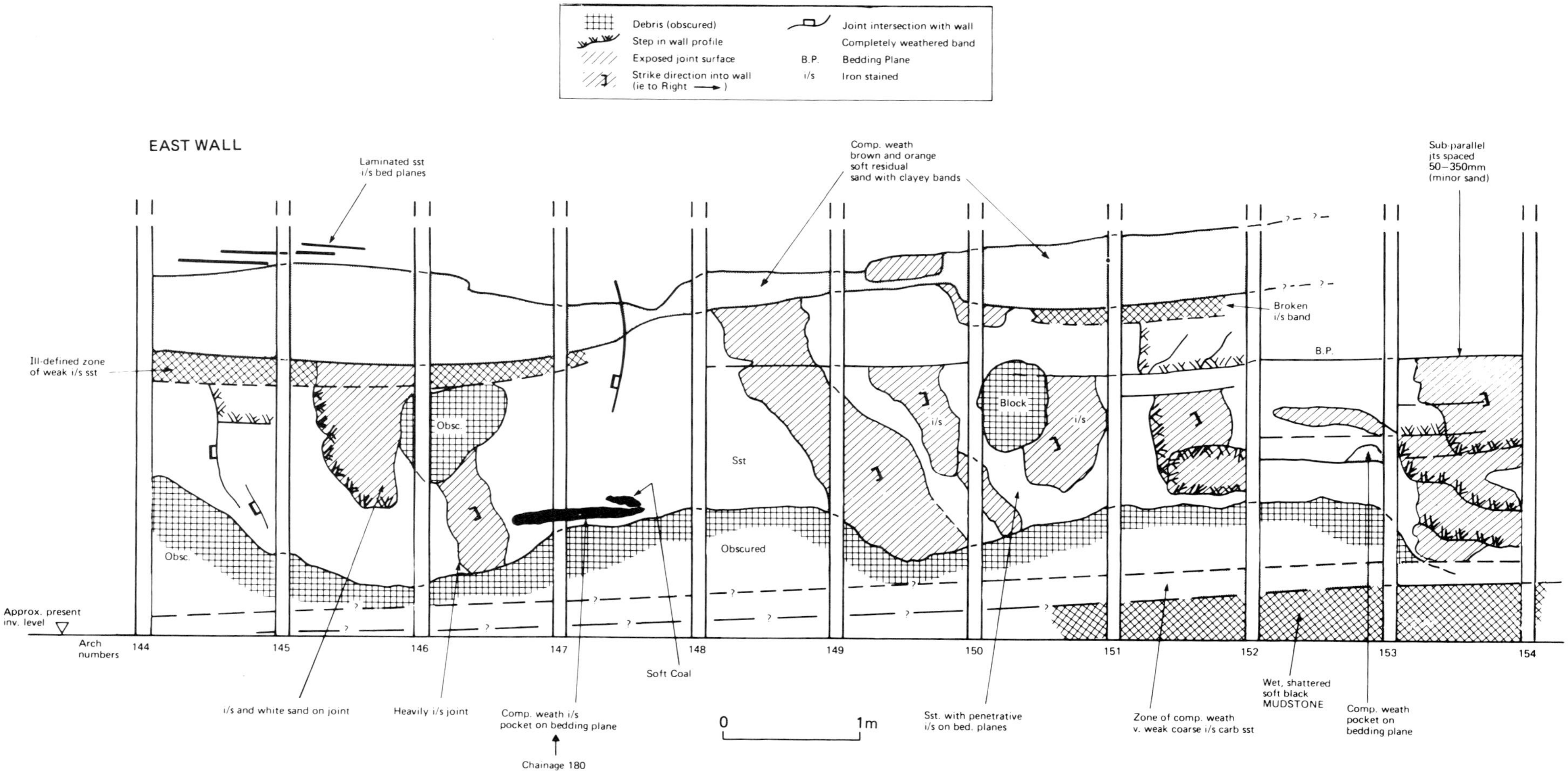

Figure 11.82 Specimen tunnel wall maps to show weathering, strata levels and joints for a section of the Letch House–Derwent tunnel, Kielder aqueduct, North-east England (From Davies *et al.*, 1981, Fig. 9, by permission)

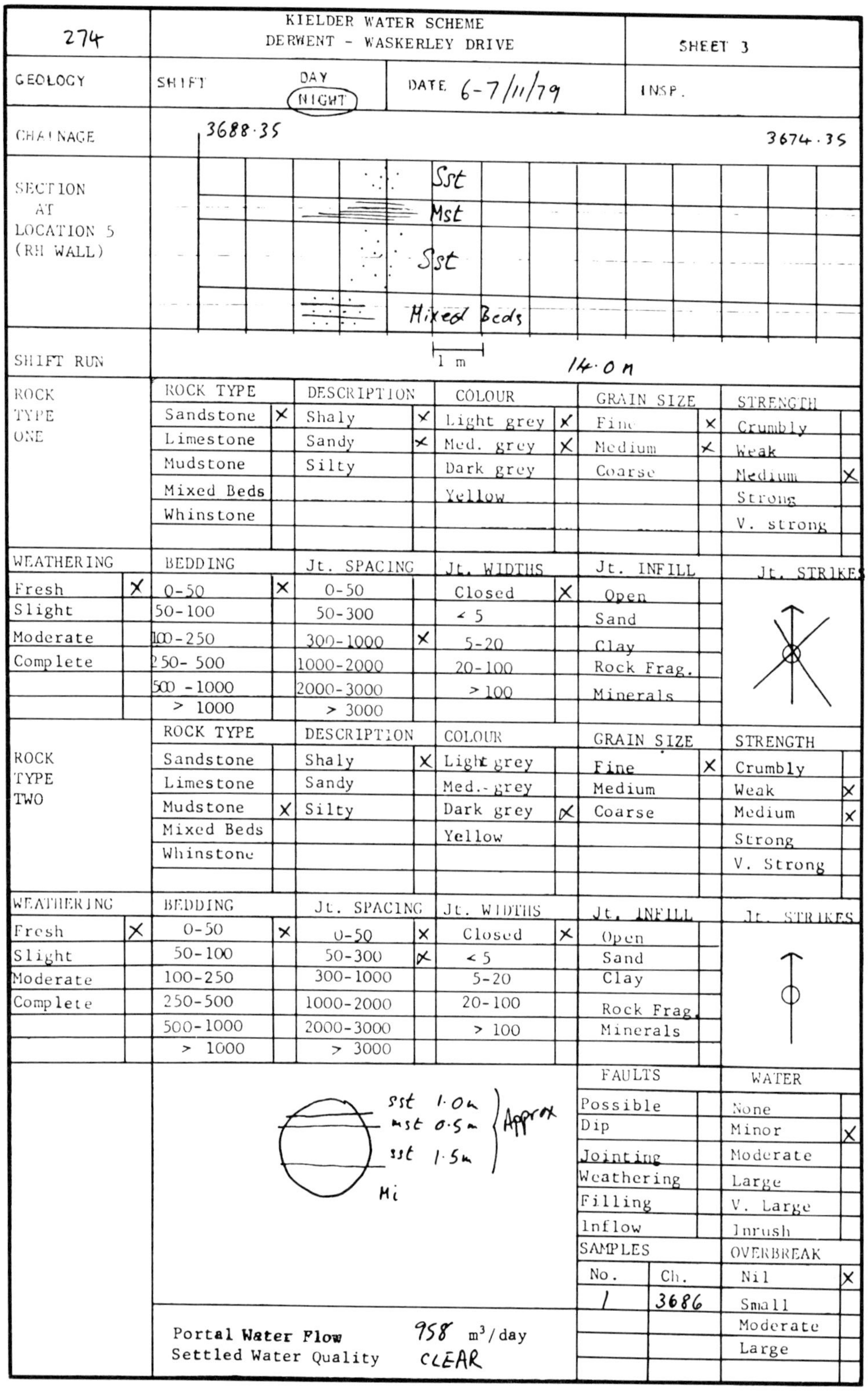

274	KIELDER WATER SCHEME DERWENT - WASKERLEY DRIVE		SHEET 3
GEOLOGY	SHIFT DAY (NIGHT)	DATE 6-7/11/79	INSP.
CHAINAGE	3688·35		3674·35
SECTION AT LOCATION 5 (RH WALL)	Sst / Mst / Sst / Mixed Beds		
SHIFT RUN	1 m	14·0 m	

ROCK TYPE ONE

ROCK TYPE		DESCRIPTION		COLOUR		GRAIN SIZE		STRENGTH	
Sandstone	X	Shaly	X	Light grey	X	Fine	X	Crumbly	
Limestone		Sandy	X	Med. grey	X	Medium	X	Weak	
Mudstone		Silty		Dark grey		Coarse		Medium	X
Mixed Beds				Yellow				Strong	
Whinstone								V. strong	

WEATHERING		BEDDING		Jt. SPACING		Jt. WIDTHS		Jt. INFILL		Jt. STRIKES
Fresh	X	0-50	X	0-50		Closed	X	Open		
Slight		50-100		50-300		< 5		Sand		
Moderate		100-250		300-1000	X	5-20		Clay		
Complete		250-500		1000-2000		20-100		Rock Frag.		
		500-1000		2000-3000		> 100		Minerals		
		> 1000		> 3000						

ROCK TYPE TWO

ROCK TYPE		DESCRIPTION		COLOUR		GRAIN SIZE		STRENGTH	
Sandstone		Shaly	X	Light grey		Fine	X	Crumbly	
Limestone		Sandy		Med.-grey		Medium		Weak	X
Mudstone	X	Silty		Dark grey	X	Coarse		Medium	X
Mixed Beds				Yellow				Strong	
Whinstone								V. Strong	

WEATHERING		BEDDING		Jt. SPACING		Jt. WIDTHS		Jt. INFILL		Jt. STRIKES
Fresh	X	0-50	X	0-50	X	Closed	X	Open		
Slight		50-100		50-300	X	< 5		Sand		
Moderate		100-250		300-1000		5-20		Clay		
Complete		250-500		1000-2000		20-100		Rock Frag.		
		500-1000		2000-3000		> 100		Minerals		
		> 1000		> 3000						

sst 1.0m, mst 0.5m, sst 1.5m (Approx), Mi

FAULTS		WATER	
Possible		None	
Dip		Minor	X
Jointing		Moderate	
Weathering		Large	
Filling		V. Large	
Inflow		Inrush	

SAMPLES		OVERBREAK	
No.	Ch.	Nil	X
1	3686	Small	
		Moderate	
		Large	

Portal Water Flow 958 m³/day
Settled Water Quality CLEAR

Figure 11.83 Specimen geology sheet for full-face machine drive from the inspector's shift report, Kielder tunnels, North-east England (From Davies *et al.*, 1981, Fig. 8, by permission)

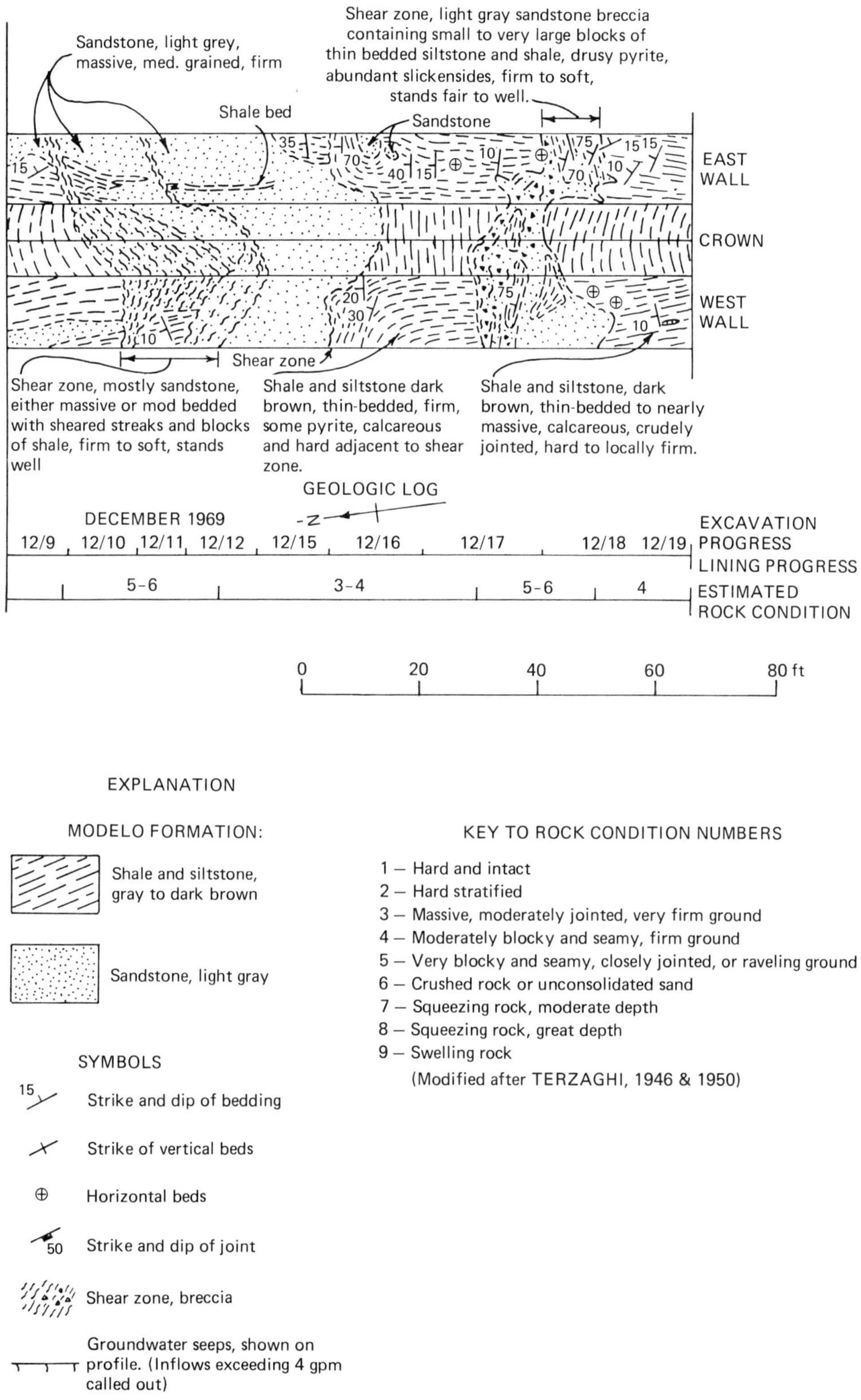

Figure 11.84 Part of an as-built tunnel log (Sepulveda tunnel, California, USA), showing the use of the Terzaghi rock condition criterion for estimation of steel arch support (After Proctor, 1971, Fig. 1)

Table 11.4 Summary of engineering geological maps and plans prepared for the High Island Scheme, Hong Kong

Stage	*Scale*	*Information shown*	*Engineering use*	*Assistance to*:
Site investigation	1:25 000	Rock types and faults	Tunnelling information	Client and consultant to finalize design. Contractor to prepare bids and plan excavations
	1:1200	Rock outcrops with weathering grade, superficial deposits, geological structures, rock types	Background to subsurface exploration. Foundation conditions at tunnel intakes and dams. Excavation characteristics	
Construction	1:100 to 1:240	Rock mass condition and weathering grade, geological structures, rock types, water inflows	Site control of temporary and permanent works in tunnels and at core base	Consultant and contractor in site operations. Consultant and client in event of future remedial treatment or claims
	1:1000	As above	Final foundation record	

For Hong Kong, a rock condition scale was devised for mapping purposes with the classes designated by Arabic numerals to distinguish them from material weathering grades. The classification (Table 11.5), based on Pender, Hosking and Mattner (1963), is similar to that devised by Terzaghi (1946) for tunnel support using steel arches. The condition scale is equally applicable to rock that has been disintegrated by weathering or faulting.

Table 11.5 Rock condition classification, High Island Scheme, Hong Kong

Class	*Rock condition*	*Description*
1	Excellent	Rock mass perfectly sound: hard rock, grade I, joints tight or cemented, tends to break through the rock on excavation, usually dry
2	Good	Rock mass perfectly sound, but somewhat loosened at joints: hard rock, grades I and II, joints often have kaolin dusting or flaky calcite or are lightly iron-stained, tends to break along joints, usually dry but drips or seeps may occur
3	Fairly good	Rock mass loosened at joints and may be seamy: mainly hard rock, grades I and II with grade III veneers at joints, many joints heavily iron-stained and thin clay seams may occur, usually wet
4	Fair	Rock mass loosely jointed and seamy: grades I–III rock with about 10% grades IV and V seams at joints, heavily iron-stained, usually wet
5	Poor	Rock mass very blocky and seamy: grades II and III rock with about 50% grades IV and V soil material, joints heavily iron-stained
6	Very poor	Soil: grades IV–VI soft material, some isolated solid blocks may occur

Types of map prepared at both the site investigation and construction stages are summarized in Table 11.4, emphasizing the importance of the plan scale at both stages.

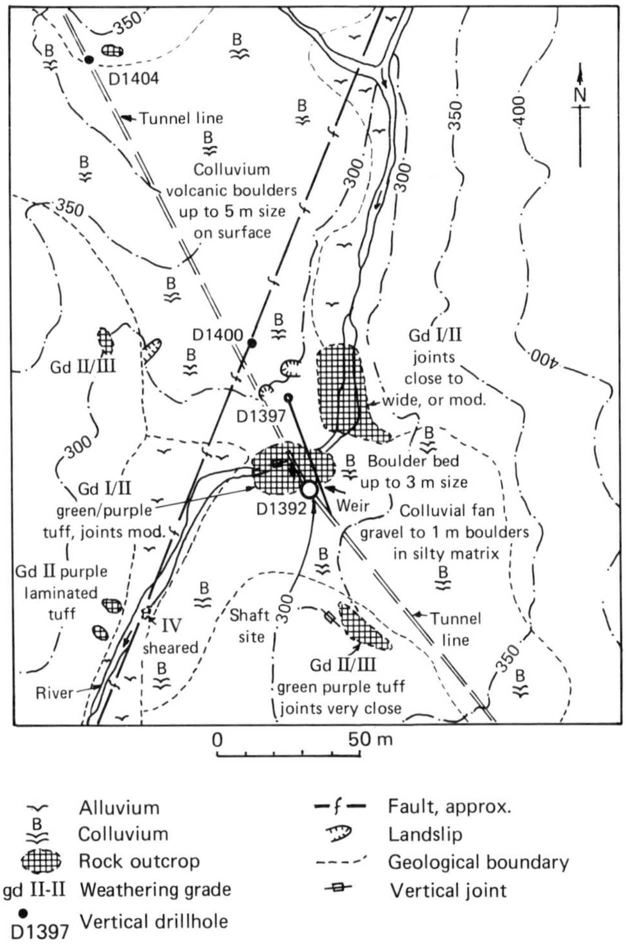

Figure 11.85 Geological plan of Intake UB, High Island Scheme, Hong Kong, at an original scale of 1:1200 (From Watkins, 1979, Fig. 3, by permission)

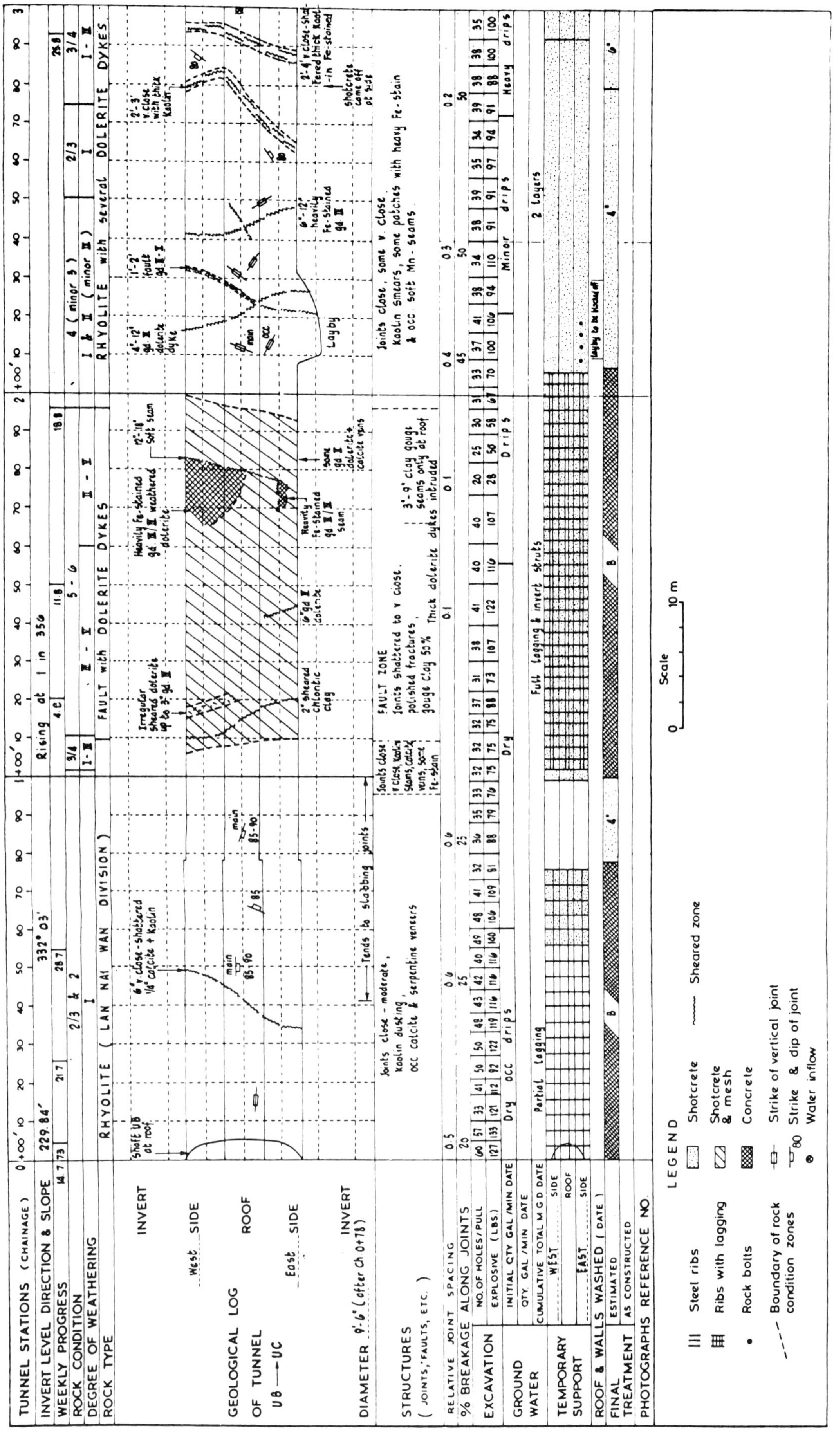

Figure 11.86 Tunnel log, High Island Scheme, Hong Kong, at an original scale of 1:24 (From Watkins, 1979, Fig. 5, by permission)

Recording at the site investigation stage

In the scheme, tunnels intercept indirect stream catchments at 19 intakes; at most of these intakes diversion is through a tunnel, with interest centring on the portal and weir areas. For each intake site, geological plans were prepared at 1:1200 scale (Figure 11.85). Fault lineaments were inserted from study of aerial photographs. At the depth of the tunnels, up to 60 m in granite, rock decomposition is mainly associated with faults. Plans conventionally show rock outcrops with an indication of weathering grade, landslips, and the distribution of alluvium and colluvium.

At the intake area in Figure 11.85, a colluvial fan has diverted the river away from a major fault towards the east bank valley side, where the river has eroded its bed into sound rock. Geological mapping confirmed the ideal conditions for a weir and shaft site, even though the cover over the tunnel was only 8 m at the river and the tunnel had to penetrate a major fault.

A reservoir was to be created by cutting off an arm of the sea with two dams. During the site investigation, areas of interest above sea level were mapped in the same style as Figure 11.85, describing rock exposures by the rock material weathering grade.

The plans produced at this stage can be called engineering geological only because they show weathering grades and joint directions; in all these aspects they are conventionally geological in approach.

Recording during construction

Tunnels and shafts

A scale of 1:240 was adopted for the geological logging of tunnels and shafts. The tunnels were mapped on both sides and roof, whereas shafts were logged over the full circumference; both were mapped over their full lengths. Figure 11.86 is an example of a tunnel log for the section north from shaft UB (Figure 11.85). It is a geotechnical plan because it shows not only the geology of the walls and roof of the tunnel but also details temporary and permanent support measures, groundwater, excavation details and joint spacing. The principal criterion used to decide support requirements was the mapping of rock condition.

11.11 Shafts

Shafts are tunnels stood on end, or inclined, ranging from man-sized site investigation boreholes and hand-dug caissons for foundations in Hong Kong, to large diameter for an exploratory shaft for an underground structure in rock, to the extremely large diameter surge shaft at Dinorwic.

11.11.1 Logging large-diameter site investigation boreholes

In Britain, periglacial conditions in the Pleistocene led to the formation of solifluction 'head' deposits on the chalk, underlain by *in situ* chalk fragmented and disturbed by first heave. These were the conditions at Mundford in Norfolk, a proposed site for the CERN proton accelerator. Extremely stringent foundation settlement limits meant that geotechnical properties had to be determined by plate loading tests, in 0.86 m diameter auger holes and a large-scale water tank loading test.

To apply the limited *in situ* test results over the whole site, the chalk profile was visually described and classified *in situ* into a series of five grades, taking into account those features likely to influence its load-deformation behaviour (Ward, Burland and Gallois, 1968). Mapping was undertaken from man-sized cylindrical cages lowered down the holes. The grades established – later extended to six in number – are now used as the standard for interpreting site investigations in the chalk of the UK.

11.11.2 Hand-dug caissons in weathered Hong Kong granite

A commonly used procedure for the formation of deep pile foundations in Hong Kong is to hand-excavate the pile, placing concrete rings as the work progresses. The work is normally carried out by a husband and wife team using hand tools and pneumatic drill as necessary. This method, known locally as 'hand-dug caissons' (Figure 11.87), is particularly suitable for advance through highly variable weathered profiles which may include substantial quantities of corestones above the sound rock. The large diameter of these piles allows detailed mapping and testing to be carried out at or above the anticipated founding level.

Figure 11.87 Surface works involved in the formation of hand-dug caissons in the Hong Kong granite

Detailed logging of three pile foundations excavated in granite was undertaken in terms of mass and material weathering grades and the properties of discontinuities (Irfan and Powell, 1985). A series of index tests, including '*N*' Schmidt hammer value, point load testing and air percussion drilling, were undertaken in these piles to establish specific correlations. The results of the index tests were compared with the detailed logging of the piles and adjacent boreholes in terms of both material and mass properties. The log of one of the advance piles, unfolded into four quadrants, is given in Figure 11.88.

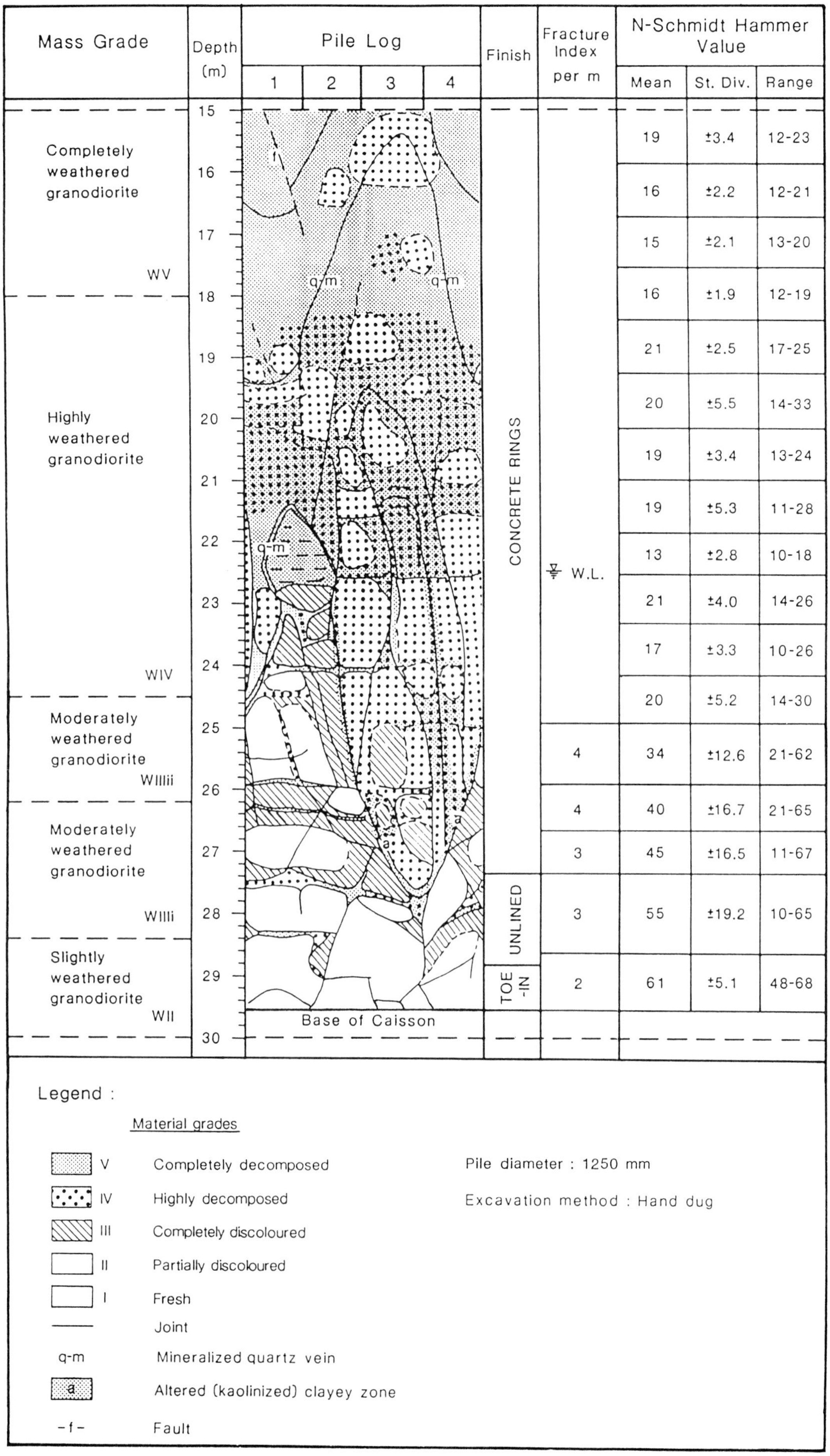

Figure 11.88 Log of part of a hand-dug caisson in Hong Kong granite (From Irfan and Powell, 1985, Fig. 3, by permission)

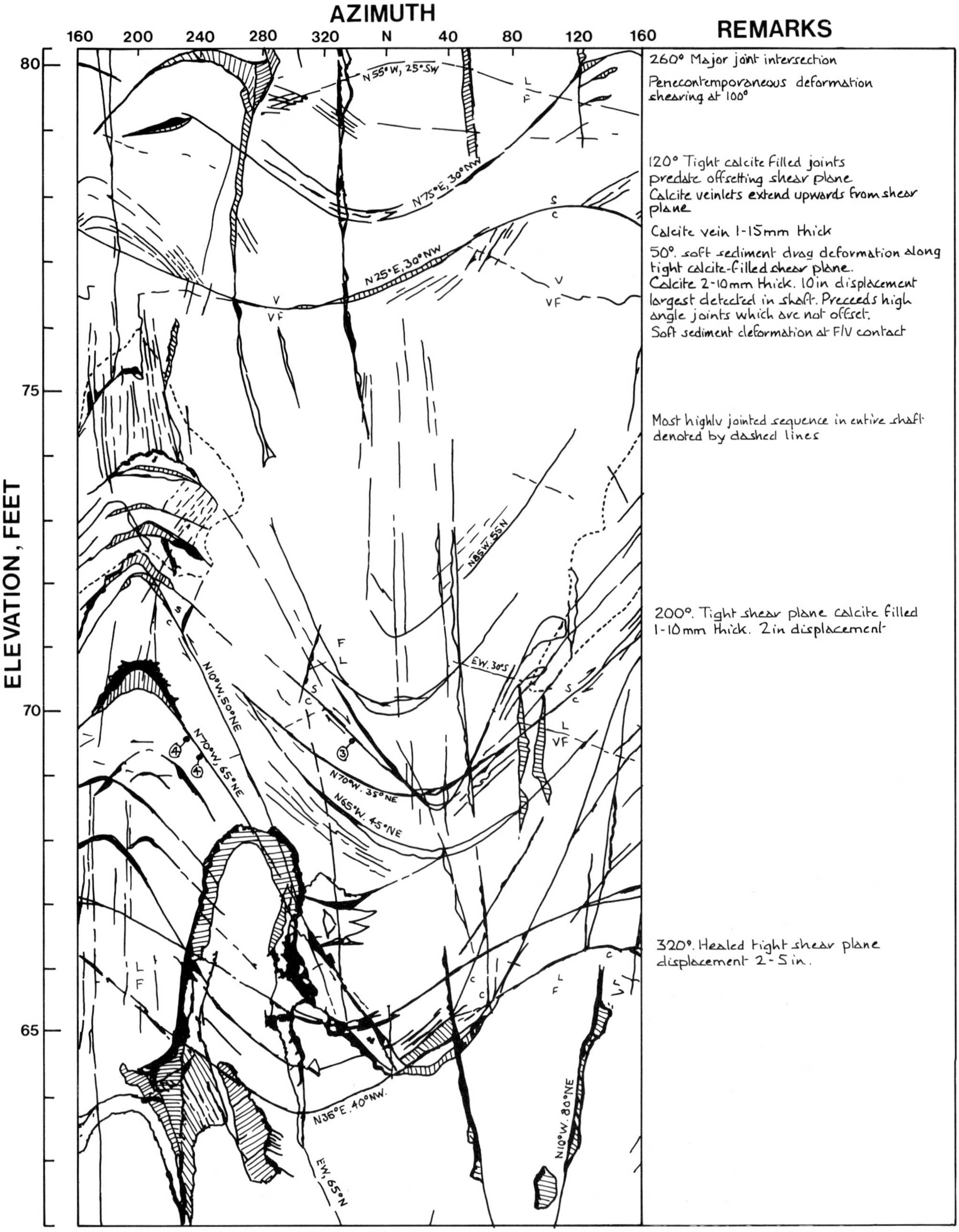

Figure 11.89 Face plan of the interior of a 0.9 m diameter exploration shaft (V and VF, degree of laminar bedding in the slates; s and c, shear planes and calcite infilling; hachured areas, over-break along joints). Mapped at 1:12 scale (Redrawn from Hatheway, 1982, Fig. 4)

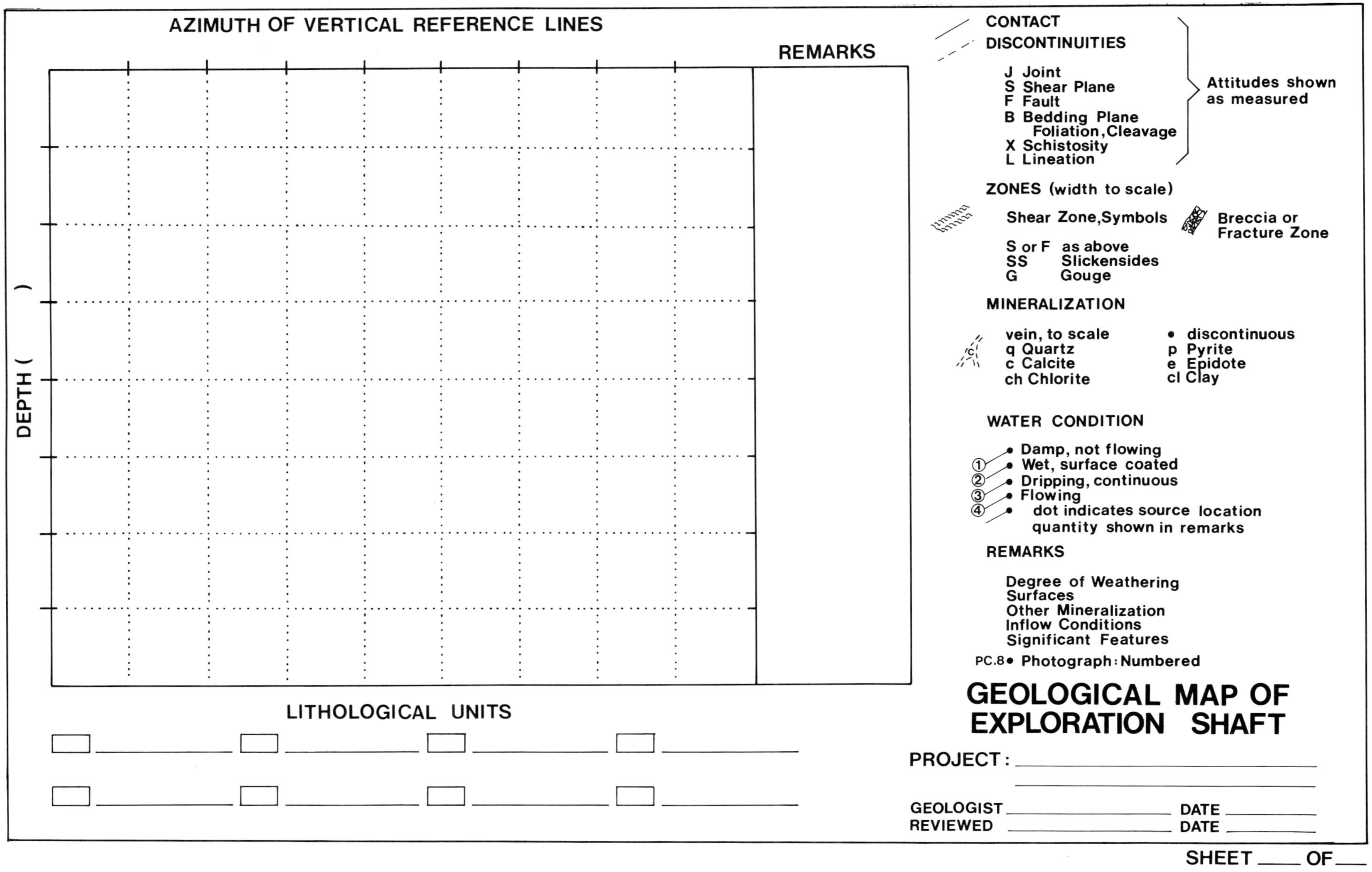

Figure 11.90 Mapping sheet used in shaft mapping by Hatheway (From Hatheway, 1982, Fig. 5, by permission)

11.11.3 Log of a 0.9 m diameter vertical exploration shaft

Hatheway (1982) has described trench, shaft and tunnel mapping, giving an excellent example of the practical problems of mapping a vertical shaft in slate. If mapping is to be effective, some kind of linear reference system is needed. Detailed geology in Figure 11.89 was mapped using a specially constructed reference cage, made up of 15.2 cm (6 in) square wire grid, reinforcing mesh. The cage was held in place on the shaft walls by a simple system of stiff guy ropes. The cage, 1.83 m (6 ft) tall, was lowered in 1.5 m (5 ft) increments, leaving a 0.3 m (1 ft) overlap for correlation between cage segments. The geologist stands in a companion man cage, specially orientated, and maps on a specially prepared, matt-surfaced, acetate grid sheet (Figure 11.90) which withstands the wet and humid conditions in the shaft. A detailed legend printed on the sheet provides both a guide to what to map – water condition, mineralization, discontinuities and so on – and standard symbols.

12

Hazard and risk assessment

12.1 Introduction

Although the effects of geological hazards are all superficial, in the sense that they occur at or near the ground surface, the hazards themselves may have an endogenetic origin, i.e. they result from internal processes, or they may be exogenetic, i.e. mainly associated with denudational processes of one kind or another. In practice, the distinction may not be clear-cut: movement of the soil or bedrock, or standing water, usually involved, may be provoked by earthquakes and volcanic activity as either a primary or secondary event. For example, in a major earthquake the ground may be fractured and displaced as a primary effect, whereas locally secondary effects such as liquefaction of sediments, landsliding and water movement may present equally great risks.

At the outset there is a need to clarify the terminology and the following definitions, which may be regarded as internationally acceptable (Varnes, 1984), have been proposed:

Natural hazard (H). The probability of occurrence within a specified period of time and within a given area of a potentially damaging phenomenon.

Vulnerability (V). The degree of loss to a given element or set of elements at risk (see below) resulting from the occurrence of a natural phenomenon of a given magnitude. It is expressed on a scale from 0 (no damage) to 1 (total loss).

Specific risk (R_s). The expected degree of loss due to a particular natural phenomenon. It may be expressed by the product of H times V.

Elements at risk (E). The population, properties, economic activities, including public services, etc., at risk in a given area.

Total risk (R_1). The expected number of lives lost, persons injured, damage to property or disruption of economic activity due to a particular natural phenomenon. It is therefore the product of specific risk (R_s) and elements at risk (E). Thus:

$$R_1 = (E)\,(R_s) = (E)\,(H \times V)$$

Unfortunately, the usage of terms by writers of English is varied, but also the equivalents in other languages may be difficult to find. In particular, the French word '*risque*', as used in the current ZERMOS programme of landslide hazard zonation, is equivalent to the English word 'hazard' as defined above for the occurrence of an event rather than consequent loss or damage. Ideally, landslide hazard maps would show where landslide processes have operated in the past, where they occur now, and the probability that a landslide will occur in the future (Brabb, 1984). In addition the maps would describe the kind of landslide movement, such as fall, slide, flow, topple, lateral spread; would indicate the rate and recurrence of landslide movement; and would provide any other information needed to judge the impact of a landslide on any person or structure, such as the anticipated direction of movement, and the thickness or height, viscosity and density of the material.

Landslide risk maps would indicate the area and/or population that is exposed to the landslide hazard, the vulnerability of this area and/or population to the landslide, and the expected consequences to people and structures if a landslide should occur.

Brabb comments that very few landslide hazard maps even come close to these standards, and very few regional landslide risk maps have been prepared.

The term 'zonation' is applied to the division of the land surface into areas which are ranked according to degrees of actual or potential hazard. Not all the maps that will be described are zoned, in particular the earlier ones which are described first; neither may they be restricted to depicting a single type of hazard. Many examples were produced in the early 1970s during the course of official geological surveys in the USA.

12.2 Hazard evaluations made in the course of geological surveys

Reports of various quadrangles prepared by the U.S. Geological Survey often comprised up to 14 different map sheets, mainly derivative from the standard geological maps.

12.2.1 Aspen Quadrangle, Colorado, USA: potential geological hazards

One map in the series (Bryant, 1972a) was made from aerial photograph interpretation and the topographic map. Actual avalanche areas are restricted to obvious gullies and steep treeless slopes. Much larger map areas are shown as slopes steep enough for small avalanches under certain conditions, and areas directly below avalanche areas. Aspen is a ski resort.

Bryant (1972b) has also outlined those areas in the mountainous Aspen Quadrangle where surface geological processes may act more vigorously and with greater frequency than elsewhere. Rockfalls, floods, mudflows and landslides have occurred in these areas during the past 100 000 years; other slopes where geological conditions are considered to be favourable for such events have also been included. Patches of colour on the map, reproduced here in black and white (Figure 12.1), outline those areas where special precautions should be taken before the natural features are disturbed.

Running water and gravity acting on earth materials are the principal active geological processes in the Aspen Quadrangle that may affect human activities. These processes work together, so areas of potentially unstable ground are more likely to move when saturated by heavy rainfall or when weakened from having been undercut by a stream in flood. Such conditions may form new landslides or reactivate old ones.

Hazard areas have a strong element of topographical control, and include the following.

Floodplains

Flat areas presenting a geological hazard may appear anomalous, although as the name implies floodplains are subject to flooding. On the west slope of the Rocky Mountains, where Aspen is located, high water occurs in late May or June during the melting of snow in the nearby high mountains. Generally such high water stays within the stream banks, but unusually heavy snowpack, combined with a late spring and sudden warming, may cause floods. In June 1921 such a combination of events accompanied by rain resulted in a flood in the Snowmass Creek and Capitol Creek drainages west of the Aspen Quadrangle.

Since the diversion of water from the head of the Roaring Fork River in 1935, the danger from floods has been greatly diminished along the river in the Aspen Quadrangle. Other creeks may have no major artificial constraints, however, and floodplains of those streams are likely to be flooded periodically.

Alluvial fans

Alluvial fans are accumulations of boulders, sand and silt at the mouths of tributary streams and gullies. This debris is deposited during floods, which may be accompanied by mudflows.

For instance, in 1965 a cloudburst on Aspen Mountain resulted in a thin mudflow which covered part of the alluvial fan at the mouth of Pioneer Gulch and caused some property damage. The small size of that alluvial fan indicates that no large mudflows have come down that gulch since the gravels on the valley floor were deposited.

Many of the fans are much larger than the one at Pioneer Gulch and could have formed in one or more major floods. A large part of the material in the fans may have been deposited soon after the retreat of glaciers from the main valleys. At that time much more unconsolidated morainal material was clinging to the valley side and could have been dislodged by melt water from the adjacent hillsides. Nevertheless, most of the fans could be reached today by floods of unusual magnitude, and in many places loose rock and other debris in the valleys of the parent streams can be washed down gullies during heavy rainfall or rapid snow-melt.

Wet meadows

In wet meadows the soil is saturated much or all of the year, and it is rich in organic matter. Typically present on highland flat benches in the mountains, possibly representing partly or completely infilled lakes, these deposits may have been mined for peat. Construction on such ground requires special care.

Potentially unstable slopes

Several types of potentially unstable slopes are outlined on the map. The most widespread type is that in which bedding in sedimentary rocks is nearly parallel to the slope. Extensive excavation could undercut the slope and promote sliding of the rock downward along the bedding planes. Where this situation is found, the steeper the slope the greater the potential for failure. Slopes in claystone or shale have more potential for failure than slopes in other rock types.

A second type of potentially unstable ground consists of steep slopes overlain by relatively thick unconsolidated debris that might slide or slump if undercut. Some large roadcuts at about 8800 ft in altitude on the north face of Aspen Mountain are difficult to maintain because of the steepness of the slope and the type of material excavated. General instability of the slope on the north side of Aspen Mountain was noted in 1888. Precise surveys of mining claims showed a downslope movement rate of 1.7 ft per year on the lower part of that slope. This movement may have been in response to the initial disturbance of the natural slope by construction of mine roads; it does not appear to have continued at that rate to the present time.

A third type of unstable slope consists of steep slopes underlain by fractured granitic rock. Saturation by rainfall, snow-melt or water diversion or undercutting by streams in flood may cause landsliding. Special care should be taken during any construction in such areas.

Landslides

Construction on mapped landslides requires special precautions even if the landslides appear to be inactive. Excavation in landslides is one of the most effective ways to reactivate them or, if they are active, to speed their downslope movement.

Rockfall areas

Areas particularly subject to rockfall are on or just below cliffs. No information is available on the frequency of rockfall, but this uncertainty makes these areas potentially hazardous building sites. If the rocks at the foot of the cliffs are lichen covered, the cliffs prob-

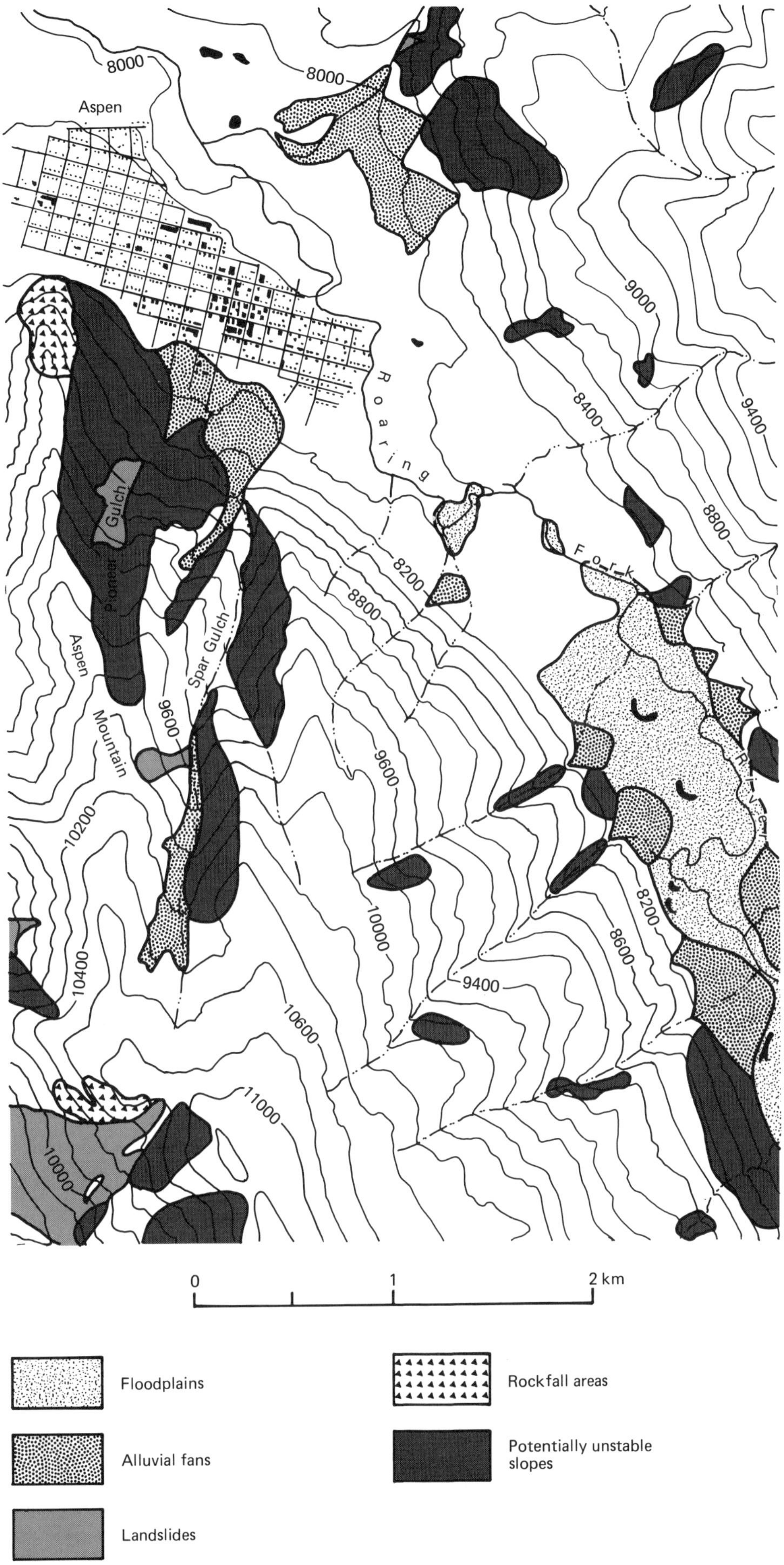

Figure 12.1 Map showing areas of potential geological hazards in Aspen Quadrangle, Colorado, USA (From Bryant, 1972b, by permission)

ably are not shedding much rock, and the area beneath them is relatively safe.

Hazards on the Aspen map are grouped by type into areas, and no attempt has been made to evaluate and classify the degree of risk, or the probability that the hazards may be activated by natural events. Some comment has been made on major, dated events, and hazards likely to be associated with ill-conceived man-made structures, particularly those on potentially unstable slopes.

An important slope map has been derived from the topographic sheet (Bryant, 1972c). The maps are printed in colour at 1:2400 scale.

12.2.2 Salina Quadrangle, Utah, USA: cloudburst floods

In the arid Salina Quadrangle, summer precipitation occurs as thunderstorms, during which an inch of rain or more may fall as a torrential downpour, in a local area. Run-off forms a cloudburst flood, eroding loose rock, soil and alluvium to form a debris-laden mudflow, which moves rapidly down gullies and canyons, eventually to spread out on the alluvial fans at the canyon mouths. Mudflows are highly erosive and destructive of man-made structures. Many towns are built on alluvial fans and consequently are at risk (see Figure 10.2).

The map (Hackman and Williams, 1972) is a simple division of drainage basins into major, secondary and minor basins. Point symbols locate the sites of historic cloudburst floods; where more than one flood is recorded this is indicated by number.

12.2.3 Parker Quadrangle, Colorado, USA

Land-use planning requires, as just one facet, a knowledge of the physical properties of rocks and soils, relative to susceptibility to erosion or relative erodibility. Usually, running water and wind are the most active agents of erosion. It is possible in the course of a detailed geological field survey to evaluate engineering characteristics, including erodibility, as was done for the Parker Quadrangle, Colorado, USA (Mayberry, 1972; Mayberry and Lindvall, 1972). Boundaries of geological units as shown on the 1:24 000 geological map have been modified for the derived map, because of the special parameters that had to be considered. These included measurement on aerial photographs of the amount of erosion from 1963 to 1970; composition of geological materials, including mineralogy, grain size, and type of binding agent or cement; depth of weathering; local steepness of slope; and relative cohesion between grains in slope-forming materials. Artificial fill and other man-made features were mainly disregarded, and the effect of vegetation cover on erodibility was not evaluated.

The geological materials in the quadrangle were grouped into three categories of relative erodibility:

1. *Highest* – most easily erodible by running water; local areas east of Parker Road are subject to severe wind erosion.
2. *Moderate* – moderately erodible by running water; local areas may be subject to moderate wind erosion.
3. *Lowest* – least readily erodible by running water or by wind action

Within each category there are wide variations in relative erodibility. For example, in the moderate category are grouped loess; parts of alluvial units that are older than modern alluvium along stream courses; and certain types of bedrock. It does not mean that loess, alluvium and bedrock all erode at the same rate, but rather that these materials erode less easily than materials in the highest category and more easily than those in the lowest category.

The usual disclaimer is made that this erodibility map is designed solely as a guide in land-use planning, and is in no way intended to supplant detailed field investigations and laboratory studies of geological materials from specific sites.

12.2.4 Henrys Lake Quadrangle, Idaho–Montana, USA: earthquakes

At a scale of 1:62 500, the map includes in its north-eastern corner a small part of Hegben Lake, and Henrys Lake overflowing by spillway into Henrys Fork. In the context of this map area, Hegben Lake is important because it was shaken by a major local earthquake in 1959. Assessment of hazard in Henrys Lake Quadrangle is based on what actually happened around Hegben Lake: a seiche occurred on the lake, and many rockfalls and one earthflow were activated.

Division of the area recognizes those parts of lake margins and the overspill valley of Henrys Lake likely to be flooded by seiches, and three zones of hazard due to rockfall and associated features (Figure 12.2).

A seiche is a periodic oscillation of a body of water in a containing basin, as is likely to happen if the area is shaken by a major local earthquake. Lake waters would almost certainly be thrown into periodic oscillations during which the lake shores would be successively inundated for brief periods.

During the seiche it is likely that some waters would top Henrys Lake outlet dam and flow south-eastward, following the Henrys Lake outlet, and spread across the basin floor. Henrys Fork might also swamp parts of its floodplain for brief periods.

If the outlet dam were to fail – a remote possibility – it is estimated that about 80 000 acre-ft of water would fan out across the southern end of the Henrys Lake basin. Although some water would seep into the porous sands and gravels which floor the basin, much of the water would escape downstream following Henrys Fork. Ultimately these waters would discharge into the Island Park reservoir.

Estimates of the shape and extent of those areas likely to be flooded by seiches are based on three assumptions. First, the earthquake will be a major local one; secondly, the lakes will be at near capacity when the earthquake occurs; and thirdly, during the first surges of the seiches the lakes will flood their shores to a level about 30 ft higher (as measured vertically, not horizontally) than their pre-earthquake levels. So, for Henrys Lake, the surface of which is at an altitude of about 6472 ft when the lake is full, the waters may extend as far inland as the 6500-ft contour (6472 ft + 30 ft = 6502 feet). For Hegben Lake, whose surface is at an altitude of 6544 ft when

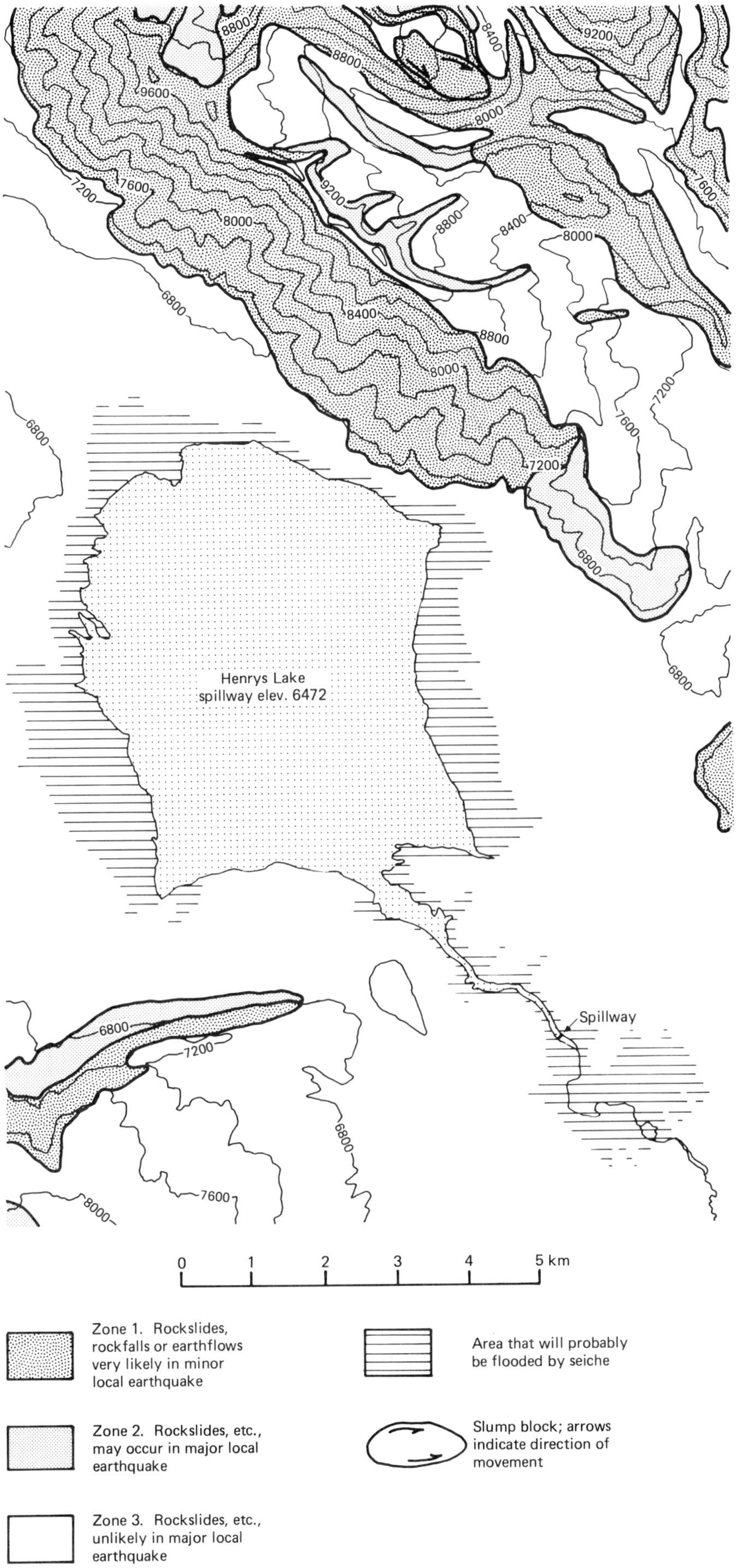

Figure 12.2 Map of Henrys Lake Quadrangle, Idaho–Montana, USA, showing seiche, rockslide, rockfall and earthflow hazards if the area is shaken by a major local earthquake (From Witkind, 1972, by permission)

the lake is near capacity, the waters may flood all ground below the 6575-ft contour (6544 ft + 30 ft = 6574 feet).

Judging by the action of Hebgen Lake during the 1959 earthquake, once the seiche has subsided, Henrys Lake will reflect the tilting of the block of the earth's crust on which it rests. Parts of its shoreline will be flooded even as the lake floor on the opposite side of the lake is exposed. The basin (in the example above) will have been tilted, and the surface of the water as it comes to rest after sloshing back and forth will act as a level to indicate the amount of tilt.

The periodic flooding that occurs during a seiche is short-lived and gradually subsides; by contrast, the lake shores that are submerged when the seiche ends will be permanently flooded. A block of the earth's crust containing the lake within its confines and possibly measuring hundreds of square miles in area will have shifted and it cannot rebound to its original position.

Rockslides, rockfalls or earthflows would occur in virtually all of the mountainous parts of the area if one of the active faults in the Henrys Lake Quadrangle were to move and so cause an earthquake. Most of these landslide phenomena would be small, but a few might be large enough to cause considerable damage.

The division of this quadrangle into three zones of hazard is based mainly on similar phenomena that occurred in the nearby Hebgen Lake area during the earthquake of August 17, 1959:

Zone 1 – area in which rockslides, rockfalls or earthflows are very likely to occur if area is shaken by a major local earthquake.
Zone 2 – area in which rockslides, rockfalls or earthflows may occur if area is shaken by a major local earthquake.
Zone 3 – area in which rockslides, rockfalls or earthflows are unlikely to occur if area is shaken by a major local earthquake.

In the Hegben Lake area, falling debris characterized practically every steep slope (45% gradient and greater). In most places the falling debris consisted of isolated boulders dislodged by the shaking of the earth, but in a few areas extensive rockslides and rockfall and debris avalanches occurred. In general, the number and intensity of the rockfalls mirrored the steepness of the slopes, and in one locality an earthflow slowly began to move a week after the earthquake. These various phenomena caused little damage to man-made structures, chiefly because few existed to be harmed.

The potential danger from such large numbers of falling boulders is great, however, and so on this map every steep slope has been placed in Zone 1. As individual falling boulders were much less common on the less steep slopes in the Hebgen Lake area, slopes of moderate gradient (25–45%) in the Henrys Lake Quadrangle have been grouped in Zone 2. Large parts of Zone 3 are virtually flat or have slopes that do not exceed 15%. Some insignificant movement of earth materials is likely on the broad flats of the Henrys Lake basin but will probably be confined to small sand and silt flows along stream embankments. A few boulders perched on some of the more gentle slopes may move short distances, but these too are likely to be of little consequence.

Specific comments are made on the map legend concerning rockfall avalanche and earthflow.

Rockfall avalanche

There is a large potential rockfall avalanche near the junction of West Targhee Creek with Targhee Creek (see Figure 12.2). Here an enormous partly fragmented slump, shown on this map by a heavy outline, is precariously perched about 1000 ft above the streams, seemingly waiting to be jostled so that it can avalanche into the valleys and effectively block them.

Earthflow

The term earthflow denotes the downslope movement of a large mass of moist to saturated unconsolidated debris. Of the earthflows in the Henrys Lake Quadrangle, one of the largest is at the head of the West Fork of Denny Creek in the north-east corner of the quadrangle. A much smaller earthflow is along the valley wall between the mouths of Tygee and Targhee Creeks. The access road for a development of summer homes crosses the earthflow, and the road is certain to be broken if the earthflow moves.

In addition to this renewed movement of the old established earthflows, new earthflows may be formed in various localities where partly saturated unstable units, mainly till, siltstone and shale, crop out.

The possibility that the events discussed will, at some unspecifiable time, occur in the Henrys Lake Quadrangle is very real, because on the Seismic Map of the USA it is within seismic risk zone 3, defined as an area in which 'major destructive earthquakes may occur'.

12.2.5 Morrison Quadrangle, Jefferson County, Colorado, USA: landslide susceptibility

This is a simple derived map showing areas of actual landslide deposits, active landslides, areas susceptible to sliding and potential rockfalls along existing highways. The map reflects the distribution of montmorillonite clays (see section 12.2.6) and steep rock slopes in sandstones (Figure 12.3).

12.2.6 Map showing areas of swelling clay in Morrison County, Colorado, USA

Focusing on hazards inherent in the mineralogical content of the rocks, the medium-scale 1:24 000 geological map of Morrison County, Colorado, USA, has been converted to a derivative map (Scott, 1972b) of built-up areas and potential development areas in which swelling-clays may cause foundation problems. Morrison County lies to the west of Denver and includes the southern part of the town of Boulder (Figure 12.4).

In so far as specialist knowledge of the clay mineralogy of the sediments is needed for map interpretation, although such information may have become available during the initial geological survey and could thus be incorporated in the map legend, this map takes derivation of maps for engineering use a significant step forward from those discussed previously. The map is backed up by specialist geotechnical laboratory testing, although the results may be held simply to quantify field observations of the performance of buildings and/or

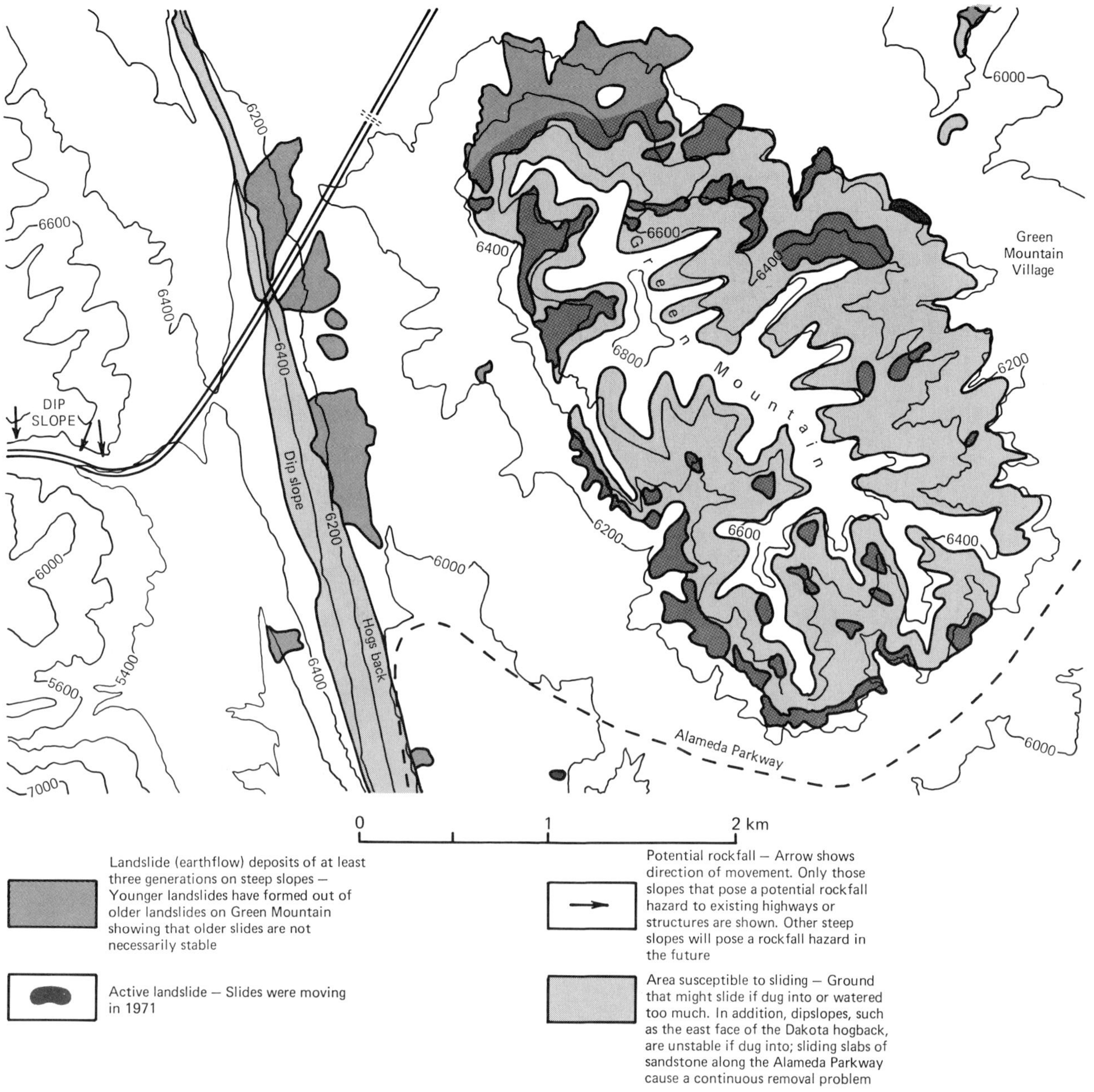

Figure 12.3 Map of landslides and areas susceptible to landsliding in the Morrison Quadrangle, Colorado, USA (From Scott, 1972a, by permission)

their foundations made during the initial survey.

There are three map units:

1. *Red.* Areas underlain by geological units that contain clays having swelling pressures higher than 2500 psf (pounds per square foot) PVC (potential volume change). Preconstruction investigations in these areas should include engineering laboratory swell-shrink tests.
2. *Pale red.* Areas underlain by 5 ft or more of non-swelling surficial deposits which in turn are underlain by geological units which may have swelling pressures higher than 2500 psf PVC. For any construction involving foundation excavation through the surficial deposits, preconstruction investigations should include engineering laboratory swell-shrink tests.
3. *Uncoloured.* Areas underlain by geological units having no or slight swelling pressures. Preconstruction swell-shrink tests unnecessary.

Four categories of swelling pressure have been recognized.

- Very critical – above 4700 psf
- Critical – 3200–4700 psf
- Marginal – 2700–3200 psf
- Non-critical – 0–2700 psf

In critical situations, swelling pressures in foundation materials would have to be determined in the laboratory.

A note on the map legend makes it clear that formations containing clays having swelling pressures higher than 2500 psf (as measured by the PVC meter) are in

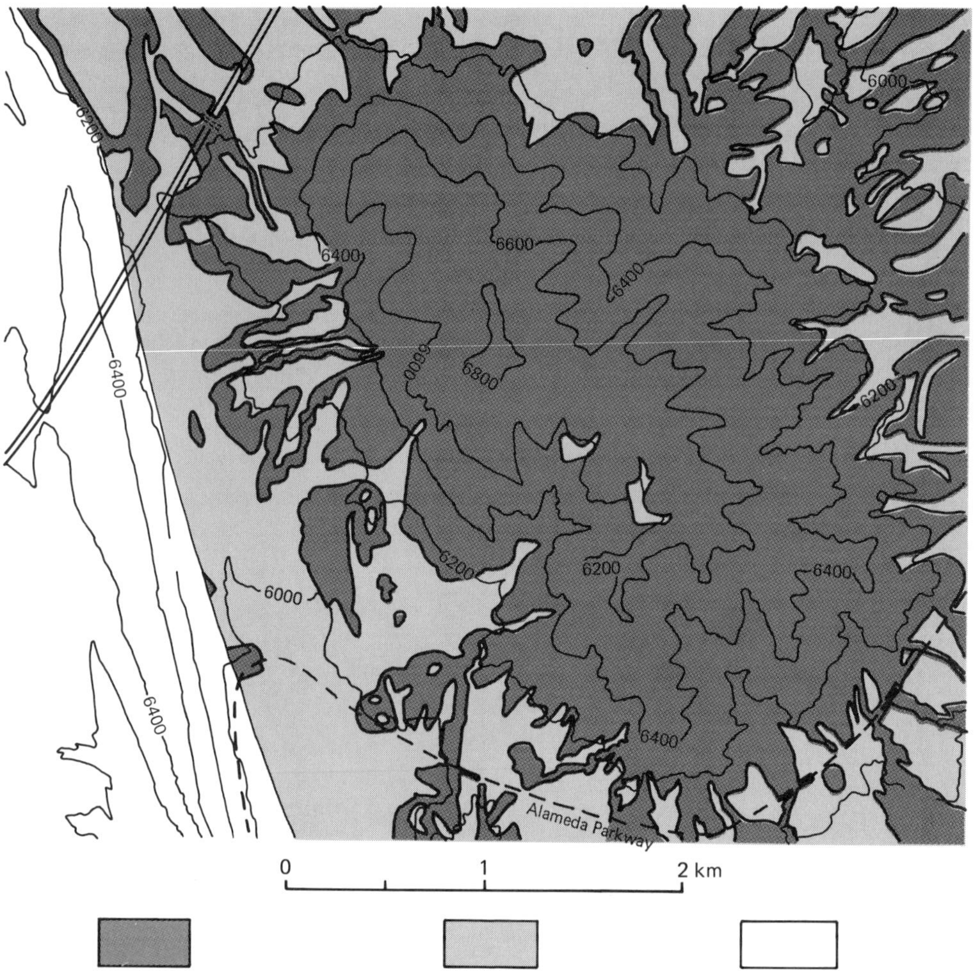

Figure 12.4 Map showing areas of swelling clay in the Morrison Quadrangle, Colorado, USA (From Scott, 1972b, by permission)

order of decreasing swelling pressure: Denver Formation, Pierre Shale, Laramie Formation, Green Mountain Conglomerate, Fox Hills, Sandstone and Arapahoe Formation. Each formation can be identified on the accompanying geological map. Some landslides derived from these formations also contain swelling clay and are mapped with the bedrock formations listed above.

Each of these bedrock formations has some beds having swelling pressures, some as high as 9000 psf. Bentonite beds in the Graneros Shale, Greenhorn Limestone, Carlile Shale, Smoky Hill Shale Member of the Niobrara Formation, and Pierre Shale have swelling pressures as high as 1000 psf, but these white clayey beds are less than 1 ft thick and are easy to remove before building. Alluvial and colluvial deposits derived from these bedrock units have inherited swelling pressures as high as 4000 psf, but generally less than one-half the swelling pressure of the original bedrock.

Montmorillonite and mixed-layer clays are the clays that swell when wetted and shrink when dried; of the two, montmorillonite has the greater swelling potential. Swell-shrink tests should be made by an engineering laboratory before construction is begun on any of the formations listed above and which constitute the group having the highest swelling pressure.

12.2.7 Comment

In the early 1970s there appears to have been a concerted effort to make the conventional products of a geological survey of more practical application. What could be selected for individual treatment obviously depended on the geological conditions, as became clear when engineering geologists became involved in the specific mapping of landslide terrains in California.

12.3 Landslide hazard and risk maps

12.3.1 Introduction

Landslides have been recorded for well over a century,

with many classic, spectacular examples. They are a widespread natural and man-induced phenomenon, and over the years may cause more damage to property than any other geological hazard. As a secondary consequence of earthquakes and intense storms (see section 12.2.2), landslides are often responsible for much of the attendant damage and loss of life. Rapid assessments can be made using engineering geomorphological mapping, recording landforms, materials and processes (Doornkamp *et al.*, 1979). Aerial photograph interpretation, at scales of 1:48 000 and larger, has major advantages – it is rapid, accurate and cost effective both before and during the mapping of a landslide on the ground. Investigation techniques are discussed in Varnes (1984), and the application to general terrain evaluation has already been illustrated (Chapter 9).

12.3.2 Landslide mapping in Czechoslovakia

In the years 1962 and 1963 all important landslide areas in Czechoslovakia were plotted on maps at a scale of 1:25 000. Active recent landslides were mapped in red, quiescent and fossil landslides in black; green was used for landslides caused by human interference. The hydrology – springs, rivers, wet ground, undrained depressions – was depicted in blue.

Basic mapping, undertaken on topographic maps or aerial photographs at various scales, has a very distinctive style (Rybar, 1968), as the two examples reproduced here demonstrate. One of the upper part of the disastrous landslide affecting the town of Handlova (Plate 16a) was mapped at a scale of 1:5000. The other, at an original scale of 1:250, is of the upper part and the toe of an active slide on the site of the Necranice dam in North-west Bohemia (Plate 16b,c).

Landslides in the Handlova area are illustrated in Varnes (1984, Figs 26–29). Maps prepared at 1:10 000 scale, both printed in colour, are of two kinds. One is a map of engineering geological conditions (Varnes's Fig. 26) with an informative explanation detailing the rocks and soils, forms of slope deformation and hydrogeological data. For active landslides the symbols are printed in red, as in Plate 16(a). The same area is also presented as a synthetic map of engineering geological zones and stability (Varnes's Fig. 28). On the map, four types of unstable area are left uncoloured and marked with distinctive symbols in red. Other zones, printed in colour, are relatively stable and stable areas. The explanation (Varnes's Fig. 29) gives schematic cross-sections for each of the nine map units, with brief statements on the character and activity of the movements and construction conditions. The maps were prepared by Malgot *et al.* (1973).

12.3.3 Landslide mapping in California, USA

The Regional Landslide Research Group of the U.S. Geological Survey, located in California, is in the Branch of Geologic Risk Assessment, formerly the Engineering Geology Branch. Early innovative work on landslide mapping began in 1970, and since 1986 an informal project on computer-generated maps of San Mateo County has received official funding. There is available, therefore, a series of maps at various scales in which the mapping methodology can be demonstrated. The order in which maps are prepared, and the methods used, may be summarized as follows.

Landslide inventories

Reconnaissance approach; simple inventories
Reconnaissance maps show those areas that appear to have failed by landslide processes. Commonly, these maps are prepared by interpreting aerial photographs, with a minimum of field checking. Old and active landslides may be distinguished in an intermediate type of simple inventory.

Detailed inventories
A detailed inventory map shows each landslide classified as to type, as well as delineation of scarps, limits of the zone of accumulation of landslide debris, and other pertinent data on depth and kind of materials involved in sliding. Active and inactive landslides are distinguished. The map would also identify excavations, trenches and boreholes used in the study of a landslide.

Although some information can be obtained from false-infrared aerial photographs, much of the data must be obtained in the field by closely spaced traverses.

Terrain analysis
Landforms associated with various kinds of landslides can be recognized and classified, and appropriate terrain units defined from small-scale (high altitude) aerial photographs. Slope processes in each terrain unit are determined by field mapping of small areas representative of each terrain unit (Chapter 9).

Landslide-hazard zonation

Slope stability maps
Slope stability maps distinguish areas that have different potentials for landsliding. Such maps predict where new landslides are likely to occur by a ranking of relative stability of slopes.

A simple form of slope stability map can be made in areas where only a single geological unit is prone to landsliding. A map of that unit therefore also shows the areas of potential landsliding, but not – it must be remembered – actual landslides. Simple slope stability maps can thus be made for large areas from an appropriate geological map. Consideration of other attributes, such as slope inclination and aspect, can improve the assessment of relative stability within specific geological units.

Actually the areal distribution of landsliding is more complex, and in such cases maps may be constructed by combining a geological map with a landslide inventory. If the landslide inventory distinguishes different types of landslides, then complex slope stability maps, showing the likelihood of different types of landslide, can be compiled.

In general, slope stability maps provide more information and include an assessment of potential for landsliding of the entire map area than do inventory maps. This increased information is, however, only gained at the expense of more field work and more extensive background information.

Landslide overview map of the conterminous USA

Three versions of the general landslide map of the USA, at a scale of 1:7 500 000, have been prepared. A preliminary landslide overview map (Radbruch-Hall *et al.*, 1976) showed landslide incidence by three categories in colour, depending on the percentage of area involved in landsliding; two categories of landslide susceptibility – high and moderate – are shown by patterns. This map was extensively modified in 1978, mainly by considering more geological units to be landslide prone (Wiggins, Slosson and Krohn, 1978). The complete map was reprinted in U.S. Geological Survey (1982, Fig. 1) and in Brabb (1984, Fig. 4). A final version of the preliminary 1976 overview map was published in colour in 1982 (Radbruch-Hall *et al.*, 1982, Plate 1).

A segment of the map, the whole of California, is presented here as Figure 12.5. The map can be used either for engineering or environmental planning to determine areas where slope stability problems may hinder construction, or where construction may exacerbate slope stability problems.

Figure 12.5 Abstract of California from the map showing relative potential of different parts of the conterminous USA to landsliding (From U.S. Geological Survey, 1982, Fig. 1, by Wiggins, Slosson and Krohn, 1978, by permission)

The map, showing both amounts of landslides and areas susceptible to landslides, was compiled primarily from existing information, by evaluating geological units shown on the geological map of the USA according to three categories of amounts of landslides and three categories of susceptibility to landsliding. The map units were evaluated primarily by consulting published information, but other criteria were also used. Aerial photograph interpretation, reference to newspaper accounts, analysis of climatic and topographic conditions, and field work were all used in compiling the map, which was also spot-checked for accuracy by ground and aerial reconnaissance after compilation. To a certain extent it is derivative, for the areas susceptible to sliding were derived from the geological map by extrapolating information generally known for only a small part of a given formation over the entire area of the formation.

An expository text (Radbruch-Hall *et al.*, 1982) accompanying the map discusses the slope stability characteristics of the physiographic regions of the USA, inasmuch as the geological factors that influence slope stability also influence the form of the land surface. To this limited extent it is also a zoning map.

A simple slope stability map of California

A small-scale map of the whole of California, published by Radbruch and Crowther (1973), shows the distribution of six units in terms of relative amounts of landslides. Slope stability units were delineated by determining from the literature, and from experience, which mappable geological units are prone to landsliding. The geological map was then used, with other attributes such as aspect and slope inclination, to outline areas in terms of least number to most landslides.

The version reproduced (Figure 12.6) has been generalized by Alfors, Burnett and Gay (1973) into four severity zones. A note on the map is a significant disclaimer.

Map of estimated abundance of landslides

An experimental map at 1:500 000 scale (Figure 12.7) was published in 1971 (Radbruch and Wentworth, 1971) to provide a first approximation of landslide abundance and potential in the San Francisco Bay region. Different parts of the region are ranked on a scale from 1 to 6 according to the estimated abundance of landslides in them. Qualitative ranking is based on the area covered by landslides estimated on three factors: slope of the ground surface, rainfall, and rock and soil conditions. Inferred relations of landsliding to these factors, with limited field knowledge of the actual abundance of landslides in some areas, have been used in the interpretation.

The region was divided into the six ranks of landslide abundance by progressively evaluating and ranking the high and low extremes. Rank 1 represents areas with a very small amount of landslides and, at the other extreme, rank 6 represents areas that contain a maximum amount of landslides in the region. Areas ranked 5 have a lesser, but still very large, amount of landslides and areas with a limited amount of landslides, but larger than range 1, are assigned as appropriate to ranks 3 and 4.

Rank 1 is defined as areas that receive less than 10 in of mean annual precipitation or have slopes of less than 5° (determined from a 1:500 000 scale topographic map with 500 ft contour interval). These criteria indicate a very small amount of landslides, because landslides are rare in areas with less than 10 in of precipitation, and a review of a limited number of reports and consultation with other geologists indicate that few landslides occur

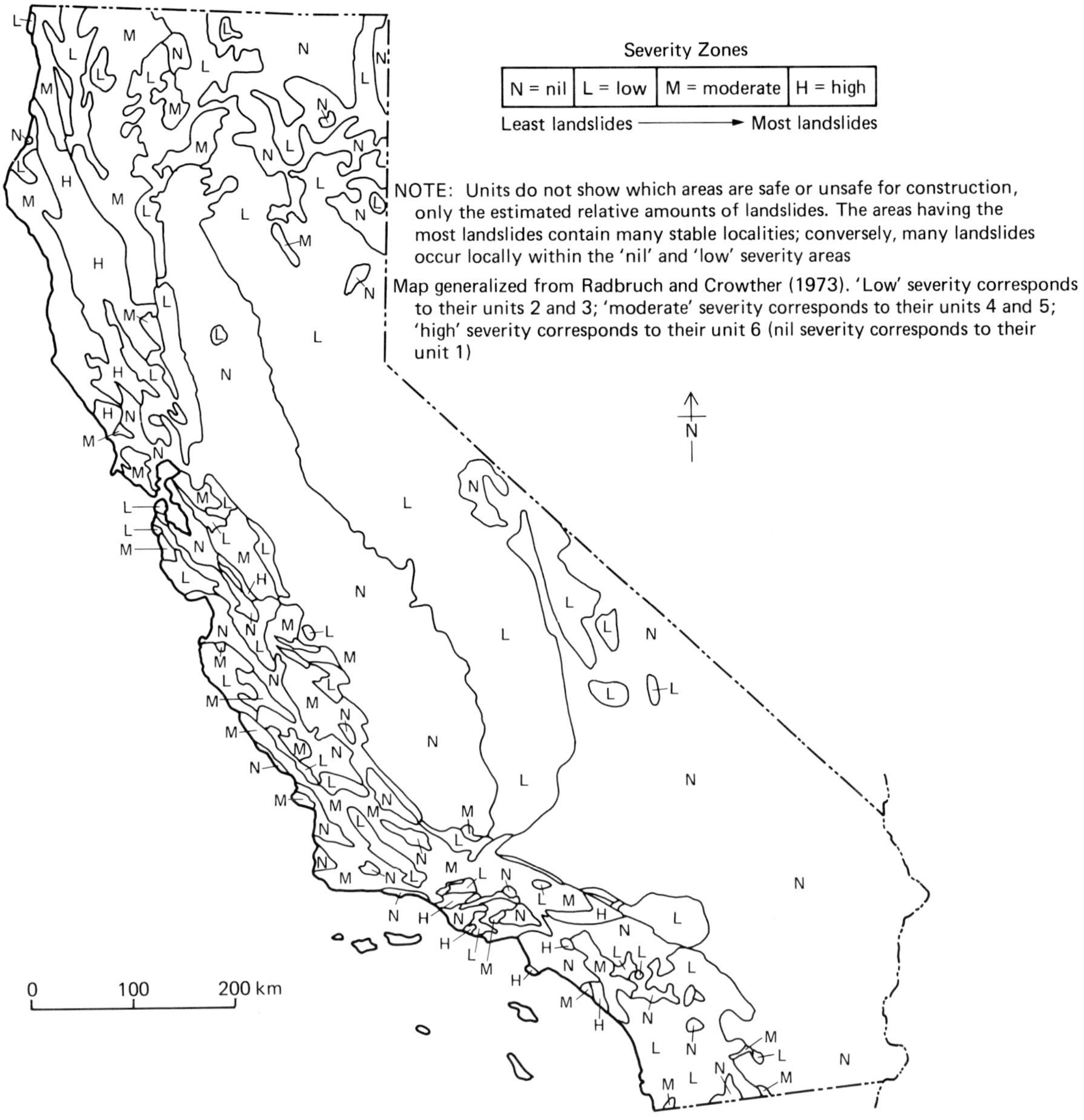

Figure 12.6 Simple slope stability map of California, originally published by Radbruch and Crowther (1973) to show relative amounts of landsliding. Shown here in generalized form with severity zones (From Alfors, Burnett and Gay, 1973, by permission)

on slopes of less than 5° (Radbruch and Crowther, 1970, pp. 4–6). Because of the large contour interval on the map from which slopes were determined, and the small scale of the final map, sea cliff areas and hilly areas with less than 500 ft relief may locally contain abundant landslides, although shown on the map as rank 1.

Ranking of the remaining region is based largely on the distribution of earth materials. For this purpose the geological units in the region that are shown on the 1:250 000 scale Jenkins edition of the Geologic Map of California were grouped into eight general classes of earth materials. The classes were selected to have as similar landslide characteristics as possible, using readily available literature on landslides in the region as a guide. Major differences in landslide characteristics within individual classes of earth materials resulting from varying topography or bedrock type of structure are used in the ranking, where the needed information is available.

Map of landslide susceptibility in San Mateo County
This map (Figure 12.8) provides a more detailed appraisal of landslide susceptibility than Figure 12.7. Constructed from three maps of basic data – a geological map, an inventory of landslides (Brabb and Pampeyan, 1972) and an experimental slope map – the map is a derivative one designed for regional planners as a simple, easily understood statement on the susceptibility of specific areas to landsliding.

The area selected as an illustration can also be located in Figure 12.7, where it can be seen that the ranking in terms of landslide abundance is 3 and 4.

A very full explanation on the map sheet includes notes on how to use the map, factors on landslide distribution, and mapping procedures. A table gives the landslide failure record for rock units cropping out in the county. Seven map units are graded from least to most susceptible.

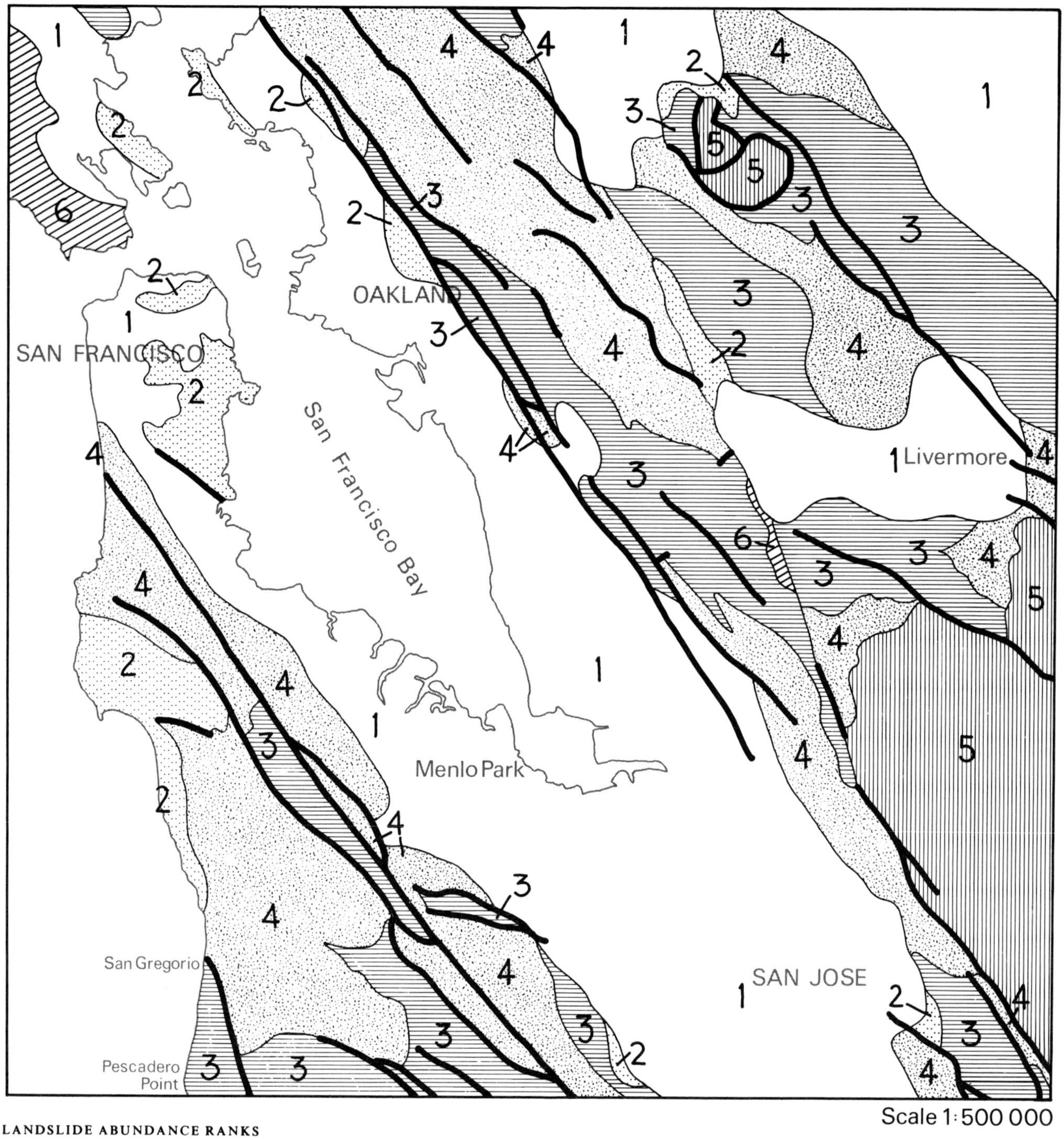

LANDSLIDE ABUNDANCE RANKS

1 — Least abundant
2
3
4
5
6 — Most abundant

Ranking is qualitative, based on estimates and extrapolation from available data. Specific safety or hazard for construction is not shown. Landslide distribution within individual map units may not be uniform: parts of the highest ranked units lack landslides, and parts of the lowest ranked units contain landslides. Hilly parts of unit 1 may contain abundant landslides

Approximate contact between map units

Fault, approximately located, marking zone of possible sheared or shattered rock not represented in the ranking but susceptible to landsliding. Not shown where concealed beneath thick overlying deposits or water

Figure 12.7 Map of estimated abundance of landslides in the San Francisco Bay Region, USA, prepared at a scale of 1:500 000 (Redrawn from Radbruch and Wentworth, 1971)

It is made abundantly clear that the map was intended primarily for solving problems involving slope stability, such as planning subdivision-size blocks of land in undeveloped areas. It indicates roughly and indirectly how extensive and how expensive the landslide problem may be in these areas, and it provides a guide to alternative areas that might be less costly. The map should not be used to determine the stability of individual building sites.

The map was adopted in 1975 as a zoning map that controls the permissible density of development on land of different susceptibility to landsliding. For areas in the highest category of susceptibility, only one structure is permitted on each 40 acres of land, provided that geological and engineering investigations show that a site exists in the parcel that can safely be developed. Since adoption of the zoning there have been no landslides on 1055 properties, about 370 of which were in hillside areas, developed under the programme (U.S. Geological Survey, 1982).

Some landslides may occur in areas I and II, where susceptibility is low, but the hazard does not appear to be great enough to warrant additional governmental restrictions at this time. From 1972 to 1982, all of the new landslides that formed were either in areas already shown as a landslide or in areas mapped as highly susceptible. In 1982, however, rarely recorded rainfall intensities induced thousands of debris flows in areas where few if any had been observed previously (Brabb, 1984). Subsequent examination of aerial photographs revealed that following heavy rains in 1955 debris flows were widespread in the county, but by 1968, the date of the photographs used for the 1972 map, most of the scars had been modified by erosion and covered with vegetation.

In mitigation of this oversight, Brabb agrees that some of them could have been recognized, but large, deep-seated landslides dominated the literature and daily concerns to such an extent in the 1960s that the debris flow process was largely overlooked.

Landslides in northern Contra Costa and adjacent counties, California

Nilsen *et al.* (1979) have provided a series of derivative maps at a scale of 1:125 000 based on analysis of the contoured topographic base map, a photo-interpretation of landslide deposits (Figure 12.9a) and the available geological map (Figure 12.9b). The landslide map is a simple inventory locating single isolated landslide deposits, and larger areas underlain by a single landslide deposit or a group of closely spaced large and small landslide deposits. From the geological map, those bedrock units and Bay mud areas that are considered to be especially susceptible to slope failure of one kind or another have been isolated. A generalized slope map, not reproduced here, breaks down the area of the map into three units on the bases of 0–5%, 5–15% and more than 15% slope.

Combine the generalized slope map, the landslide inventory map and the map of susceptible geological units, and the result is the map of relative slope stability (Figure 12.9c). Six map units are recognized: areas subject to liquefaction (a topic considered in section 12.4), and a graded sequence from stable to unstable. A point of interest is that single isolated landslide deposits are picked out on this map. The reason is clear, for most of them occur in map areas that are labelled moderately stable.

Comparing the three basic maps – slopes, landslides and susceptibility – with the interpretative map emphasizes the obvious correlations between landslide deposits and unstable areas, Bay mud deposits and areas subject to liquefaction, stable areas and flat ground. The three intermediate categories of relative slope stability result from balancing slope against susceptible bedrock units.

12.3.4 ZERMOS maps of France

National programmes relating to landslide damages exist in other countries. An example is the ZERMOS (zones exposed to hazards of movements of the soil and subsoil) plan in France, which is responsible for the production of maps at scales of 1:20 000 and smaller. The maps portray degrees of risk of various types of slope failures, and include activity, rate and potential consequences (Humbert, 1977). Recently the scope of the ZERMOS plan has been enlarged to provide guidelines for suitable locations of development and permissible land use (Porcher and Guillope, 1979).

Cartographically the maps are very attractive, and a few examples will suffice to illustrate the scope of the plan. Abstracts from some others are printed in colour in Olivier and Renet (1976) and Antoine and Pachoud (1976); these in fact represent the first essays at preparing ZERMOS maps.

Région d'Aime, La Plagne (Savoie)

The area mapped in the Savoy Alps is about 100 km^2 in extent, with about one-fifth above 2000 m. Topographically, the region shows marked effects of the last glaciation, both from the point of view of erosional features, including very late highland cirques with fresh frontal moraines, and extensive depositional moraines. Preglacial landslides on the valley sides, intensively eroded by the valley glaciers, were reactivated on their retreat and new instabilities developed. At the same time, and indeed up to the present day, boulder gravels accumulated in the valley floors, fed by alluvial fans at the mouths of lateral valleys from the mountain slopes.

Bedrock comprises Carboniferous and Triassic sediments and schists. Gypsum in the Trias, in more or less horizontal beds, underlies pastures at between 1700 and 2000 m, with solution features and characteristic erosion channels. Triassic quartzites give rise to the precipitous topography of Mount St. Jaques (Plate 17).

The map itself is printed on a brown topographic base, with the hazard zones outlined in black and printed in bold colours. Colours are graduated from red, indicating areas with the highest hazard, to orange for slight hazard, to light green for hazard free. Six degrees of hazard are used, two degrees of hazard for each colour, estimated as a function of:

1. The intensity of the dynamic phenomenon from which the hazard may be inferred. Intensity is defined by the mass in movement and the rate of movement.
2. The state of the dynamic phenomenon:
   (a) at the present time – always active or recurring frequently;

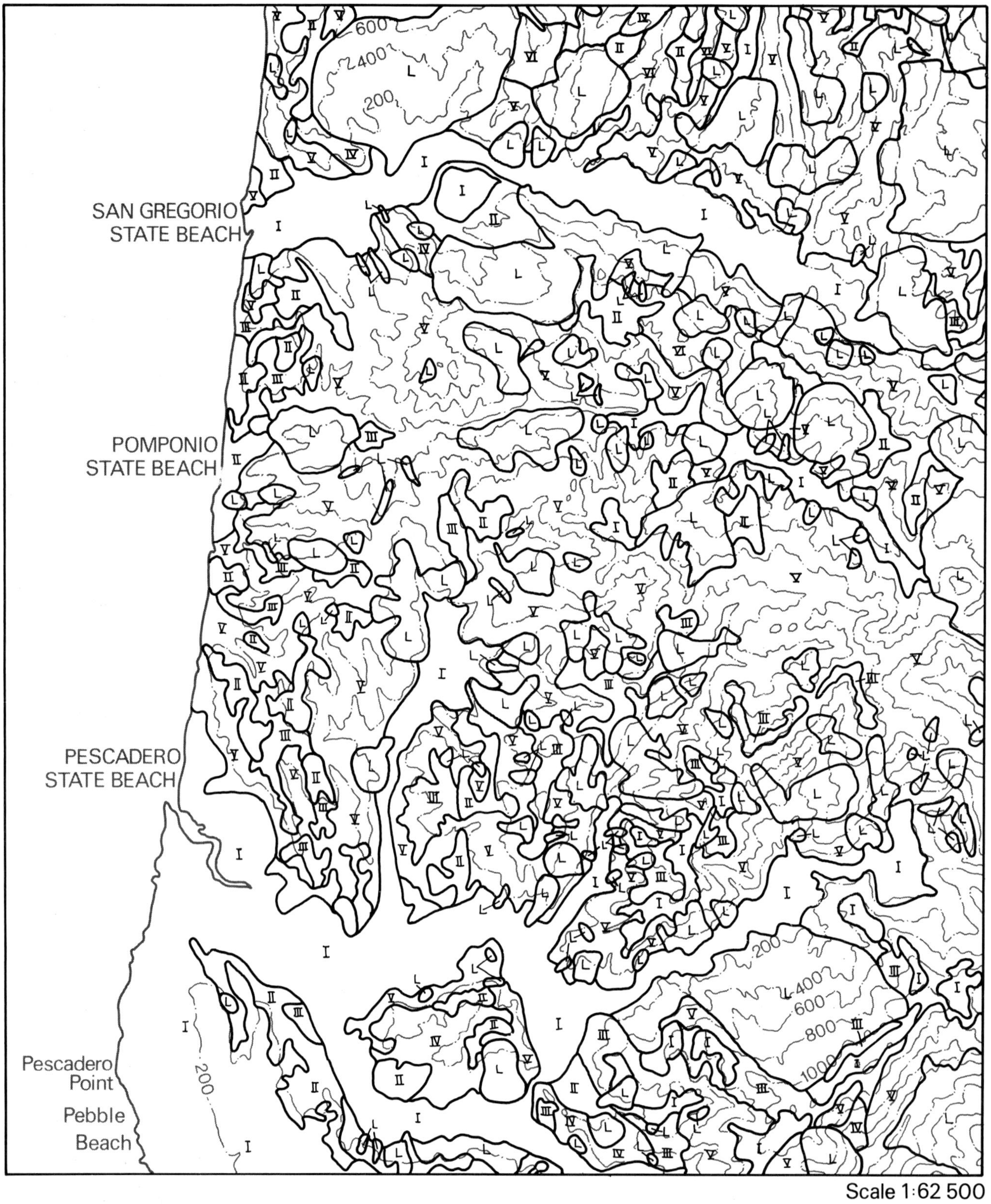

Figure 12.8 Map of landslide susceptibility, San Mateo County, California, USA, prepared at a scale of 1 : 62 500 (legend shown opposite) (From Brabb, Pampeyan and Bonilla, 1972, by permission)

Least

I Areas least susceptible to landsliding. Very few small landslides have formed in these areas. Formation of large landslides is possible but unlikely, except during earthquakes. Slopes generally less than 15 per cent, but may include small areas of steep slopes that could have higher susceptibility. Includes some areas with 30 per cent to more than 70 per cent slopes that seem to be underlain by stable rock units. Additional slope stability problems; some of the areas may be more susceptible to landsliding if they are overlain by thick deposits of soil, slopewash or ravine fill. Rockfalls may also occur on steep slopes. Also includes areas along creeks, rivers, sloughs and lakes that may fail by landsliding during earthquakes. If area is adjacent to area with higher susceptibility, a landslide may encroach into the area, or the area may fail if a landslide undercuts it, such as the flat area adjacent to sea cliffs.

II Low susceptibility to landsliding. Several small landslides have formed in these areas and some of these have caused extensive damage to homes and roads. A few large landslides may occur. Slopes vary from 5-15 per cent for unstable rock units to more than 70 per cent for rock units that seem to be stable. The statements about additional slope stability problems mentioned in I above apply in this category.

III Moderate susceptibility to landsliding. Many small landslides have formed in these areas and several of these have caused extensive damage to homes and roads. Some large landslides likely. Slopes generally greater than 30 per cent but includes some slopes 15-30 per cent in areas underlain by unstable rock units. See I for additional slope stability problems.

IV Moderately high susceptibility to landsliding. Slopes all greater than 30 per cent. These areas are mostly in undeveloped parts of the county. Several large lanslides likely. See I for additional slope stability problems.

V High susceptibility to landsliding. Slopes all greater than 30 per cent. Many large and small landslides may form. These areas are mostly in undeveloped parts of the county. See I for additional slope stability problems.

VI Very high susceptibility to landsliding. Slopes all greater than 30 per cent. Development of many large and small landslides is likely. Slopes all greater than 30 per cent. The areas are mainly in undeveloped parts of the county. See I for additional slope stability problems.

L Highest susceptibility to landsliding. Consists of landslide and possible landslide deposits. No small landslide deposits are shown. Some of these areas may be relatively stable and suitable for development, whereas others are active and causing damage to roads, houses and other cultural features.

Most

Definitions: Large landslide—more than 500 ft (150 m) in maximum dimension; small landslide—50 to 500 ft (15 to 150 m) in maximum dimension.

(b) old;
(c) potential – more or less high probability of occurrence.

3. The distribution of the dynamic phenomenon:
   (a) localized;
   (b) sporadic.

Within each hazard area on the map – a combined letter and number symbol – letters identify the dynamic phenomenon responsible for the hazard, and a number from 1 to 7 the state of the phenomenon, whether active, potential or old.

The map, at a scale of 1:20 000, is clear and easy to interpret with its six units in bold colours; the legend is simple and clear. It is altogether a good example of effective engineering geological cartography which should be easy to understand by the non-geologist.

A coloured table in the accompanying memoir relates degree of hazard to the state of the dynamic phenomenon involved. Zones of greatest risk are those in which collapses, sporadic block falls, debris flows and mining subsidence occur. The map is of interest because it does not rely on symbols to indicate the different types of phenomenon, but rather just colours with included numbers and letters.

Région de Chanonat-Sud de Clermont-Ferrand (Puy-de-Dome)

The map is published in colour at a scale of 1:25 000, with a descriptive memoir. The choice of this region south of the rapidly growing urban area of Clermont-Ferrand was motivated by the presence in the area of many natural slopes that were unstable or were considered to be of limited stability. Hills are up to 700 m high, with slopes between 10% and 30%, above the main valleys at 350 m. Present relief has resulted from Quaternary erosion, leaving hills capped by basalt flows or formed by volcanic necks.

Ground movements are essentially of five types:

- block falls;
- historic collapses;
- subsidences;
- well-defined landslides;
- diffuse movements, solifluxion.

Analysis of the region for the ZERMOS mapping programme was undertaken in the following stages:

- completion of a geological map of the solid and superficial deposits at a scale of 1:25 000 – this was not published;
- assessment of borehole records;
- aerial photograph interpretation at 1:25 000, with some areas analysed at 1:15 000;
- a geomorphological study;
- an inquiry into actual phenomena, and their development on a human time scale;
- completion of a slope map at 1:25 000, with slopes at less than 10%, 10–20%, 20–30% and more than 30%;
- assessment in the field, with consideration of the hydrogeology and compilation of the ZERMOS map with five hazard zones.

On the map (Plate 18), the zones give the location of areas where there are hazards associated with ground movements, and the degree of hazard. It is emphasized

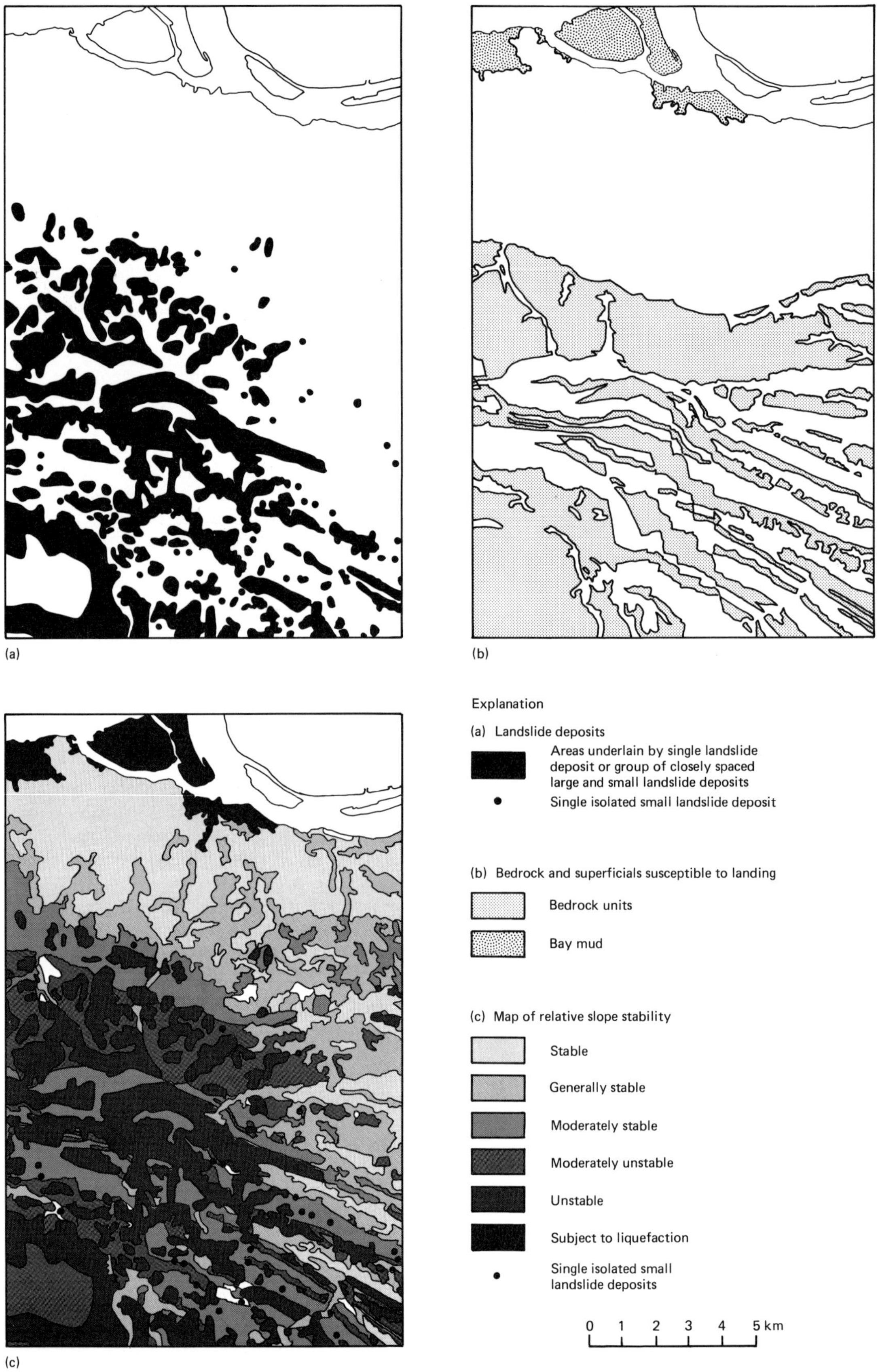

Figure 12.9 Landslides in part of north-eastern Contra Costa County, California, near San Francisco, USA: (a) generalized photo-interpretative map providing a simple inventory of landslide deposits; (b) distribution of bedrock and superficial deposits considered to be particularly susceptible to slope failures; (c) relative slope stability map (From Nilsen *et al.*, 1979, by permission)

that the map only provides information up to the date of compilation. Three basic colours are used:

Green – slight or no hazard;
Orange – moderate hazard;
Red – high hazard from an actual unstable area, likely to be very difficult to control.

The green and orange areas are subdivided.

Green V1 (map legend on Plate 18) refers to zones free from movements. This refers to most slopes of less than 10%, and to steeper slopes on massive granite.

Green V2 refers to zones apparently or actually free from movements, where nevertheless verifications or precautions, although small in number, are necessary. This zone comprises steep slopes of about or more than 30% where the rocky substratum crops out (granite, arkose, basalt), or slopes between 20% and 30% underlain by marls, or lower slopes of 10–20% where colluvium is unlikely to be thick and the hydrogeological regime presents no problems. However, in this zone precautions need to be related to the importance of the undertaking. In this context, north-facing slopes are of greater significance than those with a southern aspect. Also included in the V2 zone are those areas subject to flooding.

Orange 01, on slopes of 10–30%, indicates ground solifluxion effects, without well-defined slides, and with a generally diffuse character. Locally it is used for slopes where there is a lack of information.

Orange 02 corresponds to zones with moderate movements or with certain indications of instability capable of leading to a hazard.

Red, of limited occurrence on the map, corresponds to known zones of high hazard, or those in which the eventual scale of movements justifies inclusion as a zone in which hazard is very high.

There is the usual warning that, even in the green areas, the map should not be regarded as a substitute for a thorough site investigation.

An unusual feature of the map sheet is the inclusion of a black and white geological map at 1:100 000 and three schematic cross-sections showing hazard zones. Two of these are reproduced here (Figure 12.10) and the lines of section are marked on the map.

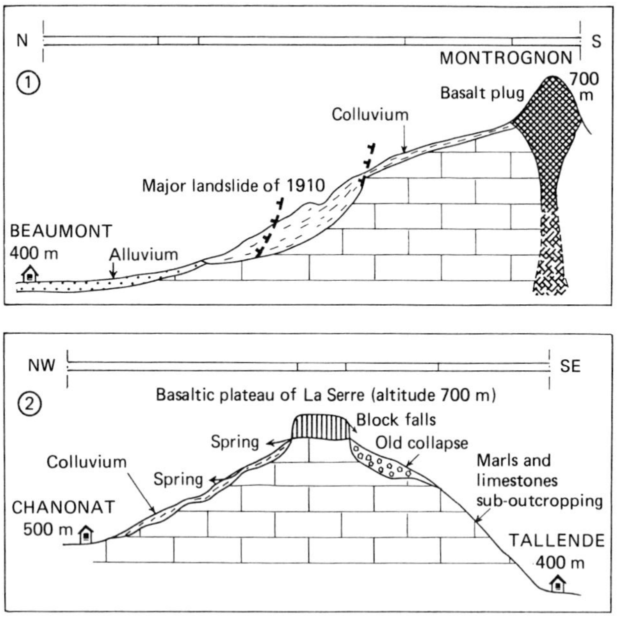

Figure 12.10 Schematic cross-sections along lines 1 and 2 on the ZERMOS map of Chanonat (see Plate 18)

Région de Pontarlier, Jura

A need became apparent to speed up the ZERMOS mapping programme, which had so far involved mapping, at 1:25 000 scale, about 30 separate map sheets, and to complete the coverage of the whole country would have taken many decades. Thus a proposal was made to map at 1:100 000, with the purpose of showing those areas that were likely to be affected by landslides, and to indicate the types of movement involved.

If the proposal was to work, it was apparent that much less checking on the ground would be possible. This in turn implied that the area to be mapped would, ideally, have a repetitive geological and geomorphological setting, and that the person making the map should have extensive knowledge of the regional geology and the stability problems likely to be encountered. The Jura region seemed likely to meet these requirements, especially as a 1:50 000 geological map was available. ZERMOS maps would be appropriate for the region – important for tourism, forestry and cattle rearing.

On the part of the map forming Figure 12.11, the background topographic features have been omitted: on the published map such details are printed in grey. Only one colour is used, light green reproduced as grey on Figure 12.11, to pick out the stable areas of modern alluvial plains with limestone or marly limestone bedrock: symbol St. The remainder of the map area is left uncoloured with three zones of marl or sandy-marl subsoil with three categories of natural slopes.

Much more restricted areas, often linear and reflecting the strike of the limestones in broad folds, are differentiated by green patterns. They comprise (see legend in Figure 12.11):

- cliffs (Ec) with associated talus accumulations (Eb);
- gorges and gullies (R) with concentrated erosion;
- karstic limestone bedrock (EF, Af, Af_1).

For each of the eleven map zones, there is a comment on the probable nature of potential movements and anticipated consequences. There are individual symbols for landslide areas, ravines, chaotic blocks and other observed movement features.

With its topographic background and detailed legend, the map can be used as a site-specific indicator of ground conditions, if due attention is paid to the limitations imposed by the small scale. The map legend is usefully supplemented by more detailed comments in the explanatory memoir. As is often the case, the point is made that there may be small areas that are unstable, for one reason or another, within areas marked on the map as stable or hazard free. This is the same, but to a lesser degree, with the 1:25 000 hazard maps. Such matters will only be resolved during the detailed investigation of a project area; the map is a guide, but not a substitute for an organized site investigation.

12.3.5 Other landslide maps

The landslide overview map of the USA has already

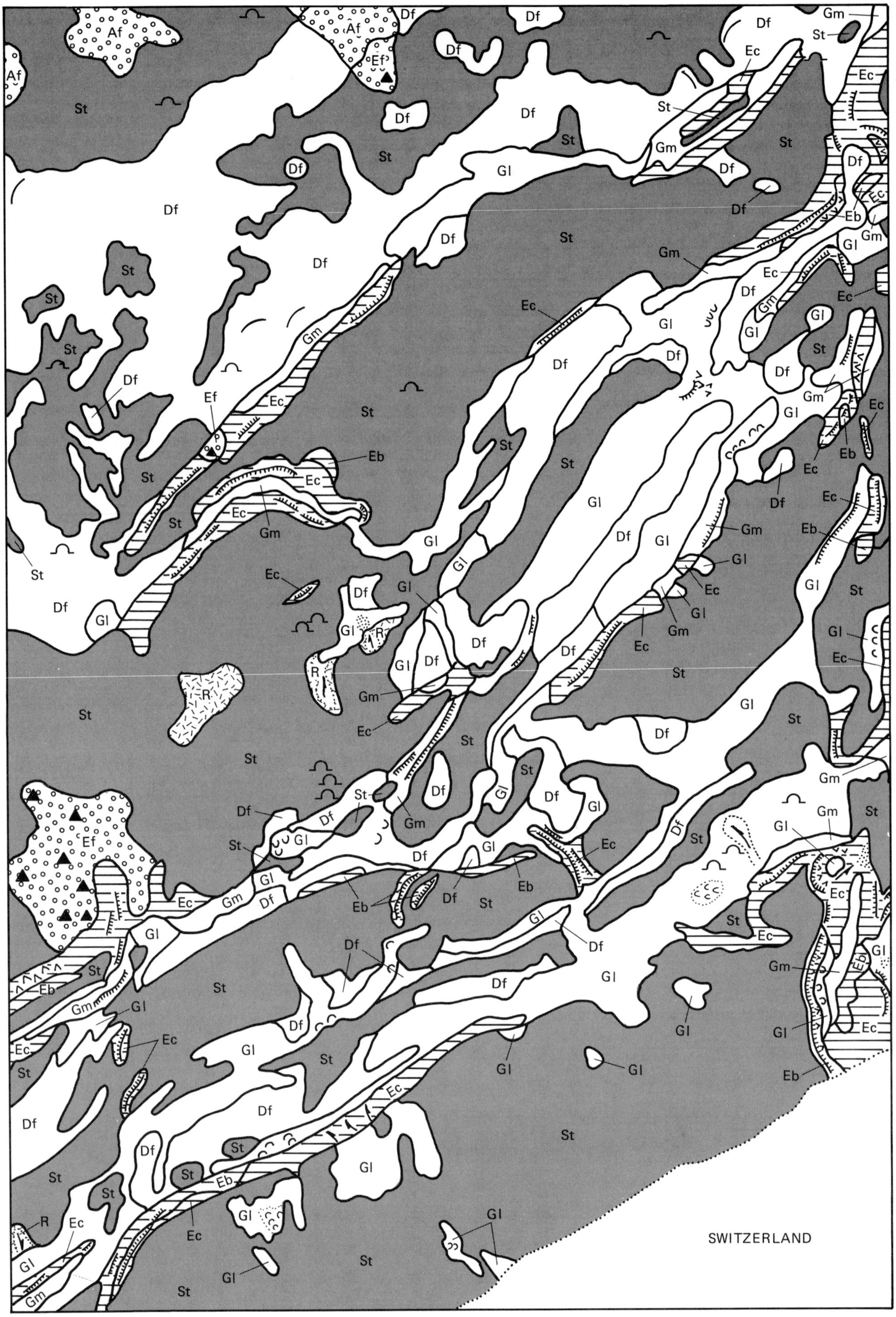
SWITZERLAND

EXPLANATION OF MAP ZONES

Symbol	Geological and morphological characteristics	Probable nature of potential movement(s). Anticipated consequences
St	Limestone or marly limestone bedrock, slightly or non-karstified, natural slopes slight or nil. Modern alluvial plains	Zone assumed to be stable, actually free from ground movements
Df	Subsoil predominantly marl or marly-sand. Natural slopes slight (0 to 10%)	Deformation slow, slight and of limited extent. Direct effects hardly visible. Slight flow and settlement
Gl	Subsoil predominantly marl or sandy-marl. Moderate natural slopes (10 to 20%)	Superficial movement affecting only weathered marl. Small extent. Direct effect: hummocky ground. Indirect effects: inclination of fences, deformation of roads, cracks in buildings
Gm	Subsoil predominantly marl or sandy-marl. Steep natural slopes (more than 20%)	Mass movement, eventually of large extent. Disruption of ground to a considerable depth. Collapse zone. Long scar
Ec	Cliffs, highly fractured with open discontinuities, or overhanging, or fissured and weathered	Collapse of cliff face, or fall of stones and blocks, collecting at the foot of the cliff
Eb	Talus, undifferentiated (old or actively accumulating), comprising blocks and stones, with or without clay matrix. Steep natural slopes (more than 20%)	Movement possible (landsliding, collapse) after heavy rain or by man's activities
R	Gorges, gullies with concentrated erosion. Steep natural slopes (more than 20%)	Gullying. Rapid removal of material from the top of the gully, and accumulation at the gully mouth (alluvial cones)
Ef	Karstic limestone bedrock with swallow holes and caves	Collapse. Marked soil settlement. Sudden appearance of swallow holes and caves
Af	Limestone bedrock with many dolines. Clay cover of variable thickness	Subsidence. Slow, progressive settlement of the soil, forming circular depressions. Effects hardly visible on the human time scale
Af_1	As Af but with thicker clay cover	
Af-Ef	Karstic limestone bedrock with thick clay cover and caves	Uncertainty on collapse and subsidence

MOVEMENT FEATURES OBSERVED

- Landslide lobes
- Slid mass (stabilized)
- Individual unstable blocks
- Ravine, localized erosion gully
- Pothole, swallow hole
- Landslide
- Chaotic blocks
- Potentially unstable cliff
- Alluvial cone
- Active sinkholes

ADDITIONAL LEGEND

- Terminal moraine
- Quarry in rock or gravel, active or disused

0 2 4 6 8 10 km

Figure 12.11 Small-scale ZERMOS map of the Région de Pontarlier (Jura), France, published at 1 : 100 000 (From Landry, 1980, by permission) been discussed in the context of California (Figure 12.5). A similar but more detailed map of France (Delaunay, 1981) at a scale of 1:1 000 000 shows areas susceptible to landslides, rock falls, subsidences and collapses, areas underlain by underground quarries, mines and saltworks, and the location by point symbols of well-known historic landslides and other ground failures. An interesting insert on the map sheet is a map showing, by 'départements', the percentage of the land surface susceptible to landslides and other ground movements (Figure 12.12). Resulting from the ZERMOS inquiry instigated by the B.R.G.M. (Champetier de Ribes, Hubert and Monition, 1974), the result, although incomplete, is a form of ground movement hazard zoning map.

An alternative approach is that by Blondeau (1976), with a synoptic map of France showing the main areas of unstable slopes with an indication of the stratigraphical age of the solid deposits involved (Figure 12.13).

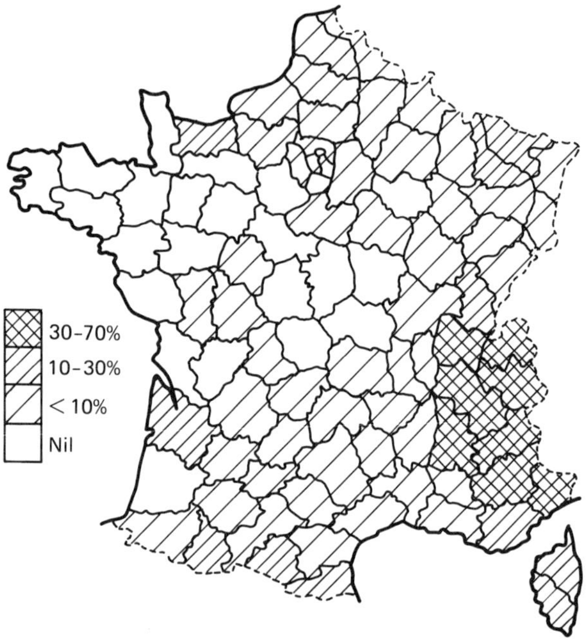

Figure 12.12 ZERMOS inventory of land surface susceptibility to landslides and other ground movements by 'départements' in France (From Delaunay, 1981, by permission)

12.4 Earthquake hazard and risk maps

Earthquake hazards, and the attendant risks, are usually related to the nature of the ground response during an earthquake. Ground shaking, for example, may provoke new landslides or mobilize old ones, or lead to liquefaction of susceptible superficial sediments. Both may damage man-made structures. The response of standing bodies of water may lead to seiches (see Figure 12.2) or tsunamis in the oceans. Earthquake maps are therefore of two kinds – those showing active fault traces and associated epicentres, and those dealing with ground effects.

12.4.1 Earthquake hazards in New Zealand

New Zealand is a seismically active area at the boundary of the Pacific and Australia–India plate. The main seismic region includes most of North Island and the northern half of South Island. Traces of the principal active faults and the epicentres of larger earthquakes since 1840 are plotted in Figure 12.14, from Dowrick (1983) and Bell (1987). Bell, reviewing the seismicity of New Zealand, quotes Eiby and Reilly (1976): 'a shock of magnitude 7 or more can be expected about once a decade, one of magnitude 6 or more once a year, and one above magnitude 5 some ten times a year.'

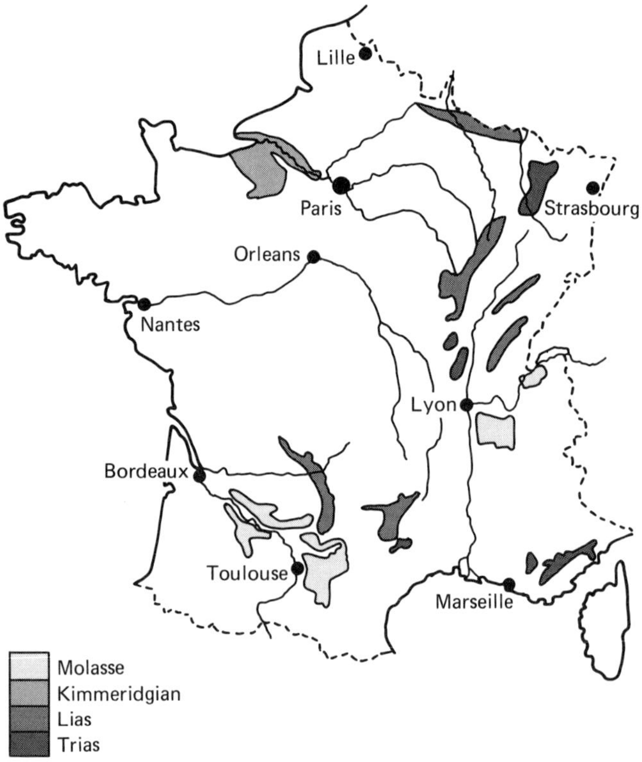

Figure 12.13 Synoptic map of France showing areas affected by natural unstable slopes (From Blondeau, 1976, Fig. 5, by permission)

A seismic hazard classification for urban areas is given in Table 12.1. Typical seismic hazard data available for urban planning are given in the contoured maps of Figure 12.14. They include a map of the observed and theoretical isoseismals for the Hawkes Bay earthquake of 2 February 1931 which largely destroyed the town of Napier (Figure 12.14b). The other two contoured maps are statistical estimations in which return periods (or probabilities) have been estimated for earthquakes of differing intensities based on historical isoseismal patterns for which there are reliable patterns since 1840.

Other examples of earthquake intensity maps

These are types of hazard maps that can be produced for any seismic area for which reliable data are available, and there are many examples in the literature.

A very detailed account of the historic ground failures in Northern California associated with earthquakes, and in particular for the 1906 San Francisco event (Youd and Hoose, 1978), provides maps of earthquake intensities (their Figs. 5 and 6) supplemented by maps of the location of ground failures.

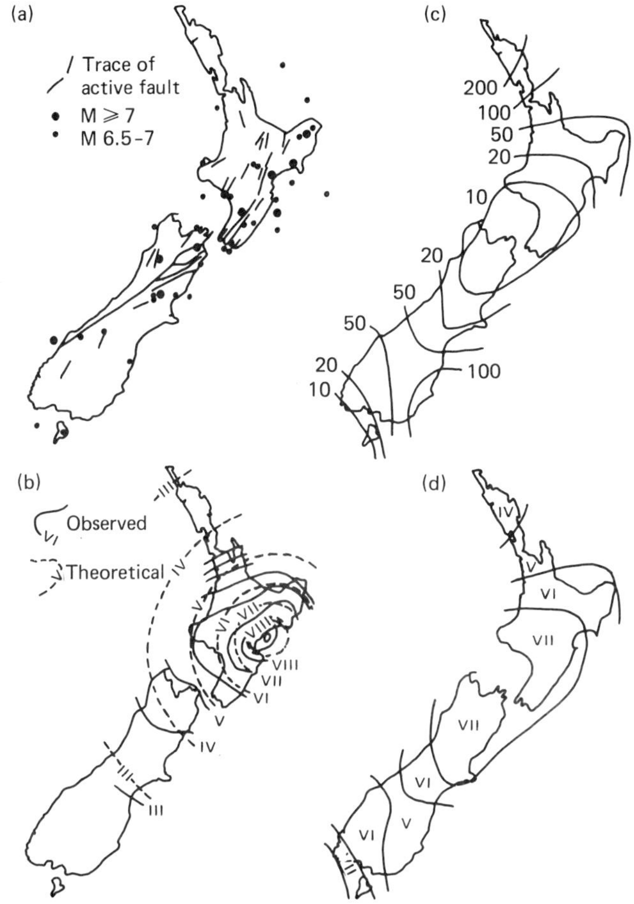

Figure 12.14 Synoptic maps of the seismicity of New Zealand. (a) Principal active faults and larger earthquakes (After Dowrick, 1983). (b) Observed and theoretical isoseismals for the Hawkes Bay earthquake of 2 February 1931 (From Smith, 1978a, by permission). (c) Return period in years for earthquake intensity of MM6 and greater (From Smith, 1978b, by permission). (d) Earthquake intensities with a return period of 50 years (From Smith, 1978b, by permission) (After Bell, 1987, Fig. 3)

Table 12.1 Seismic hazards in New Zealand urban areas

Geotechnical cause	*Scale*	*Frequency*	*Comments*
A. Ground shaking	m vl	L M	Potential for damage over large areas; aseismic design and appropriate construction codes
B. Ground displacment	s m	L M	Scarp development and associated ground warping confined to fault traces; appropriate planning codes and/or building restriction

Notes:
1. 'Scale' is represented by the area likely to be affected:
   very small (vs) $< 10^2$ m^2
   small (s) 10^2–10^4 m^2
   medium (m) 10^4–10^6 m^2
   large (l) 1–100 km^2
   very large (vl) > 100 km^2
2. 'Frequency' defined in terms of likely recurrence interval between damaging effects:
   negligible (N) > 1000 yr
   low (L) 100–1000 yr
   moderate (M) 10–100 yr
   high (H) 1–10 yr
   extreme (E) < 1 yr

12.4.2 Ground failure maps for the 1906 San Francisco earthquake, USA

At scales varying from 1:500 000 to 1:24 000 for San Francisco, five maps were prepared for the coastal zone north and south of the city affected by earthquake. The topographic maps are marked by point symbols for different failure types which include:

- hillside landslides including rotational slumps, block glides, debris avalanches and rockfalls;
- stream bank landslides including rotational slumps and soil falls;
- lateral spread;
- ground settlement;
- ground cracks not clearly associated with landslides, lateral spreads, settlement or primary fault movements;
- sand boils;
- absence of ground failure noted;
- miscellaneous effects;
- cracks in streets and pipeline breaks;
- arrows showing extent of area affected;
- symbols showing failure type;
- location number assigned to each reported ground failure type.

Five tables give specific descriptions of each ground failure noted. No attempt has been made to produce zoning maps of the different types of failure.

12.4.3 Seismic zoning and microzoning maps

Grant-Taylor *et al.* (1974), reviewing seismicity and urban planning in New Zealand, made the point that although seismologists and geologists can place broad macrozones of frequency and intensity across the country (Figure 12.14), these are not of great assistance to the architect or engineer who is designing structures against these earthquakes, or at least against the moderate ones. He must take into account the nature of the foundation rocks and soils because the interplay between earthquake vibrations and foundation materials is as important as the earthquake itself, since it controls the pattern of damage.

Earthquake microzoning is intended to determine, and display on maps, differences in ground response to earthquakes. An example from the Wellington area (Figure 12.15) recognizes three zones – basement rock, compact sediment and high porosity sediment – in which differing amounts of amplification can be expected. Also outlined are probable topographic failure zones, including areas of steep-land soils, areas of thick windblown sand and silt and areas of oversteep basement rock.

Small-scale maps of the Canadian Cordillera have been published by Eisbacher (1979) showing three seismic risk zones based on three classes of annual seismic acceleration probabilities (Whitham, 1975).

Although the probability of a major earthquake in central USA is only one-tenth of the Californian ones, and significantly less than New Zealand, the likely

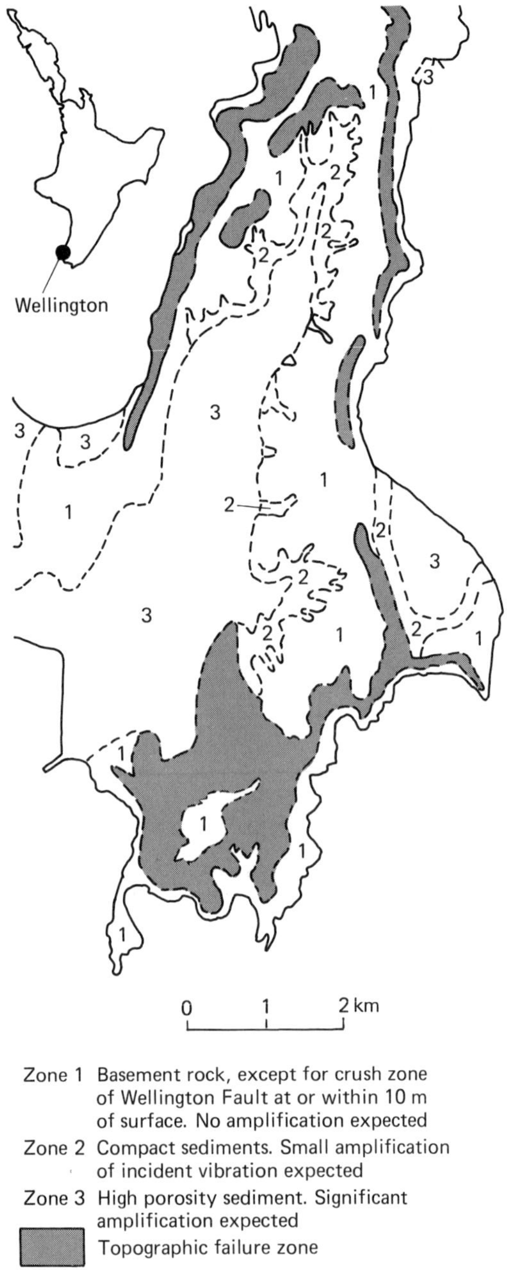

Figure 12.15 Microzoning for earthquake effects in Wellington, New Zealand (After Grant-Taylor *et al.*, 1974)

damage is expected to be 10 times greater (Sharma, 1980). The city of Memphis is situated very close to the inferred epicentre of one of the three major 1811–1812 earthquakes in the New Madrid area. Recognizing this, the seismicity of Memphis was evaluated in terms of both the maximum credible earthquake and design earthquakes of lower intensity and lower recurrence rates. Response analysis of the soils at a depth of 45 m was transformed into microzonation maps showing, among other aspects, contours of natural frequency of soils (Sharma's Fig. 7.7), and areas where soils may be susceptible to liquefaction from earthquakes with a Modified Mercalli Intensity greater than VII (Sharma's Fig. 7.11).

Within 322 km of the New Madrid fault zone is the Creve Coeur quadrangle in St Louis County, Missouri, already used as an example of unitized mapping in Chapter 7. Thick lacustrine and alluvial deposits along the Mississippi and Missouri rivers are of the type which are highly susceptible to ground failure by liquefaction. Seismic hazard mapping, based on geology, seismicity and dynamic response of the soils, used computer assessment of assumed ground motion parameters digitized to form the basic data files, including the maximum horizontal acceleration response, the velocity response spectrum, the horizontal accelerations that would cause liquefaction of deposits – yield accelerations – and, incidentally, yield accelerations that would cause slope failure. The relevant magnitude of the predominant condition was considered for each 122 × 122 m cell over the map area.

The data images created for the field acceleration data for liquefaction and for slope failure were superimposed with the maximum horizontal acceleration response data image on the screen. Where accelerations equalled or exceeded those calculated to cause, in the case of Figure 12.16, liquefaction, the area was delineated as having that hazard. The computer-generated map is compiled from the cell-squares on a yes–no basis, and as such is a very simple map. If sufficient geotechnical data are available, a next step would be to map in terms of relative liquefaction susceptibility, and this has been done by computer analysis for San Mateo County, California.

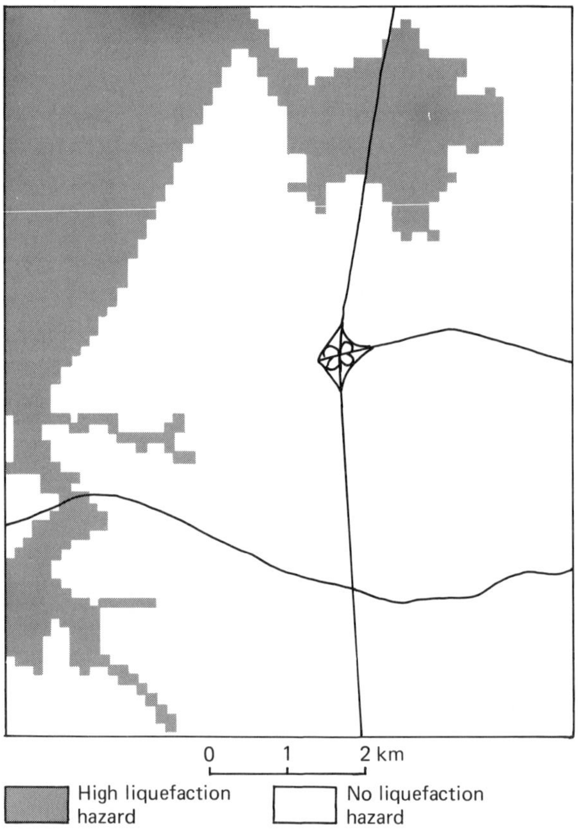

Figure 12.16 Computer-generated liquefaction hazards map for a New Madrid Region earthquake, Creve Coeur Quadrangle, Missouri, USA (After Higgins and Rockaway, 1986, Fig. 8)

12.4.4 Map of liquefaction susceptibility of San Mateo County, California, USA

Available geological maps by Lajoie *et al.* (1974) and

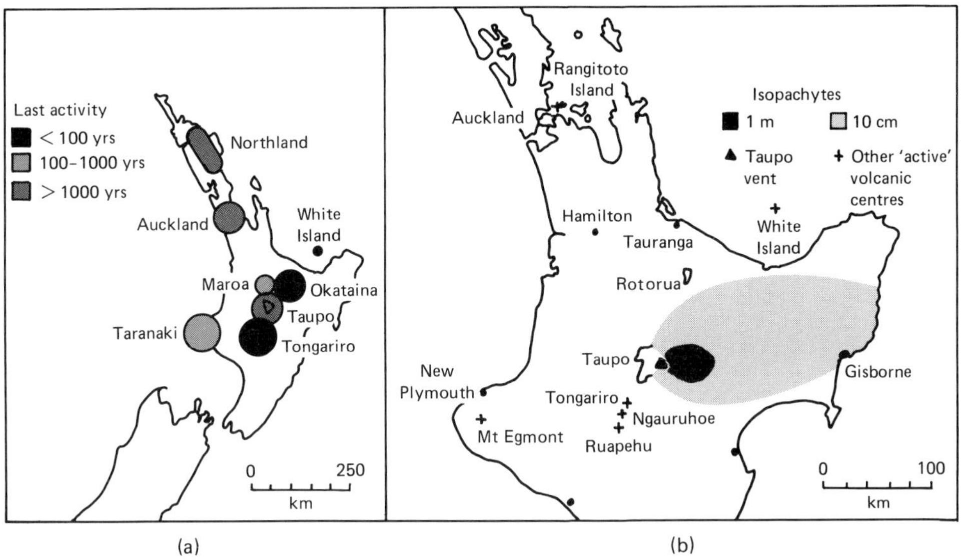

Figure 12.17 (a) Volcanic hazard areas in New Zealand; (b) Taupo pumice eruption (1800 years BP) (From Bell, 1987, by permission)

Brabb and Pampeyan (1983), which have been digitized with data assigned to 1 ha cells by the staff of the Association of Bay Area Governments (ABAG), were used to estimate the probability that sand or silt layers might lie beneath any point on the map on the basis of sedimentary materials mapped at that locality and bore-hole logs.

Three factors that control liquefaction susceptibility are (a) lithology – the presence of sand or silt layers containing less than 15% clay; (b) physical properties of sand or silt – the presence of loosely packed materials; and (c) depth to free groundwater – liquefaction occurs only in saturated or nearly saturated materials. For each map unit, the authors (Youd and Perkins, 1987) estimated the probability that each of these three factors was favourable for the development of liquefaction. The probability of liquefiable material lying beneath any point on the map was calculated by taking the product of the three component probabilities. This value was used to quantify and map the distribution of the seven liquefaction susceptibility units (Plate 19).

The map shows susceptibility to liquefaction rather than susceptibility to ground failure, which is the actual damage-producing consequence of liquefaction. Under some conditions, liquefaction can occur without generating ground failure, but the map still serves as a general guide to ground-failure susceptibility.

12.5 Volcanic hazard and risk maps

Mapping of active volcanic areas is similar to mapping seismic hazards inasmuch as the loci of volcanic eruptions can be plotted in terms of last recorded activity (Figure 12.17). In addition, it is possible to record the extent and thickness variations in the ash deposits from a known historic eruption. If future events parallel those in the past, it is then possible to predict the extent of volcanic hazards in the future, but it is quite impossible to put a time scale on those events. The past is the key to the future indefinite, and investigators such as Westercamp (1987) have attempted a zonation of volcanic hazards associated with Mt. Pelee. Analysis of four types of data are involved:

- the known eruptive history of the volcano during the past 5000 years;
- the meteorological data, influencing ash distribution and the incidence of mud flows;
- present morphology of the volcano and the barrier effect;
- the progress of certain eruptions including those on Mt. Pelee.

Westercamp produced five volcanic hazard maps for the mountain.

12.6 Maps of other geological hazards

A significant group of hazards is associated with extraction of subsurface resources: coal, water, gas and oil, and other mineral deposits. Surface effects may be represented by maps (Larson and Pewe, 1986).

12.7 Risk maps

Risk, as defined earlier, is the expected degree of loss due to a particular natural phenomenon. Although the concept of risk is often confused with hazard, none of the examples used so far attempts to assess 'risk' as such. Maps are rarely produced, although in the USA (Brabb, 1984) a substantial project to assess the extent of losses associated with landslides, earthquakes, expansive soils, floods and other natural hazards has recently been completed (Wiggins, Slosson and Krohn, 1978; Petak, Atkisson and Gleye, 1978; Legg, Eguchi and Wiggins, 1980; Petak and Atkisson, 1982). One of the maps by Wiggins, Slosson and Krohn (1978) is reproduced here as Figure 12.5.

True risk maps are uncommon, but one by Mears (1977) comes close to fulfilling basic requirements. The map (Figure 12.18) is of an area in Colorado where debris-flow basins debouch onto urbanized lower

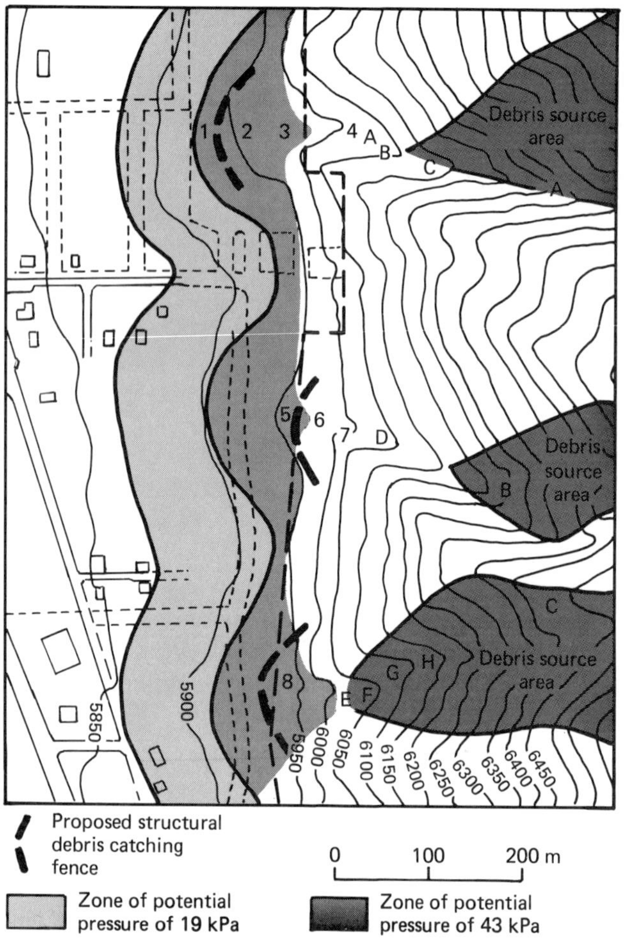

Figure 12.18 Map of Glenwood Springs, Colorado, USA, showing debris-flow drainage basins; debris source areas; boulder-sized sample localities, numbered; and locations marked by letters used in calculating velocities and discharges, debris-flow impact-pressure zones, and location of suggested debris-catching structure fences (From Mears, 1977, by permission)

ground. A debris-flow velocity of 4.5 m/sec was assumed for the area from 0 to 90 m below the mouth of the canyons, and a lower velocity of 3 m/sec from 90 to 180 m. Using these velocities and an assumed unit weight of 2 g/cm^3 for the debris flow, two zones of potential impact pressure were drawn on the map. Rainfall statistics and information on historic debris flows suggest a return interval of 50 years, or a 2% chance each year. The cost of catchment structures to arrest or slow down the debris flows would be about 10 000 US dollars per year, well below the average annual cost of 40 000 US dollars calculated for damage caused by a single 1977 event.

Thus the map and report indicate the probability that a debris-flow event will occur, the area that is likely to be affected, the impact pressures of boulders and other debris in these areas, and a cost analysis indicating that favourable cost benefits would arise from constructing the catchment structures. Missing is an assessment of the vulnerability of people and structures within the areas if a debris flow does occur.

13

Land and water management: environmental geology mapping

13.1 Introduction

As population increases and urban settlements continue to expand, there has been increased pressure on the use of land and the Earth's resources. Man's activities have, in the main, a deleterious effect on natural systems and the need to overcome problems that arise, or may arise in the future, in order to mitigate lasting damage has led to the development of environmental geology.

Environmental geology deals with the interrelationships of geological processes, the Earth's resources, and the impact of man on these, offering a tool for planning a more harmonious use of the environment. As a branch of geology, environmental geology involves a more intra- and interdisciplinary approach, calling on studies in engineering geology, hydrogeology, economic geology and the land surface as a logical extension of more conventional geological studies.

A plea is always made that the geological and other data should be presented in a form readily understandable and usable by those who are unfamiliar with geological concepts and lack a working knowledge of the subject. Environmental geology maps, as a logical extension of engineering geology maps, are directed specifically towards planners, administrators, and the interested layman – in fact all those making, or concerned about, decisions on land-use.

Internationally there is no general agreement on what topics should be included on environmental geology maps, although engineering properties of the ground, slope stability, hydrogeology, waste disposal and mineral resources are usually included. Investigations may be presented as single factor maps, but at some stage selected aspects must be interpreted into single maps from which conflicts may be resolved and specific recommendations for land-use made. Examples are given later focusing attention on the application of environmental geology maps to regional planning and to more localized planning of extensions to existing urban areas. Maps are made for a specific *purpose* and thus have a clearly defined *content*. A medium *scale* seems to be most appropriate for planning purposes, depending on the nature of the project. Contrast the requirements for a major transportation corridor with urban housing extensions or the use of sites for major industrial developments.

13.2 Environmental geology mapping in the UK

The first in a series of investigations into environmental geological mapping, sponsored by the Department of the Environment, was published in 1982. Environmental geology maps have been produced by engineering geologists and general earth scientists to meet the perceived need to present earth science data in a form that can readily be understood and used either by the planner or by the engineer at the feasibility stage of a project. British experiments in environmental geology mapping have largely remained unpublished, but recently a number of papers summarizing the results of thematic geological mapping have appeared (Bell, 1987). It is sensible to treat the publications in chronological order so that the evolution of ideas can be appreciated.

First, however, it is pertinent to recall the comments of Brook and Marker (1987) on the use of geology in planning in the UK. Their paper presents a fair and balanced critique of the sometimes conflicting attitudes of professional geologists and planners. Relatively few planners are aware of the significance of geological factors in their day-to-day tasks. The potential for the use of geological information can be highlighted by the excellent joint efforts in areas with special problems such as the subsidence of abandoned limestone mines in parts of the West Midlands of England. In areas with less spectacular but still costly problems, it can be rather more difficult to promote the value of adequate geological information.

Turning to applied earth scientists, whether they are geologists, engineering geologists or other specialized geologists, it is equally the case that their professional languages are as opaque to planners as they are in many cases between the individual disciplines. Similarly, planning terminology is often not fully understood by earth scientists.

Applied earth scientists, including engineering geologists, mainly provide information through:

- the results of scientific research;
- basic geological, geomorphological, soil and other mapping on a regional or smaller scale;
- reports on site investigation and evaluation for specific development proposals.

A planner with no geological expertise may not be aware that geological conditions might be such as to prevent the safe implementation of a planning permission.

For some years effort has been diverted to investigate the best means of collecting, collating, interpreting and presenting, in sets of maps and reports, geological results of direct use to land-use planning. Hence the development of thematic geological mapping.

13.2.1 Glenrothes district, Fife Region, Scotland

This is the first published example of environmental geology mapping at a scale of 1:25 000 in the UK (Nickless, 1982). In early 1980 the Institute of Geological Sciences (IGS), as it then was, participated in a pilot study to assess the applicability of environmental geology mapping to land-use plan-making in the UK. The study area included the rapidily expanding town of Glenrothes where coal was mined until the 1930s (Figure 10.21).

For most of lowland Britain, particularly the industrial and urbanized areas, a wealth of geological data is usually available. Glenrothes was no exception and for the topographic 1:25 000 sheet NO 20 there were:

- comprehensive geological mapping at 1:10 560 scale from 1858 to 1867, 1926 to 1946 and 1966 to 1969;
- borehole and mining data in the IGS archive;
- a geological survey of the drift deposits in 1979–1980;
- numerous site investigation reports for the new-town of Glenrothes;
- for the rural area north of Glenrothes town a wide scatter of data points arising from an assessment of sand and gravel resources in 1980;
- the results of a hydrogeological investigation in 1980.

The Glenrothes district, as well as having this wealth of information, was also selected because

- it has a varied solid geology including workable coals and potential resources of limestone and hard rock aggregate, permitting an extensive range of conditions to be studied;
- it may be regarded as representative of much of Central Scotland, both in the complexity of its geology and the uneven spread of information;
- the availability of sand and gravel and hydrogeological information allowed an assessment of the way in which data are presented in the reports and maps of these newer specialist surveys as well as in the traditional geological map.

In summary, the study area offered the opportunity to explore the potential of the environmental geology mapping method fully and to experiment in an area unhampered by current disputes on land-use.

The aim of the study has been to collate, interpret and present available geological data in a simple cartographic form, each aspect of the geology being a separate theme on an element or basic map. No new information has been collected, and one outcome of the survey was expected to be revelation of the deficiencies of the existing data base to provide answers to specialized enquiries, thus indicating the scope and scale of future revisions of the geological maps.

For Glenrothes (Table 13.1) the basic data are contained in 18 environment maps. This large number arose from an attempt to demonstrate the versatility of the techniques used to show how the same information could be presented in different ways. A greater degree of selectivity was likely to be applied to any future surveys.

Table 13.1 Environment, derived and environmental potential maps compiled for sheet NO 20, Glenrothes, Scotland

ENVIRONMENT MAPS (basic data)	
1.	Boresites
2.	Unconsolidated deposits
3.	Lithology of unconsolidated deposits
4.	Engineering properties of unconsolidated deposits
5.	Thickness of unconsolidated deposits
6.	Depth to water in unconsolidated deposits
7.	Sand and gravel thickness
8.	Bedrock geology
9.	Bedrock lithology
10.	Rockhead contours
11.	Shallow undermining
12.	Natural landslip potential
13.	Open-cast workings
14.	Hard rock aggregate resources
15.	Limestone resources
16.	Brick and tile clay
17.	Mudstone for brick making
18.	Hydrogeology
DERIVED MAPS (combining two or more elements)	
19.	Underground storage potential within 100 m of surface
20.	Sand and gravel potential
21.	Foundation conditions
22.	Groundwater resources
ENVIRONMENTAL POTENTIAL MAPS (summary maps based on element and derived maps)	
A.	Development potential
B.	Priority areas for on-site investigation
C_1.	Resources at or near surface which might be won by open-cast working
C_2.	Buried resources which might be won by open-cast working
C_3.	Buried resources which might be won by pumping or mining

A demonstration of the potential of the system in response to hypothetical requirements is given in the four 'derived' maps that have been prepared; for example, the potential for underground storage.

For the Glenrothes experiment, three main aspects were subject to special attention and form the Environmental Potential or summary maps: (a) conditions favourable to building and, (b) conversely, priority areas for site investigation; (c) resources including water.

In the event, to maintain clarity of presentation it was not possible to summarize resources on one sheet and as a result three have been drawn. Resources at or near the surface which might be won by open-cast working are the subject of the first resource map; buried resources which might be taken by open-cast working constitute the second; the third shows buried resources which might be exploited by pumping or mining.

The Environmental Potential maps draw on the

appropriate element maps and those used are listed in the sheet explanation. These summary maps do not rule out different and perhaps more specialized use of the individual element maps, nor preclude the preparation of different analyses.

For most maps the working scale has been 1:25 000, exceptions being thickness of the unconsolidated deposits (Sheet 5) and rockhead contours (Sheet 10) which were initially prepared at 1:10 000 scale. Part of the analysis for the shallow undermining map (Sheet 11) was also conducted at the larger scale, as, of course, was the geological survey on which many of the maps are based.

Environment Map, Sheet 14, Hard Rock Aggregate Resources, has been selected for illustration (Figure 13.1).

Of the various solid rocks in the Glenrothes district, only the quartz-dolerite, teschenite and basanite intrusions have a foreseeable use as hard rock aggregate and as such are judged a resource: the reasons for discounting the other rocks are given in the map legend.

Because the rock would be worked from open pits, ornament is used on the map to distinguish the proximity of the resource to the surface: the presence and nature of any overburden is also shown together with relevant borehole data.

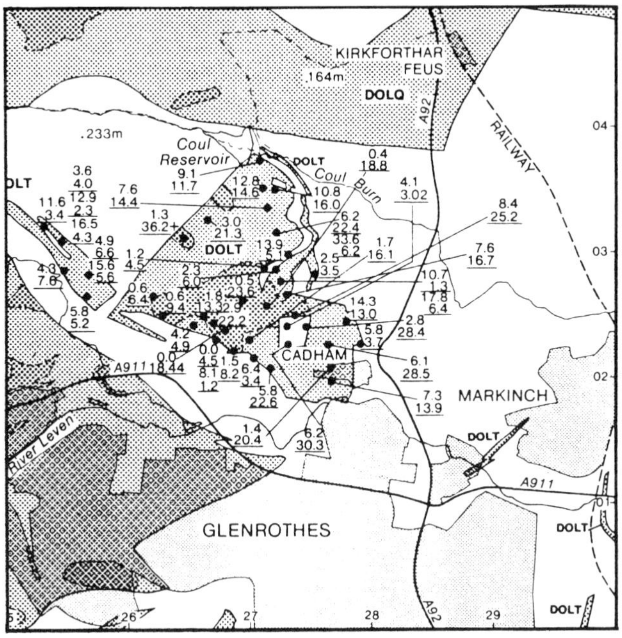

ENVIRONMENTAL GEOLOGY MAP (GLENROTHES) SHEET 14
HARD ROCK AGGREGATE RESOURCES

Resource at or near surface

Resource concealed by unconsolidated deposits

Resource overlain by rock

Resource generally absent

——— Boundary of hard rock aggregate resource

- - - - - Boundary of unconsolidated deposits overlying rock

	Rock	Comments
DOLQ	Quartz-dolerite	Medium to coarse grained, moderately jointed, horizontal or sub-horizontal intrusion
DOLT	Teschenite	Medium to fine grained, moderately to well jointed, horizontal or sub-horizontal intrusion
BASA	Basanite	Fine grained with scattered coarse grained crystals, moderately to well jointed, vertical or steep-sided intrusion

Borehole log

• bore site
3.6 overburden
18.2+ hard rock aggregate

Note
(i) the figures are thicknesses in metres
(ii) the + sign indicates that the full thickness was not proved
(iii) thicknesses underlined are hard rock aggregate: other figures refer to overburden

Comment

Within the sheet, quartz-dolerite, teschenite and basanite comprise the hard rock aggregate resource. Quartz-dolerite is extensively used as crushed rock aggregate and for road dressing, its mechanical and physical properties being superior to many other rocks. Teschenite and basanite are likely to have similar properties and end-uses. Additionally, the rocks might be worked locally for blockstone.

Sandstones, although present over large parts of the area are considered generally unsuitable for use as aggregate. Limestones which do not usually average more than 3m in thickness are most unlikely to be worked for aggregate and have been discounted.

Figure 13.1 Example of a thematic map from Glenrothes, Scotland, UK. Part of Sheet 14, scale 1:25 000, showing Hardrock Aggregate Resources and the legend (From Nickless, 1982, Fig. 16, by permission)

Criticism can be made of the retention of unusual geological names for the rock types shown on the map. Surely it would have been preferable to use the name of a 'Trade Group' (Collis and Fox, 1985, p. 78) for an aggregate resource or to select a single simple name, e.g. one advocated in Table 4.1, if the physical and mechanical properties of the rocks make them all equally suitable for the proposed purpose. An engineering geologist, and perhaps even an informed planner, would know what a dolerite is, but there is only a remote possibility of 'basanite' or 'teschenite' raising even a flicker of recognition. It is clear in this respect that the map has been compiled by geologists without the end-user always firmly in mind.

13.2.2 *Thematic maps of Glasgow, Scotland*

In Glasgow, another coal-mining area, there are additional problems associated with the large-scale underground working of sandstones for building purposes. As overburden at the quarries became too thick to make quarrying economic, so the workings were extended underground using the room-and-pillar method.

Problems now arise because the early quarries have been infilled, and the location and even knowledge of the former presence of underground extensions may not be known.

Glasgow affords another well-documented example of the spread of urban development (Price, 1971). Earliest building, north and south of the River Clyde, took place before 1773 on the relatively low-lying areas of alluvium and river terraces adjacent to the river (Figure 13.2). Later expansion was still along the river bank, with incursions into the drumlinized terrain beyond. The pattern of development reflects the drumlin array, with roads winding through the drumlins, and housing following the contours around them.

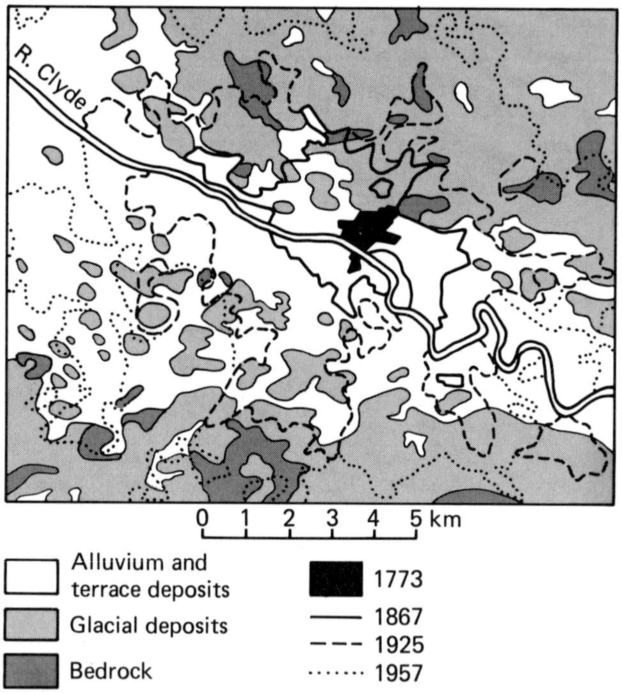

Figure 13.2 Spread of urban development in Glasgow, Scotland, UK, from 1773 to 1957, and the underlying geological features (From Price, 1971, Fig. 1, by permission)

Expansion of Glasgow over both subsurface and surface workings had led to severe engineering and related planning problems, and at a time of redevelopment and reconstruction of the inner city area, there was a need for up-to-date geological information and interpretation for future planning. The British Geological Survey was commissioned by the Department of the Environment to undertake a city-wide environmental geology assessment (Forsythe *et al.*, 1983, 1984; Browne, Forsythe and McMillan, 1986).

Geological input to the study took account of the existence and location of relevant records (mine plans, borehole logs and other archival data); the sites of quarries and their backfill; the subsurface level and extent of working of the various minerals; the nature, disposition and geotechnical properties of both superficial deposits and solid rocks; and the groundwater regime. To assist in the manipulation of existing data, a computerized borehole data base was created and is operated by an appropriate data base management system (McMillan, Browne and Robson, 1984).

As at Glenrothes, it was intended from the outset that the results of the study would be presented in map form. One example, from the 17 presented, is illustrated in Figure 13.3, a map showing mining within 30 m of rockhead. All of the maps are thematic, and no derived or summary maps were produced. From the distribution of sales of the maps to engineering consultants in the private sector, and distribution to the council's planners, it was concluded that the maps most useful to planners are: contoured thickness of superficial deposits; lithology of the superficial deposits; solid geology; 'shallow' versus deep mining; and generalized engineering properties.

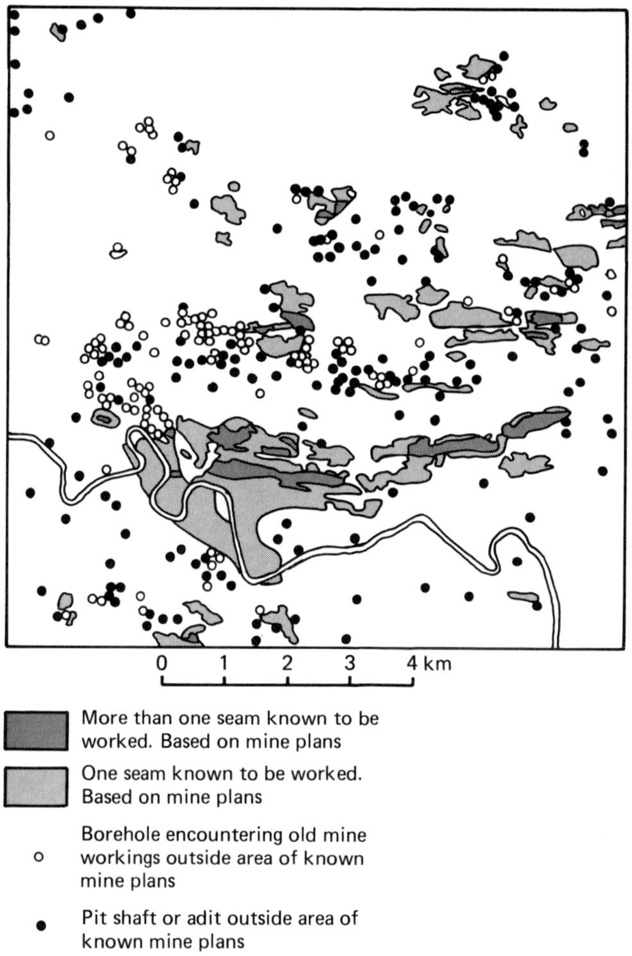

Figure 13.3 Thematic map of part of Glasgow, Scotland, UK, showing mining within 30 m of rockhead (From Browne, Forsyth and McMillan, 1986, Fig. 11, by permission)

However, it is possible that the thematic facets of particular interest to those concerned with Glasgow's problems could be summarized in one map. This might show four categories of ground concerned with the presence or absence of soft or loose superficial deposits, combined with the presence or absence of mine workings within 30 m of rockhead. These subdivisions have implications for development within each area and will vary from one extreme of thin drift on unmined solid rocks, to the other extreme of thick soft drift above shallow mining. In planning terms, such a map is about cost-effective options and not hazard assessment. Its use, therefore, might be appropriate at the strategic rather than local planning level. For an engineering assessment of ground stability, however, such a summary would be too generalized (Browne, Forsyth and McMillan, 1986, p. 520).

13.2.3 Other thematic environmental geology maps

Apart from Milton Keynes (see Figure 10.23), where Cratchley and Denness (1972) pointed out the limited use of engineering geological evaluation in both the general and detailed siting of existing new towns, and particularly the late stage at which it was started at Milton Keynes, the more favourable trend initiated at Glenrothes and Glasgow is continuing. A number of urban and urban fringe studies have been commissioned, and the results are now being published or their existence at least reported in the scientific literature.

A very detailed study of engineering conditions in the soft sediments of the upper Forth Estuary, and problems associated with the effects of past mining, was presented as a general zoning map of the suitability of the ground in the region for foundations (Gostelow and Browne, 1981, 1986, Sheet 8). Floyd *et al.* (1983) have also reported on environmental geology maps of south-east Edinburgh.

In Britain, environmental geology, reviewed in the context of British Geological Survey activities in this field by Monro and Hull (1986), is of practical use to planners and engineers concerned both with sensible conservation and the development of land. Ground conditions of particular concern in relation to land-use are:

- ground conditions for general development;
- geological hazards to development (see Chapter 12);
- ground resources and their planning implications.

Hazards can include:

- the presence of geotechnically soft and highly compressible superficial deposits;
- the existence of mine workings close to rockhead;
- the potential for the pollution of groundwater;
- natural seismicity.

For the Glenrothes experiment (section 13.2.1), three aspects were subjected to special attention and form so-called environmental potential or summary maps:

- conditions favourable to development, but excluding heavy industrial structures;
- priority areas for site investigation in areas of proposed development;
- mineral resources, including water.

In 1986, some 25 environmental geology mapping projects were in various stages of completion, with 12 topics for mapping, not all being applied in every area. Many of the areas, and especially the nine mapping areas in Scotland, are associated with problems of ground stability where there has been multi-seam workings in coal and ironstone over many centuries. Following approval by the customer, map portfolios have been released on open file. Other users may not be adequately aware of the limitations of geological maps, or of the various thematic maps at 1:10 000 or 1:10 560 scale. Perhaps the potentially most acute difficulties arising from 'misuse' have been stated by McMillan and Browne (1987) to be:

1. In cases where multiple thematic maps are available for the same area, there is a tendency to regard the shallow mining map as the most important and to use it without reference to others in the portfolio, such as bedrock geology. The moral here is that too many maps may be as bad as too few.
2. A crucial aspect of such suites of thematic maps is the use to which they may be put by finance houses and building societies. There is a genuine concern that conclusions will be drawn on ground stability, especially in urban areas where properties which have been adequately constructed to take into account ground conditions as known at the time are 'blighted' because it is later revealed that they are situated in undermined areas. This last aspect illustrates how carefully the geologist must tread when preparing maps of 'inferred' or 'suspected' mining. The best attempt must be made to prepare maps which cannot be misconstrued. It may be that a 'derived' map containing a summary drift and solid conditions, undermining and quarrying (Figure 13.4) is as useful as several contributory 'element' maps which, when taken in isolation and out of context, can be misleading.

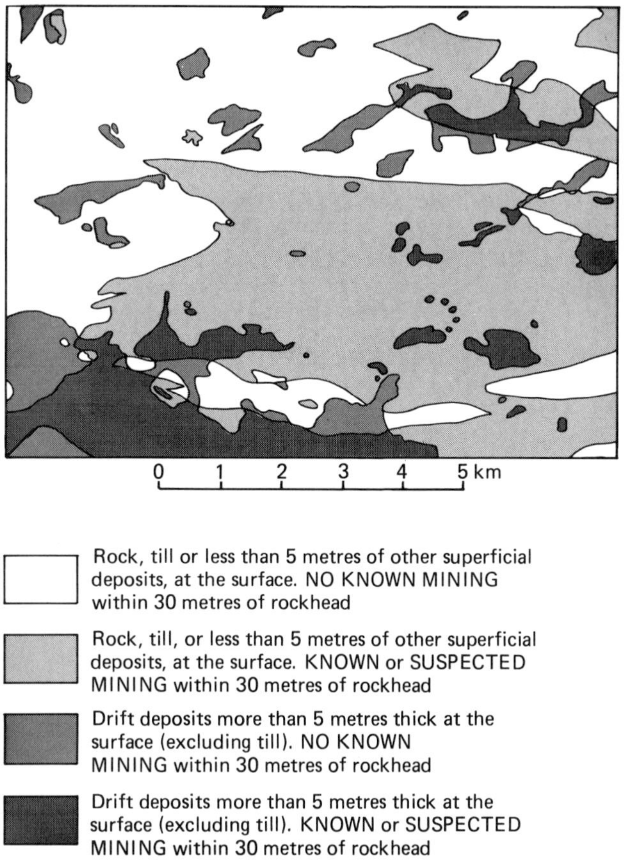

Figure 13.4 Hypothetical planning map of an area underlain by mine workings (From McMillan and Browne, 1987, Fig. 8, by permission)

13.3 Environmental geology maps in the USA

Developments in Britain paralleled those in North America where the emphasis had for long been on the production of folios of maps for use by professionals in disciplines other than geology. A similar general definition of environmental geology has been adopted which is the application of geological knowledge to the protection of human health and safety, to preserve the quality of the human environment and to facilitate the wise use of the land and its resources (McKelvey, in Tank, 1983). A practical definition, of more direct application to mapping, is the application of geological information to environmental problems (Kurston, 1982). In the USA there has been intense activity on the part of State Geological Surveys and the U.S. Geological Survey in the design of innovative maps and reports to communicate the results to users (Nichols, 1982).

A review of early work in North America (Turner and Coffman, 1973) suggests that an environmental geology map should show at least six classes of information; these are:

1. Physical properties of the soil and rock units, including seismic stability, potential as building materials and acceptability for waste disposal.
2. Slope stability, including potential for landslides, rockfalls, soil creep and avalanche.
3. Presence of active or possibly active faults, and epicentres of historical earthquakes.
4. Contours of the groundwater table or the depth to groundwater contours.
5. Thickness of the soil or depth to bedrock.
6. Extent of any floodplains present.

Some of these have already been considered in the context of the hazards they present (Chapter 12).

Only a small number of examples from the USA are given in the following.

13.3.1 Preliminary environmental map of Greater Los Angeles, California

An early example of a preliminary geological environmental map was compiled at a scale of 1:250 000 for southern California in 1970 (Wentworth, Ziony and Buchanan, 1970). Fault movement, seismic shaking and certain other geological processes, it was considered, might seriously affect nuclear reactors whose safe and reliable operation is critical to the public interest. Information scattered through the literature was presented '... to permit early identification of possible severe problems that warrant careful investigation for the siting of critical structures', as well as being useful for regional land-use planning. An interesting feature from the geological–cartographic point of view is the treatment of faults, including location, the age of most recent known or inferred displacement, and limiting stratigraphical control for that age. Additionally, selected basement faults, the location and amount of known historic surface subsidence, large mapped landslides and distribution of various rock and sediment units are also shown.

The map was compiled from available geological information, supplemented by limited field checking by the authors, after completion. Because knowledge of the geology is invariably incomplete, a reliability diagram and various line symbols for faults are included to indicate the degree of uncertainty of the information used.

This map, and the following one from Oregon, indicate how difficult it is to recommend standard symbols for use on engineering–environmental geological maps. All that can be suggested are certain minimum or desirable usages, but new ways to show special data should always be sought outside this recommended framework. The variety of maps used as illustration in this present publication will give some idea of the breadth and variety of approaches that have been taken.

13.3.2 Oregon water pollution problems

Several interdisciplinary environmental studies have been made in the State of Oregon by teams of geologists, biologists and engineers. These studies, which were primarily for water pollution control, involved geological and biological studies of water basins in different physiographic–climatic regions. In one of them (Rickert *et al.*, 1978a,b), geological factors were evaluated for areas that were classified according to whether erosion or landsliding is the predominant slope process. Biologists then evaluated the effects of erosion and landsliding on the channel properties and the fish population of streams. Both the geological and biological data were plotted on a single map. Fifteen terrane types were thus identified, and the effects of various land-use activities, including engineering construction, that would cause increased disturbance to the natural environment in each of these terranes were shown in a matrix.

A state-wide study was also made that covered six major water pollution problems (Rickert *et al.*, 1978a). The distribution of these problems, which included streambank erosion, sedimentation, excessive debris, water withdrawals causing stream-quality problems, elevated water temperatures, and nuisance algae or aquatic plant growths, was shown on maps. A composite problem map was prepared in which each stream segment or water body was assigned one point for each moderate problem and two points for each severe problem. The individual points were summed and the segments were then colour coded. Segments with 9–12 points were red; 7–8 points purple; 4–6 points orange; 1–3 points green. A small part of the map is reproduced in Figure 13.5, with patterns used instead of colours.

13.3.3 Alaska gas pipeline

A multidisciplinary environmental impact study that was made for a proposed gas pipeline across northern Alaska (E.I.S. Task Force, 1976) contained detailed discussions of the effects of geological factors on construction, and the effects of construction on the geological environment as well as related effects on other aspects of the environment. Possible related effects included damage to vegetation and wildlife if ground water is contaminated by methanol used for pipeline testing, or water for construction is withdrawn from streams necessary for fish-spawning and survival. Only small-scale, generalized maps of geological factors were shown in this report. A report on the extension of this same proposed gas pipeline across Canada (Environment Protection Board, 1974) showed on maps various engineering geological conditions (ice-wedge polygons, riverbank

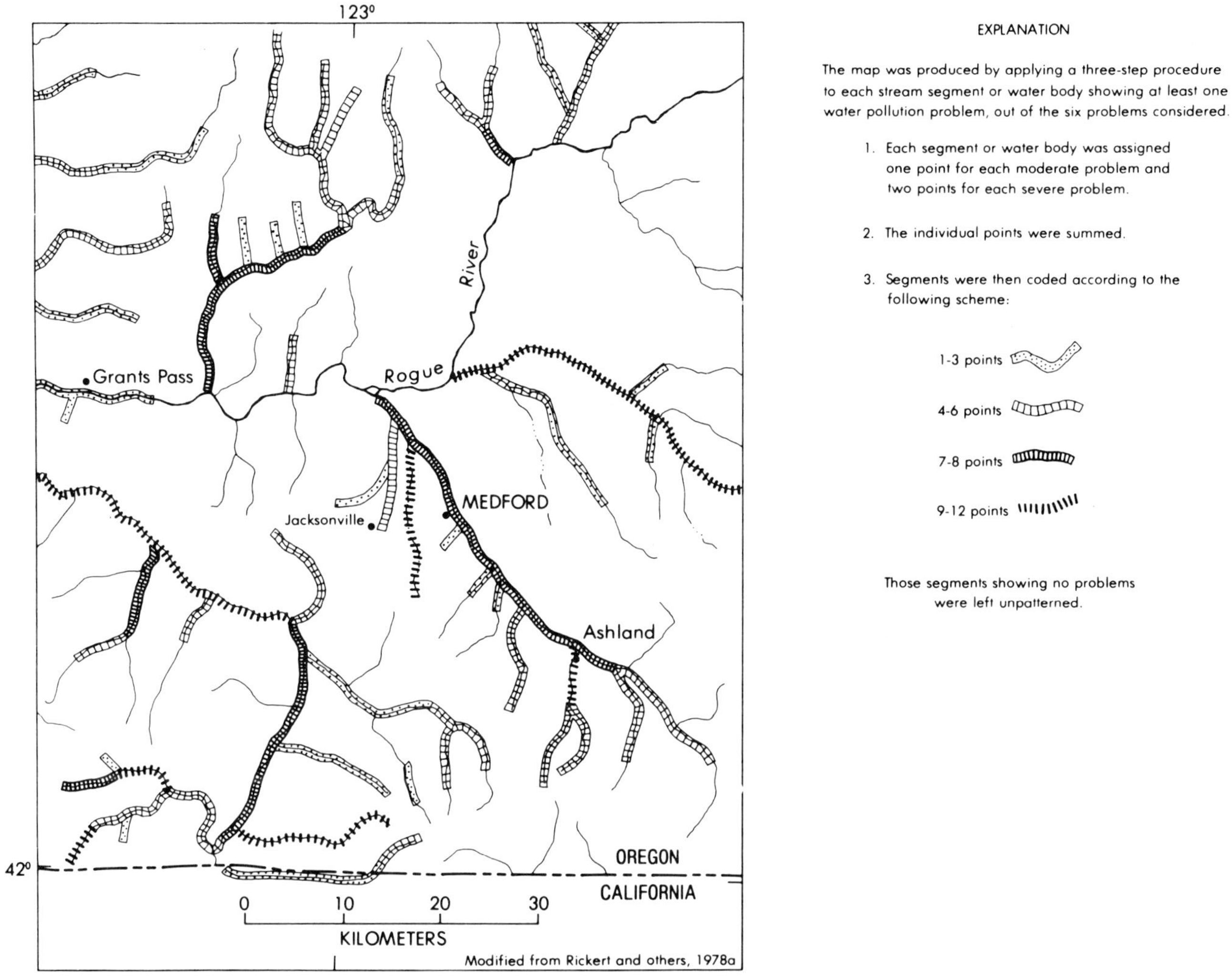

Figure 13.5 Composite map showing summation of water pollution problems in a part of south-western Oregon, USA

slumps or thaw bank erosion, solifluction) as well as vegetation and wildlife factors, and made recommendations for construction practices that would avoid damage to the total environment. General recommendations were made in a written text, and specific site problems were shown graphically by means of symbols and diagrams on maps.

Extensive studies of the possible engineering and environmental problems related to geology and other aspects of the environment were made before the Alaska pipeline was constructed to carry oil southward from northern Alaska. Examples are engineering geological maps by Kachadoorian (1971), by Weber (1971) and by Ferrians (1971). Construction was planned and modified on the basis of many of the factors shown on these maps. The pipeline and associated roads and structures will be inspected periodically to observe what the effects of certain construction practices have been on the environment. Because the pipeline was completed in 1976, not enough time has elapsed for these data to be available for use in current studies for construction in Alaska, such as those being made for the National Petroleum Reserve.

13.3.4 National Petroleum Reserve in Alaska

An example of separate studies made for a single area by scientists from many disciplines is a massive report made for the National Petroleum Reserve in Alaska (NPR-A), a large area comprising 37 000 square miles of Arctic land. A study to determine the best uses for the lands of the reserve was made by the Bureau of Land Management of the U.S. Department of the Interior, with the assistance and co-operation of numerous Federal, State and private organizations. These groups evaluated many aspects of the total physical environment, including archaeology, geology, scenic areas, possible transportation corridors, social and economic values, and the fish and wildlife.

Numerous maps were produced showing various aspects of the geology, as well as maps of wildlife habitats, archaeological sites and distribution of native peoples. The individual maps were not combined, but could be to determine how development of one resource might influence another part of the environmental system.

An illustration of how some of these elements might

be superimposed to show the effects of mineral exploitation on some of the other factors is given in Figure 13.6. The map combines maps of the generalized superficial geology, mineral deposits, wolf and fox habitats, and unique ecological areas. Most of the wolf habitats, as well as some unique ecological areas, fall within areas of metallic mineral deposits (Region II), and much of Region II is, in addition, covered with glacial till, which may make mineral exploration and exploitation difficult. These interrelationships, as well as many others on numerous additional geological and environmental maps of the NPR-A, should be considered in any use of the Petroleum Reserve. Other combinations of map data can of course be made as the need arises. The figure shown here illustrates only one possible application of such combinations.

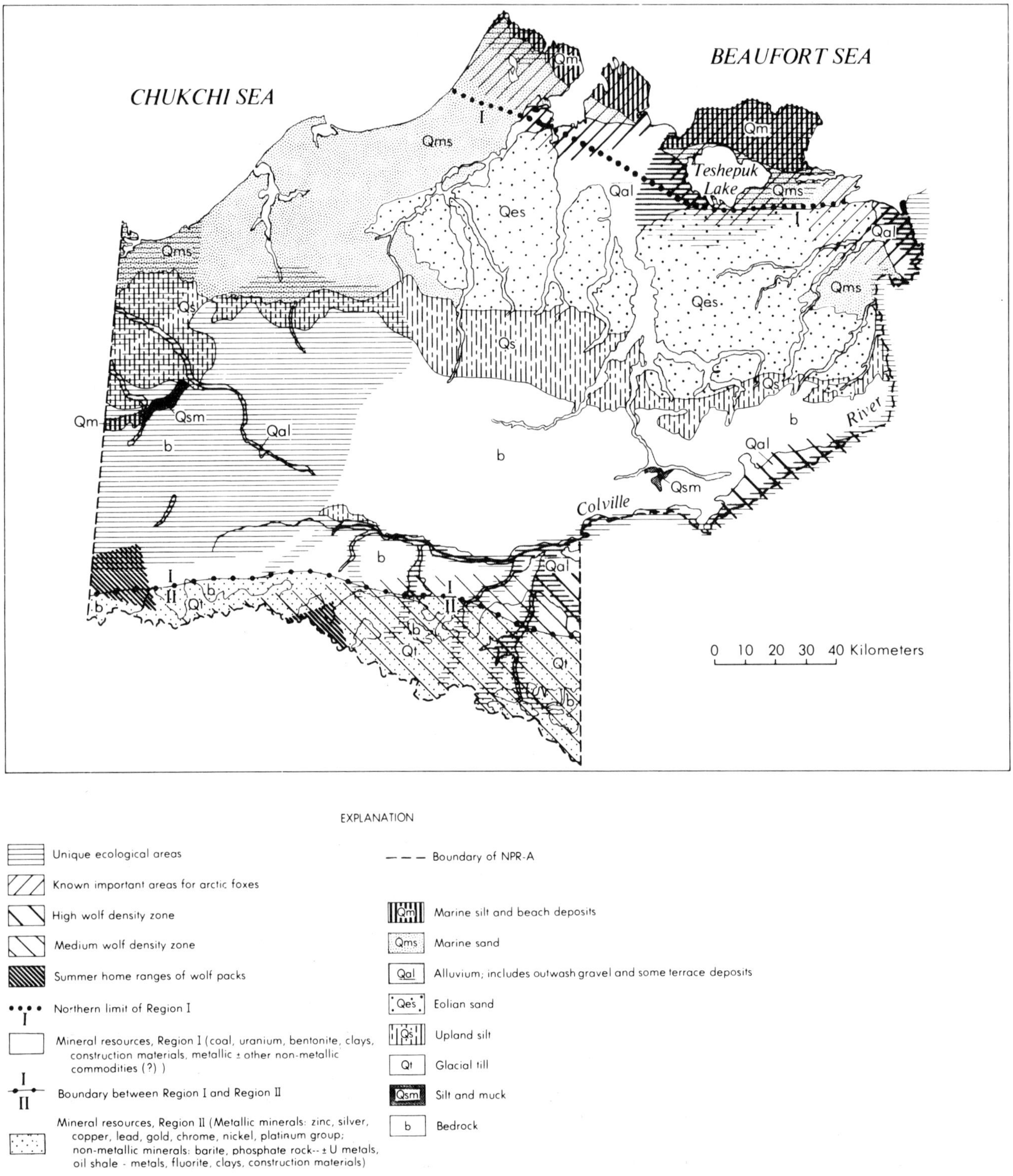

Figure 13.6 Map showing relations of some physical conditions and resources in the National Petroleum Reserve, Alaska, USA. Modified from maps by the U.S. Department of the Interior, NPR-A studies (From Radbruch-Hall, 1979, Fig. 4, by permission)

13.4 Land capability analysis

In any area the existing natural features and processes present a range of advantages and disadvantages for different uses of land. Consequently, an important part of land-use planning involved matching land-uses with the appropriate physical characteristics of the land. Evaluating these physical features with regard to different types of land-use is termed a 'land capability analysis'.

A land capability analysis attempts to make earth science information more accessible and useful to local governments and their decision makers – the politicians – and makes planning more responsive to this information. As used here, therefore, 'land capability' means the ability of land to accommodate a particular land-use with an acceptable level of costs (ABAG, 1976) relative to geological and hydrogeological conditions. The information can be used to relate geological factors to other environmental, social and economic factors that can be expressed in a dollar form (or other suitable currency) and that can contribute to a land-development decision (Laird *et al.*, 1979).

The effects of geology on land-use lie in two broad categories: constraints and resources. Constraints include hazards which only become evident when they affect people, and the type and intensity of land-use affect their importance. Constraints may be listed as those associated with:

- Earthquake problems:
  ground shaking;
  fault rupture;
  tsunamis and seiches.
- Slope stability problems:
  landslides;
  creep;
  avalanches.
- Flooding:
  stream flooding;
  dam, levee, and dyke failure.
- Bearing material problems:
  subsidence;
  liquefaction;
  settlement;
  shrink/swell potential.
- Erosion and sedimentation.
- Volcanic activity

Resources include:

- Minerals;
- Construction materials;
- Energy;
- Water;
- Soil;
- Scientific and education sites.

The geological or hydrological setting of an area can either create opportunities for, or impose constraints on, development. Its importance depends upon the costs that arise from specific constraints or resources that affect the type of land-use.

13.4.1 A land capability study from Palo Alto, California, USA

A study that considers economic, social and political factors, in addition to land capability factors, is called a 'land suitability' study. An example of a land capability study integrated within a land suitability study is contained in a report on Palo Alto, California (Livingston and Blaney and others, 1971). Without too much detail, in order to illustrate the basic concept, the following five steps are usually involved in a land capability study:

1. Identify the types of land-use for which land capability is to be determined.
2. Determine the natural factors having a significant effect on the capability of the land to accommodate each use.
3. Develop a scale of values for rating each natural factor in relation to its effect on land capability.
4. Assign a weight to each natural factor indicating its importance relative to the other factors as a determinant of land capability.
5. Establish land units, rate each land unit for each factor, calculate the weighted ratings for each factor, and aggregate the weighted ratings for each land unit.

In the Palo Alto study referred to, the area was divided by a grid into 330 cells each of 20 acres. Each cell was evaluated and rated for each factor. Twenty-three factors were selected: 10 factors related to geology and soils; and there were 4 ecological, 4 visual and recreational and 5 planning and market factors. Most of the capability factors involving earth science information were included in the 10 geology and soils factors, each of which was rated (Step 3), assigned a weighting (Step 4), and had a weighted rating determined (Step 5). The rating system used is set out in Table 13.2.

Table 13.2 Rating system for land capability factors in the Palo Alto study, California, USA (After Livingston *et al.*, 1971, in William Spangle and Associates, 1976)

1 *Geological and soils factors*	2 *Rating*	3 *Weight*	4 *Weighted rating*
Average slope:			
Over 50%	1	10	10
31–50%	2		20
16–30%	4		40
0–15%	5		50
San Andreas fault line:			
within zone	1	7	7
not within zone	5		35
Landslides:			
within slide area	1	6	6
not within slide area	5		30
Natural slope stability:			
poor	1	5	5
fair	3		15
good	5		25
Cut-slope stability:			
poor	1	4	4
fair	3		12
good	5		20
Soil suitability as fill:			
poor	1	2	2
fair	3		6
good	5		10
Soil erosion:			
severe	1	6	6
moderate	5		30
Soil expansion:			
high	1	3	3
moderate	3		9
low	5		15

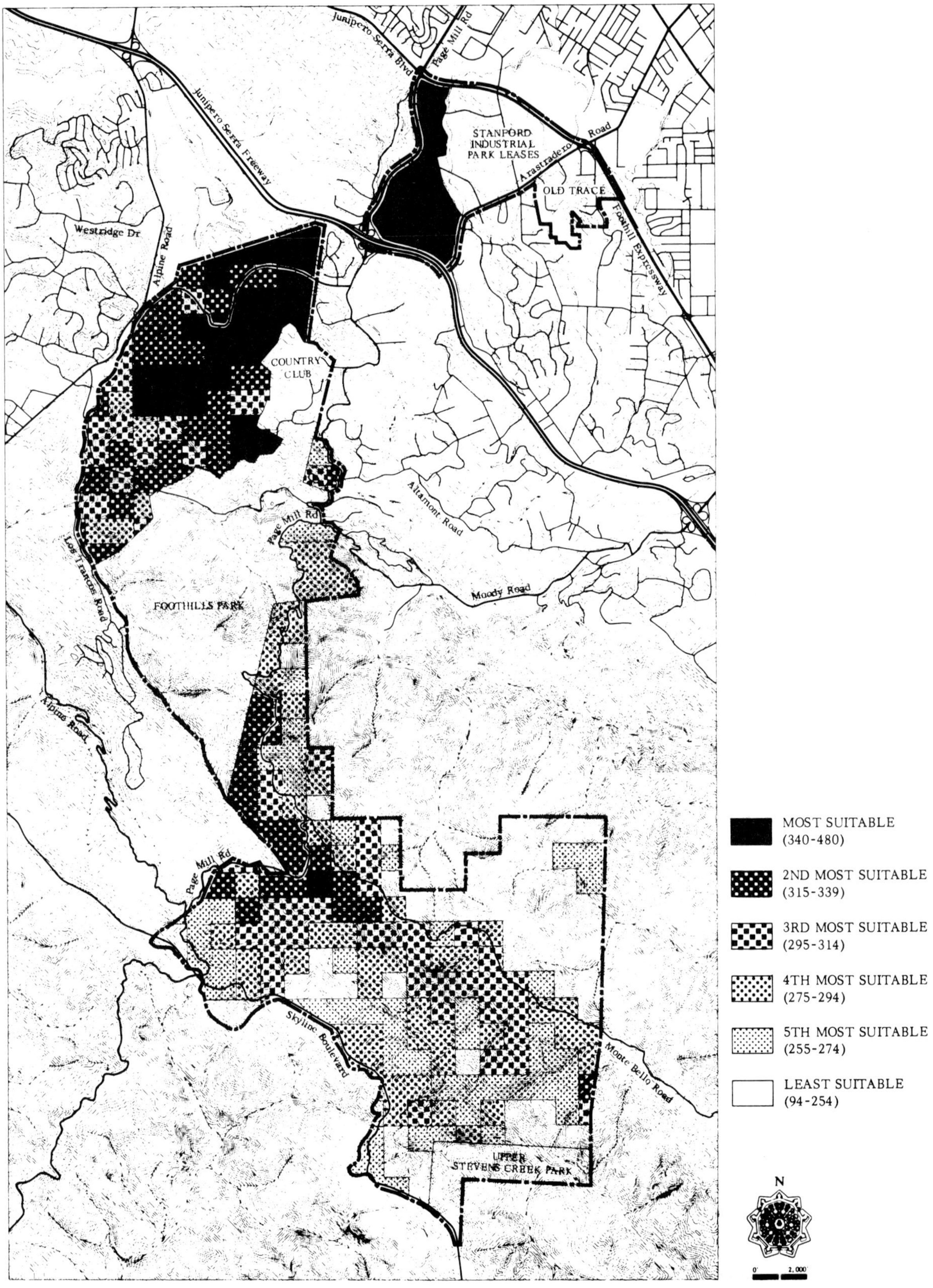

Figure 13.7 Map of part of Palo Alto, California, USA, showing land suitability for development ratings (Prepared by Livingston and Blaney and others, 1971; taken from William Spangle and Associates and others, 1976, Fig. 12, by permission)

The weighted ratings for all factors were then added for each cell to produce a total score. This combination of suitability and capability factors emphasized in a very marked way the importance of capability factors related to earth science in the determination of land suitability. The total scores for all factors ranged from 96 to 480, with the range for each factor:

- planning and market, 19–95;
- ecological, 18–90;
- visual and recreational, 11–55;
- geological and soils, 48–240.

These total scores were divided into six groups and shown on a map (Figure 13.7).

13.4.2 Land capability in Fairfax County, Virginia, USA

Published in 1978, the U.S. Geological Survey Professional Paper 'Nature to be commanded . . .', with the qualification in the quotation from Francis Bacon 'must be obeyed', sets out to demonstrate in a readily assimilated fashion the value of earth sciences information in land and water management by showing how earth science maps have been used in a variety of urban settings (Robinson and Spieker, 1978).

The background to this case study was an application for development of more than 1000 acres of undeveloped land into a high density new town. Several geological conditions would affect development of this area. Because the upland area lies at the headwaters of three drainage basins, the probability of downstream flooding and siltation, a common problem where urbanization increases the area of impervious surface, would require a comprehensive stormwater management programme. The area also has many slopes underlain by expansive clay that has a recurrent history of failure by landsliding and creep. Even level terrain underlain by susceptible clay may have foundation stability problems.

Basic data maps of the Franconia area include topography, geology and hydrology. They provide the basis for derivative maps. For example, the topographic map provides the information needed for a derivative landforms map which shows the distribution of critical surface slope categories. Combining the map of slope categories with a derivative map of engineering properties of superficial materials generates an interpretative map of slope stability.

The reviewing process of the proposed development plan starts with an 'inventory' (Figure 13.8) of basic data maps, proceeds to an 'analysis' resulting in derivative and interpretative maps, and finally a 'synthesis' – the maps showing the capability of the area for planned development housing.

As part of the inventory process, the planners consider not only the earth science capability data but such other aspects as air quality, noise, wildlife habitats and vegetation to determine land-use suitability for the area.

Because of the number of factors involved, the synthesis needed for the generation of a capability map was carried out by computer composite mapping. For the Franconia area, the computer stored the data in the form of three factor maps: distribution of clay beneath the surface gravel cap, a hydrological map showing the location of natural drainage ways and surface water, and a slope stability map showing the distribution of unstable slopes (Figure 13.9), which in turn yielded the composite capability map (Figure 13.10).

In compiling the computer composite map, the mapping technique again involves dividing the source maps into cells of preselected size (Figure 13.11a) and map units are stored as numeric codes, represented diagrammatically by shading (Figure 13.11b). Weighted numeric values are assigned to the digitized map units, based on their relative importance (Figure 13.11c), and the computer sums these values to produce a 'score' for each cell (Figure 13.11d). The composite map (Figure 13.11e) is produced on a computer line printer, with a different symbol to represent each score, or the combination of factors in each cell.

Bearing no obvious resemblance to a geological map, the composite capability map is readily interpreted in terms of favourability for planned development housing. The lighter toned areas are the more favourable. Favourable areas are on the upland in zones of moderately high and high slope stability that are underlain by sand and gravel. Potential sites for stormwater retention ponds are in low-lying areas underlain by clay.

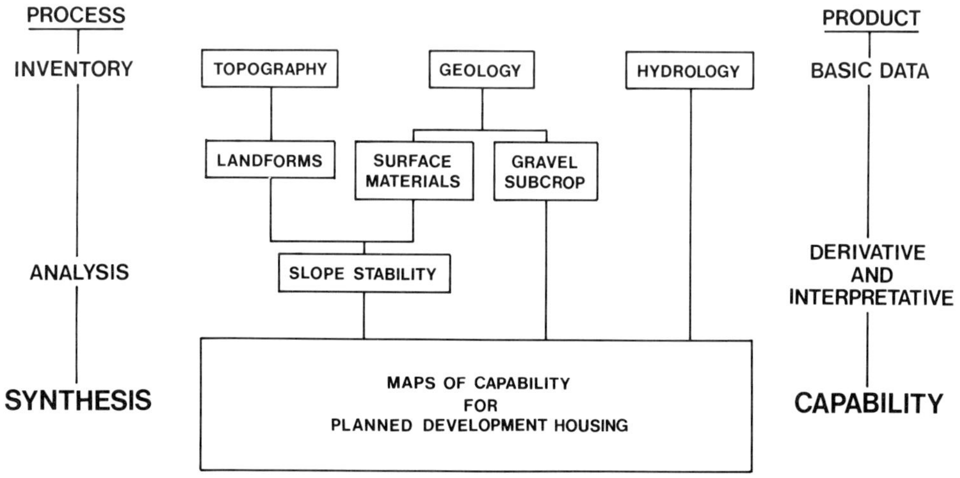

Figure 13.8 Flow chart showing the use of earth science data in the Franconia area, Virginia, USA (After Froelich, Garnaas and Van Driel, 1978)

Figure 13.9 Computer maps of three earth science factors for the composite map shown in Figure 13.10: (a) distribution of clay beneath gravel; (b) hydrological map – natural drainage and surface water; (c) slope stability map showing distribution of unstable slopes (From Froelich, Garnaas and Van Driel, 1978, by permission)

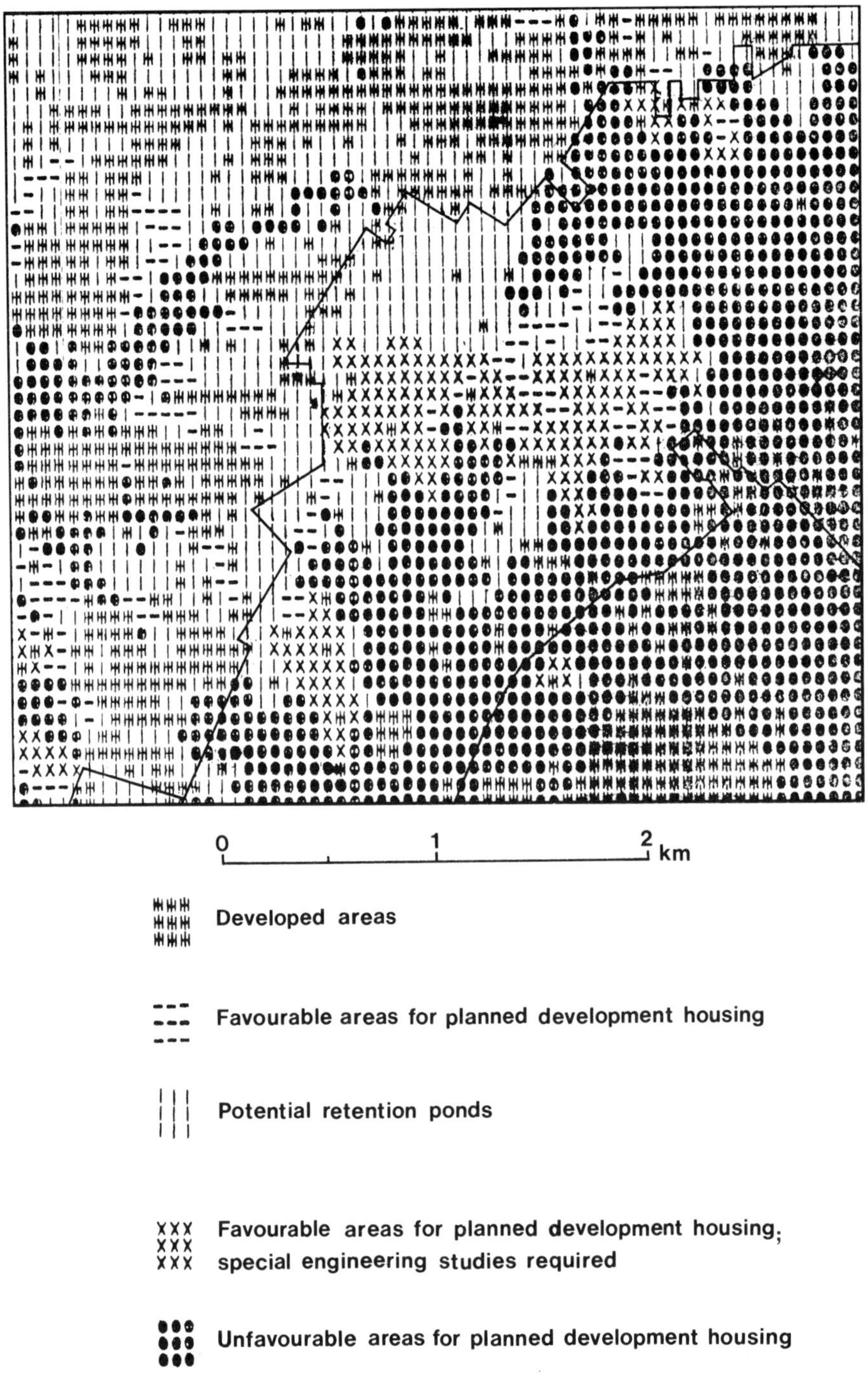

Figure 13.10 Composite capability map for planned development housing in the Franconia area, Virginia, USA (After Froelich, Garnaas and Van Driel, 1978)

This interpretive map nevertheless is easy to understand by potential users who have no geological background, and, moreover, do not need it.

The claim is made that the purpose of such maps is to inform planners, decision makers or owners so that they can:

- forestall or relocate new developments in areas where lives and property might be imperilled;
- propose appropriate design precautions in developments that cannot be placed elsewhere;
- alert inhabitants of imperilled developments to seek protection through engineering or insurance.

The approach in 'Nature to be commanded . . .' (Robinson and Spieker, 1978) to the non-geologist is admirable in the clarity and excellence of the diagrams. Particularly effective are the block diagrams in colour, giving exploded views of the surface and subsurface geology. An introduction and eight case histories, abundantly illustrated, underline the usefulness of colour in putting across the implications of the geological setting to land-use planners.

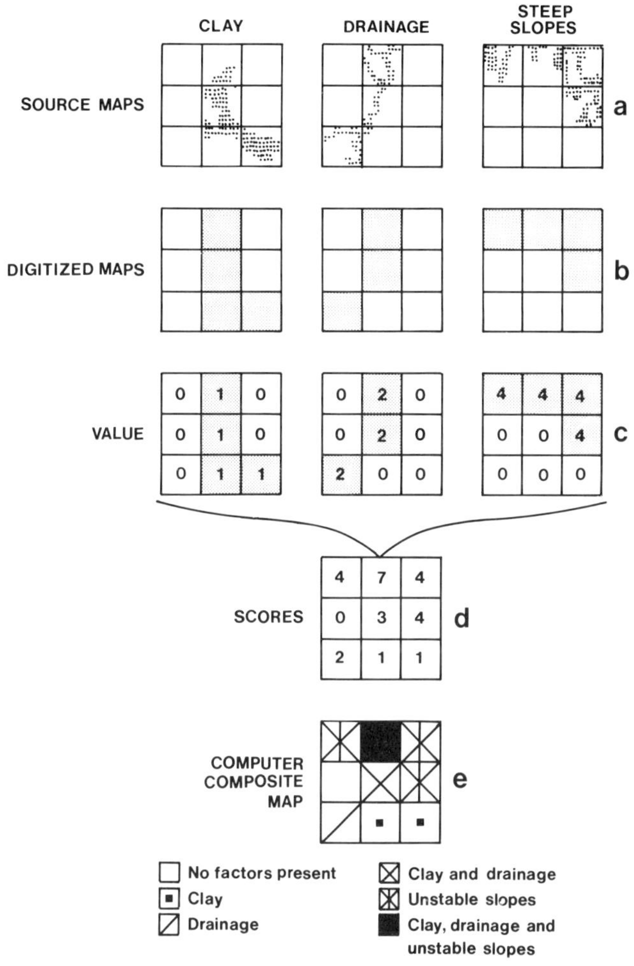

Figure 13.11 Diagram of the computer composite mapping technique for a hypothetical area (From Froelich, Garnaas and Van Driel, 1978, by permission)

13.5 Quantitative land capability analysis

In the San Francisco Bay area the Santa Clara Valley has been used as a demonstration area for quantitative land capability analysis, where the term 'quantitative' refers specifically to the assessed cost in dollars associated with each geological hazard, constraint or resource (Laird *et al.*, 1979). The relative importance of some of these land capability factors has already been set out for the Palo Alto study (see Table 13.2) as a quantitative means of comparison. Since development has to be paid for, calculating dollar costs associated with geological conditions is another possible solution to assisting the decision-making process.

Such estimates are often subject to great uncertainties. However, a technique has been worked out for evaluating the cost of several different types of construction and land development in terms of the cost of a number of kinds of earth science factors. Costs related to surface rupture and ground shaking due to fault movement, flooding, dam and dyke failure, shrinking and swelling soils, settlement, liquefaction, subsidence, landslides, soil creep, erosion and sedimentation, and to depletion of sand and gravel resources and other mineral resources, were evaluated according to land use – agricultural, residential, shopping centres, commercial, industrial and major highways. The land capability for each of these uses, based on costs due to geological factors, is shown on maps.

The method for land capability assessment involves five general steps. The first step involves collecting earth science information and preparing basic maps, once it has been decided which problems – for instance, landslides or floods – might result in significant costs for characteristic land-uses in the given region. In step 2, interpretative maps are developed for each problem, either directly from an appropriate basic map or by combining maps. These interpretative maps can then be used to identify specific problems.

In step 3, social costs (in dollars) are calculated for each type of development and each geological condition. 'Social cost', as used here, means the sum of all costs attributable to a problem regardless of who pays. There are three basic types of cost. First, study, engineering, design and mitigation costs. These are usually incurred before or immediately upon construction of a project. Secondly, disaster or damage costs. These must be paid (usually at some future time) when damage occurs. Thirdly, opportunity costs. These costs are the revenues or benefits that would have resulted from a type of land-use and are forgone if the land is used for other purposes.

Costs that accrue at different times are made commensurable by calculating the present value of future costs, using an interest, discount rate. Costs that may occur at some unknown future time are calculated by finding the average, or expected, value of the costs.

In step 4, all the expected costs for all the conditions for each land-use are totalled. This total is used as an indicator of the capability of the land to accommodate each use. Then, when the sums of these costs are displayed on a map, the result is a capability map for each land-use. This is step 5 in the procedure, and the capability assessment can now be used in the planning and decision-making processes.

13.5.1 Mapping and displaying land capability data

After interpretative maps have been compiled and costs assessed, graphically acceptable capability maps must be produced.

Mapping and information storage can be done by hand, but lack of flexibility can be a major problem in mapping of areas where change, improvement and evolution are anticipated. Various software packages are available for computer storage and processing. Data can be collected and represented by either grid cells or polygons. Grid cells produce maps which generalize areas into boxes, as has already been demonstrated in outline for the Fairfax County maps (Figure 13.11); the method of grid cell representation is illustrated in Figure 13.12.

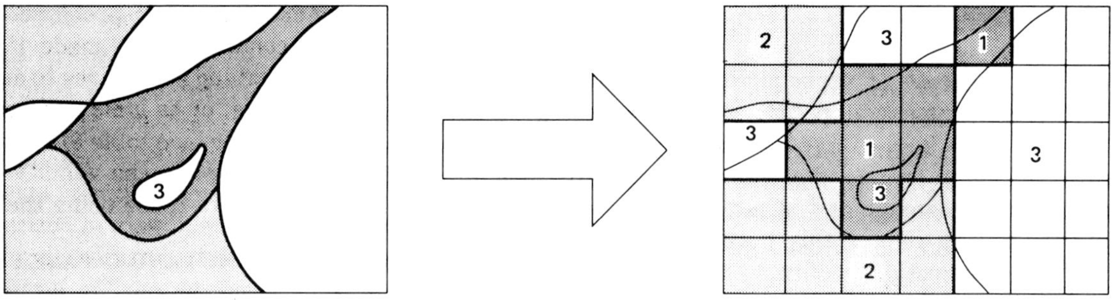

Figure 13.12 Grid cell representation of a geological map (From Laird *et al.*, 1979, Fig. 1, by permission)

As is apparent, grid cells can miss small details unless the cell size is made exceedingly small, which increases the number of cells to be collected by the square of the dimension changes (i.e. one-half width = 4 times as many cells). The smallest size that can be easily used is approximately $\frac{1}{10}$ in square. Such a cell size lends itself well to collection of mapped data as well as graphical presentation using readily available high speed line printers.

Map presentation

After the basic maps are digitized and the costs are summed, the resulting information must be displayed in a manner that is easily understood by the users. Display may be done in a number of ways, each having different advantages and disadvantages. The total dollar costs can be printed directly on the map for certain uses, such as checking a test area. In most cases, however, some sort of aggregation into different categories or levels will be more useful. The exact type needed will depend on the eventual use of the study. Several different ways of categorizing the costs may be desirable to satisfy all potential users.

Then the problem becomes one of graphic display. Hand-produced maps can represent categories either with patterns or colours (Figure 13.13a). Many interesting methods of producing computer maps are possible, depending on the software/hardware combinations available. Line printers, square or rectangular, with simple characters or with overprinting (Figure 13.13b), and graphic symbols (Figure 13.14) can provide fairly acceptable visual separation. This visual separation can be improved by photographic reduction. As the individual characters, as well as the spaces between characters, become small, the human eye visually integrates the elements into a pattern or shade of grey. A pen plotter can be used equally well to draw perspectives (Figure 13.15a), produce density maps (Figure 13.15b) or plot contours (Figure 13.15c). A pen plotter can be used with a scriber to prepare negatives directly. Stable base material can be used with a pen plotter. However, this type of material cannot be used with conventional line printers.

Another display technique now available used cathode ray tubes (CRTs) either in colour or black and white. These TV-type displays can be photographed to prepare copy. This type of display is particularly useful if the system is interactive and users want to test different assumptions rapidly.

Colour is valuable, since no more than seven categories can be represented with shades of grey on a map without loss of visual differentiation. However, shades of colour are also subject to a similar category limit.

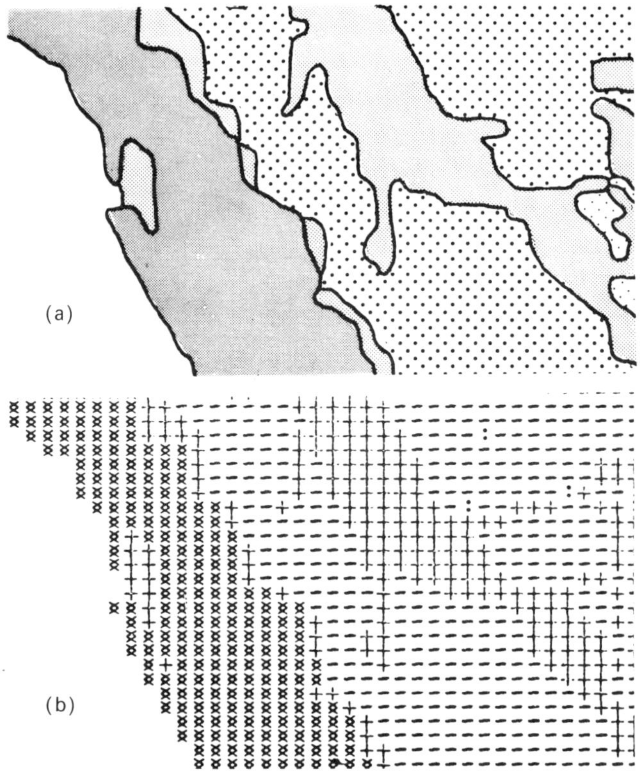

Figure 13.13 Maps produced by hand (a) and by computer (b) (From Laird *et al.*, 1979, Fig. 4, by permission)

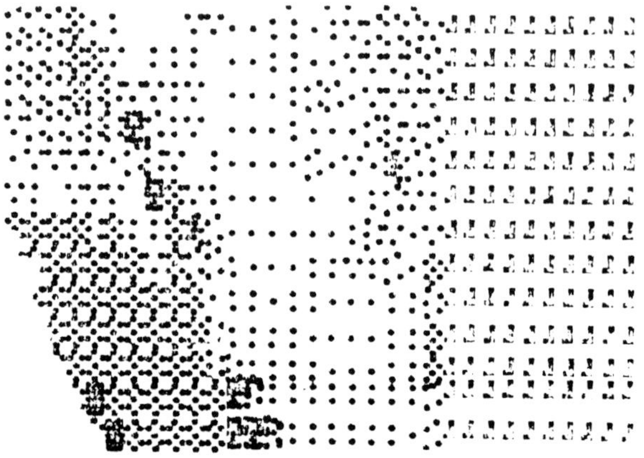

Figure 13.14 Map produced using a line printer with graphic symbols (From Laird *et al.*, 1979, Fig. 5, by permission)

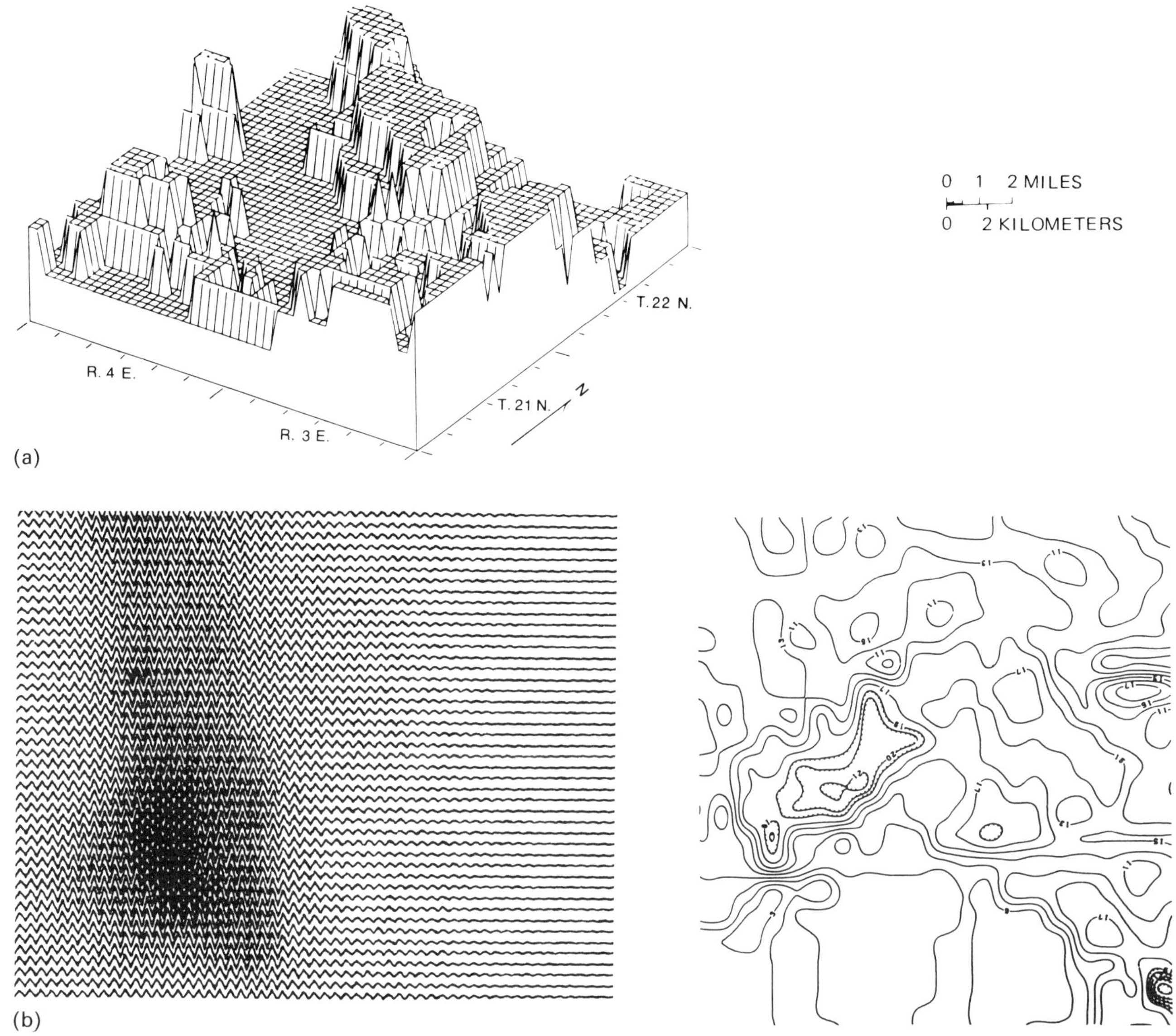

Figure 13.15 Maps from a pen plotter: (a) perspective; (b) density; (c) contours (From Laird *et al.*, 1979, Fig. 6, by permission)

Table 13.3 Standard response to geological problems in the Santa Clara Valley, California, USA*

Earth-science constraint or opportunity	*Land-use type*†							
	Rural agricultural	*Semi-rural*	*Single-family residential*	*Multi-family residential*	*Regional shopping centres*	*Downtown commercial*	*Industrial*	*Freeways*
Surface rupture	B	B	A	A	A	A	A	A,B
Ground shaking	B	B	B	B	B	B	B	B,D
Stream flooding	B	B	B	B	B	B	B	D
Dyke failure	B	B	B	B	B	B	B	D
Dam failure	A,D	A,D	A,D	A,D	A,D	A,D	A,D	A,D
Landslides	A	A	A	A	A	A	A	A
Soil creep	A	A	A	A	A	A	A	D
Shrink/swell	A	A	A	A	A	A	A	D
Liquefaction	B	B	B	B	B	B	B	D
Settlement	B	B	A	A	A	A	A	D
Erosion and sedimentation	C	C	C	C	C	C	C	C
Septic-tank limitations	D	D	D	D	D	D	D	D
Sand and gravel	D	C	C	C	C	C	C	C
Mercury extraction	D	D	D	D	D	D	D	D
Agriculture	D	C	C	C	C	C	C	C

* This is an average response based on a sample of practices in the demonstration area. Some individuals and some governmental bodies may respond differently, but the table is believed to represent a realistic overall summary. It shows whether damage, study and mitigation, or opportunity costs, are calculated in the cost analyses used in this study. These responses may not be typical of other regions and they may change with time in the demonstration area.
† A, study and mitigate; B, suffer damage; C, lose resource opportunity; D, no response required.

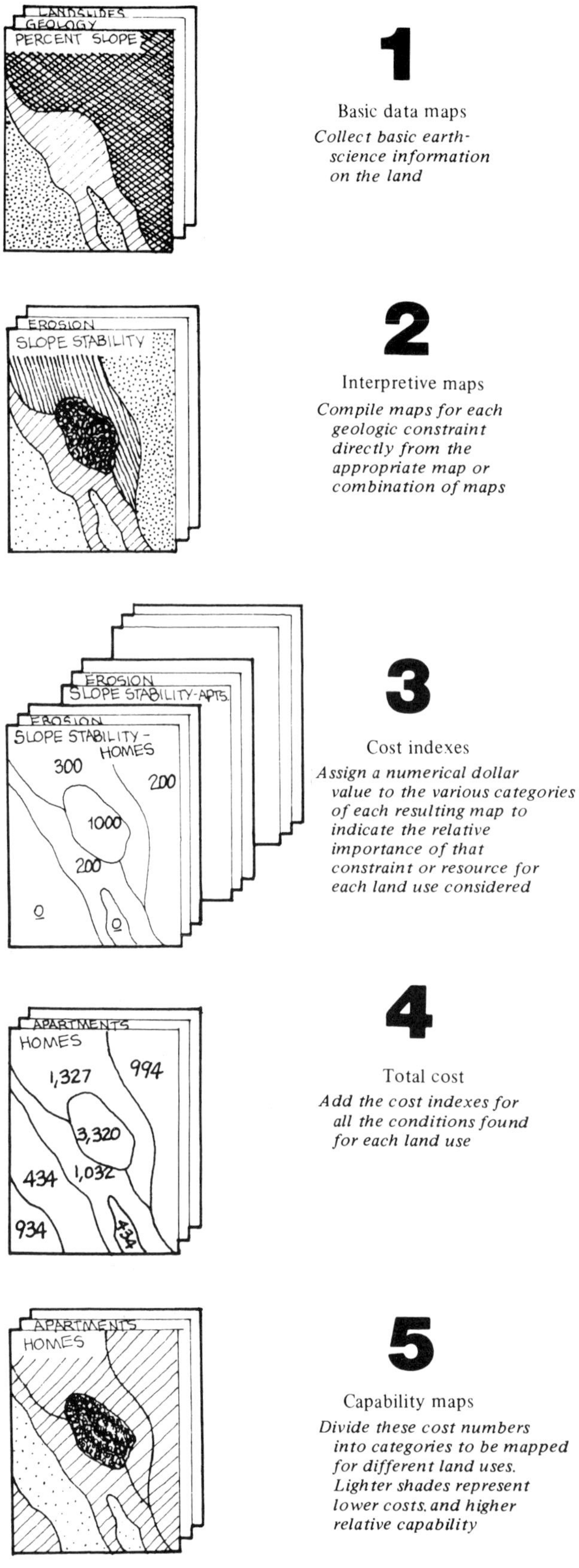

Figure 13.16 Steps in land capability analysis (From Laird *et al.*, 1979, Fig. 7, by permission)

The importance of clarity and visual impact cannot be overemphasized, but the particular problems depend on use.

A summary of the capability analysis procedure is illustrated in Figure 13.16.

13.5.2 The Santa Clara Valley study, California, USA

Fourteen source maps were compiled at a uniform scale of 1:125 000, and some of these are composites of other maps. For example, the earthquake intensity maps are based on the materials map and the fault map. The slope-stability map is based on a map showing percentage of slope, a map of photo-interpreted landslides and the materials map (Figure 13.17). For the Santa Clara Valley, the standard response to geological problems is set out in Table 13.3, and these have been taken into account in making the cost assessments.

Figure 13.18 is an example of the type of map produced, in red over a grey topographic background – the land capability map for single-family residential use. A summary of the costs for a range of constraints is set out in Table 13.4. It is interesting to compare the weightings of costs in Figure 13.18 with, as just one example, the map of relative slope stability (Figure 13.17).

13.6 Water resources in environmental planning

Hydrological maps illustrate the occurrence of water on and below the ground surface, although subsurface water is usually considered to be in the sphere of hydrogeology. Such maps are useful to planners and engineers in evaluating water supply; assessing the effect of groundwater extraction, and water-land recovery once abstraction has ceased, on civil engineering structures and ground surface levels; determining water-related hazards such as flooding. The range of maps available for any area may be limited, but among them may be general hydrogeological maps showing surface drainage areas, depth to groundwater, thickness of the water-bearing formations and groundwater quality. All of these aspects may be shown on a single comprehensive hydrogeological map, although it is often easier for the user if significant aspects are dealt with separately in interpretative maps. Block diagrams provide a particularly expressive medium for conveying the structure, form and functioning of a groundwater reservoir, in a general way suitable for planning. Such conceptual models may be replaced by mathematical models when quantitative aspects of groundwater management have to be considered.

Many examples of earth science maps applied to water management will be found in Robinson and Spieker (1978), illustrating water-related problems and water resources in Helley and Lajoie (1979), and flood-prone areas and flood-hazard maps in Waananen *et al.* (1977).

One significant contribution to the International Hydrological Decade (IHD) 1965–1975 was the publication of an international legend for hydrogeological maps (UNESCO, 1970), which included a model map (Plate 20) of an imaginary area. The purpose of hydrogeological maps is to enable various areas to be distinguished according to their hydrological character in relation to the geology. They should indicate, on a regional basis, such aspects as the extent of the principal groundwater bodies, and equally important the scarcity of groundwater elsewhere, the known or possible

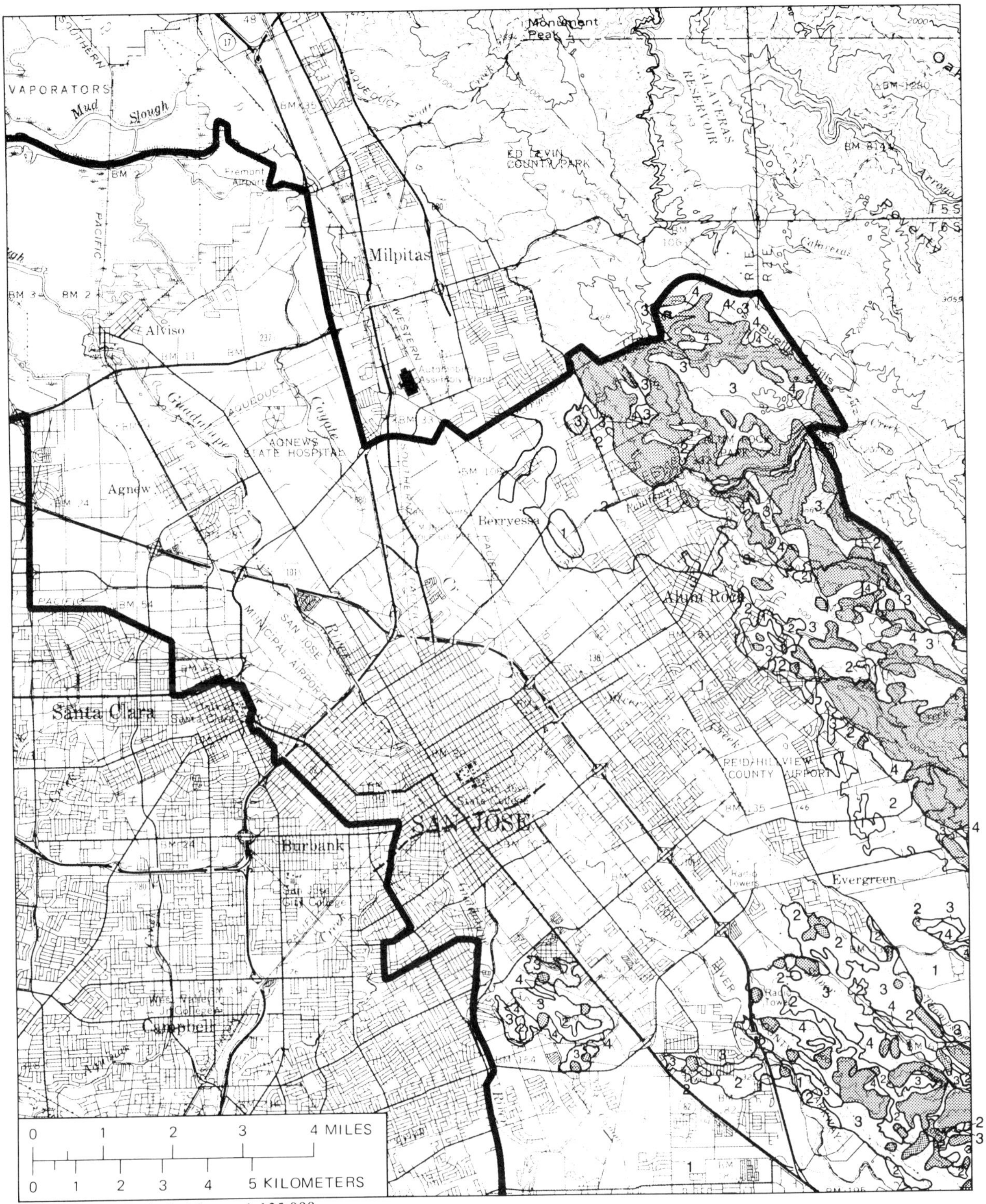

Base from U.S. Geological Survey 1:125,000

EXPLANATION

Slope-stability categories

1. Stable areas of 0-5 percent slope
2. Generally stable areas of 5-15 percent slope
3. Generally stable to marginally stable areas of >15 percent slope
4. Moderately unstable areas of >15 percent slope

Unstable areas underlain by, or immediately adjacent to, landslide deposits

Figure 13.17 Map of relative slope stability in a part of the Santa Clara Valley demonstration area, San Francisco, USA (From Laird *et al.*, 1979, Fig. 18, by permission)

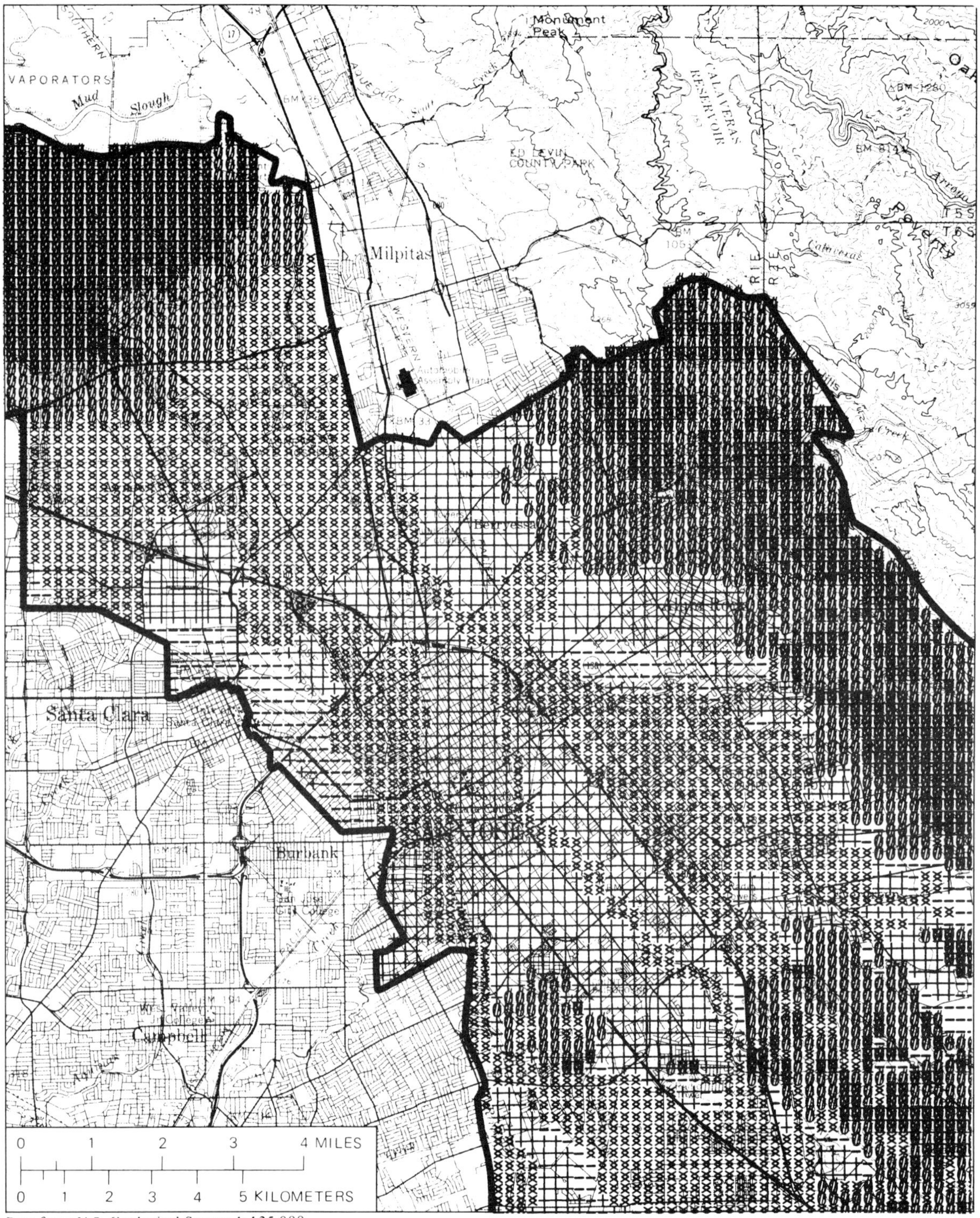

Base from U.S. Geological Survey 1:125,000

EXPLANATION

Level	Symbol	Total cost range (in dollars per acre)	
1		0.01	3,162.00
2		3,162.01	10,000.00
3	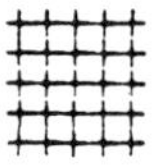	10,000.01	17,783.00
4		17,783.01	31,623.00
5		31,623.01	56,234.00
6		56,234.01	100,000.00

Figure 13.18 Land capability map for single-family residential use in the Santa Clara Valley demonstration area, San Francisco, USA (From Laird *et al.*, 1979, Fig. 33, by permission)

Table 13.4 Summary of costs for single-family residential use, in US$ per acre (Santa Clara Valley study, San Francisco Bay, USA)

Hazard, constraint or resource	*Severe* ←				→ *Slight*
Surface rupture	500	500	0	—	—
Ground shaking—San Andreas, Hayward	4 000	3 000	1 000	500	50
Ground shaking—Southern Hayward	400	300	100	50	5
Ground shaking—Calaveras	3 000	1 000	500	50	—
Stream flooding	9 000	0	—	—	—
Dam failure	0	—	—	—	—
Dyke failure	20 000	0	—	—	—
Shrink/swell soils	6 000	2 000	0	0	—
Settlement	30 000	30 000	20 000	2 000	—
Liquefaction	1 000	900	90	6	0
Subsidence	0	—	—	—	—
Landslides	40 000	20 000	8 000	2 000	0
Soil creep	10 000	4 000	0	0	—
Erosion and sedimentation	200	30	10	0	—
Septic tanks	0	—	—	—	—
Sand and gravel	20 000	0	—	—	—
Mercury	0	—	—	—	—
Agricultural land	5 000	0	—	—	—

location and extent of artesian basins, areas of saline groundwater and the potability of groundwater.

The symbols, which enable all these aspects to be shown on the map sheet, were particularly recommended for small-scale maps. They supplement those given in Chapter 5 for general engineering geological use.

13.6.1 Hydrogeological maps prepared by Geological Surveys

Since 1970, official Geological Surveys have published a range of hydrogeological maps, both general and thematic, at a variety of scales.

United Kingdom

A general hydrogeological map of England and Wales was published at a scale of 1:625 000 by the Institute of Geological Sciences (1977). Geological units are differentiated into aquifers in which flow is dominantly intergranular, and those in which flow is dominantly through fissures and other discontinuities. In turn, they are differentiated into extensive and highly productive aquifers, restricted to six in number, and locally important aquifers. A third, and very extensive unit, is the group of concealed aquifers, aquifers of limited potential, and regions without significant groundwater.

The map is based on the smaller scale Sheet B4 (London) of the International Hydrogeological Map of Europe (UNESCO, 1970).

Larger scale maps have also appeared, with the first in 1967 of North and East Lincolnshire at a scale of 126 720.

France

A general view of the hydrogeology of France appears on the 1:1 500 000 hydrogeological map (Margat, 1980). An informative, short memoir comments on the map legend and the ground hydrogeology of France. The bibliography provides an introduction to modern maps at larger scales.

The analysis of subsoil permeability conditions and the hydrogeologial characteristics of groundwater make it possible to classify aquifers with respect to their vulnerability to pollution. There are now maps of vulnerability published at various scales from 1:250 000 to 1:500 000 (BRGM, 1980), with an overview map at 1:1 000 000. Comprehensive syntheses of the chemical quality of groundwater, and the average cost of tapping groundwater in France, are examples of other thematic hydrogeological maps at 1:100 000 scale.

14

Prospect

Although it is a matter of opinion, it is apparent that there is a clear distinction in the broader aspects of engineering geological mapping between (a) the production of engineering geological plans, at appropriately large scales, of ground conditions at actual engineering sites, and (b) small-scale mapping for preliminary planning of engineering works and environmental assessments for land capability and land suitability studies with parallel consideration of natural hazards and the risks associated with them.

Large-scale engineering mapping, which really should be referred to as geotechnical mapping because the resultant plans usually incorporate the engineering solution to a problem to the basic record of geological conditions, will undoubtedly continue along the already well-established lines that have proved so effective in the past. The very act of recording in detail, for example, a rock face for which there is concern over short-term or long-term stability, focuses attention on all the significant features of the distribution of rocks and soils, geological structures, particularly discontinuities, and groundwater conditions. There can be no substitute for this close attention to detail, selectivity of those details particularly relevant to understanding the problem, and the relation of this detail to a geotechnical engineering solution. Of course, much mapping is carried out for purely record purposes, and on such plans the details of ground treatment will also be recorded. The medium of such mapping will remain the gridded sheet of paper, or plastic, to record what can be seen. Photographs, with sketched overlays, are a commonly used substitute, and may be supplemented by stereo-pairs of photographs, in black and white or preferably colour. The same techniques are used in engineering geological research.

At small scales the hands-on approach is strictly limited to spot-checks on the ground. Remote sensing, either aerial photographs or satellite images, enables the mapping to be done in the office either manually or indirectly by computer analysis with automatic cartography. Possibilities of the manipulation of data and quantitative assessments are manifold, limited only by the resolution of photographs used or the pixel size of satellite images. The rate of advance in the automatic manipulation of such data is incredible to the past generation of field mappers (Greeman, 1988), with the possibility of producing finished, hard copy maps and drawings in variety. Sophisticated versions, in colour, of the perspective maps in Figure 13.15(a) can have details of the image-interpreted geology overlain on them, and other details such as streams, roads and location of boreholes.

In most countries, with highly developed urban areas and active official Geology Surveys, the vast amounts of information on ground conditions available from large-scale geological maps and thousands of site investigation reports have been codified into geotechnical computer data banks. Experimental plans have been produced from the Tyne and Wear Data Bank in North-east England by Reekie, Coffey and Marsden (1979) and the whole system lends itself to automatic cartography which can be designed to respond to customer demand. The customer in this case may be civil engineering consultants and local planning authorities. What can be achieved has been demonstrated recently with computer maps of Madrid, generated from the Geo-Madrid Geotechnical Data Bank (Lopez Palancar and Garcia Yague, 1986).

Natural hazard mapping, and in particular slope hazards and risk assessments, lends itself to the computer-assisted statistical analysis of numerical factors, combined with automated plotting of a grid of cells to build up a map. Carrara *et al.* (1978) have brought the method to an advanced form in Calabria, Italy, which is extremely prone to slope instability. The project, as an indicator of the trend of future developments, had led to the developments of techniques for the collection, encoding, storing and processing terrain attributes in an efficient and flexible way. Of the 33 attributes stored in the computer file, 25 were examined statistically and interrelated. Their contributions to slope instability were determined and from them the slope instability map was derived (Varnes, 1984). The landslide hazard and man-made structures map (Varnes's Fig. 32) relates landslide hazard in high and low degrees to various types of man-made structures including villages, country houses, roads, bridges and pipelines. Varnes commented that as such it appears to be a unique step forward towards a formal quantitative areal assessment of actual potential loss or risk. In 1984, the whole work appeared to be one of the more advanced and accessible state-of-the-art analyses of land attributes for production of landslide-hazard maps, utilizing computer processing.

In the same way, the report by Laird *et al.* (1979), discussed in section 13.5, on land capability analysis by computer makes it easy to compare the costs associated with geological constraints on land-use, and compels the user to recognize and state his assumptions and identify the information needed to define costs.

Use of the stop light colour coding system of red, orange, green, with intermediate shades as used in the ZERMOS maps or a spectrum range of shades, produces attractive maps which have a strong impact and are easy to interpret. The colours or patterns for such maps can be generated by computer.

With sufficiently detailed topographic, geological and thematic maps it is now possible to compile engineering environmental maps of whole countries, or large parts of continents, that take the early approach of Ershova and Sergeev (1979), Sergeev and Ershova (1989) and Figures 8.1–8.3 an important step forward.

Combinations of this sort can now be very conveniently made by computer, using recently developed techniques. Computer methods were used to prepare a set of engineering environmental maps of the entire conterminous USA. A number of thematic maps showing various individual aspects of the geology, including for example landslides, karst and volcanic hazards, formed much of the basic data for the preparation of the engineering environmental maps. The map units on the component maps were digitized and converted to raster format for manipulation and display. A program assigned a specific attribute code to all image elements enclosed by lines (such as the area of a given geological formation), so that the resulting data base (in this case a map) could be manipulated with a digital image-processing system. Several maps were easily added together. Colours and patterns were generated in the computer and combined for making final coloured images in traditional map form (Edwards and Batson, 1980; Radbruch-Hall, Edwards and Batson, 1987).

A relatively conventional geological map was compiled that also served as the base map for one plate of the computer-generated engineering geological map. This map utilized the principles of zoning, for it delineates with heavy lines the major tectonic provinces of the conterminous USA, defines physical subdivisions within the provinces by medium-weight lines, and outlines the geologic units with finer lines. Because of the small scale of the map, 1:7 500 000, a segment of an early modified version of which is shown in Figure 12.5 in black and white, specific formations could not be shown.

The basic division of geological units is genetic and so by implication broadly lithologic. Sedimentary, plutonic and volcanic rocks are delineated, as is Quaternary alluvium. Continental glacial deposits are shown in the northern part of the USA. Colluvium and saprolite are indicated in the Appalachian region, with underlying rocks differentiated by symbol, because these thick surficial deposits are considered to be as critical for engineering construction as the underlying rocks.

Sedimentary rocks are divided by age, for the engineering geological properties of the rocks in general differ depending upon their age, the younger rocks commonly being less consolidated and less deformed than the older. Structural complexity of the rocks is also shown by pattern, or by letter symbol where bedrock is overlain by glacial debris, saprolite or colluvium. Complexity is classified in four grades: undeformed or mildly deformed; moderately deformed; severely deformed; and regionally metamorphosed.

The map shown in Figure 12.5 is patterned; in the final published map, colours for the geological units were generated by computer, as were the patterns for the complexity.

All of these features shown on the geological map have some relation to the physical properties of the rock units, and therefore to their engineering behaviour. They do not specifically indicate engineering or environmental problems, however. Such problems are shown on other maps, one of which – the engineering geological map – is superposed on a simplified version of the geological map with outlines of the geological units in black. The engineering geological conditions may pose constraints on construction. Thematic maps of landslides, karst, volcanic hazards, seismic probability, standing water, steep slopes and expansive materials were combined by computer to prepare the engineering geological map.

Simply to add the number of component maps for each point on the final map would have implied that each factor posing constraints on construction was of equal severity, which was not true. It was therefore necessary to evaluate and compare the maps before they could be superposed. Each numerically evaluated area on each component map was digitized and converted to raster format, using an array of numbers that contained approximately 3000 numbers east–west and 2200 numbers north–south across the USA. Colours were assigned to the resultant number on the final map. The combined map, because it is superposed on the geological base, also shows the relation of the restrictive conditions to tectonic provinces, physical subdivisions, geological complexity of the rocks, and rock and soil units.

An environmental map was prepared from the same data by evaluating the same map areas on the basis of the influence construction might have on the geological and topographical conditions shown on the component maps. The environmental map that was produced from the stored data shows where construction might cause damage to the environment, either directly, by worsening the conditions, or indirectly, by damaging other aspects of the environment.

This is admittedly a very elaborate map, prepared by currently sophisticated techniques; maps of such complexity are seldom needed. It does, however, illustrate the types of technique that are available for displaying very complex sets of geological and related data.

The basic geological map is a small-scale, comprehensive, general-purpose map; the engineering geological map and the environmental geological map are derivative. They show areas where there may be construction problems or where construction may cause environmental problems, respectively, such information being derived from the thematic maps which are themselves to some extent derivative. More detailed information on the computer techniques used in preparing the maps has been published by Edwards and Batson (1980).

Geological mapping in the field is an art, a highly

selective process – because information provided by natural exposures or by the boreholes put down during a site investigation is limited – which attempts to be predictive. Where does the boundary between two rock groups get to in the topography and if it isn't where it has been predicted, what is the reason? As an essential aid to mapping, landforms often provided a vital clue where exposures were absent, or at best few on the ground. A relatively recent innovation in engineering geological mapping, effectively dating from about 1975 (Brunsden *et al.*, 1975), has been the application of well-established techniques of geomorphological terrain mapping to engineering situations, especially highway engineering. Now called engineering geomorphological mapping, this powerful tool will continue to be used as a guide and complement to site investigation, providing for example large-scale plans that are site specific for the understanding and mitigation of hazards.

One aspect of mapping that initially (Dearman and Fookes, 1974) attracted attention was considered to be the need to develop maps showing the distribution and variation of quantitative data of direct relevance to engineering design problems. Experimental maps usually made use of site investigation boreholes for which rock and soil classification could be related to *in situ* and laboratory determinations of engineering problems. The work of Taylor (1971, Fig. 4), Cratchley and Denness (1972) and Burnett and Fookes (1974) are well documented. More recently Gostelow and Browne (1986) have published a 1:50 000 scale planning map of the upper Forth Estuary, Scotland, summarizing a vast amount of geotechnical information gathered during assessment of ground conditions in a proposed development area where difficult ground conditions would impose restraints on construction. The map is a 'Geotechnical planning map for heavy structures' on which six basic foundation zones are defined, and are described comparatively as 'very good' to 'unpredictable' for heavy structures. The descriptions are based primarily on assessments of allowable bearing pressure values.

Between the early and later works mentioned above, a distinction should be drawn between what might be called 'pure' engineering geological research, and 'applied' engineering geological research. The former satisfies scientific curiosity, the latter is a type of applied research that has an obvious and immediate practical value and as such probably establishes a pattern for the future. Geotechnical mapping offshore, requiring vast resources, is clearly applied research, from which there is likely to be a significant spin-off of scientific information.

To return briefly, and finally, to land-use planning, the natural environment has influenced the location and form of human settlement. In the UK, Glasgow, Newcastle and London were sited in the first point upstream on a major river that could easily be forded. The land and its features, including the availability of natural resources, present both opportunities and constraints for human activities. Robert F. Legget provides a particularly apt postcript:

> When planning starts, for urban community or region, the area to be developed is not the equivalent . . . of a piece of blank paper for the free materialization of the ideas of the designer, but it is rather an environment that has been exposed for a very long period to the effects of many natural modifying factors. . . . Development of new communities and the charting of regional development must, therefore, take account of this fundamental organic and dynamic character of Nature so that the works of man may fit as harmoniously as possible into the environment . . . (Legget, 1973, pp. 70–71).

Appendix 1 Alternative symbols for soils and rocks

The Mapping Commission of the International Association of Engineering Geology considered the question of symbols for soils and rocks and published alternatives, and some additions, to those set out in Chapter 5. These symbols (Figure A1.1), covering more fully the rock and soil names used in the classification (Table 4.1), have the advantage that they are based as far as possible on the recommendations made by the International Standards Organization (Anon., 1981c).

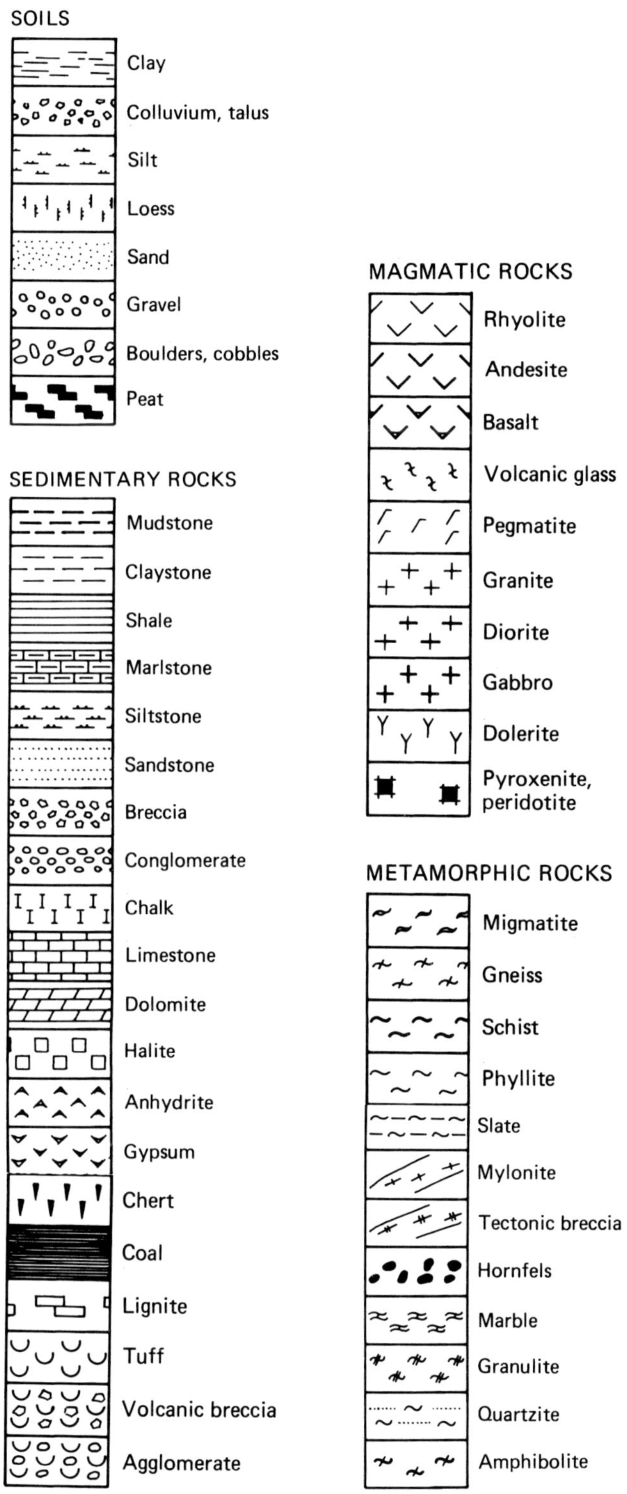

Figure A1.1 Alternative symbols for soils and rocks as recommended by the Mapping Commission of the International Association of Engineering Geology (From Anon., 1981c, by permission)

Appendix 2 Rock mass description and discontinuity data sheets

Standard forms are very useful as an aid, indeed as an *aide-mémoire*, in recording the collection of geological information when mapping rock exposures. Such forms are widely used, and two typical examples are given in Figures A2.1 and A2.2 (Anon., 1977).

ROCK MASS DESCRIPTION DATA SHEET

GENERAL INFORMATION

Seq. No. | Site | Date: Day, Month, Year | Operator

Method of location
1. By co-ordinates
2. Chainage
3. On attached map/ drawing/photograph

Co-ordinates or chainage (metres): Northings or chainage | Eastings | Elevation

Locality type
1. Natural exposure
2. Construction excavation
3. Trial pit
4. Trench
5. Adit
6. Tunnel

Size of locality
1. > $10m^2$
2. $5-10m^2$
3. $1-5m^2$
4. < $1m^2$
5. Line survey

No. of supplementary sheets of discontinuity data

Sketch

Photograph
0. No
1. Yes

Field tests
Specify type

Remarks

ROCK MATERIAL INFORMATION

Colour
1. Light
2. Dark

1. pinkish
2. reddish
3. yellowish
4. brownish
5. olive
6. greenish
7. bluish
8. greyish

1. pink
2. red
3. yellow
4. brown
5. olive
6. green
7. blue
8. white
9. grey
0. black

Grain size
1. Very coarse (>60mm)
2. Coarse (2-60mm)
3. Medium (60μ-2mm)
4. Fine (2-60μ)
5. Very fine (<2μ)

Compressive strength
1. Very strong (>100MN/m^2)
2. Strong (50-100MN/m^2)
3. Mod. strong (12.5-50MN/m^2)
4. Mod. weak (5-12.5MN/m^2)
5. Weak (1.25-5MN/m^2)
6. V. weak/hard (600-1250kN/m^2)
7. Very stiff (300-600kN/m^2)
8. Stiff (150-300kN/m^2)
9. Firm (80-150kN/m^2)
0. Soft (40-80kN/m^2)

Method of determining compressive strength
1. Measured
2. Assessed

Rock type

Qualifying terms to describe rock

ROCK MASS INFORMATION

Fabric
1. Blocky
2. Tabular
3. Columnar

Block size
1. Very large (>$8m^3$)
2. Large (0.2-$8m^3$)
3. Medium (0.008-0.2m^3)
4. Small (0.0002-0.008m^3)
5. Very small (<0.0002m^3)

State of weathering
1. Fresh
2. Slightly
3. Moderately
4. Highly
5. Completely
6. Residual soil

No. of major discontinuity sets

LINE SURVEYS TO DETERMINE DISCONTINUITY SPACINGS

	Plunge of line	Trend of line	Length of line (metres)	No. of fractures	Spacing	Remarks
Line 1						
Line 2						
Line 3						

Discontinuity spacing
1. Ext. wide (<2m)
2. Very wide (600mm-2m)
3. Wide (200-600mm)
4. Mod. wide (60-200mm)
5. Mod. narrow (20-60mm)
6. Narrow (6-20mm)
7. Very narrow (6mm)

Figure A2.1 Rock mass description data sheet as proposed by the Geological Society Engineering Group Working Party on 'The description of rock masses for engineering purposes' (From Anon., 1977, by permission)

DISCONTINUITY SURVEY DATA SHEET

GENERAL INFORMATION

Seq. No. [ ] Site [ ] Date (Day / Month / Year) [ ] Operator [ ] Discontinuity data sheet No. [ ] of [ ]

NATURE AND ORIENTATION OF DISCONTINUITY

Chainage or No.	Type	Dip	Dip direction	Persistence	Aperture	Nature of infilling	Consistency of infilling	Surface roughness	Trend of lineation	Waviness wavelength	Waviness amplitude	Water/Flow	Remarks

Type
0. Fault zone
1. Fault
2. Joint
3. Cleavage
4. Schistosity
5. Shear
6. Fissure
7. Tension crack
8. Foliation
9. Bedding

Dip, Dip direction and Trend of lineation
(Expressed in degrees)

Persistence
(Expressed in metres).

Aperture
1. Wide (>200mm)
2. Mod. wide (60-200mm)
3. Mod. narrow (20-60mm)
4. Narrow (6-20mm)
5. Very narrow (2-6mm)
6. Ext. narrow (<2mm)
7. Tight

Nature of infilling
1. Clean
2. Surface staining
3. Non-cohesive
4. Inactive clay or clay matrix
5. Swelling clay or clay matrix
6. Cemented
7. Chlorite, talc or gypsum
8. Others - specify

Compressive strength of infilling
1. Very soft (<40kN/m^2)
2. Soft (40-80kN/m^2)
3. Firm (80-150kN/m^2)
4. Stiff (150-300kN/m^2)
5. Very stiff (300-600kN/m^2)
6. Hard/v.weak (600-1250kN/m^2)
7. Weak (1.25-5MN/m^2)
8. Mod. weak (5-12.5MN/m^2)
9. Mod. strong (12.5-50MN/m^2)
10. Strong (50-100MN/m^2)
11. Very strong (100-200MN/m^2)
12. Ext. strong (>200MN/m^2)

Roughness
1. Polished
2. Slickensided
3. Smooth
4. Rough
5. Defined ridges
6. Small steps
7. Very rough

Waviness
Express wavelength & amplitude in metres

Water
1. Dry
2. Seepage

Flow
3. <10ml/sec
4. 10-100ml/sec
5. 0.1-1 l/sec
6. 1-10 l/sec
7. 10-100 l/sec
8. >100 l/sec

Figure A2.2 Discontinuity survey data sheet as proposed by the Geological Society Engineering Group Working Party on 'The description of rock masses for engineering purposes' (From Anon., 1977, by permission)

References

ABAG (1976) *Land Capability Analysis for Planning and Decision Making*, Association of Bay Area Governments, California, 30 pp

Adams, F.D. (1938) *The Birth and Development of the Geological Sciences*, Dover, New York, 506 pp

Aitchison, G.D. and Grant, K. (1967) The P.U.C.E. programme of terrain description, evaluation, and interpretation for engineering purposes. In *Proceedings of the Fourth Regional Conference in Africa on Soil Mechanics and Foundation Engineering* (Cape Town), Vol. 1, pp. 1–8

Aitchison, G.D. and Grant, K. (1968a) Proposals for the application of the P.U.C.E. programme of terrain classification and evaluation to some engineering problems. Paper No. 420 T, *Symposium on Terrain Evaluation for Engineering, Fourth Conference of the Australian Road Research Board*, Vol. 4, pp. 1648–1660

Aitchison, G.D. and Grant, K. (1968b) Terrain evaluation for engineering. In *Land Evaluation* (ed. G.A. Stewart), Macmillan, Melbourne, pp. 125–146

Alfors, J.T., Burnett, J.L. and Gay, T.E. Jr (1973) Urban geology master plan for California; the nature, magnitude, and costs of geologic hazards in California and recommendations for their mitigation. *California Division of Mines and Geology Bulletin*, **198**, 112

Anderson, J.G.C. (1974) The buried channels, rock floors and rock basins, and overlying deposits of the South Wales valleys from Wye to Neath. *Proceedings of the South Wales Institute of Engineers*, **88**, 3–17

Anderson, J.G.C., Arthur, L.J.A. and Powell, D.B. (1977) The engineering geology of the Dinorwic underground complex and its approach tunnels. In *Proceedings of the Conference on Rock Engineering* (Newcastle upon Tyne), pp. 491–510

Anon. (1921) *Manual of Map Reading and Field Sketching*, HMSO, London, 13 plates, 233 pp

Anon. (1957) British Standard Code of Practice CP 2001: 1957 Site investigations, British Standards Institution, London, 123 pp

Anon. (1962) *Dictionary of Geological Terms*, American Geological Institute/Dolphin Books, New York, 545 pp

Anon. (1970a) *International Legend for Hydrogeological Maps*, UNESCO/IASH, IAH, Paris, and Institute of Geological Sciences, London, 101 pp

Anon. (1970b) The logging of rock cores for engineering purposes. *Quarterly Journal of Engineering Geology*, **3**, 1–24

Anon. (1971) *Recommendations pour la cartographie géotechnique appliquée aux problèmes d'aménagement et d'urbanisme*, Ministère de l'équipement et du logement, Direction de l'aménagement foncier et de l'urbanisme, Paris, 15 pp

Anon. (1972) The preparation of maps and plans in terms of engineering geology. *Quarterly Journal of Engineering Geology*, **5**, 293–381

Anon. (1976) *Engineering Geological Maps. A Guide to their Preparation*, The UNESCO Press, Paris, 79 pp

Anon. (1977) The description of rock masses for engineering purposes. *Quarterly Journal of Engineering Geology*, **10**, 355–388

Anon. (1979) *Classification of Rocks and Soils for Engineering Geological Mapping. Part 1: Rock and Soil Materials. Bulletin of the International Association of Engineering Geology*, no. 19, pp. 364–371

Anon. (1981a) *Rock and Soil Description and Classification for Engineering Geological Mapping*, Report by the IAEG Commission on Engineering Geological Mapping. *Bulletin of the International Association of Engineering Geology*, no. 24, pp. 235–274

Anon. (1981b) BS 5930: 1981: Code of practice for site investigations (formerly CP 2001), British Standards Institution, London, 147 pp

Anon. (1981c) *Recommended Symbols for Engineering Geological Mapping*, Report by the IAEG Commission on Engineering Geological Mapping. *Bulletin of the International Association of Engineering Geology*, no. 24, pp. 227–234

Anon. (1982) *Land Surface Evaluation for Engineering Practice*, Report by a Working Party under the auspices of the Geological Society. *Quarterly Journal of Engineering Geology*, **15**, 265–316

Anson, W.W. and Sharp, J.I. (1960) *Surface and Rockhead Relief Features in the Northern Part of the Northumberland Coalfield.* Research Series No. 2, Department of Geography, University of Newcastle upon Tyne, 23 pp

Antoine, P. and Pachoud, A. (1976) Enseignements tirés à deux essais de cartographie systématiques de glissements de terrain. Stabilité des talus. 1—Versants naturels. 11. *Bulletin de Liaison des Laboratoires des Ponts et Chaussées*, Paris. Special No. 11, pp. 31–39

Arnould, M., Broquet, J.F., Deveughele, M. and Usseglio Polatera, J.M. (1979) Cartographie géotechnique de la Ville de Paris—Premières réalisations (13 ème, 19 ème et 20 ème arrondissements). *Bulletin of the International Association of Engineering Geology*, no. 19, 109–115

Arrowsmith, E.J. (1971) Slope stability. *Journal of the Institution of Highway Engineers*, pt 12, 13–15

Arrowsmith, E.J., Morris, A.G. and Bradley, L. (1972) Discussion on 'Geology of motorways and the role of the Institute of Geological Sciences, W.G. Evans'. *Quarterly Journal of Engineering Geology*, **5**, 71–75

Asquith, D.O. (1985) Characteristics and alternative mechanisms of Late Quaternary Faulting, Proposed LNG Terminal Site, Point Conception, California. *Bulletin of the Association of Engineering Geologists*, **22**(2), 171–192

Bachmann, G., Girowe, H. Helmerich, K., Reuter, F. and Thomas, A. (1967) Instruktion für die Anfertigung einheitlicher ingenieurgeologischer Gründkarten. *Zentralblatt für die Geologisches Instituts Abhandlungen, Berlin*, **9**, 35 pp, 9 supplements

Bachmann, G. and Reuter, F. (1968) Modern engineering-geological maps of the German Democratic Republic. *Abhandlungen des Zentralen Geologisches Instituts, Berlin*, **14**, 143–148

Barnes, J.W. (1981) *Basic Geological Mapping*, Geological Society of London, Handbook Series No. 1, 112 pp

Barton, N., Lien, R. and Lunde, J. (1974) Engineering classification of rock masses for the design of tunnel support. *Rock Mechanics*, Vol. 6, Springer-Verlag, Berlin, pp. 189–236. (Also Norwegian Geotechnical Institute, Publication No. 106, Oslo)

Bazley, R.A.B. (1971) A map of Belfast for the engineering geologist. *Quarterly Journal of Engineering Geology*, **4**, 313–314

Beavan, P.J. and Lawrence, C.J. (1982) *Terrain Evaluation for Highway Planning and Design*, Transport and Road Research Laboratory, Supplementary Report No. 725, 24 pp

Bell, D.H. (1987) Urban development practices in New Zealand. In *The Role of Geology in Urban Development*, Bulletin No. 3, pp. 43–65, Geological Society of Hong Kong, Hong Kong

Berkman, D.A. (1982) *Field Geologists' Manual*, Australasian Institute of Mining and Metallurgy Monograph Series No. 9, 2nd edn, Parkville, Victoria, Australia, 301 pp

Bickmore, D.P. and Shaw, M.A. (1963) *The Atlas of Britain and Northern Ireland*, Clarendon Press, Oxford

Bieniawski, Z.T. (1973) Engineering classification of jointed rock masses. *The Civil Engineer in South Africa*, pp. 335–343

Bieniawski, Z.T. (1974) Geomechanics classification of rock masses and its application in tunneling. In *Proceedings of the Third Congress of the International Society of Rock Mechanics* (Denver), Vol. 2A, pp. 27–32

Bieniawski, Z.T. (1976) Rock mass classification in rock engineering. In *Proceedings of the Symposium on Exploration for Rock Engineering* (Johannesburg), Vol. 1, pp. 97–106

Binnie, G.M., Clark, J.F.F. and Skempton, A.W. (1967) The effect of discontinuities in clay bedrock on the design of dams in the Mangla Project. In *Transactions of the Ninth International Congress on Large Dams* (Istanbul), Vol. 1, pp. 165–183

Binnie, G.M., Gerrard, R.T., Eldridge, J.G. *et al.* (1967) Mangla. *Proceedings of the Institution of Civil Engineers*, **38**, 337–575

Boulton, G.S. and Paul, M.A. (1976) The influence of genetic processes on some geotechnical properties of glacial tills. *Quarterly Journal of Engineering Geology*, **9**, 159–174

Brabb, E.E. (1984) Innovative approaches to landslide hazard and risk mapping. International Symposium on Landslides (Toronto, Canada), Vol. 1, pp. 307–323

Brabb, E.E. and Pampeyan, E.H. (1972) Preliminary map of landslides in San Mateo County, California. Map MF-344, Scale 1:62,500. U.S. Geological Survey, Menlo Park, California

Brabb, E.E. and Pampeyan, E.H. (1983) Geologic map of the San Mateo County, California. U.S. Map I-1257-A, Scale 1:62,500. U.S. Geological Survey, Menlo Park, California

Brabb, E.E., Pampeyan, E.H. and Bonilla, M.G. (1972) Landslide susceptibility in San Mateo County, California. Map MF-360 (San Francisco Region Environment and Resources Planning Study. Basic Data Contribution 43). U.S. Geological Survey, Menlo Park, California

BRGM (1968) Carte d'interprétation géotechnique de la zone littorale Nord-Pas-de-Calais. Rapport inédit, Orleans, France

BRGM (1969) Carte d'orientation géotechnique de la région de Tigery-Lieusaint (Essonne, Seine-et-Marne). Rapport inédit, Orleans, France

BRGM (1978) Carte Géologique de la France à 1/50 000. Sheet 230, Nancy, 2nd edn. Rapport inédit, Orleans, France

BRGM (1980) *Research and Development*. Bureau de Recherches Géologiques et Minières, Orleans, France, 111 pp

Brink, A.B., Mabbutt, J.A., Webster, R. and Beckett, P.H.T. (1966) *Report of the Working Group on Land Classification and Data Storage*. Military Experimental Establishment Report No. 940, Christchurch, Hampshire, UK, 97 pp

British Geological Survey (1970) Geological Sheet SP 83, Special Sheet, Milton Keynes (Solid and Drift), Scale 1:25,000, HMSO, London

British Standards Institution (1957) CP 2001: 1957 Site investigations, BSI, London

British Standards Institution (1981) Code of Practice for Site Investigations (formerly CP 2001), BS 5930: 1981. BSI, London

Broch, E. and Franklin, J.A. (1972) The point load strength test *International Journal of Rock Mechanics and Mineral Science*, **9**, 669–697

Brook, D. and Marker, B.R. (1987) Thematic geological mapping as an essential tool in land-use planning. In *Planning and Engineering Geology* (eds M.G. Culshaw, F.G. Bell, J.C. Cripps and M. O'Hara), Geological Society Engineering Geology Special Publications No. 4, Geological Society, London

Brown, E.T. (ed) (1981) *Rock Characterization Testing and Monitoring. ISRM Suggested Methods*, Pergamon Press, Oxford, 211 pp

Browne, M.A.E., Forsyth, I.H. and McMillan, A.A. (1986) Glasgow, a case study in urban geology. *Journal of the Geological Society, London*, **143**, 509–520

Bruning, K. (1939) Bodenkundliche Kartenwerke der Provinz Hannover. *Zeitschrift der Deutschen Geologischen Gesellschaft*, **91**, pt 8, 604–611

Brunsden, D. and Allison, J.R. (1986) Mountains and highlands. In *A Handbook of Engineering Geomorphology* (eds P.G. Fookes and P.R. Vaughan), Surrey University Press, Guildford, UK

Brunsden, D., Doornkamp, J.C., Fookes, P.G., Jones, D.K.C. and Kelly, J.M.H. (1975) Large scale geomorphological mapping and highway engineering design. *Quarterly Journal of Engineering Geology*, **8**, 227–253

Brunsden, D., Doornkamp, J.C., Green, C.P. and Jones, D.K.C. (1976) Tertiary and Cretaceous sediments in solution pipes in the Devonian Limestone of South Devon, England. *Geological Magazine*, **113**, 441–447

Brunsden, D., Jones, K.D.C., Martin, R.P. and Doornkamp, J.C. (1981) The geomorphological character of part of the Low Himalaya of Eastern Nepal. *Zeitschrift für Geomorphologie*, Suppl. Bd., **37**, 25–72

Bryant, B. (1972a) Map showing avalanche areas in the Aspen Quadrangle, Pitkin County, Colorado. Map I-786-G, U.S. Geological Survey, Washington

Bryant, B. (1972b) Map showing areas of selected potential geologic hazards in the Aspen Quadrangle, Pitkin County, Colorado. Map I-758-A, U.S. Geological Survey, Washington

Bryant, B. (1972c) Slope map of the Aspen Quadrangle, Pitkin County, Colorado. Map I-785-E, U.S. Geological Survey, Washington

Burnett, A.D., Brand, E.W. and Styles, K.A. (1985) Terrain classification mapping for a landslide inventory in Hong Kong. In *Proceedings of the Fourth International Conference and Field Workshop on Landslides* (Tokyo), pp. 63–68

Burnett, A.D. and Fookes, P.G. (1974) A regional engineering geological study of the London Clay in the London and Hampshire basins. *Quarterly Journal of Engineering Geology*, **7**, 257–295

Burnett, A.D. and Styles, K.A. (1982) An approach to urban engineering geological mapping as used in Hong Kong. In *Proceedings of the Fourth Congress of the International Association for Engineering Geology* (New Delhi), Vol. 1, pp. 167–176

Busk, H.G. (1929) *Earth Flexures*. Cambridge University Press, Cambridge, 106 pp

Carrara, A., Catalano, E., Sorriso Valvo, M., Reali, C. and Osso, I. (1978). Digital terrain analysis for land evaluation. *Geologia Applicata e Idroggeologia*, **13**, 69–127

Carter, P. and Mills, D. (1976) Engineering geological investigations for the Kielder Tunnels. *Quarterly Journal of Engineering Geology*, **9**(2), 125–141

Casagrande, A. (1962) Control of seepage through foundation and abutments of dams. *Géotechnique*, **11**, 159

Champetier de Ribes, G., Humbert, M. and Monition, L. (1974). Enquête préliminaire sur l'existence des zones exposées à des risques liés aux mouvements du sol et du sous-sol (enquête ZERMOS). Rapp. BRGM 74 SGN 244 AME, Orléans, France

Chandler, R.J. (1969) The effect of weathering on the shear strength properties of Keuper marl. *Géotechnique*, **19**, 321–334

Chandler, R.J. (1972) Lias clay. Weathering processes and their effect on shear strength. *Géotechnique*, **22**, 403–431

Chandler, R.J., Kellaway, G.A. Skempton, A.W. and Wyatt, R.J. (1976) Valley slope sections in Jurassic strata near Bath, Somerset. Pages 527–556, in 'A discussion on valley slopes and cliffs in Southern England: morphology, mechanics and Quaternary history' (by Skempton, A.W. and Hutchinson, J.N.). *Philosophical Transactions of the Royal Society of London* **A283**, 421–631

Cheffins, O.W. and Rushton, J.E.M. (1970). Edinburgh Castle Rock: a survey of the north face by terrestrial photogrammetry. *Photogrammetric Record*, **6**(35), 417–433

Clayton, K.M. (1963) A map of the drift geology of Great Britain and Northern Ireland. *Geological Journal*, **129**, 75–81

Clifton-Taylor, A. (1972). *The Pattern of English Building*, Faber and Faber, London, 466 pp

Coates, D.F. (1964a) Classification of rocks for rock mechanics. *International Journal of Rock Mechanics and Mineral Science*, **1**, 421–429

Coates, D.F. (1964b) (revised 1970) *Rock Mechanics Principles*, 3rd edn, Department of Energy, Mines and Resources, Mines Branch, Monograph 874, Ottawa, Canada, pp. A-1 to H-16

Coates, D.F. and Parsons, R.C. (1966) Experimental criteria for classification of rock substances. *International Journal of Rock Mechanics and Mineral Science*, **3**, 181–189

Collis, L. and Fox, R.A. (eds) (1985) *Aggregates: Sand, Gravel and Crushed Rock Aggregates for Construction Purposes*. Engineering Geology Special Publication No. 1, The Geological Society, London, 220 pp

Cottis, G.I., Dowell, R.W. and Franklin, J.A. (1971) A rock classification system applied to civil engineering. *Civil Engineering and Public Works Review*, June, 611–614, July, 736–738

Cratchley, C.R. and Denness, B. (1972) Engineering geology in urban planning with an example from the new city of Milton Keynes. In *Proceedings of the International Geological Congress (Montreal)*, 24th Session, Section 13, pp. 13–22

Cratchley, C.R. and Lake, R.D. (1977) *South Essex Geological and Geotechnical Survey*, Part 12: Summary. Natural Environment Research Council/Institute of Geological Sciences, Keyworth, Nottinghamshire, UK, 13 pp

Davies, T.P., Carter, P.G., Mills, D.A.C. and West, G. (1981) *Kielder Aqueduct Tunnels: predicted and actual geology*, Supplementary Report 676, Transport and Road Research Laboratory, Crowthorne, Berkshire, UK

Deacon, G.C. (1896) The Vyrnwy works for the water supply of Liverpool. *Journal of the Institution of Civil Engineers, London*, **126**, 24–67

Dearman, W.R. (1959) The structure of the Culm Measures at Meldon, near Okehampton, North Devon. *Quarterly Journal of the Geological Society, London*, **115**, 65–106

Dearman, W.R. (1974a) The characterization of rock for civil engineering practice in Britain. In *La Géologie de l'Ingénieur*. Société Géologique de Belgique, Liege, Belgium, pp. 1–75

Dearman, W.R. (1974b) Weathering classification in the characterization of rock for engineering purposes in British practice. *Bulletin of the International Association of Engineering Geology*, no. 9, 33–42

Dearman, W.R. (1976) Weathering classification in the characterization of rock: a revision. *Bulletin of the International Association of Engineering Geology*, no. 13, 123–127

Dearman, W.R. (1979) The Engineering Geological Mapping Commission of the International Association of Engineering Geology. *Bulletin of the International Association of Engineering Geology*, no. 19, 346–348

Dearman, W.R. (1983) Classification systems, design of underground structures based on classification systems. General Report Theme II. In *International Symposium on Engineering Geology and Underground Construction* (Lisboa, Portugal), Proceedings Vol. 2, pp. II.5–II.29

Dearman, W.R. (1986) *State of Weathering: The Search for a Rational Approach*, Engineering Geology Special Publication No. 2, The Geological Society, London, pp. 193–198

Dearman, W.R., Baynes, F.J. and Pearson, R. (1977) Geophysical detection of disused mineshafts in the Newcastle upon Tyne area, N.E. England. *Quarterly Journal of Engineering Geology*, **10**, 257–269

Dearman, W.R. and Coffey, R.G. (1981) An engineering zoning map of the Permian limestones of N.E. England. *Quarterly Journal of Engineering Geology*, **14**, 41–57

Dearman, W.R. and Fookes, P.G. (1974) Engineering geological mapping for civil engineering practice in the United Kingdom. *Quarterly Journal of Engineering Geology*, **7**, 223–256

Dearman, W.R. and Matula, M. (1976) Environmental aspects of engineering geological mapping. *Bulletin of the International Association of Engineering Geology*, no. 14, 141–146

Dearman, W.R., Money, M.S., Coffey, R.J., Scott, P. and Wheeler, M. (1977) Engineering geological mapping of the Tyne and Wear conurbation, North-east England. *Quarterly Journal of Engineering Geology*, **10**, 134–168

Dearman, W.R., Money, M.S., Strachan, A.D., Coffey, J.R. and Marsden, A. (1979) A regional engineering geological map of the Tyne and Wear County, N.E. England. *Bulletin of the International Association of Engineering Geology*, no. 19, 5–17

Dearman, W.R., Sergeev, E.M. and Shibakova, V.S. (eds) (1989) *Engineering Geology of the Earth*, NAUKA Publishers, Moscow

Dearman, W.R. and Strachan, A. (1983) Engineering geological plans of Tyne and Wear County, N.E. England. *Bulletin of the International Association of Engineering Geology*, no. 26, 31–41

Debaille, G. and Ghiste, S. (1969) *Carte géotechnique de la région de Mons*, Institut Reine Astrid, Mons, 44 pp and 4 maps

Deere, D.U. (1963) Technical description of rock cores. *Rock Mechanics and Engineering Geology*, **1**, 16–22

Deere, D.U., Merritt, A.H. and Coon, R.F. (1969) *Engineering Classification of in situ Rock*, Report AFWL-67-144, Air Force Systems Command, Kirtland Air Force Base, New Mexico, 272 pp

Deere, D.U. and Miller, R.P. (1966) *Engineering Classification and Index Properties for Intact Rock*, Report AFWL-TR-65-116, Air Force Weapons Laboratory (WLDC), Kirtland Air Force Base, New Mexico, 300 pp

De la Beche, H.T. (1832) *Geological Manual*, 2nd edn, Treuttel and Würtz, Treuttel Jun. and Richter, London, 564 pp

Delaunay, J. (1981) Carte de zones exposées à des glissements, écroulements, effondrements et affaissements de terrain de France, Échelle 1/1 000 000, Memoire du B.R.G.M. No. 124

Derbyshire, E. (1975) Distribution of glacial soils in Great Britain. In *The Engineering Behaviour of Glacial Materials*, Geological Abstracts, Norwich, UK, pp. 6–17

Deveughele, M. and Usseglio Polatera, J.M. (1979) Cartographie des zones exposées aux risque liés aux dissolutions de la fraction gypseuse du marno-calcaire de Saint-Ouen dans les 19 ème et 20 ème arrondissements de la ville de Paris. *Bulletin of the International Association of Engineering Geology*, no. 19, 121–125

Dobson, J. (1836) *General Plan for the Rebuilding and Extension of the Quay, Newcastle upon Tyne*, Tyne and Wear Archives, UK

Donaldson, G. *et al.* (1978) *Forestry, a Sector Policy Paper*, World Bank, Washington, DC, 65 pp

Doornkamp, J.C., Brunsden, D., Jones, K.D.C., Cooke, R.U. and Bush, P.R. (1979) Rapid geomorphological assessments for engineering. *Quarterly Journal of Engineering Geology*, **12**, 189–204

Douglas, T.H., Richards, L.R. and O'Neill, D. (1977) Site investigation for main underground complex—Dinorwic pumped storage scheme. In *Field Measurements in Rock Mechanics*, Zurich

Dowling, J.W.F. (1968) The *Classification of Terrain for Road Engineering Purposes*, Conference on Civil Engineering Problems, Overseas Session V, Communications Paper 16, Institution of Civil Engineers, London

Dowrick, D.J. (1983) An earthquake catastrophe damage assessment model with particular reference to central New Zealand. *Bulletin of the New Zealand National Society of Earthquake Engineering*, **16**(3), 213–221

Dumbleton, M.J. (1968) *The Classification and Description of Soils for Engineering Purposes: A Suggested Revision of the British System*, Road Research Laboratory Report LR 182, Ministry of Transport, Crowthorne, UK.

Dumbleton, M.J. and West, G. (1974) *Guidance on Planning, Directing and Reporting Site Investigations*, TRRL Report LR 625, Transport and Road Research Laboratory, Crowthorne, Berkshire, UK

Duncan, N. (1969) *Engineering Geology and Rock Mechanics*, 2 vols, Leonard Hill, London, pp 252, 270

Eastwood, T. (1935) *British Regional Geology, Northern England*, 1st edn, HMSO, London, 75 pp

Eckel, E.B. (1951) Interpreting geologic maps for engineers. In *Symposium on Surface and Subsurface Reconnaissance*, Special Technical Publication No. 122, American Society for Testing Materials, Washington, pp 5–15

Edwards, K. and Batson, R.M. (1980) Preparation of digital maps in raster format. *The American Cartographer*, **7**, 39–49

Eiby, G.A. and Reilly, W.I. (1976) Gravity, magnetism and seismicity. In *New Zealand Atlas* (ed. I. Wards), Government Printer, Wellington, pp. 76–79

Eisbacher, G.H. (1979) First-order regionalization of landslide characteristics in the Canadian Cordillera. *Geoscience Canada*, **6**(2), 69–79

E.I.S. Task Force, Alasca Natural Gas Transportation System, Bureau of Land Management (1976) *Alasca Natural Gas Transportation System*, Final Environmental Impact Statement, Alaska. U.S. Department of the Interior, Washington, 778 pp

Eldridge, J.G. and Little, A.L. (1967) The seismic design of earth dams of the Mangla Project. In *Transactions of the Ninth Congress on Large Dams* (Istanbul), Vol. 4, pp. Q35, R3

Environment Protection Board (Canada) (1974) *Environmental Impact Assessment of the Portion of the Mackenzie Gas Pipeline from Alasca to Alberta*, Vol. III, *Environmental Atlas*, Environment Protection Board, Ottawa, Canada

Ershova, S.B. and Sergee, E.M. (1979) Principles of the typological engineering geological zoning of the Earth. *Bulletin of the International Association of Engineering Geology*, no. 19, 18–92

Eyles, N. and Dearman, W.R. (1981) A glacial terrain map of Britain for engineering purposes. *Bulletin of the International Association for Engineering Geology*, no. 24, 173–184

Eyles, N. and Sladen, J.A. (1981) Stratigraphy and geotechnical properties of weathered lodgement till in Northumberland, England. *Quarterly Journal of Engineering Geology*, **14**, 129–141

Eyre, W.A. (1973) The revetment of rock slopes in the Clevedon Hills for the M5 motorway. *Quarterly Journal of Engineering Geology*, **6**, 223–229

Ferrians O.J. Jr (1971) *Preliminary Engineering Geologic Maps of the Proposed trans-Alaska Pipeline Route, Philip Smith Mountains Quadrangle*, Open-File Report, U.S. Geological Survey, Washington

Finlayson, A.A. (1982) *Terrain Analysis, Classification and an Engineering Geological Assessment of the Sydney Area, New South Wales*. 1, *Terrain Analysis and Classification*, 258 pp.; 2, *Terrain Assessment*, 199 pp., Technical Paper 32, CSIRO Australian Division of Applied Geomechanics, Victoria, Australia

Finlayson, A.A. (1984) Land surface evaluation for engineering practice: applications of the Australian PUCE system for terrain analysis. *Quarterly Journal of Engineering Geology*, **17**, 149–158

Floyd, J.D., Aitken, A.M., Ball, D.F., Laxton, J.L. and Long, D. (1983) *Environmental Geology Maps of South-east Edinburgh*, Open File Report SL 83/5, British Geological Survey, Keyworth, Nottinghamshire, UK

Fookes, P.G. (1965) Orientation of fissures in stiff overconsolidated clay of the Siwalik system. *Géotechnique*, **15**, 195–206

Fookes, P.G. (1967) Planning and stages of site investigation. *Engineering Geology*, **2**, 81–106

Fookes, P.G. (1969) Geotechnical mapping of soils and sedimentary rock for engineering purposes with examples of practice from the Mangla Dam Project. *Géotechnique*, **19**, 52–74

Fookes, P.G., Dearman, W.R. and Franklin, J.A. (1971) Some engineering aspects of rock weathering with field examples from Dartmoor and elsewhere. *Quarterly Journal of Engineering Geology*, **4**, 139–185

Fookes, P.G., Gordon, D.L. and Higginbottom, I.E. (1978) Glacial landforms, their deposits and engineering characteristics. In *The Engineering Behaviour of Glacial Materials*, 2nd edn, revised. Geological Abstracts, Norwich, UK, pp. 18–51

Fookes, P.G., Hinch, L.W. and Dixon, J.C. (1972) Geotechnical considerations of the site investigation for Stage IV of the Taff Vale Trunk Road to South Wales. In *Second British Regional Congress* (Cardiff), pp. 1–25. British National Committee, Permanent International Association of Road Congresses/Rendel, Palmer and Tritton, London

Fookes, P.G., Hinch, L.W., Huxley, M.A. and Simons, N.E. (1978) Some soil properties in glacial terrain in the Taff Valley, South Wales. In *The Engineering Behaviour of Glacial Materials*, Geological Abstracts, Norwich, UK, pp. 93–116

Fookes, P.G., Reeves, B.J. and Dearman, W.R. (1977) The design and construction of a rock slope in weathered slate at Fowey, Southwest England. *Géotechnique*, **27**(4), 533–556

Fookes, P.G., Sweeny, M., Manby, C.N.D. and Martin, R.P. (1985) Geological and geotechnical engineering aspects of low-cost roads in mountainous terrain. *Engineering Geology*, **21**, 1–152

Fookes, P.G. and Vaughan, P.R. (1986) *A Handbook of Engineering Geomorphology*, Chapman and Hall, New York, 343 pp

Fookes, P.G. and Wilson, D.D. (1966) The geometry of discontinuities and slope failures in Siwalik clay. *Géotechnique*, **16**(4), 305–320

Forsyth, I.H., McMillan, A.A., Browne, M.A.E. and Ball, D.F. (1983) Account accompanying *Environmental Geology Maps of Glasgow* (National Grid Sheet NS 66), Open File Report NL 83/1, British Geological Survey, Keyworth, Nottinghamshire, UK

Forsyth, I.H., McMillan, A.A., Browne, M.A.E. and Ball, D.F. (1984) Account accompanying *Environmental Geology Maps of Glasgow* (National Grid Sheet NS 56), Open File Report NL 84/3, British Geological Survey, Keyworth, Nottinghamshire, UK

Franklin, J.A. (1970) Observations and tests for engineering description and mapping of rocks. In *Proceedings of the Second Congress of the International Society of Rock Mechanics* (Beograd, Jugoslavia), Vol. I, pp. 11–16

Franklin, J.A., Broch, E. and Walton, G. (1971) Logging the mechanical character of rock. *Transactions of the Institution of Mining and Metallurgy*, **A80**, A1–A9, Discussion A43–A51

Froelich, A.J., Garnaas, A.D. and Van Driel, J.N. (1978) Franconia area, Fairfax County, Virginia. Planning a new community in an urban setting: Lehigh. In *Nature to be Commanded...* (eds G.D. Robinson and A.M. Spieker), U.S. Geological Survey Professional Paper 950, Washington DC, pp. 69–89

Garcia Yague, A. (1986) Geotechnical maps 1:10,000 of Madrid, Spain. In *Proceedings of the Fifth International Congress of the International Association of Engineering Geology* (Buenos Aires), Vol. 3, pp. 1781–1787

Gary, M., McAfee, R. Jr and Wolfe, C.L. (eds) (1972) *Glossary of Geology*, American Geological Institute, Washington DC

Gazel, J. and Peter A. (1969) Essais de cartographie géotechniques. *Annales des Mines*, Dec., 41–60. Bureau de Recherches Géologiques et Minières, Département Géotechnique, Orléans, France

Geological Society of America (1963) *Rock-color Chart*, GSA, Boulder, Colorado, USA

Glossop, R. (1968) The rise of geotechnology and its influence on engineering practice. *Géotechnique*, **18**, 105–150

Glossop, R. (1969) Engineering geology and soil mechanics. *Quarterly Journal of Engineering Geology*, **2**, 1–5

Glossop, R. and Skempton, A.W. (1945) Particle-size in silts and sands. *Journal of the Institution of Civil Engineers*, **25**, 81–105

Gojgic, D. and Lazic, M. (1971) Engineering Geological Map of the Urban Area of Herceg Novy, Jugoslavia, Scale 1:25,000. Institute for Geological and Geophysical Research, Beograd

Golodkovskaja, G.A. and Demidyuk, L.M. (1970) The problems of the engineering and geological mapping of deposits of mineral resources in the area of eternal frost. In *Proceedings of the First International Congress of the International Association of Engineering Geology* (Paris), Vol. 2, pp. 1049–1068

Gostelow, T.P. and Browne, M.A.E. (1981) *Engineering Geology of the Upper Forth Estuary*, Open File Report DG 482/87, British Geological Survey, Keyworth, Nottinghamshire, UK

Gostelow, T.P. and Browne, M.A.E. (1986) *Engineering Geology of the Upper Forth Estuary*, Report, Vol. 16, No. 8, British Geological Survey, Keyworth, Nottinghamshire, UK

Grant, K. (ed) (1968) *Proceedings of Study Tour and Symposium on Terrain Evaluation for Engineering*, Division of Applied Geomechanics, CSIRO, Victoria, Australia

Grant, K. (1972) *Terrain Classification for Engineering Purposes of the Melbourne Area, Victoria*, Technical Paper No. 11, Division of Applied Geomechanics, CSIRO, Victoria, Australia, 209 pp

Grant, K. (1975a) *The PUCE Programme for Terrain Evaluation for Engineering Purposes. Part I—Principles*, Technical Paper No. 15, 2nd edn, Division of Applied Geomechanics, CSIRO, Victoria, Australia, 32 pp

Grant, K. (1975b) *The PUCE Programme for Terrain Evaluation for Engineering Purposes. Part II—Procedures for Terrain Classification*, Technical Paper No. 19, 2nd edn, CSIRO Division of Applied Geomechanics, CSIRO, Victoria, Australia, 68 pp

Grant, K. (1976) *Terrain Classification and Evaluation for Engineering Purposes of the Canberra Area, Australian Capital Territory and New South Wales*, Technical Paper No. 22, Division of Applied Geomechamics, CSIRO, Victoria, Australia, 266 pp

Grant, K. and Ferguson, T.G. (1979) *Terrain Analysis, Classification and Assessment for Road and Railway Route Location, Construction Material and Water Resources of the Carpentaria Area, Queensland.* 1, *Terrain Analysis and Classification*, 214 pp.; 2, *Terrain Assessment*, 27 pp., Technical Report 100, CSIRO Australian Division of Applied Geomechanics, Victoria, Australia

Grant, K., Ferguson, T.G., Finlayson, A.A. and Richards, B.G. (1981) *Terrain Analysis, Classification, Assessment and Evaluation for Regional Developmental Purposes of the Albury–Wodonga Area, New South Wales and Victoria*, Technical Paper No. 30, Division of Applied Geomechanics, CSIRO, Victoria, Australia, 103 pp

Grant, K. and Finlayson, A.A. (1978a) The application of terrain analysis to urban and regional planning. In *Proceedings of the Third Congress of the International Association of Engineering Geology*, Sec. I, Vol. 1, pp. 79–91, Canberra, Australia

Grant, K. and Finlayson, A.A. (1978b) The assessment and evaluation of geotechnical resources in urban and regional environments. *Engineering Geology*, **12**, 219–293

Grant, K., Finlayson, A.A., Richards, B.G. and Pappin, J.W. (1982) *Terrain Analysis, Classification, Assessment and Evaluation for Regional Developmental Purposes of the Moreton Region, Queensland.* 1, *Terrain Analysis and Classification* (by A.A. Finlayson and K. Grant), 336 pp.; 2, *Terrain Assessment and Evaluation* (by J.W. Pappin, B.G. Richards, A.A. Finlayson and K. Grant), 203 pp., Technical Paper 23, CSIRO Australian Division of Applied Geomechanics, Victoria, Australia

Grant, K., Finlayson, A.A., Spate, A.P. and Ferguson, T.G. (1979) *Terrain Analysis and Classification for Engineering and Conservation Purposes of the Port Clinton Area, Queensland, Including the Shoalwater Bay Military Training Area.* Technical Paper 29, CSIRO Australian Division of Applied Geomechanics, Victoria, Australia, 185 pp

Grant-Taylor, T.L., Adams, R.D., Hatherton, T., Milne, J.D.G., Northey, R.D. and Stephenson, W.R. (1974) *Microzoning for Earthquake Effects in Wellington, N.Z.*, Bulletin 213, New Zealand Department of Scientific and Industrial Research, Wellington

Greeman, A. (1988) Computer makes real sense of maps. *New Civil Engineer*, 7 Jan, 18–20

Greffier, G. (1979) *Notice explicative de la Carte des Zones exposées à des risques liés aux mouvements du Sol et Sous-sol à 1/25,000. Région de Chanonat-Sud de Clermont-Ferrand (Puy-de-Dome)*, Bureau de Recherches Géologiques et Minières, Orléans, France

Hackman, R.J. and Williams, P.L. (1972) Map showing drainage basins and historic cloudburst floods in the Salina Quadrangle, Utah. Map I-591-N, U.S. Geological Survey, Washington

Hageman, B.P. (1963) A new method of representation in mapping alluvial areas. *Verhandelingen van het koninklijk Nederlands geologisch-mijnbouwkundig Genootschap*, Geol. Ser. 21-2. Jub. Conv. pt 2, pp. 211–219

Hair, T.H. (1839) *Sketches of the Coal Mines in Northumberland and Durham*, Facsimile Reprint 1969, Frank Graham, Newcastle upon Tyne, UK, 52 pp

Handel, D. (1963) Die ingeneurgeologische Detailkarte der Stadt Leipzig. *Zeitschrift für Angewandte Geologie*, **9**, pt 8, 421–423

Hatheway, A.W. (1982) Trench shaft and tunnel mapping. *Bulletin of the Association of Engineering Geologists*, **19**(2), 173–180

Hedberg, H.D. (ed.) (1972) An international guide to stratigraphic classification, terminology, and usage. *Lethaia*, **5**, 297–323 (International Subcommission on Stratigraphic Classification, Report No. 7B)

Helley, E.J. and Lajoie, R.A. (1979) *Flatland Deposits of the San Francisco Bay Region, California—Their Geology and Engineering Properties, and Their Importance to Comprehensive Planning*, Professional Paper 943, U.S. Geological Survey, Washington, 88 pp

Higginbottom, I.E. and Fookes, P.G. (1970) Engineering aspects of periglacial features in Britain. *Quarterly Journal of Engineering Geology*, **3**, 85–117

Higgins, J.D. and Rockaway, J.D. (1986) A graphics system for seismic response mapping. *Bulletin of the Association of Engineering Geologists*, **23**, 77–91

Holmes, S.C.A. (1972) Geological applications of early large-scale cartography. *Proceedings of the Geological Association*, **83**, 121–138

Horswill, P. and Horton, A. (1976) Cambering and valley bulging in the Gwash valley in Empingham, Rutland. Pages 423–462 in 'A discussion on valley slopes and cliffs in Southern England: morphology, mechanics and Quaternary history' (by Skempton, A.W. and Hutchinson, J.N.). *Philosophical Transactions of the Royal Society of London*, **A 283**, 421–631

Horton, A., Shephard-Thorn, E.R. and Thurrell, R.G. (1974) *The Geology of the New Town of Milton Keynes: Explanation of 1:25,000 Special Geological Sheet SP 83 With Parts of SP 73, 74, 84, 93 and 94*, Report, Institute of Geological Sciences

Hucka, V. (1965) A rapid method of determination of the strength of rocks *in situ*. *International Journal of Rock Mechanics and Mineral Science*, **2**(2), 127–134

Humbert, M. (1977) La cartographie en France des zones exposées à des risques liés aux mouvements du sol—Cartes ZERMOS. *Bulletin of the International Association of Engineering Geology*, no. 16, 80–82

Hvorslev, M.J. (1948) *Subsurface Exploration and Sampling of Soils for Civil Engineering Purposes*, Waterways Experimental Station, Vicksburg, Miss., USA

Institute of Geological Sciences (1971) *Geology of Belfast and District—Special Engineering Geology Sheet, Solid and Drift*, Scale 1:21120, IGS, Keyworth, Nottinghamshire, UK

Institute of Geological Sciences (1977a) Hydrogeological map of England and Wales, Scale 1:625000, IGS, Keyworth, Nottinghamshire, UK

Institute of Geological Sciences (1977b) Quaternary map of the United Kingdom, 2 Sheets, 1st edn, Scale 1:625 000, IGS, Keyworth, Nottinghamshire, UK

International Geographical Union (1968) *The Unified Key to the Detailed Geomorphological Map of the World, 1:25,000 to 1:50,000 scale*

Irfan, T.Y. and Dearman, W.R. (1978) Engineering classification and index properties of weathered granite. *Bulletin of the International Association of Engineering Geology*, no. 17, 79–90

Irfan, T.Y. and Powell, G.E. (1985) Engineering geological investigations for pile foundations on a deeply weathered granitic rock in Hong Kong. *Bulletin of the International Association of Engineering Geology*, no. 32, 67–79

ISRM (1977) *Suggested Methods for Quantitative Description of Rock Masses and Discontinuities*, Document No. 2, Final Draft, Committee on Field Tests, Norwegian Geotechnical Institute, Oslo, 147 pp

John, K.W. (1962) An approach to rock mechanics. *Proceedings of the American Society of Civil Engineers*, **88** (SM4), 1–30

Johnson, G.A.L. (compiler) (1972) Geology of Durham County. *Transactions of the Natural History Society of Newcastle*, **41**(1), 1–158 (2nd edn, revised)

Jones, D.K.C., Brunsden, D. and Goudie, A.S. (1983) A preliminary geomorphological assessment of part of the Karakoram Highway. *Quarterly Journal of Engineering Geology*, **16**, 331–355

Kabel, H. (1968) Collaboration results of different countries with reference to engineering geological mapping. *Abhandlungen der Zentralen Geologischen Instituts, Berlin*, **14**, 1–11

Kachadoorian, R. (1971) *Preliminary Engineering Geologic Maps of the Proposed trans-Alaska Pipeline Route, Wiseman and Chandalar Quadrangles*, Open-File Report, U.S. Geological Survey, Washington

Kemp, J.F. (1887) The geology of Manhattan Island. *New York Academy of Sciences Transactions*, **7**, 49–64

Knill, J.L. and Jones, K.S. (1965) The recording and interpretation of geological conditions in the foundations of the Roseires, Kariba and Latiyan Dams. *Géotechnique*, **15**, 94–124

Kurston, M.O.C. (1982) Environmental problems and geoscientific co-operation. In *Resources for the Twenty-First Century* (eds Whitmore and Williams), United States Geological Survey Professional Paper No. 1193, Washington, pp. 283–291

Lahee, F.H. (1961) *Field Geology*, 6th edn, McGraw-Hill, New York, 926 pp

Laird, R.T. *et al.* (1976) *Land Capability Analysis—A Method of Applying Earth Science Information to Planning and Decision-making*, Association of Bay Area Governments, 225 pp. and appendices

Laird R.T., Perkins, J.B., Bainbridge, D.A. *et al.* (1979) *Quantitative Land-Capability Analysis*, Professional Paper 945, U.S. Geological Survey, Washington, 115 pp

Lajoie, K.R., Helley, E.J., Nichols, D.R. and Burke, D.B. (1974) Geologic map of unconsolidated and moderately consolidated deposits in San Mateo County, California, U.S. Geological Survey Map MF-575, Scale 1:62,500.

Landry, J. (1980) *Notice explicative de la Carte des Zones Exposées à des Mouvements du Sol et du Sous-sol a 1/100,000. Région de Pontarlier (Jura)*, Bureau de Recherches Géologiques et Minières, Orléans, France

Lane, R.G.T. (1964) Rock foundations. Diagnosis of mechanical problems and treatment. *Proceedings of the Eighth Congress on Large Dams* (Edinburgh), Vol. 1, pp. 141–166

Lapworth, H. (1911) Geology of dam trenches. *Transactions of the Institution of Water Engineers, London*, **16**, 25–66

Larson, M.K. and Pewe, T.L. (1986) Origin of land subsidence and earth fissuring, Northeast Phoenix, Arizona. *Bulletin of the Association of Engineering Geologists*, **33**, 139–165

Lawrence, C.J. (1972) *Terrain Evaluation in West Malaysia*, Part 1, TRRL LR 506, Transport and Road Research Laboratory, Crowthorne, Berkshire, UK

Lawrence, C.J. (1978) *Terrain Evaluation in West Malaysia*, Part 2, TRRL SR 378, Transport and Road Research Laboratory, Crowthorne, Berkshire, UK

Legg, M.R., Eguchi, R.T. and Wiggins, I. H. (1980) *U.S. Building Exposure, County Data File—User's Manual*, J.H. Wiggins Co., Redondo Beach, California

Legget, R.F. (1973) *Cities and Geology*, New York, McGraw-Hill, New York, 552 pp

Linton, D.L. (1959) Morphological contrasts of Eastern and Western Scotland. In *Geographical Essays in Memory of Alan G. Ogilvie*, (eds R. Miller and J.W. Watson), Scottish Academic Press, Edinburgh, pp. 16–45

Little, A.L. (1967) Laterites. *Proceedings of the Asian Regional Conference on Soil Mechanics and Foundation Engineering*, 3rd Haifa, **2**, 61–71

Livingston, L. Jr and Blayney, J. (1971) *Open Space vs. Development*, Final Report to the City of Palo Alto, California, San Francisco, 190 pp

Lopez Palancar, J.J. and Garcia Yague, A. (1986) The 'Geo-Madrid geotechnical data bank. In *Proceedings of the Fifth International Congress of the International Association of Engineering Geology* (Buenos Aires), Vol. 3, pp. 1851–1860

Lovegrove, G.W. and Fookes, P.G. (1972) The planning and implementation of a site investigation for a highway in tropical conditions in Fiji. *Quarterly Journal of Engineering Geology* **5**(1,2), 43–68

Low, J.W. (1957) *Geological Field Methods*, Harper, New York, 489 pp

Lukey, M.E. (1974) Milton Keynes new city—a site survey challenge. *Ground Engineering*. **7.1**, 34–37

Lutzen, E.E. and Rockaway, J.D. Jr (1971) *Engineering Geology of St. Louis County, Missouri (U.S.A.)*, Engineering Geology Series No. 4, Missouri Geological Survey and Water Resources, Rolla, Missouri, USA

Lyell, C. (1852) *A Manual of Elementary Geology*, 4th edn, London, John Murray, 512 pp

McCave, I.N., Caston, V.N.D. and Fannin, N.G.T. (1977) The Quaternary of the North Sea. In *British Quaternary Studies* (ed. F.W. Shotton), Clarendon Press, Oxford, pp. 187–204

McClay, P.L. (1985) The use of photography to record geologic data. *Bulletin of the Association of Engineering Geologists*, **22**(2), 143–151

McCrea, R.W., Powell, D.B. and Yu, H.T. (1979) Engineering geological mapping of large caverns at Dinorwic Pumped Storage Scheme, North Wales. *Bulletin of the International Association of Engineering Geology*, no. 19, pp. 182–190

McGown, A., Anderson, W.F. and Radwan, A.M. (1978) *Geotechnical Properties of Tills in West Central Scotland*, In *Symposium of the Engineering Behaviour of Glacial Materials*, The Midland Soil Mechanics and Foundation Engineering Society, Birmingham, Geological Abstracts, pp. 89–99

McGown, A. and Derbyshire, E. (1977) Genetic influences on the properties of tills. *Quarterly Journal of Engineering Geology*, **19**, 389–410

McMillan, A.A. and Browne, M.A.E. (1987) The use and abuse of thematic mining information maps. In *Planning and Engineering Geology*, Geological Society, London, Engineering Geology Special Publication No. 4, pp. 237–245

McMillan, A.A., Browne, M.A.E. and Robson, P.G. (1984) The BGS Scottish land survey computer database—practice and use. *British Geologist*, **10**, 120–125

Madgett, P.A. and Catt, J.A. (1978) Petrography, stratigraphy and weathering of late Pleistocene tills in east Yorkshire, Lincolnshire and North Norfolk. *Proceedings of the Yorkshire Geological Society*, **42**, 55–108

Mahr, T. and Malgot, J. (1978) Zoning maps for regional and urban development based on slope stability. In *Proceedings of the Third International Congress of the I.A.E.G.* (Madrid), Sec. 1, Vol. 1, pp. 124–137

Mahr, T. and Malgot, J. (1985) Devastation of the environment by landslides activated by construction. *Bulletin of the International Association of Engineering Geology*, no. 31, pp. 81–88

Malgot, J., Baliak, F., Čabalova, D., Mahr, T. and Nemčok, A. (1973) Mapa svahovych deformacii Handlovskej kotliny: A Základná mapal; B. Mapa ingeniersko-geologickeho rajonavania, Prague

Malgot, J. and Mahr, T. (1979) Engineering geological mapping of the West Carpathian landslide area. *Bulletin of the International Association of Engineering Geology*, no. 19, pp. 116–121

Margat, J. (1980) Carte hydrogéologique de la France à l'échelle de 1/150 000. Systèmes aquifères. Notice explicative, Orléans, France

Marsland, A. (1977) The evaluation of the engineering design parameters for glacial clays. *Quarterly Journal of Engineering Geology*, **10**, 1–26

Marsland, A. (1979) The interpretation of in *situ tests* in glacial clays. In *Proceedings of the International Conference on Offshore Site Investigation*, Session 4, Paper 20, Society for Underwater Technology, London

Matson, C.R., Choy, H.H. and Gibson, A.M. (1986) Rock support prediction in the deep basement of Causeway Bay East Concourse for the Mass Transit Railway, Hong Kong. In *Proceedings of the Conference on Rock Engineering and Excavation in an Urban Environment* (Hong Kong), Institute of Mining and Metallurgy, London, pp. 285–297

Matula, M. (1961) Zakladna inziniersko-geologicka mapa, 1:25,000. *Acta geologica et geographica Universitatis Commenianae*, Geologica No. 7, Bratislava

Matula, M. (1969) *Regional Engineering Geology of Czechoslovak Carpathians*, Publishing House of the Slovak Academy of Sciences, Bratislava, 225 pp. and separate map folder

Matula, M. (1971) Engineering geologic mapping and evaluation in urban planning. In *Environmental Planning and Geology* (eds D.R. Nichols and C.C. Campbell), pp. 144–153, Geological Survey, United States Department of the Interior and the Office of Research and Technology, United States Department of Housing and Urban Development, Washington

Matula, M. (1976) Principles and types of engineering geological zoning. *Memorie della Societa Geologica Italiana, Pisa*, 327–336

Matula, M. (1979) Regional engineering geological evaluation for planning purposes. *Bulletin of the International Association of*

Engineering Geology, no. 19, pp. 18–24
Matula, M. and Holzer, R. (1978) *Engineering Geological Typology of Rock Masses*, Grundlagen und Anwendung der Felsmechanik, Felsmechanik Kolloquium Karlsruhe, Trans. Tech. Publications, Clausthal, pp. 107–121
Matula, M. and Letko, V. (1980) Engineering geology in planning the metropolitan region of Bratislava. *Bulletin of the International Association of Engineering Geology*, no. 22, pp. 139–145
Matula, M., Ondrasik, R., Wagner, P., Holzer, R. and Hyankova, A. (1985) Regional evaluation of rock mass conditions for pumped-storage plants. *Bulletin of the International Association of Engineering Geology*, no. 31, pp. 89–94
Matula. M. and Pasek, J. (1966) Zasady inzenyrskogeologickeho mapovani. *Sbor.geol. ved, HIG*, **5**, Praha
Mayberry, J.O. and Lindvall, R.M. (1972) Geologic map of the Parker Quadrangle, Arapahoe and Douglas Counties, Colorado, Map I-770-A, U.S. Geological Survey, Miscellaneous Geologic Investigations, Washington
Mayberry, J.O. (1972) Map showing relative erodibility of geologic materials in the Parker Quadrangle, Arapahoe and Douglas Counties, Colorado, Map I-770-C, U.S. Geological Survey, Miscellaneous Geologic Investigations, Washington
Mears A.J. (1977) *Debris Flow Hazard Analysis and Mitigation: an Example from Glenwood Springs, Colorado*, Colorado Geological Survey Information Series No. 8, 48 pp., Map scale 1:4800, Colorado, USA
Mitchell, C. (1973) *Terrain Evaluation*. Longman, London, 221 pp
Monro, S.K. and Hull, J.H. (1986) Environmental geology in Great Britain. In *Proceedings of the Fourth International Symposium on Geo-Resources and Environment* (Hannover), pp. 107–124
Moore, J.F.A. (1974) Mapping major joints in the Lower Oxford Clay using terrestrial photogrammetry. *Quarterly Journal of Engineering Geology*, **7**, 57–67
Moye, G.D. (1955) Engineering geology for the Snowy Mountains Scheme. *Journal of the Institution of Engineers of Australia*, **27**, 281–299
Moye, G.D. (1967) Diamond drilling for foundation exploration. *Journal of the Institution of Engineers of Australia*, **CE9**, 95–100
Muir, T.R.C., Smethurst, B.K. and Finn, R.P. (1986) Design and construction of high rock cuts for the Kornhill Development, Hong Kong. In *Proceedings of the Conference on Rock Engineering and Excavation in an Urban Environment* (Hong Kong), IMM, Hong Kong, pp. 309–329
Müller, L. (1958) Geomechanische Auswertung gefugekundlicher Details. *Geologie und Bauwesen*, **24**, 4
Müller, L. (1959) The European approach to slope stability problems in open-pit mines. *Quarterly Journal of the Colorado School of Mines*, **54**, 114–133
Müller, L. (1960) Die Geomechanik in der Praxis des Ingenieur – und Bergbaues. *Geologie und Bauwesen*, **25**, 203
Munsell, A.H. (1941) A color notation. In *Munsell Soil Color Charts, Baltimore, 1954*, Munsell Color Co. Inc., Baltimore
Newbery, J. (1970) Engineering geology in the investigation and construction of the Batang Padong hydroelectric scheme, Malaysia. *Quarterly Journal of Engineering Geology*, **3**, 151–181
Nichols, D.R. (1982) Application of Earth Sciences to land-use problems in the United States with emphasis on the role of the U.S. Geological Survey. In *Resources for the Twenty-First Century* (eds Whitmore and Williams), United States Geological Survey Professional Paper No. 1193, pp. 289–291
Nickless, E.F.P. (1982) *Environmental Geology of the Glenrothes District, Fife Region. Description of 1:25 000 Sheet NO2O*, Report of the Institute of Geological Sciences No. 83/15, p. 54
Nilsen, T.H., Wright, R.H., Vlasic, T.C. and Spangle, W. (1979) *Relative Slope Stability and Land-use Planning in the San Francisco Bay Region, California*. Professional Paper 994, U.S. Geological Survey, Washington, 96 pp
Northcote, K.H. (1965) *A Factual Key for the Recognition of Australian Soils*, 2nd edn, CSIRO Australian Division of Soils, Report No. 2/65. CSIRO, Melbourne, Australia
Northcote, K.H. (1974) *A Factual Key for the Recognition of Australian Soils*, 3rd edn, Rellim, Adelaide, 123 pp
Olivier, G. and Bernard P. (1976) *Notice explicative de la Carte des Zones Exposées à des Risques liés aux mouvements du Sol et du Sous-sol a 1/20,000. Région d'Aime, La Plagne (Savoie)*, Bureau de Recherches Géologiques et Minières, Orléans, France
Olivier, G. and Renet, J.P. (1976) *Essai de cartographie des risques liés à des mouvements de terrain dans la région de Saint-Martin-de-Belleville*, Stabilité des talus, 1-Versants naturels, Special Number 11, Bulletin Liaison des Laboratoires des Ponts et Chaussées, Paris, pp. 40-55.
Onadera, T.F. (1963) Dynamic investigation of foundation rocks *in situ*. In *Proceedings of the Fifth Symposium on Rock Mechanics* (Minnesota), Pergamon Press, New York, pp. 517–533
Ordnance Survey (1965) Gravel including associated sands. England and Wales—2 sheets, Scale 1:625,000. Compiled by Ministry of Housing and Local Government, London
Pacher, F. (1957) Über die Berechnung von Felssicherungen, verankerten Stützmauern und Füttermauern. *Geologie und Bauwesen*, **23**, 41
Pasek, J. (1968) The development of engineering geological maps in Czechoslovakia. *Abhandlungen der Zentralen Geologischen Instituts Berlin*, **14**, 75–85
Pasek, J. and Rybar, J. (1961) Die Darstellung ingenieurgeologischer Verhaltnisse in der Karte 1:25,000. *Zeitschrift für Angewandte Geologie*, **7**, pt 3, 136–139
Pender, E.B., Hosking, A.D. and Mattner, R.H. (1963) Grouted rock bolts for permanent support of major underground works. *Journal of the Institution of Engineers of Australia*, **35**, 129–150
Petak, W.J. and Atkisson, A.A. (1982) *Natural Hazard Risk Assessment and Public Policy*, New York, Springer-Verlag, 489 pp
Petak, W.J., Atkisson, A.A. and Gleye, P.H. (1978) *Natural Hazards, a Public Policy Assessment*, J.H. Wiggins Co., Redondo Beach, California
Peter, A. (1966) Essai de carte géotechnique. *Sols-Soils, Paris*, no. 16, 13–28
Pettijohn, F.J. (1957) *Sedimentary Rocks*, 2nd edn, Harper, New York
Pfannschmidt, M. (1939) Die Bodenkartierung in Stadtebau und Landesplanung. *Zeitschrift der Deutschen Geologischen Gessellschaft*, **91**, pt 8, 601–604
Phemister, T.C. and Simpson S. (1949) Pleistocene deep weathering in northeast Scotland. *Nature, London*, **164**, 318–319
Phillips, J. (1844) *Memoirs of William Smith, LL.D.*, John Murray, London, 149 pp
Pichal, Z. (1979) *Praha a Inzenyrska Geologie* (Prague and Engineering Geology). Czechoslovak Scientific–Technical Society, Prague, 131 pp with summaries in English and German
Piteau, D.R. (1970) Geological factors significant to the stability of slopes in cut rock. In *Symposium on Planning Open Pit Mines* (Johannesburg), pp. 33–53
Popov, I.V., Kats, R.S., Korikovskaia, A.K. and Lazareva, V.P. (1950) *Metodika sostavlenia inzhenernogeologicheskikhikart* (The techniques of compiling engineering-geological maps). Gosgeolizdat, Moscow
Porcher, M. and Guillope, P. (1979) Cartographie des risques Zermos appliquée à des plans d'occupation des sols en Normandie (Introduction par Ph. Gaudemer et M. Le Campion). Laboratoire Central des Ponts et Chaussées, Paris, *Bulletin de liaison*, No. 99, 43–54
Powell, D.B. and Mercer, R. (1979) The application of terrestrial photogrammetry at Tremadoc, North Wales, and an estimate of its cost-effectiveness in relation to a slope stability study. *Bulletin of the International Association of Engineering Geology*, no. 19, pp. 159–165
Price, D.G. (1971) Engineering geology in the urban environment. *Quarterly Journal of Engineering Geology* **4**, 191–208
Proctor, R.J. (1971) Mapping geological conditions in tunnels. *Bulletin of the Association of Engineering Geologists*, **8**(1), 1–43
Pupo, G., Vêncio, F.N.C. and Trinidade, T.S. (1979) Geological mapping for projects on underground works in crystalline rocks in Brazil. *Bulletin of the International Association of Engineering Geology*, no. 19, pp. 200–205

Radbruch, D.H. (1969) Areal and engineering geology of the Oakland East Quadrangle, California. U.S. Geological Survey Geology Quadrangle Map GQ-769, Washington

Radbruch, D.H. and Crowther, K.C. (1970) Map showing areas of relative amounts of landslides in California. U.S. Geological Survey Open File Report, Washington, 36 pp

Radbruch, D.H. and Crowther, K.C. (1973) Map showing areas of estimated relative amounts of landslides in California. U.S. Geological Survey, Map I-747, Scale 1:1,000,000, Washington

Radbruch, D.H. and Wentworth, C.M. (1971) Estimated Relative Abundance of Landslides in the San Francisco Bay Region, California, United States Department of the Interior, Geological Survey (San Francisco Bay Region Environment and Resources Planning Study, Basic Data Contribution 11)

Radbruch-Hall, D.H. (1979) Environmental aspects of engineering geological mapping in the United States. *Bulletin of the International Association of Engineering Geology*, no. 19, 351–358

Radbruch-Hall, D.H., Colton, R.B., Davies, W.E., Lucchitta, I., Skipp, B.A. and Varnes, D.J. (1982) Landslide overview map of the conterminous United States. Geological Survey Professional Paper 1183, Washington, 25 pp

Radbruch-Hall, D.H., Colton, R.B., Davies, W.E., Skipp, B.A., Lucchitta, I. and Varnes, D.J. (1976) Preliminary landslide overview map of the conterminous United States. U.S. Geological Survey, Map MF-771, Scale 1:7,500,000

Radbruch-Hall, D.H., Edwards, K. and Batson, R.M. (1979) Experimental engineering-geologic and environmental geologic maps of the conterminous United States prepared using computer techniques. *Bulletin of the International Association of Engineering Geology*, no. 19, 358–363

Radbruch-Hall, D.H., Edwards, K. and Batson, R.H. (1987) Experimental engineering-geologic and environmental geologic maps of the conterminous United States. U.S. Geological Survey Bulletin 1610, Washington, 7 pp

Reekie, C.F., Coffey, J.R. and Marsden, A.E. (1979) Computer aided techniques in urban engineering geological mapping. *Bulletin of the International Association of Engineering Geology*, no. 19, 322–330

Reik, G. and Soetomo, S. (1986) Influence of geological condition on design and construction of Cirata Powerhouse Cavern. In *Large Rock Caverns* (ed. K. Saari), Proceedings of an International Symposium (Helsinki), Vol. 1, Pergamon Press, Oxford, UK, 864 pp

Reuter, F. (1968) Engineering-geological mapping in Germany. *Abhandlungen des Zentralen Geologischen Instituts, Berlin*, **14**, 19–36

Rickert, D.A. *et al.* (1978a) *Oregon's Statewide Assessment of Nonpoint Source Problems.* State of Oregon, Department of Environmental Quality, Portland, Oregon, USA

Rickert, D.A. *et al.* (1978b) *Oregon's Procedure for Assessing the Impacts of Land Management Activities on Erosion-related Nonpoint Source Problems.* State of Oregon Department of Environmental Quality, Portland, Oregon, USA

Robinson, G.D. and Spieker, A.M. (eds) (1978) '*Nature to be Commanded*...', U.S. Geological Survey Professional Paper 950, Washington DC, 95pp

Robling, I and Tamm, G. (1961) Die ingenieurgeologische Kart der Stadt Berlin. Stadtbezirk Mitte, Maßstab 1:4000. *Zeitschrift für Angewandte Geologie*, **7**, pt 4, 198–201

Rockaway, J.D. and Lutzen, E.E. (1970) *Engineering Geology of the Creve Coeur Quadrangle, St. Louis County, Missouri.* Missouri Division of Geological Survey and Water Resources Engineering, Geology Series No. 2, 19 pp

Roed, M.A. (1979) *New Liskeard Area (NTS 31M/NW), District of Timiskaming; Ontario Geological Survey*, Northern Ontario Engineering Geology Terrain Study 894, 28 pp, accompanied by Maps 5021 and 5025, Scale 1:100,000

Roed, M.A. and Hallett, D.R. (1979) *Biscotasing Area (NTS 410/SE), Districts of Algoma and Sudbury*, Ontario Geological Survey, Northern Ontario Engineering Geology Terrain Study 87, accompanied by Map 5017, scale 1 : 100 000, Ontario, Canada, 15 pp

Ross-Brown, D.M., Wickens, E.H. and Markland, J.T. (1973) Terrestrial photogrammetry in open pits: 2—an aid to geological mapping. *Transactions of the Institution of Mining and Metallurgy*, **82**, A115–A130

Rudwick, M.J.S. (1976) A visual language for geology. *History of Science*, **14**(25), 149–195

Rummelen, F.F.F.E. van (1965) *Bladen Zeeuwsch-Vlaanderen West en Oost, Toelichtingen bij de Geologische Kaart van Nederland 1:50,000.* Geologische Dienst, Haarlem, 79 pp

Ruxton, B.P. and Berry, L. (1957) Weathering of granite and associated erosional features in Hong Kong. *Bulletin of the Geological Society of America*, **68**, 1263

Rybar, J. (1968) Engineering geological maps of landslide areas. *Abhandlungen der Zentralen Geologischen Instituts, Berlin*, **14**, 109–114

Saari, K. and Ylinen, A. (1979) Mapping of joints in Pre-cambrian granitic rock masses. *Bulletin of the International Association of Engineering Geology*, no. 19, pp. 190–195

Sandeman, E. (1901) The Burrator works for the water supply of Plymouth. *Proceedings of the Institution of Civil Engineers*, **156**, 4–41

Sanejouand, R. (1972) *La Cartographie Géotechnique en France*, Ministère de l'Équipement et du Logement, L.C.P.C., A.R.M.I.N.E.S., 96 pp

Scott, G.R. (1972a) Map showing landslides and area susceptible to landsliding in the Morrison Quadrangle, Jefferson County, Colorado, Map I-790-B, U.S. Geological Survey, Washington

Scott, G.R. (1972b) Map showing areas containing swelling clay in the Morrison Quadrangle, Jefferson County, Colorado, Map I-790-C, U.S. Geological Survey, Washington

Scott, P. (1976) The interpretation and presentation of urban engineering geological information: a feasibility study in the Gateshead area. *C.N.A.A. Ph.D. thesis* (unpublished), Sunderland Polytechnic, 167 pp

Sergeev, E.M. and Ershova, S.B. (1989) Classification of engineering geological conditions on the earth. In *Engineering Geology of the Earth* (eds W.R. Dearman, E.M. Sergeev and V.S. Shibakova), Nauka, Moscow

Sharma, S. (1980) *The Microzonation of the Memphis, Tennessee, Area.* Open File Report No. 80-914, U.S. Geological Survey, Washington

Simek, R. (1962) Nová inženýrsko-geologická mapa v měřítku 1:5000. *Geologicky průzkum, Praha*, **4**, pt 7, 211–214

Simek, R. (1965) Podrobna Inzenyrsko-Geologicka Mapa. A—geologicka mapa, Meritku 1:5000, Kralupy nad Vltavou 5–9

Sissons, J.B. (1967) *The Evolution of Scotland's Scenery*, Oliver and Boyd, Edinburgh, 259 pp

Skempton, A.W. and Weeks, A.G. (1976) The Quaternary history of the Lower Greensand escarpment and Weald Clay role near Sevenoaks, Kent. Pages 493–526 in Skempton, A.W. and Hutchinson, J.N. 'A discussion of valley slopes and cliffs in Southern England: morphology, mechanics and Quaternary history'. *Philosophical Transactions of the Royal Society of London*, **A283**, 421–631

Smeaton, J. (1793) *A Narrative of the Building and a Description of the Construction of the Edystone Lighthouse with Stone*, 2nd edn, G. Nicol, London

Smiles, S. (1861) *Lives of the Engineers*, Vol. II, pt VI, *Life of John Smeaton*, John Murray, London

Smith, D.B. (1972) Permian and Trias. In *Geology of Durham County*, 2nd edn (revised) (compiler: G.A.L. Johnston), Transactions of the Natural History Society of Newcastle, Vol. 41, No. 1, 158 pp

Smith, W.D. (1978a) Spatial distribution of felt intensities for New Zealand earthquakes. *New Zealand Journal of Geology and Geophysics*, **21**, 293–311

Smith, W.D. (1978b) Earthquake risk in New Zealand: statistical estimates. *New Zealand Journal of Geology and Geophysics*, **21**, 313–327

Sokal, R.R. (1966) Numerical taxonomy. *Scientific American*, **215**(6), 106–116

Sopwith, T. (1875) *Description of a Series of Elementary Geological Models*, James Tennant, London, 82 pp

Spangle and Associates, Leighton and Associates, Baxter, McDonald and Company (1976) *Earth Science Information in Land-Use Planning—Guidelines for Earth Scientists and Planners*, U.S. Geological

Survey Circular 721, 28 pp

Stamatello, H. (1965) Une carte géologique de la Ville de Varsovie, à l'usage des urbanistes. *Revue géographie physique et de géologie dynamique*, **7**(1), 3–10

Stapledon, D.H. (1976) Geological hazards and water storage. *Bulletin of the International Association of Engineering Geology*, no. 14, 249–262

Strachan, A.D. (1982) Development and application of techniques in engineering geological mapping. *M.Sc. thesis* (unpublished), University of Newcastle upon Tyne, UK

Strachan, A. and Dearman, W.R. (1982) The Tyne and Wear Data Bank, N.E. England. *Bulletin of the International Association of Engineering Geology*, no. 25, pp. 45–51

Styles, K.A., Hansen, A., Dale, M.J. and Burnett, A.D. (1984) Terrain classification methods for development planning and geotechnical appraisal: a Hong Kong case. In *Proceedings of the Fourth International Symposium on Landslides* (Toronto), Vol. 2, pp. 561–568

Sujkowski, Z. and Rosycki, S.Z. (1936) *Geologia Warszawy*. Zarzad Miejski, Warsaw

Tank, R.W. (1983) *Environmental Geology*, Oxford, 549 pp

Tarbet, M.A. (1973) Geotechnical properties and sedimentation characteristics of tills in south east Northumberland. *Ph.D. thesis*, (unpublished), University of Newcastle upon Tyne, UK (2 vols)

Taylor, B.J., Burgess, I.C., Land, D.H., Mills, D.A.C., Smith, D.B. and Warren, P.T. (1971) *British Regional Geology, Northern England*, 4th edn, HMSO, London, 121 pp

Taylor, R.K. (1971) The functions of the engineering geologist in urban development. *Quarterly Journal of Engineering Geology*, **4**, 221–240

Terzaghi, K. (1925) *Erdbaumechanik auf bodenphysicalischer Grunlage*, Franz Deuticke, Leipzig and Vienna

Terzaghi, K. (1929) *Effect of Minor Geologic Details on the Safety of Dams*, Technical Publication 215, American Institute of Mining and Metallurgical Engineering, New York

Terzaghi, K. (1946) Rock defects and loads on tunnel supports: introduction to tunnel geology. In *Rock Tunnelling with Steel Supports*, revised edition (eds M.E. Proctor and T.I. White), Commercial Shearing and Stamping Co., Ohio, 1968

Terzaghi, K. (1948) Foreword. *Géotechnique*, **1**, 3–5

Terzaghi, K. and Peck, R.B. (1967) *Soil Mechanics in Engineering Practice*, 2nd edn, Wiley, New York

Thomas, G.S.P. (1976) The Quaternary stratigraphy of the Isle of Man. *Proceedings of the Geologists' Association, London*, **87**, 56–81

Tuckwell, D.J. and Sadgove, B.M. (1977) *A Case for a National Registry of Ground Investigation Reports*, Final Draft Report (mimeographed), Construction Industry Research and Information Association, London, 37 pp

Turner, A.K. and Coffman, D.M. (1973) Geology for planning: a review of environmental geology. *Quarterly Journal of the Colorado School of Mines*, **68**, 3

UNESCO (1970) *International Legend for Hydrogeological Maps*, UNESCO, Paris, 101 pp

UNESCO/IAEG (1976) *Engineering Geological Maps. A Guide to their Preparation*, Unesco Press, Paris, 76 pp

U.S. Geological Survey (1949) *Interpreting Ground Conditions from Geologic Maps*, Geological Survey Circular 46, USGS, Washington

U.S. Geological Survey (1967) Engineering Geology of the Northeast Corridor, Washington, D.C., to Boston, Massachusetts: Coastal Plain and Surficial Deposts, Map I-514-13 (8 sheets); and Earthquake Epicentres, Geothermal Gradients and Excavations and Borings, Map I-514-C (2 sheets), USGS, Washington

U.S. Geological Survey (1982) *Goals and Tasks of the Landslide Part of a Ground-failure Hazards Reduction Program*, Circular 880, USGS, Washington, 48 pp

Varnes, D.J. (1974) *The Logic of Geological Maps, with Reference to their Interpretation and Use*, U.S. Geological Survey Professional Paper 837, Washington, 48 pp

Varnes, D.J. (1984) Landslide hazard zonation: a review of principles and practice. In *Natural Hazards*, Vol. 3, UNESCO, Paris, 63 pp

Waananen, A.O., Limerinos, J.T., Kockelman, W.J., Spangle, W.E. and Blair, M.L. (1977) *Flood-prone Areas and Land-use Planning—Selected Examples from the San Francisco Bay Region, California*. Geological Survey Professional Paper 942, USGS, Washington, 75 pp

Wakeling, T.R.M. (1970) A comparison of the results of standard site investigation methods against the results of a detailed geotechnical investigation in Middle Chalk at Mundford, Norfolk. In *Proceedings of the Conference on In Situ Investigations in Soils and Rocks*, British Geotechnical Society, London, pp. 17–22

Ward, W.H. and Burland, J.B. (1968) Assessment of the deformation properties of jointed rock in the mass. In *Proceedings of the International Symposium on Rock Mechanics* (Madrid), pp. 35–44

Ward, W.H., Burland, J.B. and Gallois, R.W. (1968) Geotechnical assessment of a site at Mundford, Norfolk, for a large proton accelerator. *Géotechnique*, **18**, 399–431

Watkins, M.D. (1979) Engineering geological mapping for the High Island Scheme in Hong Kong. *Bulletin of the International Association of Engineering Geology*, no. 19, 166–175

Weber, F.R. (1971) *Preliminary Engineering Geologic Maps of the Proposed trans-Alaska Pipeline Route, Mt. Hayes Quadrangle*, U.S. Geol. Survey Open-File Report, Washington

Webster, R. and Beckett, P.H.T. (1970) Terrain classification and evaluation using air photography: a review of recent work at Oxford. *Photogrammetria*, **26**, 51–75

Wentworth, C.M., Ziony, J.I. and Buchanan, J.M. (1970) Preliminary geologic environmental map of the greater Los Angeles area, California. U.S. Geological Survey TID–25363, issued by U.S. Atomic Energy Commission, Oak Ridge, Tennessee, USA, 41 pp

West, G. and Ewan, V.J. (1981) *Site Investigation and Construction of the Dinorwic Diversion Tunnel*, Laboratory Report 984, Transport and Road Research Laboratory, Crowthorne, Berkshire, UK

West, R.G. (1968) *Pleistocene Geology and Biology*, Longmans, London, 379 pp

Westercamp, D. (1987) Zonation of volcanic hazards at Mount Pelee, Martinique, French West Indies. In *The Role of Geology in Urban Development*, Geological Society of Hong Kong Bulletin No. 3, pp. 413–441

Whitham, K. (1975) The estimation of seismic risk in Canada. *Geoscience Canada*, **2**, 133–140

Wickens, E.H. and Barton, N.R. (1971) The application of photogrammetry to the stability of excavated rock slopes. *Photogrammetric Record*, **7**(37), 46–54

Wiggins, J.H., Slosson, J.E. and Krohn, J.P. (1978) *National hazards—Earthquake, Landslide, Expansive Soil Loss Models*, J.H. Wiggins Company Technical Report, Redondo Beach, California, 162 pp

Williams, P.L. (1972a) Map showing relative ease of excavation in the Salina Quadrangle, Utah Map I-591-J. U.S. Geological Survey, Washington

Williams, P.L. (1972b) Map showing types of bedrock and surficial deposits, Salina Quadrangle, Utah, Map I-591-M, U.S. Geological Survey, Washington

Williams, P.L. and Hackman, R.J. (1971) Geology, structure, and uranium deposits of the Salina Quadrangle, Utah, Map I-591 (2 sheets), Scale 1:250 000, U.S. Geological Survey, Washington

Wilson, H.E. (1972) The geological map and the civil engineer. In *Proceedings of the International Geological Congress, 24th Session* (Montreal), Section 13, pp. 83–86

Witkind, I.J. (1972) Map showing seiche, rockslide, rockfall, and earthflow hazards in the Henrys Lake Quadrangle, Idaho and Montana, Map I-891-C, U.S. Geological Survey, Washington

Youd, T.L. and Hoose, S.N. (1978) *Historic Ground Failures in Northern California Triggered by Earthquakes*, Geological Survey Professional Paper 993, USGS, Washington, 177 pp

Youd, T.L. and Perkins, J.B. (1987) Map showing liquefaction susceptibility of San Mateo County, California, Map I-1257-G, Scale 1:62,500, U.S. Geological Survey, Washington

Zaruba, Q. (1946) *Geologicky vyzkum mesta Turnova pro novy regulacni plan* (Geology of the town of Turnov and its significance for town planning). Turnov, Czechoslovakia, 16 pp

Zaruba, Q. (1948) Geological map of the Inner Prague Area, Scale 1:12,500,ČSR Geological Survey, Prague

Zaruba, Q. (1950) Technicko-geologicky rozbor uzemi mesta Kutne

Hory (Geologicotechnical surveying of the town of Kutna Hora) (with K. Hromada). *Geotechnica, Praha*, **9**, 45

Zaruba, Q. and Mencl, V. (1961) *Ingenieurgeologie*, Akademie Verlag, Berlin, 606 pp

Zebera, K. (1947a) Geologie v planovani oblasti a sidlist. *Geotechnica, Praha*, **3**

Zebera, K. (1947b) Technicko-geologicke pomery bustehradskeho katastru na Kladensku ve strednich Cechach. *Geotechnica, Praha*, **1**

Zeman, M. (1962) Metodika urbanisticko-geologického výzkumu se *zřetelem* k metodám podobného výzkumu v zahraniči. *Geologicky průzkum, Prague*

Bibliography

Some of the works listed here will have been referred to in the References—the list of publications cited in the text. The intention here is to give a guide to publications containing engineering and environmental maps; some are specific, others general that contain papers or references to additional maps.

Journals devoted to engineering geology

Bulletin of the Association of Engineering Geologists: From Vol. 21, No. 1, 1984, the bulletin contains good examples of engineering geological mapping.

Bulletin of the International Association of Engineering Geology: No. 1, 1970 to No. 37, 1988.

Engineering Geology: Vol. 1, 1965 to Vol. 25, 1988.

Quarterly Journal of Engineering Geology: Vol. 1, 1967 to Vol. 21, 1988.

Major publications on engineering geological mapping

The preparation of maps and plans in terms of engineering geology. Report of the Geological Society Engineering Group Working Party. *Quarterly Journal of Engineering Geology*, **5**, 295–382 (1972).

Engineering Geological Maps. A Guide to their Preparation. The UNESCO Press, Paris, 79 pp (1976).

Symposium: Engineering Geological Mapping, Newcastle upon Tyne, UK. *Bulletin of the International Association of Engineering Geology*, No. 19, 380 pp. (1979).

Papers from Sessions 1–4 of the Newcastle Symposium:

Session 1. Regional engineering geological mapping for planning purposes.

Session 2. Hazard mapping in risk evaluation for engineering structures.

Session 3. Civil engineering site mapping practice.

Session 4. Speciality sessions:

(a) Hydrogeological mapping at the engineering site scale;

(b) Land and sea floor geophysical mapping for engineering structures;

(c) Use of computers in engineering geological mapping.

Papers from the Symposium on Engineering Geological Mapping for Planning, Design and Construction in Civil Engineering. *Quarterly Journal of Engineering Geology*, **12**, 137–241 (1979).

Papers from Sessions 5 and 6 of the Newcastle Symposium:

Session 5. Terrain evaluation and remote sensing.

Session 6. Engineering geomorphological mapping.

Working Party Report on Land Surface Evaluation for Engineering Purposes. *Quarterly Journal of Engineering Geology*, *London*, **15**, 265–316 (Appendix 1, Sources of Reference, contains extensive bibliographies).

Verstappen, H. Th. and Van Zuidam, R.A. (1968) *ITC Textbook of Photo-interpretation* (contains extensive lists of symbols and maps in colour). International Institute for Aerial Survey and Earth Sciences, Delft, Netherlands.

Ronai, A. (1975) *Engineering Geological Mapping. International Post-Graduate Course on the Principles and Methods of Engineering Geology*, UNESCO/Hungarian Geological Institute, Budapest, 103 pp.

Fodor, P. (ed) (1975) *Directives for Engineering-Geological Mapping on the Scale of 1 : 10,000*. UNESCO/Hungarian Geological Institute, Budapest, 158 pp.

Ronai, R. (1985) Az alfold negyedidoszaski foldtana (The Quaternary of the Great Hungarian Plain). *Geologica Hungarica*, Series Geologica, Vol. 21, Institutum Geologicum Hungaricum, Budapest.

Hungarian Geological Institute (1969) Engineering-geological map series (Scale 1:10 000) of the Environs of Lake Balaton. HGI Budapest.

Hungarian Geological Institute (1969) The Geological Atlas of the Great Hungarian Plain, Scale 1:100 000. HGI, Budapest.

Hungarian Geological Institute (1974) The Geological Atlas of the Great Hungarian Plain, Scale 1:200 000. HGI, Budapest.

Publications with a significant mapping content

Environmental Planning and Geology. *Proceedings of the Symposium on Engineering Geology in the Urban Environment*, arranged by the Association of Engineering Geologists, the U.S. Department of Housing and Urban Development, and the U.S. Department of the Interior (eds D.R. Nichols and C.C. Campbell), Washington, 204 pp (1971).

Culshaw, M.G., Bell, F.G., Cripps, J.C. and O'Hara, M. (eds) (1987) *Planning and Engineering Geology*. Geological Society Engineering Geology Special Publication No 4. Session 4 was devoted to Environmental Geology in Planning.

Symposium: Landslides and Other Mass Movements, Prague. *Bulletin of the International Association of Engineering Geology*, No. 16, 256 pp. (1977).

Special No. II 'Stabilité des talus' 1—Versants naturels, 2—Deblais et Remblais. *Bulletin de Liaison des Laboratoires des Ponts et Chaussées, Paris* (1976), pp. I-188, II-196. Examples of maps of landslides, many in colour including ZERMOS extracts.

Études d'environnement et d'impact. *Bulletin de Liaison des Laboratoires des Ponts et Chaussées, Paris* (1981), No. 112, 353 pp.

Rock Engineering and Excavation in an Urban Environment, Institution of Mining and Metallurgy, London (1986), 552 pp.

The Role of Geology in Urban Development. Geological Society of Hong Kong, Bulletin No. 3, 601 pp. (1987).

Research and Development: *French Geological Survey*, Bureau de Recherches Géologiques et Minières, Orleans, France, 111 pp. (1980). An illustrated account in colour of the work of the Survey, with many examples of geological and thematic mapping.

Individual publications

Varnes, D.J. (1974) *The Logic of Geological Maps, with Reference to their Interpretation and Use for Engineering Purposes*, Professional Paper 837, U.S. Geological Survey, Washington, 48 pp. A philosophical discussion of the definition and classification of map units, with emphasis on the problems presented by maps intended for use in civil engineering. With 3 plates of engineering geological maps in colour.

Sanejouand, R. (1972) *La Cartographie Géotechnique en France*, Ministère de l'Équipement et du Logement. L.C.P.C., A.R.M.I.N.E.S., Paris, 96 pp.

van Zuidam, R.A. (1978) A geomorphological survey of the Ezulwini Valley, Swaziland. *ITC Journal*, 559–572 (with full-colour geomorphological map of the area at 1:30 000).

Soeters, R. and Rengers, N. (1981) An engineering geological map from large-scale photography. *ITC Journal*, 140–150 (with full-colour map at 1:10 000).

Robinson, G.D. and Spieker, A.M. (eds) (1978) *'Nature to be Commanded...'*, U.S. Geological Survey Professional Paper 950, Washington DC, 95 pp. Earth science maps applied to land and water management.

Institute of Geological Sciences, Report Series, including:

Rhind, D.W. and Sissons, J.B. (1971) *Data Banking of Drift Borehole Records for the Edinburgh area*, Report No. 71/15, 19 pp.

Rhind, D.W. (1973) *Computer Mapping of Drift Lithology from Borehole Records*, Report No. 73/6, 12 pp. with colour map.

Nickless, E.F.P. (1982) *Environmental Geology of the Glenrothes District, Fife Region*, Description of 1:25,000 sheet NO 20. Report No. 82/15, 54 pp. with 27 maps.

British Geological Survey, Report Series, including:

Gostelow, T.P. and Browne, M.A.E. (1986) *Engineering Geology of the Upper Forth Estuary*, Report 16, No. 8, 56 pp. with 8 maps at 1:50 000.

Institute of Geological Sciences (British Geological Survey, Keyworth, Nottinghamshire, UK, and South East Essex Geological and Geotechnical Survey) including:

Cratchley, C.R. and Lake, R.D. (1975) Part 1, *Introduction*, 4 pp.

Northmore, K.J., Denness, B. and Conway, B.W. (1977) Part 3, Engineering Geology. (Part 2 comprises a folio of 27 maps at 1:25 000 scale.)

Proceedings of congresses and other meetings with a significant engineering geological mapping content

International Congresses of the International Association of Engineering Geology: First (1970) to Fifth (1986).

International Geological Congresses: Engineering Geology, Section 12, Prague, 1968; Engineering Geology, Section 13, Montreal, 1972.

Bibliographies

Geofond, Prague (1970) *Bibliography and Index of Engineering Geological Mapping*. Part 1: Czechoslovakia, German Democratic Republic, Hungary, Poland, 66 pp.

Parry, R.B. and Perkins, C.R. (1987) *World Mapping Today*, Butterworths, London, 583 pp. Catalogue of available maps including geological, geophysical and environmental. Sections dealing with individual countries, each with an introduction on which organizations, including commercial publishers, produce the maps and a list of addresses.

Index